Science of Ashwagandha: Preventive and Therapeutic Potentials

Sunil C. Kaul • Renu Wadhwa
Editors

Science of Ashwagandha: Preventive and Therapeutic Potentials

Springer

Editors
Sunil C. Kaul
DBT-AIST International Laboratory for Advanced Biomedicine (DAILAB), Biomedical Research Institute
National Institute of Advanced Industrial Science & Technology (AIST)
Tsukuba, Ibaraki, Japan

Renu Wadhwa
DBT-AIST International Laboratory for Advanced Biomedicine (DAILAB), Biomedical Research Institute
National Institute of Advanced Industrial Science & Technology (AIST)
Tsukuba, Ibaraki, Japan

ISBN 978-3-319-59191-9 ISBN 978-3-319-59192-6 (eBook)
DOI 10.1007/978-3-319-59192-6

Library of Congress Control Number: 2017945748

Printed on acid-free paper

This Springer imprint is published by Springer Nature
The registered company is Springer International Publishing AG
The registered company address is: Gewerbestrasse 11, 6330 Cham, Switzerland

Foreword

Ayurveda – the traditional medicine of India – traces its roots to Atharvaveda of the fifteenth century BCE or earlier. Though the medicine of the Atharvaveda was faith-based and relied on chants and rituals for the treatment of diseases, the practice of medicine witnessed a major change during the subsequent centuries when Buddhism became the central faith of India. The change was conspicuous by the virtual disappearance of chants and rituals from medical practice, which, according to Caraka of the first century CE, had become reason-based. Ayurveda, which flourished in Buddhist India became systematized during the golden period from first to sixth centuries CE, when the great texts of Caraka and Suśruta were redacted in the presently available format and the two texts of Vāgbhaṭa – Saṅgraha and Hṛdaya – were composed. Systematization involved the division of Ayurveda into eight branches, refinement of basic concepts such as tridoṣa, standardization of medical procedures such as panchakarma, establishment of an elaborate training protocol for Ayurvedic physicians and the development of medical therapy based on a massive formulary, much of which was composed of medicinal plants.

Medicinal plants constitute the principal weapons in the therapeutic arsenal of Ayurveda. They greatly outnumber animal and mineral products. While *Caraka Samhita* mentions nearly 2000 plants for the making of drugs, the earlier Buddhist literature refers to less than 500, indicating the growth of a herbal formulary in Ayurveda over the centuries.

The European interest in India's medicinal plants and traditional medicine was triggered by the publication of the *Colloquies on the Simples and Drugs of India* in 1563 by Garcia de Orta, a Portuguese physician from Goa where he practiced medicine for 36 years. This was followed by the endeavour of the Dutch Governor of Kochi – Van Rheede – who organized in the seventeenth century a study of the medicinal plants of the Malabar Coast over 30 years and published 12 volumes of *Hortus Malabaricus* in Latin from Amsterdam. The interest in the taxonomy of medicinal plants grew rapidly among Europeans, and studies were carried out in Bengal, Madras and elsewhere by scientists such as Roxburgh and Ainslie. In the nineteenth century, 70% of the drugs of plant origin in the *British Pharmacopoeia* were based on Indian plants. The taxonomy-dominated phase was followed in the early twentieth century by the pioneering work of RN Chopra in pharmacology and the development of herbal drugs. His stated aim was "to discover remedies from the claims of Ayurvedic, Tibbi and other indigenous sources, suitable to be employed by the exponents of western medicine". His contributions included botanical identification, chemical analysis, pharmacologic studies and clinical trials of a large number of commonly used herbal drugs including *Rauwolfia serpentina*. His endeavour was the forerunner of the still vigorous and extensive research in natural products chemistry where India became a world leader.

As Ayurveda laid great emphasis on preventive and promotive aspects of health (Svasthavṛtta), a number of medicinal plants were identified as anti-ageing (vayasthapana), health promoting (jivaniya) and strengthening (brmhaniya). Many of these were studied in detail, and an extensive literature exists on their botanical, chemical and pharmacological characteristics. Ashwagandha (*Withania somnifera*), highly valued among Indian medicinal plants, occurs throughout the drier and subtropical parts of India and is cultivated widely in Madhya Pradesh. In this remarkable book on the biology of Ashwagandha, Drs. Renu Wadhwa and Sunil Kaul have drawn upon their lifetime experience in Ashwagandha research in Japan and that of scientists elsewhere in the world, who have studied diverse aspects of this remarkable "plant for all seasons". The research contributions from experts cover a wide variety of topics with reference to Ashwagandha such as phytogenomics and pharmacology, cancer treatment, brain function and biotechnology-based propagation. The papers are notable not only for their comprehensive coverage of Ashwagandha but also for their equal appeal to biological scientists and clinicians. This volume of *Science of Ashwagandha: Preventive and Therapeutic Potentials* is of much topical interest and is, at the same time, a call for accelerated search for molecular drugs from Ashwagandha in medical treatment, especially of cancer and dementia.

National Research Professor
Manipal University
Manipal, 576 104, India
msvaliathan@gmail.com

M.S. Valiathan

Preface

Ayurveda, the world's oldest home medicine system and a living tradition in India, defines that our whole-body wellness is determined by a delicate balance of three elements, the doshas (named vata, pitta and kapha), and a variety of herbs can be used for correcting, improving and innovating such balance for health care and cure. With 3000 years of experience passed through several generations, principles of Ayurveda are greatly trusted and recruited for supporting health, disease prevention and therapeutics. Owing to improved living, food and medical standards, human average lifespan has increased over the years and in effect has raised concerns on rapidly increasing old age populations in several developing parts of the world. Allied outcomes of this are the new challenges to deal with a variety of old age pathologies including neurodegenerative/cardiovascular diseases and cancers. In addition, the current scenario of industrialized lifestyle wherein we are using a large amount of chemicals and environmental pollution in a variety of forms has widely escalated incidence of these diseases.

Inspired by our own nearly three decades of research on understanding the molecular mechanism(s) of stress, ageing and cancer that has revealed their multifactorial basis and complexities in their intervention, we comprehended the need of NEW (**n**atural, **e**fficient and **w**elfare) multi-targeted and affordable drugs for supporting quality of life (QOL) under normal and diseased circumstances. By integrating traditional knowledge with the modern bioinformatics and experimental strengths for dissecting the activities and functional mechanisms of several herbs, Ashwagandha was selected as champion for further research and supporting human health and QOL. With an aim to apprehend the mechanism(s) of various well-documented activities and explore more, the book is organized into four parts: (1) "Ashwagandha – Phytogenomics and Pharmacology", (2) "Ashwagandha for Cancer", (3) "Ashwagandha for the Brain" and (4) "Active Ingredient-Enriched Ashwagandha – Biotechnologies". We hope that the book will encourage not only students at graduate and advanced undergraduate levels but also general readership and medical practitioners. We hope that this portrait of *Science of Ashwagandha: Preventive and Therapeutic Potentials* by the team of experts would help in understanding its potential as a NEW drug and asking new questions to advance knowledge for its effective use for health care and cure.

We are deeply obliged to all the authors for their interest, enthusiasm and dedication to Ashwagandha research that made this book necessary and feasible. Without their commitment to contribute chapters, it was not possible to accomplish this volume suitable for general and specialized reading.

Tsukuba, Japan

Sunil C. Kaul
Renu Wadhwa

Contents

Contributors

Luna Acharya Icahn School of Medicine at Mount Sinai (Bronx) Program, James J. Peters VA Medical Center, NY, USA

Sara S. Alhakeem Department of Microbiology, Immunology and Molecular Genetics, Markey Cancer Center, University of Kentucky, Lexington, KY, USA

Priyanshu Bhargava Graduate School of Life & Environmental Sciences, University of Tsukuba, Ibaraki, Japan

Subbarao Bondada Department of Microbiology, Immunology and Molecular Genetics, Markey Cancer Center, University of Kentucky, Lexington, KY, USA

Anupama Chaudhary DBT-AIST International Laboratory for Advanced Biomedicine (DAILAB), Biomedical Research Institute, National Institute of Advanced Industrial Science & Technology (AIST), Tsukuba, Ibaraki, Japan

Amitabha Dey Neuropharmacology Research Laboratory, Department of Pharmaceutics, Indian Institute of Technology (Banaras Hindu University), Varanasi, India

Kallol Dutta CERVO Brain Research Centre, Université Laval, Québec, Québec, Canada

Beth W. Gachuki Department of Microbiology, Immunology and Molecular Genetics, Markey Cancer Center, University of Kentucky, Lexington, KY, USA

Jessica M. Gannon Western Psychiatric Institute and Clinic, University of Pittsburgh School of Medicine, Pittsburgh, PA, USA

Sukant Garg School of Integrative & Global Majors, University of Tsukuba, Tsukuba, Ibaraki, Japan

Sukriti Goyal Department of Bioscience and Biotechnology, Banasthali University, Tonk, Rajasthan, India

Abhinav Grover School of Biotechnology, Jawaharlal Nehru University, New Delhi, India

Rupam Guleri Department of Biotechnology, Guru Nanak Dev University, Amritsar, Punjab, India

Muskan Gupta Department of Biotechnology, Guru Nanak Dev University, Amritsar, Punjab, India

Ramesh C. Gupta Department of Pharmacology and Toxicology, and James Graham Brown Cancer Center, University of Louisville, Louisville, KY, USA

Eun-Ryeong Hahm Department of Pharmacology and Chemical Biology, University of Pittsburgh Cancer Institute, University of Pittsburgh School of Medicine, Pittsburgh, PA, USA

Babli Halder Proteomics and Cell Signaling, Lab E409, Centre for Cellular & Molecular Biology, Habsiguda, Hyderabad, Telengana, India

Genji Imokawa Research Institute for Biological Functions, Chubu University, Kasugai, Aichi, Japan

Anita Jagota Neurobiology and Molecular Chronobiology laboratory, Department of Animal Biology, School of Life Sciences, University of Hyderabad, Hyderabad, India

Salma Jamal Department of Bioscience and Biotechnology, Banasthali University, Tonk, Rajasthan, India

Jean-Pierre Julien CERVO Brain Research Centre, Université Laval, Québec, Québec, Canada

Rajkumar S. Kalra DBT-AIST International Laboratory for Advanced Biomedicine (DAILAB), Biomedical Research Institute, National Institute of Advanced Industrial Science & Technology (AIST), Tsukuba, Ibaraki, Japan

Shashank P. Katiyar DAILAB, Department of Biochemical Engineering & Biotechnology, DBT-AIST International Laboratory for Advanced Biomedicine (DAILAB), Indian Institute of Technology (IIT) Delhi, Hauz Khas, New Delhi, India

Ashish Kaul School of Integrative & Global Majors, University of Tsukuba, Tsukuba, Ibaraki, Japan

Sunil C. Kaul DBT-AIST International Laboratory for Advanced Biomedicine (DAILAB), Biomedical Research Institute, National Institute of Advanced Industrial Science & Technology (AIST), Tsukuba, Ibaraki, Japan

Zeenia Kaul Department of Molecular Virology, Immunology & Medical Genetics, The Ohio State University, Columbus, USA

Gurcharan Kaur Department of Biotechnology, Guru Nanak Dev University, Amritsar, Punjab, India

Kuldeep Kaur Department of Biotechnology, Guru Nanak Dev University, Amritsar, Punjab, India

Kulwinder Kaur Department of Biotechnology, Guru Nanak Dev University, Amritsar, Punjab, India

Mandeep Kaur Department of Botanical and Environmental Sciences, Guru Nanak Dev University, Amritsar, Punjab, India

Jatinder Kaur Katnoria Department of Botanical and Environmental Sciences, Guru Nanak Dev University, Amritsar, Punjab, India

Avinash Kaur Nagpal Department of Botanical and Environmental Sciences, Guru Nanak Dev University, Amritsar, Punjab, India

Rajneet Kour Soodan Department of Botanical and Environmental Sciences, Guru Nanak Dev University, Amritsar, Punjab, India

Su-Hyeong Kim Department of Pharmacology and Chemical Biology, University of Pittsburgh Cancer Institute, University of Pittsburgh School of Medicine, Pittsburgh, PA, USA

Arpita Konar CSIR-Institute of Genomics & Integrative Biology, New Delhi, India

K. Kowshik Neurobiology and Molecular Chronobiology laboratory, Department of Animal Biology, School of Life Sciences, University of Hyderabad, Hyderabad, India

Vikas Kumar National Center of Natural Products Research, School of Pharmacy, University of Mississippi, USA

Anjani Kumari DAILAB, Department of Biochemical Engineering and Biotechnology, Indian Institute of Technology (IIT) Delhi, Hauz Khas, New Delhi, India

Fayaz Malik Department of Cancer Pharmacology, Indian Institute of Integrative Medicine, Jammu, Jammu and Kashmir, India

Vidhi Malik DAILAB, Department of Biochemical Engineering & Biotechnology, DBT-AIST International Laboratory for Advanced Biomedicine (DAILAB), Indian Institute of Technology (IIT) Delhi, Hauz Khas, New Delhi, India

Shaffi Manchanda Department of Biotechnology, Guru Nanak Dev University, Amritsar, Punjab, India

Radheshyam Maurya Department of Animal Biology, University of Hyderabad, Hyderabad, India

Mary K. McKenna Department of Microbiology, Immunology and Molecular Genetics, Markey Cancer Center, University of Kentucky, Lexington, KY, USA

Bhawana Mishra CSIR-Central Institute of Medicinal and Aromatic Plants (CSIR-CIMAP), Lucknow, UP, India

Natarajan Muthusamy Department of Internal Medicine, Ohio State University, Columbus, OH, USA

Sunil K. Noothi Department of Microbiology, Immunology and Molecular Genetics, Markey Cancer Center, University of Kentucky, Lexington, KY, USA

Karine Z. Oben Department of Microbiology, Immunology and Molecular Genetics, Markey Cancer Center, University of Kentucky, Lexington, KY, USA

Hriday S. Pandey Cellular and Molecular Neuroscience, National Brain Research Centre (NBRC), Gurgaon, Haryana, India

Neha Pandey Department of Metabolic and Structural Biology, CSIR- Central Institute of Medicinal and Aromatic Plants, Lucknow, India

Pratap Kumar Pati Department of Biotechnology, Guru Nanak Dev University, Amritsar, Punjab, India

Abhishek Rai University of Pittsburgh Medical Center – Northwest, Seneca, PA, USA

Preeti Rana Department of Biotechnology, TERI University, Vasant Kunj, New Delhi, India

Vivek M. Rangnekar Department of Radiation Medicine, Markey Cancer Center, University of Kentucky, Lexington, KY, USA

Sheikh Rayees Department of Pharmacology, University of Illinois at Chicago, Chicago, IL, USA

K.N. Roy Chengappa Western Psychiatric Institute and Clinic, University of Pittsburgh School of Medicine, Pittsburgh, PA, USA

Rajat Sandhir Department of Biochemistry, Panjab University, Chandigarh, India

Suman K. Samanta Department of Pharmacology and Chemical Biology, University of Pittsburgh Cancer Institute, University of Pittsburgh School of Medicine, Pittsburgh, PA, USA

Neelam S. Sangwan CSIR-Central Institute of Medicinal and Aromatic Plants (CSIR-CIMAP), Lucknow, UP, India

Anuradha Sehrawat Department of Pharmacology and Chemical Biology, University of Pittsburgh Cancer Institute, University of Pittsburgh School of Medicine, Pittsburgh, PA, USA

Pankaj Seth Cellular and Molecular Neuroscience, National Brain Research Centre (NBRC), Gurgaon, Haryana, India

Ashita Sharma Department of Botanical and Environmental Sciences, Guru Nanak Dev University, Amritsar, Punjab, India

Aditi Singh Department of Biotechnology, TERI University, Vasant Kunj, New Delhi, India

Baldev Singh Department of Biotechnolgy, Guru Nanak Dev University, Amritsar, Punjab, India

Pritika Singh Department of Biotechnology, Guru Nanak Dev University, Amritsar, Punjab, India

Shivendra V. Singh Department of Pharmacology and Chemical Biology, University of Pittsburgh Cancer Institute, University of Pittsburgh School of Medicine, Pittsburgh, PA, USA

Taranjeet Kaur Department of Biotechnology, Guru Nanak Dev University, Amritsar, Punjab, India

Varinder Singh Department of Biotechnology, Guru Nanak Dev University, Amritsar, Punjab, India

Deepika Soni School of Biotechnology, Jawaharlal Nehru University, New Delhi, India

Abhilasha Sood Department of Biochemistry, Panjab University, Chandigarh, India

Yashdeep Srivastava CSIR-Central Institute of Medicinal and Aromatic Plants (CSIR-CIMAP), Lucknow, UP, India

Durai Sundar DAILAB, Department of Biochemical Engineering & Biotechnology, DBT-AIST International Laboratory for Advanced Biomedicine (DAILAB), Indian Institute of Technology (IIT) Delhi, Hauz Khas, New Delhi, India

Shyam Sunder Chatterjee Retired Head of Pharmacology Research Laboratories, Karlsruhe, Germany

Vivek Swarup CERVO Brain Research Centre, Université Laval, Québec, Québec, Canada

Kazuya Tamura DAI-DAN Co., Ltd., 390 Kitanagai, Miyoshi-machi, Iruma-gun, Saitama, Japan

Manju Tewari Cellular and Molecular Neuroscience, National Brain Research Centre (NBRC), Gurgaon, Haryana, India

Mahendra K. Thakur Biochemistry and Molecular Biology Laboratory, Brain Research Centre, Department of Zoology, Banaras Hindu University, Varanasi, India

Suman S. Thakur Proteomics and Cell Signaling, Lab E409, Centre for Cellular & Molecular Biology, Habsiguda, Telengana, India

Sandhya Tripathi CSIR-Central Institute of Medicinal and Aromatic Plants (CSIR-CIMAP), Lucknow, UP, India

Teruo Wada Osaka Prefecture University, Osaka, Japan

Renu Wadhwa DBT-AIST International Laboratory for Advanced Biomedicine (DAILAB), Biomedical Research Institute, National Institute of Advanced Industrial Science & Technology (AIST), Tsukuba, Ibaraki, Japan

Yue Yu DBT-AIST International Laboratory for Advanced Biomedicine (DAILAB), Biomedical Research Institute, National Institute of Advanced Industrial Science & Technology (AIST), Tsukuba, Ibaraki, Japan

About the Editors

Dr. Sunil C. Kaul was born and brought up in Kalimpong (Darjeeling). He obtained his MPhil and PhD degrees from the University of Delhi, India. After initial postdoctoral training, he was appointed as a researcher at the National Institute of Advanced Industrial Science and Technology (AIST) in Japan and has been working there for last 28 years. He is now a chief senior research scientist. His major research interest is to understand the molecular mechanism of stress, ageing and cancer. He has been merging the traditional knowledge with modern technologies like gene silencing and imaging to understand the mechanism of action of Ayurvedic herb, Ashwagandha and others. He is a coordinator of the DBT (India)-AIST (Japan) International Laboratory for Advanced Biomedicine at AIST, Japan. With more than 200 research publications in international peer-reviewed journals and several invited talks internationally, he has been in the editorial board of several scientific journals. He is a fellow of the Geriatric Society of India (FGSI), overseas fellow of the Biotech Research Society of India (FBRSI), fellow of the Indian Academy of Neurosciences (FIAN) and foreign fellow of the National Academy of Sciences, India (FNASc).

Sunil C. Kaul, MPhil, PhD, FGSI, FBRSI, FIAN, FNASc
Chief Senior Research Scientist
National Institute of Advanced Industrial Science & Technology (AIST)
Central 5-1, 1-1-1 Higashi
Tsukuba - 305 8565, Japan
E-mail: s-kaul@aist.go.jp

Dr. Renu Wadhwa was born and brought up in Amritsar. She obtained her first PhD from the Guru Nanak Dev University, India and the second from the University of Tsukuba, Japan. She had her post-doctoral trainings at the University of Newcastle Upon Tyne, England; RIKEN, Japan and Children's Medical Research Institute, Australia. She has been working in Japan for last 27 years and leading a research team working on the mechanisms of cell proliferation controls at the Biomedical Research Institute, AIST, Japan. Her major research interest is to understand the molecular mechanism of aging and cancer using normal and cancer cells as model systems. She had originally cloned two proteins; a novel member of hsp70 family protein in 1993 and named it "mortalin", and a novel DNA damage regulatory protein in 2001 and named it CARF. Both proteins have been established as essential regulators of cell proliferation and contribute to the process of carcinogenesis. She has more than 200 publications in International peer reviewed journals with many invited/plenary talks in international conferences. She is a leader of DBT (India)-AIST (Japan) International Laboratory for Advanced Biomedicine (DAILAB) at AIST, Japan and is currently focusing on research on Natural, Efficient and Welfare (NEW) drugs for health supporting and disease therapeutic functions. She has been in Editorial board of several scientific journals including Journal of Gerontology: Biological Sciences and Mechanism of Ageing and Development. She is a Fellow of Geriatric Society of India (FGSI), Indian Academy of Neurosciences (FIAN) and Biotech Research Society, India (FBRSI).

Renu Wadhwa, PhD (India), PhD (Japan)
FGSI, FIAN, FBRSI
Primer Senior Researcher
Biomedical Research Institute Leader, DBT-AIST International Laboratory for Advanced Biomedicine (DAILAB), National Institute of Advanced Industrial Science & Technology (AIST)
Tsukuba - 305 8565, Japan
E-mail: renu-wadhwa@aist.go.jp

Professor, School of Integrative and Global Majors, Tsukuba Life Science Innovations (TLSI), University of Tsukuba, Japan

Part I
Ashwagandha – Phytogenomics and Pharmacology

Chapter 1
Phytochemical Genomics of Ashwagandha

Neelam S. Sangwan, Sandhya Tripathi, Yashdeep Srivastava, Bhawana Mishra, and Neha Pandey

Abstract *Withania somnifera* due to its characteristic secondary metabolite content is of valuable medicinal significance and thus various omics approaches have been recently gaining lots of interest. Present updates had revealed various key enzymes involved in withanolide biosynthetic pathway. Molecular markers (SSRs) resources have also been established and have provided leads to study intra and inter-specific gene diversity. Various *in-vitro* studies have been conducted to improve the withanolide content in this important plant. Further, emerging metabolomic area like phytochemical genomics has also been applied in case of *Withania somnifera* for metabolite detection and its structural validation.

Keywords *Withania somnifera* • Secondary metabolites • Biosynthetic pathway • Phytochemical genomics • Metabolite detection

1.1 Introduction

Plants present a rich source of naturally existing medicines in Indian literature having medicinal importance. These medicinal plants are responsible for producing various phytochemicals which providing immunity to combat against plant disease. Phytochemical study of medicinal plants is indispensable to identify the presence of active constituents in them. Phytochemicals also termed as secondary metabolites comprise of alkaloids, flavonoids, steroidal lactones, saponins, tannins, terpenoids, etc. which exhibit medicinal as well as pharmacological activities such as antibacterial activity, anti-inflammatory effects. Due to emerging safety records of herbal medicines in comparison to chemical drugs, medicinal plants-based drugs have

N.S. Sangwan (✉) • S. Tripathi • Y. Srivastava • B. Mishra
CSIR-Central Institute of Medicinal and Aromatic Plants (CSIR-CIMAP), Lucknow 226015, UP, India
e-mail: sangwan.neelam@gmail.com

N. Pandey
Department of Metabolic and Structural Biology, CSIR-Central Institute of Medicinal and Aromatic Plants, PO CIMAP, Lucknow 226015, India

S.C. Kaul, R. Wadhwa (eds.), *Science of Ashwagandha: Preventive and Therapeutic Potentials*, DOI 10.1007/978-3-319-59192-6_1

regained popularity for treating even the chronic human illness. The oldest written evidence of herbal medicine usage has been found on a Sumerian clay slab (approximately 5000 years old) from Nagpur, India, comprising 12 herbal preparations by using around 250 different plants (Kelly et al. 2009; Petrovska 2012). Due to Recent discoveries stating modern therapeutic usages of bioactive compounds from medicinal plants, researchers are giving immense attention to the medicinal plant biology. Out of around 18,000 flowering plants on our planet earth, 44% are of medicinal importance and are the rich source of phytochemicals (Handa et al. 2006).

1.1.1 About **Withania somnifera**

Withania somnifera is one of the most reputed Indian medicinal plant in Indian system of medicine known for its pharmaceutical and nutraceutical properties also considered as Indian ginseng and in Sanskrit it is called (Ashwagandha). *Withania somnifera*, commonly known as 'Ashwagandha, Indian Ginseng or Indian winter cherry' is a wonder herb that belongs to the family Solanaceae. In ancient system of medicines such as Ayurveda, Unani and Siddha, the root of this plant was an important ingredient of over 300 formulations for treating several physiological disorders (Tavhare 2015). *Withania somnifera* is an erect, evergreen branched, tomentose shrub. Leaves are simple, ovate and glabrous. Flowers are greenish or pale yellow born in axillary, umbellate cymes. Fruits are globose, berries, orange red when mature enclosed in membranous persistent calyx. Seeds are yellow, and reniform. The plant has wide geographic distribution and is mainly adapted to the warmer/drier parts of the tropical and subtropical regions of the world ranging from Mediterranean regions, tropical Africa, South Africa, warmer parts of Europe to Southwest Asia. The plant is native to Tropical Asia especially India, Pakistan, and Srilanka (Van Wyk and Wink 2004).

1.1.2 Pharmacological Importance of **Withania somnifera**

Withania have always been a center of attraction of plant scientific research community due to its various pharmacological activities such as anti-microbial, anti-inflammatory, anti-arthritic, anti-diabetic, anti-stress, cardio-protective, neuroprotective, anti-leishmaniasis, anti-ischemic and anti-hypoxic (Fig. 1.1) (Tripathi and Verma 2014; Dhar et al. 2015). Additionally, it has emerged as a potent anticancer agent against a wide range of cancers such as breast, cervical, lung, colon, prostrate and medullary thyroid cancer and it acts by utilizing multiple modes of action, for example by inducing cell cycle arrest, apoptosis, autophagy and suppressing various oncogenic pathways (Palliyaguru et al. 2016). The plant posses various pharmacologically important constituents in its roots, leaves, fruits and stem.

Various extracts of leaf prominently posses anti-microbial, anti-cancer, anti-diabetic, and anti-proliferative activity. For example, the leaves of Withania somnifera have been shown to possess anti-microbial activity against both gram-positive

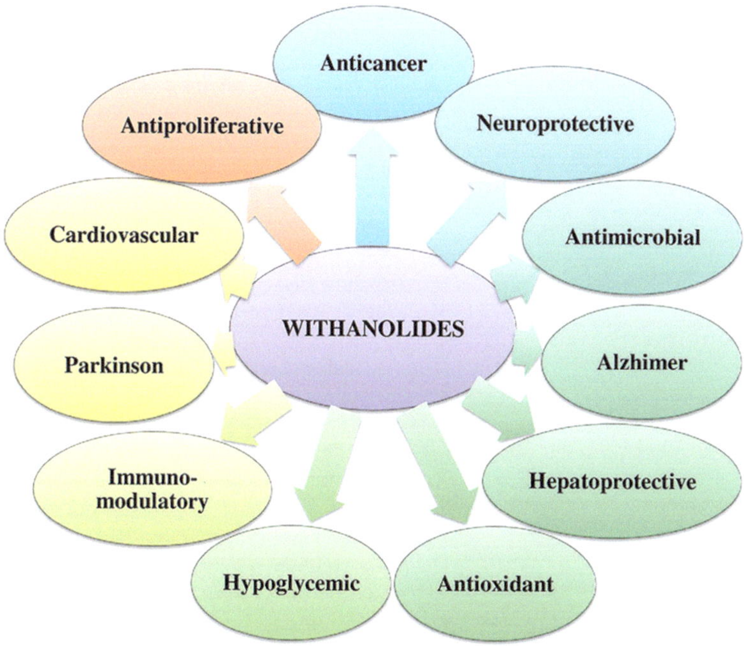

Fig. 1.1 Major pharmacological properties of withanolides from *Withania somnifera*

(methicillin-resistant *Staphylococcus aureus* and *Enterococcus spp*) (Bisht and Rawat 2014) and gram-negative bacteria (*Salmonella typhi*, *Escherichia coli*, *Pseudomonas aeruginosa*, *Proteus mirabilis*, *Klebsiella pneumonia* and *Citrobacter freundii*) (Singh et al. 2006; Alam et al. 2012). In addition, the leaf of *Withania somnifera* are also known to be effective against oral bacterial infection caused by *Streptococcus mutans*, and *Streptococcus sobrinus* (Pandit et al. 2013). The root extracts is found to have anti-cancer, anti-tumor, neuroprotective, immunomodulatory, and anti-stress activity. It is also found to be effective against Alzheimer's and Parkinson's disease. The roots of *Withania somnifera* have been used for centuries as a general tonic for vitality and longevity. The roots of *Withania somnifera* have shown to inhibit several markers of inflammation such as cytokines (Interleukin-6, Tumor necrosis factor-α), reactive oxygen species (ROS) and Nitric oxide (NO). Additionally extract from stem contains anticancer activity while berry extract is effective against Alzheimer's` disease (Table 1.1).

1.1.3 Phytochemical Constituents

As plants are sessile organisms, they possess a rich repository of metabolites around 200,000 to 1 million. Plants produce vast array of phytochemicals, among these millions of metabolites expected in plants three major groups comprises: terpenoids

Table 1.1 Pharmacological actions of bioactive constituents of *Withania somnifera*

Plant part	Extract type	Action	References
Leaf	Methanolic	Antimicrobial	
		Anticancer [against cancer cells (TIG1, U2OS, and HT1080) by activating p53, apoptosis pathway & cell cycle arrest]	Widodo et al. (2008)
		Anti-diabetic [Alloxan-induced diabetes mellitus in rats]	
	Methanolic	Hypoglycemic [Increased uptake of glucose in myotubes and adipocytes]	Gorelick et al. (2015)
		Neuroprotection through activation of neuronal proteins, oxidative stress and DNA damage	Konar et al. (2011)
		Anticancer [Neuroblastoma]	Kuboyama et al. (2014)
		Antimicrobial (against methicillin resistant *Stephylococcus aureus* and *Enterococcus sps*.)	Bisht and Rawat (2014)
	Methanolic	Antimicrobial [oral infections by *Streptococcus mutans* and *S. sobrinus*]	
	Methanolic	Antiproliferative [against MCF-7, HCT116 and HepH2 cell lines]	Alfaifi et al. (2016)
	Methanolic	Antimicrobial activity	Dhiman et al. (2016)
	Methanolic	Antibacterial [against *Salmonella typhi*] and Antioxidant	Alam et al. (2012)
	Methanolic	Cytotoxic activity against Breast cancer (MCF7), Hepatocellular carcinoma (HEPG2) & Cervix cancer (HeLa cells)	
	Ethanolic exract	Antimicrobial activity	Dhiman et al. (2016)
	Hydroalcoholic	Anti-cancer [Breast Cancer]	Nema et al. (2013)
	$CHCl_3$		Siriwardane et al. (2013)
	Aqueous	Anti-cancer	Wadhwa et al. (2013)

(continued)

Table 1.1 (continued)

Plant part	Extract type	Action	References
Root	Chloroform	Anti-cancer (Liver, breast, colon)	Siddique et al. (2014)
		Prostate cancer	
		Against chemically induced fatigue	Biswal et al. (2013)
	Alcoholic	Anti-tumor	Khazal et al. (2013)
	Ethanolic	Anti-cancer [Breast cancer]	Maliyakkal et al. (2013)
	Ethanolic of N-118	Neuroprotectant (Against cerebral stroke)	Ahmad et al. (2015)
	Ethanolic of 101R, 118R and 128R	Immunomodulatory	Kushwaha et al. (2012a), Kushwaha et al. (2012b),
	Ethanolic	Anti-cancer [Cervical]	Jha et al. (2014)
		Anti-diabetic [Stabilized blood glucose levels]	
		Anti-diabetic [Non-insulin-dependent Diabetes mellitus in rats]	
		Anti-diabetic [Alloxan-induced Diabetes mellitus in rats, normalized the urine sugar, blood glucose, glucose-6-phosphatase and tissue glycogen levels)	
	Aqueous	Antistress [reduced in T-cell population and up-regulated Th1 Cytokines in chronically stressed mouse)	Khan et al. (2006)
		Anti-stress	Jain and Saxena (2009)
		Alzheimer's disease	Sehgal et al. (2012)
		Neuroprotective (Enhanced memory and attenuated hippocampal neurodegeneration through inducing glutathione biosynthesis)	Baitharu et al. (2013), (2014)
	Methanolic	Hepatoprotective role in acetaminophen-intoxicated rats	Devkar et al. (2016)
	Methanolic	Parkinson's disease	De Rose et al. (2016)
	Methanolic	Neuroprotective effects	Bhattarai and Han (2014)
			Bhattarai et al. (2013)

(continued)

Table 1.1 (continued)

Plant part	Extract type	Action	References
Stem	Methanolic	Cytotoxic activity against Cancer cell lines MDA-MB-231 [Human breast cancer cell lines]	Srivastava et al. (2015)
	Ethanolic	Cytotoxic activity against Cancer cell lines MDA-MB-231 [Human breast cancer cell lines]	Srivastava et al. (2015)
Fruit/berry		Alzheimer's disease	Jayaprakasam et al. (2010)

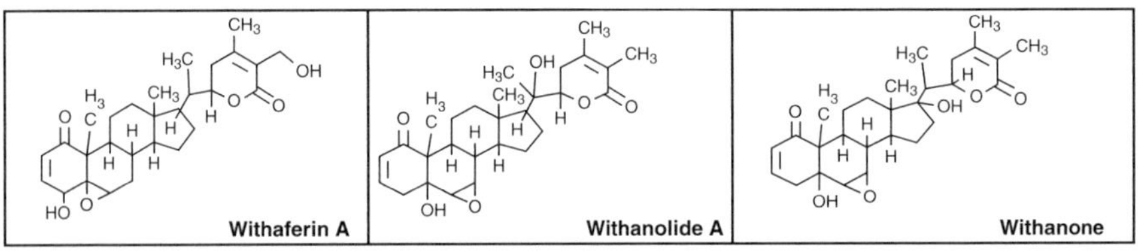

Fig. 1.2 Chemical structure of various withanolides

>40,000 different molecules, alkaloids >12,000 alkaloids and phenolic compounds > 8000 which are present in mainly all forms of life (Kabera et al. 2014; Rober 2007; Ziegler and Facchini 2008). *Withania somnifera* is tremendously rich reservoir of a wide range of secondary metabolites. The population of biologically active compounds includes around 40 withanolides (steroidal lactones). Among various withanolides, withanolide A, withaferin A, withanone and withanolide D are the most abundant having various activities (Fig. 1.2, Table 1.2). Withanolides are C-28 steroidal lactone triterpenoids having ergostane structure backbone. In this structure, six-membered lactone ring was formed after the oxidation of C-22 and C-26 atoms. These structures were nomenclatured as 22-hydroxy ergostane-26-oic acid 26, 22-lactones. Six or five membered lactone or lactol ring is the characteristic feature of withanolides at C-8 or C-9 side chain. 27-hydroxy withanolide A, iso-withanone and 6,7-epoxy-1,3,5-trihydroxy-witha-24-enolide were also found in fruits of Withania (Chatteriee et al. 2010; Dhar et al. 2015; Lal et al. 2006). More than 12 alkaloids, several steroids, ashwagandhine, cuscohygrine, anahygrine, tropine various alkaloids, salts, flavonoids, nitrogen containing compounds, steroids and steroidal lactones (withanolides) are also found to be present. Other abundant phytochemicals includes withanamides (A, B, C, D, E), withanosides, withanolide glycosides, steroidal saponins, and lignanamides, flavonoids, coagulins (C, H, L), tropane alkaloids (calistegins, pseudotropine and tropine) and nitrogen-containing compounds distributed among different plant parts (Mirjalili et al. 2009; Chatterjee et al. 2010; Singh et al. 2015; Jadaun et al. 2016; Kushwaha et al. 2013a and b). The earthy odour and flavour of Ashwagandha is attributed to the presence of steroidal

Table 1.2 Active withanolides with therapeutic potential

S.N.	Activities	Bioactive withanolides	References
1	Anti cancer	Withaferin A, Withanolide D	Mandal et al. (2008); Subramanian et al. (2014); Mondal et al. (2010, 2012a, b)
		Withalongolide A	
		Withalongolide A-4,19,27-triacetate	
		Withalongolide B-4,19-diacetate	
2	Hypoglycemic	4 β -hydroxywithanolide E	Takimoto et al. (2014)
		Withaferin A	Gorelick et al. (2015)
3	Anti microbial	17β-hydroxywithanolide K	Choudhary et al. (2010)
		Withaferin A	Subramanian and Sethi (1969)
		14,15β-epoxywithanolide I	Choudhary et al. (2010)
		Withanolide F	
		Withanolide D	
4	Anti inflammatory	Withaferin A	Noh et al. (2016)
		3β-hydroxy-2,3-dihydrowithanolide F	Budhiraja et al. (1984)
	Antioxidant	WIthaferin A	Bhattacharya et al. (1997)
	Neuroprotection	Withanolide A	Baitharu et al. (2014)
5	Immunosuppressive	Withaferin A	Shohat et al. (1978)
		Withanolide E	Shohat et al. (1978)
		5,20α(R)-dihydroxy-6α,7α-epoxy-1-oxo-(5α)-with a-2,24dienolide	Bähr and Hänsel (1982)
6	CNS related	3β-hydroxy-2,3-dihydrowithanolide F	Budhiraja et al. (1984)
7	Alzheimer's disease	Withanolide A	Sehgal et al. (2012)
8	Cardiovascular	3β-hydroxy-2,3-dihydrowithanolide F	Budhiraja et al. (1984)
		Withaferin A	Ravindran al. (2015)
9	Hepatoprotective	3β-hydroxy-2,3-dihydrowithanolide F	Budhiraja et al. (1986)

lactones or withanolides. Maximum amount of withaferin A is present in shoot tips, followed by leaves, nodes, internodes, roots, flowers and seeds (Praveen and Murthy 2010). Our group for the first time succeeded in showing the differential biosynthesis of withanolides during different morphogenetic stages of the plant *in-vitro* (Sabir et al. 2012). The previous beliefs that the withanolides are synthesized in leaves and transported to roots got faded after we had demonstrated the de-novo synthesis of withanolide A in the roots (Sangwan et al. 2008). The unique composition of different withanolides along with other secondary metabolites is the key determinant of varied medicinal significance of different plant parts. We have reported an efficient method for isolation of withaferin A (Sangwan and Sangwan 2013). These withanolides have been reported to be present in various plant parts and different developmental stages, however, their concentration/abundance varies greatly.

For maximum economic benefit it is important to identify a specific genotype of plant that synthesize bioactive metabolites in more quantity with much better quality. Also, there is considerable variations in both qualitative and quantitative patterns of secondary phytochemicals in plants belonging to different agro-climatic regions. This becomes highly pertinent to study as this factor will impact the efficacy of the pharmaceutical preparations prepared from these plants. Mir et al. (2014b) found variations in the level of withanolides viz. withanolide A, withanone and withaferin investigated from *Withania somnifera* samples collected from different regions of India. These regions included Neemuch, Mandsaur, Manasa and Hyderabad. Withanone was detected only in plants of Manasa population (1.312% in the leaves) and 0.083% withanolide A was present in the roots. Highest content of withanolides was found in mature roots and leaves, showing a correlation between the developmental stages of the plant and withanolide accumulation (Mir et al. 2014b).

1.1.3.1 Phytochemical Variations with Different Accessions

Fascinating researches were actively pursued by different research laboratories around the world to understand the phytochemical variability present in different parts as well as varieties of *W. somnifera* plant. Pharmacological properties of *Withania somnifera* are mainly because of its secondary metabolite content that is present in the parts like leaf, stem, root, seed or some times in the whole plant.

Withania somnifera produces specific bunch of withanolides in different parts as well as in different variety with independent biogenesis capability (Sangwan et al. 2007; Chaurasiya et al. 2007; Misra et al. 2005; Chatterjee et al. 2010; Chaurasiya et al. 2008, 2009). There exist variations in the metabolome of plants under different conditions. These include nutritional stress, physical stress mutation or environmental changes. Further phytochemical variations in different accessions constitute a drawback for the quality control in herbal formulations as they lead to inconsistency in pharmacological properties. There is a need to identify these phytochemical variations among different accessions of the same plant. Investigations showed that there is a correlation between phytochemical data of different chemotypes of *Withania somnifera* with genetic factors (AFLP markers, RAPD, isozymes) for the presence of secondary metabolites (Dhar et al. 2006; Sangwan et al. 2014). Withanolide biogenesis is linked with the morphogenic transitions in the plant. The level of total withanolide content was higher in field grown roots (1.9 mg/g DW) in comparison to *in-vitro* developed roots (0.31 mg/g DW). Withanolide A content was two times higher with equal amount of withanone and absence of withaferin A in 15 days old green callus as compared to the mother explants, however compared to the non green callus derived from the multiple shoots which were found to contain high propotions of withanolides in best optimized media conditions (Sabir et al. 2013).

1.1.4 Omics and In Silico Approaches to Study Plant Metabolism

Newly emerging fields of omics and bioinformatics are nowadays mandate to realize the significance of molecular mechanisms underlying different plant activities. Current game-changing next generation sequencing (NGS) approaches have revolutionized the sequencing protocols in genomics thus providing a platform to novel analytical applications. Influenced by these technological advancements new omics areas are emerging for example interactome, epigenome and hormonome etc. These in turn address several biological properties of different plant species model or non-model. Combination of information from these omics approaches is a rising area of research to investigate significant biological issues and promotion of translational research. Technological advancements provide a backbone for plant scientists to understand the metabolic pathways involved in the production of various secondary metabolites. These mainly includes ultra-high-throughput DNA sequencing, or 'next-generation DNA sequencing', which made the availability of DNA sequence not only limited to model plants like *Arabidopsis* and rice but also for crop plants as well as medicinal plants for which no genome sequencing data is available. The datasets available from genome or transcriptomic sequencing can provide us a relationship between the expressions of a gene under particular conditions in relation to its metabolite contents.

The present chapter thus describes in detail the various omics approaches (metabolomics, transcriptomics and genomics) to decipher the withanolide identification, its significance, biosynthesis and its multilevel regulation.

1.2 Withanolide Biosynthesis Under Different Growth Conditions

The diverse chemical nature of the plant, which is helpful in such a wide array of health conditions, has puzzled plant biologists and chemists equally. The National Medicinal Plant Board of India has listed *Withania somnifera* among the 32 most demanding medicinal plant around the world that drives its huge demand (Prajapati et al. 2007). It has been estimated that the yearly global demand (around 12,120 tonnes) (Ramawat and Goyal 2008) of withanolides exceeds the annual supply (5905 tonnes) (Sharada et al. 2007). This leads to need for enhanced production of these phytochemicals in *Withania somnifera*. Chemical synthesis of these compounds has also been reported but due to complex structure of withanolides (Neumann et al. 2008) the chemical synthesis is uneconomical. Many approaches have been screened for their potential for increased production of withanolides. Among them *in-vitro* approaches gained much attention. In addition, various stress signals have also been reported to enhance the withanolide biosynthesis in *Withania somnifera*.

1.2.1 Regulation of Withanolide Biosynthesis Under In-vitro Conditions

In-vitro culture technique for secondary metabolite production has gained much attention during last two decades. Further, alterations in culture's growth conditions have been variously used for improving the yield of secondary metabolites. Successful callus cultures of *Withania somnifera* were established long back in 1979 (Vishoni et al. 1979) by inoculating anthers as explants onto the MS medium supplemented with 10^{-6} M 6-benzyladenine (BAP). *In-vitro* cultures of *Withania somnifera* have been extensively and variably used for withanolides production. Roja et al. (1991) have established the protocol for the *in-vitro* shoot generation from axillary meristems of *Withania somnifera* using benzyladenine (BA). Various combinations of BA, 2,4-dichlorophenoxyacetic acid (2,4-D) and indolebutyric acid (IBA) were used for optimizing the generation of rooted shoot cultures of *Withania somnifera* (Sen and Sharma 1991). Various studies have been conducted for the optimization of *in-vitro* propagation of *Withania somnifera*. Our group has developed a protocol for producing large numbers of plant *via* sterile lines having stable production of withanolides. For this study six accession of *Withania somnifera* (RSS-8, RSS118, RSS-27, RSS 28, RSS41 and RSS44) were taken. Nodal segment of these accession were used for the generation of sterile lines with various concentration of BAP and kinetin in combination or alone. After shoot induction, IBA and IAA were used in combination or alone for the rooting of shootlets, which were further successfully transplanted in field under various optimized conditions such as fertilizer dose etc. (Sabir et al. 2008a, b, Tables 1.3 and 1.4).

Manipulation in the growth medium provides an easier way for varied degree of improved withanolide production. MS medium supplemented with 5.0 μM BAP and 1.0 μM Kinetin has been shown to be appropriate for withaferin A production. Wherein, withaferin A production in the liquid cultures was higher as compared to the static composition of the same media (Mir et al. 2014a). Surprisingly, withanolide-A, which was hardly detectable in the aerial plant parts of field-grown *Withania somnifera* was found to be present in the *in-vitro* shoot cultures of the plant (Sangwan et al. 2007). In another study, changes in composition of macro elements (addition of KNO_3 and removal of $NH4^+$) in the medium of adventitious root cultures resulted in increased withanolide-A production (Murthy and Praveen 2012). In addition, when adventitious root cultures were given methyl jasmonate and salicylic acid treatments, withanolides contents were increased. However, comparatively salicylic acid show better performance and has been shown to boost accumulation of withanone, withaferin A, withanolide A and withanolide B along with minor constituents likely 12-deoxy withastramonolide, withanoside V and withanoside IV (Sivanandhan et al. 2012a, b).

Agrobacterium rhizogenes-mediated hairy root production is also an efficient tool for withanolide biosynthesis. Our group has developed the protocol for the efficient *Agrobacterium*-mediated genetic transformation using leaf and node of *Withania somnifera*. By this method, we can modify the biosynthesis of secondary metabolite content by altering the enzymes activity involved in their biosynthesis pathway (Mishra et al. 2013; Mishra et al. 2016a, b).

Table 1.3 List of the developed methods for the *in-vitro* propagation of *Withania somnifera*

Objective of the study	Hormone used for propagation	References (column shifted to extreme right)
In-vitro propagation by organogenesis	Various concentration of Bap, NAA and IAA and TDZ in combination or alone.	Kulrkarni et al. (1996)
Callus cultures were initiated from axillary leaves, axillary shoots, hypocotyls, and root segments	Fir axiilary leaves different concentrations of 2,4-D and KN,either alone or in combination for axillary shoots different concentrations of BA and KN (alone or in combination) for hypocotyl and root segment 2,4-D and KN (alone or in combination) for plant regenration combinations of 2,4-D, KN, BA, and 2-iP [N6-(2-isopentenyl) adenosine]	Rani and Grover (1999)
Rapid micropropagation of selected chemotypes using nodes, internodes, hypocotyls and embryo explants	Various concentration of BA and TDZ in combination or alone; Fatima et al. 2016.	Kulkarni et al. (2000)
Plant regeneration from encapsulated shoot tips	Various concentration of IBA	Singh et al. (2006)
Large scale propagation using seed as an explant	Various concentration of BAP, IAA and TDZ in combination or alone.	Supe et al. (2006)
Direct rooting from leaf explants	Various concentration of IBA, NAA and IAA and TDZ in combination or alone.	Wadegaonkar et al. (2006)
In-vitro mass propagation	Various combination of kinetin	Shukla et al. (2010)
In-vitro plant regeneration from cotyledonary nodes	Various concentration of BAP, IBA and KN in combination or alone.	Nayak et al. (2013)

Recently, modified *Agrobacterium rhizogenes*-mediated genetic transformation of *Withania somnifera* has been developed with leaf tissue explants of *in-vitro* raised plantlets involving sonication and heat treatments. These techniques may be useful for large-scale production of bioactive phytochemicals by metabolic engineering approaches of *Withania somnifera* (Chandrasekaran et al. 2015).

Withanolides can be produced efficiently by *in-vitro* cultures of *Withania somnifera*. *In-vitro* callus, root, multiple shoots and cell suspension cultures are used as a source for bioproduction of withanolides *in-vitro*. Genetic transformation in this plant by different *Agrobacterium rhizogenes* strains leads to an increase in secondary metabolites like withaferin A which was produced up to 2.25 and 1.77 times more over control hairy roots, also the metabolites was found to be released into the culture media (Varghese et al. 2014). Various types of elicitors have effect on secondary metabolites production in *Withania somnifera*. When *Withania somnifera* was transformed by fungal elicitor protein β-cryptogein, it was observed that the hairy root cultures possess more withaferin A and withanolide A content with an increase in wall-bound ferulic acid (Sil et al. 2015). When elicitors like cadmium chloride, aluminium chloride and chitosan were inspected in different concentrations with the precursor feeding (cholesterol, mevalonic acid and squalene), the

Table 1.4 *In-vitro* studies to improve the withanolide production in *Withania somnifera*

Used material	Method	Results	References (column shifted to extreme right)
Callus culture	Transformation with opaline and octopine strains of agrobacterium tumefaciens.	Increase withanolide D and withaferin	Ray and Jha (2001)
In-vitro culture	At different developmental stage	Differential pattern of withanolide content	Sharada et al. (2007)
Callus culture	Agrobacterium transfomation/and biotic and abitoic elicitaion strategy	Enhancement in Withaferin accumulation	Baldi et al. (2008)
Root line, callus line, rooty callus line	Transformation by Agrobacterium rhizogenes,	Presence of withasteroids in undifferentiated callus cultures	Bandyopadhyay et al. (2007)
Callus culture	Effect of different growth regulators (auxins, combination of auxin and cytokinin)	2,4-D in combination with kinetin induce the withanolide production kinetin was suitable for withanolide A production	Nagella and Murthy (2010)
Root cultures	Induce adventitious root culture	Induce withanolide production	Praveen and Murthy (2010)
Leaves, stems, and roots	Addition of polyamine, spermidine (20 mg L^{-1})	Increase in the quantities of withanolide A, withanolide B,	Sivanandhan et al. (2011)
Root cultures	Aluminium chloride at 10 mg L^{-1}, chitosan at 100 mg L^{-1}	Enhance withanolide content	Sivanandhan et al. (2012a, b)
Hairy root culture	Effects of carbon source [sucrose, glucose, fructose, maltose) and ph change pH (4.0, 4.5, 5.0, 5.5, 5.8, 6.0 and 6.5) on	4% sucrose favoured the production of withanolide A withanolide A production was highest in the medium pH set at 6.0 (13.84 mg g^{-1} DW).	Praveen and Murthy (2012)

study reveals that the presence of total withanolides (withanolide A withanolide B withanone, withaferin A, 12 deoxy withanstramonolide, withanoside IV and withanoside V) were highest in the combined treatment of chitosan and squalene when applied along with picloram, kinetin, L-glutamine and 5% sucrose in culture at an interval of 4 h and 48 h respectively on 28th day of culture in bioreactor. Higher concentrations of total withanolides (2.13-fold) were observed in shake-flask culture as well as in bioreactor (1.66-fold) in comparison to control treatments (Sivanandhan et al. 2014).

In comparison to methyl jasmonate salicylic acid was proved to be the best elicitor to enhance the production of withanolides, when examined on adventitious root culture of *Withania somnifera*. Salicylic acid improved the level of major withanolides namely withanone, withaferin A, withanolide A and withanolide B along with minor constituents likely 12-deoxy withastramonolide, withanoside V and withanoside IV. Treatment of root biomass on 30-day-old adventitious root cultures with SA in 150 μM concentration for 4 h exposure period resulted in the production of high dry weight of withanolide A with an increase of 48-fold, withanolide B with an increase of 29-fold and withaferin A which shows 20-fold increase, withanone with 37 fold increment, 12-deoxy withastramonolide with nine fold increase, withanoside V with seven fold increase, and withanoside IV which was a nine fold increment after a elicitation of 10 days (40th day of culture) in comparison to untreated culture was observed (Sivanandhan et al. 2012a, b).

Withaferin A production was obtained on MS medium supplemented with a concentration of 5.0 μM of BAP and 1.0 μM of kinetin, In this study Withaferin A production in the liquid cultures was higher as compared to the static composition of the same media (Mir et al. 2014a). Also withanolide-A, which was hardly detectable in the aerial plant parts of field-grown *Withania somnifera* was found to be present in the *in-vitro* shoot cultures of the plant (Sangwan et al. 2007).

Genetic transformation of *Withania somnifera* using two strains of *A. rhizogenes* (strains LBA 9402 and A4) shows accumulation of withaferin A in the transformed root line to a maximum (0.44%dry weight) while callus lines produced both withaferin A ranging from 0.15 to 0.21% in dry weight and withanolide D (ranging from 0.08 to 0.11% in DW (Bandyopadhyay et al. 2007).

Inter-conversion of withanolides takes place when some withanolides *viz.* withanolide A, withaferin A, and withanone were added as precursor substrates, these get converted to some unknown compounds, and further released to the culture media, withanolide A is converted to withanone, and *vice versa* though at a lower level (Sabir et al. 2011).

Plant cell culture has been used successfully for the production of various useful secondary metabolites, which are useful in pharmaceutical applications, pigments and other fine chemicals. Optimization studies were carried out in *Withania somnifera* cell suspension cultures to enhance the withanolides production. The effects of macro elements for the optimization of the adventitious root suspension culture of ashwagandha plant were studied (Murthy and Praveen 2012).

For optimization of different factors concentrations of auxins, cytokinins and their combinations, agitation speed, carbon sources, organic additives and seaweeds were altered in cell suspension culture. The study reveals that production of withanolides namely withanolide A, withanolide B, withaferin and withanone in higher quantities (in DW) were recorded after treatment of *Gracilaria edulis* at 40%. Organic additives like L-glutamine when used in combination with picloram and kinetin, promoted growth characteristics up to 11.87 g in formula weight and 2.96 g in dry weight with an increase in withanolides biosynthesis, likely, withanolide A to 5.04 mg/g DW, withanolide B to 2.59 mg/g DW, withaferin A to 2.36 mg/g DW and withanone up to 4.32 mg/g DW). Sucrose at 5% level revolved out to be a superior carbon source and yielded the highest withanolides production likely withanolide A to 2.88 mg/g DW, withanolide B to 1.48 mg/g DW, withaferin A to 1.35 mg/g DW and withanone to 2.47 mg/g DW) (Sivanandhan et al. 2013).

The detected level of Withanolide A in transformed roots was 2.7-fold more in comparison to non-transformed cultured roots (Murthy et al. 2008). They also used adventitious roots from generated from leaf explants of Withania to assess the level of production of withanolide-A (Praveen and Murthy 2010). With increase in age of plant the concentration of withaferin A was increased and withanolide A decreased in the leaves of *Withania somnifera.* It was observed that carbohydrate and elicitor treatment in hairy root cultures of *Withania somnifera* shows that there was increase in withaferin A with increasing glucose levels peaked at 5% and withanolide A was detected only at 5% glucose concentration while both withaferin A and withanolide A accumulated at 3% sucrose levels. Treatment of chitosan and nitric acid increased the levels of withaferin A while jasmonic acid decreased it (Doma et al. 2012). In addition, aeroponics studies have shown production of unusual withanolides from W. somnifera (Xu et al. 2011). These new molecules were derivatives of withaferin and thus aeroponics offer induction of novel molecules not present in plants.

1.2.2 Regulation of Withanolide Biosynthesis Through Various Stress-Signals

Secondary metabolites have been variously shown to be regulated by different biotic and abiotic stress-signals due to their protective effects under adverse environmental conditions. Withanolides, being an important member of steroidal lactones, have also been shown to be regulated by many abiotic and biotic stresses. Drought stress has been reported to alter withanolide content in *Withania somnifera.* While observing the trend of various withanolides accumulation under drought stress, it was found that withaferin-A content was increased by 42.7% with concomitant decrease of withanolide-A and 12-withastromonolides by 78 and 71% respectively as compared to control (Sanchita et al. 2015). Similarly, cold stress triggered differential accumulation of different withanolides in *Withania somnifera.* While a significant elevated level of leaf withaferin-A was found throughout the cold stress treatment (from day 1 to day 7), withanolide-A and withanone contents were decreased (Mir et al. 2015). Effect of light was also studied for withanolide biosynthesis in *Withania somnifera* and the results suggested the inducible effect of light on withanolide accumulation when 75% shade conditions were applied to 5 months old *Withania* plant for 30 days (Jacob et al. 2014). When *Withania somnifera* plants were subjected to supplemental ultraviolet-B radiations (UV-B) exposure, withaferin-A content was induced while reduction in withanolide-A was observed. Thus, s-UVB stress can be a potential tool for withanolide production (Kalidhasan et al. 2013; Takshak and Agarwal 2014). In addition to the abiotic stresses, effects of biotic factors on withanolides have also been reported, though less explored. In a report addressing the impact of leaf spot disease on *Withania somnifera*, significantly higher accumulation of withanolide contents by 15.4% (withaferin-A) and 76.3% (withanolide-A) and total alkaloids by 49.3% was observed (Pati et al. 2008). Secondary metabolites such alkaloids accumulation were also enhanced against the microbial consortium which also induced the plant growth activity (Rajasekar and Elango 2011).

Based on various reports regarding impact of stress-signals on withanolides, it was observed that most of the abiotic and biotic stress-signals caused differential accumulation of withanolides. Under most of the abiotic stresses, only withaferin-A was significantly induced however other withanolides were more or less decreased. This suggests probable role of withaferin-A in conferring abiotic stress tolerance in *Withania somnifera*. However this needs further validation through various experiments that may provide a valid mechanistic evidence for the hypothesis.

1.3 Omics of the Withanolide Biosynthetic Pathway Genes in *Withania somnifera*

Due to enormous medicinal applications of the plant it has been chosen for sequencing project namely Sol Genomics Network (http://solgenomics.net/organism/sol100/view) in 100 Solanaceae genomes, which have been proposed to be sequenced. Consecutively various omics studies have opened up the gateway for some more studies to be conducted through *in silico* studies with risen interest of plant biologists in NGS generated data from the plant for different sets of experiments under different situations.

1.3.1 Genomics of the Withanolide Biosynthetic Pathway Genes in Withania somnifera

The major goal of genomics is to reveal the structure, function, and evolution of genomes of an organism. Although the whole genome sequencing of *Withania somnifera* has not been carried out yet but the characterization of some major pathway genes related to secondary metabolites, which helps us to understand the function and effect on secondary metabolite production was done. Phytochemical genomics provides a platform for investigating biosynthesis, function and regulation of plant secondary metabolites. This provides ample information for altering and tweaking with the biosynthetic pathway for improved secondary metabolite production. The withanolide biosynthetic pathway has been almost completely elucidated, however, their regulation at molecular level has still not been completely explored. The withanolide biosynthetic pathway genes that have been manipulated (Table 1.5). Besides these enzymes, others enzymes such as several glycosyltransferases 27β-hydroxy glucosyltransferase, 3β-hydroxy specific sterol glucosyltransferase, tryptophan decarboxylase, tropine reductase were also stuided (Madina et al. 2007; Sharma et al. 2007; Kushwaha et al. 2013c; Jadaun et al. 2016) Withanolides are synthesized in genus Withania in most prominent and diversified amounts. In view of the pharmacological importance of this plant, *Withania somnifera* has been extensively studied. Previous studies revealed the chemotypic variation, tissue specific abundance of withanolide in *Withania somnifera*. The withanolide content was also affected by various factors such as stress, growth rate, environments and geographical conditions (Srivastava et al. 2015; Mishra et al. 2014, 2016b; Sabir et al. 2012; Dhar et al. 2013; Chaurasiya et al. 2007; Sangwan et al. 2007, 2008). These withanolides are synthesized from triterpenoid pathway intermediates, which imply the

Table 1.5 List of withanolide biosynthetic pathway gene manipulation studies

Gene	Manipulation strategy	Remarks	References
WsHMGR	Expressed in *E. coli*	Elevated expression in response to salicylic acid, methyl jasmonate and mechanical injury	Akhtar et al. (2013)
WsSGTL1	Transformed in *Arabidopsis thaliana*	Elevated level of *WsSGTL1*	Mishra et al. (2013)
		Conferred abiotic stress tolerance	
	RNAi-mediated gene silencing	Downregulation of SGTs	Singh et al. (2016)
		Increase in withanolide A, Withaferin A	
		Increased expression of *WsHMGR, WsDXR, WsFPPS, WsCYP710A1, WsSTE1* & *WsDWF5* genes	
		Reduced biotic tolerance	
	Ectopic overexpression	Increased glycowithanolide	Saema et al. (2016)
		Provided tolerance to biotic and abiotic stress	
WsDXS		Differentially expressed under salicylic acid, methyl jasmonate and mechanical	Gupta et al. (2013)
WsDXR		Differentially expressed under salicylic acid, methyl jasmonate and mechanical	Gupta et al. (2013)
WsCYP98A	Expressed in *E. coli*	Elevated expression under methyl jasmonate and salicylic acid while reduced expression under gibberellic acid treatment.	Rana et al. (2014)
WsCYP76A	Expressed in *E. coli*	Elevated expression under methyl jasmonate and salicylic acid while reduced expression under gibberellic acid treatment.	Rana et al. (2014)
WSCYP93Id	Expressed in *E. coli*	Elevated expression under methyl jasmonate and salicylic acid	Srivastava et al. (2015)
WsCPR1& *WsCPR2*	Expressed in *E. coli*	Expression of *WsCPR1* remained unchanged while *WsCPR2* expression induced under Salicylic acid and methyl jasmonate treatment	Rana et al. (2013)
WsSQS	Expressed in *E. coli*	Positively regulate withanolide content	Bhat et al. (2012)

(continued)

Table 1.5 (continued)

Gene	Manipulation strategy	Remarks	References
WsSQE	Expressed in *E. coli*	Positively regulate withanolide content	Razdan et al. (2013)
OSCs (WsOSC/ LS,	Expressed in *Schizosaccharomyces pombe*	Differentially expressed under methyl jasmonate, salicylic acid and yeast treatments	Dhar et al. (2014)
WsOSC/BS,			
WsOSC/CS)			
WsDWF1	Expressed in *E. coli*	Overexpressed under methyl jasmonate, 2,4-D and gibberellic acid treatments	Razdan et al. (2016)
WsDWF5	Virus induced gene silencing	Confirmed its role in withanolide biosynthetic pathway	Gupta et al. (2015)
WsFPPS		Upregulation under mechanical injury, methyl jasmonate and salicylic acid treatment	Gupta et al. (2011)
WsCAS	Overexpressing lines	Enhanced withanolide content	Mishra et al. (2016b)
	RNA i-mediated gene silencing	Reduced withanolide content	Mishra et al. (2016b)

isoprenogenesis followed by withanolide production. Chaurasiya et al. (2012) concluded that two pathways, mevalonate (MVA) in cytosol and non-mevalonate or DOXP/MEP in plastids involved in withanolide biosynthesis. Isopentenylnpyrophosphate (IPP) and dimethyl allyl pyrophosphate (DMAPP) serve as the precursor for the withanolide biosynthesis. Seven enzymes are involved in the Mevalonate Pathway. First is AcAc-CoA thiolase which catalyses the synthesis of acetoacetyl-CoA by the condensation of two molecules of acetyl-CoA. Second HMG-CoA synthase, synthesizes 3-hydroxy-3-methylglutaryl-coenzyme (HMG-CoA) after the condensation of AcAc-CoA with one molecule of acetyl-CoA. Mevonalate is synthesized from HMG-CoA by the HMG-CoA reductase (HMGR) enzyme. The IPP unit is formed after various sequential reactions including two phosphorylations and decarboxylation by mevalonate kinase, phosphomevalonate kinase, and mevalonate diphosphate decarboxylase enzymes. Final step of this biosynthesis is carried out by isopentenyl diphosphate isomerase which catalyses the conversion of IPP into DMAPP (Sangwan et al. 2004; Srivastava et al. 2015; Mishra et al. 2016b; Chaurasiya et al. 2012).

The first step of the MEP pathway is the synthesis of glyceraldehyde 3-phosphate into 1-deoxy-D-xylulose 5-phosphate (DXP) by 1-deoxy-D-xylulose 5-phosphate synthase (DXS) from pyruvate and glyceraldehyde 3-phosphate. After several consecutive enzymatic reactions IPP unit is synthesized in cytosol via DXP reductoisomerase, 2-C-methyl-D-erythritol 4-phosphate cytidylyltransferase, 4-diphosphocytidyl-2-C-methyl-D-erythritol kinase, 2-C-methyl-D-erythritol 2,4-cyclodiphosphate synthase, and (E)-4-hydroxy-3-methylbut-2-enyl diphosphate synthase. (E)-4-hydroxy-3-methyl but-2-enyl diphosphate reductase (HDR) catalyses the last step of the reaction, which is the conversion of the IPP into the DMAPP. Further, farnesyl diphosphate synthase (FPPS) enzyme catalyzes the head to tail condensation of IPP units to from FPP (main precursor for triterpenoids). Condensation of two FPP molecules synthesize squalene (squalene synthase) which converts into squalene 2,3- epoxide (squalene epoxidase). Squalene 2,3- epoxide converted in to ring structure cycloartenol by cycloartenol synthase (CAS). Cycloartenol further converted into 24-methylenecholestro (an immediate precursor for different withanolides). The δ-lactonization between C-22/C-26 and hydroxylation of C-22 of 24-methylenecholestrol, lead to the biosynthesis of withanolides. After various sequential reactions various withanolides synthesized.

1.3.2 Transcriptomics of Withania somnifera

Various transcriptomic studies have been conducted on *Withania somnifera* in recent years (Table 1.6). As far as the first reports on application of transcriptomic approach is concerned, it was reported in 2010 by Senthil and co-workers through construction of cDNA library from 2 months old *in-vitro* leaf and root cultures. The plasmid DNA templates were subjected to sequencing reactions and the product was then sequenced on ABI3730XL DNA Analyzer. As a result there was generation of 1047 leaf (48.5% unique sequences) and 1034 root (61.5% unique sequences) sequences, which were clustered into 239 leaf identities and 230 root identities to represent 22.8 and 22.2% respectively out of total repertoire of sequences. E -value $> = 10^{-14}$ non-redundant database search against annotated and characterized proteins returned 70% similarity of encoded proteins from the dataset. Annotated expressed sequence tags (ESTs) were also assigned to various biological processes and molecular functions on the basis of gene ontology (GO) classification. Genes like cytochrome P-450 having significant role in photosynthesis, arginine decarboxylase and chitinase with potential roles in pathogenesis and squalene epoxidase and CDP-ME kinase involved in withanolide biosynthesis were identified. The transcript with highest abundance in leaf (0.85%) was reported for cytochrome-P450. In addition to this, the sequences were also subjected for *CpG* island search and Pfam analysis for identification of functional domains. In conclusion the study presented a comparative analysis in two vital sites (i.e., leaf and root) of withanolide biosynthesis using transcriptomic approach (Senthil et al. 2010).

Next to this after 4 years in 2013, NGS was used to establish transcriptomic resource for leaf and root tissues of NMITLI-101 chemotype of *Withania* through 454 GSFLX Titanium platform. Here, 34,068 and 7,21,755 HQ reads for leaf and root tissues respectively were generated resulting in 89,548 unigenes in 101 L and 1, 14,814 unigenes in 101R. Further annotations were also provided by comparison to TAIR 10, non-redundant database, tomato and potato databases. To look for an elaborated view of enzymes putatively involved in biosynthesis of withanolides, Gene Ontology assignment and KEGG analysis was performed on the generated unigenes. Tissue specific biosynthesis of withanolides was predicted on the basis of

Table 1.6 Fundamental findings of transcriptomic approach studies relevant to *Withania somnifera*

Senthil et al. (2010)		
Unique contigs (unassembled + clustered)	*In-vitro* leaf	*In-vitro* root
No. of contigs	508	636
ESTs identified for genes related to Withanolide biosynthetic pathway		
Gene name	*In-vitro* leaf EST ID	*In-vitro* root EST ID
Squalene epoxidase	–	K-Root-1-1a-N13
Sesquiterpene cyclase	–	K-Root-2-1a-I03
CDP-ME kinase	–	K-Root-1-3a-F11
HMG CoA synthase	–	K-Root-1-3a-D01
Cytochrome p-450	K-SL-1-3a-T3-E05	K-Root-1-3a-N21
Glucosyltransferase	K-SL-1-4a-T3-A03	K-Root-1-4a-D10
Genes highly expressed	Total no. of ESTs	
Universal stress protein	11	
Zinc finger protein	9	
UDP-arabinose-4-epimerase	9	
Methionine synthase	9	
Cytochrome p-450	9	
UDP-apiose/xylose synthase	7	
Phosphoglucomutase	7	
Elongation factor-1-gamma chain	6	
Aspartyl protease	6	
Heat shock protein	6	
Xylulose-6-phosphate-1-dehydrogenase	6	

Gupta et al. (2013)		
Chemotype	101 L	101R
Total (contigs + singletons)	89,548	114,814
Unigenes related to secondary metabolites	1068	1034
Unigenes involved in triterpenoid biosynthesis		
MVA pathway (step 1)	41	41
MEP pathway (step 1)	23	25
Step 2	49	60
Step 3	116	72
Total SSRs identified	2553	3262

Parmar et al. (2015)				
Source species used for cross-species-transfer of ESTs	Tomato	Chili	Tobacco	Egg plant
No. of primer pairs used in study	9	7	3	1

Dasgupta et al. (2014)		
	Relative fold expression after SA treatment	
Pathogenesis related gene name	17 h	36 h
Pathogenesis- related protein PR-1 (WsPR1)	3.02	21.76
b-1,3-glucanase (WsB13G)	21.49	287.74
Class I chitinase (WsCHTN1)	0.3	1.23

(continued)

Table 1.6 (continued)

Dasgupta et al. (2014)		
	Relative fold expression after SA treatment	
Class II chitinase (WsCHTN2)	83.97	5.42
Class IV chitinase (WsCHTN4)	33.58	190.62
Thaumatin-like protein (WsTHAU)	1.58	7
Cystatin-like protein (WsCYST)	7.14	149.24
Serine protease inhibitor like protein (WsSPI)	1.51	9.06
Class III chitinase (WsCHTN3)	75.03	46.29
Lignin-forming anionic peroxidase (WsL-PRX)	12.12	377.42
Suberization-associated anionic peroxidase (WsS-PRX)	49.07	6532.89
PR-10 type pathogenesis- related protein (WsPR10)	0.12	0.07
Chitinase, class V (WsCHTN5)	2.26	13.39
Defensin (WsDFSN)	0.83	5.75
Nonspecific lipid transfer protein 2 (WsLTPa)	3.57	5.4
Non-specific lipid-transfer protein-like protein At2g13820-like (WsLTPb)	0.8	4.85
Germin-like protein subfamily 1 member 20 (WsGER1–20)	1	19.87

Senthil et al. (2015)			
Differentially expressed transcripts in major secondary metabolite biosynthetic pathways			
No. of pathways assigned	Total no. of differentially expressed transcripts	No. of unique transcripts in adventitious roots	No. of unique transcripts in adventitious leaves
25	801	389	1
Details of transcripts involved in withanolide biosynthesis			
No. of enzymes for which transcripts were found	Total transcripts number	Transcripts involved in adventitious roots + leaves	Transcripts involved in adventitious roots
30	778	686	82

Gupta et al. (2015)				
	NMITLI-118	NMITLI-135	NMITLI-101	
Leaf (contig + singletons above 200 bases)	66,807	67,005	71,630	
Root (contig + singletons above 200 bases)	78,790	74,282	87,216	
Total SSRs (leaf + root) in contigs	1150	884	1432	
Total SSRs (leaf + root) in singletons	2282	2127	4383	
Gene families	Cytochrome P450s	Glycosyl-transferase	Methyltransferase	Transcription factors
Total unigenes	228	259	419	1210
Differentially expressed genes	143	144	143	624

differential expression and uniquely present gene members from cytochrome P450, glycosyltransferase and methyltransferase in leaf and root tissues. Detection of simple sequence repeats (SSRs) also provided a platform to various genetic studies in near future. In this study attempt was made to assist the fundamental learning about the biosynthesis of withanolide in *Withania somnifera* chemotype 101 L specifically involved in biosynthesis of withaferin A and chemotype 101R, specifically involved in the biosynthesis of withanolide A through pyrosequencing (Gupta et al. 2013).

Utilizing the data from previously mentioned resources cross-species transferability of EST-SSR markers from *Withania* and other solanaceous species was validated. Conclusively 30 SSR markers were valuably reported to be important to analyze intra and inter-specific gene diversity in *Withania somnifera* and established an important resource for SSR related studies (Parmar et al. 2015). Similarly salicylic acid induced leaf transcriptome of *Withania somnifera* was characterized to study expression of gene related to pathogenesis. The study resulted in generation of 45.6 million reads which were further assembled *de novo* to produce 73, 523 transcript contig, each having an average length of 1620 bp. On annotation 53,424 transcripts were assigned with GO terms. Presence of 182 pathways revealed on mapping of transcripts to biological pathways. Seventeen genes belonging to 12 pathogenesis-related gene family were studied from the transcriptome and documentation of expression pattern after 17 and 36 h of salicylic acid (SA) treatment was carried out. Pattern of accumulation of major secondary metabolites of *Withania e.g.*, withanoside V, withaferin A and withanolide A on SA treatment was analyzed and increase in corresponding metabolite content was observed (Dasgupta et al. 2014).

For example, in 2015 Senthil and group again proposed *in-vitro* leaf and root cultures as resources for targeted withanolide biosynthesis by pyrosequencing approach using illumina platform. Here 1,77,156 assembled transcripts were generated with each unigene having an average length of 1033 bp. Specifically 13% of overall transcripts were observed in case of *in-vitro* adventitious root and no unique proportion of transcripts was found for *in-vitro* leaf tissues. In terms of transcripts, 1,52,839 sequences have shown their presence in both the *in-vitro* tissues while 24,319 were specific to be expressed in root. Amongst 12,822 differentially expressed transcripts, 8013 were found to get up regulated whereas 4809 were showing down regulation. Mapping of the assembled transcripts to KEGG database resulted in deduction of putative biosynthetic pathway for withanolide biosynthesis. A total of 505 transcripts were related to phenylpropanoid biosynthesis, 174 were found to be involved in monoterpenoid biosynthesis and 155 members were putatively present in biosynthesis of terpenoid backbone. Also the expression of candidate genes of withanolide biosynthesis was subjected to qRT PCR validation. Additionally, 13 specific beta amyrin synthase transcripts were observed for *in-vitro* root tissues as against absence of corresponding transcripts in field grown root tissues. Variation was observed in withaferin A and withanolide A accumulation pattern based on tissue type and culture period. Further anticancer activity of *in-vitro* leafs have been demonstrated against human gastric adenocarcinoma cancer cell line (Senthil et al. 2015).

Recently, comparative analysis of transcriptome from differnt chemotypes of *Withania somnifera* was carried out to elucidate biosynthesis of withanolides. Variability in different chemotypes with respect to nature and content of withanolides was observed on chemoprofiling of leaf and root tissues. Leaf and root transcriptomes specific to different chemotypes were established for identification of genes in particular chemotype for tissue specific biosynthesis of withanolides. Total (20,621contigs and 63,910 singletons in leaf), (22,438 contigs and 78,645 singletons in root) for NMITLI-118, (20,135 contigs and 66,176 singletons in leaf), (18,413 contigs and 79,737 singletons in root) for NMITLI-135 and (21,445 contigs and 68,103 singletons in leaf) and (20,797 contigs and 94,017 singletons in root) for NMITLI-101 were generated in the study. Combined assembly of the two tissues from different chemotypes generated 43,287 contigs 1,78, 956 singletons and 18,317 transcripts, which were differentially expressed. Moreover on differential comparison 142 contigs were matched against non-redundant database, 943 were showing matches with Tomato genome and 6 were found similar against TAIR database.

Genes corresponding to enzymes of intermediary steps of terpenoid backbone biosynthesis were identified along with paralogous and alternatively spliced isoforms. For example seven enzymes namely AACT having 1 unigene, HMGS having 2 unigenes, HMGR with 5 unigenes, MK comprising 4 unigenes, pmk (5 unigenes) were found for Step1 MVA pathway in addition to MDD and IPP having 3 and 1 unigenes respectively. Similarly, for MEP pathway also seven genes were characterized as *DXS, DXR, CDP-MES, CDP-MEK, HDS, HDR* containing 2, 1, 3, 1, 2, 3 unigenes respectively, in addition to partial gene sequence for MECPS. Additionally, 12 full-length unigenes were identified in Step 2 and one partial unigene for cycloeucalenol cycloisomerase (CEC1) was found. Among full length unigenes 1 was found for (*GPPS*), (*CAS*), Δ14-sterol reductase (*FK*), C-7,8 sterol isomerase (*HYD1*), sterol-4α-methyl oxidase 2 (*SMO2*), C-5 sterol desaturase (*STE1*) and sterol δ7 reductase (*DWF5*) each, 4 for (*FPPS*), 2 for (*SqS*), (*SMT1*), (*SMO1*), 2 for (*CYP51G1*) and sterol δ7 reductase (*DWF5*) were found. Moreover, 228 CYPs, 259 GTs, 419 MTs and 1210 transcription factors unigenes were involved in secondary metabolism of which 143, 144, 143 and 624 unigenes were differentially expressed respectively in the four gene families. It was proposed that this differential expression might account for chemodiversity in *Withania somnifera*. In the same study nucleotide sequence of *WsDWF5* was mined and constructs for virus-induced gene silencing (VIGS) analysis were prepared to observe transcript accumulation and its role in withanolide biosynthesis (Gupta et al. 2015).

1.3.3 Metabolomics of **Withania somnifera**

The term 'metabolomics' was introduced by Oliver Fiehn in 2002 for holistic analysis for quantification and identification of whole of the metabolites in an organism. Metabolomics is an important tool for understanding the metabolism of plant including both primary and secondary (Tugizimana et al. 2013). It is like a fingerprint of an organism, which represents the sum total of compounds present in a particular organism at a given time point. Search is going on to find out the phytochemicals, a major investigation in which done metabolic fingerprinting of crude extracts of leaf and root of *Withania somnifera* by using Nuclear magnetic resonance (NMR) and chromatographic (HPLC and GC-MS), 62 major and minor primary and secondary metabolites from leaves and 48 from roots were identified (Singh et al. 2015). There are various applications of NMR in metabolomics of plant as it can be used as tool for functional genomics and to differentiate between plants from different origin and after different treatments. Using NMR and IR spectroscopy and chemical transformations some unusually sulfated containing the rare 3-O-sulfate group with the saturation in A ring and oxygenated unusual 1,4-dien-3-one group steroids along with 5,6-deepoxy- 5-en-7-one-17-hydroxy withaferin A (Siddique et al. 2014) were isolated from leaves of *Withania somnifera* (Misra et al. 2005). Chlorinated withanolides like 6a–chloro-5b,17a–dihydroxywithaferin A is also reported from *Withania somnifera* dried plant by using spectroscopic techniques like (IR, HRESIMS, 1D/2D NMR) and further validated by X-ray crystallography (Tong et al. 2011). Leaves of *Withania somnifera* are the best source of withaferin A patents were developed for the ion of withaferin A from leaves of *Withania somnifera* (Sangwan et al. 2006).

1.3.4 Phytochemical Genomics for Identification of Withanolides

Phytochemical genomics is an emerging field of plant science that integrates biosynthetic mechanisms and genomics to the plants metabolomics. This area emerges as a combinatorial approach of high-throughput massive DNA sequencing technology and metabolomics. This current field is based on advancements in metabolomics and allied 'omics' like transcriptomics, proteomics and genomics to find the biosynthetic mechanism, function evolution and regulation of plant metabolites (Saito 2013; Muranaka and Saito 2013).

There are two cornerstone techniques used for the detection of metabolites in a plant: NMR and MS. Other correlated techniques for single-separation of a metabolite, based on physicochemical properties are: gas chromatography-MS, capillary electrophoresis-MS, reversed phase liquid chromatography (LC)-MS, collision-induced dissociation (CID) in tandem mass spectrometry (MS/MS), LC-fourier transform-ion cyclotron resonance (FT-ICR)-MS, LC-NMR and LC-CD (Nakabayashi and Saito 2013).

Now a days' various software tools are available that makes it possible to project both metabolome and transcriptome data onto maps for systematic analyses (Joung et al. 2009; Thimm et al. 2004; Tokimatsu et al. 2005). Transcriptome and metabolome data provide us a clue about expression pattern of genes along with pattern of metabolite accumulation. These data also helps us a way to find out the functions of genes. Since many of the model plants like tomato, potato, pepper and tobacco belong to Solanaceae family, transcriptome and metabolome research is much advanced in solanaceous plants during their different developmental phases (Alba et al. 2005; Carrari et al. 2006; Urbanczyk et al. 2003, 2005). The Sol Genomics Network (SGN) is dedicated to provide us a useful genomics data for solanaceous plants. At present, complete genome data is available for tomato, potato, wild tomato and pepper with draft genome available for *Nicotiana benthamiana*, *Nicotiana tabacum*, *Solanum pimpinellifolium*. Metabolite analysis (metabolome) using GC-MS and comprehensive gene expression analysis in different accessions or introgression lines of wild tomato performed to recognize the genes involved in production of volatile metabolites (Fridman et al. 2005; Tieman et al. 2006). The AtMetExpress is a detailed database about metabolites detected in *A. thaliana*, which is based on liquid chromatography-mass spectrometry and contains 167 metabolite structures from different growth stages and organs (Matsuda et al. 2010). Using omics approaches (transcriptomics and metabolomics) genes and metabolites responsible for biosynthesis of terpenoids, glucosinolates, flavonoids, phenylpropanoids and lipids in *A. thaliana* are recognized (Higashi and Saito 2013; Okazaki et al. 2013). RIKEN tandem mass spectral database (Re Spect) is a database based on literature and tandem mass spectrometry (MS/MS) data that covers up to 9017 spectra records of different plants (Sawada et al. 2012). Some other important databases of MS and MS/MS spectra are Madison Metabolomics Consortium Database (MMCD, http://mmcd.nmrfam.wisc.edu/), METLIN and MassBank databases (Nakabayashi and Saito 2013). KNApSAcK family databases contains 101,500 species metabolite relationship encompassing 20,741 species and 50,048 metabolites searched with their name, molecular mass or molecular formula, or mass spectra in several ionization modes (Afendi et al. 2012). For figuring out the genetic background of plant metabolites metabolomics is used in combination with a QTL analysis of natural variants of some plants like Rice and tomato and their inbred line (Schauer et al. 2006; Okazaki and Saito 2016).

Advanced analytical tools like GC–MS, HPLC and NMR creates new opportunities to explore the rapid identification of various metabolites qualitatively as well as quantitatively. Recently, 62 major and minor primary and secondary metabolites from leaves and 48 from roots were identified by using NMR and chromatographic (HPLC and GC-MS) techniques (Singh et al. 2015).

1.3.4.1 Various Techniques for Structural Identification

Metabolomics approaches using GC-MS, LC-MS, or NMR spectroscopy served as efficient tools for quality control of medicinal plants and their herbal products. Metabolomic studies based on NMR have been widely used to decipher problems in functional genomics. NMR has very wider applications in plant sciences including the comparison studies of metabolites in different plant accessions collected from different sites as well as to check the effect of various known biological factors on the metabolome of a plant (Kim et al. 2011).

Metabolic investigations from roots of *Withania somnifera* by using ^{1}H NMR, ^{13}C NMR and chemical transformation reveals the presence of two new withanolides with rare 16 b-acetoxy-17(20)-ene and 6a–hydroxy-5,7a–epoxy functional groups in the withasteroid skeleton and also ergosterol (Misra et al. 2008, 2012).

Further by using techniques like column chromatography, normal-phase silica gel column chromatography, reversed-phase silica gel, column chromatography and HPLC, seven new withanolide glycosides called withanosides I, II, III, IV, V, VI, and VII along with withaferin A, 5a,20aF(R)-dihydroxy-6a,7a–epoxy-1-oxowitha-2,24-dienolide, physagulin D, and coagulin Q were identified from methanolic root extracts of *Withania somnifera* (Matsuda et al. 2009).

By using NMR, HPLC and GC-MS techniques, metabolite profiling of *Withania somnifera* was carried out (Chatterjee et al. 2010) that revealed a total of 62 and 48 secondary metabolite from leaves and roots respectively. It was analyzed that out of four fractions of different polarities used for ion, quantities of metabolite in leaf (228.57 ± 5.2 mg/g of DW) was much higher than that in roots (15.00 ± 1.6 mg/g of DW), especially in the aqueous-methanolic fraction. NMR spectrum of *Withania somnifera* leaf indicated the presence of withaferin-A and withanone as the major metabolite present in the leaves while withanolide A and withanone are major metabolites in the roots. The HPLC-PDA analysis of butanolic fraction of leaf indicates the presence of physagulin, withanoside IV and withanoside VI.

Applying HR-MASNMR spectroscopy for metabolite analysis of root and leaf tissues led to the identification of a total of 41 metabolites including both primary and secondary metabolites. This technique is advantageous as it provides a complete characterization of metabolites from the whole tissue without any ion procedure (Bharti et al. 2011).

Another technique of metabolite identification is ^{1}H NMR. Being a fast method, ^{1}H NMR has many applications like classification and characterization based on metabolites (chemotaxonomy), quality control, to check the difference in genetically modified plants and to check the changes in plants after interaction with other organism or environment. NMR alone has a lower sensitivity than MS therefore it is used in association with MS where high sensitivity is required (Kim et al. 2011). In *Withania somnifera*, this technique was used for determining the phytochemical variability among *Withania somnifera* plants collected from six different locations in Pune (India). Surprisingly, leaf metabolites showed highest degree of phytochemical variability among them. Besides the common withanolide, the technique led to the identification of two other withanolides such as 4-OH-5,6-epoxy withanolides (withaferin A-like steroids) and 5-OH-6,7-epoxy withanolides (Namdeo et al. 2011).

Similar to leaves and roots, fruits of *Withania somnifera* are also loaded with a rich population of secondary metabolites. Application of liquid chromatography (LC) coupled with high-resolution mass spectrometry (HRMS) along with data-dependent and targeted MS/MS enabled identification of a total of 62 secondary metabolites and their structures from fruits of *Withania somnifera*. As per studies of Bolleddula et al. (2012) several novel phytomolecules including withanamides (32), and withanolides (22), saponins (3), and other tyramine based phytomolecules were also reported, in addition to several knoen compounds (Bolleddula et al. 2012). Further the variations in metabolites of fruits from four different chemotypes (NMITLI-101, NMITLI-108, NMITLI-118 and NBRI-WS) were detected using GC–MS and ^{1}H NMR. In this study 82 metabolites including both primary and secondary metabolites were identified namely, fatty acids, aliphatic organic acids and aromatic amino acids, sugars and sugar alcohols, polyols, tocopherols, sterols, phenolic acids, and withanamides (Bhatia et al. 2013).

1.4 Conclusion

Withania somnifera and its most important bioactive constituents withanolides have gained significant attention from the researchers belonging to both plant biology and medical science fields. However, the information gathered from this plant till date is still very little. Researchers from both the areas plant biology and medical sciences are engaged in bringing depth of information to an open platform for more specific utilization of *Withania somnifera* and withanolides for betterment of mankind. Although various approaches have been adopted to enhance the withanolide production, it is still a challenge to find out the most appropriate way to bring global withanolide production equivalent to its demand.

References

Afendi FM, Okada T, Yamazaki M, Hirai-Morita A, Nakamura Y, Nakamura K, Ikeda S, Takahashi H, Altaf-Ul-Amin M, Darusman LK, Saito K, Kanaya S (2012) KNApSAcK family databases: integrated metabolite – plant species databases for multifaceted plant research. Plant Cell Physiol 53:e1

Ahmad H, Khandelwal K, Samuel SS, Tripathi S, Mitra K, Sangwan RS, Shukla R, Dwivedi AK (2015) Neuro-protective potential of a vesicular system of a standardized of a new chemotype of Withania somnifera Dunal (NMITLI118RT+) against cerebral stroke in rats. Drug Deliv 23:2630–2641

Akhtar N, Gupta P, Sangwan NS, Sangwan RS, Trivedi PK (2013) Cloning and functional characterization of 3-hydroxy-3-methylglutaryl coenzyme A reductase gene from Withania somnifera: an important medicinal plant. Protoplasma 250(2):613–622

Alam N, Hossain M, Mottalib MA, Sulaiman SA, Gan SH, Khalil MI (2012) Methanolic extracts of Withania somnifera leaves, fruits and roots possess antioxidant properties and antibacterial activities. BMC Complement Altern Med 12(1):175

Alba R, Payton P, Fei Z, McQuinn R, Debbie P, Martin GB, Tanksley SD, Giovannoni JJ (2005) Transcriptome and selected metabolite analyses reveal multiple points of ethylene control during tomato fruit development. Plant Cell 17:2954–2965

Alfaifi MY, Saleh KA, El-Boushnak MA, Elbehairi SEI, Alshehri MA, Shati AA (2016) Antiproliferative activity of the Methanolic extract of Withania somnifera leaves from Faifa Mountains, Southwest Saudi Arabia, against several human cancer cell lines. Asian Pac J Cancer Prev 17(5):2723–2726

Bähr V, Hänsel R (1982) Immunomodulating Properties of 5, 20α (R)-Dihydroxy-6α, 7α-epoxy-1-oxo-(5α)-witha-2, 24-dienolide and Solasodine. Planta Med 44(01):32–33

Baitharu I, Jain V, Deep SN, Hota KB, Hota SK, Prasad D, Ilavazhagan G (2013) Withania somnifera root extract ameliorates hypobaric hypoxia induced memory impairment in rats. J ethnopharmacol 145(2):431–441

Baitharu I, Jain V, Deep SN, Shroff S, Sahu JK, Naik PK, Ilavazhagan G (2014) Withanolide A prevents neurodegeneration by modulating hippocampal glutathione biosynthesis during hypoxia. PLoS One 9(10):e105311

Baldi A, Singh D, Dixit VK (2008) Dual elicitation for improved production of withaferin A by cell suspension cultures of Withania somnifera. Appl Biochem Biotechnol 151:556–564

Bandyopadhyay M, Jha S, Tepfer D (2007) Changes in morphological phenotypes and withanolide composition of Ri-transformed roots of Withania somnifera. Plant Cell Rep 26(5):599–609

Bharti SK, Bhatia A, Tewari SK, Sidhu OP, Roy R (2011) Application of HR-MAS NMR spectroscopy for studying chemotype variations of Withania somnifera (L) Dunal. Magn Reson Chem 49:659–667

Bhat WW, Lattoo SK, Razdan S, Dhar N, Rana S, Dhar RS et al (2012) Molecular cloning, bacterial expression and promoter analysis of squalene synthase from Withania somnifera (L.) Dunal. Gene 499(1):25–36

Bhatia A, Bharti SK, Tewari SK, Sidhu OP, Roy R (2013) Metabolic profiling for studying chemotype variations in Withania somnifera (L) Dunal fruits using GC–MS and NMR spectroscopy. Phytochemistry 93:105–115

Bhattacharya SK, Satyan KS, Ghosal S (1997) Antioxidant activity of glycowithanolides from Withania somnifera. Indian J Exp Biol 35:236–239

Bhattarai JP, Han SK (2014) Phasic and tonic type A γ-Aminobutryic acid receptor mediated effect of Withania somnifera on mice hippocampal CA1 pyramidal Neurons. J Ayurveda and Integrative Med 5(4):216

Bhattarai JP, Park SJ, Han SK (2013) Potentiation of NMDA receptors by Withania somnifera on hippocampal CA1 pyramidal neurons. Am J Chin Med 41(03):503–513

Bisht P, Rawat V (2014) Antibacterial activity of Withania somnifera against Gram-positive isolates from pus samples. Ayu 35(3):330

Biswal BM, Sulaiman SA, Ismail HC, Zakaria H, Musa KI (2013) Effect of Withania somnifera (Ashwagandha) on the development of chemotherapy-induced fatigue and quality of life in breast cancer patients. Integrative Cancer Therapies 12(4):312–322

Bolleddula JFW, Vareed SK, Nair MG (2012) Identification of metabolites in Withania sominfera fruits by liquid chromatography and high-resolution mass spectrometry. Rapid Commun Mass Spectrom 26:1277–1290

Budhiraja RD, Sudhir S, Garg KN (1984) Antiinflammatory activity of 3 β -Hydroxy-2, 3-dihydrowithanolide F. Planta Med 50(02):134–136

Budhiraja RD, Garg KN, Sudhir S, Arora B (1986) Protective effect of 3-ss-hydroxy-2, 3-dihydrowithanolide F against CCl4-induced hepatotoxicity. Planta Med 52(01):28–29

Carrari F, Baxter C, Usadel B, Urbanczyk-Wochniak E, Zanor MI, Nunes-Nesi A, Nikiforova V, Centero D, Ratzka A, Pauly M, Sweetlove LJ, Fernie AR (2006) Integrated analysis of metabolite and transcript levels reveals the metabolic shifts that underlie tomato fruit development and highlight regulatory aspects of metabolic network behavior. Plant Physiol 142:1380–1396

Chatterjee S, Srivastava S, Khalid A, Singh N, Sangwan RS, Sidhu OP, Roy R, Khetrapal CL, Tuli R (2010) Comprehensive metabolic fingerprinting of Withania somnifera leaf and roots. Phytochemistry 71:1085–1094

Chaurasiya ND, Gupta VK, Sangwan RS (2007) Leaf ontogenic phase-related dynamics of withaferin A and withanone biogenesis in Ashwagandha (Withania somnifera Dunal) – an important medicinal herb. J Plant Biol 50:508–513

Chaurasiya ND, Uniyal GC, Lal P, Misra L, Sangwan NS, Tuli R, Sangwan RS (2008) Analysis of withanolides in root and leaf of Withania somnifera by HPLC with photodiode array and evaporative light scattering detection. Phytochem Anal 19:148–154

Chaurasiya ND, Sangwan RS, Misra LN, Tuli R, Sangwan NS (2009) Metabolic clustering of a core collection of Indian ginseng Withania somnifera Dunal through DNA isoenzyme polypeptide and withanolide profile diversity. Fitoterapia 80:496–505

Chaurasiya ND, Sangwan NS, Sabir F, Misra L, Sangwan RS (2012) Withanolide biosynthesis recruits both mevalonate and DOXP pathways of isoprenogenesis in Ashwagandha Withania somnifera L (Dunal). Plant Cell Rep 31:1889–1897

Choudhary MI, Hussain S, Yousuf S, Dar A, Mudassar A-u-R (2010) Chlorinated and diepoxy withanolides from Withania somnifera and their cytotoxic effects against human lung cancer cell line. Phytochemistry 71:2205–2209

Dasgupta MG, George BS, Bhatia A, Sidhu OP (2014) Characterization of Withania somnifera leaf transcriptome and expression analysis of pathogenesis-related genes during salicylic acid signaling. PLoS One 9:e94803

De Rose F, Marotta R, Poddighe S, Talani G, Catelani T, Setzu MD, Acquas E (2016) Functional and morphological correlates in the drosophila LRRK2 loss-of-function model of Parkinson's disease: drug effects of Withania somnifera (Dunal) administration. PLoS One 11(1):e0146140

Devkar ST, Kandhare AD, Zanwar AA, Jagtap SD, Katyare SS, Bodhankar SL, Hegde MV (2016) Hepatoprotective effect of withanolide-rich fraction in acetaminophen-intoxicated rat: decisive role of TNF-α, IL-1β , COX-II and iNOS. Pharm Biol 54(11):2394–2403

Dhar RS, Verma V, Suri KA, Sangwan RS, Satti NK, Kumar A, Tuli R, Qazi GN (2006) Phytochemical and genetic analysis in selected chemotypes of Withania somnifera. Phytochemistry 67:2269–2276

Dhar N, Rana S, Bhat WW, Razdan S, Pandith SA, Khan S et al (2013) Dynamics of withanolide biosynthesis in relation to temporal expression pattern of metabolic genes in Withania somnifera (L.) Dunal: a comparative study in two morpho-chemovariants. Mol Biol Rep 40(12):7007–7016

Dhar N, Rana S, Razdan S, Bhat WW, Hussain A, Dhar RS et al (2014) Cloning and functional characterization of three branch point oxidosqualene cyclases from Withania somnifera (L.) dunal. J Biol Chem 289(24):17249–17267

Dhar N, Razdan S, Rana S, Bhat WW, Vishwakarma R, Lattoo SK (2015) A decade of molecular understanding of withanolide biosynthesis and in vitro studies in Withania somnifera (L) Dunal: prospects and perspectives for pathway engineering. Front Plant Sci 6:1031

Dhiman R, Aggarwal N, Aneja KR, Kaur M (2016) In vitro antimicrobial activity of spices and medicinal herbs against selected microbes associated with juices. Int J microbiol 2016:9015802

Doma M, Abayankar G, Reddy VD, Kishor PBK (2012) Carbohydrate and elicitor enhanced withanolide (withaferin A and withanolide A) accumulation in hairy root cultures of Withania somnifera (L). Indian J Exp Biol 50:484–490

Fatima N, Ahmad N, Anis M (2016) In vitro propagation and conservation of Withania somnifera (Dunal) L. Methods Mol Biol 1391:303–315

Fridman E, Wang J, Iijima Y, Froehlich JE, Gang DR, Ohlrogge J, Pichersky E (2005) Metabolic genomic and biochemical analyses of glandular trichomes from the wild tomato species Lycopersicon hirsutum identify a key enzyme in the biosynthesis of methylketones. Plant Cell 17:1252–1267

Gorelick J, Rosenberg R, Smotrich A, Hanuš L, Bernstein N (2015) Hypoglycemic activity of withanolides and elicitated Withania somnifera. Phytochem 116:283–289

Gupta P, Akhtar N, Tewari SK, Sangwan RS, Trivedi PK (2011) Differential expression of farnesyl diphosphate synthase gene from Withania somnifera in different chemotypes and in response to elicitors. Plant Growth Regul 65(1):93–100

Gupta P, Goel R, Pathak S, Srivastava A, Singh SP, Sangwan RS, Asif MH, Trivedi PK (2013) De novo assembly functional annotation and comparative analysis of Withania somnifera leaf and root transcriptomes to identify putative genes involved in the withanolides biosynthesis. PLoS One 8:e62714

Gupta P, Goel R, Agarwal AV, Asif MH, Sangwan NS, Sangwan RS, Trivedi PK (2015) Comparative transcriptome analysis of different chemotypes elucidates withanolide biosynthesis pathway from medicinal plant Withania somnifera. Sci Rep 5:18611

Handa SS, Rakesh DD, Vasisht K (2006) Compendium of medicinal and aromatic plants ASIA. ICS UNIDO. Asia, 2, 305

Higashi Y, Saito K (2013) Network analysis for gene discovery in plant specialized metabolism. Plant Cell Environ 36:1597–1606

Jacob L, Manju RV, Stephen R, Edison LK (2014) Alterations in withanolide production in Withania somnifera (l) Dunal under low light stress. J Plant Sci Res 30:121–123

Jadaun JS, Sangwan NS, Narnoliya LK, Tripathi S, Sangwan RS (2016) Withania coagulans tryptophan decarboxylase gene cloning heterologous expression and catalytic characteristics of the recombinant enzyme. Protoplasma 254:181–192

Jain SM, Saxena P (2009) Protocols for in vitro cultures and secondary metabolite analysis of aromatic and medicinal plants. In: Methods in molecular biology, vol 1391, 2nd edn. Humana Press, New York, pp 303–315

Jayaprakasam B, Padmanabhan K, Nair MG (2010) Withanamides in Withania somnifera fruit protect PC-12 cells from β -amyloid responsible for Alzheimer's disease. Phytother Res 24(6):859–863

Jha AK, Nikbakht M, Capalash N, Kaur J (2014) Demethylation of RARβ 2 gene promoter by Withania somnifera in HeLa cell line. European J Med Plants 4(5):503–510

Joung JG, Corbett AM, Fellman SM, Tieman DM, Klee HJ, Giovannoni JJ, Fei Z (2009) Plant MetGenMAP: an integrative analysis system for plant systems biology. Plant Physiol 151:1758–1768

Kabera JN, Semana E, Mussa AR, He X (2014) Plant secondary metabolites: biosynthesis classification function and pharmacological properties. J Pharm Pharmacol 2:377–392
Kalidhasan N, Bhagavan NB, Kannan ND (2013) Ultraviolet-B (280–320 nm) enhanced radiation induced changes in secondary metabolites and photosystem-II of medicinal plant Withania somnifera Dunal. J Med Plants Res 7:3112–3120
Kelly-Pieper K, Patil SP, Busse P et al (2009) Safety and tolerability of an Antiasthma Herbal Formula (ASHMI™) in adult subjects with asthma: a randomized, double-blinded, placebo-controlled, dose-escalation Phase I study. J Altern Complement Med 15(7):735–743
Khan B, Ahmad SF, Bani S, Kaul A, Suri KA, Satti NK, Qazi GN (2006) Augmentation and proliferation of T lymphocytes and Th-1 cytokines by Withania somnifera in stressed mice. Int Immunopharmacol 6(9):1394–1403
Khazal KF, Samuel T, Hill DL, Grubbs CJ (2013) Effect of an extract of Withania somnifera root on estrogen receptor-positive mammary carcinomas. Anticancer Res 33(4):1519–1523
Kim HK, Choi YH, Verpoorte R (2011) NMR-based plant metabolomics: where do we stand, where do we go? Trends Biotechnol 29:267–275
Konar A, Shah N, Singh R, Saxena N, Kaul SC, Wadhwa R, Thakur MK (2011) Protective role of Ashwagandha leaf extract and its component withanone on scopolamine-induced changes in the brain and brain-derived cells. PLoS One 6(11):e27265
Kuboyama T, Tohda C, Komatsu K (2014) Effects of Ashwagandha (roots of Withania somnifera) on neurodegenerative diseases. Biol Pharm Bull 37(6):892–897
Kulkarni AA, Thengane SR, Krishnamurthy K (1996) Direct in vitro regeneration of leaf explants of Withania somnifera (L) Dunal. Plant Sci 119:163–168
Kulkarni AA, Thengane SR, Krishnamurthy KV (2000) Direct shoot regeneration from node, internode, hypocotyl and embryo explants of Withania somnifera plant cell. Tiss Org 62:203–209
Kushwaha S, Roy S, Maity R, Mallick A, Soni VK, Singh PK, Mandal C (2012a) Chemotypical variations in Withania somnifera lead to differentially modulated immune response in BALB/c mice. Vaccine 30:1083–1093
Kushwaha S, Soni VK, Singh PK, Bano N, Kumar A, Sangwan RS, Misra-Bhattacharya S (2012b) Withania somnifera chemotypes NMITLI 101R, NMITLI 118R, NMITLI 128R and withaferin A protect Mastomys coucha from Brugia malayi infection. Parasite Immunol 34:199–209
Kushwaha AK, Sangwan NS, Tripathi S, Sangwan RS (2013a) Molecular cloning and catalytic characterization of a recombinant tropine biosynthetic tropinone reductase from Withania coagulans leaf. Gene 516:238–247
Kushwaha AK, Sangwan NS, Trivedi PK, Negi AS, Misra L, Sangwan RS (2013b) Tropine forming tropinone reductase gene from Withania somnifera (Ashwagandha): biochemical characteristics of the recombinant enzyme and novel physiological overtones of tissue-wide gene expression patterns. PLoS One 8:e74777
Kushwaha S, Roy S, Maity R, Mallick A, Soni VK, Singh PK et al (2013c) Chemotypical variations in Withania somnifera lead to differentially modulated immune response in BALB/c mice. Vaccine 30(6):1083–1093
Lal P, Misra L, Sangwan RS, Tuli R (2006) New withanolides from fresh berries of Withania somnifera. Zeitschrift für Naturforschung B 61:1143–1147
Madina BR, Sharma LK, Chaturvedi P, Sangwan RS, Tuli R (2007) Purification and physico-kinetic characterization of 3β-hydroxy specific sterol glucosyltransferase from Withania somnifera (L) and its stress response. Biochim Biophys Acta 1774:392–402
Maliyakkal N, Udupa N, Pai KS, Rangarajan A (2013) Cytotoxic and apoptotic activities of extracts of Withania somnifera and Tinospora cordifolia in human breast cancer cells. Inter J Appl Res in Nat Prod 6(4):1–10
Mandal C, Dutta A, Mallick A, Chandra S, Misra L, Sangwan RS, Mandal C (2008) Withaferin A induces apoptosis by activating p38 mitogen-activated protein kinase signaling cascade in leukemic cells of lymphoid and myeloid origin through mitochondrial death cascade. Apoptosis 13:1450–1464
Matsuda H, Murakami T, Kishi A, Yoshikawa M (2009) Structures of withanosides i, ii, iii, iv, v, vi, and vii, new withanolide glycosides, from the roots of Indian Withania somnifera Dunal and inhibitory activity for tachyphylaxis to clonidine in isolated Guinea-Pig ileum. Bioorg Med Chem 9:1499–1507

Matsuda F, Hirai MY, Sasaki E, Akiyama K, Yonekura-Sakakibara K, Provart NJ, Sakurai T, Shimada Y, Saito K (2010) AtMetExpress development: a phytochemical atlas of Arabidopsis development. Plant Physiol 152:566–578

Mir BA, Mir SA, Koul S (2014a) In vitro propagation and withaferin A production in Withania Ashwagandha, a rare medicinal plant of India. Physiol Mol Biol Plants 20:357–364

Mir BA, Khazirb J, Hakeemc KR, Kumara A, Koul S (2014b) Withanolides array of Withania Ashwagandha sp. novo populations from India. Indus Crops Prod 59:9–13

Mir BA, Mir SA, Khazir J, Tonfack LB, Cowan DA, Vyas D, Koul S (2015) Cold stress affects antioxidative response and accumulation of medicinally important withanolides in Withania somnifera (L) Dunal. Indus Crops Prod 74:1008–1016

Mirjalili MH, Moyano E, Bonfill M, Cusido RM, Palazón J (2009) Steroidal lactones from Withania somnifera, an ancient plant for novel medicine. Molecules 14:2373–2393

Mishra S, Sangwan RS, Bansal S, Sangwan NS (2013) Efficient genetic transformation of Withania coagulans (stocks) Dunal mediated by Agrobacterium tumefaciens from leaf explants of in vitro multiple shoot culture. Protoplasma 250:451–458

Mishra B, Sangwan RS, Mishra S, Jadaun JS, Sabir F, Sangwan NS (2014) Effect of cadmium stress on inductive enzymatic and nonenzymatic responses of ROS and sugar metabolism in multiple shoot cultures of Ashwagandha (Withania somnifera Dunal). Protoplasma 251:1031–1045

Mishra S, Bansal S, Mishra B, Sangwan RS, Jadaun JS, Sangwan NS (2016a) RNAi and homologous over-expression based functional approaches reveal triterpenoid synthase gene-cycloartenol synthase is involved in downstream withanolide biosynthesis in Withania somnifera. PLoS One 11:e0149691

Mishra S, Bansal S, Sangwan RS, Sangwan NS (2016b) Genotype independent and efficient Agrobacterium-mediated genetic transformation of the medicinal plant Withania somnifera Dunal. J Plant Biochem Biotechnol 25(2):191–198

Misra L, Lal P, Sangwan RS, Sangwan NS, Uniyal GC, Tuli R (2005) Unusually sulfated and oxygenated steroids from Withania somnifera. Phytochemistry 66:2702–2707

Misra L, Mishra P, Pandey A, Sangwan RS, Sangwan NS, Tuli R (2008) Withanolides from Withania somnifera roots. Phytochemistry 69:1000–1004

Misra L, Mishra P, Pandey A, Sangwan RS, Sangwan NS (2012) 1,4-Dioxane and ergosterol derivatives from Withania somnifera roots. J Asian Nat Prod Res 14:39–45

Mondal S, Mandal C, Sangwan R, Chandra S, Mandal C (2010) Withanolide D induces apoptosis in leukemia by targeting the activation of neutral sphingomyelinase-ceramide cascade mediated by synergistic activation of c-Jun N-terminal kinase and p38 mitogen-activated protein kinase. Mol Cancer 9:239

Mondal S, Bhattacharya K, Mallick A, Sangwan R, Mandal C (2012a) Bak compensated for Bax in p53-null cells to release cytochrome c for the initiation in Withania somnifera. PLoS One 11:e0149691

Mondal S, Roy S, Maity R, Mallick A, Sangwan R, Misra-Bhattacharya S, Mandal C (2012b) Withanolide D, carrying the baton of Indian rasayana herb as a lead candidate of antileukemic agent in modern medicine. In: Biochem roles Eukar cell surface Macromol. Springer, New York, pp 295–312

Muranaka T, Saito K (2013) Phytochemical genomics on the way. Plant Cell Physiol 54:645–646

Murthy HN, Praveen N (2012) Influence of macro elements and nitrogen source on adventitious root growth and withanolide-A production in Withania somnifera (L) Dunal. Nat Product Res 26:466–473

Murthy HN, Dijkstra C, Anthony P, White DA, Davey MR, Power JB, Hahn EJ, Paek KY (2008) Establishment of Withania somnifera hairy root cultures for the production of withanolide. J Integ Plant Biol 50:975–981

Nagella P, Murthy HN (2010) Establishment of cell suspension cultures of Withania somnifera for the production of withanolide. A Biores Technol 101:6735–6739

Nakabayashi R, Saito K (2013) Metabolomics for unknown plant metabolites. Anal Bioanal Chem 405:5005–5011

Namdeo AG, Sharma A, Yadav KN, Gawande R, Mahadik KR, Lopez-Gresa MP, Kim HK, Choi YH, Verpoorte R (2011) Metabolic characterization of Withania somnifera from different regions of India using NMR spectroscopy. Planta Med 77:1958–1964

Nayak S, Kumar S, Satapathy K, Moharana A, Behera B, Barik DP, Acharya L, Mohaptra PK, Jena PK, Naik SK (2013) In vitro plant regeneration from cotyledonary nodes of Withania somnifera (L.) Dunal and assessment of clonal fidelity using RAPD and ISSR markers. Acta Physiol Plant 35:195–203

Nema R, Khare S, Jain P, Pradhan A (2013) Anticancer activity of Withania somnifera (leaves) flavonoids compound. Int J Pharm Sci Rev Res 19(1):103–106

Noh EJ, Kang MJ, Jeong YJ, Lee JY, Park JH, Choi HJ et al (2016) Withaferin A inhibits inflammatory responses induced by Fusobacterium nucleatum and Aggregatibacter actinomycetemcomitans in macrophages. Mol Med Rep 14(1):983–988

Okazaki Y, Saito K (2016) Integrated metabolomics and phytochemical genomics approaches for studies on rice. GigaScience 5:11

Okazaki Y, Otsuki H, Narisawa T, Kobayashi M, Sawai S, Kamide Y, Kusano M, Aoki T, Hirai MY, Saito K (2013) A new class of plant lipid is essential for protection against phosphorus depletion. Nat Commun 4:1510

Palliyaguru DL, Singh SV, Kensler TW (2016) Withania somnifera: from prevention to treatment of cancer. Mol Nutr Food Res 60(6):1342–1353

Pandit S, Chang KW, Jeon JG (2013) Effects of Withania somnifera on the growth and virulence properties of Streptococcus mutans and Streptococcus sobrinus at sub-MIC levels. Anaerobe 19:1–8

Parmar EK, Fougat RS, Patel CB, Zala HN, Patel MA, Patel SK, Kumar S (2015) Validation of dbEST-SSRs and transferability of some other solanaceous species SSR in ashwagandha [Withania somnifera (L.) Dunal]. 3 Biotech 5(6):933

Pati PK, Sharma M, Salar RK, Sharma A, Gupta AP, Singh B (2008) Studies on leaf spot disease of Withania somnifera and its impact on secondary metabolites. Indian J Microbiol 48:432–437

Petrovska BB (2012) Historical review of medicinal plants? Usage. Pharmacogn Rev 6(11):1

Praveen N, Murthy HN (2010) Production of withanolide-A from adventitious root cultures of Withania somnifera. Acta Physiol Plant 32:1017–1022

Praveen N, Murthy HN (2012) Synthesis of withanolide a depends on carbon source and medium pH in hairy root cultures of Withania somnifera. Ind Crop Prod 35:241–243

Rajasekar S, Elango R (2011) Effect of microbial consortium on plant growth and improvement of alkaloid content in Withania somnifera (Ashwagandha). Curr. Bot. 2:27–30

Rana S, Lattoo SK, Dhar N, Razdan S, Bhat WW, Dhar RS, Vishwakarma R (2013) NADPH-cytochrome P450 reductase: molecular cloning and functional characterization of two paralogs from Withania somnifera (L.) Dunal. PLoS One 8(2):e57068

Rana S, Bhat WW, Dhar N, Pandith SA, Razdan S, Vishwakarma R, Lattoo SK (2014) Molecular characterization of two A-type P450s, WsCYP98A and WsCYP76A from Withania somnifera (L.) Dunal: expression analysis and withanolide accumulation in response to exogenous elicitations. BMC Biotechnol 14(1):89

Rani G, Grover IS (1999) In vitro callus induction and regeneration studies in Withania somnifera. Plant Cell Tissue Org Cult 57:23–27

Ravindran R, Sharma N, Roy S, Thakur AR, Ganesh S, Kumar S et al (2015) Interaction studies of Withania somnifera 's key metabolite Withaferin A with different receptors associated with cardiovascular disease. Current Computer-Aided Drug Design 11(3):212–221

Ray S, Jha S (2001) Production of withaferin A in shoot cultures of Withania somnifera. Planta Med 67:432–436

Razdan S, Bhat WW, Rana S, Dhar N, Lattoo SK, Dhar RS,Vishwakarma RA (2013) Molecular characterization and promoter analysis of squalene epoxidase gene from Withania somnifera (L.) Dunal. Mol Biol Rep:1–12

Razdan S, Bhat WW, Dhar N, Rana S, Pandith SA, Wani TA et al (2016) Molecular characterization of DWF1 from Withania somnifera (L.) Dunal: its implications in withanolide biosynthesis. J Plant Biochem Biotechnology 1(26):52–63

Rober SC (2007) Production and engineering of terpenoids in plant cell culture. Nat Chem Biol 3:387–395

Roja G, Heble M, Sipahimalani A (1991) Tissue cultures of Withania somnifera: morphogenesis and withanolide synthesis. Phytother Res 5:185–187

Sabir F, Sangwan NS, Chaurasiya ND, Misra LN, Sangwan RS (2008a) In vitro Withanolide production by Withania somnifera L cultures. Z Naturforsch C 63:409–412

Sabir F, Sangwan NS, Chaurasiya ND, Misra LN, Tuli R, Sangwan RS (2008b) Rapid micropropagation of Withania somnifera L accessions from axillary meristems. J Herbs Spices Med Plant 13:123–133

Sabir F, Sangwan RS, Jyoti SJ, Misra LN, Pathak N, Sangwan NS (2011) Biotransformation of withanolides by cell suspension cultures of Withania somnifera (Dunal). Plant Biotechnol Rep 5:127–134

Sabir, F., Sangwan, R. S., Kumar, R., & Sangwan, N. S. (2012). Salt stress-induced responses in growth and metabolism in callus cultures and differentiating in vitro shoots of Indian ginseng (Withania somnifera Dunal). Journal of plant growth regulation, 31(4), 537–548

Sabir F, Mishra S, Sangwan RS, Jadaun JS, Sangwan NS (2013) Qualitative and quantitative variations in withanolides and expression of some pathway genes during different stages of morphogenesis in Withania somnifera Dunal. Protoplasma 250:539–549

Saema S, Rahman LU, Singh R, Niranjan A, Ahmad IZ, Misra P (2016) Ectopic overexpression of WsSGTL1, a sterol glucosyltransferase gene in Withania somnifera, promotes growth, enhances glycowithanolide and provides tolerance to abiotic and biotic stresses. Plant Cell Rep 35(1):195

Saito K (2013) Phytochemical genomics – a new trend. Curr Opin Plant Biol 16:373–380

Sanchita SR, Mishra A, Dhawan SS, Shirke PA, Gupta MM, Sharma A (2015) Physiological performance secondary metabolite and expression profiling of genes associated with drought tolerance in Withania somnifera. Protoplasma 252:1439–1450

Sangwan NS, Sangwan RS (2013) Applied plant cell biology secondary metabolites of traditional medical plants: a case study of Ashwagandha (Withania somnifera). Plant Cell Monogr 22:325–367

Sangwan NS, Sangwan RS (2014) Secondary metabolites of traditional medical plants: a case study of Ashwagandha (Withania somnifera). Appl Plant Cell Biol 22:325–367

Sangwan RS, Chaurasia ND, Misra LN, Lal P, Uniyal GC, Sharma R, Sangwan NS, Suri KA, Qazi GN, Tuli R (2004) Phytochemical variability in commercial herbal products and preperations of Withania somnifera (Ashwagandha). Curr Sci 86:461–465

Sangwan RS, Chaurasiya ND, Misra LN, Lal P, Uniyal GC, Sangwan NS, Srivastava AK, Suri KA, Qazi GN, Tuli R (2006) Process for isolation of withaferin-a from plant materials and products there from. US Patent Patent 7,108,870, 19 Sept 2006

Sangwan RS, Chaurasiya ND, Lal P (2007) Withanolide a biogeneration in in vitro shoot cultures of Ashwagandha (Withania somnifera Dunal) a main medicinal plant in Ayurveda. Chem Pharm Bull 55:1371–1375

Sangwan RS, Das Chaurasiya N, Lal P, Misra L, Tuli R, Sangwan NS (2008) Withanolide A is inherently de novo biosynthesized in roots of the medicinal plant Ashwagandha (Withania somnifera). Physiol Plant 133:278–287

Sawada Y, Nakabayashi R, Yamada Y, Suzuki M, Sato M, Sakata A, Akiyama K, Sakurai T, Matsuda F, Aoki T, Hirai MY, Saito K (2012) RIKEN tandem mass spectral database (ReSpect) for phytochemicals A plant-specific MS/MS-based data resource and database. Phytochemistry 82:38–45

Schauer N, Semel Y, Roessner U (2006) Comprehensive metabolic profiling and phenotyping of interspecific introgression lines for tomato improvement. Nat Biotechnol 24:447–454

Sehgal N, Gupta A, Valli RK, Joshi SD, Mills JT, Hamel E et al (2012) Withania somnifera reverses Alzheimer's disease pathology by enhancing low-density lipoprotein receptor-related protein in liver. Proc Natl Acad Sci 109(9):3510–3515

Sen J, Sharma AK (1991) Micropropagation of Withania somnifera from germinating seeds and shoot tips. Plant Cell Tissue Organ Cult 26:71–73

Senthil K, Wasnik NG, Kim YJ, Yang DC (2010) Generation and analysis of expressed sequence tags from leaf and root of Withania somnifera (Ashwgandha). Mol Biol Rep 37:893–902

Senthil K, Jayakodi M, Thirugnanasambantham P, Lee SC, Duraisamy P, Purushotham PM, Rajasekaran K, Nancy Charles S, Mariam Roy I, Nagappan AK, Kim GS, Lee YS, Natesan S, Min TS, Yang TJ (2015) Transcriptome analysis reveals in vitro cultured Withania som-

nifera leaf and root tissues as a promising source for targeted withanolide biosynthesis. BMC Genomics 16:14
Sharada M, Ahuja A, Suri K, Vij SP, Khajuria RK, Verma V, Kumar A (2007) Withanolide production by in vitro cultures of Withania somnifera and its association with differentiation. Biol Plant 51:161–164
Sharma H, Chandola HM, Singh G, Basisht G (2007) Utilization of Ayurveda in health care: an approach for prevention, health promotion, and treatment of disease. Part 2—Ayurveda in primary health care. J Altern Complement Med 13(10):1135–1150
Shohat B, Kirson I, Lavie D (1978) Immunosuppressive activity of two plant steroidal lactones withaferin A and withanolide E. Biomedicine 28(1):18–24
Shukla DD, Bhattarai N, Pant B (2010) In-vitro mass propagation of Withania somnifera (L) Dunal. Nepal J Sci Technol 11:101–106
Siddique AA, Joshi P, Misra L, Sangwan NS, Darokar MP (2014) 5 6-De-epoxy-5-en-7-one-17-hydroxy withaferin A a new cytotoxic steroid from Withania somnifera L Dunal leaves. Nat Prod Res 28:392–398
Sil B, Mukherjee C, Jha S, Mitra A (2015) Metabolic shift from withasteroid formation to phenylpropanoid accumulation in cryptogein-cotransformed hairy roots of Withania somnifera (L) Dunal. Protoplasma 252:1097–1110
Singh AK, Varshney R, Sharma M (2006) Regeneration of plants from alginate-encapsulated shoot tips of Withania somnifera (L) Dunal a medicinally important plant species. J Plant Physiol 163:220–223
Singh P, Guleri R, Singh V, Kaur G, Kataria H, Singh B, Kaur G, Kaul SC, Wadhwa R, Pati PK (2015) Biotechnological interventions in Withania somnifera (L) Dunal. Biotechnol Genet Eng Rev 31:1–20
Singh G, Tiwari M, Singh SP, Singh S, Trivedi PK, Misra P (2016) Silencing of sterol glycosyltransferases modulates the withanolide biosynthesis and leads to compromised basal immunity of Withania somnifera. Sci Rep 6:25562
Sivanandhan G, Mariashibu TS, Arun M, Rajesh M, Kasthurirengan S, Selvaraj N, Ganapathi A (2011) The effect of polyamines on the efficiency of multiplication and rooting of Withania somnifera (L) Dunal and content of some withanolides in obtained plants. Acta Physiol Plant 33:2279–2288
Sivanandhan G, Arun M, Mayavan S, Rajesh M, Jeyaraj M, Dev GK, Manickavasagam M, Selvaraj N, Ganapathi A (2012a) Optimization of elicitation conditions with methyl jasmonate and salicylic acid to improve the productivity of withanolides in the adventitious root culture of Withania somnifera (L) Dunal. Appl Biochem Biotechnol 168:681–696
Sivanandhan G, Arun M, Mayavan S, Rajesh M, Mariashibu TS, Manickavasagam M, Selvaraj N, Ganapathi A (2012b) Chitosan enhances withanolides production in adventitious root cultures of Withania somnifera (L) Dunal. Ind. Crops Prod 37:124–129
Sivanandhan G, Dev GK, Jeyaraj M (2013) A promising approach on biomass accumulation and withanolides production in cell suspension culture of Withania somnifera (L) Dunal. Protoplasma 250:885–898
Sivanandhan G, Selvaraj N, Ganapathi A (2014) Enhanced biosynthesis of withanolides by elicitation and precursor feeding in cell suspension culture of Withania somnifera (L) Dunal in shake-flask culture and bioreactor. PLoS One 9:e104005
Siriwardane AS, Dharmadasa RM, Samarasinghe K (2013) Varieties of Withania sommfera (L.) Dunal. Grown in Sri Lanka. Pakistan journal of Bbiological Sciences 16(3):141–144
Srivastava S, Sangwan RS, Tripathi S, Mishra B, Narnoliya LK, Misra LN, Sangwan NS (2015) Light and auxin responsive cytochrome P450s from Withania somnifera Dunal: cloning, expression and molecular modelling of two pairs of homologue genes with differential regulation. Protoplasma 252(6):1421–1437
Subramanian SS, Sethi PD (1969) Withaferin–A from Withania somnifera coagulants roots. Current science (India) 38:267–268
Subramanian C, Zhang H, Gallagher R, Hammer G, Timmermann B, Cohen M (2014) Withanolides are potent novel targeted therapeutic agents against adrenocortical carcinomas. World J Surg 38(6):1343

Supe U, Dhote F, Roymon MG (2006) In vitro plant regeneration of Withania somnifera. Plant Tissue Cult Biotech 16:111–115

Takimoto T, Kanbayashi Y, Toyoda T, Adachi Y, Furuta C, Suzuki K et al (2014) 4β -Hydroxywithanolide E isolated from Physalis pruinosa calyx decreases inflammatory responses by inhibiting the NF-κB signaling in diabetic mouse adipose tissue. Int J Obes 38(11):1432–1439

Takshak S, Agrawal SB (2014) Secondary metabolites and phenylpropanoid pathway enzymes as influenced under supplemental ultraviolet-B radiation in Withania somnifera Dunal an indigenous medicinal plant. J Photochem Photobiol 140:332–343

Tavhare SD (2015) Collection practices of medicinal plants vedic ayurvedic and modern perspectives. Int J Pharm Biol Arch 5(5)

Thimm O, Blaesing O, Gibon Y (2004) MAPMAN: a user-driven tool to display genomics data sets onto diagrams of metabolic pathways and other biological processes. Plant J 37:914–939

Tieman D, Taylor M, Schauer N, Fernie AR, Hanson AD, Klee HJ (2006) Tomato aromatic amino acid decarboxylases participate in synthesis of the flavor volatiles 2-phenylethanol and 2-phenylacetaldehyde. Proc Natl Acad Sci U S A 103:8287–8292

Tokimatsu T, Sakurai N, Suzuki H, Ohta H, Nishitani K, Koyama T, Umezawa T, Misawa N, Saito K, Shibata D (2005) KaPPA-view a web-based analysis tool for integration of transcript and metabolite data on plant metabolic pathway maps. Plant Physiol 138:1289–1300

Tong X, Zhang H, Timmermann BN (2011) Chlorinated withanolides from Withania somnifera. Phytochem Lett 4:411–414

Tripathi V, Verma J (2014) Current updates of indian antidiabetic medicinal plants. Int J Res Pharm Chem 4(1):114–118

Tugizimana F, Piater L, Dubery (2013) Plant metabolomics: a new frontier in phytochemical analysis. S Afr J Sci 109(5/6):1–11

Urbanczyk-Wochniak E, Luedemann A, Kopka J, Selbig J, Roessner-Tunali U, Willmitzer L, Fernie AR (2003) Parallel analysis of transcript and metabolic profiles: a new approach in systems biology. EMBO Rep 4:989–993

Urbanczyk-Wochniak E, Baxter C, Kolbe A, Kopka J, Sweetlove LJ, Fernie AR (2005) Profiling of diurnal patterns of metabolite and transcript abundance in potato (Solanum tuberosum) leaves. Planta 221:891–903

Van Wyk BE, Wink M (2004) Medicinal plants of the world: an illustrated scientific guide to important medicinal plants and their uses. Timber Press

Varghese S, Keshavachandran R, Baby B, Nazeem PA (2014) Genetic transformation in ashwagandha Withania somnifera (L) Dunal for hairy root induction and enhancement of secondary metabolites. J Trop Agri 52:39–46

Wadegaonkar PA, Bhagwat KA, Rai MK (2006) Direct rhizogenesis and establishment of fast growing normal root organ culture of Withania somnifera Dunal. Plant Cell Tissue Organ Cult 84:223–225

Wadhwa R, Singh R, Gao R, Shah N, Widodo N, Nakamoto T, Kaul S (2013) Water extract of Ashwagandha leaves has anticancer activity: identification of an active component and its mechanism of action. PLoS One 8(10):e77189

Widodo N, Takagi Y, Shrestha BG, Ishii T, Kaul SC, Wadhwa R (2008) Selective killing of cancer cells by leaf extract of Ashwagandha: components, activity and pathway analyses. Cancer Lett 262(1):37–47

Xu YM, Gao S, Bunting DP (2011) Unusual withanolides from aeroponically grown Withania somnifera. Phytochemistry 72:518–522

Ziegler J, Facchini PJ (2008) Alkaloid biosynthesis: metabolism and trafficking. Annu Rev Plant Biol 59:735–769

Chapter 2
Phytopharmacology of Ashwagandha as an Anti-Diabetic Herb

Vikas Kumar, Amitabha Dey, and Shyam Sunder Chatterjee

Abstract Ashwagandha (*Withania somnifera*) extracts and several pharmaceutical formulations containing them are currently often used as tonics useful for prevention and cure of mental health problems, including sleep disturbances, accompanying or caused by diverse slowly progressing chronic diseases. The possibility that *W. somnifera* could be used for treatments of diabetes and associated metabolic disturbances were first suggested by the results of an exploratory clinical study conducted with its root powdered in diabetic patients and published in 2000. Since then, numerous preclinical and a randomized, double blind and placebo controlled clinical study with extracts of the plant have continued to add experimental evidences in favor of the convictions of the scholars and practitioners of Ayurvedic and other traditionally known systems of medicine that the plant could also be used for prevention and cure of diabetes and other metabolic disorders associated physical and mental health problems. Currently available information suggesting such possibilities are summarized and critically analyzed in this chapter. Potential uses of our current knowledge on phytopharmaclogy and medicinal phytochemistry of the plant and its bioactive constituents for obtaining more sustainable and reproducible health benefits from the plant in patients suffering from, or at risk to, metabolic disorders associated mental health problems, or for discovering novel therapeutic lead against such health problems of the twenty-first century are also discussed.

Keywords *Withania somnifera* • Hyperglycemia • Diabetic complications • Oxidative stress • Psychosomatic disorders • Multi-targeted therapy

V. Kumar (✉)
National Center of Natural Products Research, School of Pharmacy, University of Mississippi, Oxford MS 38677, USA
e-mail: vkumar@olemiss.edu

A. Dey
Neuropharmacology Research Laboratory, Department of Pharmaceutics, Indian Institute of Technology (Banaras Hindu University), Varanasi 221 005, India

S.S. Chatterjee
Retired Head of Pharmacology Research Laboratories, Stettiner Straße 1, D-76139 Karlsruhe, Germany

S.C. Kaul, R. Wadhwa (eds.), *Science of Ashwagandha: Preventive and Therapeutic Potentials*, DOI 10.1007/978-3-319-59192-6_2

2.1 Introduction

Diabetes is a slowly and silently progressing metabolic disorder, which ultimately leads to structural and functional deteriorations of almost all bodily organs including those of circulating fluids and central and peripheral nervous systems. It is one of the major health problems now affecting almost all countries, irrespective of their socioeconomic, cultural, and genetic backgrounds (Shaw et al. 2010). According to recent estimates, more than 85% of diabetic patients live in India, China and others developing countries where herbal remedies and advices of the practitioners of traditionally known systems of medicine still continue to be the only health care and therapeutic possibilities for a vast majority of their population. Most probably, prevalence of diabetes in those countries is also continuing to increase during the next two decades (Guariguata et al. 2014).

Co-morbidities of anxiety, depression and other mental health problems are more often encountered in diabetic patients than in normal population (Mendenhall et al. 2014; Balhara and Sagar 2011; Snoek et al. 2015; Petrak et al. 2015; Bajor et al. 2015), and it is now well recognized that serious psychological distress has a significant negative impact on the initiation and progression of diabetes and associated comorbidities (Egede and Dismuke 2012; Egede et al. 2014). However, cardiovascular diseases are the leading causes of death and other health problems of patients suffering from, or at risk to, type 2 diabetes. While lowering blood glucose levels in such patients have major beneficial effects on microvasculature complications, as yet no very definitive statements on beneficial effects of available psychoactive drugs and other treatments on mental health problems and cardiovascular outcome are possible (Huqi 2015; McMurray et al. 2014).

The most ancient text of traditionally known system of medicine, i.e. Ayurveda, mention that metabolic abnormalities leading to sweet tasting urine are associated with abnormal body weight changes and sedentary behaviour (Sharma and Chandola 2011a, b; Puranik and Patwardhan 2012). The Ayurvedic concept that proper choices of food and eating habits in combination with physical and mental exercises and medicine is useful for prevention and cure of such disorders, now referred to as diabetes or diabesity, have now been well accepted by all other traditional known or modern systems of medicine (Ahmed 2002; Henschen 1969; White 2014; Zajac et al. 2010; Zhang et al. 2010). Specific combinations of diverse edible and psychoactive plants with anti-stress or adaptogenic properties are often used as therapeutics in Ayurvedic and other traditionally known systems of medicine for prevention and cure of physical and mental health problems accompanying almost all chronic diseases. Amongst them, *W. somnifera* (Ashwagandha, or Winter cherry) is one of the clinically and preclinically more extensively scrutinized ones now also attracting considerable attention of modern drug designers. It is one of the few plants of the *Withania* genus of flowering plants of the Solanacea family with 23 species that are native to many Afro-Asian countries (Mirjalili et al. 2009). The two other medicinally used ones in India and other Afro-asiatic countries are *W. coagulans* and *W. obtusifolia* (Kumar et al. 2011; Alali et al. 2014), some scattered reports on their beneficial effects against diabetes have also appeared during more recent years.

A group of naturally occurring poly-oxygenated C-28 estrogen-type steroids commonly known as Withanolides (Misico et al. 2011; Singh et al. 2010; Choudhary and Yousuf 2013) are chemotaxonomic markers of the plants of the *Withania* genus (Alali et al. 2014; Chen et al. 2011; Zhang et al. 2014; Samadi 2015). Since anti-hyperglycemic and diverse other therapeutically interesting bioactivities of structurally diverse Withanolides have been reported (Gorelick et al. 2015; Khodaei et al. 2012), they are now often considered to be their quantitatively major bioactive secondary metabolites. It cannot be ignored though, that numerous other extractable secondary metabolites of *W. somnifera* with Withanolides like bioactivity profiles in preclinical models are now known also (Singh et al. 2001, 2003; Wadhwa et al. 2013). Therefore, it is apparent that like for all other medicinal plant extracts, bioactivity profiles or therapeutic benefits of *W. somnifera* extracts do not also depend entirely on their contents of Withanolides, or any other chemotaxonomic marker of the plant, only. Since our current knowledge on biological interactions between the structurally diverse extractable bioactive constituent of the plant still remain far from being satisfactory, as yet no very definitive statements on therapeutic potentials of diverse types of medicinally used *W. somnifera* extracts, or on their bioactive constituents involved in their such potentials, can yet be made.

Critical analysis of currently available preclinical and clinical information on such potentials of *W. somnifera* extracts and their quantitatively major bioactive constituents strongly suggest though, that appropriately processed and standardized extracts of the plant and several bioactive metabolite of the plant could be useful therapeutic leads against diabetes mellitus and associated co-morbidities. Therefore, efforts are now being made in several laboratories, including ours, to better define the medicinal phytochemistry and phytopharmaclogy of the plant. Currently available preclinical and clinical information on such extracts and their bioactive constituents potentially involved in their glucose and insulin homeostasis regulating effects are summarized and critically analyzed in this chapter.

2.2 Why Should *Withania somnifera* Be Considered as an Anti-Diabetic Plant?

Together with sedentary behaviour, malnutrition (both over and under-nutrition) is a major risk factor for diabetes mellitus (Zimmet et al. 2014; Schmidt and Duncan 2003), defined as a disease characterized by hyperglycemia and secretion and excretion of excessive amount of urine. Available therapeutic options for prevention and cure of diabetes (Colagiuri 2010) either do not meet the therapeutic demands of patients, or are not affordable or available to a vast majority of global population. Diabesity, *i.e.* obesity associated type-2 diabetes, is the spreading epidemic of the twenty-first century and is also major economic burden in many countries including

USA and other economically developed ones (Farag and Gaballa 2010). Amongst all currently available antidiabetic drugs, metformin is the only one highly recommended for prevention and cure of type-2 diabetes, which is the most prevalent type (ca. 90%) encountered amongst all diabetic patients (Liu et al. 2010; Anwer et al. 2008). The development of type-2 diabetes begins with an impairment of glucose tolerance and insulin resistance, i.e. a condition when the cells in the body are unable to use insulin secreted from the β-pancreatic and other cells, and which eventually leads to hyperglycemia and hyperinsulinemia (Anwer et al. 2008; Zimmet and Thomas 2003; Robertson and Harmon 2006). The symptoms of diabetes include high blood sugar, polydypsia, polyuria, unusual thirst, polyphagia, weakness and tiredness, intense hunger and sudden weight loss. In addition to high blood sugar, other factors including dyslipidemia or hyperlipidemia also play important roles in the development of micro and macro-vascular complications of Type-2 diabetes, which ultimately leads to cardiovascular disorders and other co-morbidities and death (Tiwari et al. 2014). Cognitive deficits and diverse spectrums of mental health problems, including anxiety and depression, are more often encountered in diabetic patients than in normal population, and it is now well documented that serious psychological suffering has a major negative impact on diabetes process and outcome (Thakur et al. 2015; Balhara and Sagar 2011; Sarkar and Balhara 2014).

Like for numerous other lifestyles associated medical conditions, sleep disturbances has now also been well recognized to be another major risk factor for diabetes development (Anothaisintawee et al. 2015; Kowall et al. 2016). Changes in diet, physical activity, body eight, and obesity (as judged by body mass index) do not seem to be effective in decreasing diabesity risk of sleep disturbances (Cespedes et al. 2016; Lee et al. 2016). Due to their diabetogenic and obesogenic effects, numerous currently available psychoactive and sleep inducing or tranquilizing drugs are often contraindicated in patients suffering from or at risk to diabetes (Flanagan 2008). Environmental and metabolic stress triggered anxiety and depression are the root causes of sleep disturbances (Spoormaker and van den Bout 2005; Leroith and Vinik 2008), and *W. somnifera* has since long been well recognized to be a sleep inducing plant without sedative or hypnotic effects (Archana et al. 2015; Pingali et al. 2013; Choudhary et al. 2015). It has recently been pointed out indeed, that while most herbal adaptogens possess stimulating effects, *W. somnifera* is a calming adaptogenic herb (Winston and Maimes 2007). Numerous preclinical and some clinical studies have consistently revealed and reconfirmed beneficial effects of diverse types of *W. somnifera* extracts against stress triggered psychopathologies leading to sleep disturbance (Chandrasekhar et al. 2012; Ashok and Shende 2015; Archana et al. 2015; Pratte et al. 2014; Dar et al. 2015; Wadhwa et al. 2015). Therefore, *W. somnifera* seems to be a particularly well-suited adaptogenic herb for prevention and cure of sleep disturbances triggered metabolic disorders like diabetes and diabesity.

2.3 Pathophysiology of Diabetes Comorbidities

It has since long been well recognized that diabetes is not a single disease entity, but rather a group of metabolic disorders sharing the common underlying feature of hyperglycemia (Anderson and Auslander 1980). Insulin is the principle hormone that regulates the uptake of glucose from the blood into most other cells. Therefore, deficiency of insulin or insensitivity of its receptors plays a crucial role in all forms of diabetes. The body can obtains glucose either from food, digested and processed inside the gastrointestinal tract, or by breakdown of from stored glucose precursor glycogen or from gluconeogenesis i.e., the generation of glucose from non-carbohydrate substrates in the body. Glucose is used by about two thirds of the body's cells for maintaining their energy balance and for conversion of glucose to other needed molecules or for storage and other functions.

Lower glucose levels results in decreased insulin release from pancreatic cells and in breakdown of glycogen to glucose. This process is mainly controlled by the hormone glucagon, which acts in opposite manner to insulin (Gardner and Shoback 2011). When the glucose concentration in blood remains high for a long time, the kidney reach a threshold of glucose reabsorption, and glucose is excreted in the urine causing glucosuria. This increases the osmotic pressure of the urine and inhibits water reabsorption by kidney, resulting in increased urine production (polyuria) and increased fluid loss. Resulting loss of blood volume is then replaced osmotically from water held in the body cells and other body compartments, causing dehydration and increased thirst.

Apart from insulin and glucagon, numerous other hormones of the gastrointestinal tract are also involved in pathogenesis of obesity and type-2 diabetes (Vincent et al. 2008; Adamska et al. 2014; Thomas et al. 2003; Scheen and Paquot 2015). It is now becoming increasingly apparent that gut microbial ecology also plays a crucial role in regulating the homeostasis of gastrointestinal hormones, and that dysbiosis or dysbacteriosis inside the gastrointestinal tracts often leads to numerous chronic diseases, including diabetes and obesity (Nicholson et al. 2012; Miele et al. 2015; Durg et al. 2015; Choudhury et al. 2016). Diverse gastrointestinal hormones are encountered or even biosynthesized inside the brain, and that they are also involved in the pathogenesis and progression of diabesity associated co-morbid psychopathologies (Torres-Fuentes et al. 2015; Thakur et al. 2014). Since like structurally diverse plant derived molecules and other products, *W. somnifera* extracts and substances derived from them also possess bactericidal activities (Kapoor et al. 2015; Singh and Kumar 2011; Girish et al. 2006), it could as well be that modulation of gut microbial ecology are also involved in broad spectrums of beneficial effects of the plant observed in diabetic patients and animals. Efforts to verify this possibility could not only be useful for better understanding of therapeutic potential of the plant, but also for obtaining novel therapeutic leads against diabesity and associated co-morbidities.

2.4 Preclinical Pharmacology of *Withania somnifera*

The very first reports dealing with preclinical pharmacology of *W. somnifera* started appearing in English language journals during 1980s, and dealt mainly with adaptogenic or stress response regulating effects of the herb. Since several observations have revealed Ginseng like effects of the plant in diverse animal models, it is now often referred to as "Indian ginseng" (Kulkarni and Dhir 2008). It was not until the year 2000 when the very first reports on beneficial effects of the plant in diabetic patients had appeared (Andallu and Radhika 2000). This clinical trial was conducted to corroborate an earlier preclinical finding made with an Ayurvedic formulation of the plant (Trasina) in diabetic rats (Bhattacharya et al. 1997a). Currently available experimental evidences reaffirming that *W. somnifera* is indeed another anti-diabetic, or anti-hyperglycemic, or glucose homeostasis regulating plant are summarized in the following. For more detailed currently available preclinical and clinical information on neuro-psycho-pharmacology (Durg et al. 2015), and other therapeutically interesting bioactivities of the plant, more recent reviews mentioned before and other chapters of this book has to be consulted.

2.4.1 Anti-Hyperglycemic Activities in Rodent Models

Prevention of diabetes is the primary goal of almost all health care authorities around the globe (WHO 2006). The biomarker or diagnostic criteria now regularly used in all epidemiological or clinical studies for identifying patients prone to, or suffering from, diabetes are blood glucose responses to meals and to diverse versions of glucose-tolerance and insulin sensitivity tests in relation to body weights or body mass index of a person (Bentley-Lewis et al. 2008; Marrero et al. 2014; Hostalek et al. 2015). Several reports on beneficial effects on body weight changes, blood glucose and insulin levels, and glucose or insulin tolerance in stressed, or obese, or diabetic rodents have appeared during the past decade. Some of the most cited and more recent reports revealed such effects of diverse types of *W. somnifera* extracts are summarized in Table 2.1. One such report dealing with effects of *W. somnifera* extracts in alloxan-induced diabetic rats attributed the observed effects of the extract to the flavonoids encountered in roots of the plant, and suggested that this is due to their ability to stimulate insulin release from pancreatic cells (Udayakumar et al. 2009). Such and numerous other *W. somnifera* extracts like beneficial effects of pure flavonoids and their metabolites have often been reported (Dall'Asta et al. 2015). It cannot be ignored though, that numerous other phytochemicals and other substances capable of modulating glucose homeostasis and with antidiabetic activity in animal models are also encountered in such extracts, and that pancreatic cells are not the only insulin producing and secreting cells involved in etiology and pathogenesis diabetes (Kojima et al. 2006; Lehner et al. 2012).

Table 2.1 Some often cited and more recent reports on anti-diabetic (anti-hyperglycemic) potential of *Withania somnifera*

Part of plant	Type of extract/formulation	Dose, duration and route of administration	Mechanism of action	References
Leaves and roots	Ethanolic	100 and 200 mg/kg/day, for 56 days, p.o.	Increased insulin secretion from pancreatic β cells	Udayakumar et al. (2009)
Roots	Aqueous	200 and 400 mg/kg/day, for 35 days, p.o.	Increased insulin sensitivity and inhibited insulin resistance	Anwer et al. (2008)
Leaves and roots	Ethanolic	200 mg/kg/day, for 56 days, p.o.	Insulin mimetic and increased insulin secretion from pancreatic β cells	Sarangi et al. (2013)
Leaves	Ethanolic	1 ml/rat/day (30 and 200% concentration), for 10 days, i.p.	Increased hepatic metabolism of glucose and increased insulin secretion	Navinder et al. (2013)
Roots	Dried powder	3 g/day (500 mg/capsule) for 30 days, p.o.	Increased serum level of insulin, decreased serum level of lipids	Andallu and Radhika (2000)
Leaves and roots	Methanolic	100 μg/ml in cellular model of diabetes	Increased insulin secretion and increased insulin sensitivity	Gorelick et al. (2015)
Roots	Aqueous	200 and 400 mg/kg/day for 35 days, p.o.	Increased glucose metabolism, increased insulin sensitivity	Safhi and Anwer (2011)
Roots	Hydromethanolic	25, 50 and 100 mg/kg/day for 10 days, p.o.	Insulin mimetic and anti-oxidant action	Thakur et al. (2015)
Roots	Ethanolic	1.4 g/kg/day for 15 days, p.o.	Increased insulin secretion and insulin sensitivity	Jatwa and Kar (2009)
Root	Dried powder	5, 10, 20 mg/kg, p.o. for in vivo assay and 0.3, 0.6, 0.9, 1.2 and 1.5 mg/ml for in vitro assay	Increased transport f glucose, increased insulin sensitivity and inhibition of α- amylase enzyme	Nirupama et al. (2014)
Roots	Acetone	20 mg/kg/day for 1 month, p.o.	Antioxidant action and increased insulin sensitivity	Parihar et al. (2004)

(continued)

Table 2.1 (continued)

Part of plant	Type of extract/formulation	Dose, duration and route of administration	Mechanism of action	References
Leaves	Aqueous and methanolic	1 ml of extract used for in vitro assay	Inhibition of α- amylase enzyme	Prabhakar et al. (2013)
Whole plant	Methanolic	Withanolides (IC_{50}–38.20 μg/ml) isolated were used for in vitro assay	Inhibition of α-glucosidase enzyme	Khan et al. (2014)
Roots	Aqueous	0.58 mg/ml for in vitro assay	Inhibition of both α- amylase and α-glucosidase enzyme	Balaji et al. (2015)
Roots	Standardized extract	500 mg/kg/day for 15 days, p.o.	Antioxidant action and increased insulin secretion	Kyathanahalli et al. (2014)
Roots	Aqueous	200 and 400 mg/kg/day for 35 days, p.o.	Antioxidant action and protected pancreatic β cells from oxidative damage	Anwer et al. (2012)
Roots	Hydoalcoholic	25 and 50 mg/kg/day for 21 days, p.o.	Antioxidant and adaptogenic action	Bhattacharya and Muruganandam (2003)
Fruits of *W. coagulans*	Aqueous and choroform	1 g/kg/day for 14 days, p.o.	Hypoglycemic, antioxidant and immunomodulation activities	Hoda et al. (2010)
Roots	Powder	62.5 mg/g WS powder with standard pellet diet for 56 days	Increased insulin sensitivity and decreased insulin resistance	Samadi Noshahr et al. (2015)
Roots and leaves	Ethanolic	100 mg/kg/day for 56 days, p.o.	Antioxidant action, and rejuvenation of β-cells leading to increased insulin production and secretion	Udayakumar et al. (2010)
Roots	Ethanolic	1.4 g/kg/day for 15 days, p.o.	Reduced oxidative stress and increased glucose metabolism	Jatwa and Kar (2009)
Trasina (Ayurvedic formulation)	Powder containing *W. somnifera*	100 and 200 mg/kg/day for 28 days, p.o.	Anti-oxidant action and increased insulin sensitivity	Bhattacharya et al. (1997a)

Extra-pancreatic insulin production and secretion have been observed in streptozotocin-induced diabetic rats (Cunha et al. 2007), and antidiabetic activity of repeated daily treatments with *W. somnifera* extracts have also been reported in such animals as well (Sarangi et al. 2013; Safhi and Anwer, 2011). In one of the reports dealing with anti-diabetic activity of *W. somnifera* (Anwer et al. 2008), no effects of even 400 mg/kg daily oral doses of an extract on blood glucose and glycosylated haemoglobin (HbA1C) levels, or in glucose and insulin tolerance tests, or on hepatic insulin resistance, were observed, whereas 200 mg/kg daily doses of the same extract was effective in affording protections against all these altered parameters in diabetic rats. Others using other animal models for hyperglycemia and in non-diabetic animals have also reported analogous observations. Such observations strongly suggest that beneficial effects of *W. somnifera* in diabetic rodents is not necessarily due to its effects on insulin secretion, and that its observed bioactivity profile in normal animals is not very predictive for its potential therapeutic benefits observed in diabetic and other animals with disturbed glucose homeostasis.

The alternative possibility that stimulatory and enhancing effects of Withanolides on glucose transporters and cellular glucose transport are involved in anti-diabetic activity of the plant is suggested by the observations made in some cellular as well as in animal models (Gorelick et al. 2015; Kumar et al. 2015a, b; Safhi and Anwar 2011). Plant metabolites structurally analogous to Withanolides (Coagunolides) encountered in *W. coagulans* have also been reported to be effective in suppressing postprandial hyperglycemia and anti-hyperglycemic activity in streptozotocin diabetic rats (Maurya et al. 2008). Since diverse steroidal lactones are encountered in varying concentrations in roots and leaves of *W. somnifera*, and antidiabetic activities of diverse parts of this and other plants of *Withania* family has been reported, it seems logical that they are also involved in traditionally known medicinal uses in diabetic patients. However, since several other fairly common plant metabolites like fumaric, succinic, 4-hydroxybenzoic, nicotinic and several other organic acids and other bioactive molecules with antidiabetic and other glucose homeostasis influencing bio-activities are also found in *W. somnifera* (Chatterjee et al. 2010), they could also be involved in observed effects of its crude extracts and root powders in animal models and clinical trials. Moreover, since depending on cultivations and extraction procedure used, the relative contents of Withanolides and other bioactive phytochemical in different parts and types of *W. somnifera* extracts vary considerably (Bharti et al. 2011; Dhanani et al. 2017; Fernando et al. 2013), observations made with one type of extract in a study cannot be generalized for all types of extracts.

In any case, there is now more than sufficient preclinical evidences suggesting preventive and curative potentials of diverse types of extracts from different parts of *W. somnifera* against diabetes and other chronic diseases caused by or associated with disturbances of glucose homeostasis. Recent observations in our laboratories and elsewhere (Thakur et al. 2015; Nirupama et al. 2014; Bhattacharya and Muruganandam 2003) have revealed that fairly low daily oral dose (25 mg/kg/day or lower) of *W. somnifera* root extracts are highly effective in regulating body weights and glucose metabolism in diabetic as well as in stressed non-diabetic rodents. These observations, taken together with the fact that such low oral doses of

W. somnifera extracts also modulate brain functions, strongly suggest that their observed anti-diabetic activities are most probably due to their modulating effects against environmental and metabolic stress triggered alterations in brain functions and that their regulating effects on biological oxidative processes are involved in their anti-hyperglycemic activities observed after their repeated daily doses. This inference is well supported by currently available evidences on the neurobehavioral activity profiles of *W. somnifera* (Durg et al. 2015; Dar et al. 2015).

2.4.2 Mechanistic Studies

The initial report (Bhattacharya et al. 1997a) appearing in 1997, and leading to the very first exploratory clinical study with *Withania somnifera* root powder in diabetic patients (Andallu and Radhika 2000), had indicated that hypoglycemic effect of the *Withania somnifera* containing Ayurvedic formulation (Trasina) could be due to its protective effects against oxidative free radicals triggered pathologies in pancreatic β-cells. A report suggesting that Glycowithanolides (Sitoindosides VII to X) and Withaferin A are involved in the antioxidative and stress response modulating or adaptogenic activity of *Withania somnifera* extracts had also appeared during the same year (Bhattacharya et al. 1997b). Another study revealing protective effects of fairly low daily oral doses (25 mg/kg/day) of a *Withania somnifera* extract against chronic stress triggered alteration in insulin sensitivity and glucose homeostasis had appeared after a few years thereafter (Bhattacharya and Muruganandam 2003). Since then the numbers of reports reaffirming protective effects of diverse types of *Withania somnifera* derived products and their bioactive constituents against diverse oxidative stress triggered pathologies have continued to increase (Mirjalili et al. 2009; Rai et al. 2016). Although it is now almost certain that Glycowithanolides and their aglycones are some of the bioactive anti-diabetic and oxidative stress protective constituents of *W. somnifera* extracts, the fact that they also contain numerous other bioactive molecules with anti-oxidative and other activities similar to those of Withanolides in numerous *in vitro* and completely animal models cannot be neglected.

In patients suffering from diabetes mellitus, oxidative stress induced by dysregulation of glucose homeostasis accompanies chronic inflammation, which eventually leads to tissue damages, and the crucial role of oxidative stress in the development of diabetic endothelial dysfunctions is also underlined in numerous studies (Rochette et al. 2014; Domingueti et al. 2015; Fiorentino et al. 2013; Hasnain et al. 2016). Diverse cellular heat shock proteins are involved in regulation of stress triggered alterations in glucose homeostasis and insulin sensitivity (Hooper and Hooper 2005; Hooper et al. 2014). It has been suggested that manipulation these cellular chaperons in metabolically relevant tissues represent a therapeutic avenue for prevention and cure of obesity associated diabetes and other metabolic disorders (Hooper 2014; Henstridge et al. 2014; Molina et al. 2016). *W. somnifera* root extracts and Withanolides, especially Withaferin A and Withanone, modulate the functions of

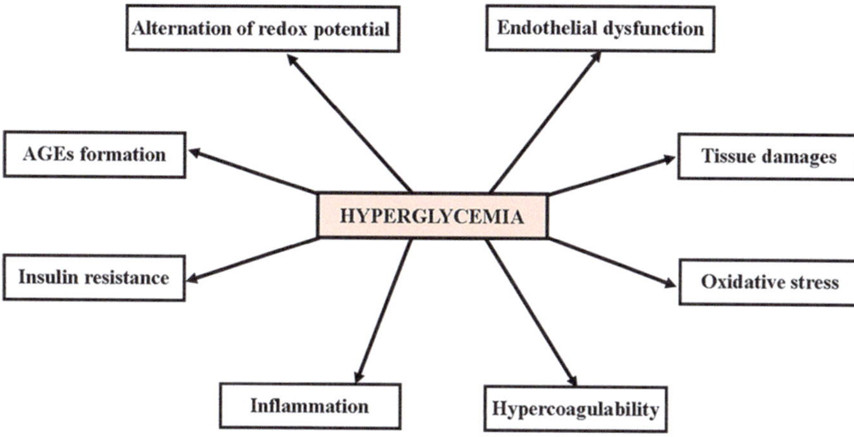

Fig. 2.1 The major pathological mechanisms involved in the development of vascular and other complications in diabetes mellitus (Adapted from: Domingueti et al. 2015)

such cellular chaperons and other nuclear processes regulating oxidative stress and cellular stress responses (Vanden Berghe et al. 2012; Gao et al. 2014; Misico et al. 2002; Mohan and Bargagna-Mohan 2016; Heyninck et al. 2014; Vaishnavi et al. 2012; Cui et al. 2014). Covalent or irreversible binding to sulfhydryl groups of proteins and other biological targets have been implicated in their broad spectrums of activities observed against diverse cellular metabolic processes (Antony et al. 2014). However, yet little concentrated efforts have been made to identify such mechanistic possibilities for other *W. somnifera* metabolites with anti-oxidative and anti-diabetic activities in animal and cellular models. Available information on such bioactivities of numerous phytochemicals encountered in *W. somnifera* (Chatterjee et al. 2010) strongly suggest though, that many of them can modulate the functions of diverse biological targets and processes involved in diverse hyperglycemia triggered pathological changes involved in diabesity associated vascular and other complications. Some such major mechanisms and processes involved in hyperglycemia-triggered pathologies are summarized in Fig. 2.1.

Apart from intracellular, several extracellular sites of actions for *W. somnifera* extracts have been suggested during more recent years (Fig. 2.2). Glucose transporters and other cellular processes regulating glucose uptake in skeletal muscle myotubes and adipocytes and some others involved in insulin secretion from basal pancreatic β cells are just a few of them (Jonathan et al. 2015; Kumar et al. 2015a; Safhi and Anwer 2011; Nirupama et al. 2014). Inhibition of enzymatic activities of α-amylase and α-glucosidase involved in starch degradation and digestive processes regulating glucose homeostasis has also been implicated in anti-hyperglycemic activities of *Withania somnifera* extracts and their bioactive constituents (Balaji et al. 2015; Prabhakar et al. 2013; Khan et al. 2014). In one of these reports (Khan et al. 2014), α-glucosidase inhibitory IC_{50} value of *W. somnifera* derived Withanolides in isolated cells was estimated to be 38.20 μg/ml. In another study (Khan et al.

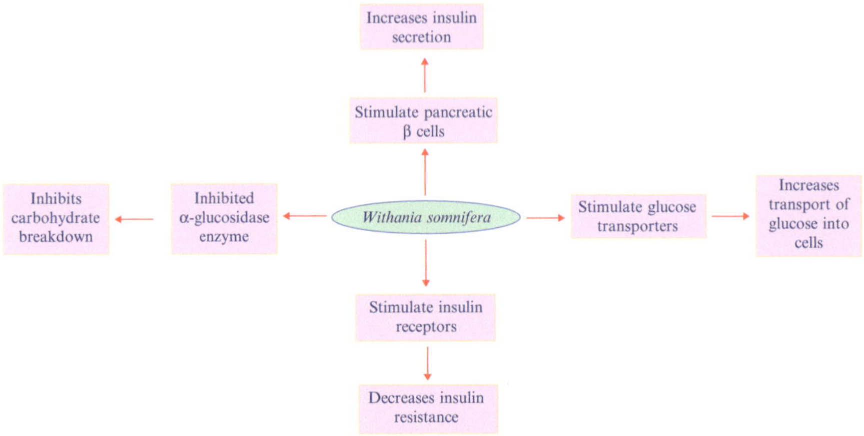

Fig. 2.2 Diverse proposed mechanisms that could be involved in glucose and insulin homeostasis influencing effects of *Withania somnifera* and its bioactive constituents. All such mechanisms could be influenced diverse *Withania somnifera* extract constituents by reversible as well as irreversible binding to diverse biological targets regulating glucose homeostasis

2014a) the same authors have reported that a roots extract of *W. somnifera* at 0.1 and 0.2 mg/ml dose levels possess anti-glycating activity (*in vitro*) via inhibition and reduction of cross-link breaker of glycated proteins or of extra cellular matrix proteins. Since after oral intake, the high concentrations can be expected only inside the gastrointestinal tract, and both α-amylase and α-glucosidase are involved in digestive process and other functions of the gastrointestinal tract, it could as well be that their anti-hyperglycemic and diverse other effects observed after their repeated daily oral doses in metabolically or otherwise stressed animals are also due to their modulating effects on the physiological functions of the digestive system inside this tract. Considering the facts that gut microbial ecology influences the functions of almost all innervated bodily organs (Fig. 2.3), and numerous *W. somnifera* constituents possess bactericidal activities as well as inhibitory effects on enzymes hydrolyzing carbohydrate polymers, this possibility is theoretically a very plausible one also.

Although many questions concerning bioactive constituents, modes and sites of actions still remain unanswered, it is now certain that Glycowithanolides and their aglycones are some of the quantitatively major bioactive constituents of *W. somnifera*, and that like numerous other secondary plant metabolites, they are also pluripotent molecules with therapeutically interesting effects on homeostatic processes regulating blood glucose levels, insulin sensitivity, and physiological functions of the peripheral as well as of the central nervous system. Since numerous therapeutically interesting effects of *W. somnifera* derived products become apparent, or more pronounced, in metabolically abnormal or stressed animals only, it seems reasonable to assume that their ability to facilitate adaptability of diverse bodily organs against noxious stress responses are also involved in their anti-diabetic or anti-hyperglycemic activities. In any case, they do not seem to be carbon copies of any

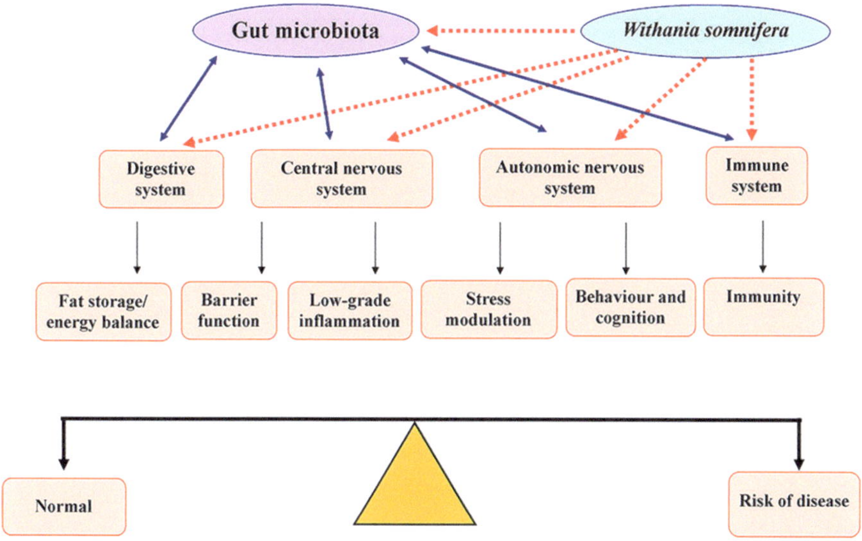

Fig. 2.3 Role of gut microbiota in regulating glucose homeostasis (Adapted from: Foster and Neufeld 2013). *Withania somnifera* extracts and many of their bioactive constituents posses bactericidal activity, and in addition can also modulate the functions of all organs functions regulating glucose homeostasis

currently known and therapeutically used anti-diabetic and other drugs. Since fairly high daily oral *W. somnifera* extract doses (400 mg/kg/day for several weeks) are well tolerated by diabetic or stressed rodents, and their estimated minimal daily stress response modulating and antidiabetic activities in rodent models are 25 mg/kg/day or lower (Thakur et al. 2015; Bhattacharya and Muruganandam 2003; Nirupama et al. 2014), it can also be said that safety range of products derived from the plant are very big. Thus, depending on the severity of diabetes in a given patient, daily oral doses of *W. somnifera* extracts could be fairly low or even very high.

2.5 *Withania somnifera* and Diabetic Complications

Hyperlipidemia is one of the major risk factor for cardiovascular disorders and premature deaths in patients suffering from diabetes mellitus, and it is now becoming increasingly apparent that dyslipidemia accompanying hyperglycemia can also eventually lead Alzheimer's disease and other dementia disorders (Saxena 2010; Tiwari et al. 2014; Aulston et al. 2013). Reports revealing anti-hyperlipidemic and other beneficial effects of *W. somnifera* and Withanolides against altered lipid metabolism in diabetic or hyperlipidemic animals are summarized in Table 2.2. In one of these reports (Visavadiya and Narasimhacharya 2007), enzymatic activity of HMG-CoA (the enzyme regulating cholesterol synthesis) in liver of cholesterol fed

Table 2.2 Some often cited and more recent reports on anti-hyperlipidemic potential of *Withania somnifera*

Part of plant	Type of extract/formulation	Dose, duration and route of administration	Mechanism of action	References
Leaves and roots	Ethanolic	100 and 200 mg/kg/day, for 56 days, p.o.	Increased high density lipoprotein-bound cholesterol and decreased very low density lipoprotein-bound cholesterol and low-density lipoprotein-bound cholesterol	Udayakumar et al. (2009)
Roots	Powder	Root powder combined with high fat diet and administered for 4 weeks	Increased hepatic antioxidant status and activity of hepatic HMG-CoA reductase enzyme	Visavadiya and Narasimhacharya (2011)
Roots	Powder	Root powder combined with diet at 0.75 and 1.5 gm/rat/day was administered for 4 weeks	Increased antioxidant activity and HMG-CoA reductase activity	Visavadiya and Narasimhacharya (2007)
Fruits of *Withaia coagulans*	Hydroalcoholic (60%)	1000 mg/kg/day for 4 weeks	Inhibited HMG-CoA reductase enzyme	Datta et al. (2013)
Leaves and roots	Ethanolic	200 mg/kg/day, for 56 days, p.o.	Increased high density lipoprotein-bound cholesterol and decreased very low density lipoprotein-bound cholesterol and low-density lipoprotein-bound cholesterol	Sarangi et al. (2013)
Roots	Dried powder	3 g/day (500 mg/capsule) for 30 days, p.o.	Decreased serum level of cholesterol, decreased serum level of lipids	Andallu and Radhika (2000)
Ambrex (a poly herbal formulation)	Powder containing *W. somnifera* (100 mg/g)	Ambrex (40 mg/kg/day) for 15 days, p.o.	Antioxidant action, increase HDL cholesterol and reduce LDL cholesterol	Devi and Rajkumar (2013)
Caps HT2 (herbal Ayurvedic formulation)	Capsule containing roots of *W. somnifera* (100 mg/g)	100–400 mg/kg/day for 30 days, p.o.	Antioxidant action, increased HDL cholesterol levels	Mary et al. (2003)

Trasina (Ayurvedic formulation)	Powder containing *W. somnifera*	100 and 200 mg/kg/day for 28 days, p.o.	Antioxidant and hypocholesterolemic activity	Bhattacharya et al. (1997)
Fruits of *Withania coagulans*	Aqueous	1 g/kg/day for 49 days, p.o.	Decresed elevated serum cholesterol, triglycerides and lipoprotein levels	Hemalatha et al. (2006)
Fruits of *Withania coagulans*	Chloroform	1 g/kg/day for 14 days, p.o.	Decrease in the blood triglyceride, total cholesterol, LDL and VLDL levels	Hoda et al. (2010)
Fruits of *Withania coagulans*	Coagulanolide isolated from aqueous decoctions	100 mg/kg for 10 days, p.o.	Lowered the level of plasma triglycerides, total cholesterol, low-density lipoprotein cholesterol and very low-density lipoprotein cholesterol	Maurya et al. (2008)
Leaves and roots	Methanolic and isolated Withanolides	100 μg/ml in cellular model of diabetes	Increased insulin secretion and increased insulin sensitivity	Gorelick et al. (2015)
Fruits of *Withania coagulans*	Aqueous	1000 mg/kg/day for 28 days, p.o.	Increased glucose metabolism and decreased elevated cholesterol levels	Saxena, (2010)
Roots	Ethanolic	1.4 g/kg/day for 15 days, p.o.	Increased antioxidant enzymes and decreased elevated cholesterol levels	Jatwa and Kar (2009)
Roots	Ethanolic (5%)	50 mg/kg/day for 56 days, p.o.	Antioxidant activity and decreased elevated cholesterol levels	Kumar et al. (2013)
Roots	Standardized	750 mg/day for 10 days followed by 1000 mg/day for next 10 days and 1250 mg/day for last 10 days, p.o.	Decreased lipid levels and increased fat metabolism	Raut et al. (2012)

rats was reduced by incorporating *W. somnifera* root powder in the hypercholesteremic diet. Increased bile acid synthesis and improvements in antioxidative status of the animals treated with *W. somnifera* were also observed in that study. Antihyperglycemic and anti-hyperlipidemic effects of some Withanolides isolated from *W. coagulans* (Coagulin: 50 mg/kg/day) have also been reported (Maurya et al. 2008). These observations suggest that appropriate structure activity studies and structural modification of Withanolides could eventually lead to structurally and functionally novel drug leads for prevention and cure of diabetes associated hyperlipidemia also. However, till now all such efforts have been made for obtaining potential anticancer drugs, or for obtaining structurally novel inducers of heat shock proteins only (Zhang et al. 2012; Wijeratne et al. 2014).

Functions and structures of all parts of the nervous system are eventually disrupted in diabetic patients causing encephalopathy and myelopathy in the central nervous system, and peripheral neuropathy in all divisions of peripheral nervous system. Autonomic neuropathy is most commonly associated with dysregulation of gastrointestinal (Drewes et al. 2016) and cardiovascular functions (Serhiyenko and Serhiyenko 2015), while peripheral sensory neuropathy manifests as pain, dysesthesia, and/or loss of sensation. Peripheral neuropathy ultimately affects more than 50% of all diabetic patients. It is most prevalent in poorly controlled diabetic patients and its complications start arising immediately after sudden improvement of glycaemic control (Lee-Kubli et al. 2014; Hosseini and Abdollahi 2013). Since currently available drugs are not effective in relieving pain in all patients, and all of them have their own drawbacks, efforts are now being made in several laboratories to identify novel therapeutic leads against diabetic neuropathy associated pain.

The very first report suggesting *Withania somnifera* roots could have anti-nociceptive activities had appeared also in 1997 (Kulkarni and Ninan 1997), and a very recent one has reconfirmed and further extended those observations (Orrù et al. 2014). However, efforts to identify the pharmacological receptors possibly involved in anti-nociceptive activity of diverse types of *W. somnifera* extracts and numerous of their known bioactive constituents using *in vitro* radio-ligand binding experiments were unsuccessful in pinpointing any specific neurotransmitter or opioid receptors for the extracts of for any of their bioactive constituents studied (Sonar et al. 2015). In any case, the observations summarized in this report strongly suggest that opioid, cannabinoid, GABAergic and glutamatergic receptors are not the primary pharmacological targets involved in anti-nociceptive and other brain function modulating activities of *W. somnifera* extracts and many of their bioactive constituents observed in animal models after their repeated daily fairly low oral doses.

Several other reports have continued to reconfirm anti-nociceptive effects of *W. somnifera* against the second phase of formalin-induced pain in diabetic rodents (Khalili 2009; Pradeep et al. 2010; Orrù et al. 2016; Roughani et al. 2007). Reported results of one of these studies (Pradeep et al. 2010) suggest that such effects of the tested dose of the extract (100 mg/kg/day for several weeks) is due to its anti-depressant activities, and that protective effects of the extract against noxious stimuli triggered oxidative stress is involved in its such effects. They reveal also that antidepressant drugs like effects of the extract increases with increasing numbers of treatment days, and that such effects of the extracts is qualitatively similar that to that of the flavonoid quercetin (10 mg/kg/day). In another study, similarities between the effects of *W. somnifera* and salicylic acid in formalin induced pain after their repeated daily doses were observed (Khalili 2009). Since both salicylic acid and quercetin are also encountered in *W. somnifera* extracts, they could also contribute to the observed effects of the extracts in pain models. Available preclinical and clinical information on therapeutic potentials salicylic acid (Rainsford 2013; Ugurlucan et al. 2012; Rumore and Kim 2010; Berk et al. 2013; Khan et al. 2016) and quercetin (Nabavi et al. 2015; Gormaz et al. 2015) are in agreement with this possibility.

Salicylates and the anti-hyperlipidemic drug simvastatin are currently often prescribed for prevention of diabetes and metabolic syndromes associated with or caused by environmental stress and psychiatric disorders. Although, their potential uses as antidepressants or for treatments of psychiatric disorders are now controversially discussed (Rahola 2012), it cannot be denied that psychiatric disorders often accompany or are caused by metabolic disorders, and that prevalence of morbidity and mortality are higher in diabetes with psychosomatic abnormalities (Balhara and Sagar 2011; Sarkar and Balhara 2014; Maia et al. 2012). Several reports during more recent years have consistently revealed antidepressive, anxiolytic and other beneficial effects of *W. somnifera* and its metabolites against diverse neurological disorders in diabetic and stressed or intoxicated rodents after their repeated daily oral doses (Parihar et al. 2004; Xu et al. 2013; Thakur et al. 2015; Kumar et al. 2015a; Roughani et al. 2006; Parihar et al. 2015; Wadhwa et al. 2015; Sehgal et al. 2012; Kuboyama et al. 2014). Several of these reports, have reconfirmed that such effects of the extracts of the plant are due to their protective effects against hyperglycemic stress triggered alterations in oxidative defense systems in brain and other bodily organs. However, yet no very definitive statement on possible involvements of bioactive metabolites of the plant, other than Withanolides, in such effects of *W. somnifera* derived crude products can be made.

Except for a very few reports dealing with stress resistance increase or adaptogenic and anxiolytic and antidepressant other bioactivities activities of the plant, most of them deal with arbitrarily chosen extraction procedures, doses and dosing regimen for the tested extracts. However, in one such more recent report (Bharathi et al. 2015) it was inferred that 30 mg/kg daily oral doses of a commercially available *W. somnifera* extract is its minimally effective anti-depressive ones in albino mice. Dose finding studies conducted with another commercially available extract of the roots of the plant (containing only 2.6% Withanolides and their glycosides) have revealed that 25 mg/kg/day or lower doses of the extract for 10 days is high enough not only for its statistically significant antidepressant and anxiolytics like effects in diabetic and non-diabetic rats, but also for its protective effects against chronic unpredictable mild stress triggered alterations in body weight and temperature changes in such rats (Thakur et al. 2015). Such were also the more recent observations made with root, leaf, and stem extracts of *W. somnifera* extracts containing different contents of Withanolides, and the minimally effective daily oral doses of a *W. somnifera* root extract devoid of Withanolides for analogous activities was only a bit higher than that of the root extracts containing Withanolides (Dey et al. 2014, 2016a, b).

These observations reaffirm and strongly suggest that constituents of *Withania somnifera* extracts other than Withanolides also contribute to the bioactivity profiles of its extracts, and that such profiles of Withanolides present in *W. somnifera* roots and other parts of the plant are modulated by numerous other bioactive constituents of the plants, or those of other edible or medicinal plants. Salicylic and other hydroxylated aromatic acids, quercetin and other flavonoids, nicotinic, fumaric and ascorbic acids, an triethylene glycol are some such phytochemicals present in *W. somnifera* extracts, and their diverse combinations are encountered also in numerous other medicinal or edible plants often used in Ayurvedic and other traditionally systems of medicine for prevention and cure of diabetes and other metabolic disorders (Singh et al. 2014; Kumar et al. 2015a). Unfortunately, as yet only very little attention has been paid to the possibility that neurohormetic and stress response modulating properties of structurally and functionally diverse phytochemical (Son et al. 2008; Mattson and Cheng 2006; Calabrese et al. 2012) of *W. somnifera* could also be involved in antidiabetic and other therapeutically interesting bioactivities of diverse types of extracts of the plant. It is now becoming increasingly apparent though, that heat shock proteins and other cellular chaperons are also involved in their hormetic and stress response regulating activities (Dattilo et al. 2015). Diverse biomarkers of such physiological processes are useful tools for diagnosing and treatments of neurodegenerative and other comorbidities commonly encountered in chronic diseases, including those leading to or caused by insulin resistance and dysregulation of glucose homeostasis (Bhakta-Guha and Efferth 2015; Farooqui 2013). Appropriate uses of these biomarkers for better understanding of therapeutic potentials of *W. somnifera* and its metabolites could eventually lead to novel therapeutic leads and pharmacological strategies urgently needed for prevention and cure of diabetes and associated comorbidities.

2.6 Clinical Studies

The very first exploratory clinical observations suggesting that *W. somnifera* could be an antidiabetic plant with anti-hyperlipidemic activities (Andallu and Radhika 2000) was conducted with its root powder containing capsules administer for 30 consecutive days to mild non-insulin dependent mildly diabetic patients maintained on antidiabetic diabetic drug Daonil (glibenclamide), and in mildly hypercholesteremic patients not undergoing any drug treatments. Daily dose of the root powder used was 3 g/day, and the effects of *W. somnifera* treatments on hyperglycemia, hyperlipidemia and other parameters were compared with those observed in subjects not undergoing any other additional treatments. This study indicate that regular intake of *W. somnifera* root powder can induce potassium sparing diuretic effects in diabetic patients, and can also reduce serum cholesterol and triglycerides in hyperlipidemic patients. However, the magnitude of effects of root powder treatment on hyperglycemia in diabetic patients maintained on glibenclamide (12% reduction in blood glucose level) was equal to that observed in only glibenclamide treated diabetic patients. No adverse effects of treatments were observed in this study.

Report of another exploratory clinical study dealing with effects of *W. somnifera* extract on blood glucose and lipid levels have appeared 12 years after that exploratory study (Raut et al. 2012). In this study published in 2012, increasing numbers of capsules of a *W. somnifera* aqueous extract (extraction procedure used and analytical characteristics of the extract used was not mentioned) was administer to young healthy volunteers during 30 days for evaluating its dose related safety, tolerability and effectiveness against diabetes associated vascular abnormalities. Increasing doses of the extract used were equivalent to 6, 8, and 10 g of crude pulverized of *W. somnifera* roots (not mentioned whether dried or not). The extract was well tolerated by 17 of the 18 volunteers included in the study. One of the volunteers showed increased appetite, libido and hallucinogenic effects with vertigo even after the lowest dose of the extract tested (750 mg/day for 10 days) and was excluded from the study. Improvement in sleep quality was observed in 6 subjects. Although significant reduction in mean total blood cholesterol and blood urea nitrogen values were observed, extract treatments for 30 days had no statistically significant effects on blood sugar, triglyceride, and HDL-, LDL-, and VLDL-cholesterol values.

However, significant and dose dependent lowering effects of an analytically well characterized *W. somnifera* root extract on total cholesterol, triglyceride, and LDL-cholesterol levels were observed in a more recently reported randomized, double blind, and placebo controlled study in Type-2 diabetic patients maintained on 1500–2500 mg/day metformin (Usharani et al. 2015). This study was designed to evaluate the effects of the extract on endothelial functions of type-2 diabetic patients and to verify whether antioxidative effects of *W. somnifera* extracts observed in animals could be confirmed in patients or not. Daily 250 and 500 mg doses of the extract were administered for 12 weeks in this study. Significant improvements of endothelial function, as well as in biomarkers of oxidative stress, systemic inflammation, and HbA1C levels were also observed in this study, and the extract was well tolerated by the patients. The authors conclude that the extract can be used as a therapeutic adjunctive in type-2 diabetic patients. Since diabetic patients included in the study were maintained in on metformin and yet a significant reduction in the HbA1aC level was observed in *W. somnifera* treated group, the observation of the study are also in agreement with several observations made in laboratory rodents suggesting preventive effects of oxidative processes as well as glycation reactions (Fig. 2.4) are involved in the modes of action(s) of the *W. somnifera* extract tested.

The same research group had also reported similar effects of an ayurvedic polyherbal formulation (CardiPro) containing *W. somnifera* as one of its several active ingredients (Fatima et al. 2012), and also were the authors of another report revealing analgesic activity a *W. somnifera* extract currently commercialized in the USA (Sensoril®) and containing ca 15.7% Withanolides glycosides, 40.2% oligosaccharides and 0.24% Withaferin A (Usharani et al. 2013; Kumar et al. 2015b). The same author (Pingali et al. 2014) has also reported mental function improving effects of the same extract in healthy persons. Although several other reports revealing

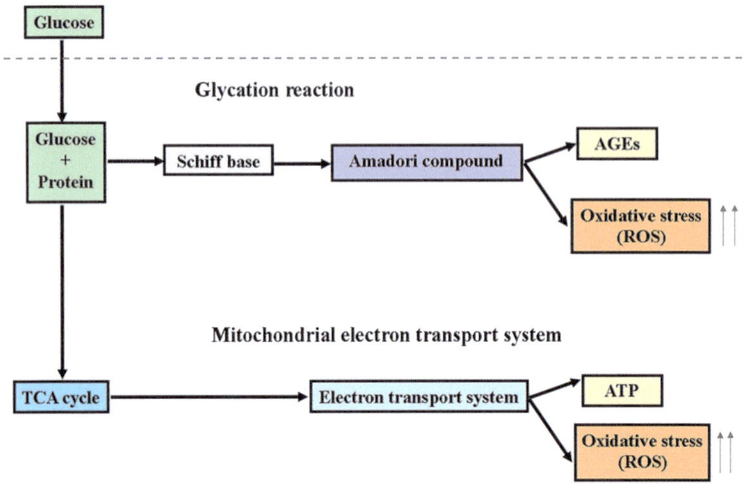

Fig. 2.4 Increase in oxidative stress in the diabetic state: Acceleration of glycation response and the intramitochondrial electron transfer system is observed in the diabetic (hyperglycemic) state, causing oxidative stress as the responses accelerated. [*Black arrows* indicate pathway; *grey arrows* indicate increase; *AGE* advanced glycosylation end products, *ROS* reactive oxygen species, *TCA cycle* tricarboxylic acid cycle, *ATP* adenosine tri-phosphate (Adapted from: Kawahito et al. 2009)]. *Withania somnifera* extracts can inhibit both glycation reactions leading to elevated HbA1c levels as well as oxidative damages

anti-stress, anxiolytic, and safety in volunteers and patients have appeared during more recent years (Chandrasekhar et al. 2012; Pratte et al. 2014), as yet only two reports (Andallu and Radhika 2000; Usharani et al. 2015) on potential health benefits of *W. somnifera* in diabetic patients have appeared in more recent English language journals. Both of them and several others made in volunteers and non-diabetic patients do reaffirm that fairly high daily *W. somnifera* extracts doses are well tolerated without any severe adverse effects and that they could be effective and safe herbal therapeutic alternatives against diabetes associated hyperlipidemia and other comorbidities. Identification of bioactive metabolites of the plant involved in their anti-hyperlipidemic and Hb1Ac level suppressing effects of its extracts could eventually lead to novel therapeutic leads and strategies urgently needed for prevention and cure of diabetes and other metabolic disorders, i.e. the most widely spreading epidemics of the twenty-first century affecting ca. 20–25% of global adult population (Alberti et al. 2006).

2.7 Conclusions and Future Perspective

Although it is now evident that *W. somnifera* is an anti-diabetic plant, many questions concerning its bioactive constituents and pharmacological interactions between them as well as on their pharmacological targets and their locations of actions in different bodily organs remain unanswered. From the preclinical and clinical information summarized in this chapter it is evident though, that homeostatic processes and mechanisms regulating circulating glucose levels and its biological functions are involved in the modes of actions of diverse types of extracts obtainable from different parts of the plant. Hereupon, modulations of oxidative processes and mechanisms regulating glucose and/or insulin homeostasis by the extracts of the plant and their numerous bioactive constituents play important roles. A few, but not all, such effects of *W. somnifera* derived products could be due their antioxidative properties, and due to their irreversible and nonspecific binding to diverse biological targets involved in regulation of biological oxidative processes. Like for numerous other plant-derived products, the so-called synergy and polyvalence paradigms (Houghton 2009) can explain their therapeutically interesting bioactivity profiles.

Although glycosylated Withanolides and their aglycones are certainly some of the quantitatively major secondary metabolites with modulating effects on glucose and insulin homeostasis, they must not necessarily be the only ones involved in such effects of extracts of the plant. However, since they are structurally and functionally the unique constituents biosynthesized by *W. somnifera* and many other plants, better understanding of their sites and modes of action are certainly be useful not only for drug discovery purposes, but also better understanding of therapeutic potentials of numerous other plants synthesizing and storing them in different concentrations in their different anatomical parts. Since medicinal values of other plants synthesizing and storing structurally and functionally diverse Withanolides for prevention of metabolic and other disorders has been well recognized by practitioners of Ayurvedic and other traditionally known systems of medicine, better understanding of their sites and modes of actions could be useful for identifying novel therapeutic leads from several such plants as well.

Since *W. somnifera* is a fairly nontoxic psychoactive plant with traditionally known and reasonably validated sleep quality improving and anti-diabetic activities in experimental and some clinical models, it seems to be particularly well suited for prevention and cure of diabesity and other metabolic disordered triggered by, or associated with sleep disorders. Proper understanding of its bioactive constituents and pharmacological interactions between them, involved in such effects of the plant is an essential prerequisite for obtaining sustainable and reproducible therapeutic benefits from the plant. To date, only very little concentrated efforts have been made to identify a convenient and validated bioassay system necessary for such purposes. It is now well recognized though, that environmental and/or metabolic disturbances triggered stress leads to sleep disturbances (Fig. 2.5), which in turn leads to diabesity and further diabetic complicates. Therefore, quantification of the extracts and their bioactive constituents in stressed or metabolically disturbed rodent models for such purposes can be warranted.

More recent observations made in our laboratories and elsewhere using stressed rodents have revealed sleep quality improving effects of *W. somnifera* extracts (Kumar and Kalonia 2007, 2008) and their antihyperglycemic activities on stress diabetic and non-diabetic rats (Thakur et al. 2015). The most sensitive and easily

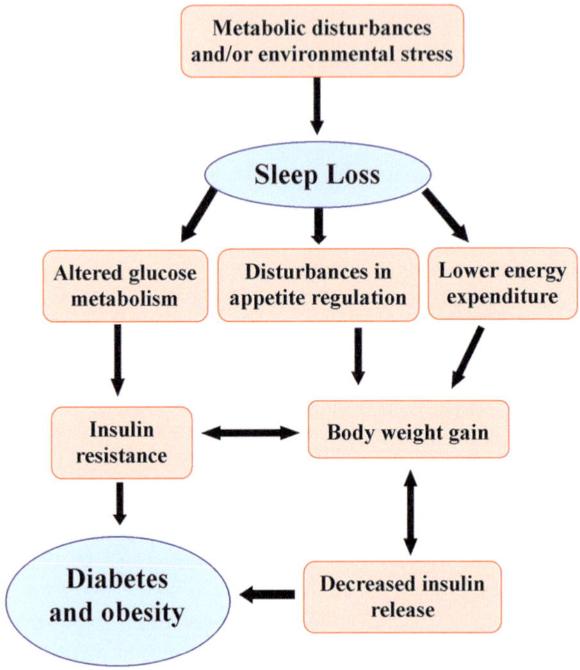

Fig. 2.5 Pathways leading to sleep loss to diabetes risk (Adapted from: Knutson et al. 2007)

quantifiable effects of such extracts identified to date in our laboratory are their protective effects against alterations in stress triggered body weight and temperature of laboratory rodents after their repeated daily fairly low oral doses. Using a bioassay procedure identified and validated during these efforts we have already led us identify several bioactive constituents of their extracts other than Withanolides with very high oral efficacies against both these biological markers of environmental or metabolic stress. Therefore, it seems reasonable to suggest that appropriate uses of this bioassay system and conventionally known activity guided fraction procedures could be used not only or better understanding of therapeutic potentials of *W. somnifera* derived products for prevention and cure of diabetes and associated co-morbidities, but also for clarifying pharmacological interactions between Withanolides and other bioactive constituents of the plant often encountered in many other edible and medicinal plants.

China, India and South American countries are the most populous countries now affected the most from diabetes, and these are the also the ones spending the most on drug discovery efforts based on the uses of medicinal plants. However, the phyto-pharmacological and other strategies often used by preclinical and clinical researches are those based on reductive and specific target based ones evolving from efforts to better understand the biological processes and mechanisms involved in modes of actions of already known drugs. Unfortunately, even diverse combinations of currently available and more costly antidiabetic and other drugs evolving from such strategies are not meeting the therapeutic needs and demands of the patients suffering from, or at risk to, diabetes and other metabolic disorders. The phytopharmacological lessons learned from *W. somnifera* strongly suggest that more realistic and holistic strategies are necessary for better understanding of therapeutic potentials, of this and other plants many of which also biosynthesize and store many stress response modulating secondary metabolites of *W. somnifera*.

A quotation cited in the introduction of recently published book on herbal adaptogens (Winston and Maimes 2007), states: "*All plants contain adaptogenic/tonic compounds, because plants have to contend with good deal of stress themselves*". Currently available information on the effects of metabolites of *W. somnifera* and other adaptogenic plants on brain functions (Kennedy 2014), have already added more than sufficient experimental evidences that numerous of them are also adaptogenic or stress resistance improving agents in all mammals including human beings. Numerous preclinical and clinical observations made with Withanolides and other metabolites synthesized and stored by *W. somnifera* have continued to add further evidences in favor of the conviction that loss of adaptability to metabolic stress and leading to diabetes can be compensated by their regular oral intake. What needs to be done now is to assemble their appropriate combination and doses in a capsule, or in foods and drinks, for obtaining reliable and reproducible therapeutic benefits from them. Judicious uses of the knowledge and knowhow evolving from the efforts of numerous phyto-pharmacologists and medicinal phytochemists during the past decade, and leading to our current understanding of therapeutic potentials of the plant, will certainly be useful for such purposes.

References

Adamska E, Ostrowska L, Górska M, Krętowski A (2014) The role of gastrointestinal hormones in the pathogenesis of obesity and type 2 diabetes. Prz Gastroenterol 9:69–76

Ahmed AM (2002) History of diabetes mellitus. Saudi Med J 23:373–378

Alali FQ, Amrine CSM, El-Elimat T, Alkofahi A, Tawaha K, Gharaibah M, Swanson SM, Falkinham JO III, Cabeza M, Sanchez A, Figueroa Oberlies NH (2014) Bioactive Withanolides from Withania obtusifolia. Phytochem Lett 9:96–101

Alberti G, Zimmet P, Shaw J (2006) The IDF consensus worldwide definition of the metabolic syndrome. In: Grundy SM (ed) International diabetes federation. Brussels, Belgium

Andallu B, Radhika B (2000) Hypoglycemic, diuretic and hypocholesterolemic effect of winter cherry (Withania somnifera, Dunal) root. Indian J Exp Biol 38:607–609

Anderson BJ, Auslander WF (1980) Research on diabetes management and the family: a critique. Diabetes Care 3:696–702

Anothaisintawee T, Reutrakul S, Van Cauter E, Thakkinstian A (2015) Sleep disturbances compared to traditional risk factors for diabetes development: systematic review and meta-analysis. Sleep Med Rev 30:11–24

Antony ML, Lee J, Hahm ER, Kim SH, Marcus AI, Kumari V, Ji X, Yang Z, Vowell CL, Wipf P, Uechi GT, Yates NA, Romero G, Sarkar SN, Singh SV (2014) Growth arrest by the antitumor steroidal lactone withaferin A in human breast cancer cells is associated with down-regulation and covalent binding at cysteine 303 of β-tubulin. J Biol Chem 289:1852–1865

Anwer T, Sharma M, Pillai KK, Iqbal M (2008) Effect of Withania somnifera on insulin sensitivity in non-insulin-dependent diabetes mellitus rats. Basic Clin Pharmacol Toxicol 102:498–503
Anwer T, Sharma M, Pillai KK, Khan G (2012) Protective effect of Withania somnifera against oxidative stress and pancreatic beta-cell damage in type 2 diabetic rats. Acta Pol Pharm 69:1095–1101
Archana SS, Mishra HS, Yadav KN (2015) Ashwagandha: solution to stress and insomnia. Indian J Ancient Med Yoga 8:153–159
Ashok GA, Shende MB (2015) A clinical evaluation of antistress activity of ashwagandha (Withania somnifera Dunal) on employees experiencing mental stress at work place. Int J Ayur Pharma Res 3:37–45
Aulston BD, Odero GL, Aboud Z, Glazner GW (2013) Alzheimer's disease and diabetes. InTech Open, Hampshire
Bajor LA, Gunzler D, Einstadter D, Thomas C, McCormick R, Perzynski AT, Kanuch SW, Cassidy KA, Dawson NV, Sajatovic M (2015) Associations between comorbid anxiety, diabetes control, and overall medical burden in patients with serious mental illness and diabetes. Int J Psychiatry Med 49:309–320
Balaji RM, Jeyaram C, Sundaram KM, Ramasamy MS (2015) Studies on antidiabetic activity of Indian medicinal plants using α–amylase and α-glucosidase inhibitory activity: a pathway to antidiabetic drugs. World J Med Sci 12:207–212
Balhara YPS, Sagar R (2011) Correlates of anxiety and depression among patients with type 2 diabetes mellitus. Indian J Endocrinol Metabol 15:S50–S54
Bentley-Lewis R, Levkoff S, Stuebe A, Seely EW (2008) Gestational diabetes mellitus: postpartum opportunities for the diagnosis and prevention of type 2 diabetes mellitus. Nat Clin Pract Endocrinol Metab 4:552–558
Berk M, Dean O, Drexhage H, McNeil JJ, Moylan S, O'Neil A, Davey CG, Sanna L, Maes M (2013) Aspirin: a review of its neurobiological properties and therapeutic potential for mental illness. BMC Med 11:74
Bhakta-Guha D, Efferth T (2015) Hormesis: decoding two sides of the same coin. Pharmaceuticals 8:865–883
Bharathi P, Seshayamma V, Jagannadharao GH, Sivakumar N (2015) Evaluation of antidepressant activity of aqueous extract of Withania somnifera [Aswagandha] roots in albino mice. IOSR J Pharma Biol Sci 10:27–29
Bharti SK, Bhatia A, Tewari SK, Sidhu OP, Roy R (2011) Application of HR-MAS NMR spectroscopy for studying chemotype variations of Withania somnifera (L.) Dunal. Magn Reson Chem 49:659–667
Bhattacharya SK, Muruganandam AV (2003) Adaptogenic activity of Withania somnifera: an experimental study using a rat model of chronic stress. Pharmacol Biochem Behav 75:547–555
Bhattacharya SK, Satyam SK, Chakrabarti A (1997a) Effect of Trasina an Ayurvedic herbal formulation, on pancreatic islet superoxide dismutase activity in hyperglycaemic rats. Indian J Exp Biol 35:297–299
Bhattacharya SK, Satyam KS, Ghosal S (1997b) Antioxidant activity of Glycowithanolides from Withania somnifera. Indian J Exp Biol 35:236–239
Calabrese V, Cornelius C, Dinkova-Kostova AT (2012) Cellular stress responses, hormetic phytochemicals and vitagenes in aging and longevity. Biochim Biophys Acta 1822:753–783
Cespedes EM, Bhupathiraju SN, Li Y, Rosner B, Redline S, Hu FB (2016) Long-term changes in sleep duration, energy balance and risk of type 2 diabetes. Diabetologia 59:101–109
Chandrasekhar K, Kapoor J, Anishetty S (2012) A prospective, randomized double-blind, placebo-controlled study of safety and efficacy of a high-concentration full-spectrum extract of ashwagandha root in reducing stress and anxiety in adults. Indian J Psychol Med 34:255–262
Chatterjee S, Srivastava S, Khalid A, Singh N, Sangwan RS, Sidhu OP, Roy R, Khetrapal CL, Tuli R (2010) Comprehensive metabolic fingerprinting of Withania somnifera leaf and root extracts. Phytochemistry 71:1085–1094
Chen LX, He H, Qiu F (2011) Natural Withanolides: an overview. Nat Prod Rep 28:705–740

Choudhary MI, Yousuf S (2013) Withanolides: chemistry and antitumor activity. In: Ramawat KG, Me'rillon JM (eds) Natural products. Springer, Berlin/Heidelberg, pp 3465–3495
Choudhary B, Shetty A, Langade DG (2015) Efficacy of Ashwagandha (Withania somnifera [L.] Dunal) in improving cardiorespiratory endurance in healthy athletic adults. Ayu 36:63–68
Choudhury SM, Tan TM, Bloom SR (2016) Gastrointestinal hormones and their role in obesity. Curr Opin Endocrinol Diabetes Obes 23:18–22
Colagiuri S (2010) Diabesity: therapeutic options. Diabetes Obes Metab 12:463–473
Cui ZG, Piao JL, Rehman MU, Ogawa R, Li P, Zhao QL, Kondo T, Inadera H (2014) Molecular mechanisms of hyperthermia-induced apoptosis enhanced by withaferin A. Eur J Pharmacol 723:99–107
Cunha DA, de Alves MC, Stoppiglia LF, Jorge AG, Módulo CM, Carneiro EM, Boschero AC, Saad MJ, Velloso LA, Rocha EM (2007) Extra-pancreatic insulin production in RAt lachrymal gland after streptozotocin-induced islet beta-cells destruction. Biochim Biophys Acta 1770:1128–1135
Dall'Asta M, Bayle M, Neasta J, Scazzina F, Renato B, Cros G, Rio DD, Oiry C (2015) Protection of pancreatic β-cell function by dietary polyphenols. Phytochem Rev 14:933–959
Dar NJ, Hamid A, Ahmad M (2015) Pharmacologic overview of Withania somnifera, the Indian Ginseng. Cell Mol Life Sci 72:4445–4460
Datta A, Bagchi C, Das S, Mitra A, Pati AD, Tripathi SK (2013) Antidiabetic and antihyperlipidemic activity of hydroalcoholic extract of Withania coagulans Dunal dried fruit in experimental rat models. J Ayurveda Integr Med 4:99–106
Dattilo S, Mancuso C, Koverech G, Di Mauro P, Ontario ML, Petralia CC, Petralia A, Maiolino L, Serra A, Calabrese EJ, Calabrese V (2015) Heat shock proteins and hormesis in the diagnosis and treatment of neurodegenerative diseases. Immun Ageing 12:20
Devi J, Rajkumar J (2013) Antihyperlipidemic effect of ambrex, a polyherbal formulation against experimentally induced hypercholesterolemia in rats. Afr J Pharm Pharmacol 7:1737–1743
Dey A, Chatterjee SS, Kumar V (2014) Comparison of the Adaptogenic efficacy of three different Withania somnifera extracts in mice. Indian J Pharm 46:S87–S87
Dey A, Chatterjee SS, Kumar V (2016a) Low dose effects of a Withania somnifera extract on marble burying behaviour in stressed mice. J Intercult Ethnopharmacol 5:274–277
Dey A, Chatterjee SS, Kumar V (2016b) Analgesic activity of a Withania somnifera extract in stressed mice. Orient Pharm Exp Med 16(4):295–302. Under review
Dhanani T, Shah S, Gajbhiye NA, Kumar S (2017) Effect of extraction methods on yield, phytochemical constituents and antioxidant activity of *Withania somnifera*. Arab J Chem 10:S1193–S1199
Domingueti CP, Dusse LM, Md C, de Sousa LP, Gomes KB, Fernandes AP (2015) Diabetes mellitus: the linkage between oxidative stress, inflammation, hypercoagulability and vascular complications. J Diabetes Complicat 30:738–745
Drewes AM, Søfteland E, Dimcevski G, Farmer AD, Brock C, Frøkjær JB, Krogh K, Drewes AM (2016) Brain changes in diabetes mellitus patients with gastrointestinal symptoms. World J Diabetes 7:14–26
Durg S, Dhadde SB, Vandal R, Shivakumar BS, Charan CS (2015) Withania somnifera (Ashwagandha) in neurobehavioural disorders induced by brain oxidative stress in rodents: a systematic review and meta-analysis. J Pharm Pharmacol 67:879–899
Egede LE, Dismuke CE (2012) Serious psychological distress and diabetes: a review of the literature. Curr Psychiatry Rep 14:15–22
Egede LE, Gebregziabher M, Zhao Y, Dismuke CE, Walker RJ, Hunt KJ, Axon RN (2014) Impact of mental health visits on healthcare cost in patients with diabetes and comorbid mental health disorders. PLoS One 9:e103804
Farag YM, Gaballa MR (2010) Diabesity: an overview of a rising epidemic. Nephrol Dial Transplant 26:28–35

Farooqui AA (2013) Effect of dietary phytochemicals on metabolic syndrome and neurological disorders. In: Metabolic syndrome: an important risk factor for stroke, alzheimer disease, and depression. Springer, New York, pp p191–p234
Fatima N, Usharani P, Muralidhar N (2012) Randomised double blind placebo controlled study to evaluate the effect of cardipro on endothelial dysfunction and biomarkers in patients with type 2 diabetes mellitus. Innovative J Med Health Sci 2:122–128
Fernando IDNS, Abeysinghe DC, Dharmadasa RM (2013) Determination of phenolic contents and antioxidant capacity of different parts of Withania somnifera (L.) Dunal. from three different growth stages. Ind Crop Prod 50:537–539
Fiorentino TV, Prioletta A, Zuo P, Folli F (2013) Hyperglycemia-induced oxidative stress and its role in diabetes mellitus related cardiovascular diseases. Curr Pharm Des 19:5695–5703
Flanagan RJ (2008) Side effects of clozapine and some other psychoactive drugs. Curr Drug Saf 3:115–122
Foster JA, Neufeld KAM (2013) Gut-brain axis: how the microbiome influences anxiety and depression. Trends Neurosci 36:305–312
Gao R, Shah N, Lee JS (2014) Withanone-rich combination of ashwagandha Withanolides restricts metastasis and angiogenesis through hnRNP-K. Mol Cancer Ther 13:2930–2940
Gardner D, Shoback D (2011) Greenspan's basic & clinical endocrinology. LANGE Clinical Medicine, McGraw-Hill Medical, New York
Girish KS, Machiah KD, Ushanandini S, Harish Kumar K, Nagaraju S, Govindappa M, Vedavathi M, Kemparaju K (2006) Antimicrobial properties of a non-toxic glycoprotein (WSG) from Withania somnifera (Ashwagandha). J Basic Microbiol 46:365–374
Gorelick J, Rosenberg R, Smotrich A, Hanuš L, Bernstein N (2015) Hypoglycemic activity of Withanolides and elicitated Withania somnifera. Phytochemistry 116:283–289
Gormaz JG, Quintremil S, Rodrigo R (2015) Cardiovascular disease: a target for the pharmacological effects of quercetin. Curr Top Med Chem 15:1735–1742
Guariguata L, Whiting DR, Hambleton I, Beagley J, Linnenkamp U, Shaw JE (2014) Global estimates of diabetes prevalence for 2013 and projections for 2035. Diabetes Res Clin Pract 103:137–149
Hasnain SZ, Prins JB, McGuckin MA (2016) Oxidative and endoplasmic reticulum stress in β-cell dysfunction in diabetes. J Mol Endocrinol 56:R33–R54
Hemalatha S, Wahi AK, Singh PN, Chansouria JP (2006) Hypolipidemic activity of aqueous extract of Withania coagulans Dunal in albino rats. Phytother Res 20:614–617
Henschen F (1969) On the term diabetes in the works of Aretaeus and Galen. Med Hist 13:190–192
Henstridge DC, Whitham M, Febbraio MA (2014) Chaperoning to the metabolic party: the emerging therapeutic role of heat-shock proteins in obesity and type 2 diabetes. Mol Metab 3:781–793
Heyninck K, Lahtela-Kakkonen M, Van der Veken P, Haegeman G, Vanden Berghe W (2014) Withaferin A inhibits NF-kappaB activation by targeting cysteine 179 in IKKβ. Biochem Pharmacol 91:501–509
Hoda Q, Ahmad S, Akhtar M, Najmi AK, Pillai KK, Ahmad SJ (2010) Antihyperglycaemic and antihyperlipidaemic effect of poly-constituents, in aqueous and chloroform extracts, of Withania coagulans Dunal in experimental type 2 diabetes mellitus in rats. Hum Exp Toxicol 29:653–658
Hooper PL (2014) Induction of heat shock proteins in the therapy of type 2 diabetes and metabolic syndrome. EBio Med 1:14–15
Hooper PL, Hooper JJ (2005) Loss of defense against stress: diabetes and heat shock proteins. Diabetes Technol Ther 7:204–208
Hooper PL, Balogh G, Rivas E (2014) The importance of the cellular stress response in the pathogenesis and treatment of type 2 diabetes. Cell Stress Chaperones 19:447–464
Hosseini A, Abdollahi M (2013) Diabetic neuropathy and oxidative stress: therapeutic perspectives. Oxidative Med Cell Longev 2013:168039
Hostalek U, Gwilt M, Hildemann S (2015) Therapeutic use of metformin in prediabetes and diabetes prevention. Drugs 75:1071–1094

Houghton PJ (2009) Synergy and polyvalence: paradigms to explain the activity of herbal products. In: Houghton PJ, Mukherjee PK (eds) Evaluation of herbal medicinal products. Pharmaceutical Press, London, pp 85–94
Huqi A (2015) Antidiabetic treatment and cardiovascular events: all a matter of perspective. Heart Metab 68:39–40
Jatwa R, Kar A (2009) Amelioration of metformin-induced hypothyroidism by Withania somnifera and Bauhinia purpurea extracts in Type 2 diabetic mice. Phytother Res 23:1140–1145
Jonathan G, Rivka R, Avinoam S et al (2015) Hypoglycemic activity of withanolides and elicitated Withania somnifera. Phytochem 116:283–289
Kapoor A, Kaur G, Kaur R (2015) Antimicrobial activity of different herbal plants extracts: a review. World J Pharm Pharm Sci 4:422–459
Kawahito S, Kitahata H, Oshita S (2009) Problems associated with glucose toxicity: role of hyperglycemia-induced oxidative stress. World J Gastroenterol 15:4137–4142
Kennedy DO (2014) Plants and the human brain. Oxford University Press, London
Khalili M (2009) The effect of oral administration of Withania somnifera root on formalin-induced pain in diabetic rats. Basic Clin Neurosci 1:29–31
Khan MA, Khan H, Ali T (2014) Withanolides isolated from Withania somnifera with α-glucosidase inhibition. Med Chem Res 23:2386–2390
Khan SA, Chatterjee SS, Kumar V (2016) Low dose aspirin like analgesic and anti-inflammatory activities of mono-hydroxybenzoic acids in stressed rodents. Life Sci 148:53–62
Khodaei M, Mehrana J, Noori M (2012) Remedial use of Withanolides from Withania coagolans (Stocks) Dunal. Adv Life Sci 2:6–19
Knutson KL, Spiegel K, Penev P, Van Cauter E (2007) The metabolic consequences of sleep deprivation. Sleep Med Rev 11:163–178
Kojima H, Fujimiya M, Terashima T, Kimura H, Chan L (2006) Extrapancreatic proinsulin/insulin-expressing cells in diabetes mellitus: is history repeating itself? Endocr J 53:715–722
Kowall B, Lehnich AT, Strucksberg KH, Führer D, Erbel R, Jankovic N, Moebus S, Jöckel KH, Stang A (2016) Associations among sleep disturbances, nocturnal sleep duration, daytime napping, and incident prediabetes and type 2 diabetes: the heinz nixdorf recall study. Sleep Med 21:35–41
Kuboyama T, Tohda C, Komatsu K (2014) Effects of Ashwagandha (roots of Withania somnifera) on neurodegenerative diseases. Biol Pharm Bull 37:892–897
Kulkarni SK, Dhir A (2008) Withania somnifera: an Indian ginseng. Prog Neuro-Psychopharmacol Biol Psychiatry 32:1093–1105
Kulkarni SK, Ninan I (1997) Inhibition of morphine tolerance and dependence by Withania somnifera in mice. J Ethnopharmacol 57:213–217
Kumar A, Kalonia H (2007) Protective effect of Withania somnifera Dunal on the behavioral and biochemical alterations in sleep-disturbed mice (Grid over water suspended method). Indian J Exp Biol 45:524–528
Kumar A, Kalonia H (2008) Effect of Withania somnifera on sleep-wake cycle in sleep-disturbed rats: possible GABAergic mechanism. Indian J Pharm Sci 70:806–810
Kumar S, Vinothkumar D, Kumar S, Aslam A, Shajahan A (2011) The phytochemical constituents of Withania somnifera and Withania obtusifolia by GCMS analysis. Int J Pharm Phytochem Res 3:31–34
Kumar R, Singh JK, Nath A, Ali M, Kumar A, Anshupriya SP (2013) Effect of Withania somnifera on estrogen, cholesterol and subcellular structure of ovary of chlorpyrifos exposed mice. Carib J SciTech 1:61–69
Kumar V, Dey A, Hadimani MB, Markovic TL, Emerald M (2015a) Chemistry and pharmacology of Withania somnifera: an update. TANG [Humanitas Medicine] 5:e1
Kumar CU, Pokuri VK, Pingali U (2015b) Evaluation of the analgesic activity of standardized aqueous extract of Terminalia chebula in healthy human participants using hot air pain model. J Clin Diagn Res 9:FC01–FC04

Kyathanahalli CN, Manjunath MJ, Muralidhara (2014) Oral supplementation of standardized extract of Withania somnifera protects against diabetes-induced testicular oxidative impairments in prepubertal rats. Protoplasma 251:1021–1029

Lee SWH, Ng KY, Chin WK (2016) The impact of sleep amount and sleep quality on glycaemic control in type 2 diabetes: a systematic review and meta-analysis. Sleep Med Rev 31:91–101

Lee-Kubli CA, Mixcoatl-Zecuatl T, Jolivalt CG, Calcutt NA (2014) Animal models of diabetes-induced neuropathic pain. In: Taylor BK, Finn D (eds) Behavioral neurobiology of chronic pain. Springer, Berlin/Heidelberg, pp p147–p170

Lehner C, Gehwolf R, Wagner A, Resch H, Hirzinger C, Augat P, Stephan D, Aigner L, Rivera FJ, Bauer HC, Tempfer H (2012) Tendons from non-diabetic humans and rats harbor a population of insulin-producing, pancreatic beta cell-like cells. Horm Metab Res 44:506–510

Leroith D, Vinik AI (2008) Controversies in treating diabetes: clinical and research aspects. (Contemporary endocrinology). Humana press, Totowa

Liu Q, Chen L, Hu L, Guo Y, Shen X (2010) Small molecules from natural sources, targeting signaling pathways in diabetes. Biochim Biophys Acta 1799:854–865

Maia AC, Braga Ade A, Brouwers A, Nardi AE, Oliveira e Silva AC (2012) Prevalence of psychiatric disorders in patients with diabetes types 1 and 2. Compr Psychiatry 53:1169–1173

Marrero D, Pan Q, Barrett-Connor E, de Groot M, Zhang P, Percy C, Florez H, Ackermann R, Montez M, Rubin RR (2014) Impact of diagnosis of diabetes on health-related quality of life among high-risk individuals: the diabetes prevention program outcomes study. Qual Life Res 23:75–88

Mary NK, Babu BH, Padikkala J (2003) Antiatherogenic effect of Caps HT2, a herbal Ayurvedic medicine formulation. Phytomedicine 10:474–482

Mattson MP, Cheng A (2006) Neurohormetic phytochemicals: low-dose toxins that induce adaptive neuronal stress responses. Trends Neurosci 29:632–639

Maurya R, Akanksha J, Singh AB, Srivastava AK (2008) Coagulanolide, a withanolide from Withania coagulans fruits and antihyperglycemic activity. Bioorg Med Chem Lett 18:6534–6537

McMurray JJ, Gerstein HC, Holman RR, Pfeffer MA (2014) Heart failure: a cardiovascular outcome in diabetes that can no longer be ignored. Lancet Diabetes Endocrinol 2:843–851

Mendenhall E, Norris SA, Shidhaye R, Prabhakaran D (2014) Depression and type 2 diabetes in low-and middle income countries: a systematic review. Diabetes Res Clin Pract 103:276–285

Miele L, Giorgio V, Alberelli MA (2015) Impact of gut microbiota on obesity, diabetes, and cardiovascular disease risk. Curr Cardiol Rep 17:1–7

Mirjalili MH, Moyano E, Bonfill M, Cusido RM, Palazón J (2009) Steroidal lactones from Withania somnifera, an ancient Plant for Novel Medicine. Molecules 14:2373–2393

Misico RI, Nicotra VE, Oberti JC, Barboza G, Gil RR, Burton G (2002) Induction of quinone reductase by Withanolides. J Nat Prod 65:677–680

Misico RI, Nicotra VE, Oberti JC et al (2011) Withanolides and related steroids. In: Kinghorn AD, Falk H, Kobayashi J (eds) Progress in the chemistry of organic natural products. Springer, Vienna, pp p127–p229

Mohan R, Bargagna-Mohan P (2016) The use of Withaferin A to study intermediate filaments. In: Handel M (ed) Methods in enzymology, vol 568. Elsevier, Amsterdam, pp p187–p218

Molina MN, Ferder L, Manucha W (2016) Emerging role of nitric oxide and heat shock proteins in insulin resistance. Curr Hypertens Rep 18:1–13

Nabavi SF, Russo GL, Daglia M, Nabavi SM (2015) Role of quercetin as an alternative for obesity treatment: you are what you eat! Food Chem 179:305–310

Navinder KM, Sehrawat R, Khatak S (2013) A comparative study: homoepathic medicine and a medicinal plant Withania somnifera for antidiabetic activity. J Pharmacogn Phytochem 2:109–112

Nicholson JK, Holmes E, Kinross J, Burcelin R, Gibson G, Jia W, Pettersson S (2012) Host-gut microbiota metabolic interactions. Science 336:1262–1267

Nirupama R, Devaki M, Nirupama M et al (2014) In vitro and in vivo studies on the hypoglycaemic potential of Ashwagandha (Withania somnifera) root. Pharma Sci Monitor 5(3):45–58

Orrù A, Marchese G, Casu G, Casu MA, Kasture S, Cottiglia F, Acquas E, Mascia MP, Anzani N, Ruiu S (2014) Withania somnifera root extract prolongs analgesia and suppresses hyperalgesia in mice treated with morphine. Phytomedicine 21:745–752

Orrù A, Casu MA, Tambaro S, Marchese G, Casu G, Ruiu S (2016) Withania somnifera (L.) Dunal root extract alleviates formalin-induced nociception in mice: involvement of the opioidergic system. Behav Pharmacol 27:57–68

Parihar MS, Chaudhary M, Shetty R, Hemnani T (2004) Susceptibility of hippocampus and cerebral cortex to oxidative damage in streptozotocin treated mice: prevention by extracts of Withania somnifera and Aloe vera. J Clin Neurosci 11:397–402

Parihar P, Shetty R, Ghafourifar P, Parihar MS (2015) Increase in oxidative stress and mitochondrial impairment in hypothalamus of streptozotocin treated diabetic rat: Antioxidative effect of Withania somnifera. Cell Mol Biol (Noisy-le-Grand) 62:73–83

Petrak F, Baumeister H, Skinner TC, Brown A, Holt RI (2015) Depression and diabetes: treatment and health-care delivery. Lancet Diabetes Endocrinol 3:472–485

Pingali U, Pilli R, Fatima N (2013) Effect of Withania somnifera extract on mental stress induced changes in hemodynamic properties and arterial wave reflections in healthy subjects. Curr Topics Nutraceuticals Res 11:151

Pingali U, Pilli R, Fatima N (2014) Effect of standardized aqueous extract of Withania somnifera on tests of cognitive and psychomotor performance in healthy human participants. Pharm Res 6:12–18

Prabhakar VK, Jaidka A, Singh R (2013) In vitro study on α -amylase inhibitory activity and phytochemical screening of few Indian medicinal plant having anti-diabetic properties. Int J Sci Res Pub 3:1–6

Pradeep S, Prem Kumar N, Khajuria DK, Rao GS (2010) Preclinical evaluation of antinociceptive effect of Withania somnifera (ashwagandha) in diabetic peripheral neuropathic rat models. Pharmacol Ther 2:283–298

Pratte MA, Nanavati KB, Young V, Morley CP (2014) An alternative treatment for anxiety: a systematic review of human trial results reported for the Ayurvedic herb ashwagandha (Withania somnifera). J Altern Complement Med 20:901–908

Puranik A, Patwardhan B (2012) Ayurveda and metabolic diseases: the whole is greater than the sum of the parts. In: Bagchi D, Sreejayan N (eds) Nutritional and therapeutic interventions for diabetes and metabolic syndrome. Elsevier Inc./Academic, Amsterdam/Boston, pp 443–456

Rahola GJ (2012) Somatic drugs for psychiatric diseases: aspirin or simvastatin for depression? Curr Neuropharmacol 10:139–158

Rai M, Jogee PS, Agarkar G, dos Santos CA (2016) Anticancer activities of Withania somnifera: current research, formulations, and future perspectives. Pharm Biol 54:189–197

Rainsford KD (2013) Aspirin and the salicylates. Elsevier, Burlington

Raut AA, Rege NN, Tadvi FM, Solanki PV, Kene KR, Shirolkar SG, Pandey SN, Vaidya RA, Vaidya AB (2012) Exploratory study to evaluate tolerability, safety, and activity of Ashwagandha (Withania somnifera) in healthy volunteers. J Ayurveda Integr Med 3:111–114

Robertson RP, Harmon JS (2006) Diabetes, glucose toxicity, and oxidative stress: a case of double jeopardy for the pancreatic islet β-cell. Free Radic Biol Med 41:177–184

Rochette L, Zeller M, Cottin Y, Vergely C (2014) Diabetes, oxidative stress and therapeutic strategies. Biochim Biophys Acta 1840:2709–2729

Roughani M, Baluchnejadmojarad T, Khalili M, Salimi SFM (2006) The effect of chronic oral administration of Withania somnifera root on learning and memory in diabetic rats using passive avoidance test. Sci J Hamadan Univ Med Sci 13:10–17

Roughani M, Baluchnejadmojarad T, Khalili M, Salimi SFM (2007) Evaluation of the effect of chronic oral administration of Withania somnifera root on pain score in male diabetic rats. Jundishapur Sci Med J:168–174

Rumore MM, Kim KS (2010) Potential role of salicylates in type 2 diabetes. Ann Pharmacother 44:1207–1221

Safhi MM, Anwer T (2011) Comparative analysis of Withania somnifera and Rhus coriaria on hyperglycemia and insulin sensitivity in type 2 diabetic rats. J Jazan University 1:39–49
Samadi AK (2015) Chapter three-potential anticancer properties and mechanisms of action of Withanolides. Enzyme 37:73–94
Samadi Noshahr Z, Shahraki MR, Ahmadvand H, Nourabadi D, Nakhaei A (2015) Protective effects of Withania somnifera root on inflammatory markers and insulin resistance in fructose-fed rats. Rep Biochem Mol Biol 3:62–67
Sarangi A, Jena S, Sarangi AK, Swain B (2013) Anti-diabetic effects of Withania somnifera root and leaf extracts on streptozotocin induced diabetic rats. J Cell Tissue Res 13:3597–3601
Sarkar S, Balhara YPS (2014) Diabetes mellitus and suicide. Indian J Endocrinol Metabol 18:468–474
Saxena B (2010) Anti-hyperlipidemic activity of Withania coagulans in streptozotocin-induced diabetes: a potent anti-atherosclerotic agent. Drug Discov Ther 4:334–340
Scheen AJ, Paquot N (2015) Obesity: a new paradigm for treating obesity and diabetes mellitus. Nat Rev Endocrinol 11:196–198
Schmidt MI, Duncan BB (2003) Diabesity: an inflammatory metabolic condition. Clin Chem Lab Med 41:1120–1130
Sehgal N, Gupta A, Valli RK, Joshi SD, Mills JT, Hamel E, Khanna P, Jain SC, Thakur SS, Ravindranath V (2012) Withania somnifera reverses Alzheimer's disease pathology by enhancing low-density lipoprotein receptor-related protein in liver. PNAS 109:3511
Serhiyenko VA, Serhiyenko AA (2015) Diabetic cardiac autonomic neuropathy: do we have any treatment perspectives? World J Diabetes 6:245–258
Sharma H, Chandola HM (2011a) Prameha in Ayurveda: correlation with obesity, metabolic syndrome, and diabetes mellitus. Part 1–Etiology, classification, and pathogenesis. J Altern Complement Med 17:491–496
Sharma H, Chandola HM (2011b) Ayurvedic concept of obesity, metabolic syndrome, and diabetes mellitus. J Altern Complement Med 17:549–552
Shaw JE, Sicree RA, Zimmet PZ (2010) Global estimates of the prevalence of diabetes for 2010 and 2030. Diabetes Res Clin Pract 87:4–14
Singh G, Kumar P (2011) Evaluation of antimicrobial efficacy of flavonoids of Withania somnifera L. Indian J Pharm Sci 73:473–478
Singh B, Saxena AK, Chandan BK, Gupta DK, Bhutani KK, Anand KK (2001) Adaptogenic activity of a novel, withanolide-free aqueous fraction from the roots of Withania somnifera Dun. Phytother Res 15:311–318
Singh B, Chandan BK, Gupta DK (2003) Adaptogenic activity of a novel withanolide-free aqueous fraction from the roots of Withania somnifera Dun. (Part II). Phytother Res 17:531–536
Singh A, Duggal S, Singh H, Singh J, Katekhaye S (2010) Withanolides: phytoconstituents with significant pharmacological activities. Int J Green Pharm 4:229–237
Singh R, Arif T, Khan I, Sharma P (2014) Phytochemicals in antidiabetic drug discovery. J Biomed Ther Sci 1:1–33
Snoek FJ, Bremmer MA, Hermanns N (2015) Constructs of depression and distress in diabetes: time for an appraisal. Lancet Diabetes Endocrinol 3:450–460
Son TG, Camandola S, Mattson MP (2008) Hormetic dietary phytochemicals. NeuroMolecular Med 10:236–246
Sonar VP, Cottiglia F, Ruiu S, Orru A, Anzani N, Floris C (2015) Withanolides and alkaloids from Withania somnifera roots with binding affinity to opioid, cannabinoids and GABAergic receptors. Planta Med 81:122
Spoormaker VI, van den Bout J (2005) Depression and anxiety complaints; relations with sleep disturbances. Eur Psychiatry 20:243–245
Thakur AK, Shakya A, Husain GM, Emerald M, Kumar V (2014) Gut-microbiota and mental health: current and future perspectives. J Pharmacol Clin Toxicol 2:1016–1031
Thakur AK, Dey A, Chatterjee SS, Kumar V (2015) Reverse ayurvedic pharmacology of Ashwagandha as an adaptogenic anti-diabetic plant: a pilot study. Curr Trad Med 1:51–61

Thomas RP, Hellmich MR, Townsend CM Jr, Evers BM (2003) Role of gastrointestinal hormones in the proliferation of normal and neoplastic tissues. Endocr Rev 24:571–599
Tiwari N, Thakur AK, Kumar V, Dey A, Kumar V (2014) Therapeutic targets for diabetes mellitus: an update. Clinic Pharmacol Biopharmaceut 3:117
Torres-Fuentes C, Schellekens H, Dinan TG, Cryan JF (2015) A natural solution for obesity: Bioactives for the prevention and treatment of weight gain. A review. Nutr Neurosci 18:49–65
Udayakumar R, Kasthurirengan S, Mariashibu TS, Rajesh M, Anbazhagan VR, Kim SC, Ganapathi A, Choi CW (2009) Hypoglycaemic and hypolipidaemic effects of Withania somnifera root and leaf extracts on alloxan-induced diabetic rats. Int J Mol Sci 10:2367–2382
Udayakumar R, Kasthurirengan S, Vasudevan A, Mariashibu TS, Rayan JJ, Choi CW, Ganapathi A, Kim SC (2010) Antioxidant effect of dietary supplement Withania somnifera L. reduce blood glucose levels in alloxan-induced diabetic rats. Plant Foods Hum Nutr 65:91–98
Ugurlucan M, Caglar IM, Caglar FN, Ziyade S, Karatepe O, Yildiz Y, Zencirci E, Ugurlucan FG, Arslan AH, Korkmaz S, Filizcan U, Cicek S (2012) Aspirin: from a historical perspective. Recent Pat Cardiovasc Drug Discov 7:71–76
Usharani P, Nalini P, Manjunath NK, Reddy KSK (2013) Evaluation of the analgesic activity of standardized aqueous extract of Withania somnifera in healthy human volunteers using hot air pain model. Res J Life Sci 1:1–6
Usharani P, Fatima N, Kumar CU, Kishan PV (2015) Evaluation of a highly standardized Withania somnifera extract on endothelial dysfunction and biomarkers of oxidative stress in patients with type 2 diabetes mellitus: a randomized, double blind, placebo controlled study. Int J Ayur Pharma Res 2:22–32
Vaishnavi K, Saxena N, Shah N, Singh R, Manjunath K, Uthayakumar M, Kanaujia SP, Kaul SC, Sekar K, Wadhwa R (2012) Differential activities of the two closely related Withanolides, Withaferin A and Withanone: bioinformatics and experimental evidences. PLoS One 7:e44419
Vanden Berghe W, Sabbe L, Kaileh M, Haegeman G, Heyninck K (2012) Molecular insight in the multifunctional activities of Withaferin A. Biochem Pharmacol 84:1282–1291
Vincent RP, Ashrafian H, le Roux CW (2008) Mechanisms of disease: the role of gastrointestinal hormones in appetite and obesity. Nat Clin Pract Gastroenterol Hepatol 5:268–277
Visavadiya NP, Narasimhacharya AV (2007) Hypocholesteremic and antioxidant effects of Withania somnifera (Dunal) in hypercholesteremic rats. Phytomedicine 14:136–142
Visavadiya NP, Narasimhacharya AV (2011) Ameliorative effects of herbal combinations in hyperlipidemia. Oxidative Med Cell Longev 2011:160408
Wadhwa R, Singh R, Gao R, Shah N, Widodo N, Nakamoto T, Ishida Y, Terao K, Kaul SC (2013) Water extract of Ashwagandha leaves has anticancer activity: identification of an active component and its mechanism of action. PLoS One 8:e77189
Wadhwa R, Konar A, Kaul SC (2015) Nootropic potential of Ashwagandha leaves: beyond traditional root extracts. Neurochem Int 95:109–118
White JR Jr (2014) A brief history of the development of diabetes medications. Diabetes Spectr 27:82–86
WHO (2006) Definition and diagnosis of diabetes mellitus and intermediate hyperglycemia: report of a WHO/IDF consultation. WHO, Geneva, p 21
Wijeratne EM, Xu YM, Scherz-Shouval R, Marron MT, Rocha DD, Liu MX, Costa-Lotufo LV, Santagata S, Lindquist S, Whitesell L, Gunatilaka AA (2014) Structure–activity relationships for Withanolides as inducers of the cellular heat-shock response. J Med Chem 57:2851–2863
Winston D, Maimes S (2007) Adaptogens: herbs for strength, stamina, and stress relief. Inner Traditions/Bear & Co/Healing Arts Press, Rochester
Xu X, Guo L, Tian G (2013) Diabetes cognitive impairments and the effect of traditional chinese herbs. Evid Based Complement Alternat Med 2013:649396
Zajac J, Shrestha A, Patel P, Poretsky L (2010) The main events in the history of diabetes mellitus. In: Principles of diabetes mellitus. Springer, New York
Zhang H, Tan C, Wang H, Xue S, Wang M (2010) Study on the history of traditional Chinese medicine to treat diabetes. Eur J Integr Med 2:41–46

Zhang H, Samadi AK, Cohen MS, Timmermann BN (2012) Antiproliferative Withanolides from the Solanaceae: a structure–activity study. Pure Appl Chem 84:1353–1367
Zhang H, Cao CM, Gallagher RJ, Timmermann BN (2014) Antiproliferative Withanolides from several solanaceous species. Nat Prod Res 28:1941–1951
Zimmet P, Thomas CR (2003) Genotype, obesity and cardiovascular disease – has technical and social advancement outstripped evolution. J Intern Med 254:114–125
Zimmet PZ, Magliano DJ, Herman WH, Shaw JE (2014) Diabetes: a 21st century challenge. Lancet Diabetes Endocrinol 2:56–64

Chapter 3
Allium cepa Root Chromosomal Aberration Assay: An Application in Assessing Anti-genotoxic Potential of Ashwagandha

Rajneet Kour Soodan, Ashita Sharma, Mandeep Kaur, Jatinder Kaur Katnoria, and Avinash Kaur Nagpal

Abstract Plant kingdom is rich in phytochemicals with high potential for medicinal and therapeutic uses. These can be particularly recruited to combat the environmental genotoxins that have increased tremendously in last few decades due to industrialization of human lifestyle. Screening of plants for this purpose involves a battery of bioassays; the mammalian or bacterial assays are used worldwide. Considering the ethical and economic reasons, the use of plant bioassays for the initial screening is advocated. Of all plant bioassays available, *Allium cepa* root chromosomal aberration assay is an efficient, most-easy and inexpensive tool that can be used for (i) assessment of genotoxicity of environmental mutagens and (ii) evaluation of anti-genotoxic potential of plant extracts. The present chapter discusses the history, application and technical aspects of this assay in assessing the geno-protective potential of various plants with special reference to Ashwagandha, the wonder herb.

Keywords *Allium cepa* • Chromosomal aberrations • Assessment of genotoxicity • Ashwagandha • Geno-protective potential

3.1 Introduction

The phenomenal growth of chemical industry in the past few decades has led to immense social and economic advancements. At the same time, it has accentuated environmental contamination leading to severe health hazards. Large-scale use of synthetic chemicals as a result of changes in lifestyle has caused degradation of environmental quality resulting in increase in magnitude and variety of ill health effects including hypertension, impairment of kidneys, cognitive disorders and even

---

R.K. Soodan • A. Sharma • M. Kaur • J.K. Katnoria • A.K. Nagpal (✉)
Department of Botanical and Environmental Sciences, Guru Nanak Dev University, Amritsar -143005, Punjab, India
e-mail: avnagpal@rediffmail.com

S.C. Kaul, R. Wadhwa (eds.), *Science of Ashwagandha: Preventive and Therapeutic Potentials*, DOI 10.1007/978-3-319-59192-6_3

death. Exposure to different environmental contaminants has caused considerable damage to the human gene pool and biodiversity at large (Weil et al. 2012; Brei et al. 2016; Kim et al. 2016).

Evidences have been accumulating to prove that environmental contaminants not only increase the genetic load but also cause cancer through induction of somatic mutations (Ames et al. 1983; Popesku et al. 2010; Xia et al. 2013). Environmental biologists throughout the world are presently concerned to safeguard human beings from unduly exposure to environmental mutagens and carcinogens. To frame out programs and policies to protect the human beings from adverse effects of various environmental chemicals, it is important to determine the degree of health risk to humans by an environmental agent under a specified level of exposure. Since, the direct risk assessment in humans is not feasible due to logistic, ethical and practical considerations so a variety of tests using different experimental models ranging from viruses, bacteria, insects, animals, plants, human cell cultures and also intact mammals have been employed to assess the genotoxic effects of environmental contaminants (Kong and Ma 1999; Matsumoto et al. 2006; Mouchet et al. 2006; Yıldız et al. 2009; Pohren et al. 2013).

A number of international organizations including World Health Organization recommended the utility of higher plant bioassays such as *Allium cepa* root chromosome aberration assay (Grant 1982); *Hordeum vulgare* chromosomal aberration assay (Constantin and Nilan 1982); *Tradescantia* staminal hair mutation assay, *Tradescantia* micronucleus assay (Ma et al. 1983); *Zea mays* specific locus mutation assay (Plewa 1982) etc. for the detection of different types of environmental genotoxins/mutagens. The cytotoxic response in the meristematic tissues of plants is similar to the embryogenic or spermatogenic tissues in animals (Kristen 1997).

Allium cepa root chromosomal aberration assay is the most widely used assay among higher plant bioassays. The popularity of this assay is because of cost benefit factors and simple methodology involving minimum use of expensive laboratory equipment. The less number (2n = 16) and appropriate size (8–16 μm) of chromosomes in *Allium cepa* further adds to ease aspects of this assay (Bolle et al. 2004). The results obtained have been largely demonstrated to be in good correlation with results obtained from other test systems using eukaryotic and prokaryotic cells (Grant 1978; Hazarika and Sarkar 2001; Ozkara et al. 2015). Also, unlike *Ames* assay (which requires addition of S9), the plant system possesses the important enzymes, which can activate the promutagens to mutagens (Fiskesjo 1985). Thus, it is widely used for the assessment of genotoxicity of environmental chemicals such as pesticides (Nagpal and Grover 1994; Asita and Makhobo 2013); chemical mutagens (Rank and Neilson 1997); heavy metals etc. (Fiskesjo 1988; Soodan et al. 2009a; Eleftheriou et al. 2012; Cigerci et al. 2014) and environmental mixtures such as air, water and soil (Grover and Kaur 1999; Katnoria et al. 2011; Souza et al. 2013; Kaur et al. 2014). Though, this assay was introduced to assess the geno-toxicity but later it was successfully used to assess anti-genotoxicity of plant extracts rich in various kinds of phyto-chemicals which can further help in drug discovery (Fiskesjo 1969; Rani et al. 2005; Fedel-Miyasato et al. 2014; Mekki et al. 2015; Prajitha and Thoppil 2016; Roberto et al. 2016).

3.2 *Allium* Assay: History and Applications

The *Allium cepa* test was first introduced by Levan (1938) while studying effects of colchicine in the root tip cells of *Allium fistulosum* and *Allium cepa*. Colchicine inhibited the formation of spindle, delayed the division of centromeres and led to the scattering of chromosomes in the cell. This study was followed by a number of other studies where some chemicals like organic hydrocarbons, pesticides, metal ions, etc. showed colchicine like effect in this test system (Quidet and Hitier 1948; Scholes 1955; Wilson et al. 1956). Later in 1969, Fiskesjo reported other kinds of aberrations like chromosomal stickiness, breakage, acentric fragments, anaphase bridges, multipolar mitosis, c-mitosis and c-tumors as a result of treatment with five organic mercury halogenides *viz.* betoxin (technical preparation); methyl, ethyl and methoxyethyl mercuric chloride; and butyl mercuric bromide. Grant (1978) observed that chromosomal aberrations in plants could serve as a reliable monitoring system for detection of genotoxic nature of environmental chemicals. He presented examples of some pesticides, which showed parallel effects in induction of chromosomal aberrations in plants and mammalian systems in different studies, advocating use of plant systems in the first-tier bioassay systems for detection of environmental genotoxins. Grant (1979) studied genotoxicity of herbicide 2, 4, 5-trichlorophenoxyacetic acid (2, 4, 5-T) using *Allium cepa* test. He discussed molecular activity, toxicology, morphological and cytological effects; mutagenicity; carcinogenicity; and teratogenicity of 2, 4, 5-T. Also, Fiskesjo (1988) recommended *Allium* test as an alternative method for testing the relative toxicity of heavy metals viz., Hg, Cu, Ni, Cd, Be, A1, Mn and Li. Induction of c-tumors, c-mitosis, stickiness, chromosome breaks, 'banded' or 'fragmented' chromosomes was reported. The study strongly advocated the use of *Allium* assay as an important tool for evaluation of environmental chemicals like heavy metals and their ranking with reference to their toxicity. The use of this assay for unknown natural/environmental samples was also proposed. Rank and Nielsen (1993) studied the genotoxicity of complex mixtures like non-diluted wastewater from slaughterhouse and dye-house using this assay. The genotoxicity of samples was estimated by observing the chromosomal aberrations in metaphasic and anaphasic cells. Various aberrations including bridges, fragments, vagrants, c-mitosis and stickiness were reported. Rank (2003) carried out a study and advocated use of *Allium* anaphase-telophase chromosome aberration assay as a modification of *Allium* test as it was simpler and faster.

The slight modifications in the *Allium cepa* assay were proposed by Fiskesjo (1979) to use it as a tool to assess anti-gentoxicity. Later, the assay was widely and efficiently used to assess anti-genotoxicity of plant extracts and/or phytoconstituents.

3.3 Methodology

The widely used protocol proposed through various studies for evaluation of genotoxicity of environmental chemicals and anti-genotoxicity of plant extracts is given in Fig. 3.1. Young and fresh onion bulbs of uniform size are used. For genotoxicity

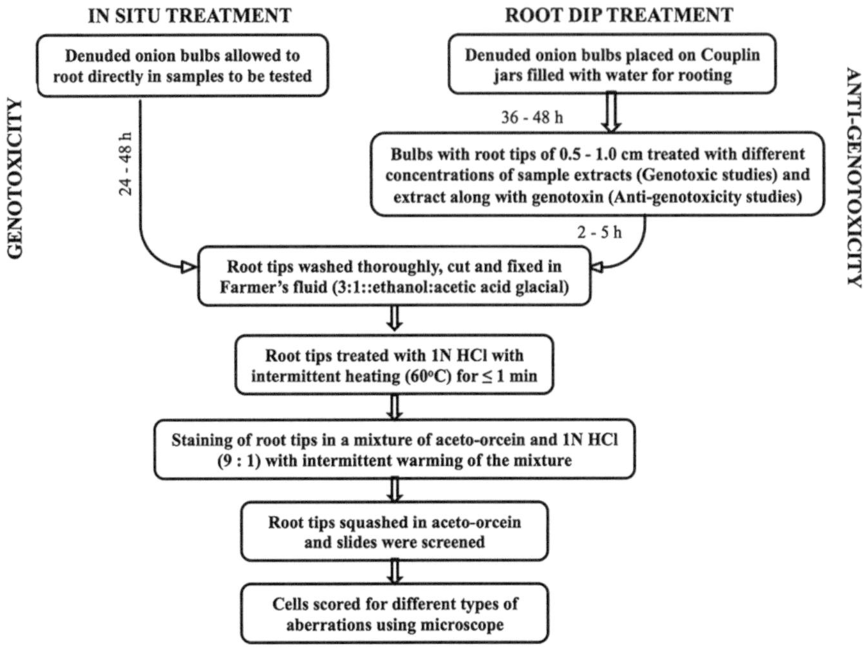

Fig. 3.1 Protocol for assessment of genotoxicity and anti-genotoxicity using *Allium cepa* root chromosomal aberration assay

studies, the assay can be carried out with two different modes of treatment, *in situ* and root dip. In both modes of treatment, the primary roots are first removed. During *in situ* treatment, the denuded bulbs are placed directly on solution to root while in root dip treatment, the bulbs are placed on Couplin jars filled with water for 24–36 h for rooting. After 24–36 h, bulbs with freshly emerged roots of size 0.5–2 cm are treated with different concentrations of test sample for 2–5 h (mostly 3 h). Assessment of anti-genotoxicity of a plant extract can be carried by treatment of onion root tips with genotoxin along-with plant extract. There are three modes of treatment for the assessment of anti-genotoxicity – pre-treatment, post-treatment and simultaneous treatment. For pre-treatment, root tips are first treated with the plant extract and then with the genotoxin while for post treatment it's the other way round. For simultaneous treatment, root tips of onion bulbs are treated with both genotoxin and plant extract simultaneously. After treatment, the bulbs are thoroughly washed, and root tips are plucked and fixed in Farmer's fluid (ethanol and glacial acetic acid; 3:1). At least, nine root tips are squashed in Aceto-Orcein to prepare slides and the slides are screened under microscope to score different types of chromosomal aberrations.

When a genotoxic agent attacks a meristematic cell, it causes damage to the genetic compliment of the cell; either this damage gets repaired through the cellular repair mechanisms or if unrepaired this leads to various kinds of chromosomal aberrations (Fig. 3.2). Chromosomal aberrations are considered as end result of

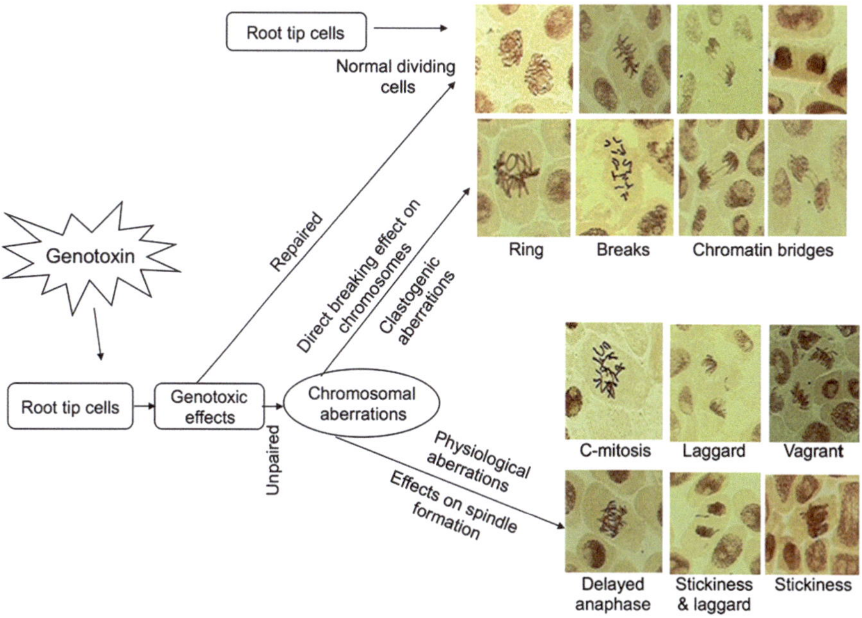

Fig. 3.2 Representation of effects of a genotoxin on the root tip cells of *Allium cepa*

genotoxic effects of various physical and chemical agents and are also estimates of exposure of various organisms to these agents that impair human health (Pohren et al. 2013). Chromosomal aberrations are caused due to effect on DNA (synthesis or replication) or nucleoproteins, which can lead to either direct breaking effect on chromosomes or abnormal segregation of chromosomes. The aberrations can broadly be classified as clastogenic aberrations (direct breaking effect on chromosomes) and physiological (effect on spindle proteins). The clastogenic effects are observed in the form of chromosomal/chromatid breaks, rings and bridges. The chromosomal break is considered to be result of unfinished or misrepair of DNA involved in the linear continuity of chromosomes (Evans 1977). The broken ends of chromosomes stick together to form ring chromosomes. The loss of telomeres results in the sticky broken ends which join together to form ring (Peacock et al. 1973; Raghuvanshi and Singh 1976). The formation of bridge is attributed to "end to end" fusion of uncapped chromosomes resulting in the formation of dicentric chromosomes (Tusell et al. 2010). Also, formation of bridges is attributed to unequal exchange leading to formation of dicentric chromosomes, which are pulled equally to both poles resulting in bridge like structure (Sax 1940). Among physiological aberrations, when during the separation of chromosomes, one chromosome moves ahead of the group towards the pole and gets separated from the rest of group, this anomaly is called as vagrant chromosome while, laggard is a situation where during separation a chromosome lags behind the rest of the group as it doesn't get attached to the spindle fiber. Both laggards and vagrants result in unequal distribution of chromosomes in daughter cells. Increased chromosomal contraction, depolymerization of DNA or dissolution of nucleoproteins lead to the condensation of chromosomes, which gives the sticky appearance and this anomaly is labeled as stickiness. At times, the separation of chromosomes doesn't occur during anaphase and all the chromosomes lie near the equatorial plate, this is recorded as delayed anaphase. The spatial abnormality during anaphase with respect to equatorial plate is considered as abnormal anaphase.

3.4 *Allium cepa* Assay as a Tool to Screen Anti-genotoxic Products

The risk associated with exposure to genotoxic carcinogens incites scientific community to carry out rigorous research for the identification of compounds of plant kingdom with anti-genotoxic potential, which can be used either directly for medicinal purposes or for drug development. The secondary metabolites present in plants are known to have anti-genotoxic and anti-carcinogenic properties. These phytoconstituents can scavenge reactive oxygen species that contribute to DNA damage and also prevent cell proliferation. *Allium cepa* assay, as discussed above is also used to assess the anti-genotoxicity of natural products. Fiskesjo (1979) suggested the modifications in the original protocol to use this assay as a tool to assess anti-genotoxic potential. He studied the protective effects of selenium against mercury poisoning in plants. This study reported that treatment with mercury induced stickiness, delayed anaphases, vagrant chromosomes and c-mitosis in root tip cells, and the frequency of chromosomal aberrations was reduced by the treatment of root tips with selenium.

This assay is widely used as an efficient tool to screen the anti-genotoxic potential of plant extracts against various toxic compounds. Extracts of plants like *Brassica juncea* (Sharma et al. 2012); *Coriandrum sativum, Cuminum cyminum* (Sharma et al. 2011) and *Emblica officinalis* (Samanta and Bandyopadhyay 2011) have shown genoprotective effects against heavy metals using *Allium cepa* assay. Also, extracts of plants like, *Adhota vasica, Carica papya* and *Cinacarthus nutans* (Rathnasamy et al. 2013); *Erythrina velutina* (Silva et al. 2013) and *Schinus terebinthifolitus* (Fedel-Miyasato et al. 2014) have been reported to decrease the percentage of chromosomal aberrations induced by MMS in root-tip cells of *Allium cepa*. Anti-genotoxic potential of *Calotrpis gigantea* (Ravi et al. 2011) and *Terminalia chebula* (Rathore and Choubey 2005) against various drugs was also reported using this assay.

3.5 Anti-genotoxicity Potential of Ashwagandha: A Wonder Herb of India

Withania somnifera Dunal (family Solanaceae), also known as Ashwagandha, winter cherry, wonder herb of India and queen of Ayurveda is amongst the plants, which are widely used as traditional medicine. In Vedas, it is referred as herbal tonic and a healthy food. The plant is known to possess anti-inflammatory, anti-tumor, anti-stress, anti-bacterial, anti-fungal, anti-diabetic, anti-oxidant, immunomodulatory, cardioprotective, neuroprotective, hemopoetic and rejuvenating properties (Arora et al. 2004; Kaur et al. 2004, 2007; Widodo et al. 2007; Hamza et al. 2008; Shah et al. 2015). Roots and leaves of Ashwagandha are known to have number of bioactive withanolides, which have various medicinal properties (Buddhiraja and Sudhir 1987; Ziauddin et al. 1996). Rani et al. (2005) investigated anti-genotoxic effects of leaf extract of Ashwagandha (LASH) against MNNG (N-methyl-N′-nitro-N-nitrosoguanidine) induced chromosomal aberrations in root tip cells of *Allium cepa*. For preparation of LASH, finely ground leaf powder (40 g) was exhaustively extracted in soxhlet apparatus using methanol at 60 °C for 100–110 h. Methanol extract was made aqueous by addition of distilled water and then hexane was added to remove chlorophyll and other pigments. Diethyl ether was added to the residual aqueous methanolic extract and diethyl ether extract (DEE) was separated using a separatory funnel. Diethyl ether extract was concentrated, dried and was called as LASH. 1 g of LASH was solubilized in DMSO (1 ml) and 99 ml of distilled water was added to make 1.0% LASH. This extract (1%) was further diluted to get different concentrations (0.25, 0.50, and 0.75%) of LASH. To study anti-genotoxic potential of LASH, freshly emerged onion root tip cells were subjected to three different kinds of treatment *viz.*, pre-, post- and simultaneous with different concentrations of LASH. For pre-treatment, the onion roots were first treated for 2 h with different concentrations of LASH followed by 2 h treatment with 0.1% MNNG. During post-treatment, 2 h MNNG (0.1%) treatment was given first followed by LASH concentrations for 2 h. For simultaneous treatment, onion roots were treated simultaneously with 0.1% MNNG and each concentration of LASH. 1.0% DMSO (2 h) and 0.1% MNNG (2 h) served as negative and positive controls, respectively. After treatment, the root tips were thoroughly washed and squashed in aceto-orcein and slides were screened for different types of chromosomal aberrations. Treatment of onion roots with MNNG (0.1%) alone induced different kinds of chromosomal aberrations, which were classified as physiological aberrations (c-mitosis, stickiness, delayed anaphases, laggards and vagrant chromosomes) and clastogenic aberrations (chromatid bridges and chromosomal breaks). Different kinds of treatment (pre-, post- and simultaneous) with different concentrations of LASH imparted a dose dependent protective effect on MNNG-induced chromosomal aberrations in root tip cells of *A. cepa* (Fig. 3.3). About 91–92% inhibition in the induction of chromosomal aberrations due to MNNG treatment was observed at the highest concentration (1.0%) of LASH tested. In another study, LASH showed anti-proliferative activity in human osteosarcoma (U2OS) and breast carcinoma (MCF-7) cancer cell lines (Kaur et al. 2004). Soodan et al. (2009b) determined the anti-genotoxic potential of an aqueous extract of Ashwagandha leaves against lead induced genotoxicity in *Allium cepa* root chromosomal aberration assay and micronucleus test (MNT) in bone marrow cells of male albino rats. Treatment of *Allium* roots with lead (0.5 ppm) led to different kinds of physiological and clastogenic aberrations. A dose dependent decrease in percent chromosomal aberrations was observed in RTCs of *Allium cepa* treated with lead (0.5 ppm) and different concentration of aqueous extract of Ashwagandha

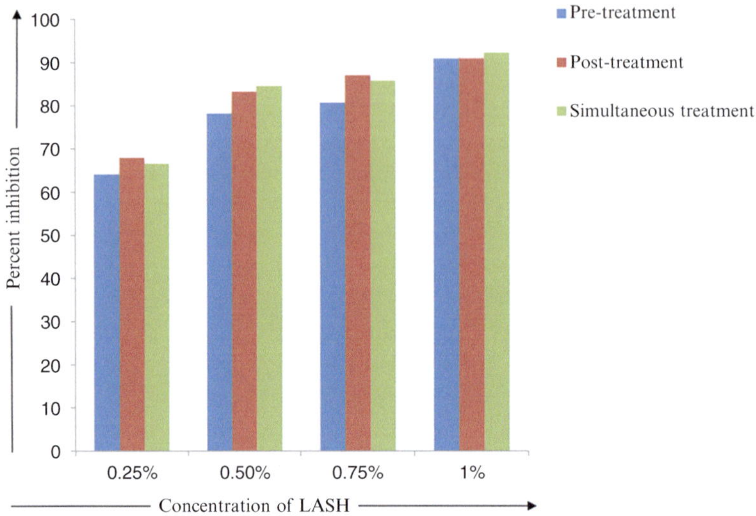

Fig. 3.3 Effect of pre-, post- and simultaneous LASH treatments on MNNG-induced genotoxic effects in root tip cells of *Allium cepa* represented as % inhibition of chromosomal aberrations

in three mode of treatment as described by Rani et al. (2005). Maximum genoprotective effect was observed with 1% Ashwagandha leaf extract (aqueous) in all the three modes of treatment. Treatment of male albino rats with Ashwagandha leaf extract along with lead resulted in a decrease in number of micronucleated polychromatic erythrocytes as compared to when rats were treated with lead alone. This study demonstrated strong correlation between *Allium* test and mammalian micronucleus test.

3.6 Conclusion

Plant kingdom is rich in compounds that can protect the cells from effects of various genotoxins released in the environment through accelerated use of chemicals. Screening of these plants to determine their anti-genotoxic potential can prove a boon for mankind. *Allium cepa* root chromosomal aberration assay is validated by various international agencies like WHO, UNEP as an efficient and easy tool for screening of genotoxic agents. This assay not only assesses genotoxic compounds but also used for the screening of anti-genotoxicity of plant extracts. The meristematic cells of onion roots depict cytotoxic effects similar to those observed in mammalian cells. Genoprotective substances can support human beings by providing protection against genotoxic carcinogens. Screening of Ashwagandha, for anti-genotoxic properties depicts that this wonder herb can prove to be nectar against carcinogens.

References

Ames BN, Lee FD, Durston WE (1983) An improved bacterial test system for the detection and classification of mutagens and carcinogens. Proc Natl Acad Sci U S A 70:782–786

Arora S, Dhillon S, Rani G, Nagpal A (2004) The *in vitro* antibacterial and synergistic activities of *Withania somnifera* extracts. Fitoterapia 7:385–388

Asita AO, Makhobo MM (2013) Clastogenic and cytotoxic effects of four pesticides used to control insect pests of stored products on root meristems of *Allium cepa*. Environ Nat Resour Res 3:133–145

Bolle P, Mastrangelo S, Tucci P, Evandri MG (2004) Clastogenicity of atrazine assessed with the *Allium cepa* test. Environ Mol Mutagen 43:137–141

Brei M, Perez-Barahona A, Strobl E (2016) Environmental pollution and biodiversity: light pollution and sea turtles in the Caribbean. J Environ Econ Manag 77:95–116

Buddhiraja RD, Sudhir S (1987) Review of biological activity of withanolides. J Sci Ind Res Ind 46:488–491

Cigerci IH, Liman R, Ozgul E, Konuk M (2014) Genotoxicity of indium tin oxide by *Allium* and Comet tests. Cytotechnology 67:157–163

Constantin MJ, Nilan RA (1982) Chromosomal aberration assays in barley (*Hordeum vulgare*) a report of the US environmental protection agency gene tox program. Mutat Res 99:13–36

Eleftheriou EP, Adamakis I-DS, Melissa P (2012) Effects of hexavalent chromium on microtubule organization, ER distribution and callose deposition in root tip cells of *Allium cepa* L. Protoplasma 249:401–416

Evans HJ (1977) Molecular mechanism in the induction of chromosome aberrations. In: Scott D, Bridges BA, Sobels FH (eds) Progress in Genetic Toxicology. Elsevier/North-Holland and Biomedical Press, Amsterdam, p 57

Fedel-Miyasato LES, Formagio ASN, Auharek SA, Kassuya CAL, Navarrol SD, Cunha-Laura AL, Monreal ACD, Vieira MC, Oliveira RJ (2014) Antigenotoxic and antimutagenic effects of *Schinus terebinthifolius* Raddi in *Allium cepa* and Swiss mice: a comparative study. Genet Mol Res 13:3411–3425

Fiskesjo G (1969) Some results from *Allium* tests with organic mercury halogenides. Hereditas 62:314–322

Fiskesjo G (1979) Mercury and selenium in a modified *Allium* test. Hereditas 9:169–178

Fiskesjo G (1985) The *Allium* test as a standard in environmental monitoring. Hereditas 102:99–112

Fiskesjo G (1988) The *Allium* test-an alternative in environmental studies: the relative toxicity of metal ions. Mutat Res 197:243–260

Grant WF (1978) Chromosome aberrations in plants as a monitoring system. Environ Health Perspect 27:37–43

Grant WF (1979) The genotoxic effects of 2,4,5-T. Mutat Res 65:83–119

Grant WF (1982) Cytogenitic studies of agricultural chemicals in plants. In: Fleck RA, Hollander A (eds) Genetic toxicology, An agricultural Perspective, vol 21. Plenum Press, New York and London, pp 353–378

Grover IS, Kaur S (1999) Genotoxicity of wastewater samples from sewage and industrial effluent detected by the *Allium cepa* root anaphase aberration and micronucleus assays. Mutat Res 426:183–188

Hamza A, Amin A, Daoud S (2008) The protective effect of a purified extract of *Withania somnifera* against doxorubicin-induced cardiac toxicity in rats. Cell Biol Toxicol 24:63–73

Hazarika A, Sarkar SN (2001) Effect of isoproturon pretreatment on the biochemical toxicodynamics of anilofos in male rats. Toxicology 165:87–95

Katnoria JK, Arora S, Bhardwaj R, Nagpal A (2011) Evaluation of genotoxic potential of industrial waste contaminated soil extracts of Amritsar, India. J Environ Biol 32:363–367

Kaur K, Rani G, Widodo N, Nagpal A, Taira K, Kaul SC, Wadhwa R (2004) Evaluation of the anti-proliferative and antioxidative activities of leaf extract from *in vivo* and *in vitro* raised *Ashwagandha*. Food Chem Toxicol 42:2015–2020

Kaur K, Widodo N, Nagpal A, Kaul SC, Wadhwa R (2007) Sensitization of human cancer cells to anticancer drugs by leaf extract of *Ashwagandha* (LASH). Tissue Cult Res Commun 26:193–199

Kaur M, Soodan RK, Katnoria JK, Bhardwaj R, Pakade YB, Nagpal AK (2014) Analysis of physico-chemical parameters, genotoxicity and oxidative stress inducing potential of soils of some agricultural fields under rice cultivation. Trop Plant Res 1:49–61

Kim K, Kabir E, Jahan SA (2016) A review on the distribution of Hg in the environment and its human health impacts. J Hazard Mater 306:376–385

Kong MS, Ma TH (1999) Genotoxicity of contaminated soil and shallow well water detected by plant bioassays. Mutat Res 426:221–228

Kristen U (1997) Use of higher plants as screens for toxicity assessment. Toxicol In Vitro 11:181–191

Levan A (1938) The effect of colchicine on root mitoses in *Allium*. Hereditas 24:471–486

Ma TH, Anderson VA, Harris MM, Bare JL (1983) Tradescantia-micronucleus (Trad-MCN) test on the genotoxicity of malathion. Environ Mutagen 5:127–137

Matsumoto ST, Mantovani MS, Malagutti MIA, Dias AL, Fonseca IC, Marin-Morales MA (2006) Genotoxicity and mutagenicity of water contaminated with tannery, as evaluated by the micronucleus test and comet assay using the fish *Oreochromis niloticus* and chromosome aberration in onion root-tips. Genet Mol Biol 29:148–158

Mekki LL, Mansour H, Eldean EG, Nasser AA (2015) Evaluation of anti-genotoxicity of the ethanolic plant extract of *Beta vulgaris* Maritima using *Allium cepa* root assay. Egypt J Exp Biol 11:147–153

Mouchet F, Gauthier L, Mailhes C, Jourdain MJ, Ferrier V, Triffault G, Devaux A (2006) Biomonitoring of the genotoxic potential of aqueous extracts of soils and bottom ash resulting from municipal solid waste incineration, using the comet and micronucleus tests on amphibian (*Xenopus laevis*) larvae and bacterial assays (Mutatox and Ames tests). Sci Total Environ 15:232–246

Nagpal A, Grover IS (1994) Genotoxic evaluation of some systemic pesticides in *Allium cepa* following in situ and direct treatments I mitotic effects. Nucleus 37:99–105

Ozkara A, Akyl D, Eren Y, Erdogmus SF (2015) Potential cytotoxic effect of Anilofos by using *Allium cepa* assay. Cytotechnology 67:783–791

Peacock WJ, Wolff S, Lindley DL (1973) Continuity of chromosome subunits. Chromosom Today 4:85–100

Plewa MJ (1982) Specific-locus mutation assays in *Zea mays* a report of the U.S. Environmental Protection Agency Gene-Tox program. Mutat Res 99:317–337

Pohren P, Thatiana C, Vargas VMF (2013) Investigation of sensitivity of the *Allium cepa* test as an alert system to evaluate the genotoxic potential of soil contaminated by heavy metals. Water Air Soil Pollut 224:1460–1470

Popesku JT, Tan EY, Martel PH, Kovacs TG, Rowan-Carroll A, Williams A, Yauk C, Trudeau VL (2010) Gene expression profiling of the fathead minnow (*Pimephales promelas*) neuroendocrine brain in response to pulp and paper mill effluents. Aquat Toxicol 99:379–388

Prajitha V, Thoppil JE (2016) Genotoxic and antigenotoxic potential of the aqueous leaf extracts of *Amaranthus spinosus* Linn. Using *Allium cepa* assay. S Afr J Bot 102:18–25

Quidet P, Hitier H (1948) The production of polyploidy plants by treatment with hexachlorocyclohexane sulphide. C R Acad Sci 226:833–835

Raghuvanshi SS, Singh AK (1976) Effect of gamma rays on growth and Karyokinetic activity in *Trigonella foenum-oraceum* L. Cytologia 4:177–186

Rani G, Kaur K, Wadhwa R, Kaul SC, Nagpal A (2005) Evaluation of the anti-genotoxicity of leaf extract of *Ashwagandha*. Food Chem Toxicol 43:95–98

Rank J (2003) The method of *Allium* anaphase-telophase chromosome aberration assay. Ekologija (Vilnius) 1:38–42

Rank J, Neilson MH (1997) *Allium cepa* anaphase-telophase root tip chromosome aberration assay on *N*-methyl *N*-nitrosourea, maleic hydrazide, sodium azide and ethylmethane sulfonate. Mutat Res 390:121–127

Rank J, Nielsen MH (1993) A modified *Allium* test as a tool in the screening of the genotoxicity of complex mixtures. Hereditas 118:49–53

Rathnasamy S, Mohamed KB, Sulaiman SF, Akinboro A (2013) Evaluation of cytotoxic, mutagenic and antimutagenic potential of leaf extracts of three medicinal plants using *Allium cepa* chromosome assay. Int Curr Pharm J 2:131–140

Rathore HS, Choubey P (2005) Prevention of acetaminophen-induced Mitodepression with Myrobalan (fruit of *Terminalia chebula*) in *Allium cepa* model. Iran J Pharmacol Ther 4:100–104

Ravi RG, Harikesh D, Chandrasekhar TR, Pramod YG, Angad PM (2011) Cytotoxic activity of ethanolic root extract of *Calotropis gigantea* Linn. Int J Drug Dev Res 3:101–108

Roberto MM, Jamal CM, Malaspina O, Marin-Morales MA (2016) Antigenotoxicity and antimutagenicity of ethanolic extracts of Brazilian green propolis and its main botanical source determined by the *Allium cepa* test system. Genet Mol Biol 39:257–269

Samanta A, Bandyopadhyay B (2011) Prevention of cadmium induced genotoxicity with *Emblica officinalis* L. (Amla) in *Allium* test. Res J Pharm, Biol Chem Sci 3:890–897

Sax K (1940) An analysis of X-ray induced chromosomal aberrations in *Tradescantia*. Genetics 25:41–68

Scholes ME (1955) The effects of aldrin, dieldrin, isodrin, endrin and DDT on mitosis of the onion (*Allium cepa* L.) J Hortic Sci 30:181–187

Shah N, Singh R, Sarangi U, Saxena N, Chaudhary A, Kaur G, Kaul SC, Wadhwa R (2015) Combinations of *Ashwagandha* leaf extracts protect brain –derived cells against oxidative stress and induce differentiation. PLoS One 10:1–14

Sharma A, Kumar M, Kaur S (2011) *Cuminum cyminum* Linn. And *Coriandrum sativum* Linn. Extracts modulate chromium genotoxicity in *Allium cepa* chromosomal aberration assay. Nucleus 54:99–105

Sharma S, Nagpal A, Vig AP (2012) Genoprotective potential of *Brassica juncea* (L.) Czern. Against mercury-induced genotoxicity in *Allium cepa* L. Turk J Biol 36:622–629

Silva DSBS, Garcia ACFS, Mata SS, de Oliveira B, Estevam CS, Scher R, Pantaleao SM (2013) Genotoxicity and cytotoxicity of *Erythrina velutina* Willd., *Fabaceae*, on the root meristem cells of *Allium cepa*. Braz J Pharm 21:92–97

Soodan RK, Katnoria JK, Nagpal A (2009a) Evaluation of genotoxic effects of lead nitrate in *Allium cepa* root chromosomal aberration assay. Int J Biosci Rep 7:245–249

Soodan RK, Katnoria JK, Kaur G, Nagpal A (2009b) Genoprotective potential of an aqueous extract of leaves of *Withania somnifera* against lead induced genotoxicity in plant and mammalian bioassays. Perspect Cytol Genet 15:283–296

Souza TS, Hencklein FA, Angelis DF, Fontanetti CS (2013) Clastogenicity of landfarming soil treated with sugar cane vinasse. Environ Monit Assess 185:1627–1636

Tusell L, Pampalona J, Soler D, Frias C, Genesca A (2010) Different outcomes of telomere-dependent anaphase bridges. Biochem Soc Trans 38:1698–1703

Weil RE, Spade DJ, Knoebl I, Hemming JM, Tongue ML, Szabo NJ, Kroll KJ, Tate WB, Denslow ND (2012) Evaluation of water quality threats to the endangered Okaloosa darter (*Etheostoma okaloosae*) in East Turkey Creek on Eglin Air Force Base. Aquat Toxicol 110–111:177–186

Widodo N, Kaur K, Shrestha BG, Takagi Y, Ishii T, Wadhwa R, Kaul SC (2007) Selective killing of cancer cells by leaf extract of *Ashwagandha*: identification of tumor-inhibitory factor and the first molecular insights to its effects. Cancer Prev Res 13:2298–2306

Wilson SM, Daniel A, Wilson GB (1956) Cytological and genetical effects of the defoliant endothal. J Hered 47:151–155

Xia Z, Duan X, Tao S, Qiu W, Liu D, Wang Y, Wei S, Wang Y, Wei S, Wang B, Jiang Q, Lu B, Song Y, Hu X (2013) Pollution level, inhalation exposure and lung cancer risk of ambient atmospheric polycyclic aromatic hydrocarbons (PAHs) in Taiyuan, China. Environ Pollut 173:150–156

Yıldız M, Cigerci IH, Konuk M, Fidan AF, Terzi H (2009) Determination of genotoxic effects of copper sulphate and cobalt chloride in *Allium cepa* root cells by chromosome aberration and comet assays. Chemosphere 75:934–938

Ziauddin M, Phansalka N, Patki P, Diwanay S, Patwardhan B (1996) Studies on the immunomodulatory effects of *Ashwagandha*. J Ethnopharmacol 50:69–76

Chapter 4
Withania somnifera: From Traditional Use to Evidence Based Medicinal Prominence

Sheikh Rayees and Fayaz Malik

Abstract *Withania somnifera* is wonder shrub used traditionally as a folk medicine for several remedies in the Indian subcontinent. During last few decades numerous scientific studies have shown the potential therapeutic prospects of this herb and of its various constituents in different disease models. Several classes of compounds including withanolides, sitoindosides and other useful alkaloids have been isolated from this plant with promising medicinal value. It has been demonstrated that different parts of plant including roots, leaves and fruits exhibit unique biological activities that actually are manifested to the presence and the abundance of specific constituent/s in the respective parts of the plant. It was also established that same plant grown at different locations under varied environmental conditions influences the synthesis of the individual constituents in different parts of the plant, hence their biological activities. Studies conducted in cellular and murine models have shown that extracts made of roots bear immune activating properties, whereas leaf extracts showed anticancer activities. It has been observed that root and leaf extracts of plant characterised with about a dozen markers displayed that root extracts demonstrated Th1 specific immunomodulatory and anti-inflamatory activities, whereas leaf extracts showed anticancer properties, respectively. It was further explored that anticancer potential of leaf extracts was mainly due to Withaferin A, a potent cytotoxic withanoloid, existing in abundance and a major constituent in the leaves of the plant. Similarly, anti-inflammatory and Th1 immune skewing properties of the root extract was conferred to the higher amount of Withanolide A, present in the roots of the plant along with other constituents. Meanwhile several studies have deciphered the role of individual constituents in various biological activities with extensive mechanism of action discussed in this review. This review while summarises the potential medicinal benefits of the *W.somnifera*, it also emphasizes that marketed product of the plant extracts for human consumption should scientifically validated for bioactive constitutents.

S. Rayees
Department of Pharmacology, University of Illinois at Chicago, Chicago, IL, USA

F. Malik (✉)
Department of Cancer Pharmacology, Indian Institute of Integrative Medicine, Canal Road, Jammu, Jammu and Kashmir 180001, India
e-mail: fmalik@iiim.ac.in

S.C. Kaul, R. Wadhwa (eds.), *Science of Ashwagandha: Preventive and Therapeutic Potentials*, DOI 10.1007/978-3-319-59192-6_4

Keywords *Withania somnifera* • Root extract • Anti-inflamatory activity • Leaf extract • Anticancer

4.1 Introduction

Since the origin of human civilisation, human beings have been using plants, available in their habitat, both for therapeutic purposes and as a source of food. Despite all the advancement in synthetic chemistry and drug development processes, plants are still a key source of modern medicinal preparations. Numerous species of plants are known for their medicinal values and are used to treat various human ailments and diseases. Medicinal properties of the plants are associated with their roots, seeds, flowers, leaves, fruits or as a whole plant itself. Traditional system of medicine, which is based on beliefs and practices has always been a remedy to every kind of human ailments. This system of medicine dates back to hundreds of centuries ago when there was hardly any drug based on authentic scientific validation (Winters 2006; Mirjalili 2009a; Kulkarni and Dhir 2008; Rayees et al. 2012, 2013). More importantly this pharmacological treatment of diseases through traditional system of has proven to be successful in curing several human ailments and wounds, though most of the industrialized countries have reduced the use of plants in therapeutics. Some of the most powerful and useful medicinal plants in the history of mankind include *Emblica officinalis, Saraca Asoca, Withania Somnifera, Aegle marmelous, Bacopa monnieri, Azardirchata indica,* etc. There are more than 120 plant based chemical agents that were development into drugs and used in modern medicine to treat diseases related to cardiovascular system, nervous system, reproductive system, respiratory disorders, etc. More importantly, these drugs have been very successful in treating human illnesses of every class. Some of the best known drugs of plant origin include Paclitaxel, Etoposide, Vinblastine, Acetyldigoxin, Aescin, Atropine, etc. The current strategies employed for the development of new drugs include the development of a novel chemical moity or modification of the existing drug structures. Later has proven to be more successful and economical. Modern science and medicine has also started developing links with traditional medicine (homeopathic) which has resulted in exploration of traditionally used herbs/plants to underdstand and explore the individual constitutents for any particular biological activity (Dhar et al. 2012; Gauttam and Kalia 2013; Sukanya et al. 2010; Singh and Kumar 1998; Linder 1996; Murthy et al. 2008; Rayees et al. 2014; Bhandari 1970). *W. somnifera*, widely used herb in Indian traditional medicine has been scientifically explored for it's medicinal properties and efforts have been made to identify the individual constituents present in various parts of the plant by using *in vitro* and animal models. In this chapter, we will try to highlight the recent scientific advances made in the exploration of therapeutic potential of the plant extracts and indicidual constitutuents.

4.1.1 *Withania somnifera, a Widely Grown Shrub*

Withania somnifera (L.) Dunal, commonly known as Ashwagandha or Indian Ginseng is a valuable medicinal plant used since ancient times (Winters 2006). It is an evergreen shrub belonging to the solanaceae family. It is widely used in traditional system of medicine in India (Unani and Ayurvedic system). *W. somnifera* grows in most parts of the world mainly in tropical and subtropical zones. However, it is abudnatly found in sub-Himalayan (1000 m) tracts in India, Pakistan, Sri Lanka, Afghanistan, South Africa Jordan, Egypt and Morocco. Among the 23 species of *Withania* genus, this plant is economically more valuable because of its medicinal importance (Mirjalili 2009a; Kulkarni and Dhir 2008).

Various pharmacological activities have been documented to be associated with *W. somnifera*. It is regarded as "Rasayana" in Ayurveda which means potent rejuvenator, as it enhances haemoglobin count, hair melanin pigmentation (Singh et al. 2011) physiological endurance and general health (Dhar et al. 2012). All parts of this plant have been reported to possess medicinal values e.g. its roots are used to treat various physiological disorders. Their paste or powder is used to cure ulcers, rheumatic swellings and as a nutrient or health restorative in pregnant or old age people. Similarly, the leaves of this plant are used as antihelmantic, anti-pyretic, anti-cancer, anti-tuberculosis and anti-ulcer.

Furthermore, various parts of this plant are constituents of more than 200 formulations in Indian system of traditional medicine and have been used to cure various health disorders (Gauttam and Kalia 2013; Sukanya et al. 2010).

4.1.2 *Geographical Distribution of* W. somnifera

W. somnifera is widely distributed across the globe. Its distribution area extends from the Canary Islands and the Mediterranean region through India, the Middle East, Africa and Sri Lanka to China. It also occurs in Australia, Pakistan, Afghanistan, Jordan, Morocco and Spain. In Africa, it occurs wild or naturally, mostly on drier parts like South Africa and Indian Ocean islands. However, *W. somnifera* is found almost on all drier parts and sub-tropical India. It is widely distributed in Punjab, Madhya Pradesh, Mumbai, Uttar Pradesh, Rajasthan, Jammu and Himachal Pradesh (Singh and Kumar 1998; Linder 1996; Murthy et al. 2008).

4.1.3 *Classical Uses of* W. somnifera

W. somnifera has been widely used as Rasayana in Ayurvedic system of medicine, practiced in India which can be traced back to 6000 BC (Singh et al. 2011). Rasayana has been described in Ayurveda as a herbal or metallic preparation that upholds a youthful state of physical and mental health. Different parts of this plant have been used for different activities. The roots of this plant are used as anthelmintic, diuretic, narcotic, astringent, thermogenic and aphrodisiac.

Further, the roots are used for rheumatism, leucoderma, constipation, insomnia, nervous breakdown, goitre, joint inflammation (Singh et al. 2011; Bhandari 1970), ulcers and painful swellings (Kritikar and Basu 1935) pimples, worms and piles (Mishra 2004). Because of horse like smell of roots, this plant is also called as Ashwagandha (on consuming it gives the power of a horse). The leaves of this plant are bitter and are used for fever, anthelmintic and for painful swellings. The flowers are aphrodisiac and diuretic. The seeds are anthelmintic, however when mixed with rock salt, can remove white spots from cornea. These are also used in hysteria, memory loss, anxiety and increasing the sperm count (Govardhan 1938).

4.2 Pharmacological Properties Associated with Whole Plant, Parts of the Plant and Individual Constituents

4.2.1 Anti-stress

W. somnifera has traditionally been used for mood soothing in patients with behavioural disorders and as a tranquiliser. Studies were conducted on different rodent models to explore its anti-stress and anti-depressant potential. In a study conducted on a rodent model, Imipramine, an anti-depressant drug and Lorazepam, anti-anxiety drug were compared for their respective pharmacological effects with the extract of *W. somnifera* and the results obtained for *W. somnifera* were found quit promising (Archana and Namasivayam 1999). In a similar study conducted on a Wistar rat chronic stress model, *W. somnifera* root extract was compared with Panax ginseng for its anti-stress properties. In this study, the extracts of the both the herbs were compared and contrasted for their potential to relieve some adverse effects of chronic stress. It was observed that both *W. somnifera* and Panax ginseng reduced the severity and frequency of induced stress, reversed stress induced hampering of sexual behaviour and showed a positive effect on learning and memory tasks (Bhattacharya and Muruganandam 2003). In yet another study, conducted on rodents, chronic stress was induced in animals using electric shock method which resulted in different adverse effects like mental depression, glucose intolerance, sexual dysfunction, immuno suppression and gastric ulcerations The study concluded that glycowithanolides isolated from the whole plant *W. somnifera* root extract has a significant anti-stress and adaptogenic effect (Bhattacharya et al. 2001). Further, another study done at University of Texas, USA, found that anti-anxiety effects of *W. somnifera* are possibly due to its GABA (Gamma Amino-butyric acid)-like activity. GABA, is an inhibitory neurotransmitter which functions by attenuating the neuron activity and suppresses the nerve cells from over firing (Mehta et al. 1991).

4.2.2 Anticancer

W. somnifera has been reported to show anti-cancer activity both *in vitro* and *in vivo* in various studies. Muralikrishnan et al. (1996) reported that *W. somnifera* was able to normalise the count of immune cells and immunoglobins in an experimental colon cancer model induced by Azoxymethane in mice (Muralikrishnan et al. 2010). Several studies reported that root extract of *W. somnifera* suppressed stomach and skin cancer and tumour development in mice (Singh et al. 1984; Chandra et al. 2012; Kuttan 1996). The root extract of *W. somnifera* was also documented to inhibit methylcholanthrene induced sarcoma development in mice. This study also reported that the extract increased the life span of tumour bearing animals (Ziauddin et al. 1996). *W. somnifera* was also found to reduce tumour growth in urethane-induced lung tumours in adult male mice (Sharma et al. 2011) and Sarcoma 180 (S-180), a transplantable mouse tumour in BALB/c mice (Devi et al. 1992). *W. somnifera* has also demonstrated anti-cancer properties in several *in vitro* studies performed on various human cancer cells of different origin like breast cancer, lung cancer, colon cancer, blood cancer and pancreatic cancer (Kritikar and Basu 1935; Muralikrishnan et al. 2010).

4.2.3 Anti-inflammatory

W. somnifera has been explored for its anti-inflammatory effects in arthritic and other inflammatory conditions and found to be a good anti-arthritic agent in different studies (Singh et al. 1984). In various other studies conducted on Wistar rats, *W. somnifera* was observed to reduce inflammation induced by Freund's complete adjuvant or Formaline and Carragenaan (Chandra et al. 2012; Singh et al. 2007). In a similar study, *W. somnifera* was reported to reduce inflammation by inhibiting cyclooxygenase (Gorain et al. 2012; Mishra et al. 2000). *W. somnifera* has been reported to have high steroidal content which is higher than that of hydrocortisone, a commonly used anti-inflammatory agent which may be a possible reasons of its significant anti-inflammatory potential (Kaur et al. 2007).

4.2.4 Anti-oxidant

W. somnifera has been documented to show anti-oxidant properties by up regulating some natural anti-oxidants and inhibiting free radical production. In an *in vivo* study conducted on rat brains, it was observed that *W. somnifera* up regulated the levels of various natural anti-oxidants like glutathione peroxidise, superoxide dismutase and catalase. In another study conducted on mice and rabbits, *W. somnifera* treated animals showed a significant inhibitory effect on stress induced lipid peroxidation (Dhuley Dhuley 1998a, b). The anti-oxidant effect of *W. somnifera* may also explain the reasons of its reported anti-aging, anti-inflammatory and anti-stress effects.

4.2.5 Cardio Protective Activity

W. somnifera has been reported to demonstrate cardio protective effects both in clinical and preclinical studies. In a study conducted in dogs, *W. somnifera* showed a prolonged hypotensive, bradycardiac and respiratory-stimulant action. This study concluded that hypotensive effect was due to autonomic ganglion blocking action of *W. somnifera*. This study also found that the alkaloids present in this botanical stimulate the vasomotor and respiratory centres in the brain stem of experimental animals (Mishra et al. 2000; Ojha and Arya 2009). In another study conducted on human subjects, hypocholesterolemic, diuretic and hypoglycaemic effects of hydro-alcoholic extract of *W. somnifera* were evaluated. It was observed that *W. somnifera* treatment increased urine volume, and urine sodium and decreased triglycerides, serum cholesterol and low-density lipoproteins (Mohanty et al. 2004).

4.2.6 Immuno Modulatory Activity

W. somnifera has been documented as an effective immune regulator in several animal studies. Various animal studies demonstrated that methanolic extract *W. somnifera* enhances the production of white blood cells (Kuttan 1996; Ziauddin et al. 1996). Similarly, in a study performed on Swiss mice and Wistar rats, whole plant extract of *W. somnifera* caused the augmentation of peritoneal macrophages and increased the activity of the lysosomal enzymes.

Furthermore, this extract was found to augment the phagocytic activity of macrophages, hamper delayed-type hypersensitivity reactions in mice and stimulate the generation of cytotoxic T lymphocytes (Davis and Kuttan 2000; Davis and Kuttan 2002). Malik et al. (2007) has shown that both marker based chemically standardised root and leaf extract are able to stimulate the Th1 cell mediated and humoral immune response in BALB/c mice. The leaf extract of *W. somnifera* was shown to activate immune system by selectively augmenting Th1 immune response *in vitro*. In another study, the root extract of *W.somnifera* was found to enhance cell mediated immunity by predominantly enhancing Th1 immunity. In the same study, the authors found that root extract with Withanolide-A as a major constituent, was responsible for this Th1 polarisation of immune response (Malik et al. 2007). In yet another study, same group in 2009, reported that a chemically standardized herbal mixture of standardised root and leaf extract of *W.somnifera* (WSF) possesses both anti-cancer and Th1 immune up-regulatory activities.WSF which produced cytotoxicity in a panel of human cancer cell lines also showed marked reduction in in tumor growth in different mouse tumor models.The study also demonstrated that WSF enhanced Th1 tumor reactive immune response in tumor bearing mice with simultaneous induction of apoptosis in cancer cells (Malik et al. 2009).

4.2.7 Other Documented Pharmacological Actions

In several *in vitro* or *in vivo* studies, the whole plant extract or extracts of different plant parts of *W. somnifera* were found to be effective againt different disease conditions e.g. root extract of this plant was demonstrated to be an effective treatment against osteoarthritis (Kulkarni et al. 1991) and stroke (Anbalagan and Sadique 1981), hydroalcoholic extract of the whole plant or its constituents were found to be effective against tardive dyskinesia (Chaudhary et al. 2003) and microbial and bacterial infections (Choudhary et al. 1995; Ali et al. 2001). Similarly, the ethanolic extract of *W. somnifera* was found to be effective against Alzheimer's disease (Dickson and Vickers 2001).

4.3 Chemical Characterisation of *W. somnifera* Visa-Vis Pharmacological Activities

4.3.1 Chemical Characterisation

Phytochemical profile of *W. somnifera* has been extensively studied for the first time isolated Dimethyltryptamine ($C_{12}H_{16}N_2$), an amorphous alkaloid from a South African strain of *W. somnifera* Majumdar (1955). The presence of this alkaloid was later confirmed by Majumdar and Guha. They also isolated several other alkaloids from the roots of this plant which were named as withanaminine, pseudowithamine, somniferinine, withanmine, somniferine, somnine and withamine; however no structural elucidation of these alkaloids was provided. Several other groups of chemical constituents were identified, extracted and isolated from time to time including steroidal lactones, flavonoids, saponins, tannins and alkaloids (Murthy et al. 2008; Eastwood et al. 1980; Rahman et al. 1993).

However, steroidal lactones are the major constituents found in the leaves of *W.somnifera* and responsible for most of the therapeutic properties of this plant (Rahman et al. 1993; Iqbal et al. 1996; Dhar et al. 2012). These steroidal Lactones, called as withanolides are a group of naturally occurring C28- steroidal lactones built on an intact or rearranged ergostane framework, in which C-22 and C-26 are appropriately oxidized to form a six-membered lactone ring. The basic structure (Fig. 4.1) of withanolides (withanolide skeleton) is given below (Mirjalili 2009b).

Withanolides are also present in other plant species like members of Taccaceae (Leet et al. 1982), Fabaceae (Leguminosae) (Khan et al. 1999) and Lamiaceae

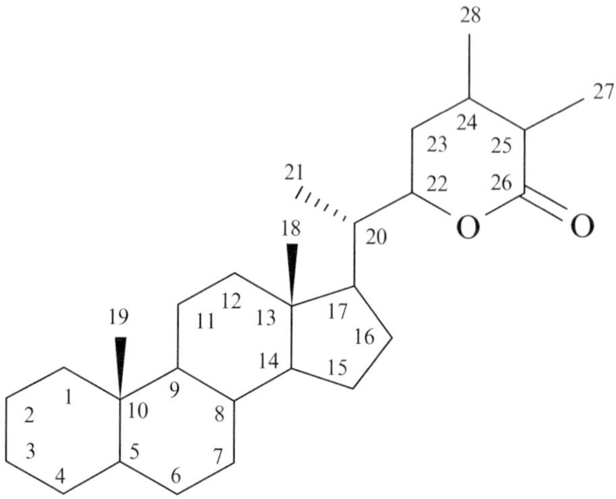

Fig. 4.1 The basic structure of withanolides

(Labiatae) (Sultanbawa et al. 1985) families and fifteen Solanaceous genera i.e. *Acnistus*, *Datura*, *Deprea*, *Dunalis*, *Iochroma*, *Jaborosa*, *Lycium*, *Nicandra*, *Physalis*, *Salpichroa*, *Tubocapsicum*, *Discopodium*, *Trechonaetes*, *Withania* and *Witheringia*. Further, withanolides have also been isolated from marine organisms (Singh et al. 2011). Presently, more than 12 alkaloids and 40 withanolides have been reported and isolated from different parts of *W. somnifera* plant. The concentration of major withanolides in the dry weight of this plant ranges from 0.001 to 1.5% (Kumar et al. 2007; Kapur 2001). Various bioactive compounds isolated of *W. somnifera* are depicted in Table 4.1. The most bioactive molecules isolated from this plant along with the description of their pharmacological actions and chemical structures, are given in Table 4.1.

4.3.2 Pharmacological Significance of Individual Constituents of W. somnifera

Till date various active molecules have been isolated from *W. somnifera* (Figs. 4.2, 4.3, 4.4, 4.5, 4.6 and 4.7). Given below are the pharmacological properties of the active constituents isolated from the plant.

4.3.2.1 Withaferin A

Withaferin A is an oxygenated steroidal lactone, isolated from *W.somnifera*. It was the first pharmologically active withanolide isolated from this plant. Further, it is the most profuse and primary pharmacological agent of *W. somnifera*. Withaferin A

Table 4.1 Chemical constituents of pharmacological importance identified in various parts of *W. somnifera* (Mir et al. 2012)

Plant part	Chemical constituent
Roots	Sitoindosides VII (Acylsteryl-glucoside)
	Sitoindosides VIII (Acylsteryl-glucoside)
	Sitoindosides IX (Glycowithanolide)
	Sitoindosides X (Glycowithanolide)
	Withanine (Alkaloid)
	Withananine (Alkaloid)
	Ashwagandhanolide
Leaves	Withaferin (Steroidal lactone)
	Withaferin A (Steroidal lactone)
	Withanolie D (Steroidal lactone)
	Withanolie E (Steroidal lactone)
	Withanone (Steroidal lactone)
	Withanolide Z (Novel)
	Withanolide B
	7-hydroxywithanolide
	3α-methoxy-2, 3-dihydro-
	27-deoxywithaferin A (Steroidal lactone)
	4β, 17α-dihydroxy-1-1oxo-
	5β, 6β-epoxy-22R-witha-
	2, 24-dienolide (steroidal lactone)
	4β-dihydroxy-5β, 6β-epoxy-
	1-oxo-22R-witha-2, 14-24-
	Trienolide (steroidal lactone)
Seeds	Withanolide –WS 2 (aliphatic ester), Withanolide –WS 1 (aliphatic ketone)

(Fig. 4.2) has been documented to possess diverse pharmacological activities. However, the primary molecular target of Withaferin A is unknown.

Anti-cancer Activity

Withaferin A has been documented as a versatile pharmacological agent with diverse pharmacological activities. More importantly, this important derivative of *W.somnifera* has been reported to possess a significant anti-cancer activity both *in vivo* and *in vitro*.

Withaferin A was shown to cause a concentration dependent cell cycle arrest of estrogen-independent (MDA-MB-231) and estrogen-responsive (MCF-7) human breast cancer cell lines in their G2-M phase, decreasing levels of cyclin-dependent kinase 1 (Cdk1), Cdc-25C (cell division cycle-25C) and Cdc-25C proteins (Stan et al. 2008). It also exhibited a strong growth inhibitory effect on several other human cell lines like human myeloma (U266), adipocytes (3 T3-L1), endothelial cell (HUVEC), pancreatic cells (Panc-1, MiaPaCa2 and BxPc3) and primary cells

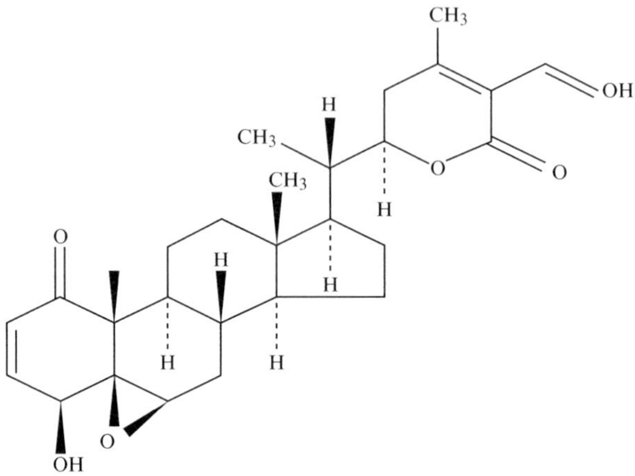

Fig. 4.2 Chemical structure of Withaferin A (Molecular formula: $C_{28}H_{38}O_6$; Molecular mass: 470.60 g mol^{-1})

Fig. 4.3 Chemical structure of Withanolide A (Molecular formula: $C_{28}H_{38}O_6$; Molecular mass: 470.60 g mol^{-1})

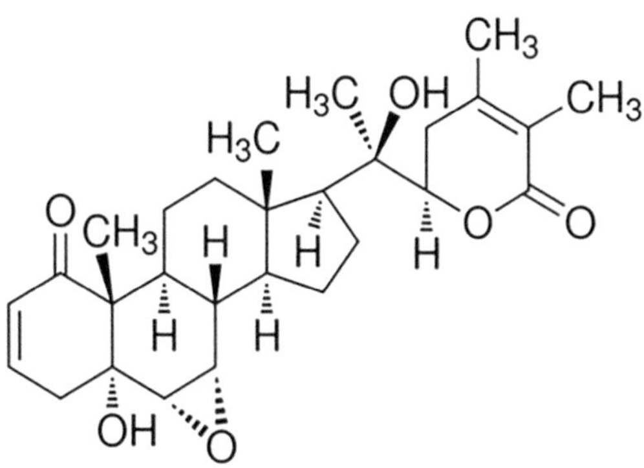

from patients with myeloid or lymphoblastic leukemia and oral squamous cell carcinoma (Mandal et al. 2008; Malara et al. 2008; Park et al. 2008; Panjamurthy et al. 2009).

Withaferin A also showed a significant potential to combat tumors *in vivo*. It demonstrated a potent anti- angiogenic activity in mice by interfering with the ubiquitin-mediated proteasome pathway (Mohan et al. 2004). It also suppressed tumour growth in pancreatic xenografts via an ATP-independent mechanism. Furthermore, Withaferin A also demonstrated a significant anti-tumour potential using a stringent orthotopic human glioma xenograft model in (Santagata et al. 2012).

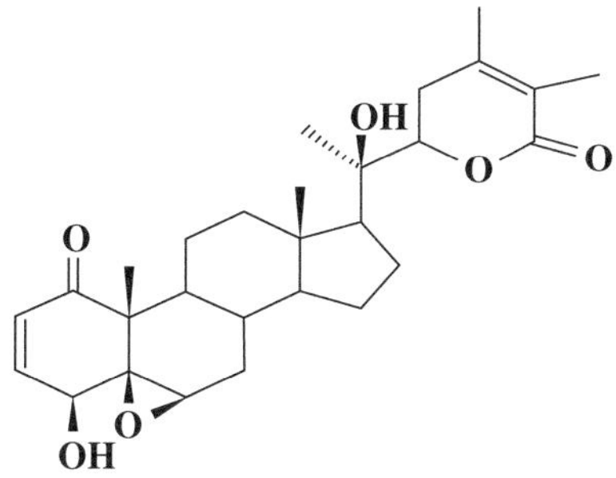

Fig. 4.4 Chemical structure of Withanolide D (Molecular formula: $C_{28}H_{38}O_6$; Molecular mass: 470.59 g mol^{-1})

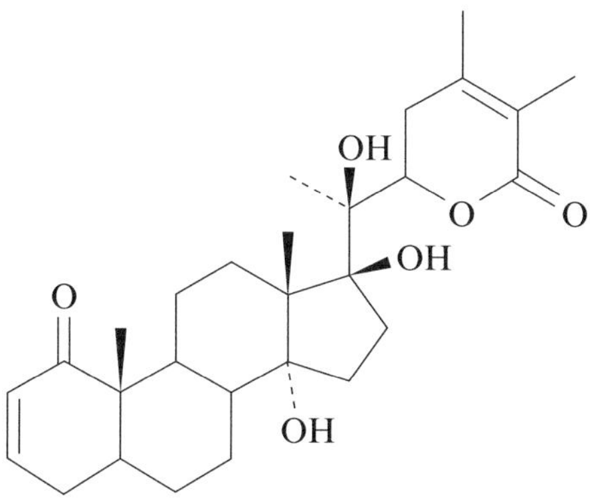

Fig. 4.5 Chemical structure of Withanolide E (Molecular Formula: $C_{28}H_{38}O_{7;}$ Molecular mass: 486.59712 $gmol^{-1}$)

Anti-inflammatory Effect

Withaferin A has been widely studied for its anti-inflammatory potential. In pre-clinical studies it has shown a notable effect against inflammatory disorders like rheumatoid arthritis and airway inflammation. Withaferin A was found to attenuate collagen production and inflammation in rabbit articular chondrocytes by up regulating the cyclooxygenase-2 expression. It was also found to down regulate intercellular adhesion molecule-1 and vascular cell adhesion molecule-1 (VCAM-1) expression in TNF-alpha stimulated human lung epithelial (A549) cells. Furthermore, it also showed a notable decrease in LPS induced nitric oxide production in RAW 264.7 cells, a mouse macrophage cell line (Mohan et al. 2004; Singh et al. 2007; Oh and Kwon 2009; Yu and Kim 2013).

4.3.2.2 Withanolide A

Withanolide A, a well oxygenated steroidal lactone, is isolated from roots of *W. somnifera* (Yu and Kim 2013). It is has been documented to possess a strong pharmacological profile especially related to nervous system (Jain et al. 2001; Kuboyama et al. 2002, 2005). It was found to target and block β-amyloid deposition in rat brain

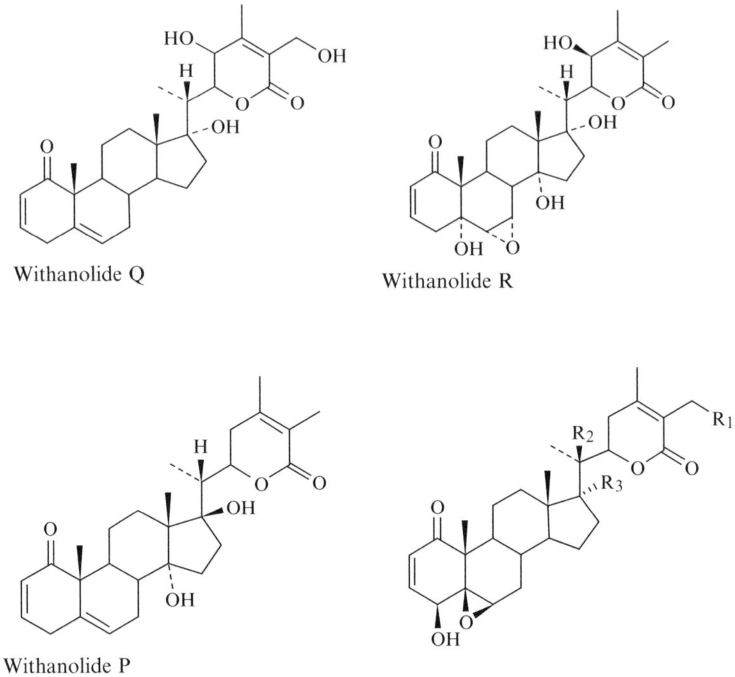

Withanolide G: 14α-OH, 14=OH, R1=OH, R2=H; **Withanolide H:** 14α-OH, R1=R2=OH; **Withanolide J:** 7α-OH, R1=H, R2=OH

Fig. 4.6 Structure of different withanolides (Mirjalil 2009a)

by targeting multiple proteins like BACE1, ADAM10, IDE, and NEP (Patil et al. 2010). Withanolide A (Fig. 4.3) has also been documented to possess a neuro protective ability showing cholinesterase inhibition potential, reducing the epileptic seizures in rats and ameliorating the impaired memory in rats (Grover et al. 2012; Soman et al. 2012). Further, Withanolide A improved neuronal atrophy and synaptic loss in mice brain (Kuboyama et al. 2005). Thus, Withanolide A seems to have a potential to combat various nervous system pathologies especially related to Alzheimer's disease. Also, Withanolide A has been reported to cause a notable recovery of immune system functioning with regard to normalising the T cell population, depleted as a result of stress in mice (Kour et al. 2009).

4.3.2.3 Withanolide D

Withanolide D is a steroidal lactone and a pure herbal compound isolated from *W. somnifera (*Kour et al. 2009). It has been reported to possess diverse pharmacological activities like anti fungal (Roumy et al. 2010), anti-bacterial, anti-oxidant

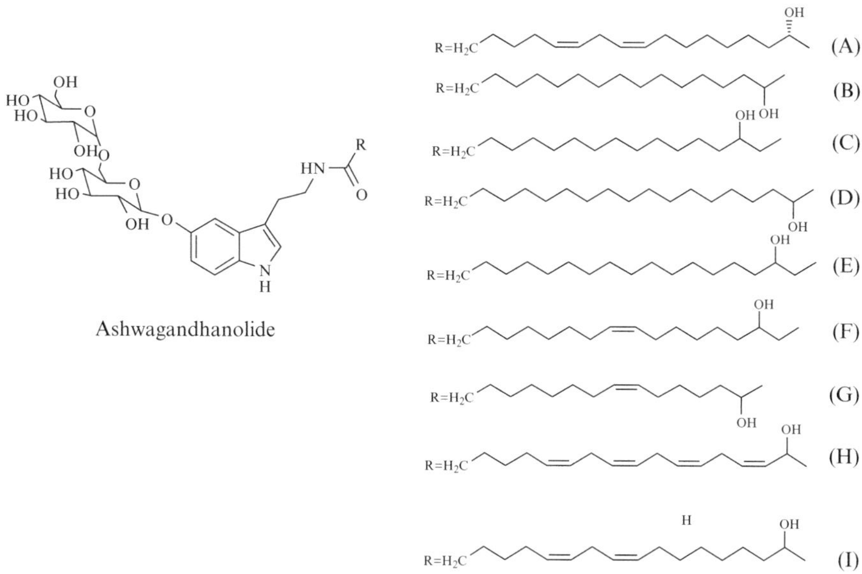

Fig. 4.7 Chemical structure of Ashwagandhanolide and Withanamides A-I

hepatoprotective, adaptogenic, anti-inflammatory and anti-stress (Budhiraja et al. 2000). However, its anti-cancer activity, both *in vitro and in vivo*, is a hot topic in preclinical research. Withanolide D (Fig. 4.4) has been reported to block cell cycle at G2/M phase in pancreatic ductal adenocarcinoma cells and induce apoptosis (Sarkar et al. 2014). It also caused a significant apoptosis of estrogen-independent (MDA-MB-231) and estrogen-responsive (MCF-7) human breast cancer cells (Wang et al. 2012). Further, this withanolide showed a significant anti-apoptotic potential in other cell types also like human myeloid (K562), lymphoid (MOLT-4) cells and human epidermoid carcinoma of the nasopharynx (KB). It also showed a notable *in vivo* anti-tumour activity against Sarcoma 180 and Ehrlich ascites carcinoma (Mondal et al. 2010; Wall and Wani 1977).

4.3.2.4 Withanolide E

Withanolide E is a naturally occurring steroidal lactone isolated from *W. somnifera*. This withanolide has been reported to significantly halt cell proliferation of human pancreatic cancer (Panc-1) and breast cancer cells (MDA-MB-231 and MCF-7). The anti-proliferative activity of Withanolide E in Panc-1 cells was found to be due to its binding to Hsp90, causing Hsp90 aggregation and hence inhibiting its chaperone activity to induce degradation of Hsp90 client proteins Akt and Cdk4 through proteasome-dependent pathway (Wang et al. 2012; Gu et al. 2014) Further, withanolide E was demonstrated to possess a notable immunosuppressive activity. It was shown to suppress the proliferation of T and B lymphocytes and affect antigen recognition (Shohat et al. 1978).

4.3.2.5 Other Withanolides

There are several other withanolides, isolated from *W. somnifera* that have been documented to possess different pharmacological activities, e.g., 5, 2O-α dihydroxy-6α 7α-epoxy-1- OXO, a withanolide isolated from *W. somnifera* possesses significant immuno modulatory properties (Bahr and Hansel 1970). Another steroildal lactone 3-β-hydroxy-2, 3-hydroxy Withanloide was found to possess anti-inflammatory properties, evaluated in Wistar rats and antibacterial activity against different microbial strains (Mir et al. 2012; Budhiraja et al. 1984). Similarly, another withanolide namely Withaferinil was demonstrated to possess anti-proliferative activity against HeLa cells (Palyi 1969). The structures of several other withanolides isolated from *W. somnifera* is given below (Fig. 4.6):

Further examination of *W. somnifera* roots resulted in the isolation of a series of compounds named as Withanamides A-I. Later, another dimeric thiowithanolide was also isolated from this plant, named Ashwagandhanolide (Fig. 4.7). All these compounds were isolated by a bioassay-guided purification of the methanolic extract of the plant. However, their pharmacological actions have not been explored yet (Mirjalili 2009b) (Fig. 4.7).

4.4 Combitorial Studies with the Constuients of *W. somnifera* to Deter Pathobiology of Diseases

Most of the chemical constituents of *W. somnifera* possess various pharmacological activities, as demonstrated by various studies. However, in several studies, a combinational approach has been demonstrated where the effect of combination of two chemical constituents of *W. somnifera* was observed on a particular pharmacological activity and the results so obtained where comparatively promising e.g. in a study performed on glioma (C6 and YKG1) cell lines, the effect of Withaferin A, Withanone, Withanolide A and their different combinations, was evaluated on the growth inhibition and differentiation of these cells. It was observed that Withaferin A combined with Withanone and Withaferin A combined with Withanolide A showed better inhibitory action rather than evaluated individually (Shah et al. 2009). In another study, Withaferin A in combination with Sorafenib, a clinically approved drug for the treatment of renal cell carcinoma, demonstrated synergistic efficacy in apoptosis induction in papillary and anaplastic cancer cells. This combination achieved potent anti-cancer activity at lower doses of Sorafenib, which may decrease its toxicity for future translational studies (Cohen et al. 2012). In another study performed on nude mice, Withaferin A in combination with Cisplatin, a chemotherapy

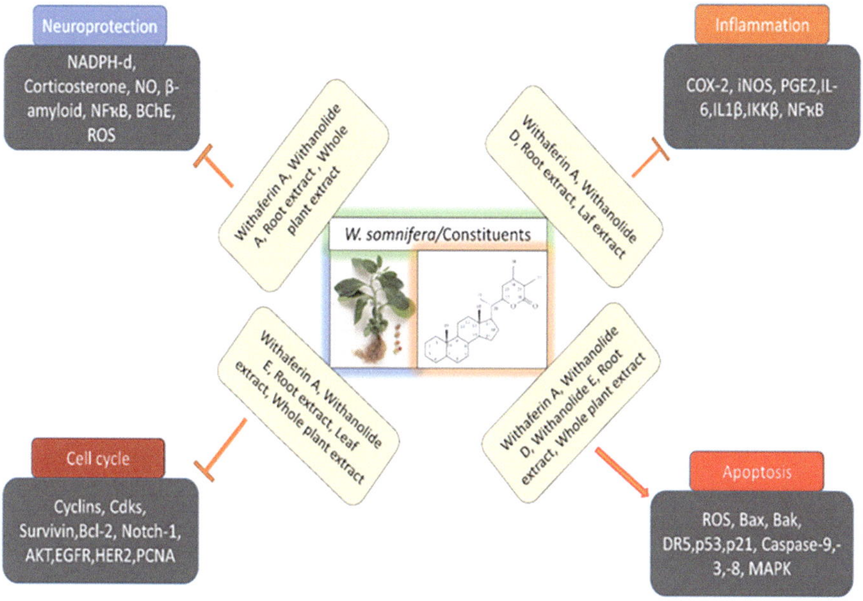

Fig. 4.8 Overview of potential targets of *W. somnifera* and its constituents

drug, reduced tumour growth up to 70 to 80% and completely inhibited the metastasis of other organs compared to untreated controls and Cisplatin group (Kakar et al. 2014). In a different study, Withaferin A effectively enhanced the ionizing radiation induced apoptosis in human lymphoma U937 cells, so the presumption of using Withaferin A as radio sensitizers to enhance the therapeutic efficiency of radiotherapy cannot be underestimated (Yang et al. 2011) (Fig. 4.8).

4.5 Clinical Studies with *W. somnifera*

Apart from the various preclinical studies performed with *W. somnifera* or its constituents, several clinical studies have been also done for different pharmacological actions:

1. *W.somnifera* was found to recover the seminal plasma levels of antioxidant enzymes, vitamins A, C, and E and corrected fructose in infertile men. Moreover, treatment also significantly increased serum T and LH and reduced the levels of FSH and PRL, which is a good indication of healthy semen (Ahmad et al. 2010). In a similar study, *W. somnifera* repaired the distorted concentrations of lactate, alanine, citrate, GPC, histidine, and phenyl alanine in seminal plasma and recovered the quality of semen of post-treated compared to pre-treated infertile men.

Furthermore, serum biochemistry was also improved over post-therapy in infertile men (Gupta et al. 2013).

2. In another study performed on infertile men of different age groups, it was demonstrated that frequency of sperm apoptosis and intracellular ROS levels were significantly higher in infertile men compared with control subjects. Similarly, the seminal plasma levels of the essential metal ions Cu^{2+}, Zn^{2+}, Fe^{2+} and Au^{2+} were low. Treatment with *Withania somnifera* significantly reduced the sperm apoptosis, ROS levels and improved serum metal ion levels of infertile men of all the age groups under study (Shukla et al. 2011).
3. In a double-blind, placebo-controlled study performed on patients with ICD-10 anxiety disorders. It was observed that by the end of the first month of treatment, 72% of the patients showed moderate to excellent improvement and in about half of these cases, benefits were observed within the first fortnight. These 72% patients were continued on the drug for a maximum of 18 months. It was observed that the drug satisfactorily attenuated anxiety symptoms and did not occasion any adverse effects (Andrade et al. 2000).
4. In a double-blind, multi-dose, placebo-controlled study, 20 healthy male controls were randomized to receive 250 mg two capsules twice daily of an encapsulated dried aqueous extract of roots and leaves of *W. somnifera* or a matching placebo for a period of 14 days. Cognitive and psychomotor performance was assessed pre-dose (day 1) and at 3-h post-dose on day 15 using a battery of computerized psychometric tests. After a washout period of 14 days, the subjects crossed-over to receive the other treatment for a further period of 14 days as per prior randomization schedule. Same battery of test procedures was performed to assess cognitive and psychomotor performance. Significant improvements were observed in reaction times with simple reaction, choice discrimination, digit symbol substitution, digit vigilance, and card sorting tests with *W. somnifera* extract compared to placebo. All the results indicate that *W. somnifera* extract can improve cognitive and psychomotor performance (Pingali et al. 2014).
5. In a double-blind, randomized, placebo-controlled trial performed on patients with chronic stress, a 300 mg capsule was given for 60 days. It was observed that the patients who received 300 mg capsule of *W. somnifera* showed a significant improvement towards stress which improved their self-assessed quality of life. Further, the consumption of *W. somnifera* was found to be safe health wise (Chandrasekhar et al. 2012).
6. In another study conducted on subjects with moderate to severe anxiety, the group of patients who received naturopathic therapy involving deep breathing relaxation technique, a standard multi-vitamin and *W. somnifera* (300 mg b.i.d.), were found to recover effectively, with no toxic manifestation (Cooley et al. 2009).
7. In another study conducted on post myocardial infarction patients, *W. somnifera* extract, administered in the form of a herbal cocktail, was able to reduce the levels of cholesterol and triglycerides and augment high density lipoprotein cholesterol, very effectively (Dwivedi et al. 2000).

8. Study performed on hypoglycaemic and hypo-cholesterolemic subjects, *W. somnifera* was able to increase urine volume, urine sodium, decrease serum cholesterol and triglyceride levels significantly (Dhuley Dhuley 1998a, b).
9. Study conducted on breast cancer patients undergoing combination chemotherapy with oral *W. somnifera*(2 g every 8 h) showed a notable and remarkable anticancer activity. The consumption was found to be clinically safe (Biswal et al. 2013).

4.6 Marketed Preparations of *W. somnifera*

W. somnifera has been used in various herbal preparations for treating different ailments in countries where Ayurvedic or Unani system of medicine is officially recognised, such as India, Sri Lanka, Nepal, Malaysia and Bangladesh. Given below are listed some marketed products (Sangwan et al. 2004) of *W. somnifera* with their therapeutic properties (Table 4.2):

Table 4.2 Various marketed preprations of *W. somnifera* and their health benefits

Name of the medicine	Manufacturer	Application
Ashwagandharista	Baidyanath Ayurvedic Bhawan	Used for improvement of memory and cognition, sleep induction and general nervous weakness
Himalaya Ashwagandha	The Himalaya Drug Company	Used for stress management
Stresswin	Baidyanath Ayurved Bhawan	Used for relieving stress and physical exertion, memory enhancement, sleep induction and for general body weakness during menopause
Stresscom	Dabur India Ltd	Used as an anti-depressant and relieves physical and mental stress and general body weakness
Himalaya massage oil	The Himalaya Drug Company	Used for stress relief and against insomnia, backache and as a muscle relaxant
Lovemax	BACFO Pharmaceuticals Ltd	Used to induce vigour and cure sexual dysfunctioning
Vigomax	Charak Pharmaceuticals Pvt Ltd	Used to cure sexual dysfunctioning
Vital plus	Mukhti Pharma	Used to cure sexual dysfunctioning, oligospermia and for general body weakness
Amrukha Kasthuri	Pankajakasthuri Herbals India Ltd	Used to treat Neurasthenia and for convalescence
Brento	Zandu Pharmaceutical Works Ltd	Used as a nervine tonic

4.7 Conclusion

The traditional plant derived medicinal products used in disease therapies are a rich source of medicines. In this context, *W. somnifera*, one of the most useful medicinal plants, has been used for centuries to cure various disease conditions like inflammation, immune-modulation and several nervous disorders. This palnt is chemically enriched with various active constituents such as withanolides, sitoindosides and many useful alkaloids that have been explored through valid scientific analysis. These active isolated constituents have been documented to be beneficial for curing a wide range of human ailments. Currently, several studies are being conducted globally to explore the pharmacologic benefits of the plant and its constituents. Although several individual constitunets have been isolated and characterised for their health benefits, there is still need to explore the treasure of small molecules in this plant and to identify bioactive constituents. A more robust scientific evidence vis-à-vis pharmacological evalauation including drug like properties needs to be performed on bioactive constitunets to promote them as drug like candidates.

References

Ahmad MK, Mahdi AA, Shukla KK, Islam N, Rajender S, Madhukar D, Shankhwar SN, Ahmad S (2010) Withania somnifera improves semen quality by regulating reproductive hormone levels and oxidative stress in seminal plasma of infertile males. Fertil Steril 94(3):989–996

Ali NA, Julich WD, Kusnick C, Lindequist U (2001) Screening of Yemeni medicinal plants for antibacterial and cytotoxic activities. J Ethnopharmacol 74:173–179

Anbalagan K, Sadique J (1981) Influence of an Indian medicine (Ashwagandha) on acute-phase reactants in inflammation. Indian J Exp Biol 19:245–249

Andrade C, Aswath A, Chaturvedi SK, Srinivasa M, Raguram R (2000) A double-blind, placebo-controlled evaluation of the anxiolytic efficacy of an ethanolic extract of Withania Somnifera. Ind J Psychaitry 42:295–301

Archana R, Namasivayam A (1999) Antistressor effect of Withania somnifera. J Ethnopharmacol 64:91–39

Bahr V, Hansel R (1970) Immunomodulating properties of 5,20 alpha (R)-dihydroxy-6 alpha,7 alpha-epoxy-1-oxo-(5 alpha)-whitha-2,24-dienolide and solasodine. Planta Med 44:32–33

Bhandari C (1970) Ashwagandha (Withania somnifera)"Vanaushadhi Chandroday"(An Encyclopedia of Indian Herbs), vol 1. CS Series of Varanasi Vidyavilas Press, Varanasi, pp 96–97

Bhattacharya SK, Muruganandam AV (2003) Adaptogenic activity of Withania somnifera: an experimental study using a rat model of chronic stress. Pharmacol Biochem Behav 75:547–555

Bhattacharya A, Ghosal S, Bhattacharya SK (2001) Anti-oxidant effect of Withania somnifera glycowithanolides in chronic footshock stress-induced perturbations of oxidative free radical scavenging enzymes and lipid peroxidation in rat frontal cortex and striatum. J Ethnopharmacol 74:1–6

Biswal BM, Sulaiman SA, Ismail HC, Zakaria H, Musa KI (2013) Effect of Withania somnifera (Ashwagandha) on the development of chemotherapy-induced fatigue and quality of life in breast cancer patients. Integr Cancer Ther 12:312–322

Budhiraja RD, Sudhir S, Garg KN (1984) Anti-inflammatory activity of 3 beta-hydroxy-2,3-dihydro-withanolide F. Planta Med 50:134–136

Budhiraja R, Krishan P, Sudhir S (2000) Biological activity of withanolides. J Sci Ind Res 59:904–911

Chandra S, Chatterjee P, Dey P, Bhattacharya S (2012) Evaluation of anti-inflammatory effect of Ashwagandha: a preliminary study in vitro. Phcog J 4:47–49

Chandrasekhar K, Kapoor J, Anishetty S (2012) A prospective, randomized double-blind, placebo-controlled study of safety and efficacy of a high-concentration full-spectrum extract of Ashwagandha root in reducing stress and anxiety in adults. Indian J Psychol Med 34:255–262

Chaudhary G, Sharma U, Jagannathan NR, Gupta YK (2003) Evaluation of Withania somnifera in a middle cerebral artery occlusion model of stroke in rats. Clin Exp Pharmacol Physiol 30:399–404

Choudhary MI, Dur ES, Parveen Z, Jabbar A, Ali I, Atta-ur R (1995) Antifungal steroidal lactones from Withania coagulance. Phytochemistry 40:1243–1246

Cohen SM, Mukerji R, Timmermann BN, Samadi AK, Cohen MS (2012) A novel combination of withaferin A and sorafenib shows synergistic efficacy against both papillary and anaplastic thyroid cancers. Am J Surg 204: 895–900; discussion -1

Cooley K, Szczurko O, Perri D, Mills EJ, Bernhardt B, Zhou Q, Seely D (2009) Naturopathic care for anxiety: a randomized controlled trial ISRCTN78958974. PLoS One 4:e6628

Davis L, Kuttan G (2000) Immunomodulatory activity of Withania somnifera. J Ethnopharmacol 71:193–200

Davis L, Kuttan G (2002) Effect of Withania somnifera on CTL activity. J Exp Clin Cancer Res 21:115–118

Devi PU, Sharada AC, Solomon FE, Kamath MS (1992) In vivo growth inhibitory effect of Withania somnifera (Ashwagandha) on a transplantable mouse tumor, sarcoma 180. Indian J Exp Biol 30:169–172

Dhar RS, Gupta SB, Singh PP, Razdan S, Bhat WW, Rana S, Lattoo SK, Khan S (2012) Identification and characterization of protein composition in Withania somnifera-an Indian ginseng. J Plant Biochem Biotechnol 21:77–87

Dhuley JN (1998a) Adaptogenic and cardioprotective action of ashwagandha in rats and frogs. J Ethnopharmacol 70:57–63

Dhuley JN (1998b) Effect of ashwagandha on lipid peroxidation in stress-induced animals. J Ethnopharmacol 60:173–178

Dickson TC, Vickers JC (2001) The morphological phenotype of beta-amyloid plaques and associated neuritic changes in Alzheimer's disease. Neuroscience 105:99–107

Dwivedi S, Gupta D, Sharma K, Kumar S, Kukreja A, Dwivedi S, Singh AK (2000) Modification of coronary risk factors by medicinal plants. J Med Aroma Plant Sci 22:616–620

Eastwood FW, Kirson I, Lavie D, Abraham A (1980) New withanolides from a cross of a South African chemotype by chemotype II (Israel) in Withania somnifera. Phytochemistry 19:1503–1507

Gauttam VK, Kalia AN (2013) Development of polyherbal antidiabetic formulation encapsulated in the phospholipids vesicle system. J Adv Pharm Technol Res 4:108–117

Gorain M, Ahmed S, Bhandari R, Sharma N (2012) Biology of Phalantha Phalantha Drary (Nymphalidae: Lepidoptera)-an injurious pest of well-known medicinal plant Withania Somnifera Dunal. Biotech Res 2:4–6

Govardhan S (1938) Ashwagandharishta – Rastantra Sar Evam Sidhyaprayog Sangrah. Krishna–Gopal Ayurveda Bhawan (Dharmarth Trust), Nagpur, pp 743–744

Grover A, Shandilya A, Agrawal V, Bisaria VS, Sundar D (2012) Computational evidence to inhibition of human acetyl cholinesterase by withanolide a for Alzheimer treatment. J Biomol Struct Dyn 29:651–662

Gu M, Yu Y, Gunaherath GM, Gunatilaka AA, Li D, Sun D (2014) Structure-activity relationship (SAR) of withanolides to inhibit Hsp90 for its activity in pancreatic cancer cells. Investig New Drugs 32:68–74

Gupta A, Mahdi AA, Shukla KK, Ahmad MK, Bansal N, Sankhwar P, Sankhwar SN (2013) Efficacy of Withania somnifera on seminal plasma metabolites of infertile males: a proton NMR study at 800 MHz. J Ethnopharmacol 149:208–214
Iqbal CM, Abbas S, Arshad JS (1996) Withania somnifera: a source of exotic withanolides. Heterocycles 42:555–563
Jain S, Shukla SD, Sharma K, Bhatnagar M (2001) Neuroprotective effects of Withania somnifera Dunn. in hippocampal sub-regions of female albino rat. Phytother Res 15:544–548
Kakar SS, Ratajczak MZ, Powell KS, Moghadamfalahi M, Miller DM, Batra SK, Singh SK (2014) Withaferin a alone and in combination with Cisplatin suppresses growth and metastasis of ovarian cancer by targeting putative cancer stem cells. PLoS One 9:e107596
Kapur LD (2001) Handbook of ayurvedic medicinal plants. CRC Press, London
Kaur K, Widodo N, Nagpal A, Kaul SC, Wadhwa R (2007) Sensitization of human cancer cells to anti-cancer drugs by leaf extract of Ashwagandha (LASH). Tiss Cult Res Commun 26:193–199
Khan PM, Ahmad S, Rubnawaz H, Malik A (1999) The first report of a withanolide from the family Labiatae. Phytochemistry 51:669–671
Kour K, Pandey A, Suri KA, Satti NK, Gupta KK, Bani S (2009) Restoration of stress-induced altered T cell function and corresponding cytokines patterns by Withanolide A. Int Immunopharmacol 9:1137–1144
Kritikar KR, Basu BD (1935) Withania somnifera. In: Lalit Mohan Basu A (ed) Indian medicinal plants, pp 1774–1776
Kuboyama T, Tohda C, Zhao J, Nakamura N, Hattori M, Komatsu K (2002) Axon- or dendrite-predominant outgrowth induced by constituents from Ashwagandha. Neuro Report 13:1715–1720
Kuboyama T, Tohda C, Komatsu K (2005) Neuritic regeneration and synaptic reconstruction induced by withanolide A. Br J Pharmacol 144:961–971
Kulkarni SK, Dhir A (2008) Withania somnifera: an Indian ginseng. Prog Neuro-Psychopharmacol Biol Psychiatry 32:1093–1105
Kulkarni RR, Patki PS, Jog VP, Gandage SG, Patwardhan B (1991) Treatment of osteoarthritis with a herbomineral formulation: a double-blind, placebo-controlled, cross-over study. J Ethnopharmacol 33:91–95
Kumar A, Kaul M, Bhan M, Khanna PK, Suri K (2007) Morphological and chemical variation in 25 collections of the Indian medicinal plant, Withania somnifera (L.) Dunal (Solanaceae). Genet Resour Crop Evol 54:655–660
Kuttan G (1996) Use of Withania somnifera Dunal as an adjuvant during radiation therapy. Indian J Exp Biol 34:854–856
Leet J, Hussain SF, Minard R, Shamma M (1982) Sindamine, Punjabine and Gilgitine-3 New secobisbenzylisoquinoline alkaloids. Heterocycles 19:2355–2360
Linder S (1996) Withania somnifera: winter cherry, Indian ginseng, ashwagandha. Aust J Med Herb 8:78
Majumdar D (1955) Withania somnifera Dunal, part II: alkaloidal constituents and their chemical characterisation. Indian J Pharm 17:158–168
Malara N, Foca D, Casadonte F, Sesto MF, Macrina L, Santoro L, Scaramuzzino M, Terracciano R, Savino R (2008) Simultaneous inhibition of the constitutively activated nuclear factor kappaB and of the interleukin-6 pathways is necessary and sufficient to completely overcome apoptosis resistance of human U266 myeloma cells. Cell Cycle 7:3235–3245
Malik F, Singh J, Khajuria A, Suri KA, Satti NK, Singh S, Kaul MK, Kumar A, Bhatia A, Qazi GN (2007) A standardized root extract of Withania somnifera and its major constituent withanolide-A elicit humoral and cell-mediated immune responses by up regulation of Th1-dominant polarization in BALB/c mice. Life Sci 80:1525–1538
Malik F, Kumar A, Bhushan S, Mondhe DM, Pal HC, Sharma R, Khajuria A, Singh S, Singh G, Saxena AK, Suri KA, Qazi GN, Singh J (2009) Immune modulation and apoptosis induction: two sides of antitumoural activity of a standardised herbal formulation of Withania somnifera. Eur J Cancer 45:1494–1509

Mandal C, Dutta A, Mallick A, Chandra S, Misra L, Sangwan RS (2008) Withaferin A induces apoptosis by activating p38 mitogen-activated protein kinase signaling cascade in leukemic cells of lymphoid and myeloid origin through mitochondrial death cascade. Apoptosis 13:1450–1464

Mehta AK, Binkley P, Gandhi SS, Ticku MK (1991) Pharmacological effects of Withania somnifera root extract on GABAA receptor complex. Indian J Med Res 94:312–315

Mir BA, Khazir J, Mir NA, T-u H, Koul S (2012) Botanical, chemical and pharmacological review of Withania somnifera (Indian ginseng): an ayurvedic medicinal plant. Indian J Drugs and Dis 1:147–160

Mirjalili MH, Fakhr-Tabatabaei SM, Alizadeh H, Ghassempour A, Mirzajani F (2009a) Genetic and Withaferin A analysis of Iranian natural populations of Withania somnifera and W. coagulans by RAPD and HPTLC. Nat Prod Commun 4:337–346

Mirjalili MH, Moyano E, Bonfill M, Cusido RM, Palazón J (2009b) Steroidal lactones from Withania somnifera, an ancient plant for novel medicine. Molecules 14:2373–2393

Mishra B (2004) Ashwagandha – Bhavprakash Nigantu (Indian Materia Medica). Chaukhambha Bharti Academy, Varanasi, pp 393–394

Mishra L-C, Singh BB, Dagenais S (2000) Scientific basis for the therapeutic use of Withania somnifera (ashwagandha): a review. Altern Med Rev 5:334–346

Mohan R, Hammers HJ, Bargagna-Mohan P, Zhan XH, Herbstritt CJ, Ruiz A, Zhang L, Hanson AD, Conner BP, Rougas J, Pribluda VS (2004) Withaferin A is a potent inhibitor of angiogenesis. Angiogenesis 7:115–122

Mohanty I, Arya DS, Dinda A, Talwar KK, Joshi S, Gupta SK (2004) Mechanisms of cardioprotective effect of Withania somnifera in experimentally induced myocardial infarction. Basic Clin Pharmacol Toxicol 94:184–190

Mondal S, Mandal C, Sangwan R, Chandra S (2010) Withanolide D induces apoptosis in leukemia by targeting the activation of neutral sphingomyelinase-ceramide cascade mediated by synergistic activation of c-Jun N-terminal kinase and p38 mitogen-activated protein kinase. Mol Cancer 9:239

Muralikrishnan G, Dinda AK, Shakeel F (2010) Immunomodulatory effects of Withania somnifera on azoxymethane induced experimental colon cancer in mice. Immunol Investig 39:688–698

Murthy HN, Dijkstra C, Anthony P, White DA, Davey MR, Power JB, Hahn EJ, Paek KY (2008) Establishment of Withania somnifera hairy root cultures for the production of withanolide A. J Integr Plant Biol 50:975–981

Oh JH, Kwon TK (2009) Withaferin A inhibits tumor necrosis factor-alpha-induced expression of cell adhesion molecules by inactivation of Akt and NF-kappaB in human pulmonary epithelial cells. Int Immunopharmacol 9:614–619

Ojha SK, Arya DS (2009) Withania somnifera Dunal (Ashwagandha): a promising remedy for cardiovascular diseases. World J Med Sci 4:156–158

Palyi I (1969) Phagocytosis in HeLa cell cultures induced by anti-cancer drugs. Arch Geschwulstforsch 33:345–349

Panjamurthy K, Manoharan S, Nirmal MR, Vellaichamy L (2009) Protective role of Withaferin-A on immunoexpression of p53 and bcl-2 in 7,12-dimethylbenz(a)anthracene-induced experimental oral carcinogenesis. Investig New Drugs 27:447–452

Park HJ, Rayalam S, Della-Fera MA, Ambati S, Yang JY, Baile CA (2008) Withaferin A induces apoptosis and inhibits adipogenesis in 3T3-L1 adipocytes. Biofactors 33:137–148

Patil SP, Maki S, Khedkar SA, Rigby AC, Chan C (2010) Withanolide A and asiatic acid modulate multiple targets associated with amyloid-beta precursor protein processing and amyloid-beta protein clearance. J Nat Prod 73:1196–1202

Pingali U, Pilli R, Fatima N (2014) Effect of standardized aqueous extract of Withania somnifera tests of cognitive and psychomotor performance in healthy human participants. Pharm Res 6:12–18

Rahman AU, Abbas S, Shahwar DE, Jamal S, Choudhary M (1993) New withanolides from Withania spp. J Nat Prod 56:1000–1006

Rayees S, Johri RK, Singh G, Tikoo MK, Singh S, Sharma SC, Satti NK, Gupta VK, Bedi YS (2012) Sitopaladi, a poly-herbal formulation inhibits IgE mediated allergic reactions. Spatula DD 2:75–82

Rayees S, Kumar A, Rasool S, Kaiser P, Satti NK, Sangwan PL, Singh S, Johri RK, Singh G (2013) Ethanolic extract of Alternanthera sessilis (AS-1) inhibits IgE-mediated allergic response in RBL-2H3 cells. Immunol Investig 42:470–480

Rayees S, Satti NK, Mehra R, Nargotra A, Rasool S, Sharma A, Sahu PK, Rajnikant GVK, Nepali K, Singh G (2014) Anti-asthmatic activity of azepino [2, 1-b] quinazolones, synthetic analogues of vasicine, an alkaloid from *Adhatoda vasica*. Med Chem Res 23:1–11

Roumy V, Biabiany M, Hennebelle T, Aliouat el M, Pottier M, Joseph H, Joha S, Quesnel B, Alkhatib R, Sahpaz S, Bailleul F (2010) Antifungal and cytotoxic activity of withanolides from Acnistus arborescens. J Nat Prod 73:1313–1317

Sangwan R, Chaurasiya N, Misra L, Lal P, Uniyal G, Sharma R, Sangwan NS, Suri KA, Qazi GN (2004) Phytochemical variability in commercial herbal products and preparations of Withania somnifera (Ashwagandha). Curr Sci 86:461–465

Santagata S, Xu YM, Wijeratne EM, Kontnik R, Rooney C, Perley CC, Kwon H, Clardy J, Kesari S, Whitesell L, Lindquist S, Gunatilaka AA (2012) Using the heat-shock response to discover anticancer compounds that target protein homeostasis. ACS Chem Biol 7:340–349

Sarkar S, Mandal C, Sangwan R (2014) Coupling G2/M arrest to the Wnt/beta-catenin pathway restrains pancreatic adenocarcinoma. Endocr Relat Cancer 21:113–125

Shah N, Kataria H, Kaul SC, Ishii T, Kaur G, Wadhwa R (2009) Effect of the alcoholic extract of Ashwagandha leaves and its components on proliferation, migration, and differentiation of glioblastoma cells: combinational approach for enhanced differentiation. Cancer Sci 100:1740–1747

Sharma H, Parihar L, Parihar P (2011) Review on cancer and anticancerous properties of some medicinal plants. J Med Plant Res 5:1818–1835

Shohat B, Kirson I, Lavie D (1978) Immunosuppressive activity of two plant steroidal lactones withaferin A and withanolide E. Biomedica 28:18–24

Shukla KK, Mahdi AA, Mishra V, Rajender S, Sankhwar SN, Patel D, Das M (2011) Withania somnifera Improves semen quality by combating oxidative stress and cell death and improving essential metal concentrations. Reprod Bio Med Online 22:421–427

Singh S, Kumar S (1998) Withania somnifera: the Indian ginseng ashwagandha. Central Institute of Medicinal and Aromatic Plants, Lucknow

Singh N, Singh S, Nath C, Kohli R, Bhargava K (1984) Anti-stress plants as anti-rheumatic agents. 5th sepal congress of rheumatology, Bangkok

Singh D, Aggarwal A, Maurya R, Naik S (2007) Withania somnifera Inhibits NF-κB and AP-1 transcription factors in human peripheral blood and synovial fluid mononuclear cells. Phytother Res 21:905–913

Singh N, Bhalla M, de Jager P, Gilca M (2011) An overview on ashwagandha: a Rasayana (rejuvenator) of Ayurveda. Afr J Tradit Complement Altern Med 8:208–213

Soman S, Korah PK, Jayanarayanan S, Mathew J, Paulose CS (2012) Oxidative stress induced NMDA receptor alteration leads to spatial memory deficits in temporal lobe epilepsy: ameliorative effects of Withania somnifera and Withanolide A. Neurochem Res 37:1915–1927

Stan SD, Zeng Y, Singh SV (2008) Ayurvedic medicine constituent withaferin a causes G2 and M phase cell cycle arrest in human breast cancer cells. Nutr Cancer 60(Suppl 1):51–60

Sukanya DH, Lokesha AN, Datta G, Himabindu K (2010) Phytochemical diversity in ashwagandha (Withania somnifera). J Med Aromat Plants 1(2):TS2-p26

Sultanbawa M, Sotheeswaran S, Balasubramaniam S, Moustafa A-K, Slatkin D, Schiff P Jr (1985) Pachyovatamine, a bisbenzylisoquinoline alkaloid, and other alkaloids from Pachygone ovata. Phytochemistry 24:589–592

Wall ME, Wani M (1977) Antineoplastic agents from plants. Annu Rev Pharmacol Toxicol 17:117–132

Wang HC, Tsai YL, Wu YC, Chang FR, Liu MH, Chen WY, Wu CC (2012) Withanolides-induced breast cancer cell death is correlated with their ability to inhibit heat protein 90. PLoS One 7:e37764

Winters M (2006) Ancient medicine, modern use: Withania somnifera and its potential role in integrative oncology. Altern Med Rev 11:269–277

Yang ES, Choi MJ, Kim JH, Choi KS, Kwon TK (2011) Combination of withaferin A and X-ray irradiation enhances apoptosis in U937 cells. Toxicol In Vitro 25:1803–1810

Yu SM, Kim SJ (2013) Production of reactive oxygen species by withaferin A causes loss of type collagen expression and COX-2 expression through the PI3K/Akt, p38, and JNK pathways in rabbit articular chondrocytes. Exp Cell Res 319:2822–2834

Ziauddin M, Phansalkar N, Patki P, Diwanay S, Patwardhan B (1996) Studies on the immunomodulatory effects of Ashwagandha. J Ethnopharmacol 50:69–76

Chapter 5
Withanolides: A Prospective Drug for Infectious and Tropical Diseases

Radheshyam Maurya

Abstract *Withania somnifera* (Ashwagandha) is a shrub plant from the family of Solanaceae and widely used by practitioners in the folk medicines as immune boosting agent. The medicinal properties of plants have been attributed due to the presence of steroidal lactones called as withanolides. Withanolides are a group of highly oxygenated steroids derived from a C28 ergostane skeleton and have attracted significant scientific interest due to their complex structural features and multiple bioactivities. Last one decade's more than 130 new natural withanolides were isolated from *Withania spp.*, and the scientist from the different field demonstrated that Ashwagandha possesses antioxidant, anti-inflammatory, antitumor, antimicrobial properties. This chapter briefly reviews the drug-likeness activity of withanolides against antibacterial, antifungal, antiparasitic diseases and more focussed on anti-leishmanial and immunomodulatory activities of withanolides.

Keywords Withanolides • Withaferin-A • Antibacterial • Antifungal • Antiparasitic • Anti-leishmanial • Immunomodulatory

5.1 Introduction

Withanolides are a group of steroidal lactones, isolated from various plants of the family, concluding *W. somnifera*, distributed in the South Asian sub-continent (Subramanian and Sethi 1971). *W. somnifera* (WS) has been safely used for centuries for the treatment of different ailments is one of the examples for Medhya Rasayana. In Sanskrit "Ashwa" means horse and "Gandha" means odor, which resembles the sweaty horse. The name "somnifera" in Latin means "sleep-inducer" which relates its activity of relieving stress. WS is an erect, grayish, slightly hairy evergreen shrub with relatively long tuberous roots and also known as winter cherry,

R. Maurya (✉)
Department of Animal Biology, University of Hyderabad,
Prof. C.R. Rao Road, Gachibowli, Hyderabad 500046, India
e-mail: rmusl@uohyd.ernet.in; radhemauryabhu@gmail.com

S.C. Kaul, R. Wadhwa (eds.), *Science of Ashwagandha: Preventive and Therapeutic Potentials*, DOI 10.1007/978-3-319-59192-6_5

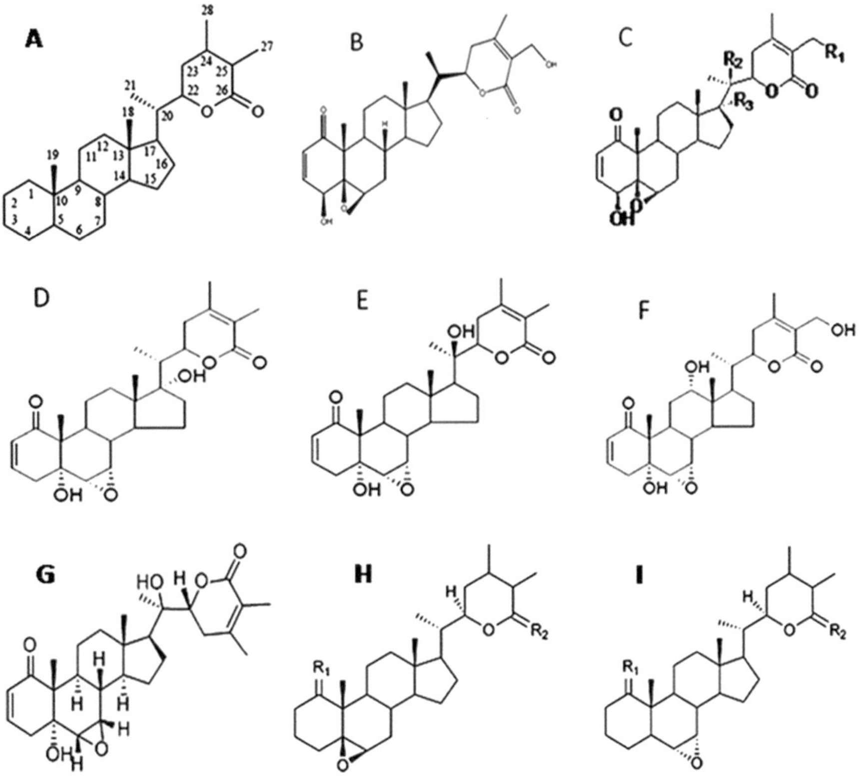

Fig. 5.1 Some bioactive molecules and their chemical structures isolated from Withania somnifera. (**a**) The basic structure of Withanolide, (**b**) Withaferin-A, (**c**) Withanolide-D, (**d**) Withanone (**e**) Withanolide-A, (**f**) Withastramonolide (**g**) Withanolides E (**h**) 5, 6-epoxy withanolides (**i**) 6, 7-epoxy withanolides

or Indian ginseng (Fig. 5.1). The genus *Withania* includes 26 species out of which two occur in India. These are readily distinguishable as *W. somnifera* and *W. coagulans* (Singh and Kumar 1998). The pharmacological and therapeutic properties of the plant have been attributed to the presence of withanolides, the most important of which is Withaferin-A, a steroidal lactone first isolated by Kurup (1956) from the leaves of the Indian plant and later isolated from the Israeli variety of the plant and characterized by Lavie et al. (1965).

WS leaves consist of steroidal lactone triterpenoids compounds called as withanolides. Withanolides are a group of highly oxygenated steroids derived from a C28 ergostane skeleton as the major chemical constituents (Lavie et al. 1972; Menssen and Stapel 1973). It is functionalized at carbons 1, 22 and 26, commonly known as the withanolide skeleton, in which C22 and C26 are oxidized to form a six-membered lactone ring (Fig. 5.1). Withaferin-A, the first withanolide-type, was reported in 1965 (Kurup) More than 1000 withanolides have been discovered over the past 50 years from the families of Solanaceae, Leguminosae, Labiatae,

Table 5.1 Antibacterial, antifungal and antiparasitic activities of some Withanolides isolated from *Withania somnifera*

Chemical constituent	Compounds	Biological activity
Withanolides	Withaferin A	Antibacterial and Antifungal Antibiotic and Antimitotic Antiparasitic and Antitumour
	Withanolide E	Antibacterial, Immunosuppressive, Insectantifeedant
	(3-β-hydroxy-2, 3-hydroxy Withanloide)	Antibacterial , Antitumour, Antiinflammatory, Insect antifeedant
	Withanolide D	Antitumour
	17β-hydroxywithanolide K	Antifungal
	5, 20-α dihydroxy-6α 7a-epoxy-1- OXO (5α-1 with a-2, 24-dienolide)	Immunolomodulatory

Myrtaceae, and Taccaceae. Most of these compounds are 1-oxo steroids and/or oxidized at C-26 and C-22, or C-26 and C-23 to form a β- or α-lactone (Kurup 1958; Sethi et al. 1974; Kirson et al. 1969, 1970, 1971; Jaffer et al. 1988). Some of these (Table 5.1) are 14 a-hydroxy steroids, 4-OH and 5, 6-epoxy withanolides (Withaferin-A like steroids) and 5-OH and 6, 7-epoxy withanolides (withanolides A and D like steroids), 12-deoxywithastramonolide, (Kirson and Glotter 1980; Namdeo et al. 2011; Srivastava et al. 2008; Pretorius et al. 2009; Chatterjee et al. 2010) and 14, 15β-epoxywithanolide I and 17β-hydroxywithanolide K (Choudhary et al. 1995). Ashwagandhanolide, a bioactive dimeric thiowithanolide has been isolated from the roots of WS (Subbaraju et al. 2006). Withanolides are synthesized de novo in different parts of the plant (root and leave) and possess diverse biological and pharmacological activities (Mirjalili et al. 2009; Chen et al. 2011).

The antimicrobial activity of the roots, as well as leaves, has been shown experimentally (Kurup 1958; Kohlumnezer and Krupinska 1963). Withaferin-A possess antibiotics activity (10 μg/ml) against the growth of various Gram-positive bacteria, acid-fast aerobic bacilli, pathogenic fungi, Micrococcus pyogenes var. aureus and partially inhibited the activity of glucose-6-phosphate dehydrogenase in Bacillus subtilis (Atta-ur-Rahman and Choudhary 1993). Withaferin-A inhibited Ranikhet virus. The shrub's extract is active against Vaccinia virus and Entamoeba histolytica (Wealth of India 1982). Antibiotic activity of Withaferin-A is due to the presence of the unsaturated lactone ring. The lactone showed a potent therapeutic activity in experimentally induced sores in rabbits and the being slightly stronger than that of Penicillin. It substantiates the reliability of the leaves as a cure for abscesses and carbuncles in the indigenous system of medicine (Wealth of India 1982).

The comprehensive study has reported the diverse activities of WS (Winters 2006) and whole plant extract and its constituents were associated anti-inflammatory, immunomodulatory and anti-stress activities. WS is also considered an adaptogen, a preventive medication that normalizes physiological functions which are disturbed

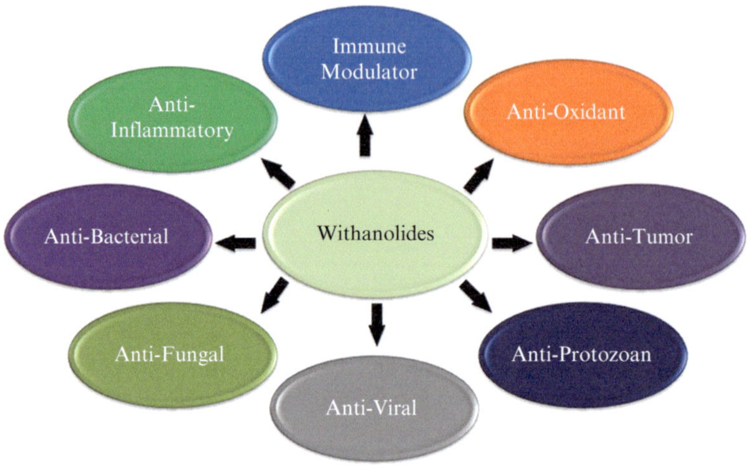

Fig. 5.2 Schematic flow chart represents the Withanolides as a potential drug for human diseases

by due to chronic stress, through correction of imbalances in the neuroendocrine and immune systems (Ven Murthy et al. 2010). Antitumor mechanisms of WS reveal its potential to upregulate phase II liver enzymes, cell cycle proliferation, tumor apoptosis, inhibition of angiogenesis; NF-KB suppression enhanced immune system (Oh et al. 2008). Withaferin-A and Withanolide-E exhibited a specific immunosuppressive effect on human B and T lymphocytes and mice thymocytes. Withanolide-E had a particular effect on T lymphocytes whereas Withaferin-A affected both B and T lymphocytes (Aggarwal et al. 2012). WS has been used in different forms, such as dried powder of root and leaf.

In recent years, withanolides have attracted significant scientific interest due to their complex structural and pharmaceuticals properties, such as their antibacterial, antiviral, antifungal, anti protozoan, antitumor, immunosuppressive, antifeedant, antistress, and anti-inflammatory capacities (Fig. 5.2). This chapter provides a comprehensive review on withanolides in the treatment of various infectious and tropical diseases last one decade.

5.1.1 Anti-bacterial Activity

The antibacterial properties of this multipronged medicinal plant were for the first time reported by Kurup (1956) against *Salmonella aureus*. Withaferin-A and 3-b-hydroxy-2,3-dihydrowithanolide-F isolated from WS show promising antibacterial, antitumoral, immunomodulating and anti-inflammatory properties (Kurup 1958; Sethi et al. 1974; Budhiraja and Sudhir 1987; Ziauddin et al. 1996; Mishra et al. 2000). The methanolic and hexane extracts of both leaves and roots of WS were demonstrated as a potent antibacterial activity (Shanazbanu et al. 2006). A synergistic increase in the antibacterial effect with Tibrim was noticed in agar plate disc-diffusion assay against *Salmonella typhimurium* and *Escherichia coli* (Arora et al. 2004). Recently, the antibacterial activity of WS root and leaf of both aqueous as well as alcoholic extracts of the plant were found to possess potent antibacterial activity against a range of bacteria including *Salmonella typhimurium, E.coli* as revealed by in vitro Agar Well Diffusion Method (Owais et al. 2005; El-Boshy et al. 2013). Further, the oral administration of the aqueous fruit extracts successfully obliterated Salmonella infection in mice as revealed by increased survival rate, as well as less bacterial load in various vital organs of the treated animals (Owais et al. 2005; Heyman et al. 2009). WS leaf extract indirectly modulates immune activity and probably disengages *Listeria- monocyte* genes induced suppression by provoking a higher reserve of myeloid progenitors in the bone marrow, proliferation of lymphocytes and increased INF-γ levels in infected mice (Teixeira et al. 2006).

Antimicrobial activity of alkaloids of different parts (root, stem, leaf and fruits) of WS was evaluated against four bacteria (*Escherichia coli, Pseudomonas aeruginosa, Proteus mirabilis, Staphylococcus aureus*) and four fungi (*Aspergillus flavus, Aspergillus niger, Candida albicans* and *Trichophyton mentagrophytes*). The alkaloid extracts from the different source of the plant were found to possess effective antimicrobial activity against test pathogens, as revealed by Disc Diffusion Assay. All the microorganisms were found to be sensitive against the extracts tested except *A. flavus, A. niger*. The results of these studies advocate the exploitation of alkaloid extracts of WS for future antimicrobial drugs (Rajendran and Ramakrishnan 2009; Singh and Kumar 2011). The methanolic extract of WS (MEW) was restraining physiological activities of carcinogenic bacteria and virulence properties of *Streptococcus mutans* and *Streptococcus*. Moreover, MEW inhibited acid production, acid tolerance, and biofilm formation of S. mutans and S. sobrinus at sub-MIC levels and showed a broad antibacterial range against oral bacteria (Pandit et al. 2013). A Recent study demonstrated that the herbal extracts of WS showed the highest MIC against drug resistant *E. coli* (erythromycin, tetracycline, ceftazidime, cefixime, penicillin, ampicillin and nalidixic acid) was 200 ppm and lowest MIC was 50 ppm. These results indicated that WS extracts had antibacterial effects on drug resistant *E. coli* strains isolated from clinical samples (Bokaeian et al. 2015).

Antibacterial activities of WS leaf extracts displayed the highest activity against *S. typhi* (32.00 ± 0.75 mm zone of inhibition), although the lowest activity was against *K. pneumoniae* (19.00 ± 1.48 mm zone of inhibition). The lowest minimum inhibitory concentration valuc was 6.25 mg/ml, which was against *S. typhi*, followed by 12.5 mg/ml against *E. coli*. Finally, WS exhibited significant antibacterial activities against Gram-negative bacteria, particularly *S. Typhi* (Alam et al. 2012). The root extracts of WS were found to be effective against all multi-drug resistant (MDR) *Staphylococcus aureus* strains isolated from local and patient sources. The antimicrobials activities of different root extracts of WS showed a different degree of effectiveness against the strains and may raise an alternative therapy for MDR staphylococcal infections in the near future (Jaffer et al. 1988; Datta et al. 2011). The different fraction of WS extract was tested on clinically isolated bacterial pathogens from skin, upper respiratory tract, gastrointestinal and urinogenital tract infection (*Staphylococcus aureus, E. coli, Pseudomonas aeruginosa and Bacillus subtilis*) and possess high antibacterial activity for gram negative as well as gram positive bacteria. (Mahesh and Satish 2008; Sundaram et al. 2011). The acetone extract of WS leaves possessed promising antimicrobial activity against the growth of six pathogenic bacteria viz., *Bacillus cereus, B. subtilis, Staphylococcus epidermis, S. aureus, Streptococcus pneumoniae and S. pyogenes and five bifidobacteria viz., Bifidobacterium animalis, B. breve, B. catenulatum, B. infantis and B. longum* (Polesny et al. 2008; Halamova et al. 2013). The above finding also demonstrated that the withanolides are extracted from ethanol, methanol, ethyl acetate and acetone, the range of polar solvents which are a potent inhibitor of bacterial growth (Table 5.2).

5.1.2 Anti-fungal Activity

Therapeutic efficacy of *W. somnifera* was demonstrated to possess potent antifungal activity and was found effective in the treatment of murine aspergillosis. Ashwagandha showed the protective action against systemic *Aspergillus* infection. This protective activity was probably related to the activation of the macrophage function revealed by the observed increases in phagocytosis and intracellular killing of peritoneal macrophages induced by Ashwagandha treatment in mice (Dhuley 1998). Moreover, Dhuley (1997) demonstrated that this plant stimulates chemotaxis and the production of Interleukin-1 and TNF-α in macrophages from immunosuppressed mice. Another study showed that extracts of WS with 100 μg concentration in each disk, made an inhibitory zone (15 ± 0.5 mm) against *Aspergillus fumigatus* (Bansod and Rai 2008). 14, 15β-epoxywithanolide I and 17β-hydroxywithanolide K isolated from *W.cogulance* exhibited antifungal activity against human pathogens *Nigrospora oryzae, Aspergillus niger, Curvularia lunata, Stachybotrys atra, Allescheria boydii, Drechslera rostrata, Microsporum canis* and *Epidermophyton floceosum* and plant pathogen *Pleurotus ostreatus* (Choudhary et al. 1995) Table 5.2.

5.1.3 Anti-parasitic Activity

In vitro study has been shown that chloroform extract from *W. somnifera* posse's antimalarial activity against *P. falciparum* and the IC50 of the extract is 2.04 μg/ml (Bogale and Petros 1996). The anti-malarial activities of WS also have been reported in vivo model of malaria (Dikasso et al. 2006; Teklemariam 2005). Results of a 4-day malaria suppressive test in *P. berghei*-infected mice also revealed that the methanolic extract fractions of WS leaves showed a significant reduction of parasitemia at lower doses compared to the crude extracts. The decrease of parasitemia by

Table 5.2 Chemical constituents of pharmaceutical importance identified in *Withania somnifera* (Mir BA et al. 2012)

Plants parts	Chemical constituent	References
Roots	Sitoindosides VII (Acylsteryl-glucoside)	Bhattacharya et al. (1987)
	Sitoindosides VIH (Acylsteryl-glucoside)	Bhattacharya et al. (1987)
	Sitoindosides IX (Glycowithanolide)	Ghosal et al. (1988)
	Sitoindosides X (Glycowithanolide)	Ghosal et al. (1988)
	Withanine (Alkaloid)	Majumdar (1955)
	Withananine (Alkaloid)	Majumdar (1955)
	Ashwagandhanolide	Subaraju et al. (2006) and Mirjalili et al. (2009)
Leaves	Withaferin (Steroidal lactone)	Anjaneyulu and Satyanarayana Rao (1997)
	Withaferin A (Steroidal lactone)	Kirson et al. (1970) and Lavie et al. (1965, 1972)
	Withanolie D (Steroidal lactone)	Kirson et al. (1970) and Lavie et al. (1975)
	Withanolie E (Steroidal lactone)	Glotter et al. (1973)
	Withanone (Steroidal lactone)	Dhalla et al. (1961) and Kirson et al.(1971)
	Withanolide Z (Novel)	Pramanick et al. (2008)
	Withanolide B	Pramanick et al. (2008)
	7-hydroxywithanolide	Pramanick et al. (2008)
	3α-methoxy-2, 3-dihydro-	Anjaneyulu and Satyanarayana Rao (1997)
	27-deoxywithaferin A (Steroidal lactone)	Kirson et al. (1970)
	4β, 17α-dihydroxy-l-loxo-	Lavie et al. (1965)
	5β, β-epoxy-22R-witha-	Lavie et al. (1965)
	2,24-dienolide (steroidal lactone)	Kirson et al. (1970) and Lavie et al. (1968)
	β-dihydroxy-5β, 6β-epoxy-	Glotter et al. (1973)
	l-oxo-22R-witha-2, 14-24-	Dhalla et al. (1961)
	Trienolide (steroidal lactone)	Kirson et al. (1971)
	5,20α (R)-dihydroxy-6α, 7α-epoxy-l-oxo- (5α) -Witha-2, 24-dienolide (steroidal laetone) 2,3-dihydrowithaferin A-3beta-0-sulfate	Menben Von and Stapel (1973), Xu et al. (2011)
Seeds	Withanolide —WS 2 (aliphatic ester) Withanolide —WS 1 (aliphatic ketone)	Kundu et al. (1976) and Khan et al. (1993)

44% and 57% at doses of 200 and 300 mg/kg body weight, respectively, by fractions of crude methanolic extract (Dame et al. 2013; Amuamuta and Na-Bangchang 2015).

The methanolic extracts from W. somnifera (Ashwagandha) possess antileishmanial activity showed in-vitro (Sharma et al. 2009; Sen et al. 2007). A recent study also revealed that the antileishmanial activity of the herbal drug from WS is exhibited through blocking of the protein kinase C signaling pathway (Grover et al. 2012). The treatment of infected mice with a combination of Asparagus racemosus and WS extracts not only resulted in the successful reduction of parasite level but also, generated protective Th1 type of immune responses with normalization of biochemical and hematological tests suggesting their role as an important antileishmanial agent (Kaur et al. 2014).

Withaferin-A has been identified as one of Ashwagandha's prominent compounds from WS and has shown antileishmanial activities with possible inhibitory action on DNA topoisomerase-I (Pramanick et al. 2008; Sharma et al. 2009). A previous study with L. donovani showed that Withaferin-A inhibits protein kinase, which induces apoptosis through apoptotic topoisomerase I- DNA complex (Sen et al. 2007). Withaferin-A was collectively known as withanolides and exhibits other therapeutic activities like anticancer (Grover et al. 2010, 2012), anti-herpetic property (Grover et al. 2011). It is a cell permeable steroidal lactone having many other pharmacological properties (Sharma et al. 2009). Recently, our lab demonstrated that the mode of action of Withaferin-A, an abundant withanolide in the leaves of WS inhibits Pteridine Reductase-1 enzymes from the Pteridine salvage pathway of the parasite (Chandrasekaran et al. 2015), exclusively present in trypanosomatids (Table 5.2).

5.1.4 Withanolides Controls the **L. donovani** *Infection*

Recently we have demonstrated that Withanolides isolated from WS leaves using bioassay-guided fractionation (F5 and F6 of ethanolic fractions) showed potent antileishmanial activity on the promastigote stage of the *L. donovani* parasites in-vitro (Chandrasekaran et al. 2013). The study revealed that the withanolides were altered the morphology of parasites from spindle to round shape and loss of flagella/cell integrity in promastigote stage of parasites. Moreover, it also induced DNA cleavage, inhibition of cell cycle and externalization of phosphatidylserine in dose and time-dependent manner via an increase in reactive oxygen species (ROS) and decreased in mitochondrial membrane potentials (Ψm) in parasites. Further, in-vivo study revealed a drastic reduction in Th2 cytokine (IL-10, IL-4, TGF-β) expression and increased IgG2a levels when compared with IgG1 in treated mice suggesting its immunomodulatry properties (Chandrasekaran et al. 2017) (Fig. 5.3). In conclusion, withanolides induce apoptotic-like death through the production of ROS from mitochondria and disruption of Ψm in promastigotes of *L. donovani* and shown potent immunomodulatory properties.

An ex-vivo study of fractions F5, F6, and Withaferin-A have shown the drastic reduction of amastigotes count in infected macrophage compared to control through the induction of reactive oxygen species upon treatment. BALB/c mice infected with L. donovani and treated for ten consecutive days showed a drastic reduction in parasite burden in both spleen and liver. The treatment resulted in increased expression of Th1 cytokines (IFN-γ, IL-12) and reduced expression of Th2 cytokines (IL-10, IL-4, and TGF-β). Besides, the IgG2a levels were significantly increased as compared to IgG1 in the sera of treated mice, indicating the induction of Th1 type

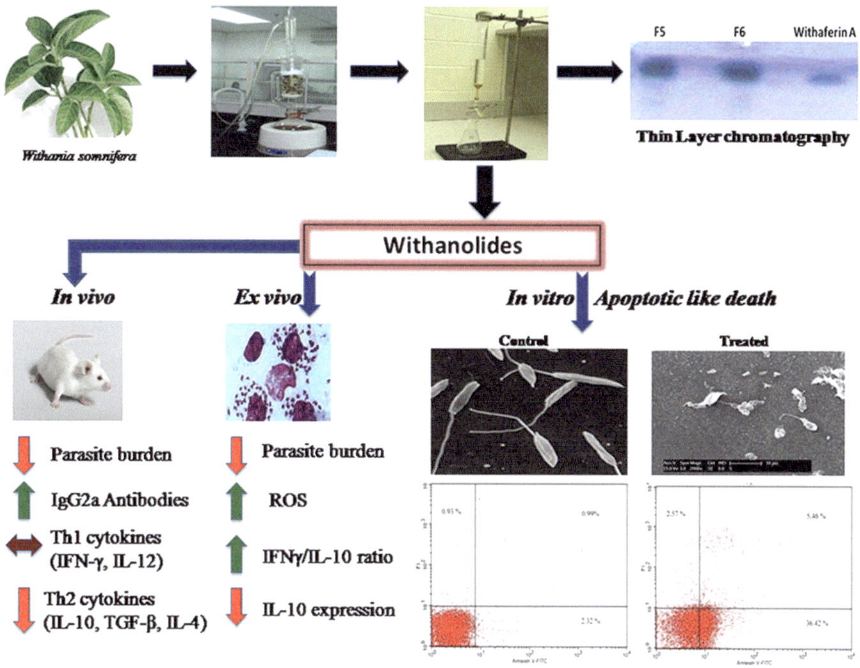

Fig. 5.3 Schematic representation of *Withania somnifera* leaf extract (F5 & F6) and Withaferin-A activities in experimental visceral leishmaniasis. Withanolides have shown potential antileishmanial and immunomodulatory properties against *Leishmania donovani* parasites in *in vitro, ex vivo* and *in vivo* studies

of immunity (Fig. 5.3). In conclusion, our results suggest that withanolides from W. somnifera have potent antileishmanial and immunomodulatory activities in experimental VL mice.

Furthermore, our lab has also demonstrated the Withaferin-A can alone inhibit the parasite-specific Pteridine reductase 1 (PTR1) enzymes. This is a crucial enzyme involved in pteridine salvage pathway and an essential for the growth and survival of the parasite. Hence, we employed molecular docking studies to identify the binding mode of Withaferin-A with PTR1 in-silico. The enzymes kinetic assayed with purified PTR-1 protein and whole parasites lysates revealed that Withaferin-A inhibits PTR-1 enzymes (Fig. 5.4) through the uncompetitive mode of inhibition in the parasites (Chandrasekaran et al. 2015).

5.1.5 Immune-Modulation in Infectious Diseases

The immune response elicited by the treatment as well as an effect of therapy on the immune response. Since, the Th1 and Th2 responses play a prominent role in recovery from leishmaniasis and shift from Th2 to Th1 response helps in providing

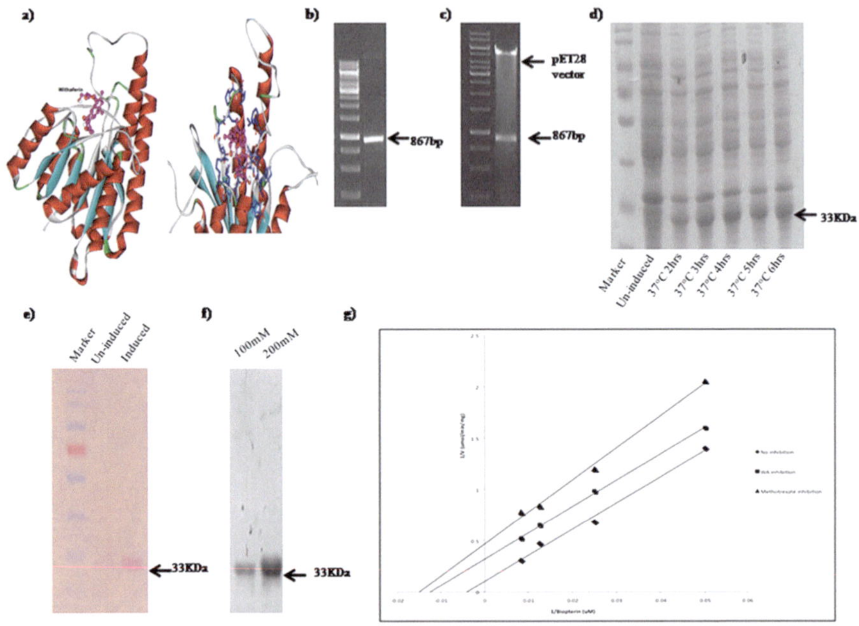

Fig. 5.4 Docking, Cloning, Expression and Purification of Pteridine reductase 1 (PTR 1) protein. The enzyme assayed with Withaferin-A has shown the inhibition of PTR1 enzyme through the uncompetitive mode of enzyme inhibition suggesting its antileishmanial activity. (**a**) Withaferin-A Docking with Pteridine Reductase 1 enzymes, (**b**) Amplified PCR product of PTR1 gene (**c**) Cloning & confirmation of PTR1 gene, (**d**) Expression of PTR1 proyein, (**e**) PTR1 protein confirmation by Western Blot and (**f**) PTR1 Enzymes assay with substrate, Withaferin-A and PTR1 Inhibitors

resistance to infection (Chowdhury et al. 2012; Chandrasekaran et al. 2013; Yamamoto et al. 2014). Some studies have demonstrated the benefits of combining conventional treatment with other compounds that can control immune response, or they have immunomodulatory property (Calla-Magarinos et al. 2013; Gamboa-Leon et al. 2014). An in vivo study on anti-leishmanial and immunomodulatory activity of withanolides against *L. donovani* has shown that it is more effective with the combination of cisplatin compared to the drugs alone (Sachdeva et al. 2013). This inhibitory effect of WS has been attributed to the presence of Withaferin-A (steroidal lactone) and withanolide-A. Withaferin-A can induce *Leishmania*-affected cells apoptosis via inhibition of protein kinase C affecting apoptosis and pteridine reductase-1 activity in parasite (Sachdeva et al. 2013; Chandrasekaran et al. 2013). Additionally, withanolides having its hepatoprotective and antioxidant properties can eliminate hepatotoxicity and the liver dysfunction either induced by cisplatin used for the treatment of VL or by the parasite itself (Sachdeva et al. 2013). Plant extracts, secondary metabolites or biomolecules can exert immunostimulatory properties (Chouhan et al. 2014). It is believed that the most common mechanism for exerting immunomodulatory activity is the stimulation of NO or ROS production that is principle effector molecule for killing of *Leishmania* amastigotes (Chowdhury et al. 2012; Chandrasekaran et al. 2013).

W. somnifera show an immuno-potentiating and myeloprotective effects know roots by enhancing the levels of interferon (IFN)-γ, interleukin (IL)-2 and granulocyte macrophage colony stimulating factor in normal and cyclophosphamide-treated mice (Malik et al. 2009; Davis and Kuttan 1998). The active compound (withanolide-A) in the roots of WS significantly increases the expression levels of T-helper 1 (Th1) cytokines, as well as CD4 and CD8 T cells counts. It also enhances natural killer (NK) cell activity in a dose-dependent manner with a prompt recovery of CD4+ T cells in immune suppressed animals (Malik et al. 2009; Davis and Kuttan 2000; Khan et al. 2006; Bani et al. 2006; Khan et al. 2009). Apart from this, the activated macrophage functioning indicated by enhanced secretion of nitric oxides, IL-2, and TNF-2, decreases moderately IL-4 with no effect on IL-10 suggesting that it only influenced Th1 profile of the cytokines (Chandrasekaran et al. 2013 and 2017; Khan et al. 2009). The root powder is also reported to stimulate the cell-mediated immunity, IgM, and IgG and a noticeable improvement in proliferation and differentiation of lymphocytes as indicated by lymphocyte surface markers of T cells (CD3+, CD4+, and CD8+) and B cells (CD19+) (Bani et al. 2006; Khan et al. 2009). Furthermore, the aqueous suspension of WS also showed anti-inflammatory and immunosuppressive effects by inhibiting the complement system, mitogen-induced lymphocyte proliferation and delayed type hypersensitivity (DTH) in rats (Rasool and Varalakshmi 2006). Though, no effect on the humoral response was seen while others have reported that elevated levels of IgG2a over lgGl in the withanolides treated BALBc mice (Bani et al. 2006). However, the effect of Withanolides on the immunoregulatory cellular response has not been tested yet. Hence, natural compounds isolated from WS are the future drugs candidates that might address the antibacterial and antifungal and antiparasitic activities together with induced immune response to control the infectious diseases.

5.2 Conclusion

The future perspective of basic and translational research with *Withania* somnifera must enlist the investigation for immunoregulatory molecules induced by withanolides in infectious diseases. The involvement of immune- regulatory cells induced by withanolides molecules might have several functions: such as regulating antigen presentation and control of immunosuppressive microenvironment along with a physiological cytokine milieu for an effector T cell function. Finally, the investigations toward the use of withanolides, in accord with the holistic Ayurvedic approaches to treating infectious diseases may impact on several forms of bacterial, fungal and parasitic disease together with immune modulation.

References

Aggarwal R, Diwanay S, Patki P, Patwardhan B (2012) Studies on immunomodulatory activity of *Withania somnifera* (Ashwagandha). J App Pharm Sci 2:170–175

Alam N, Hossain M, Ma M, Sa S, Gan SH, Khalil MI (2012) Methanolic extracts of *Withania somnifera* leaves, fruits and roots possess antioxidant properties and antibacterial activities. BMC Complement Altern Med 12:175

Amuamuta A, Na-Bangchang K (2015) A review of ethnopharmacology of the commonly used antimalarial herbal agents for traditional medicine practice in Ethiopia. Afr J Pharm Pharmacol 9:615–627

Anjaneyulu ASR, Satyanarayana Rao D (1997) New withanolides from the roots of Withania somnifera. Indian J Chem 36:424–433

Arora S, Dhillon RG, Nagpal A (2004) The in vitro antibacterial/ synergistic activities of *Withania somnifera* extracts. Fitoterapia 75:385–388

Atta-ur-Rahman, Choudhary M (1993) New withanolides from Withania somnifera spp. J Nat Prod 56:1000–1006

Bani S, Gautam M, Sheikh FA, Khan B, Satti NK, Suri KA, Qazi GN, Patwardhan B (2006) Selective Th-1 up-regulating activity of *Withania somnifera* aqueous extract in an experimental system using flow cytometry. J Ethnopharmacol 107:107–115

Bansod S, Rai M (2008) Antifungal activity of essential oils from Indian medicinal plants against human pathogenic Aspergillus fumigatus and A. niger. World J Med Sci 3:81–88

Bhattacharya SK, Goel RK, Kaur R, Ghosal S (1987) Antistress activity of sitoindosides VII and VIII, new acylsterylglucosides from *Withania somnifera*. Phytother Res 1:32–39

Bogale M, Petros B (1996) Evaluation of the antimalarial activity of some Ethiopian traditional medicinal plants against *P. falciparum in vitro*. SINET 19:233–243

Bokaeian M, Fakheri BA, Nejad NM, Zarei H, Saeidi S, Mohasseli T (2015) The effect of *Withania somnifera* extract on drug resistant strains of Escherichia coli. Int J Inf Secur 2:e23036

Budhiraja RD, Sudhir S (1987) Review of biological activity of Withanolides. J Sci Ind Res 46:488–491

Calla-Magarinos J, Quispe T, Gimenez A, Freysdottir J, Troye-Blomberg M, Fernandez C (2013) Quinolinic alkaloids from Galipea Longiflora Krause suppress production of proinflammatory cytokines in vitro and control inflammation in vivo upon Leishmania infection in mice. Scand J Immunol 77:30–38

Chandrasekaran S, Dayakar A, Veronica J, Sundar S, Maurya R (2013) An in vitro study of apoptotic like death in Leishmania donovani promastigotes by withanolides. Parasitol Int 62:253–261

Chandrasekaran S, Veronica J, Dayakar A, Sundar S, Maurya R (2015) Exploring the inhibitory activity of Withaferin-A against Pteridine reductase-1 of *L. donovani*. J Enzyme Inhib Med Chem 25:1–9

Chandrasekaran S, Veronica J, Sundar S, Maurya R (2017) Alcoholic fractions F5 and F6 from Withania somnifera leaves show a potent Antileishmanial and Immunomodulatory activities to control experimental Visceral Leishmaniasis. Front Med 4:55

Chatterjee S, Srivastava S, Khalid A, Singh N, Sangwan RS, Sidhu OP, Raja R, Khetrapal CL, Tuli R (2010) Comprehensive metabolic fingerprinting of *Withania somnifera* leaf and root extract. Phytochemistry 71:1085–1094

Chen L, He H, Qiu F (2011) Natural withanolides: an overview. Nat Prod Rep 28:705–740

Choudhary MI, Dur-E-Shanhwar Z, Parveen A, Jabbar IA, Atta-ur-Rahman (1995) Antifungal steroidal lactones from Withania coagulance. Phytochemistry 40:1243–1246

Chouhan G, Islamuddin M, Sahal D, Afrin F (2014) Exploring the role of medicinal plant-based immunomodulators for effective therapy of leishmaniasis. Front Immunol 5:193

Chowdhury S, Mukherjee T, Mukhopadhyay R, Mukherjee B, Sengupta S, Chattopadhyay S, Jaisankar P, Roy S, Majumder HK (2012) The lignan niranthin poisons *Leishmania donovani* topoisomerase IB and favours a Th1 immune response in mice. EMBO Mol Med 4:1126–1143

Dame Z, Petros B, Mekonnen Y (2013) Evaluation of anti-*Plasmodium berghei* activity of crude and column fractions of extracts from *Withania somnifera*. Turk J Biol 37:147–150

Datta S, Kumar Pal NK, Nandy AK (2011) Inhibition of the emergence of multi drug resistant Staphylococcus aureus by *Withania somnifera* root extracts. Asian Pac J Trop Med 4:917–920

Davis L, Kuttan G (1998) Suppressive: effect of cyclophosphamide-induced toxicity by *Withania somnifera* extract in mice. J Ethnopharmacol 62:209–214

Davis L, Kuttan G (2000) Effect of *Withania somnifera* on cyclophosphamide induced urotoxicity. Cancer Lett 148:4–17

Dhalla NS, Gupta KC, Sastry MS, Malhotra CL (1961) Comparative studies of Withania somnifera. Indian J Pharmacol 23:126

Dhuley JN (1997) Effect of some Indian herbs on macrophage functions in ochratoxin A treated mice. J Ethnopharmacol 58:15–20

Dhuley JN (1998) Effect of Asgand on lipid peroxidation in stress induced animals. J Ethnopharmacol 7:173–178

Dikasso D, Mekonnen E, Debella A (2006) Antimalarial activity of *Withania somnifera* L. Dunal extracts in mice. J Ethiop Med 44:279–285

El-Boshy ME, Abdalla OM, Risha A, Moustafa F (2013) Effect of *Withania somnifera* extracts on some selective biochemical, hematological, and immunologi- cal parameters in guinea pigs experimental infected with *E. coli*. ISRN Vet Sci 2013:6

Gamboa-Leon R, Vera-Ku M, Peraza-Sanchez SR, Ku-Chulim C, Horta-Baas A, Rosado-Vallado M (2014) Antileishmanial activity of a mixture of Tridax procumbens and Allium sativum in mice. Parasite 21:15

Ghosal S, Kaur R, Srivastava RS (1988) Sitoindosides IX and X, new glycowithanolides from *W. somnifea*. Ind J Nat Pr 4:12–13

Glotter E, Kirson I, Abraham A, Lavis D (1973) Constituents of Withania somnifera (Dunal) XIII-The Withanolides of chemotype III. Tetrahedron Lett 29:1353–64

Grover A, Shandilya A, Bisaria VS, Sundar D (2010) Probing the anticancer mechanism of prospective herbal drug Withaferin A on mammals: a case study on human and bovine proteasomes. BMC Genomics 11(Suppl 4):S15

Grover A, Agrawal V, Shandilya A, Bisaria VS, Sundar D (2011) Non-nucleosidic inhibition of Herpes simplex virus DNA polymerase: mechanistic insights into the anti-herpetic mode of action of herbal drug withaferin A. BMC Bioinformatics 12(Suppl 13):S22

Grover A, Singh R, Shandilya A, Priyandoko D, Agrawal V, Bisaria VS, Wadhwa R, Kaul SC, Sundar D (2012) Ashwagandha derived withanone targets TPX2-Aurora A Complex: computational and experimental evidence to its anticancer activity. Plos One 7(1):e30890

Grover A, Katiya SP, Jeyakanthan J, Dubey VK, Sundar D (2012) Blocking protein kinase C signaling pathway: mechanistic insights into the anti-leishmanial activity of prospective herbal drugs from *Withania somnifera*. BMC Genomics 13(Suppl 7):S20

Halamova K, Kokoska L, Polesny Z, Macakova K, Flesar J, Rada V (2013) Selective in vitro growth inhibitory effect of *Withania a somnifera* on human pathogenic bacteria and bifidobacteria. Pak J Bot 45:667–670

Heyman HM, Hussein AA, Meyer JJM, Lall N (2009) Antibacterial activity of South African medicinal plants against methicillin resistant *Staphylococcus aureus*. Pharm Biol 47:67–71

Jaffer HJ, Jawad ALM, Saber HS, Al-Naib A (1988) Evaluation of antimicrobial activity of *Withania somnifera* extracts. Fitoterapia 59:497–500

Kaur S, Chauhan K, Sachdeva H (2014) Protection against experimental visceral leishmaniasis by immunostimulation with herbal drugs derived from *Withania somnifera* and Asparagus racemosus. J Med Microbiol 63:1328–1338

Khan FZ, Saeed MA, Alam M, Chaudhary AR, Ismail M (1993) Biological studies of indigenous medicinal plants III Phytochemical and antimicrobial studies on the non alkaloidal constitutes of some solanaceous fruit. Fac Pharmacy Gazi Univ 10:105–16

Khan B, Ahmad SF, Bani S, Kaul A, Suri KA, Satti NK, Athar M, Qazi GN (2006) Augmentation and proliferation of T lymphocytes and Th-1 cytokines by *Withania somnifera* in stressed mice. Int Immunopharmacol 6:1394–1403

Khan S, Malik F, Suri KA, Singh J (2009) Molecular insight into the immune up-regulatory properties of the leaf extract of Ashwagandha and identification of Th1 immunostimulatory chemical entity. Vaccine 27:6080–6087

Kirson I, Glotter E (1980) 14 a-hydroxy steroids from *W somnifera* (L) Dunal. J Chem Res Synop 10:338–339

Kirson I, Glotter E, Abraham A, Lavis D (1969) Two new constituents of *W somnifera* (L) Dunal. Indian type. Isr J Chem (proceedings) 15

Kirson I, Glotter E, Abraham A, Lavis D (1970) Constituents of *Withania somnifera*. Dunal XI the structure of three new withnolides. Tetrahedron 26:2209–2215

Kirson I, Glotter E, Lavis D, Abraham A (1971) Constituents of *Withania somnifera* Dunal XII the withanolides of an Indian Chemotype. J Chem Soc (Org) 2032–2044

Kohlumnezer S, Krupinska J (1963) Isolation of an antibiotic substance unsaturated lactine, m.P. 159- 60oC from the leaves of *W somnifera*. Disseratones Pharm 11:501–506

Kurup PA (1956) The antibacterial principle of the leaves of *Withania somnifera*. Curr Sci Bangalore 25:57–58

Kurup PA (1958) The antibacterial principle of *Withania somnifera*. I. Isolation and antibacterial activity. Antibiot Chemother 8:511–515

Kundu AB, Mukerjee A, Dey AK (1976) A new Withanolide from the seeds of Withania somnifera. Indian J Chem 14B:434–5

Lavie D, Glotter E, Shvo Y (1965) Constituents of *Withania somnifera Dun*. Part IV. The structure of withaferin A J Chem Soc 30:7517–7531

Lavie D, Kirson I, Glotter E (1968) Constituents of Dun. Part X the structure of Withanolide D. Israel J Chem 6(5):671–678

Lavie D, Kirson I, Abraham A (1972) Steroidal lactones of *W. somnifera*. Harokaech. Haivri 14:362–368

Lavie D, Glotter E, Kirson I, Abraham A (1975) Isolation of Withanolide D from W.somnifera chemotype II. Phytochemistry 14:189–94

Mahesh B, Satish S (2008) Antimicrobial activity of some important medicinal plant against plant and human pathogens. World J Agric Sci 4:839–843

Majumdar DN (1955) *Withania somnifera* Dunal, Part II, alkaloidal constituents and their chemical characterization. Ind J Pharmaco 17:158–161

Malik F, Kumar A, Bhushan S, Mondhe DM, Pal HC, Sharma R, Khajuria A, Singh S, Singh G, Saxena AK, Suri KA, Qazi GN, Singh J (2009) Immune modulation and apoptosis induction: two sides of antitumoural activity of a standardised herbal formulation of *Withania somnifera*. Eur J Cancer 45:1494–1509

Menssen HG, Stapel G (1973) A C28-steroidlacton from the roots of *Withania somnifera*. Planta Med 24:8–12

Mir BA, Koul S, Kumar A, Sushant S, Kaul MK, Soodan AS (2012) Reproductive behaviour and breeding system of wild and cultivated types of Withania somnifera (L.) Dunal. J Med Plants Res 6(5):754–762

Mirjalili MH, Moyano E, Bonfill M, Cusido RM, Palazon J (2009) Steroidal lactones from *Withania somnifera*, an ancient plant for novel medicine. Molecules 14:2373–2393

Mishra LC, Singh BB, Dagenais S (2000) Scientific basis for the therapeutic use of *Withania somnifera* (Ashwagandha): a review. Altern Med Re 5:334–346

Namdeo AG, Sharma A, Yadav KN, Gawande R, Mahadik KR, Lopez-Gresa MP, Kim HK, Choi YH, Verpoorte R (2011) Metabolic characterization of *Withania somnifera* from different regions of India using NMR spectroscopy. Planta Med 17:1958–1964

Oh JH, Lee TJ, Kim SH, Choi YH, Lee SH, Lee JM, Kim YH, Park JW, Know TK (2008) Induction of apoptosis by withaferin A in human leukaemia U937 cells through downregulation of Akt phosphorylation. Apoptosis 13:1494–1504

Owais M, Sharad KS, Shehbaz A, Saleemuddin M (2005) Antibacterial efficacy of *Withania somnifera* (Ashwagandha) an indigenous medicinal plant against experimental murine salmonellosis. Phytomedicine 12:229–235

Pandit S, Chang KW, Jeon JG (2013) Effects of *Withania somnifera* on the growth and virulence properties of Streptococcus mutans and Streptococcus sobrinus at sub-MIC levels. Anaerobe 19:1–8

Polesny Z, Flesar J, Jankovska I, Rada V, Kokoska L (2008) Selective in vitro inhibitory effect of *Withania somnifera* extract on growth of human pathogenic bacteria and bifidobacteria. First European Food Congress ´08, Ljubljana, Slovenia, Abstract p30

Pramanick S, Roy A, Ghosh S, Majumder H, Mukhopadhyay S (2008) Withanolide Z, a New Chlorinated Withanolide from Planta. Medica 74(14):1745–1748

Pretorius E, Oberholzer HM, Becker PJ (2009) Comparing the cytotoxic potential of *Withania somnifera* water and methanol extracts. Afr J Trad CAM 6:275–280

Rajendran NK, Ramakrishnan J (2009) In vitro evaluation of antimicrobial activity of crude extracts of medicinal plants against multi drug resistant pathogens. Biyoloji Bilimleri Araptyrma Dergisi 2:97–101

Rasool M, Varalakshmi P (2006) Immunomodulatory role of *Withania somnifera* root powder on experimental induced inflammation: an in vivo and in vitro study. Vasc Pharmacol 44:406–410

Sachdeva H, Sehgal R, Kaur S (2013) Studies on the protective and immunomodulatory efficacy of *Withania somnifera* along with cisplatin against experimental visceral leishmaniasis. Parasitol Res 112:2269–2280

Sen N, Banerjee B, Das BB, Ganguly A, Sen T, Pramanik S, Mukhopadhyay S, Majumder HK (2007) Apoptosis is induced in leishmanial cells by a novel protein kinase inhibitor withaferin A and is facilitated by apoptotic topoisomerase I-DNA complex. Cell Death Differ 14:358–367

Sethi PD, Sharma KB, Subramanian Sankara S (1974) Antibacterial activity of some C28 steroidal lactones. Indian J Pharm 36:122–123

Shanazbanu SS, Babu VLA, Dhanapal R (2006) Isolation of withaferin-A from Withania somnifera Dun. Leaves and its antibacterial activity. Asian J Chem 18:1243–1247

Sharma U, Velpandian T, Sharma P, Singh S (2009) Evaluation of anti-leishmanial activity of selected Indian plants known to have antimicrobial properties. Parasitol Res 105:1287–1293

Singh S, Kumar S (1998) *Withania somnifera* – the Indian ginseng Ashwagandha. Central Institute of Medicinal and Aromatic Plants, Lucknow

Singh G, Kumar P (2011) Evaluation of antimicrobial efficacy of flavonoids of *Withania somnifera* L. Indian J Pharm Sci 73:473–478

Srivastava P, Tiwari N, Yadav AK, Kumar V, Shanker K, Verma RK, Gupta MM, Gupta AK, Khanuja SP (2008) Simultaneous quantification of withano- lides in *Withania somnifera* by a validated high-performance thin-layer chromatographic method. J AOAC Int 91:1154–1161

Subbaraju GV, Vanisree M, Rao CV, Sivaramakrishna C, Sridhar P, Jayaprakasam B, Nair MG (2006) Ashwagandhanolide, a bioactive dimeric thiowithanolide isolated from the roots of *Withania somnifera*. J Nat Prod 69:1790–1792

Subramanian SS, Sethi PD (1971) Steroidal lactone of Withania ashwagandha. Indian J Pharm 33:25–26

Sundaram S, Priyanka D, Shalini P (2011) In vitro evaluation of antibacterial activities of crude extracts of *Withania somnifera* (Ashwagandha) to bacterial pathogens. Asian J Biotechnol 3:194–199

Teixeira ST, Valadares MC, Goncalves SA, de Melo A, Queiroz MLS (2006) Prophylactic administration of *Withania somnifera* extract increases host resistance in Listeria monocytogenes infected mice. Int Immunopharmacol 6:1535–1542

Teklemariam Z (2005) *In vivo* evaluation of antimalarial activities of *Moringa stenopetela, Withania somnifera* and *Vernonia amygdalina.* MSc Thesis, Department of Biology, Addis Ababa University

The Wealth of India (1982) Publications and information directorate, Council of Scientific and Industrial Research (CSIR), New Delhi, pp 580–585

Ven Murthy MR, Ranjekar PK, Ramassamy C, Deshpande M (2010) Scientic basis for the use of Indian ayurvedic medicinal plants in the treatment of neurodegenerative disorders: Ashwagandha. Cent Nerv Syst Agents Med Chem 10:238–246

Winters M (2006) Ancient medicine, modern use: *Withania somnifera* and its potential role in integrative oncology. Altern Med Rev 11:269–277

Xu Y-m, Gao S, Bunting DP, Gunatilaka AAL (2011) Unusual withanolides from aeroponically grown Withania somnifera. Phytochemistry 72:518–522

Yamamoto ES, Campos BL, Laurenti MD, Lago JH, Grecco Sdos S, Corbett CE, Passero LF (2014) Treatment with triterpenic fraction purified from Baccharis Uncinella leaves inhibits Leishmania (Leishmania) amazonensis spreading and improves Th1 immune response in infected mice. Parasitol Res 113:333–339

Ziauddin M, Phansalkar N, Patki P, Diwanay S, Patwardhan B (1996) Studies on the immunomodulatory effects of ashwagandha. J Ethnopharmacol 50:69–76

Chapter 6
Withania somnifera Extract/Withaferin A as a Prospective Anti-pigmenting Agent

Genji Imokawa

Abstract A screening test was used to assess more than 400 kinds of herbal extracts and related chemicals for their ability to interrupt the endothelin (EDN)1- and/or stem cell factor (SCF)-activated interacellular signaling that leads to the increased activity of tyrosinase in normal human melanocytes (NHMs). A *Withania somnifera* extract (WSE) and one of its active compounds withaferin A (WFA) showed distinct abrogating effects on the EDN1- and SCF- stimulation of tyrosinase activity without any direct inhibitory effect on tyrosinase, which suggests that the WSE and WFA interrupt signaling pathways upstream of tyrosinase expression. In addition, in a co-culture system using UVB-exposed normal human keratinocytes (NHKs) and NHMs, WFA has a potent ability to abrogate the UVB-stimulated tyrosinase activity in NHMs by abolishing the increased secretion of the intrinsic melanogenic cytokine EDN1 in UVB-exposed NHKs as well as by interrupting signaling pathways upstream of tyrosinase expression in NHMs. Consistently, signaling analysis with immunoblots revealed that in WSE-treated human melanoma cells or NHMs in culture, there is a marked deficiency in the EDN1-stimulated phosphorylation of Raf-1, MEK, ERK, CREB and MITF at 15 min after EDN1 treatment. The absence of an inhibitory effect on EDN1-induced intracellular calcium mobilization in NHMs suggests that the WSE inhibits EDN1-triggered PKC activity and/or abolishes its activation. In NHMs treated with SCF, the WSE also has a distinct potential to abrogate the stimulated phosphorylation of ERK, MITF and CREB, but not of Raf-1 and MEK, which indicates that the WSE attenuates the SCF-triggered phosphorylation of ERK by inhibiting MEK kinase activity. In WFA-treated NHMs, there is a marked deficiency in the SCF-stimulated series of phosphorylations of c-KIT, Shc, Raf-1, MEK, ERK, CREB and MITF. Taken together, the analysis of SCF binding and the effect of dithiothreitol (DTT) revealed that WFA attenuates the SCF-induced activation of c-KIT in NHMs by interrupting the auto-phosphorylation of c-KIT through DTT-suppressible Michael addition thioalkylation reactions without interrupting the binding of SCF to the c-KIT receptor. In human epidermal equivalents consisting of multilayered NHKs and NHMs, addition of the WSE or

G. Imokawa (✉)
Research Institute for Biological Functions, Chubu University,
1200 Matsumoto, Kasugai, Aichi 4878501, Japan
e-mail: imokawag@dream.ocn.ne.jp; imokawag@cc.utsunomiya-u.ac.jp

S.C. Kaul, R. Wadhwa (eds.), *Science of Ashwagandha: Preventive and Therapeutic Potentials*, DOI 10.1007/978-3-319-59192-6_6

WFA elicits a marked depigmenting effect on EDN1- or SCF-stimulated pigmentation accompanied by a significant decrease in eumelanin content. Real-time RT-PCR and Western blotting revealed that the stimulated expression of melanocyte-specific mRNAs and proteins, including tyrosinase and microphthalmia associated transcription factor (MITF), is significantly suppressed at days 7–10 of culture by the WSE and by WFA. The sum of these findings strongly suggests that the WSE and WFA can serve as potent melanogenic signaling-interruption type anti-pigmenting agents without any risk of hypopigmentation, thus they are prospective therapeutic candidates for treating hyperpigmentary disorders such as UVB-melanosis, solar lentigo and melasma where the over-expression of SCF and/or EDN1 is involved in the hyperpigmentation mechanism.

Keywords Withania *somnifera* • Extract • Withaferin A • Interrupt endothelin and stem cell factor- activated interacellular signaling • Anti-pigmenting agents

6.1 Introduction

Medicinal plants contain antioxidants consisting of polyphenols and flavonoids, and many other active ingredients (such as alkaloids). They have several biological properties such as depleting reactive oxygen species (ROS), modulating the cell cycle and alterating intracellular signaling. *Withania somnifera* (WS), commonly known as *Ashwagandha* or Indian winter cherry, is a medicinal plant that contains *Withaferin A* (WFA), a bioactive steroidal lactone. The known pharmacologic effects of the *Withania somnifera* extract (WSE) include modulation of immune function (Agarwal et al. 1999), cardioprotection from ischemia and reperfusion injury (Gupta et al. 2004), protection from 6-hydroxydopamine–induced Parkinsonism in rats (Ahmad et al. 2005), antibacterial effects (Owais et al. 2005) and anti-inflammatory effects (Rasool and Varalakshmi 2006). As for effects associated with modulating signal transduction, cell signaling, transcription, apoptosis and the cell cycle in a variety of cell types, the WSE has been reported to suppress the lipopolysaccharide-induced production of inflammatory cytokines, including tumor necrosis factor (TNF)-α, interleukin (IL)-1 and IL-12, in peripheral blood mononuclear cells and it inhibits nuclear factor-nkB activation (Singh et al. 2007; Ichikawa et al. 2006; Kaileh et al. 2007). Treatment with the WSE also significantly down-regulates the expression of proinflammatory cytokines (IL-6, IL-1β, IL-8, Hsp70 and STAT-2) in prostate cancer cells, while a reciprocal up-regulation is observed in the expression of p38 MAPK, PI3K, caspase 6, cyclin D and c-myc (Aalinkeel et al. 2010). Furthermore, treatment with the WSE significantly modulates the JAK-STAT pathway, which regulates both the apoptosis process as well as MAP kinase (MAPK) signaling (Aalinkeel et al. 2010). WFA has been reported to inhibit Notch-1 signaling and to down-regulate prosurvival pathways, such as Akt/NF-kappaB/Bcl-2, in 3 colon cancer cell lines (HCT-116, SW-480 and SW-620)

(Koduru et al. 2010). In addition, WFA down-regulates the expression of the rapamycin signaling targets, pS6K and p4E-BP1, and activates c-Jun-NH(2)-kinase-mediated apoptosis in colon cancer cells (Koduru et al. 2010). WFA has also been shown to be a radiosensitizer and a suppressor of mouse Ehrlich ascites carcinoma growth (Devi et al. 1992, 1995, 1996, 2000). More recent studies have shown that WFA suppresses the growth of human cancer cells by causing apoptosis (Yang et al. 2007; Srinivasan et al. 2007; Malik et al. 2007). In leukemic cells of lymphoid and myeloid origin, WFA induces apoptosis by activating the p38 MAPK signaling cascade, which further activates downstream signaling by phosphorylating ATF-2 and HSP27 (Mandal et al. 2008). Further, WFA has a potent capacity to inhibit inflammation through the inhibition of NO production and iNOS expression by blocking Akt and subsequently down-regulating NF-kappaB activity in RAW 264.7 cells (Oh et al. 2008). Withanolide sulfoxide from *Aswagandha* roots completely suppresses TNF-induced NF-kappaB activation followed by the inhibition of cyclooxygenase and tumor cell proliferation (Mulabagal et al. 2009). Recently, the bioactive properties of WFA have been reported to include an anti-tumor capacity by inhibiting proteasomal chymotrypsin-like activity (Yang et al. 2007) and apoptosis induction through the inhibition of protein kinase C (Sen et al. 2007). WFA also inhibits cell adhesion through the inhibition of ICAM-1 and VCAM-1 expression by blocking Akt and down-regulating NF-kappaB activity in human pulmonary epithelial cells (Oh and Kwon 2009). Despite abundant reports on the biological activities of the WSE and WFA derivatives, there have been no reports elucidating their effects on melanogenesis in melanocytic cells. The abundant evidence for the WSE and WFA derivatives to modulate intracellular signaling and apoptosis prompted us to examine their effects on the endothelin 1 (EDN1)- and stem cell factor (SCF)-stimulated melanogenesis and pigmentation in melanocytic cells including normal human melanocytes (NHMs) and human melanoma cells, and human epidermal equivalents (HEEs) consisting of multilayered normal human keratinocytes (NHKs) and NHMs.

6.2 Hyperpigmentation Mechanisms

Ultraviolet B (UVB) irradiation is a harmful stimulus for human skin and elicits various types of alterations and degenerations especially in epidermal cells, with one of the most frequently occurring phenomenon being the accentuation of skin pigmentation such as UVB-melanosis and solar lentigo. As depicted in Fig. 6.1, the biological mechanism underlying UVB-melanosis has been extensively characterized (Hachiya et al. 2001, 2004; Imokawa et al. 1992, 1995, 1996, 1997; Yada et al. 1991; Yang et al. 2007; Yokota et al. 2006). The mechanism involves a sequence of cellular events that predominantly occurs in the epidermis, where the production and secretion of several cytokines, such as EDN1, SCF and α-melanocyte stimulating hormone (αMSH), is accentuated in an autocrine fashion by the initial secretion of the primary inflammatory cytokine IL-1α in UVB-exposed NHKs.

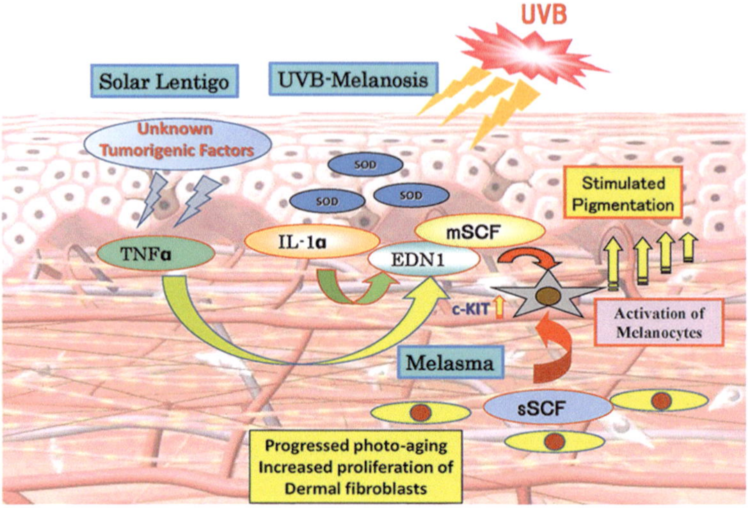

Fig. 6.1 Paracrine cytokine mechanisms underlying the hyperpigmentation in UVB-melanosis, solar lentigo and melasma. *TNF* tumor necrosis factor, *IL-1* interleukin-1, *EDN1* endothelin-1, *mSCF* membrane-bound stem cell factor, *sSCF* soluble stem cell factor (Imokawa and Ishida 2014)

Those cytokines trigger neighboring melanocytes to activate intracellular signaling cascades that are initiated by the binding of EDN1, SCF and/or αMSH to their specific receptors (Abdel-Malek et al. 1995; Imokawa et al. 1992, 1995, 1996, 1997, 2000; Suzuki et al. 1996). This activation of intracellular signaling pathways eventually leads to cAMP response element binding protein (CREB) activation (phosphorylation) at the terminal point of corresponding intracellular signaling, which results in an accentuated expression of the melanocyte master transcription factor microphthalmia-associated transcription factor (MITF). This then up-regulates the expression of genes encoding melanocyte-specific proteins, including the major melanin synthetic enzymes tyrosinase, tyrosinase-related protein 1 (TYRP1) and dopachrome tautomerase (DCT) as well as melanosomal structural proteins (PMEL17) and the corresponding receptors for EDN1 (EDNRB) and SCF (c-KIT). This results in increased melanization within melanosomes in melanocytes and subsequently an accumulation of melanized melanosome complexes in keratinocytes after they phagocytize them from the dendritic tips of melanocytes. On the other hand, solar lentigos, which are the most common type of pigmented spots in Asians and frequently occur in sun-exposed skin areas, have been implicated in a similar but distinct biological mechanism, as depicted in Fig. 6.1. In that mechanism, the secretion of TNFα by lesional keratinocytes is up-regulated independent of UV, which triggers keratinocytes to stimulate the production of EDN1 and SCF in an autocrine fashion (Hattori et al. 2004; Kadono et al. 2001; Hachiya et al. 2001, 2004). Those same cytokines cause melanocytes to activate several intracellular signaling pathways, which lead to the accentuated expression of melanocyte-specific proteins including tyrosinase, via the up-regulated levels of MITF in the same fashion as UVB-melanosis. In melasma, the secretion of soluble SCF (sSCF) by dermal fibroblasts (Imokawa et al. 1998) is upregulated in the lesional dermis, probably due at least in part to the photoaging process (Kang et al. 2006). This leads to the penetration of sSCF into the epidermis through the basement membrane and then to the activation of epidermal melanocytes via the SCF signaling cascade, all of which results in the stimulation of epidermal pigmentation.

Few anti-pigmenting agents have been designed and developed according to these known hyperpigmentation mechanisms and corresponding cytokine-triggered intracellular signaling cascades. In most cases, anti-pigmenting agents have been shown to inhibit tyrosinase activity in vitro (Noh et al. 2007; An et al. 2010; Cha et al. 2011; Chung et al. 2012; Emami et al. 2012; Vontzalidou et al. 2012), to stimulate the proteolytic degradation of tyrosinase via the proteasome system (Ando et al. 1999) and/or to disrupt the transfer of melanin granules to keratinocytes from melanocytes (Paine et al. 2001). Thus, those anti-pigmenting agents are mechanistically involved in the interruption of natural melanogenic mechanisms by which constitutive skin color is maintained at a normal and unstimulated level. These mechanistic approaches to search for appropriate therapeutic and anti-melanogenic agents face a serious problem as follows: Even in normal skin conditions, tyrosinase functions to synthesize melanin in melanosomes to a specifically regulated level and the melanin produced is transferred to keratinocytes, which maintains the constitutive level of skin color. Thus, because of the difficulty of topically applying agents limited to a hyperpigmented skin area, some potent anti-pigmenting agents capable of suppressing these natural non-stimulated pigmentation processes may lead to hypopigmentation, which hampers further research into potent anti-pigmenting agents without the risk of hypopigmentation. Such a skin problem occured in 2013 in Japan among approximately 20,000 consumers (corresponding to 2% of the total users) who used the tyrosinase inhibitor rhododendrol containing whitening cosmetics (Imokawa 2014). In this respect, the general notion that hyperpigmentation should be diminished at most to the normal level of constitutive pigmentation by anti-pigmenting agents would be ideal. From this viewpoint, suppressing the over-activation of melanogenesis, including these melanogenic intracellular signaling pathways, would be an appropriate approach to develop new anti-pigmenting agents with an effective potential at nM levels but without the risk of causing hypopigmentation because the intracellular signaling pathway does not function substantially in normal skin conditions and is activated only by environmental stimuli such as UVB to contribute to the hyperpigmentation.

6.3 Abrogating Effect of WFA on Stimulated Tyrosinase Activity in Co-culture

We first characterized the paracrine cytokine interactions between UVB-exposed NHKs and NHMs, which lead to the increased activity of the key melanin synthetic enzyme tyrosinase using co-cultures of NHMs and UVB-exposed NHKs. Co-cultures with cell culture inserts were performed as previously reported (Nakajima et al. 2012c; Niwano et al. 2015).

Thus, in the co-cultue system with cell culture inserts, we found that NHMs co-cultured with UVB-exposed NHKs had significantly increased tyrosinase activity as measured using 3-methyl-2-benzothiazolinone hydrazine (MBTH) as reported by Winder and Harris (1991) in a UVB dose-dependent manner (Niwano et al. 2015). Real-time RT-PCR analysis revealed that co-culture with UVB-exposed NHKs stimulated mRNA levels of MITF, tyrosinase and TYRP1 at 48 h post-irradiation in NHMs. In the co-culture medium, the secretion of IL-1α, TNFα, Il-6, IL-8, EDN1, prostaglandin (PG) E_2 and granulocyte macrophage colony stimulatory factor (GM-CSF), but not αMSH, was significantly increased during 24–72 h post-irradiation (Niwano et al. 2015). The addition of an EDN1 neutralizing antibody to the co-culture significantly abrogated the stimulation of tyrosinase activity elicited by UVB-exposed NHKs (Niwano et al. 2015). This indicates that among the several soluble cytokines secreted, EDN1 is the one responsible for stimulating tyrosinase activity in the co-culture, which is consistent with our previous reports concerning intrinsic soluble cytokines involved in UVB pigmentation mechanisms (Imokawa et al. 1992, 1995). The sum of our results indicates that paracrine factors, including EDN1, are secreted by UVB-exposed NHKs, which pass through the membrane of the cell insert well and trigger NHMs in cell culture plate to increase tyrosinase activity. Thus, this co-culture system with cell culture inserts serves as an appropriate new epidermal model for UVB-induced hyperpigmentation. We have determined using the co-culture system whether WFA can interrupt the paracrine interaction between UVB-exposed keratinocytes and melanocytes to abolish the increased tyrosinase activity. The results revealed that while UVB-exposure of NHKs stimulated tyrosinase activity in NHMs, treatment with WFA (at 12.5, 25 and 50 nM) at 2 h pre-irradiation abrogated the stimulatory effect at 96 h post-irradiation (Fig. 6.2).

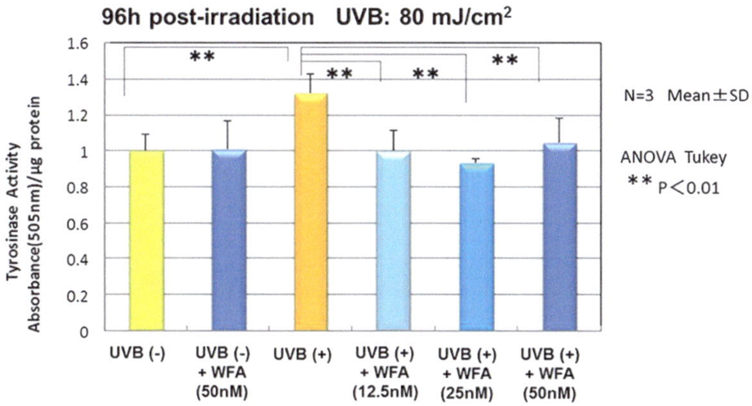

Fig. 6.2 Abrogation of the stimulated tyrosinase activity by WFA in co-cultures with UVB-exposed NHKs (Niwano et al. 2015)

6.4 Abrogating Effects of WFA on the Increased Secretion of Cytokines in UVB-Exposed NHKs

To elucidate the mechanism(s) involved in the abrogating effect of WFA on the increased activity of tyrosinase in NHMs in the co-culture system, we used ELISA kits to measure the levels of IL-6, IL-8 and EDN1 in UVB-exposed NHKs after 2 h treatment with WFA at 50 nM (Niwano et al. 2015). The ELISA assays reveal that pre-treatment with WFA significantly attenuates the increased secretion of IL-6 and IL-8 at 24, 36, 48 and 72 h post-irradiation and of EDN1 at 24, 36 and 48 h post-irradiation (Fig. 6.3) (Niwano et al. 2015). The facts reported above show that EDN1 is a major factor that stimulates tyrosinase expression by NHMs in the co-culture with UVB-exposed NHKs and that WFA has the potential to suppress the secretion of EDN1 by UVB-exposed NHKs. Thus, the finding that WFA abrogates the increased tyrosinase activity suggests that it may serve as an inhibitor of EDN1 secretion and/or the EDN1-activated intracellular signaling pathway, either of which would prevent the stimulated tyrosinase activity in EDN1-activated NHMs.

6.5 Abrogating Effects on the Stimulated Tyrosinase Activity in EDN1- or SCF-Treated NHMs

To determine which herbal extracts have suppressive effects on EDN1- or SCF-activated intracellular signaling which leads to the increased activity of tyrosinase in NHMs, we screened more than 400 kinds of herbal extracts and found that the WSE has a distinct abrogating effect on the stimulated tyrosinase activity without any direct inhibitory effect on tyrosinase. Thus, NHMs were cultured for 72 h after

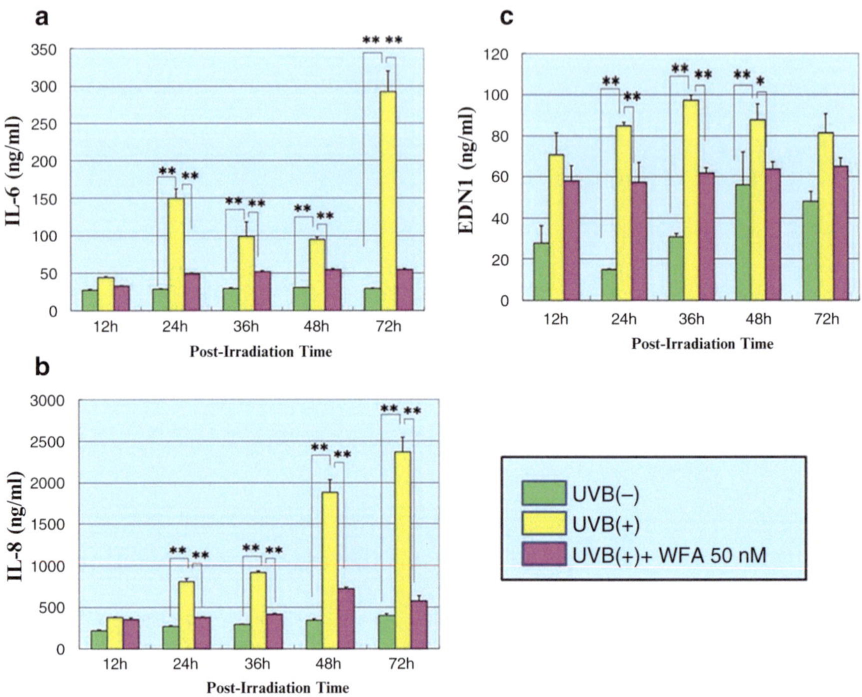

Fig. 6.3 Abrogating effects of WFA on the increased secretion of cytokines in UVB-exposed NHKs (Niwano et al. 2015)

EDN1 or SCF stimulation together with 3 h pre-incubation with the WSE or with WFA at concentrations of 10 μg/ml or 50 nM, respectively, after which cell viability was evaluated by cellular morphology and by the MTT assay. The cell lysates were measured for tyrosinase activity in the WSE- or WFA-treated NHMs (Nakajima et al. 2011, 2012b; Terazawa et al. 2015; Niwano et al. 2015). In separate experiments, lysates of NHMs cultured in the absence of the WSE for 72 h after EDN1 or SCF stimulation were directly incubated with the WSE or with WFA at a concentration of 10 μg/ml or 50 nM, respectively, after which tyrosinase activity was measured according to the method of Winder and Harris (1991). The results showed that there were no cytotoxic changes caused by the WSE in NHM morphology and the cells are 110.2% viable (Nakajima et al. 2011). Tyrosinase activity in cell lysates of NHMs cultured for 72 h was significantly suppressed by 3 h pre-incubation with the WSE or with WFA before EDN1 or SCF stimulation. In contrast, tyrosinase activity was not affected by the direct addition of the WSE or WFA (Nakajima et al. 2011b; 2012b), indicating that they have no melano-cytotoxic effect and that tyrosinase activity is not directly inhibited by the WSE or by WFA. Thus, it is likely that the WSE or WFA interrupts an upstream pathway that leads to the increased tyrosinase expression.

6.6 Melanogenic Intracellular Signaling Pathways Activated by SCF and EDN1

Fortunately, the melanogenic intracellular signaling pathways that are activated by SCF and by EDN1 have been elucidated in detail. In NHMs, as summarized schematically in Fig. 6.4, SCF binding to the c-KIT receptor elicits its dimerization and activation of its intrinsic tyrosine kinase activity, which results in its autophosphorylation (Blume-Jensen et al. 1991). The activated c-KIT receptor then phosphorylates various substrates and associates with a number of different signaling molecules, including phosphatidylinositol 3-kinase (PI 3-kinase), the Shc and Grb2 adaptor proteins and the guanine nucleotide exchange factor, SOS, the latter two of which are linked to activation of the Ras-MAPK pathway (Cutler et al. 1993; Imokawa et al. 2000; Lennartsson et al. 1999; Liu et al. 1994). Activated SOS converts Ras GDP to Ras GTP, which is required for the activation of Raf-1 kinase through its series of phosphorylations (Troppmair et al. 1992; Yang et al. 2007). Raf-1 then phosphorylates and activates MAPK kinase (MAPKK), also known as MEK1 (Imokawa et al. 2000), which initiates a protein kinase cascade that leads to the phosphorylation and activation of the extracellular signal-regulated MAPK isoforms ERK1 and ERK2 (Imokawa et al. 2000). Upon activation, MAPK is translocated from the cytoplasm into the nucleus, where it phosphorylates several

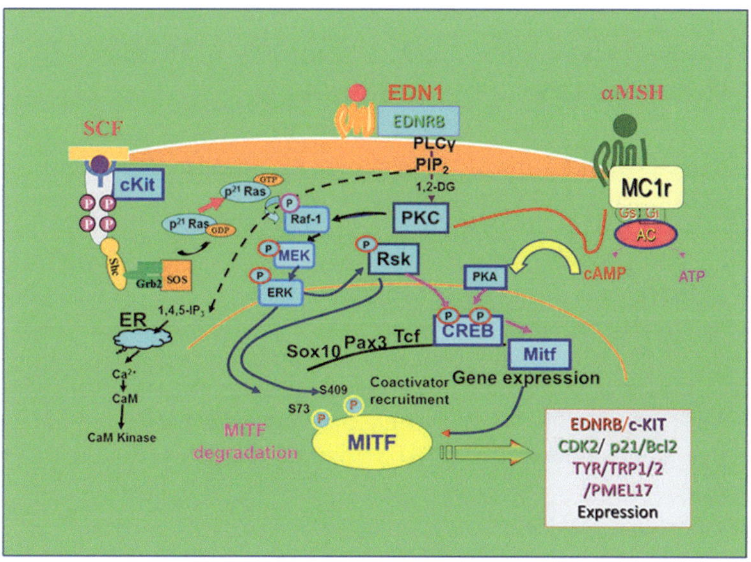

Fig. 6.4 Intracellular signaling mechanisms associated with EDN1 and SCF. *EDN1* endothelin-1, *EDNRB* endothelin B receptor, *MITF* microphthalmia associated transcription factor, *PKA* protein kinase A, *PKC* protein kinase C, *SCF* stem cell factor, *TYRP-1* tyrosinase-related protein-1, *DCT* dopachrome tautomerase, *TYK* tyrosine kinase, *αMSH* alpha melanocyte stimulating hormone (Imokawa and Ishida 2014)

transcription factors, including CREB and MITF (Sato-Jin et al. 2008), all of which culminates in stimulating the proliferation and/or melanization of melanocytes.

Furthermore, in NHMs, the intracellular signaling mechanisms by which ligands such as EDN1 activate melanogenic cellular events have been well characterized. After the binding of EDN1 to its receptor EDNRB, phospholipase Cγ is activated via G protein to stimulate the hydrolysis of polyphosphoinositide. That hydrolysis yields equal amounts of inositol triphosphate (IP3) and diacylglycerol, which results in intracellular Ca^{2+} mobilization and the translocation of PKC from the cytosol to the cell surface membrane, respectively (Yada et al. 1991; Imokawa et al. 1996). The translocated and then activated PKC directly phosphorylates RAF proto-oncogene serine/threonine-protein kinase (Raf-1) at many serine residues, or the Raf-1 inhibitory protein that activates Raf-1 via complex mechanisms (Kolch et al. 1993; Schonwasser et al. 1998; Marais et al. 1998; Mason et al. 1999). The Raf-1 activation then leads to the activation of the MAPK cascade, which includes the sequence of protein phosphorylations elicited by the kinase activities of Raf-1 on MEK, of MEK on ERK1/2, of ERK on ribosomal s6 kinase (RSK), of RSK on CREB and of ERK/RSK on MITF (Imokawa et al. 2000; Wan et al. 2011). By a cross-talk mechanism, the activation of PKC also increases levels of cyclic AMP to activate PKA leading to the phosphorylation of CREB, which is also mediated by the action of RSK following the activation of ERK (Imokawa et al. 2000; Wan et al. 2011).

6.7 Effect of the WSE on EDN1 Signaling in Human Melanoma Cells

We next asked whether the WSE interrupts the EDN1-induced phosphorylation of ERK, CREB and/or MITF in acral lentigo malignant (ALM) melanoma cells (Nakajima et al. 2011b). When the WSE was added at 10 μg/ml 3 h prior to EDN1 stimulation, the EDN1-stimulated phosphorylation of ERK, CREB and MITF was significantly suppressed at 15 min post-EDN1 incubation. To determine which signaling factor(s) through the MAPK pathway is affected by the WSE, we examined the effects of the WSE on the phosphorylation of MEK and Raf-1, which occur upstream of ERK at 15–30 min after EDN1 incubation. Western blotting analysis revealed that the WSE at 10 μg/ml significantly suppressed MEK and Raf-1 phosphorylation at 15–30 min after EDN1 incubation (Nakajima et al. 2011b). This result indicates that the WSE interrupts the EDN-signaling pathway upstream of Raf-1 activation, which results in the down-regulated phosphorylation of Raf-1 and MEK. To determine which signaling factor(s) upstream of Raf-1 is affected by the WSE, we next examined their effects on the calcium mobilization that occurs upstream of PKC at 0–300 sec after EDN1 incubation. Calcium mobilization analysis revealed that the WSE at 10 μg/ml did not suppress the EDN1-induced calcium mobilization starting at 15 sec after EDN1 incubation (Nakajima et al. 2011b).

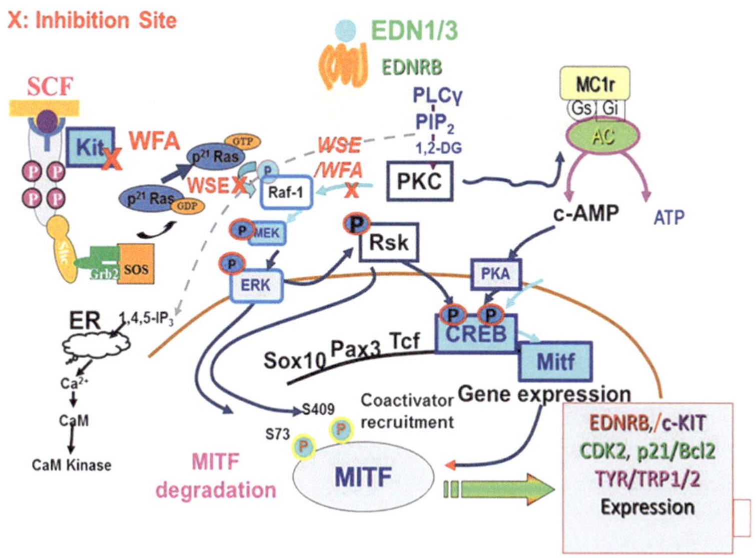

Fig. 6.5 Intracellular signaling pathways regulating the expression of melanocyte-specific proteins in NHMs and inhibition sites of the WSE and WFA (Nakajima et al. 2011b, 2012b; Imokawa and Ishida 2014; Terazawa et al. 2015)

This result indicates the possibility that, as depicted in Fig. 6.5, the WSE abrogates EDN1-induced PKC activation or directly inhibits PKC activity, resulting in the down-regulated phosphorylation of Raf-1 that occurs downstream of PKC.

6.8 Effect of the WSE on SCF Signaling in ALM Human Melanoma Cells or NHMs

We next asked whether the WSE interrupts the SCF-triggered intracellular signaling including the phosphorylation of ERK, CREB and/or MITF in ALM human melanoma cells or in NHMs (Nakajima et al. 2012b). When the WSE was added at 10 μg/ml 3 h prior to SCF treatment, the SCF-stimulated phosphorylation of ERK, CREB and MITF in ALM melanoma cells was significantly suppressed at 15 min post-SCF treatment. Similarly, the SCF-stimulated phosphorylation of ERK, CREB and MITF but not Raf-1 and MEK in NHMs was distinctly suppressed by treatment with the WSE, especially at 10 μg/ml, at 3 h prior to SCF addition in NHMs (Nakajima et al. 2012b). This result indicates the possibility that, as depicted in Fig. 6.5, the WSE abrogates SCF-induced MEK activation or directly inhibits MEK activity, resulting in the down-regulated phosphorylation of ERK which occurs downstream of MEK.

6.9 Effects of WFA on SCF Signaling in NHMs

We next asked whether WFA interrupts the SCF-triggered intracellular signaling pathway where the phosphorylation/activation of CREB occurs at the terminal step, being mainly responsible for the increased gene expression of MITF (Nakajima et al. 2012b). When WFA was added to NHMs at concentrations of 12.5, 25 and 50 nM prior to SCF stimulation, the SCF-stimulated phosphorylation of CREB was distinctly attenuated at 15 min post-SCF treatment by WFA at concentrations of 25 and 50 nM (Fig. 6.6a) (Terazawa et al. 2015). The abrogating effect of WFA at 50 nM on the stimulated phosphorylation of CREB was accompanied by a similar suppressive effect on the stimulated phosphorylation of MITF (Fig. 6.6b) and of ERK (Fig. 6.6c), which indicates that the down-regulated phosphorylation of ERK is attributable to the diminished phosphorylation levels of CREB and MITF. Western blotting analysis of the phosphorylation of SCF-triggered signaling molecules upstream of ERK demonstrates that WFA at 50 nM elicits an attenuating effect on the stimulated phosphorylation of MEK (Fig. 6.6d), Raf-1 (Ser338) (Fig. 6.6e), Shc (Fig. 6.6f) and c-KIT (Fig. 6.6g) (Terazawa et al. 2015). This indicates that the signaling step initially affected by WFA occurs at the auto-phosphorylation of c-KIT owing to SCF binding to c-KIT, followed by the suppressed phosphorylation of Shc, which leads to the abrogated series of phosphorylations of Raf-1/MEK/ERK, and results in the diminished phosphorylation of CREB and MITF.

As for the biochemical mechanism(s) by which WFA elicits the interruption of c-KIT auto-phosphorylation induced by its binding to SCF and subsequent dimerization, several reports have described the regulatory actions elicited by WFA or its derivatives on protein kinases or protein phosphatases involved in a variety of

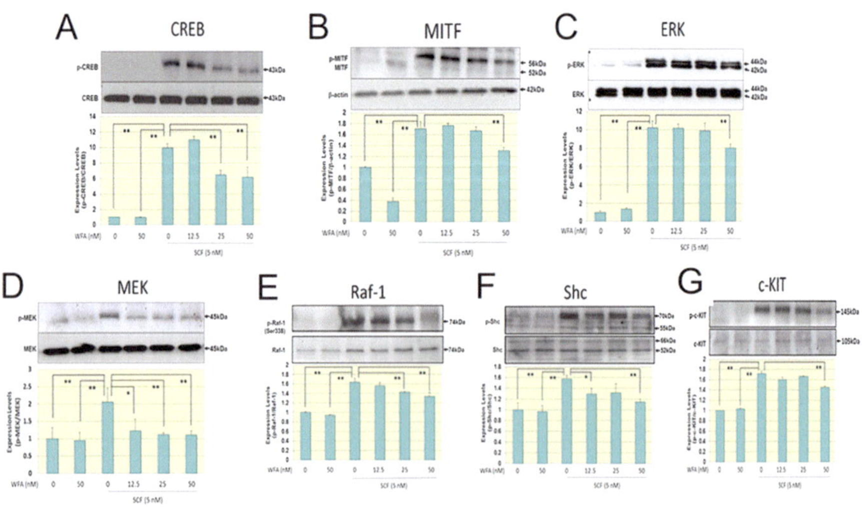

Fig. 6.6 Effects of WFA on SCF-stimulated phosphorylation ofCREB (**a**), MITF (**b**), ERK (**c**), MEK (**d**), Raf-1 (**e**), Shc (**f**) and c-KIT (**g**) (Terazawa et al. 2015)

intracellular signaling pathways. The major action by WFA is implicated in Michael addition thioalkylation reactions through either its epoxide or lactone ring (Rosazza et al. 1978; Yokota et al. 2006), which are involved with directly suppressing IKK kinase activity upon attack of Cys-179 in the kinase domain activation loop or of Cys-662/716, that affects IKK/ complex formation (Bernier et al. 2006; Hehner et al. 1999; Liang et al. 2003, 2006; Na and Surh 2006). Also, those dual specificity tyrosine phosphatases have been shown to be susceptible to thioalkylation due to a conserved "XHCXXGXSRS" motif in the catalytic domain, which can result in sustained MEK/ERK signaling (Ueda et al. 2002; Foley et al. 2004; Peyregne et al. 2005; Vasudevan et al. 2005; Rahmouni et al. 2006; Shin et al. 2006). WFA can also target various cysteine residues of multiple protein kinases/protein phosphatases (Adams et al. 2002; Prajapati et al. 2004; Kray et al. 2005; Humphries et al. 2005; Chi et al. 2006; Codreanu et al. 2006; Reynaert et al. 2006; Salojin et al. 2006; Seth and Rudolph 2006) that affect the phosphorylation of p38, MEK/ERK, JNK and IKK, ultimately resulting in the inhibition of IKK kinase activity upon attack of critical cysteine residues involved in kinase activation and/or complex formation.

Based on the above evidence, there are three mechanistic possibilities for the inhibition of c-KIT phosphorylation by WFA, as follows: (1) WFA may stimulate c-KIT conjugated protein tyrosine phosphatase via a thioalkylation-sensitive redox mechanism (Kaileh et al. 2007). (2) WFA may directly inhibit c-KIT auto-phosphorylation by inhibiting its tyrosine kinase activity via a similar redox mechanism, and/or (3) WFA may interrupt the binding of SCF to the c-KIT receptor. Since ERK hyperphosphorylation by treatment with WFA in concert with IKK kinase inhibition (Kaileh et al. 2007) was not observed in the present study using NHMs treated with WFA and since the known effects of WFA on tyrosine phosphatase would lead to the up-regulation of c-KIT phosphorylation, the first possibility can be ruled out.

As for the second possibility, WFA can manipulate intracellular signaling cascades by down- or up-regulating protein kinases or protein phosphatases, respectively, by Michael addition thioalkylation reactions using its epoxide or lactone ring (Rosazza et al. 1978; Yokota et al. 2006). DTT is known to interrupt Michael addition thioalkylation reactions by WFA to several protein kinases or protein phosphatases with an SH domain in their catalytic sites, which leads to the down-regulation of protein kinase activity or sustained phosphorylation of many signaling intermediates (Kaileh et al. 2007). Therefore, we determined whether pretreatment with DTT can abrogate the WFA-suppressed phosphorylation of c-KIT and the subsequent series of phosphorylations of signaling intermediates in SCF-treated NHMs. Western blotting using anti-phospho-antibodies to c-KIT or other signaling intermediates revealed that DTT distinctly abrogates the suppressive effect of WFA on the SCF-stimulated phosphorylation of c-KIT/ MITF/CREB at 5 min post-SCF treatment (Terazawa et al. 2015). Thus, treatment with DTT was found to significantly abrogate the attenuated phosphorylation of c-KIT by WFA.

To address the third possibility, we determined whether or not WFA interrupts the binding of SCF to the c-KIT receptor, resulting in the interruption of the c-KIT phosphorylation induced by SCF treatment. NHMs were incubated at 4 °C for 2 h with SCF at 5 nM, followed by removal of free unbound SCF by washing with fresh medium and raising the temperature to 37 °C to start the signaling reaction. c-KIT was distinctly phosphorylated by incubation for 5 min at 37 °C followed by phosphorylation of MITF and CREB to a similar extent by treatment with SCF only and with SCF + WFA (Terazawa et al. 2015), which suggests that WFA does not interrupt the binding of SCF to the c-KIT receptor. Taken together, these findings strongly suggest that WFA attenuates the SCF binding-induced c-KIT activation in NHMs by interrupting the auto-phosphorylation of c-KIT tyrosine kinase through Michael addition thioalkylation reactions.

Concerning the mode of action of WFA, the sum of our findings indicates that while other depigmenting agents generally inhibit tyrosinase, WFA functions in a distinct manner by interrupting the phosphorylation of c-KIT, the initial step of the SCF-triggered intracellular signaling cascade, which mainly consists of the MAPK pathway, that leads to down-regulation of the expression and function of the melanocyte master transcription factor MITF. The decrease in MITF function then suppresses the expression and function of its downstream targets, including most of the melanocyte-specific proteins, which in turn results in the attenuated synthesis of melanin.

6.10 Effect of WFA on EDN1 Signaling in NHMs

We next asked whether WFA interrupts the EDN1–induced phosphorylation of Raf-1, MEK, ERK, CREB and/or MITF in NHMs (Niwano et al. 2015). When WFA was added at 12.5, 25 and 50 nM 2 h prior to treatment with EDN1, the EDN1-stimulated phosphorylation of Raf-1 (at 25 and 50 nM), MEK (at 12.5, 25 and 50 nM), ERK (at 12.5, 25 and 50 nM), CREB (at 12.5, 25 and 50 nM) and MITF (at 25 and 50 nM) in NHMs was significantly suppressed at 15 min post-EDN1 treatment (Fig. 6.7). To elucidate the intracellular action site(s) affected by WFA, we next asked whether WFA interrupts the calcium mobilization induced by EDN1, which occurs upstream of PKC. Treatment with EDN1 markedly elicited calcium mobilization with a peak at 80 sec post-incubation in NHMs, which was not affected by 2 h pre-treatment with WFA (Niwano et al. 2015).

As mentioned above regarding the mode of action of WFA and DTT, we determined whether pretreatment with DTT can abrogate the WFA-suppressed phosphorylation of MITF and/or CREB in EDN1-treated NHMs because MITF phosphorylation is directly linked to ERK/RSK phosphorylation. Western blotting using antibodies to MITF and phosphorylated CREB revealed that DTT did not abrogate the suppressive effect of WFA on the EDN1-stimulated phosphorylation of MITF and CREB at 15 min post-EDN1 treatment (Niwano et al. 2015). Since WFA does not interrupt intracellular calcium mobilization in EDN1-treated NHMs, taken together, these findings indicate that, similar to the effect of the WSE on the EDN1-triggered signaling cascade, WFA inhibits the PKC activation process and/or its activity that occurs upstream of Raf-1 in EDN1-treated NHMs.

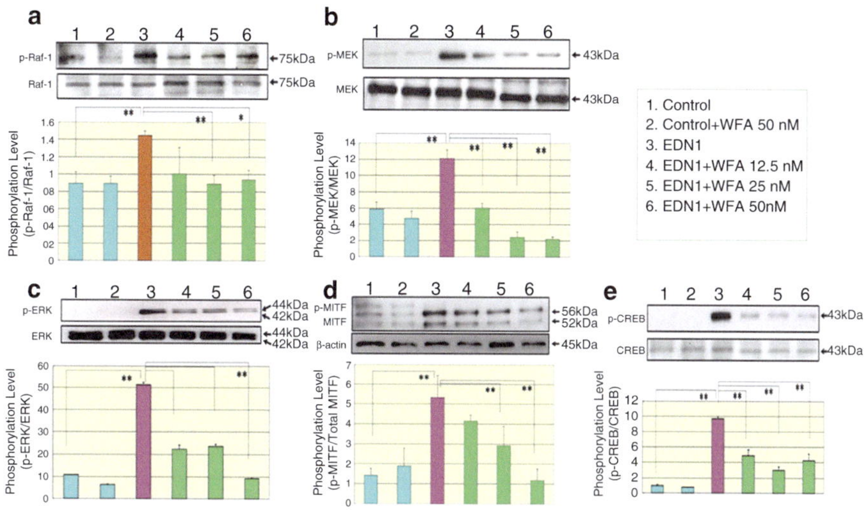

Fig. 6.7 Effects of WFA on the EDN1-stimulated phosphorylation of Raf-1 (**a**), MEK (**b**), ERK (**c**), MITF (**d**) and CREB (**e**) in NHMs (Niwano et al. 2015)

6.11 Human Epidermal Equivalents (HEEs)

There are now strict regulations that animal studies cannot be used to screen anti-pigmenting agents in the medical cosmetic field. Therefore, there is an urgent need for an alternative human skin model in which melanogenic paracrine interactions between UVB-exposed NHKs and NHMs can be precisely evaluated. For this purpose, HEEs consisting of multilayered NHKs and NHM at a cell number ratio of 10 to 1 have been developed to characterize the inhibitory effects of bioactives on visible pigmentation. This allows the levels of proteins and genes of melanocyte-specific proteins, such as tyrosinase, TYRP1, DCT, PMEL17, EDNRB and c-KIT, to be assessed in SCF- or EDN1-treated HEEs. However, the HEE model has limitations such as that UVB exposure does not necessarily increase the activity of tyrosinase and subsequently their pigmentation, probably due to the medium-mediated dilution of melanogenic cytokines secreted by UVB-exposed NHKs. In compensation for that defect, HEE models are cultured in the presence of several melanogenic cytokines, and therefore, they do not allow the evaluation of anti-melanogenic agents capable of inhibiting the secretion of melanogenic cytokines by UVB-exposed NHKs. To resolve that limitation, we described above a co-culture system using UVB-exposed NHKs and NHMs that serves as an appropriate new model for characterizing UVB-induced hyperpigmentation. Using that co-culture system, we showed the blocking effects of WFA on paracrine cytokine interactions between UVB-exposed NHKs and NHMs and characterized its mechanism of action. On the other hand, using the HEE model system, we next determined the abrogating effects of the WSE and of WFA on the SCF- and EDN1-stimulated pigmentation and the biological mechanisms involved.

6.12 Effect of the WSE on EDN1-Stimulated Pigmentation of HEEs

In HEEs, the addition of EDN1 at 10 nM gradually stimulates visible pigmentation over 14 days of culture. During the EDN1-induced stimulation of pigmentation, addition of the WSE at 5 or 10 μg/ml markedly reduced the increase in visible pigmentation over those 14 days with the most marked suppression attained at 10 μg/ml, the suppression level of which was slightly higher than the basal pigmentation level without EDN1 stimulation (Nakajima et al. 2011b). While there was no degeneration of the epidermal tissue visible at day 14 by hematoxylin and eosin (H&E) staining, melanin deposition detected by Fontana Masson (FM) staining throughout the epidermis (including the stratum corneum) was markedly reduced in the WSE-treated HEEs at day 14 compared to the untreated controls (Nakajima et al. 2011b). Chemical analysis of eumelanin (pyrrole-2,3,5-tricarboxylic acid: PTCA) content according to the method established by Ito and Fujita (1985) revealed that the WSE significantly decreased the eumelanin content compared with the untreated control (Nakajima et al. 2011b). Immunohistochemistry with anti-S-100, a protein normally present in cells derived from the neural crest such as Schwann cells and melanocytes, revealed that there was no reduction in the number of S-100 positive melanocytes in the WSE-treated HEEs at day 14 compared with the untreated controls and the non-EDN1 stimulated control (Nakajima et al. 2011b). The lack of a reduced number of S-100 positive melanocytes strongly suggests that there is no involvement of a melano-cytotoxic effect in the depigmenting effect elicited by the WSE.

6.13 Effect of the WSE on the Expression of Melanocyte-Specific Genes

To elucidate the biochemical mechanism(s) involved in the depigmenting effect of the WSE, we used real-time RT-PCR analysis to examine its effects on the expression of melanocyte-specific genes during the EDN1-stimulated pigmentation of HEEs. Since our previous time course study in the identical system with EDN1 stimulation (Nakajima et al. 2011a) demonstrated that the expression of genes encoding all the melanocyte-specific proteins tested (MITF, Tyrosinase, TYRP1, DCT, PMEL17, c-KIT and EDNRB) was gradually up-regulated over 14 days, with a peak at day 7, we used real-time RT-PCR analysis at day 7 to characterize the effects of the WSE. The results reveal that addition of the WSE at 10 μg/ml (the minimal concentration effective for the depigmentation in HEEs) significantly down-regulated the increased expression of genes encoding MITF, Tyrosinase, TYRP1, DCT, PMEL17, c-KIT and EDNRB in EDN1-treated HEEs compared with the untreated HEEs controls (Nakajima et al. 2011b). These findings are consistent with the interrupting effects on EDN1-triggered intracellular signaling, leading to MITF gene expression.

6.14 Effect of the WSE on the Expression of Melanocyte-Specific Proteins

To further elucidate the biochemical mechanism(s) involved in the depigmenting effect of the WSE, we next used Western blotting analysis to examine its effects on the expression of melanocyte-specific proteins at day 10 of culture during the EDN1-stimulated pigmentation of HEEs, at which time their protein expression reaches a peak in the identical system with EDN1 stimulation (Nakajima et al. 2011a). Western blotting analysis reveals that the addition of WSE at 10 μg/ml (the minimal concentration effective for the depigmentation in HEEs) significantly down-regulated the increased protein expression of all melanocyte-specific proteins examined, except for EDNRB and c-KIT, in EDN1-treated HEEs compared with the untreated control (Nakajima et al. 2011b).

6.15 Effect of the WSE on SCF-Stimulated Pigmentation in HEEs

In HEEs, the addition of SCF (5 nM) gradually stimulated visible pigmentation over 14 days of treatment. During the SCF-induced stimulation of pigmentation, addition of the WSE at 2, 5 or 10 μg/ml distinctly reduced the increase in visible pigmentation in a dose-dependent manner over those 14 days (Nakajima et al. 2012b). The most marked suppression attained at 10 μg/ml, the level of which was still slightly higher than the basal pigmentation levels without SCF stimulation. While there was no degeneration of the epidermal tissue visible at day 14 by HE staining, melanin deposition detected by FM staining throughout the epidermis (including the stratum corneum) was markedly reduced in the WSE-treated HEEs at day 14 compared to the untreated controls (Fig. 6.8a). Chemical analysis revealed that the WSE significantly reduced the eumelanin content compared with the untreated control (Fig. 6.8b). Immunohistochemistry with anti-S-100 revealed that there was no reduced number of S-100 positive melanocytes in the WSE-treated HEEs at day 14 compared with the untreated controls and the non-SCF stimulation control (Nakajima et al. 2012b). The lack of a reduced number of S-100 positive melanocytes strongly suggests that there is no involvement of a melano-cytotoxic effect in the anti-melanogenic effect elicited by the WSE.

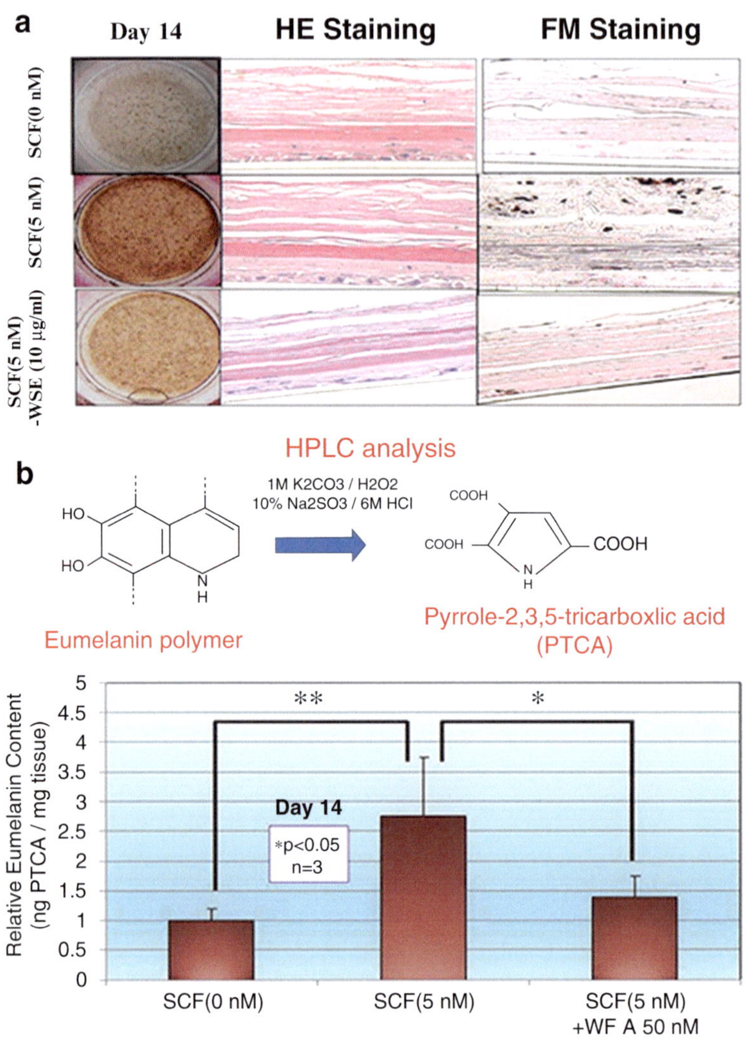

Fig. 6.8 Inhibitory effects of the WSE on the SCF-stimulated pigmentation of HEEs, a: HE and FM staining of HEEs at day 14, b: HPLC analysis for eumelanin quantitation (Nakajima et al. 2012b)

6.16 Effect of the WSE on the Expression of Melanocyte-Specific Genes

To elucidate the biochemical mechanism(s) involved in the anti-melanogenic effect of the WSE, we used real-time RT-PCR analysis to examine its effects on the expression of melanocyte-specific genes during the SCF-stimulated pigmentation of HEEs. A time course study demonstrated that the expression of genes encoding all melanocyte-specific proteins tested (MITF-M (Amae et al. 1998), Tyrosinase,

TYRP1 and PMEL17) was gradually upregulated over 14 days, with a peak at day 7 (Nakajima et al. 2012c). Real-time RT-PCR analysis of the effects of the WSE at day 7 revealed that the addition of the WSE at 10 µg/ml significantly down-regulated the increased expression of genes encoding MITF, Tyrosinase, TYRP1 and PMEL17 in SCF-treated HEEs compared with the untreated controls, the down-regulated levels of which are still slightly higher than the levels without SCF (Nakajima et al. 2012b). These findings are consistent with the interrupting effects on SCF-triggered intracellular signaling, leading to MITF gene expression.

6.17 Effect of the WSE on the Levels of Melanocyte-Specific Proteins

To elucidate the biochemical mechanisms involved in the anti-melanogenic effect of the WSE, we next used Western blotting analysis to examine its effects on the levels of melanocyte-specific proteins at day 10 at which time the protein expression peaked during the SCF-stimulated pigmentation of HEEs (Nakajima et al. 2012c). Western blotting analysis reveales that the addition of the WSE at 10 µg/ml significantly abolished the increased levels of all melanocyte-specific proteins examined compared with the untreated control at day 10, the down-regulated levels of which were still slightly higher than the levels without SCF (Nakajima et al. 2012b).

6.18 Effects of WFA on the SCF-Stimulated Pigmentation of HEEs

Because an extract of *Ashwagandha* at 10 µg/ml has already been shown to distinctly abrogate the SCF-stimulated pigmentation of HEEs (Nakajima et al. 2011b), we next examined the effects of one of its major active substances, WFA, on the same pigmentation system. The addition of WFA at 50 nM markedly abolished the increase in the visible pigmentation of HEEs elicited by SCF over 14 days, whereas WFA at 50 nM did not decrease the level of visible pigmentation in untreated HEEs (Terazawa et al. 2015). While there was no degeneration of the epidermal tissue visible at day 14 by HE staining, melanin deposition detected by FM staining throughout the epidermis (including the stratum corneum) was markedly reduced in the WFA-treated HEEs at day 14 compared to the untreated controls. To eliminate the possible influence of a chromophore other than melanin on visible pigmentation, we performed chemical analysis of eumelanin content by HPLC. Those results show that WFA at 50 nM significantly reduced the eumelanin content (expressed as relative eumelanin ratio) to the unstimulated level compared with the untreated control. We also assessed the viability of HEEs treated with or without WFA. HEEs were cultured for 14 days in the presence of WFA at 50 nM after which cell viability was evaluated for the last 24 h of culture using the MTT assay. The results showed that there was no significant reduction in cell viability of HEEs treated with 50 nM WFA compared with untreated controls (Terazawa et al. 2015).

6.19 Effects of WFA on the Expression of Melanocyte-Specific Genes

To elucidate the biochemical mechanism(s) involved in the anti-pigmenting effect of WFA, we used real-time RT-PCR analysis to examine the effects of WFA on the expression of melanocyte-specific genes at day 7 when those gene levels reached a peak during the SCF-stimulated pigmentation of HEEs (Nakajima et al. 2012c). Those analyses revealed that the addition of WFA at a concentration of 50 nM significantly down-regulated the increased expression of melanocyte-specific genes at day 7 for MITF, tyrosinase, TYRP1, DCT, PMEL17 and c-KIT in SCF-treated HEEs compared with the untreated controls (Terazawa et al. 2015). These findings are consistent with the interrupting effects on SCF-triggered intracellular signaling, leading to MITF gene expression.

6.20 Effects of WFA on the Expression of Melanocyte-Specific Proteins

We next performed Western blotting to examine the effects of WFA on the expression of melanocyte-specific proteins at day 10 when those protein levels reached a peak during the SCF-stimulated pigmentation of HEEs (Nakajima et al. 2012a, b). Those analyses revealed that the addition of WFA at 50 nM distinctly down-regulated the increased expression of melanocyte-specific proteins at day 10 for MITF, TYR, TYRP1, DCT, PMEL17 and c-KIT, compared with the untreated controls (Fig. 6.9) (Terazawa et al. 2015). Western blotting analysis revealed that the increased expression of melanocyte-specific proteins was significantly abrogated for tyrosinase (Fig. 6.9a), TYRP1 (Fig. 6.9b), DCT (Fig. 6.9c) and c-KIT (Fig. 6.9e) by WFA in a dose-dependent manner at concentrations of 12.5, 25 and 50 nM, and 12.5 and 50 nM, respectively, and for PMEL17 (Fig. 6.9d) at a concentration of 25 nM (Terawawa et al. 2015).

6.21 Conclusion

In this study, for the first time, we show that an extract of Withania somnifera (WSE) and one of its active compounds, WFA, significantly abrogates both the EDN1- and the SCF-stimulated pigmentation of HEEs at concentrations as low as 10 μg/ml and

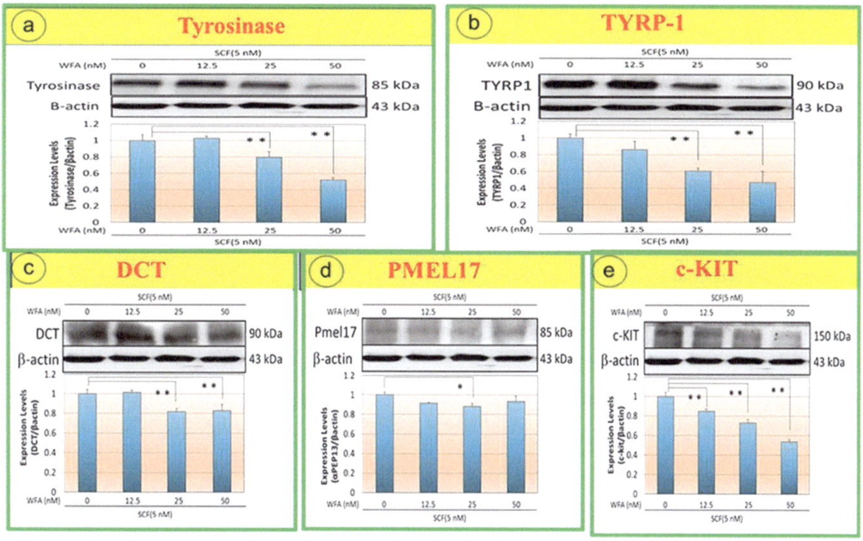

Fig. 6.9 Inhibitory effects of WFA on SCF-stimulated protein expression. a: Tyrosinase, b: TYRP-1, c: DCT, d: PMEL17, e: c-KIT (Terazawa et al. 2015)

50 nM, respectively, without affecting the normal non-stimulated levels of pigmentation. These depigmenting effects of the WSE and WFA are accompanied by the significantly down-regulated expression of major melanocyte-specific genes and proteins (MITF, tyrosinase, TYRP1, DCT and PMEL17) compared with their up-regulated expression in the EDN1 or SCF treated-HEEs. Unlike ordinary anti-pigmenting agents whose anti-melanogenic mechanisms are involved in the interruption of natural melanogenic mechanisms by which constitutive skin color is maintained at a normal and unstimulated level, those anti-melanogenic mechanisms are specifically characterized by their interrupting effect on the EDN1- or SCF-activated intracellular signaling pathways which results in the up-regulation of the melanocyte-specific master transcription factor, MITF, then leading to the up-regulated expression of all melanocyte-specific proteins. Those activation processes of the EDN1- or SCF-triggered intracellular signaling are mechanistically involved in the anti-pigmenting agent-targeted hyperpigmentary disorders such as UVB-melanosis (Hachiya et al. 2001, 2004), lentigo senilis (Kadono et al. 2001; Hattori et al. 2004) as well as melasma (Kang et al. 2006). These mechanistic characteristics provide a great advantage to avoid the risk of hypopigmentation because the intracellular signaling pathways involved are not substantially activated in normal skin conditions and are initiated to function only by environmental stimuli such as UVB to contribute to the hyperpigmentation. The sum of these results suggest that the WSE and WFA can serve as therapeutic tools for treating EDN1- and SCF-associated hyperpigmentary disorders, such as UVB-melanosis, lentigo senilis as well as melasma.

References

Aalinkeel R, Hu Z, Nair BB, Sykes DE, Reynolds JL, Mahajan SD, Schwartz SA (2010) Genomic analysis highlights the role of the JAK-STAT signaling in the anti-proliferative effects of dietary flavonoid-'Ashwagandha' in prostate cancer cells. Evid Based Complement Alternat Med 7:177–187

Abdel-Malek Z, Swope VB, Suzuki I, Akcali C, Harriger MD, Boyce ST, Urabe K, Hearing VJ (1995) Mitogenic and melanogenic stimulation of normal human melanocytes by melanotropic peptides. Proc Natl Acad Sci U S A 92:1789–1793

Adams JD Jr, Yang J, Mishra LC, Singh BB (2002) Effects of Ashwagandha in a rat model of stroke. Altern Ther Health Med 8:18–19

Agarwal R, Diwanay S, Patki P, Patwardhan B (1999) Studies on immunomodulatory activity of Withania somnifera (Ashwagandha) extract in experimental immune inflammation. J Ethnopharmacol 67:27–35

Ahmad M, Saleem S, Ahmad AS, Ansari MA, Yousuf S, Hoda MN, Islam F (2005) Neuroprotective effects of Withania somnifera on 6-hydroxydopamine induced Parkisonism in rats. Hum Exp Toxicol 24:137–147

Amae S, Fuse N, Yasumoto K, Sato S, Yajima I, Yamamoto H, Udono T, Durlu YK, Tamai M, Takahashi K, Shibahara S (1998) Identification of a novel isoform of microphthalmia-associated transcription factor that is enriched in retinal pigment epithelium. Biochem Biophys Res Commun 247:710–715

An SM, Koh JS, Boo YC (2010) p-Coumaric acid not only inhibits human tyrosinase activity in vitro but also melanogenesis in cells exposed to UVB. Phytother Res 24:1175–1180

Ando H, Funasaka Y, Oka M, Ohashi A, Furumura M, Matsunaga J, Matsunaga N, Hearing VJ, Ichihashi M (1999) Possible involvement of proteolytic degradation of tyrosinase in the regulatory effect of fatty acids on melanogenesis. J Lipid Res 40:1312–1316

Bernier M, Kwon YK, Pandey SK, Zhu TN, Zhao RJ, Maciuk A, He HJ, Decabo R, Kole S (2006) Binding of manumycin A inhibits IkappaB kinase beta activity. J Biol Chem 281:2551–2561

Blume-Jensen P, Claesson-Welsh L, Siegbahn A, Zsebo KM, Westermark B, Heldin CH (1991) Activation of the human c-kit product by ligand-induced dimerization mediates circular actin reorganization and chemotaxis. EMBO J 10:4121–4128

Cha SH, Ko SC, Kim D, Jeon YJ (2011) Screening of marine algae for potential tyrosinase inhibitor: those inhibitors reduced tyrosinase activity and melanin synthesis in zebrafish. J Dermatol 38:354–363

Chi H, Barry SP, Roth RJ, Wu JJ, Jones EA, Bennett AM, Flavell RA (2006) Dynamic regulation of pro- and anti-inflammatory cytokines by MAPK phosphatase 1 (MKP-1) in innate immune responses. Proc Natl Acad Sci U S A 103:2274–2279

Chung KW, Park YJ, Choi YJ, Park MH, Ha YM, Uehara Y, Yoon JH, Chun P, Moon HR, Chung HY (2012) Evaluation of in vitro and in vivo anti-melanogenic activity of a newly synthesized strong tyrosinase inhibitor (E)-3-(2,4 dihydroxybenzylidene)pyrrolidine- 2,5-dione (3-DBP). Biochim Biophys Acta 1820:962–969

Codreanu SG, Adams DG, Dawson ES, Wadzinski BE, Liebler DC (2006) Inhibition of protein phosphatase 2A activity by selective electrophile alkylation damage. Biochemistry 45:10020–10029

Cutler RL, Liu L, Damen JE, Krystal G (1993) Multiple cytokines induce the tyrosine phosphorylation of Shc and its association with Grb2 in hemopoietic cells. J Biol Chem 268:21463–21465

Devi PU, Sharada AC, Solomon FE, Kamath MS (1992) In vivo growth inhibitory effect of Withania somnifera (Ashwagandha) on a transplantable mouse tumor, sarcoma 180. Indian J Exp Biol 30:169–172

Devi PU, Sharada AC, Solomon FE (1995) In vivo growth inhibitory and radiosensitizing effects of withaferin A on mouse Ehrlich ascites carcinoma. Cancer Lett 95:189–193

Devi PU, Akagi K, Ostapenko V, Tanaka Y, Sugahara T (1996) Withaferin A: a new radiosentizer form the Indian medicinal plant Withania somnifera. Int J Radiat Biol 69:193–197

Devi PU, Kamath R, Rao BS (2000) Radiosensitization of a mouse melanoma by withaferin A: in vivo studies. Indian J Exp Biol 38:432–437

Emami S, Hosseinimehr SJ, Shahrbandi K, Enayati AA, Esmaeeli Z (2012) Synthesis and evaluation of 2(3H)-thiazole thiones as tyrosinase inhibitors. Arch Pharm 345:629–637

Foley TD, Armstrong JJ, Kupchak BR (2004) Identification and H_2O_2 sensitivity of the major constitutive MAPK phosphatase from rat brain. Biochem Biophys Res Commun 315:568–574

Gupta SK, Mohanty I, Talwar KK, Dinda A, Joshi S, Bansal P, Saxena A, Arya DS (2004) Cardioprotection from ischemia and reperfusion injury by Withania somnifera: a hemodynamic, biochemical and histopathological assessment. Mol Cell Biochem 260:39–47

Hachiya A, Kobayashi A, Ohuchi A, Takema Y, Imokawa G (2001) The paracrine role of stem cell factor/c-kit signaling in the activation of human melanocytes in ultraviolet-B-induced pigmentation. J Invest Dermatol 116:578–586

Hachiya A, Kobayashi A, Yoshida Y, Kitahara T, Takema Y, Imokawa G (2004) Biphasic expression of two paracrine melanogenic cytokines, stem cell factor and endothelin-1, in ultraviolet B-induced human melanogenesis. Am J Pathol 165:2099–2109

Hattori H, Kawashima M, Ichikawa Y, Imokawa G (2004) The epidermal stem cell factor is overexpressed in lentigo senilis: implication for the mechanism of hyperpigmentation. J Invest Dermatol 122:1256–1265

Hehner SP, Hofmann TG, Dröge W, Schmitz ML (1999) The antiinflammatory sesquiterpene lactone parthenolide inhibits NF-kappa B by targeting the I kappa B kinase complex. J Immunol 163:5617–5623

Humphries KM, Deal MS, Taylor SS (2005) Enhanced dephosphorylation of cAMP-dependent protein kinase by oxidation and thiol modification. J Biol Chem 280:2750–2758

Ichikawa H, Takada Y, Shishodia S, Jayaprakasam B, Nair MG, Aggarwal BB (2006) Withanolides potentiate apoptosis, inhibit invasion, and abolish osteoclastogenesis through suppression of nuclear factor-nB (NF-nB) activation and NF-nB-regulated gene expression. Mol Cancer Ther 5:1434–1445

Imokawa G (2014) Mechanistic difference and similarity compared to occupational vitiligo and vitiligo vulgaris. Fragr J 9:41–54

Imokawa G, Ishida K (2014) Inhibitors of intracellular signaling pathways that lead to stimulated epidermal pigmentation: perspective of anti-pigmenting agents. Int J Mol Sci 15:8293–8315

Imokawa G, Yada Y, Miyagishi M (1992) Endothelins secreted from human keratinocytes are intrinsic mitogens for human melanocytes. J Biol Chem 267:24675–24680

Imokawa G, Miyagishi M, Yada Y (1995) Endothelin-1 as a new melanogen: coordinated expression of its gene and the tyrosinase gene in UVB-exposed human epidermis. J Invest Dermatol 105:32–37

Imokawa G, Yada Y, Kimura M (1996) Signalling mechanisms of endothelin-induced mitogenesis and melanogenesis in human melanocytes. Biochem J 314:305–312

Imokawa G, Kobayashi T, Miyagishi M, Higashi K, Yada Y (1997) The role of endothelin-1 in epidermal hyperpigmentation and signaling mechanisms of mitogenesis and melanogenesis. Pigment Cell Res 10:218–228

Imokawa G, Yada Y, Morisaki N, Kimura M (1998) Biological characterization of human fibroblast-derived mitogenic factors for human melanocytes. Biochem J 330:1235–1239

Imokawa G, Kobayasi T, Miyagishi M (2000) Intracellular signaling mechanisms leading to synergistic effects of endothelin-1 and stem cell factor on proliferation of cultured human melanocytes cross-talk via trans-activation of the tyrosine kinase c-kit receptor. J Biol Chem 275:33321–33328

Ito S, Fujita K (1985) Microanalysis of eumelanin and pheomelanin in hair and melanomas by chemical degradation and liquid chromatography. Anal Biochem 144:527–536

Kadono S, Manak I, Kawashima M, Kobayashi T, Imokawa G (2001) The role of the epidermal endothelin cascade in the hyperpigmentation mechanism of lentigo senilis. J Invest Dermatol 116:571–577

Kaileh M, Vanden Berghe W, Heyerick A, Horion J, Piette J, Libert C, De Keukeleire D, Malik F, Kumar A, Bhushan S, Khan S, Bhatia A, Suri KA, Qazi GN, Singh J (2007) Withaferin a

strongly elicits IkappaB kinase beta hyperphosphorylation concomitant with potent inhibition of its kinase activity. J Biol Chem 82:4253–4264
Kang HY, Hwang JS, Lee JY, Ahn JH, Kim JY, Lee ES, Kang WH (2006) The dermal stem cell factor and c-kit are overexpressed in melasma. Br J Dermatol 154:1094–1099
Koduru S, Kumar R, Srinivasan S, Evers MB, Damodaran C (2010) Notch-1 inhibition by Withaferin-A: a therapeutic target against colon carcinogenesis. Mol Cancer Ther 9:202–210
Kolch W, Heidecker G, Kochs G, Hummel R, Vahidi H, Mischak H, Finkenzeller G, Marme D, Rapp UR (1993) Protein kinase Cα activates RAF-1 by direct phosphorylation. Nature 364:249–252
Kray AE, Carter RS, Pennington KN, Gomez RJ, Sanders LE, Llanes JM, Llanes JM, Khan WN, Ballard DW, Wadzinski BE (2005) Positive regulation of IkappaB kinase signaling by protein serine/threonine phosphatase 2A. J Biol Chem 280:35974–35982
Lennartsson J, Blume-Jensen P, Hermanson M, Pontén E, Carlberg M, Rönnstrand L (1999) Phosphorylation of Shc by Src family kinases is necessary for stem cell factor receptor/c-kit mediated activation of the Ras/MAP kinase pathway and c-fos induction. Oncogene 18:5546–5553
Liang MC, Bardhan S, Li C, Pace EA, Porco JA Jr, Gilmore TD (2003) Jesterone dimer, a synthetic derivative of the fungal metabolite jesterone, blocks activation of transcription factor nuclear factor kappaB by inhibiting the inhibitor of kappaB kinase. Mol Pharmacol 64:123–131
Liang MC, Bardhan S, Pace EA, Rosman D, Beutler JA, Porco JA Jr, Gilmore TD (2006) Inhibition of transcription factor NF-kappaB signaling proteins IKKbeta and p65 through specific cysteine residues by epoxyquinone A monomer: correlation with its anti-cancer cell growth activity. Biochem Pharmacol 71:634–645
Liu L, Damen JE, Cutler RL, Krystal G (1994) Multiple cytokines stimulate the binding of a common 145-kilodalton protein to Shc at the Grb2 recognition site of Shc. Mol Cell Biol 14:6926–6935
Malik F, Kumar A, Bhushan S, Khan S, Bhatia A, Suri KA, Qazi GN, Singh J (2007) Reactive oxygen species generation and mitochondrial dysfunction in the apoptotic cell death of human myeloid leukemia HL-60 cells by a dietary compound withaferin A with concomitant protection by N-acetyl cysteine. Apoptosis 12:2115–2133
Mandal C, Dutta A, Mallick A, Chandra S, Misra L, Sangwan RS, Mandal C (2008) Withaferin A induces apoptosis by activating p38 mitogen-activated protein kinase signaling cascade in leukemic cells of lymphoid and myeloid origin through mitochondrial death cascade. Apoptosis 13:1450–1464
Marais R, Light Y, Mason C, Paterson H, Olson MF, Marshall CJ (1998) Requirement of Ras-GTP-Raf complexes for activation of Raf-1 by protein kinase C. Science 280:109–112
Mason CS, Springer CJ, Cooper RG, Superti-Furga G, Marshall CJ, Marais R (1999) Serine and tyrosine phosphorylations cooperate in Raf-1, but not B-Raf activation. EMBO J 18:2137–2148
Mulabagal V, Subbaraju GV, Rao CV, Sivaramakrishna C, Dewitt DL, Holmes D, Sung B, Aggarwal BB, Tsay HS, Nair MG (2009) Withanolide sulfoxide from Aswagandha roots inhibits nuclear transcription factor-kappa-B, cyclooxygenase and tumor cell proliferation. Phytother Res 23:987–992
Na HK, Surh YJ (2006) Transcriptional regulation via cysteine thiol modification: a novel molecular strategy for chemoprevention and cytoprotection. Mol Carcinog 45:368–380
Nakajima H, Wakabayashi Y, Wakamatsu K, Imokawa G (2011a) An extract of Melia toosendan attenuates endothelin-1-stimulated pigmentation in human epidermal equivalents through the interruption of PKC activity within melanocytes. Arch Dermatol Res 303:263–276
Nakajima H, Wakabayashi Y, Wakamatsu K, Imokawa G (2011b) An extract of Withania somnifera attenuates endothelin-1-stimulated pigmentation in human epidermal equivalents through the interruption of PKC activity within melanocytes. Phytother Res 25:1398–1411
Nakajima H, Ezaki Y, Nagai T, Yoshioka R, Imokawa G (2012a) Epithelial-mesenchymal interaction during UVB-induced up-regulation of neutral endopeptidase. Biochem J 443:297–305

Nakajima H, Fukazawa K, Wakabayashi Y, Wakamatsu K, Imokawa G (2012b) Withania somnifera extract attenuates stem cell factor-stimulated pigmentation in human epidermal equivalents through interruption of ERK phosphorylation within melanocytes. J Nat Med 66:435–446

Nakajima H, Fukazawa K, Wakabayashi Y, Wakamatsu K, Senda K, Imokawa G (2012c) Abrogating effect of a xanthophyll carotenoid astaxanthin on the stem cell factor-induced stimulation of human epidermal pigmentation. Arch Dermatol Res 304:803–816

Niwano T, Terazawa S, Nakajima H, Wakabayashi Y, Imokawa G (2015) Astaxanthin and withaferin A block paracrine cytokine interactions between UVB-exposed human keratinocytes and human melanocytes via the attenuation of endothelin-1 secretion and its downstream intracellular signaling. Cytokine 73:184–197

Noh JM, Kwak SY, Kim DH, Lee YS (2007) Kojic acid-tripeptide amide as a new tyrosinase inhibitor. Biopolymers 88:300–307

Oh JH, Kwon TK (2009) Withaferin A inhibits tumor necrosis factor-alpha-induced expression of cell adhesion molecules by inactivation of Akt and NF-kappaB in human pulmonary epithelial cells. Int Immunopharmacol 9:614–619

Oh JH, Lee TJ, Park JW, Kwon TK (2008) Withaferin A inhibits iNOS expression and nitric oxide production by Akt inactivation and down-regulating LPS-induced activity of NF-kappaB in RAW 264.7 cells. Eur J Pharmacol 599:11–17

Owais M, Sharad KS, Shehbaz A, Saleemuddin M (2005) Antibacterial efficacy of Withania somnifera (Ashwagandha) an indigenous medicinal plant against experimental murine salmonellosis. Phytomedicine 12:229–235

Paine C, Sharlow E, Liebel F, Eisinger M, Shapiro S, Seiberg M (2001) An alternative approach to depigmentation by soybean extracts via inhibition of the PAR-2 pathway. J Invest Dermatol 116:587–595

Peyregne VP, Kar S, Ham SW, Wang M, Wang Z, Carr BI (2005) Novel hydroxyl naphthoquinones with potent Cdc25 antagonizing and growth inhibitory properties. Mol Cancer Ther 4:595–602

Prajapati S, Verma U, Yamamoto Y, Kwak YT, Gaynor RB (2004) Protein phosphatase 2Cbeta association with the IkappaB kinase complex is involved in regulating NF-kappaB activity. J Biol Chem 279:1739–1746

Rahmouni S, Cerignoli F, Alonso A, Tsutji T, Henkens R, Zhu C, Louis-dit-Sully C, Moutschen M, Jiang W, Mustelin T (2006) Loss of the VHR dual-specific phosphatase causes cell-cycle arrest and senescence. Nat Cell Biol 8:524–531

Rasool M, Varalakshmi P (2006) Immunomodulatory role of Withania somnifera root powder on experimental induced inflammation: an in vivo and in vitro study. Vasc Pharmacol 44:406–410

Reynaert NL, van der Vliet A, Guala AS, McGovern T, Hristova M, Pantano C, Heintz NH, Heim J, Ho YS, Matthews DE, Wouters EF, Janssen-Heininger YM (2006) Dynamic redox control of NF-kappaB through glutaredoxin-regulated S-glutathionylation of inhibitory kappaB kinase beta. Proc Natl Acad Sci U S A 103:13086–13091

Rosazza JP, Nicholas AW, Gustafson ME (1978) Microbial transformations of natural antitumor agents. 7. 14-alpha-hydroxylation of withaferin-A by Cunninghamella elegans (NRRL 1393). Steroids 31:671–679

Salojin KV, Owusu IB, Millerchip KA, Potter M, Platt KA, Oravecz T (2006) Essential role of MAPK phosphatase-1 in the negative control of innate immune responses. J Immunol 176:1899–1907

Sato-Jin K, Nishimura EK, Akasaka E, Huber W, Nakano H, Miller A, Du J, Wu M, Hanada K, Sawamura D, Fisher DE, Imokawa G (2008) Epistatic connections between MITF and endothelin signaling in Waardenburg syndrome and other pigmentary disorders. FASEB J 22:1155–1168

Schonwasser DC, Marais RM, Marshall CJ, Parker PJ (1998) Activation of the mitogen-activated protein kinase/extracellular signal-regulated kinase pathway by conventional, novel, and atypical protein kinase C isotypes. Mol Cell Biol 18:790–798

Sen N, Banerjee B, Das BB, Ganguly A, Sen T, Pramanik S, Mukhopadhyay S, Majumder HK (2007) Apoptosis is induced in leishmanial cells by a novel protein kinase inhibitor withaferin A and is facilitated by apoptotic topoisomerase I-DNA complex. Cell Death Differ 14:358–367
Seth D, Rudolph J (2006) Redox regulation of MAP kinase phosphatase 3. Biochemistry 45:8476–8487
Shin HM, Lee YR, Chang YS, Lee JY, Kim BH, Min KR, Kim Y (2006) Suppression of interleukin-6 production in macrophages by furonaphthoquinone NFD-37. Int Immunopharmacol 6:916–923
Singh D, Aggarwal A, Maurya R, Naik S (2007) Withania somnifera inhibits NF-kB and AP-1 transcription factors in human peripheral blood and synovial fluid mononuclear cells. Phytother Res 21:905–913
Srinivasan S, Ranga RS, Burikhanov R, Han SS, Chendil D (2007) Par-4-dependent apoptosis by the dietary compound withaferin A in prostate cancer cells. Cancer Res 67:246–253
Suzuki I, Cone RD, Im S, Nordlund J, Abdel-Malek ZA (1996) Binding of melanotropic hormones to the melanocortin receptor MC1R on human melanocytes stimulates proliferation and melanogenesis. Endocrinology 137:1627–1633
Terazawa S, Nakajima H, Fukasawa K, Imokawa G (2015) Withaferin A abolishes the stem cell factor-stimulated pigmentation of human epidermal equivalents by interrupting the autophosphorylation of c-KIT in human melanocytes. Arch Dermatol Res 307:73–88
Troppmair J, Bruder JT, App H, Cai H, Liptak L, Szeberényi J (1992) Ras controls coupling of growth factor receptors and protein kinase C in the membrane to Raf-1 and B-Raf protein serine kinases in the cytosol. Oncogene 7:1867–1873
Ueda K, Usui T, Nakayama H, Ueki M, Takio K, Ubukata M, Osada H (2002) 4-Isoavenaciolide covalently binds and inhibits VHR, a dual-specificity phosphatase. FEBS Lett 525:48–52
Vasudevan SA, Skoko J, Wang K, Burlingame SM, Patel PN, Lazo JS, Nuchtern JG, Yang J (2005) MKP-8, a novel MAPK phosphatase that inhibits p38 kinase. Biochem Biophys Res Commun 330:511–518
Vontzalidou A, Zoidis G, Chaita E, Makropoulou M, Aligiannis N, Lambrinidis G, Mikros E, Skaltsounis AL (2012) Design, synthesis and molecular simulation studies of dihydrostilbene derivatives as potent tyrosinase inhibitors. Bioorg Med Chem Lett 22:5523–5526
Wan P, Hu Y, He L (2011) Regulation of melanocyte pivotal transcription factor MITF by some other transcription factors. Mol Cell Biochem 354:241–246
Winder AJ, Harris H (1991) New assays for the tyrosine hydroxylase and dopa oxidase activities of tyrosinase. Eur J Biochem 198:317–326
Yada Y, Higuchi K, Imokawa G (1991) Effects of endothelins on signal transduction and proliferation in human melanocytes. J Biol Chem 266:18352–18357
Yang H, Shi G, Dou QP (2007) The tumor proteasome is a primary target for the natural anticancer compound Withaferin A isolated from Indian winter cherry. Mol Pharmacol 71:426–437
Yokota Y, Bargagna-Mohan P, Ravindranath PP, Kim KB, Mohan R (2006) Development of withaferin A analogs as probes of angiogenesis. Bioorg Med Chem Lett 16:2603–2607

Part II
Ashwagandha for Cancer

Chapter 7
Ashwagandha Bioactives for Cancer Treatment: Experimental Evidence and Their Mechanism(s) of Action

Renu Wadhwa, Sukant Garg, Ashish Kaul, and Sunil C. Kaul

Abstract Cancer is a complex disease showing rapid increase in its incidence that is often related to (i) increasing old age societies- an outcome of improved standards of living, healthcare and hygiene, and (ii) harmful impact of environment- an outcome of excessive industrialization of lifestyle and use of chemicals. In contrast to normal cells that have strictly regulated division potential controlled by tumor suppressor proteins, cancer cells escape limits on proliferation and divide indefinitely resulting in formation of tumors in the body. These are genetically unstable and yield self-propagating signals leading to progressively aggressive stages that are extremely difficult to treat. Extensive cancer research in last 2–3 decades has generated sensitive and sophisticated technologies for early diagnosis as well as better treatment. However, it still remains a scary and dreadful disease due to high rate of mortality, high toxicity and side effects of chemotherapeutic drugs. Since several herbs have been shown to possess anticancer potential, a lot of interest has hence been generated to understand their mechanism of action for cancer prevention and treatment. In this chapter, we provide a basic understanding of cancer, the use of cell culture and modern molecular biology techniques to understand its mechanism(s) and anticancer activity in leaves of Ashwagandha. We found that both alcoholic and water extract possess anticancer bioactives that operate through different signaling pathways in cells providing molecular evidence to the multi-module action and basis of the long-trusted fact that extracts work better than the individual and purified herbal components.

Keywords Cancer • Progression • Ashwagandha • Extracts • Therapy • Mechanism

---

R. Wadhwa • S.C. Kaul (✉)
DBT-AIST International Laboratory for Advanced Biomedicine (DAILAB), Biomedical Research Institute, National Institute of Advanced Industrial Science & Technology (AIST), Central 5-41, 1-1-1 Higashi, Tsukuba, Ibaraki 305-8565, Japan
e-mail: s-kaul@aist.go.jp

S. Garg • A. Kaul
School of Integrative & Global Majors, University of Tsukuba, 1-1-1 Tennodai, Tsukuba, Ibaraki 305-8577, Japan

S.C. Kaul, R. Wadhwa (eds.), *Science of Ashwagandha: Preventive and Therapeutic Potentials*, DOI 10.1007/978-3-319-59192-6_7

7.1 Introduction

Cancer is a disease of cell proliferation and most commonly defined as the process in which the normal cells exit from their proliferation control checkpoints and set to the path of infinite division. The process, called carcinogenesis, oncogenesis, or tumorigenesis, is a continuum of discrete molecular changes over time resulting in aberrant growth of cells and tissues. Although extremely complex, the process is often divided into various stages from initiation to progression to malignant transformation, and involves a series of genetic and epigenetic changes leading to altered cell physiology and disruption balancing between cell proliferation and death. In spite of their diverse etiology including genetics (mutation, DNA amplification and chromosomal rearrangement) and environmental factors (radiations, chemical and viral toxins), all cancers encompass rapidly dividing cancer cells that often loose their differentiation and specific functional characteristics, and become highly unstable genetically. Cancer cells exploit their normal neighbor cells for space and nutrients. According to the World Health Organization, global cancer burden is likely to rise from 14 million annual new cases in 2012 to beyond 22 million by 2030, due to (i) rapidly increasing aging populations world-wide, (ii) industrialization of life style that enrolls heavy use of chemicals, (iii) environmental pollution and radiations, (iv) dietary insults and carcinogen exposure leading to hormonal and neurodegenerative disorders and (v) infectious agents. Cancer can occur in any tissue or cell type in the body in the form of either benign or malignant tumors. **Benign tumors** are lumps that grow slowly, limit their spread to other tissues, and do not threaten life except when localized at complex organs such as brain or grow extremely large making uncomfortable or start pressing other organs. **Malignant tumors** are fast growing, have ability to invade and destroy neighboring tissue, and spread to other parts of the body through a process known as metastasis.

Based on the lineages, cancer can be classified into the following four categories:

- **Carcinoma** - It arises from the epithelial (the lining of cells that enclose organs) cells and often invades to the neighboring or far-off tissues by metastasizing to the lymph nodes. The most common forms of carcinomas are breast, prostate, lung and colon cancers.
- **Sarcoma** - It arises from bone or soft connective tissues including fat, muscle, blood vessels, nerves. The most common forms of sarcomas are leiomyosarcoma, liposarcoma and osteosarcoma.
- **Lymphoma** - Also called cancer of immune system, it arises from the lymphocytes, the cells that fight disease and infection. The two main forms of lymphoma include Hodgkin and non-Hodgkin lymphoma. Hodgkin's lymphoma constitutes of large abnormal lymphocytes. Non-Hodgkin lymphoma comprises of different types of white blood cells (B-cells, T-cells, NK cells). Whereas Hodgkin's lymphoma is often slow growing and curable, non-Hodgkin lymphoma are aggressive and difficult to treat.

- **Leukemia** - Leukemia, a cancer of the white blood cells and bone marrow, comprises of several subtypes of which the most common are lymphocytic leukemia and chronic lymphocytic leukemia.

Just like the multiple causes of cancer (genetics, environment, age, food and stress), its symptoms are diverse and depend on its type, lineage, niche and stage. However, escape from cellular senescence and immortalization (capacity to proliferate indefinitely) remains the common feature of most cancers.

7.2 Cellular Senescence

Senescence or irreversible growth arrest is the stage at which a cell stops dividing yet stays viable and functionally active. Normal cells have a limited lifespan (capacity to divide) that is determined by the following major in-build mechanisms –

- DNA end-replication problem of eukaryotic chromosomes- Specialized chromosome ends (telomeres) consisting of tightly held specialized and repetitive (10–15 kb) TTAGGG sequence and protein caps protect chromosomes from end to end fusion (Hayflick 1965; Harley 1990). With each division of normal cells, telomeres shorten (by 50–200 base pairs) due to the inherit incapability of DNA polymerase enzyme to initiate the replication process. Such telomere erosion triggers a normal cell into an irreversible and non-proliferative stage called replicative senescence through complex signaling cascades resulting in activation of several tumor suppressor proteins.
- Limited defense mechanisms and hence accumulation of metabolic, physiological and molecular damage leading to DNA damage and repair signaling cascades and activation of tumor suppressor proteins.

Senescence is an anticancer mechanism in normal cells, naturally designed to prevent them to replicate beyond their healthy lifespan (typically about 60–80 population doublings). Telomere shortening, DNA damage, increasing endogenous stress induced by reactive oxygen species, metabolic or molecular damage cause activation of tumor suppressor proteins and cellular senescence. At the same time, a threshold of these factors reverses the signaling resulting in inactivation of tumor suppressor mechanisms, activation of oncogenes leading to breakdown of controls on cell proliferation and escape of cells from senescence resulting in continued proliferation, immortalization. It is most often associated with activation of telomerase, a reverse transcriptase enzyme that prevents telomere erosion through semi-conservative DNA replication by adding TTAGGG repeat sequences to the chromosome ends (Blackburn et al. 2015; Hayflick 2000). Cancer cells divide uncontrollably, fail to achieve physiological differentiation, escape normal apoptotic pathways and become genetically unstable. They are resistant to extrinsic growth-check signaling pathways, and are capable of angiogenesis, invasion, and metastasis, and escape the immune and tumor suppressor regulatory mechanisms.

Here we discuss, in brief, only the main tumor suppressor mechanisms involved in regulation of cellular senescence and carcinogenesis.

7.2.1 p53 Tumor Suppressor as a Regulator of Cellular Senescence and Carcinogenesis

p53 is an established multifunctional regulator of cellular senescence and carcinogenesis. The two main activities of p53, DNA binding and transcriptional activation, have been shown to increase in cells that undergo either normal replicative senescence or stress-induced premature senescence in response to oncogenic or environmental stimuli (Hengartner 2000). p53 acts as a transcriptional activator of a variety of genes including $p21^{WAF1}$, Bax, GADD45, cyclin G, TGFα, IGF-BP3, and MDM2 (Cheung and Vousden 2010; Engelmann and Putzer 2014) and in turn gets regulated by HDM2 by proteasome mediated degradation. Several studies have shown that p53 regulates senescence by its downstream effector, cyclin dependent kinase inhibitor $p21^{WAF1}$ that causes cell cycle arrest at G1/S or G2/M stage of cell cycle. Whereas exogenous expression of $p21^{WAF1}$ induced senescence in early passage human diploid fibroblasts (Fang et al. 1999; McConnell et al. 1998; Vogt et al. 1998), its disruption caused lifespan extension (Brown et al. 1997). p53 is inactivated in more than 70% of human cancers by mechanisms involving mutations, inactivation either by DNA tumour virus oncoproteins or cellular partners/antagonists (Sharpless et al. 2002).

DNA damage and other stresses have been shown to cause site-specific phosphorylation of p53 that prevents its binding to antagonist HDM2, resulting in its stability and up regulation in stressed cells resulting in their growth arrest. Several other factors including proteins and miRNAs regulate p53. For example, ARF (Alternate Reading Frame) protein coded by INK4a locus on human chromosome 9p21 (also encodes $p16^{INK4A}$, an upstream regulator of pRb) has been shown to be a major upstream regulator of p53 and its control in replicative senescence, Stress Induced Premature Senescence (SIPS), Oncogenic Stress Induced Premature Senescence (OIPS) and immortalization of human cells (Kamijo et al. 1997; Quelle et al. 1995; Serrano et al. 1996). It is further regulated by CARF (collaborator of ARF) that activates ARF-induced activation of p53 (Cheung and Vousden 2010; Hasan et al. 2002, 2004, 2008) and regulates senescence and carcinogenesis by its dose dependent two-way regulation of DNA damage response. High level of CARF activated DNA damage response and p53 pathway causing growth arrest of cells. Super-high levels were shown to inactivate these pathways leading to malignant transformation (Cheung et al. 2014). Knockdown of CARF, on the other hand, caused apoptosis depicting that it is an essential protein for cell survival (Cheung et al. 2011, 2014). p53 is regulated by stress chaperone mortalin, enriched in cancer cells. Mortalin sequesters p53 in cell cytoplasm (Deocaris et al. 2013; Wadhwa et al. 2006) by binding to its c-terminus region in cancer cells (Kaul et al. 2001, 2005).

Small molecules and peptides that disrupted the binding of mortalin to p53 resulted in nuclear translocation and reactivation of p53 activities resulting in either growth arrest or apoptosis of cancer cells (Deocaris et al. 2007; Grover et al. 2012a; Kanai et al. 2007; Kaul et al. 2005; Lu et al. 2011a, b; Ma et al. 2006; Wadhwa et al. 2000, 2002).

7.2.2 pRb Tumor Suppressor as a Regulator of Cellular Senescence and Carcinogenesis

Retinoblastoma (Rb) protein is coded by RB1 tumor suppressor gene, and remains in an inactive state bound to E2F in the cytoplasm. It is the key regulator of cell cycle progression, apoptosis and a number of other biological processes (Korenjak and Brehm 2005). At the G_1-S checkpoint, upregulation of cyclin-Cdk kinases cause phosphorylation of pRB protein resulting in abrogation of its interaction with E2F proteins. The free E2F translocates to the nucleus and triggers DNA replication. Several studies have shown that pRB is under-phosphorylated in senescent cells, favoring pRB-E2F interactions and cell cycle arrest. Similar to p53, pRB is a target of DNA tumor virus transforming proteins and remains inactivated in a large majority of tumors. Introduction of pRB gene into p53/pRB deficient immortal tumor cells lead to induction of senescence (Xu et al. 1997). pRB phosphorylation is regulated by an inhibitor of cyclin D-dependent kinase ($p16^{INK4a}$) (Kamijo et al. 1997; Stott et al. 1998). It has been shown to maintain hypo-phosphorylated pRB in senescent human cells. Unlike $p21^{WAF1}$, $p16^{INK4A}$ remains high in late senescent cells. Introduction of exogenous $p16^{INK4A}$ into immortal and normal human cells resulted in their growth arrest, and induction of premature senescence, respectively (Lin et al. 1998; Serrano 1997). Consistent with the role of p53 and pRb in cellular senescence *in vitro*, mice with mutations in p53 and pRb or $p16^{INK4A}$ showed high incidence of tumors (Sharpless et al. 2002), suggesting the role of these proteins in organismal ageing. Of note, $p16^{INK4a}$ and p53 double knockout animals showed severely short lifespan (Sharpless et al. 2004). Stress-induced senescence in human cells is associated with increase in $p16^{INK4A}$ expression (Rayess et al. 2012). The two major tumor suppressor pathways, p53 and pRb, show independent as well as overlapping activities to trigger and maintain senescence through activities of $p21^{WAF1}$ and $p16^{INK4A}$. Since senescence is a permanent cell cycle arrest that limits tumor progression, it serves as a crucial marker for anticancer therapies. Activation of tumor suppressor mechanisms by small molecules/drugs/peptides and induction of senescence in cancer cells is a promising strategy for cancer intervention. Inhibition of telomerase activity or telomere maintenance can slow down the abnormal and uncontrolled proliferation of cancer cells.

7.3 Oncogenes and Cancer Pathogenesis

Proto-oncogenes are the cellular homologues of retroviral genes that activate into oncogenes following random and multiple mutations and stresses, and affect the cell biomechanics in a variety of ways. They promote tumor development by altering signal transduction, gene transcription, ribosomal translation, protein transportation, mitosis, or other biochemical reactions essential for physiological cell cycle. Majority of the mutated cells die owing to innate immunity and defective genomic targets. Few survive, escape and settle in nourished niches resulting in dynamic proliferation and vandalize normal cells for nourishment and space.

7.4 Apoptosis

Apoptosis, a programmed cell death, is a phenomenon in which a cell identifies its incapability to sustain growth or repair damage, and commits suicide. It is a genetically regulated inbuilt cancer-preventive mechanism in a cell. Through a cascade of pro-apoptotic signaling, the cell shrinks and degrades, and the phospholipid phosphatidylserine is exposed to attract circulating phagocytes like macrophages and dendritic cells, which then phagocytize them and secrete cytokines to inhibit inflammation in the surrounding tissue. Neighboring cells remain healthy and unaffected. Apoptosis is important to remove cells that pose threat to the integrity of a tissue.

In response to stress or damage (internal or extrinsic stimulus or pathway), the cell recruits serine/threonine kinase proteins (ATM, ATR) and phosphorylate checkpoint kinases (Chk1, Chk2) that in turn phosphorylate and activate p53 resulting in an up regulation of its downstream effector protein $p21^{WAF1}$ and inhibition of cell cycle progression by inhibiting cyclin-dependent kinases (cdc2, cdk2). As a result, the cell is forced to take up DNA repair (for minor damage), continued growth arrest (for medium damage), or undergo apoptosis (for major damage). The latter involves up regulation of pro-apoptotic BH3 family proteins (Bad, Bim, Bid, Puma, Noxa) and inhibition of anti-apoptotic proteins (Bcl-2, Bcl-xL). The apoptotic proteins control the formation of Mitochondrial Apoptosis-induced Channel-MAC (whereas pro-apoptotic proteins- BAX and BAK promote MAC, anti-apoptotic proteins, Bcl-2 or Bcl-xL prevent MAC formation) and release cytochrome c from mitochondria to cytosol, a key step in apoptotic cell death. Several pharmacological inhibitors of BAX have been shown to cause inhibition of apoptosis by depletion of MAC activity and prevention of cytochrome c release. When there is an external injury (external stimulus, extrinsic pathway), tumor necrosis factors (TNFs) and ligands (FAS of CD95, TRAIL of DR4 and DR5) cross-link and associate Fas-associated death domain (FADD) and procaspase-8 to the receptor death domain motifs. Active caspase-8 further catalyzes the process and rapidly produces large amount of the protein to lead the cell into the same fate following karyorrhexis. Apoptosis, like senescence, could be a great tool and target for anticancer therapies. Since it is a physiological mechanism and can also be promoted by external stimuli, pro-apoptotic drugs that promote apoptosis have fair potential in cancer intervention.

7.5 Cancer Treatment and Prevention

Cancer research establishes therapeutics that comply with the safety and ethical standards. Since cancer has multitudinous ways to initiate, express and operate, its treatment cannot be generalized into one or few targeted approaches. Enormous research has been done to understand and treat cancer. The fundamental established points include that cancer cells (i) escape from cellular senescence, (ii) grow exponentially and are genetically unstable, (iii) may possess multitude of molecular changes in tumor suppressor and oncogenic pathways, and (iv) show a physiologically-stressed state in which cells compete for oxygen and nutrients. These edges direct to the viewpoint, i.e. prevention of cancer is worthwhile and may be less cumbersome than the treatment. Early detection and prevention have hence been prioritized in modern cancer research.

7.6 Cancer Prevention and Treatment by Herbs

Search and development of preventive strategies against cancer began as early as 1845 when cisplatin was discovered as a soluble platinum complex generated by the electrolysis of platinum electrodes. It inhibited binary fission in *Escherichia coli* bacteria (Rosenberg et al. 1967). Years later, cisplatin was tested in rat sarcoma and extended to various other tumor cell lines to develop its medicinal applications and approvals for human use. Doxorubicin was also a result of an accidental discovery (Tan et al. 1967). A compound called Daunorubicin was found in a soil sample that showed potent anti-tumor properties in rats. However, it posed fatal cardio-toxicity. With minor changes in its chemical structure, newly termed as Adriamycin or doxorubicin, it lost its cardio-toxicity but retained its chemotherapeutic index. More than 2000 analogs of this compound are known.

Since the ancient era, plants have increasingly gained attention and importance in the prevention of cancer. Many drugs have been developed exclusively from plant extracts. However, cancer prevention remains controversial and yet to be explored. Kaefer and Milner first described scientific surprises in a number of herbs, and their probable role in cancer prevention (Kaefer and Milner 2008, 2011). They demonstrated a profound link in the establishment of cancer prevention and curative therapies using herbs like allspice, basil, clove, and garlic, etc. Active ingredients from a number of herbs have been identified in recent years that interfere with the process of carcinogenesis. Several studies have supported that autonomous or collaborative therapy with herbs could contribute to reduce the risk of developing cancer through various mechanisms. Although the pharmacodynamics and pharmacokinetics of herbal medicines have not been clearly demonstrated, being natural they are considered to be easily tolerated with lesser adverse effects or chances of chemo-resistance than the synthetic molecules. Furthermore, ingredients for natural medicines are more readily available and the manufacturing costs are relatively lower than the synthetic molecules. Therefore, switching to herbs, rather than completely on synthetic medicines, shall also be cost effective.

7.7 Ashwagandha Anticancer Bioactives and Their Mechanism of Action

Ashwagandha (*Withania somnifera)*, commonly known as winter cherry or Indian ginseng, is a shrub belonging to Solanaceae family. It is famously known in an Ayurveda health supplements and tonics that possesses a large variety of activities. In traditional Indian home medicine, its tuberous roots have been used to prepare medicinal formulations, while the flower, leaf and fruit extracts have been used for external applications on tumors, ulcers, inflamed skin rash or wounds. Ashwagandha has been professed for its anti-oxidant, analgesic, anti-inflammatory, anti-anxiety, anti-stress, anti-aging, and fertility related properties (Deocaris et al. 2008). Its adverse effects, only if consumed in heavy doses, are gastro-intestinal and gestational complications. Scientific basis of its health promoting activities have not been completely understood as yet. In this chapter, we discuss the experimental evidences and molecular mechanism(s) of action of anticancer activity in Ashwagandha extracts and their constituents.

Ashwagandha is rich in Withanolides, a group of phytochemicals that consist of steroid backbone bound to lactone and/or its derivatives. The main phytochemicals extracted from Ashwagandha roots/leaves/fruits are alkaloids, anaferine, anahygrine, beta sisterol, chlorogenic acid, cysteine, cuscohygrine, iron, pseudotropine, scopoletin, somniferiene, tropanol, withanine, steroidal lactones, and withanolides (Dar et al. 2016; Dhar et al. 2006). They are produced as secondary metabolites by oxidation of steroids. Several studies have reported anticancer activity in the root and leaf extracts of Ashwagandha (Vyas and Singh 2014; Wadhwa et al. 2016). The active anticancer Withanolides include Witahferin-A (Wi-A) and Withanone (Wi-N). Wi-A is the most extensively studied among all the Withanolides. Several studies have shown that anti-cancer activity of Wi-A is mediated by (i) inactivation of NFκB, a multifunctional transcriptional factor that regulates cytokine production and stress response (Gambhir et al. 2015), (ii) collapse of intermediate filament protein Vimentin, an important regulator of cell shape and migration (Ivaska et al. 2007; Thaiparambil et al. 2011), (iii) Oxidative stress in endoplasmic reticulum and mitochondria and c-FOS-FLIP-TRAIL induced apoptosis (Malik et al. 2007; Mayola et al. 2011; Nishikawa et al. 2015), (iv) Inhibition of proteasome mediated degradation leading to upregulation of p21, BAX, IκBα and apoptotic cell death (Yang et al. 2012), (v) Inhibition of STAT3 phosphorylation (an anti-apoptotic, pro-migratory factor in various types of cancers) (Lee et al. 2010; Spitzner et al. 2014; Yu et al. 2014), (vi) Inhibition of Hedgehog signaling pathway, an important contributor to proliferation, differentiation, and maintenance of tissue patterning (Yoneyama et al. 2015), (vii) Inhibition of c-MET, AKT and RAF-1 signaling (Grogan et al. 2013; Samadi et al. 2012; Suman et al. 2016), (viii) inhibition of snail and e-Cadherin mediating EMT signaling (Suman et al. 2016) (ix) downregulation of BRCA1 and HSF1 proteins (Zhang et al. 2012) and (x) downregulation of HPV oncoproteins in human cervical cancer cells and restoration of p53 mediated growth arrest or apoptosis (Munagala et al. 2011). Furthermore, Wi-A has been shown to sensitize cancer cells to radiotherapy (Kalthur and Pathirissery 2010) and work in a synergistic manner with other commercial anticancer compounds including sorafenib (Cohen et al. 2012), oxaliplatin (Li et al. 2015) and cisplatin (Kakar et al. 2014). Table 7.1 provides a comprehensive review on the cellular targets and molecular mechanism(s) of its anticancer activity tested on various kinds of human cancer cells *in vitro* and *in vivo*.

We initially prepared an alcoholic extract of leaves of Ashwagandha (called i-Extract) (Fig. 7.1a), and investigated its tumor suppressor activity in nude mice tumor-growth suppression assays. Significant tumor-growth suppression with i-Extract was recorded (Fig. 7.1b). *In vitro* cytotoxicity assays, using a variety of human cancer cell lines, revealed that i-Extract caused toxicity to cancer cells and was safe to normal cells at the equivalent doses (Fig. 7.1c). HPLC profiling of the i-Extract showed that (i) it contained Withaferin-A and Withanone as major constituents, (ii) the extracts with high ratio of Withanone/Wi-A possess higher anticancer activity *in vivo* and were selectively toxic to cancer cells *in vitro* (Fig. 7.1a). In order to identify the cellular targets for anticancer activity of i-Extract, we performed phenotype based loss-of-function screening using a shRNA library. In a platform of 96-well cell culture dishes, shRNA-mediated gene knockdown was carried out before the cells were subjected to i-Extract. shRNAs that rendered the cells unresponsive to the cytotoxic effect of i-Extract were identified and their gene targets were considered crucial for i-Extract mediated cancer cell death (Fig. 7.1d). By similar, but more robust and unbiased, loss-of-function screening using ribozymes involved infection of cancer cells with randomized ribozyme library prior to the treatment with i-Extract (Fig. 7.1e). Ribozymes were recovered from cells that survived the i-Extract treatment. Gene targets of the selected ribozymes (as predicted by database search) were analyzed by bioinformatics and pathway analyses. Interestingly, we identified p53 tumor suppressor in these screenings, and also found it to be activated in response to the treatment by i-Extract and its constituent (Withanone) in cancer, but not normal, cells (Fig. 7.2) (Widodo et al. 2007). Withaferin-A (Wi-A), on the other hand, that caused cytotoxicity to cancer as wells as normal cells (Widodo et al. 2007) induced p53 in both cell types endorsing that activation of p53 tumor suppressor is one of the mechanisms of anticancer activity in Ashwagandha bioactives. It has been reported that p53 binds to mortalin (mtHSP70) in cancer cells. This interaction causes cytoplasmic retention of p53 and inactivation of its transactivation function in cancer cells (Kaul et al. 2005). Pancytoplasmic subcellular distribution of mortalin has been demonstrated as a

Table 7.1 A comprehensive review on the cellular targets and molecular mechanism(s) of its anticancer activity tested on various kinds of human cancer cells *in vitro* and *in vivo*

	Disease	Target	Pathway	Result	References
1	Graft-to-host reaction	NFκB	MHC-mediated immune reaction	Delay of rejection of allogenic transplant	Gambhir et al. (2015)
2	Breast cancer	Vimentin	Vimentin ser56 phosphorylation	Invasion inhibition	Thaiparambil et al. (2011)
3	Endothelial cell inflammation	Vimentin	Vimentin-IκBα-NFκB	Anti-inflammatory	Mohan and Bargagna-Mohan (2016)
4	Prostate cancer	c-FOS	PERK-eIF2α-CHOP	Apoptosis	Nishikawa et al. (2015)
5	Prostate cancer	CYCLIN A, E1 and B1, p21, WEE1	Chromosomal condensation	Depolymerization of α-tubulin	(Roy et al. 2013)
6	Malignant pleural mesothelioma	p21, BAX, IκBα, c-MYC, c-FOS, c-JUN, TIMP-2, (CARP)-1/CCAR1	p53	Apoptosis	Yang et al. (2012)
7	Renal cancer	STAT3, JAK2	JAK-STAT3	Apoptosis	Um et al. (2012)
8	Renal cancer	STAT3, JAK2	Endoplasmic reticulum stress	Apoptosis	Choi et al. (2011)
9	Pancreatic cancer	HSP90	Hedgehog	Apoptosis, proliferation inhibition	Yu et al. (2010) and Yoneyama et al. (2015)
10	Myeloid leukemia	Caspase 3, p65/REL, NFκB	p53	Apoptosis	Malik et al. (2007)
11	Leukemia	p38MAPK, BAX	p53	Apoptosis	Mandal et al. (2008)
12	Malignant melanoma	ROS production	p53	Apoptosis	(Mayola et al. (2011)
13	Murine skin cancer	Acetyl-CoA carboxylase	Carcinogen-induced anchorage-independent growth	Proliferation inhibition	Li et al. (2016)

14	Metastatic uveal melanoma	c-MET, AKT and RAF-1	AKT-mTOR, p53	Growth arrest, apoptosis	Samadi et al. (2012)
15	Murine skin epidermal (pre-cancerous)	Isocitrate dehydrogenase	Anaerobic glycolysis	Transformation/proliferation inhibition	Li and Zhao (2013)
16	Colorectal cancer	MAD2, CDC20	Chromosomal segregation	Growth arrest, apoptosis	Das et al. (2014)
17	Colorectal cancer	Notch-1, IκKα, p70-S6 K, 4E-BP1, c-JUN, JNK, ERK1/2	Notch-1, AKT-mTOR, NFκB, p53, c-JUN-NH2-kinase	Growth arrest, apoptosis	Koduru et al. (2010)
18	Colorectal cancer	STAT3	JAK-STAT3	Apoptosis	Choi and Kim (2015)
19	Colorectal cancer	AKT	AKT-mTOR	Apoptosis	Suman et al. (2016)
20	Breast cancer	β-tubulin	Mitotic nuclear segregation	Growth arrest, apoptosis	Antony et al. (2014)
21	Breast cancer	ERK 1/2 and RSK	ERK1/2-RSK-ELK1-CHOP-DR5	Apoptosis	Nagalingam et al. (2014)
22	Breast cancer	BRCA1, HSF1	p53	Apoptosis	Zhang et al. (2012)
23	Breast cancer	STAT3	JAK-STAT3	Apoptosis	Lee et al. (2010)
24	Cervical cancer	p53, pRb, STAT3	p53	Apoptosis, Proloferation inhibition	Munagala et al. (2011)
25	Neuroblastoma	STAT3	IL-6-mediated cell death	Apoptosis	Yco et al. (2014)
26	Glioblastoma	AKT, mTOR, AMPKα	AKT-mTOR, p53	Growth arrest, apoptosis	Grogan et al. (2013)
27	Glioblastoma	–	–	Neurosphere collapse	(Chang et al. (2016)
28	Thyroid cancer	Synergism with sorafenib	–	Similar efficacy with reduced toxicity	Cohen et al. (2012)
29	Pancreatic cancer	Synergism with oxaliplatin	–	Similar efficacy with reduced toxicity	Li et al. (2015)
30	Renal cancer	Synergism with radiotherapy	p53	Apoptosis	Yang et al. (2012)

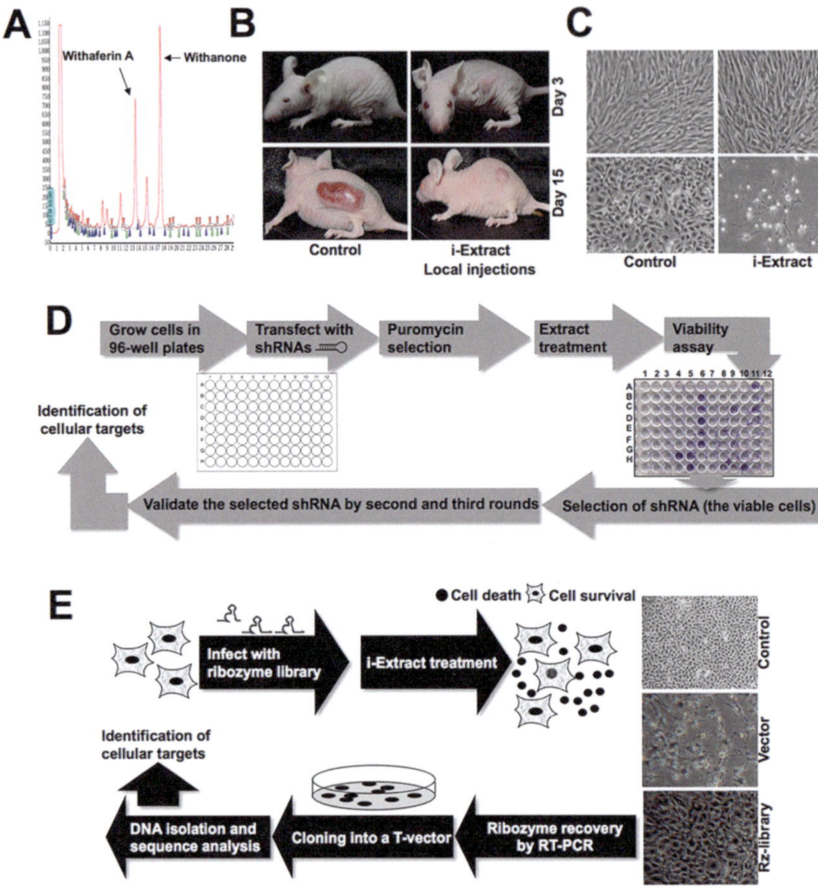

Fig. 7.1 Anticancer activity in leaves of Ashwagandha. (**a**) Chemical profile of alcoholic extract (i-Extract) of Ashwagandha showing high level of Withanone than Witahferin A. (**b**) Nude mice tumor progression assay showing that whereas, in 2 weeks, subcutaneous xenograft of HT1080 cells formed solid and big tumors in control mice. The i-Extract fed mice showed suppression of tumor growth. (**c**) Cell culture assays showed cytotoxicity of i-Extract (12 μg/ml) to MCF7 (human breast cancer cells), but not TIG-1 (normal cells). (**d** and **e**) Loss of function screening using shRNA or ribozymes for identification of cellular targets for cytotoxicity of i-Extract to cancer cells (Widodo et al. 2007)

marker for normal cells; tumor cells show perinuclear-staining pattern (Wadhwa et al. 1993). Interestingly, a shift in perinuclear mortalin staining pattern, typical of tumor cells, to pancytoplasmic pattern was observed in i-Extract treated cancer cells. The latter also showed nuclear translocation and activation of p53 (Widodo et al. 2007). Similar change in the subcellular distribution of mortalin was found when cancer cells were induced to undergo senescence-like growth arrest in response to treatment with a rhodacyanine dyes (MKT-077 and its derivatives), bromo-deoxyuridine (BrdU), single chromosomes and active component of honey

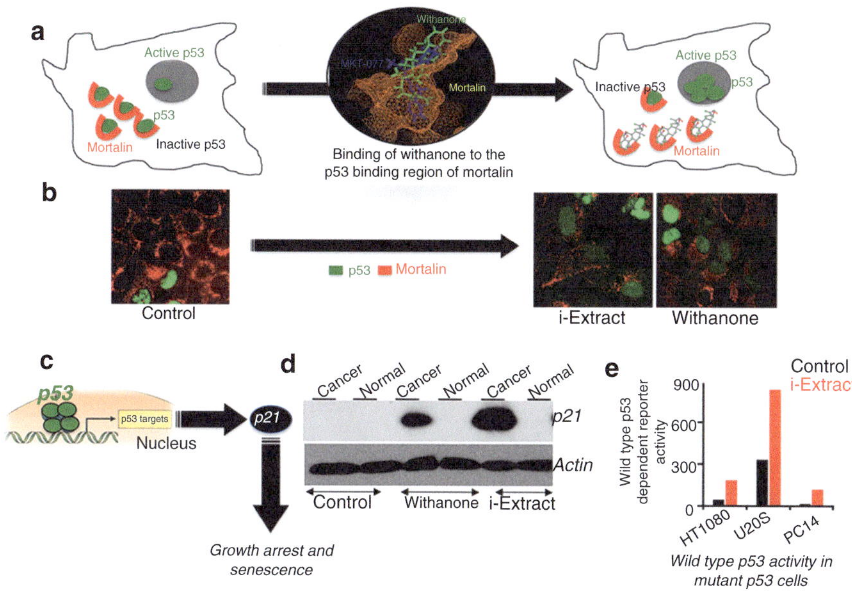

Fig. 7.2 Activation of tumor suppressor p53 in cancer cells Withanone. (**a**) A model showing inactivation of p53 by its binding to p53 in cytoplasm, docking to Withanone into mortalin-p53 complexes, resulting in nuclear translocation of p53. (**b**) i-Extract and Withanone treated cells showed increased number of cells with nuclear p53. (**c**) A model showing transcriptional activation of p21 by p53. (**d**) i-Extract and Withanone caused increase in p21 in cancer cells. (**e**) Wild type p53 reporter assay in control and i-Extract treated cells showed that the later caused increase wild type p53 activity (Widodo et al. 2008, 2010)

bee propolis-CAPE (all of these have been demonstrated to possess anticancer activity) (Kaul et al. 1995; Michishita et al. 1999; Wadhwa et al. 2000, 2016). By molecular docking studies, we further found that Wi-A and Wi-N are able to dock into the p53-mortalin complexes in which mortalin protein inactivates the tumor suppressor activities (Fig. 7.2a). By docking to the protein sites that are involved in their binding, these small molecules cause abrogation of mortalin-p53 complexes, resulting in re-translocation of p53 to the cell nucleus and activation of its transcriptional activation and growth arrest functions mediated by p21$^{WAF\text{-}1}$ protein (Fig. 7.2b). Of note, since mortain-p53 interaction occurs in cancer, not in normal, these molecular data accounted for selective toxicity of Wi-A to cancer cells. Taken together, these data suggested that the i-Extract abrogated mortalin-p53 interactions and induced senescence-like growth arrest in cancer cells through an activation of wild type p53 function that was endorsed by increase in p21 protein in i-Extract treated cancer cells (Fig. 7.2c, d). Furthermore, as discussed above, mutant p53 (a hall mark of cancer cells) was seen to decrease in i-Extract treated cells along with the appearance of wild type p53 activity, suggesting that i-Extract is useful for cancer treatment.

Cell phenotype driven molecular analyses showed that i-Extract induced growth arrest in cancer cells with wild type p53 (U2OS, MCF7, and HeLa) and apoptosis in mutant type p53 (HS578T, SK-BR3 and PC14) proteins. Apoptosis in cancer cells with mutant p53 (HT29A4) has been reported when wild type p53 protein was restored in these cells (Barberi-Heyob et al. 2004). Reactivation of wild type p53 activity was also achieved by low molecular weight peptides including PRIMA-1 (Bykov et al. 2003) and by global suppressor motif approach (Baroni et al. 2004). In view of this information, our data was suggestive that i-Extract has an ability to endow wild type p53 function in mutant p53 protein in cancer cells. Since such endowment of wild type p53 function (reactivation of the growth arrest or the apoptotic pathway) to mutant p53 protein in cancer cells is the most beneficial approach for cancer therapeutics. We undertook biochemical, reporter and imaging assays to confirm wild type p53 activity in i-Extract treated mutant p53-harboring cells. As shown in Fig. 7.2e, we detected wild type p53 activity by reporter assays as well as endogenous downstream targets ($p21^{WAF-1}$, HDM2 and Bax) in treated cells. Decrease in the level of mutant p53 and reactivation of wild type p53 in i-Extract treated cancer cells suggested that it is a useful for cancer treatment.

In order to obtain further molecular insights to the anticancer bioactives and their mechanism of action, i-Extract was fractionated by silica gel column chromatography and subjected them to cell based activity analyses. We identified seven fractions that could cause cancer cell killing; Wi-N was most selective for cancer cells. Wi-N rich Ashwagandha leaf powder was nontoxic and anti-tumorigenic in mice assays. Gene silencing and pathway analyses approach on the i-Extract and its components revealed that they cause cancer cell killing by at least by five different pathways viz., p53 signaling, GM-CFS signaling, death receptor signaling, apoptosis signaling and G2-M DNA damage regulation pathway of which p53 signaling was most common (Fig. 7.3) (Widodo et al. 2008, 2010). Imaging analyses of p53 and mortalin further revealed that i-Extract, Fractions F1 & F4, and Wi-N caused an abrogation of mortalin-p53 interactions and reactivation of p53 function while the fractions F2, F3, F5 work through other mechanisms. These analyses provided evidence on the multi-module action of i-Extract and its components Wi-A and Wi-N. Such multi-site mechanisms of action of i-Extract may offer greater success as compared to the single-site agents (Fig. 7.3). Bioinformatics on the selected gene-targets once again revealed involvement of p53, apoptosis and insulin/IGF signaling pathways linked to the ROS signaling. We verified these findings by biochemical and imaging assays and found that the selective killing of cancer cells is mediated by induction of p53 activity and oxidative stress including increase in ROS levels, DNA damage, changes in mitochondrial structure and membrane potential (Fig. 7.4). Furthermore, cellular targets, TPX2, TFAP2A, LHX3 and ING1 were identified.

TPX2 is a microtubule-associated protein that functions as an allosteric regulator of Aurora-A, an oncogene that plays essential role in centrosome maturation and chromosome segregation during mitosis (Bibby et al. 2009). It is overexpressed in multiple human cancers (Tanenbaum et al. 2009) and regulate the Aurora A kinase that plays a key role during mitosis and cytokinesis and has also been found to be up regulated in several cancer types. TPX2 and Aurora A kinase together control the

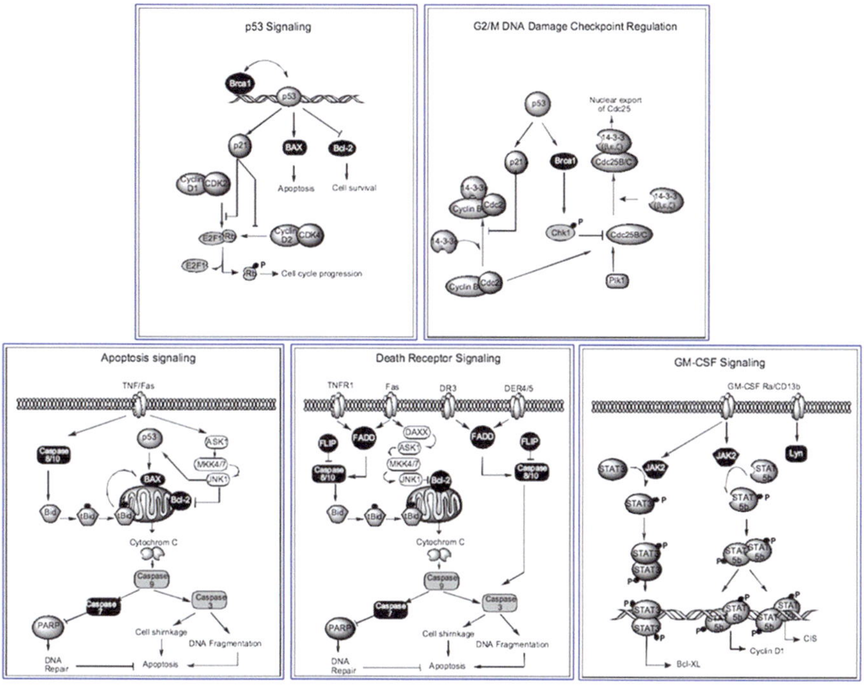

Fig. 7.3 Cellular pathways involved in cytotoxicity of i-Extract to cancer cells. Cellular targets identified by loss-of-function screening assays using shRNA and ribozymes were analyzed by bioinformatics. The identified five pathways are shown (Widodo et al. 2008)

genetic instability that results from improper segregation of sister chromatids during mitosis. They associates with the centrosome and the spindle microtubules during mitosis, and function in centrosome maturation, spindle assembly, maintenance of spindle bipolarity, and mitotic checkpoint control. Aurora A inhibitors have been proposed as anticancer drugs that work by blocking its ATP binding site. However, the ATP binding site is common to other kinases, such inhibitors would lack specificity, and hence advocate the need of some alternative inhibition route. Withanone inhibited TPX2. By computational and experimental approaches, we found that Withanone binds to TPX2-Aurora A complex and results in their dissociation causing disruption of mitotic spindle apparatus and cell death. Furthermore, phosphorylation analysis of Aurora A and its subtrate histone H3 in control and Withanonc treated cells showed significant decrease in phosphorylated histone H3 in the latter suggesting that Withanone is a candidate Aurora A targeting anticancer drug (Grover et al. 2012b).

TFAP2A has been shown to play crucial role in tumor growth and progression by regulation of E-cadherin, insulin like growth factor receptor-1, MMP-2, c-kit, HER-2, Bcl-2, and Smad signaling (Koinuma et al. 2009). LHX3 is a homeodomain transcription factor and plays key role in cell fate determination, embryonic development and oncogenesis (Dietrich et al. 2009). On the other hand, ING1, an ING family

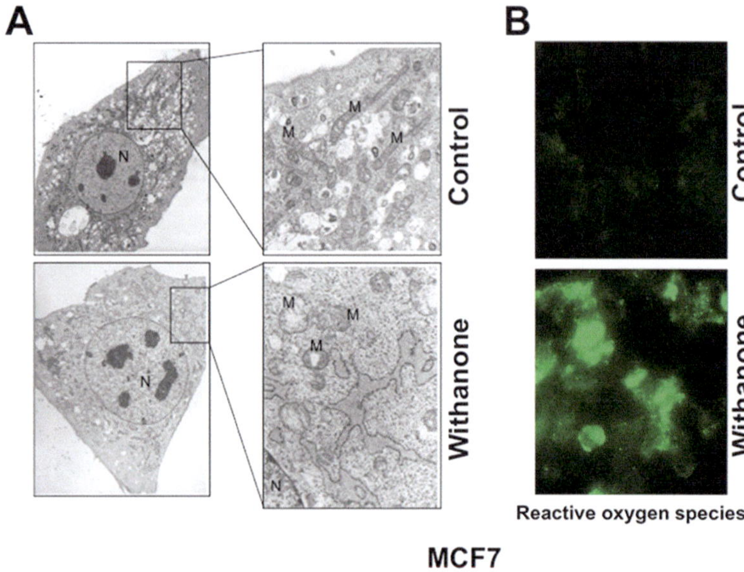

Fig. 7.4 Withanone caused oxidative stress and ROS induction in cancer cells. Electron micrographs of control and Withanone-treated cells showing induction mitochondrial damage (**a**) and reactive oxygen species (**b**) in Withanone treated cells (Widodo et al. 2007, 2008, 2010).

protein, is involved in human cellular senescence, tumor suppression and apoptosis (Menendez et al. 2009; Shah et al. 2009). ING1 has been shown to modulate p53 activity and its downstream effectors, p21^{WAF1} and Bax by acetylation and stabilization (Zhu et al. 2009). Taken together, the data suggested that the cancer cell killing by i-Extract might involve repression of TPX2, TFAP2A and LHX3, and activation of ING1 functions; supporting its multi-site action. Bioinformatics and systems biology analyses on these gene targets revealed that these gene targets are involved in several kinds of biological processes such as, oncogenesis, cell cycle, DNA repair and nucleic acid metabolism. The top two pathways were p53 tumor suppressor (gene targets-DDB2, CDKN1A, CDKN2B) and apoptosis (gene targets-IGF2R and HSPA9). In addition, Ras, insulin/IGF, angiogenesis and cytoskeleton regulation pathways that are tightly linked with apoptosis and tumor development, were also identified. Involvement of these gene clusters during i-Extract induced cytotoxicity suggested that it might be characterized as cellular responses including, stress response (HSPA9, CDKN1A), and DNA damage response (ING1, DDB2 and TFAP2A) culminating into either cell cycle arrest or apoptosis.

Based on these data it was suggestive that i-Extract causes activation of cellular stress signaling by ROS-mediated pathways initiated at two levels (i) mitochondrial stress leading to change in membrane potential and (ii) DNA damage stress leading to activation of DNA damage and repair machinery. In order to test this hypothesis, we investigated molecular markers of stress in control and i-Extract treated cells. As shown in Fig. 7.2, i-Extract caused an activation of CDKN1A-p21 in cancer cells; it remained unaltered in normal cells when treated with either i-Extract or Withanone. Of note, i-Extract induced increase in CDKN1A-p21 expression was abrogated by knockdown of each of these four (TPX2, LHX3, ING1 and TFAP2A) target genes suggesting CDKN1A-p21 is a critical mediator of i-Extract induced selective growth arrest in cancer cells. Furthermore, whereas $p53^{+/+}$ and $p53^{-/-}$ cells showed comparable sensitivity to i-Extract, the $p21^{-/-}$ cells were 2–3 times more tolerant to i-Extract treatment as compared to the $p21^{+/+}$ cells endorsing that CDKN1A-p21 is a critical mediator of i-Extract induced growth arrest in cancer cells (Widodo et al. 2008, 2010).

The involvement of DNA damage and repair-signaling pathway in i-Extract induced cancer cell killing was investigated by analysis of DNA damage-induced molecular markers. i-Extract, Wi-A and Wi-N treated cells showed increase in γH2AX (an early marker of DNA damage response) in cancer cells only (Widodo et al. 2008, 2010). Similarly, stress response initiated at mitochondria in i-Extract induced MCF7 (breast cancer) cell killing was supported by induction of ROS (chemically reactive molecules that have essential role in signal transduction, cell growth and differentiation, regulation of enzyme activities and immune response including inflammation and cytokine production). Whereas a moderate increase in ROS promotes cell proliferation and differentiation, its excessive amount causes irreversible oxidative damage to DNA, proteins and lipids leading to cell death. Cancer cells frequently exhibit high oxidative stress and increased generation of reactive oxygen species (ROS) as compared to the normal counterparts. This higher level of ROS has been considered as a selective therapeutic target for cancer treatment (Piccirillo et al. 2009). Consistent with the cytotoxicity of two components (Wi-A and Wi-N) in i-Extract, ROS induction was detected by both the components in cancer cells and only by Wi-A in normal cells (Widodo et al. 2008, 2010). These data demonstrated that the treatment with i-Extract and Withanone lead to induction of ROS selectively in cancer cells and hence could be responsible for selective cancer cell killing (Fig. 7.4) (Widodo et al. 2008, 2010). Investigations of mitochondrial membrane potential that is highly affected by ROS levels, revealed decrease in mitochondrial membrane potential in i-Extract treated cancer cells. Ultrastructure of control and i-Extract treated MCF7 cells at single-cell level showed typical mitochondrial morphology, characterized by a double membrane containing a homogeneous matrix and a system of parallel cristae. Wi-N treated cells showed swollen mitochondria with an altered morphology of which the shortening and reduction in number of cristae were apparent (Fig. 7.4) (Widodo et al. 2007).

With the experimental strategies described above, we next investigated anticancer activity in the water extract of Ashwagandha leaves (WEX) and found that it possesses considerable anticancer potential. Since water extracts are eco- as well as bio- friendly by which plant is not sacrificed and organic solvents are replaced with easy, economic, convenient and safe alternative i.e., water, the anticancer activity in water extract was deemed extremely beneficial for human consumption. We extended the study to examine (i) the toxicity assays in mice, (ii) *in vivo* anti-tumor test in nude mice subcutaneous xenograft and tail vein metastasis models (Fig. 7.5). Nude mice fed with 500 mg of WEX/Kg body weight on every alternate day for 60

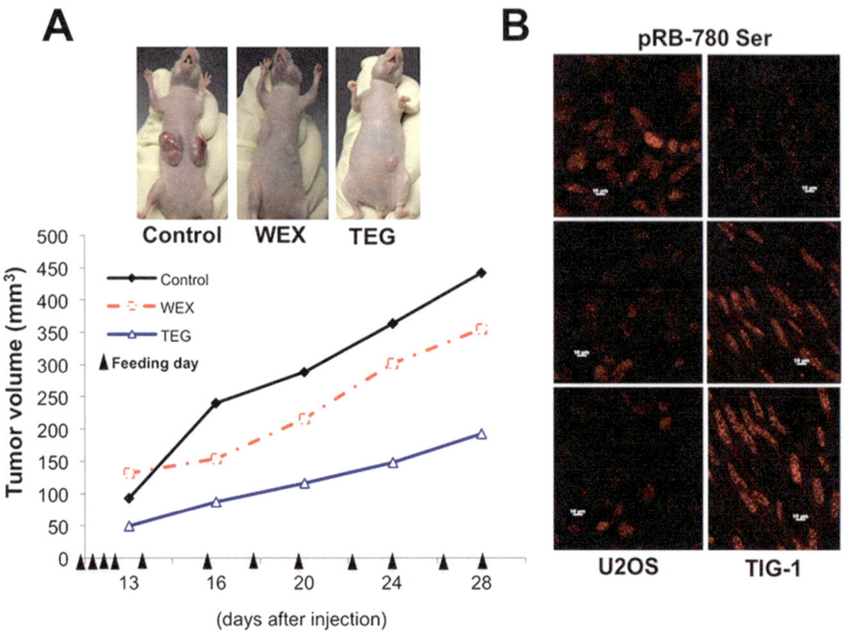

Fig. 7.5 Anticancer activity in water extract of Ashwagandha leaves and its active component Triethylene glycol. (**a**) Nude mice tumor assays showing growth suppression in mice fed either with WEX or TEG. (**b**) Cancer (U2OS), but not normal (TIG-1), cells treated with WEX or TEG showed decrease in phosphorylated pRB signifying their cell cycle arrest (Wadhwa et al. 2013)

days did not show toxicity in terms of change in body weight and physical activity. After implanting HT1080 subcutaneous xenografts and tail vein injections, the nude mice were fed with WEX 250 mg/Kg body weight every alternate day and the tumor formation was monitored. In about 20 days, out of 24 control mice, 21 showed big tumor, one showed small tumor and two showed no tumors. WEX fed mice, on the other hand, showed strong tumor suppression. In this group, out of 24 mice, 12 showed no tumors, 3 showed tumor buds, 5 showed small tumors and only 4 showed big tumors. In lung metastasis assays, WEX fed mice showed strong suppression of metastasis. There were less than 5 tumors in WEX fed mice as compared to the control mice that had more than 50 tumors. These data suggested that the WEX has considerable anticancer activity *in vitro* and *in vivo*.

Chemical analysis of WEX showed that it contains very low levels of Wi-N or Wi-A. Analysis on the cellular targets of WEX and i-Extract by shRNA library revealed that their cytotoxicity was mediated by some unique and several common gene-targets. The latter were attributed to low levels of Wi-N and Wi-A. By activity based fractionation and NMR analysis, we identified triethylene glycol (TEG) as a key component that possess tumor suppression activity in *in vivo* nude mice tumor formation and metastasis assays (Fig. 7.5). TEG is a member of dihydroxy alcohol family, known to have high ability to hold water molecules. It has been well established as a low toxicity mild disinfectant towards a variety of airborne, solution and surface bound microbes including bacteria and viruses (Rudnick et al. 2009). An old study on dogs having inoperable malignant and metastatic tonsillar epithelioma and reticulum cell sarcoma reported the regression of tumors in response to the treatment with TEG without any toxicity (Owen 1962). Liu et al. (2008) reported that the triethylene tetramine (TETA) as a novel ligand for g-quadruplex and has many kinds of biological activities, including telomerase inhibition and induction of senescence in tumor cells. Low dose of TETA caused week inhibition of the growth of tumor cells in culture. However, it could significantly enhance anti-tumor activity of traditional anti-tumor drugs *in vitro* and *in vivo* (Liu et al. 2008).

Our biochemical, molecular and imaging analyses revealed that the cytotoxicity of TEG is mediated by activation of p53 and pRb tumor suppressor pathways. However, in contrast to the selective activation of p53 in response to i-Extract treatment as discussed above, WEX caused activation of p53 in cancer as well as normal cells. On the other hand, investigation on the level of phosphorylated pRb by Western blotting and immunostaining in control, WEX and TEG-treated cells revealed that whereas pRb phosphorylation and cyclin B1 decreased in cancer cells, normal cells showed increase (Fig. 7.5) (Wadhwa et al. 2013). Cyclin D1 was increased in cancer and decreased in normal treated cells. Cyclins are a highly conserved family of proteins that act as regulators of CDK kinases and show dramatic periodicity in their expression level during the cell cycle progression. Different cyclins exhibit distinct expression and degradation patterns that contribute to the temporal coordination of each mitotic event. It has been established that cyclin D1 forms a complex with CDK4 or CDK6 and is absolutely required for G1/S transition. Cyclin D1-CDK4/CDK6 complex forms an active kinase for the retinoblastoma protein (RB) resulting in RB-phosphorylation, release of E2F from RB-E2F complex and transcription of genes required for cell cycle progression. Cyclin D1 also plays a key role in linking the extracellular signaling to cell cycle progression. Its expression level increases during G2 phase, maintains through mitosis and G1 phase, and declines in S phase when DNA synthesis begins. During the cell cycle progression in response to pro proliferative signals, cyclin D1 level gets induced once again during G2 phase and decline on entry to S phase resulting commitment to continuing proliferation. WEX and TEG caused hypophosphorylation of Rb (activation of its tumor suppressor activity) in cancer cells selectively (Fig. 7.5) (Wadhwa et al. 2013). These studies have not only demonstrated the molecular basis of the anticancer activity in the leaves of Ashwagandha but also provided compelling evidence to the fact that often herbs are seen to have better activity than their extracts (Fig. 7.6a). In view of this, we finally investigated anticancer potential of Ashwagandha leaf powder and found that mice fed with Ashwagandha (500 mg/Kg Body weight) showed no toxicity, but strong suppression of tumor progression (Fig. 7.6b). Furthermore, mixture of leaf powder with gamma cyclodextrin (γCD) showed higher efficacy that may be caused by increased stability and bioavailability of active ingredients (Fig. 7.6c) (Kaul et al. 2017).

In last one decade, large number of studies has provided experimental evidence and molecular insights to the anticancer bioactive compounds in leaves of

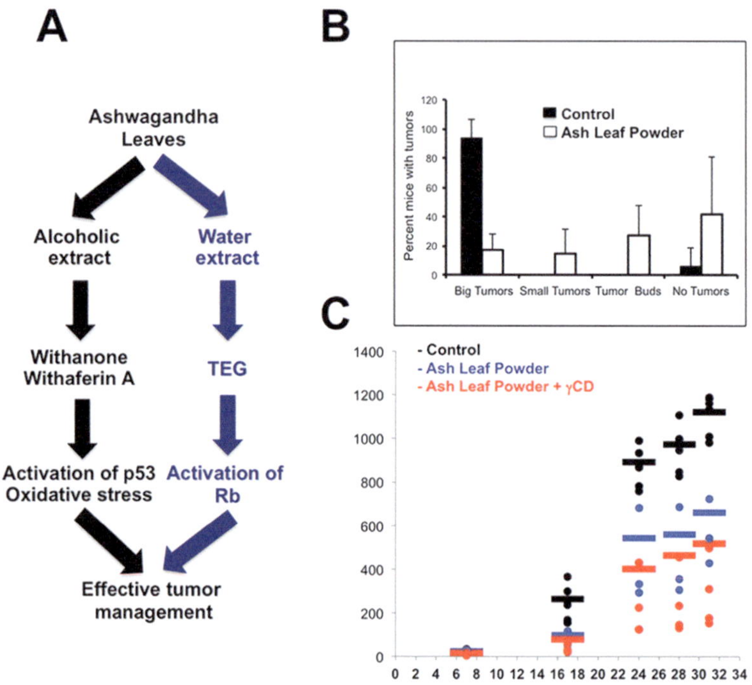

Fig. 7.6 Multimodule anticancer action of Ashwagandha leaves. (**a**) A model showing anticancer components of Ashwgandha leaf extract and their mechanism of action. (**b**) Antitumor effect of Ashwagandha leaf powder and its mixture with γCD (**c**) feeding on tumor growth in nude mice. Leaf powder was effective in suppressing the growth of tumors and γCD enhanced the antitumor efficacy

Ashwagandha. While further studies to dissect their efficacy in *in vivo* clinical models of cancers and pharmokinetics are warranted, the traditional knowledge on use of extracts in preference to the individual active components is validated by our analyses and deserves attention for anticancer treatment.

References

Antony ML, Lee J, Hahm ER, Kim SH, Marcus AI, Kumari V, Ji X, Yang Z, Vowell CL, Wipf P, Uechi GT, Yates NA, Romero G, Sarkar SN, Singh SV (2014) Growth arrest by the antitumor steroidal lactone withaferin A in human breast cancer cells is associated with down-regulation and covalent binding at cysteine 303 of beta-tubulin. J Biol Chem 289:1852–1865

Barberi-Heyob M, Vedrine PO, Merlin JL, Millon R, Abecassis J, Poupon MF, Guillemin F (2004) Wild-type p53 gene transfer into mutated p53 HT29 cells improves sensitivity to photodynamic therapy via induction of apoptosis. Int J Oncol 24:951–958

Baroni TE, Wang T, Qian H, Dearth LR, Truong LN, Zeng J, Denes AE, Chen SW, Brachmann RK (2004) A global suppressor motif for p53 cancer mutants. Proc Natl Acad Sci U S A 101:4930–4935

Bibby RA, Tang C, Faisal A, Drosopoulos K, Lubbe S, Houlston R, Bayliss R, Linardopoulos S (2009) A cancer-associated aurora A mutant is mislocalized and misregulated due to loss of interaction with TPX2. J Biol Chem 284:33177–33184

Blackburn EH, Epel ES, Lin J (2015) Human telomere biology: a contributory and interactive factor in aging, disease risks, and protection. Science 350:1193–1198

Brown JP, Wei W, Sedivy JM (1997) Bypass of senescence after disruption of p21CIP1/WAF1 gene in normal diploid human fibroblasts. Science 277(5327):831–834

Bykov VJ, Selivanova G, Wiman KG (2003) Small molecules that reactivate mutant p53. Eur J Cancer 39:1828–1834

Chang E, Pohling C, Natarajan A, Witney TH, Kaur J, Xu L, Gowrishankar G, D'Souza AL, Murty S, Schick S, Chen L, Wu N, Khaw P, Mischel P, Abbasi T, Usmani S, Mallick P, Gambhir SS (2016) AshwaMAX and Withaferin A inhibits gliomas in cellular and murine orthotopic models. J Neuro-Oncol 126:253–264

Cheung EC, Vousden KH (2010) The role of p53 in glucose metabolism. Curr Opin Cell Biol 22:186–191

Cheung CT, Singh R, Yoon AR, Hasan MK, Yaguchi T, Kaul SC, Yun CO, Wadhwa R (2011) Molecular characterization of apoptosis induced by CARF silencing in human cancer cells. Cell Death Differ 18:589–601

Cheung CT, Singh R, Kalra RS, Kaul SC, Wadhwa R (2014) Collaborator of ARF (CARF) regulates proliferative fate of human cells by dose-dependent regulation of DNA damage signaling. J Biol Chem 289:18258–18269

Choi BY, Kim BW (2015) Withaferin-A inhibits Colon cancer cell growth by blocking STAT3 transcriptional activity. J Cancer Prev 20:185–192

Choi MJ, Park EJ, Min KJ, Park JW, Kwon TK (2011) Endoplasmic reticulum stress mediates withaferin A-induced apoptosis in human renal carcinoma cells. Toxicol in Vitro 25:692–698

Cohen SM, Mukerji R, Timmermann BN, Samadi AK, Cohen MS (2012) A novel combination of withaferin A and sorafenib shows synergistic efficacy against both papillary and anaplastic thyroid cancers. Am J Surg 204:895–900

Dar PA, Singh LR, Kamal MA, Dar TA (2016) Unique medicinal properties of Withania somnifera: phytochemical constituents and protein component. Curr Pharm Des 22:535–540

Das T, Roy KS, Chakrabarti T, Mukhopadhyay S, Roychoudhury S (2014) Withaferin A modulates the spindle assembly checkpoint by degradation of Mad2-Cdc20 complex in colorectal cancer cell lines. Biochem Pharmacol 91:31–39

Deocaris CC, Widodo N, Shrestha BG, Kaur K, Ohtaka M, Yamasaki K, Kaul SC, Wadhwa R (2007) Mortalin sensitizes human cancer cells to MKT-077-induced senescence. Cancer Lett 252:259–269

Deocaris CC, Widodo N, Wadhwa R, Kaul SC (2008) Merger of ayurveda and tissue culture-based functional genomics: inspirations from systems biology. J Transl Med 6:14

Deocaris CC, Lu WJ, Kaul SC, Wadhwa R (2013) Druggability of mortalin for cancer and neuro-degenerative disorders. Curr Pharm Des 19:418–429

Dhar RS, Verma V, Suri KA, Sangwan RS, Satti NK, Kumar A, Tuli R, Qazi GN (2006) Phytochemical and genetic analysis in selected chemotypes of Withania somnifera. Phytochemistry 67:2269–2276

Dietrich D, Lesche R, Tetzner R, Krispin M, Dietrich J, Haedicke W, Schuster M, Kristiansen G (2009) Analysis of DNA methylation of multiple genes in microdissected cells from formalin-fixed and paraffin-embedded tissues. J Histochem Cytochem 57:477–489

Engelmann D, Putzer BM (2014) Emerging from the shade of p53 mutants: N-terminally truncated variants of the p53 family in EMT signaling and cancer progression. Sci Signal 7:re9

Fang L, Igarashi M, Leung J, Sugrue MM, Lee SW, Aaronson SA (1999) p21Waf1/Cip1/Sdi1 induces permanent growth arrest with markers of replicative senescence in human tumor cells lacking functional p53. Oncogene 18:2789–2797

Gambhir L, Checker R, Sharma D, Thoh M, Patil A, Degani M, Gota V, Sandur SK (2015) Thiol dependent NF-kappaB suppression and inhibition of T-cell mediated adaptive immune responses by a naturally occurring steroidal lactone Withaferin A. Toxicol Appl Pharmacol 289:297–312

Grogan PT, Sleder KD, Samadi AK, Zhang H, Timmermann BN, Cohen MS (2013) Cytotoxicity of withaferin A in glioblastomas involves induction of an oxidative stress-mediated heat shock response while altering Akt/mTOR and MAPK signaling pathways. Investig New Drugs 31:545–557

Grover A, Priyandoko D, Gao R, Shandilya A, Widodo N, Bisaria VS, Kaul SC, Wadhwa R, Sundar D (2012a) Withanone binds to mortalin and abrogates mortalin-p53 complex: computational and experimental evidence. Int J Biochem Cell Biol 44:496–504

Grover A, Singh R, Shandilya A, Priyandoko D, Agarwal D, Bisaria VS, Wadhwa R, Kaul SC, Sundar D (2012b) Ashwagandha derives withanone taregts TPX2-Aurora A complex: computational and experimental evidence to its anticancer activity. PLoS One 7:e30890

Harley CB (1990) Aging of cultured human skin fibroblasts. Methods Mol Biol 5:25–32

Hasan MK, Yaguchi T, Sugihara T, Kumar PK, Taira K, Reddel RR, Kaul SC, Wadhwa R (2002) CARF is a novel protein that cooperates with mouse p19ARF (human p14ARF) in activating p53. J Biol Chem 277:37765–37770

Hasan MK, Yaguchi T, Minoda Y, Hirano T, Taira K, Wadhwa R, Kaul SC (2004) Alternative reading frame protein (ARF)-independent function of CARF (collaborator of ARF) involves its interactions with p53: evidence for a novel p53-activation pathway and its negative feedback control. Biochem J 380:605–610

Hasan MK, Yaguchi T, Harada JI, Hirano T, Wadhwa R, Kaul SC (2008) CARF (collaborator of ARF) interacts with HDM2: evidence for a novel regulatory feedback regulation of CARF-p53-HDM2-p21WAF1 pathway. Int J Oncol 32:663–671

Hayflick L (1965) The limited in vitro lifetime of human diploid cell strains. Exp Cell Res 37:614–636

Hayflick L (2000) The future of ageing. Nature 408:267–269

Hengartner MO (2000) The biochemistry of apoptosis. Nature 407:770–776

Ivaska J, Pallari HM, Nevo J, Eriksson JE (2007) Novel functions of vimentin in cell adhesion, migration, and signaling. Exp Cell Res 313:2050–2062

Kaefer CM, Milner JA (2008) The role of herbs and spices in cancer prevention. J Nutr Biochem 19:347–361

Kaefer CM, Milner JA (2011) Herbs and spices in cancer prevention and treatment. In: IFF B, Wachtel-Galor S (eds) Herbal medicine: biomolecular and clinical aspects, 2nd edn. CRC Press/Taylor & Francis, Boca Raton

Kakar SS, Ratajczak MZ, Powell KS, Moghadamfalahi M, Miller DM, Batra SK, Singh SK (2014) Withaferin a alone and in combination with cisplatin suppresses growth and metastasis of ovarian cancer by targeting putative cancer stem cells. PLoS One 9:e107596

Kalthur G, Pathirissery UD (2010) Enhancement of the response of B16F1 melanoma to fractionated radiotherapy and prolongation of survival by withaferin A and/or hyperthermia. Integr Cancer Ther 9:370–377

Kamijo T, Zindy F, Roussel MF, Quelle DE, Downing JR, Ashmun RA, Grosveld G, Sherr CJ (1997) Tumor suppression at the mouse INK4a locus mediated by the alternative reading frame product p19ARF. Cell 91:649–659

Kanai M, Ma Z, Izumi H, Kim SH, Mattison CP, Winey M, Fukasawa K (2007) Physical and functional interaction between mortalin and Mps1 kinase. Genes Cells 12:797–810

Kaul SC, Wadhwa R, Matsuda Y, Hensler PJ, Pereira-Smith OM, Komatsu Y, Mitsui Y (1995) Mouse and human chromosomal assignments of mortalin, a novel member of the murine hsp70 family of proteins. FEBS Lett 361:269–272

Kaul SC, Reddel RR, Mitsui Y, Wadhwa R (2001) An N-terminal region of mot-2 binds to p53 in vitro. Neoplasia 3:110–114

Kaul SC, Aida S, Yaguchi T, Kaur K, Wadhwa R (2005) Activation of wild type p53 function by its mortalin-binding, cytoplasmically localizing carboxyl terminus peptides. J Biol Chem 280:39373–39379

Kaul SC, Ishida Y, Tamura K, Wada T, Iitsuka T, Garg S, Kim M, Gao R, Nakai S, Okamoto Y, Terao K, Wadhwa R (2017) Novel methods to generate active ingredient-enriched Ashwagandha leaves and extracts. PLoS One 11:e0166945

Koduru S, Kumar R, Srinivasan S, Evers MB, Damodaran C (2010) Notch-1 inhibition by Withaferin-A: a therapeutic target against colon carcinogenesis. Mol Cancer Ther 9:202–210
Koinuma D, Tsutsumi S, Kamimura N, Taniguchi H, Miyazawa K, Sunamura M, Imamura T, Miyazono K, Aburatani H (2009) Chromatin immunoprecipitation on microarray analysis of Smad2/3 binding sites reveals roles of ETS1 and TFAP2A in transforming growth factor beta signaling. Mol Cell Biol 29:172–186
Korenjak M, Brehm A (2005) E2F-Rb complexes regulating transcription of genes important for differentiation and development. Curr Opin Genet Dev 15:520–527
Lee J, Hahm ER, Singh SV (2010) Withaferin A inhibits activation of signal transducer and activator of transcription 3 in human breast cancer cells. Carcinogenesis 31:1991–1998
Li W, Zhao Y (2013) Withaferin A suppresses tumor promoter 12-O-tetradecanoylphorbol 13-acetate-induced decreases in isocitrate dehydrogenase 1 activity and mitochondrial function in skin epidermal JB6 cells. Cancer Sci 104:143–148
Li X, Zhu F, Jiang J, Sun C, Wang X, Shen M, Tian R, Shi C, Xu M, Peng F, Guo X, Wang M, Qin R (2015) Synergistic antitumor activity of withaferin A combined with oxaliplatin triggers reactive oxygen species-mediated inactivation of the PI3K/AKT pathway in human pancreatic cancer cells. Cancer Lett 357:219–230
Li W, Zhang C, Du H, Huang V, Sun B, Harris JP, Richardson Q, Shen X, Jin R, Li G, Kevil CG, Gu X, Shi R, Zhao Y (2016) Withaferin A suppresses the up-regulation of acetyl-coA carboxylase 1 and skin tumor formation in a skin carcinogenesis mouse model. Mol Carcinog 55:1739–1746
Lin AW, Barradas M, Stone JC, van Aelst L, Serrano M, Lowe SW (1998) Premature senescence involving p53 and p16 is activated in response to constitutive MEK/MAPK mitogenic signaling. Genes Dev 12:3008–3019
Liu J, Guo L, Yin F, Zheng X, Chen G, Wang Y (2008) Characterization and antitumor activity of triethylene tetramine, a novel telomerase inhibitor. Biomed Pharmacother 62:480–485
Lu WJ, Lee NP, Kaul SC, Lan F, Poon RT, Wadhwa R, Luk JM (2011a) Induction of mutant p53-dependent apoptosis in human hepatocellular carcinoma by targeting stress protein mortalin. Int J Cancer 129:1806–1814
Lu WJ, Lee NP, Kaul SC, Lan F, Poon RT, Wadhwa R, Luk JM (2011b) Mortalin-p53 interaction in cancer cells is stress dependent and constitutes a selective target for cancer therapy. Cell Death Differ 18:1046–1056
Ma Z, Izumi H, Kanai M, Kabuyama Y, Ahn NG, Fukasawa K (2006) Mortalin controls centrosome duplication via modulating centrosomal localization of p53. Oncogene 25:5377–5390
Malik F, Kumar A, Bhushan S, Khan S, Bhatia A, Suri KA, Qazi GN, Singh J (2007) Reactive oxygen species generation and mitochondrial dysfunction in the apoptotic cell death of human myeloid leukemia HL-60 cells by a dietary compound withaferin A with concomitant protection by N-acetyl cysteine. Apoptosis 12:2115–2133
Mandal C, Dutta A, Mallick A, Chandra S, Misra L, Sangwan RS, Mandal C (2008) Withaferin A induces apoptosis by activating p38 mitogen-activated protein kinase signaling cascade in leukemic cells of lymphoid and myeloid origin through mitochondrial death cascade. Apoptosis 13:1450–1464
Mayola E, Gallerne C, Esposti DD, Martel C, Pervaiz S, Larue L, Debuire B, Lemoine A, Brenner C, Lemaire C (2011) Withaferin A induces apoptosis in human melanoma cells through generation of reactive oxygen species and down-regulation of Bcl-2. Apoptosis 16:1014–1027
McConnell BB, Starborg M, Brookes S, Peters G (1998) Inhibitors of cyclin-dependent kinases induce features of replicative senescence in early passage human diploid fibroblasts. Curr Biol 8:351–354
Menendez C, Abad M, Gomez-Cabello D, Moreno A, Palmero I (2009) ING proteins in cellular senescence. Curr Drug Targets 10:406–417
Michishita E, Nakabayashi K, Suzuki T, Kaul SC, Ogino H, Fujii M, Mitsui Y, Ayusawa D (1999) 5-Bromodeoxyuridine induces senescence-like phenomena in mammalian cells regardless of cell type or species. J Biochem 126:1052–1059

Mohan R, Bargagna-Mohan P (2016) The use of Withaferin A to study intermediate filaments. Methods Enzymol 568:187–218
Munagala R, Kausar H, Munjal C, Gupta RC (2011) Withaferin A induces p53-dependent apoptosis by repression of HPV oncogenes and upregulation of tumor suppressor proteins in human cervical cancer cells. Carcinogenesis 32:1697–1705
Nagalingam A, Kuppusamy P, Singh SV, Sharma D, Saxena NK (2014) Mechanistic elucidation of the antitumor properties of withaferin a in breast cancer. Cancer Res 74:2617–2629
Nishikawa Y, Okuzaki D, Fukushima K, Mukai S, Ohno S, Ozaki Y, Yabuta N, Nojima H (2015) Withaferin A induces cell death selectively in androgen-independent prostate cancer cells but not in normal fibroblast cells. PLoS One 10:e0134137
Owen LN (1962) Treatment of 30 cases of spontaneous tumours in dogs with triethylene glycol diglycidyl ether. Br J Cancer 16:441–452
Piccirillo S, Filomeni G, Brune B, Rotilio G, Ciriolo MR (2009) Redox mechanisms involved in the selective activation of Nrf2-mediated resistance versus p53-dependent apoptosis in adenocarcinoma cells. J Biol Chem 284:27721–27733
Quelle DE, Zindy F, Ashmun RA, Sherr CJ (1995) Alternative reading frames of the INK4a tumor suppressor gene encode two unrelated proteins capable of inducing cell cycle arrest. Cell 83:993–1000
Rayess H, Wang MB, Srivatsan ES (2012) Cellular senescence and tumor suppressor gene p16. Int J Cancer 130:1715–1725
Rosenberg B, Van Camp L, Grimley EB, Thomson AJ (1967) The inhibition of growth or cell division in Escherichia coli by different ionic species of platinum(IV) complexes. J Biol Chem 242:1347–1352
Roy RV, Suman S, Das TP, Luevano JE, Damodaran C (2013) Withaferin A, a steroidal lactone from Withania somnifera, induces mitotic catastrophe and growth arrest in prostate cancer cells. J Nat Prod 76:1909–1915
Rudnick SN, McDevitt JJ, First MW, Spengler JD (2009) Inactivating influenza viruses on surfaces using hydrogen peroxide or triethylene glycol at low vapor concentrations. Am J Infect Control 37:813–819
Samadi AK, Cohen SM, Mukerji R, Chaguturu V, Zhang X, Timmermann BN, Cohen MS, Person EA (2012) Natural withanolide withaferin A induces apoptosis in uveal melanoma cells by suppression of Akt and c-MET activation. Tumour Biol 33:1179–1189
Serrano M (1997) The tumor suppressor protein p16INK4a. Exp Cell Res 237:7–13
Serrano M, Lee H, Chin L, Cordon-Cardo C, Beach D, DePinho RA (1996) Role of the INK4a locus in tumor suppression and cell mortality. Cell 85:27–37
Shah S, Smith H, Feng X, Rancourt DE, Riabowol K (2009) ING function in apoptosis in diverse model systems. Biochem Cell Biol 87:117–125
Sharpless NE, Alson S, Chan S, Silver DP, Castrillon DH, DePinho RA (2002) p16(INK4a) and p53 deficiency cooperate in tumorigenesis. Cancer Res 62:2761–2765
Sharpless NE, Ramsey MR, Balasubramanian P, Castrillon DH, DePinho RA (2004) The differential impact of p16(INK4a) or p19(ARF) deficiency on cell growth and tumorigenesis. Oncogene 23:379–385
Spitzner M, Ebner R, Wolff HA, Ghadimi BM, Wienands J, Grade M (2014) STAT3: a novel molecular mediator of resistance to Chemoradiotherapy. Cancers (Basel) 6:1986–2011
Stott FJ, Bates S, James MC, McConnell BB, Starborg M, Brookes S, Palmero I, Ryan K, Hara E, Vousden KH, Peters G (1998) The alternative product from the human CDKN2A locus, p14(ARF), participates in a regulatory feedback loop with p53 and MDM2. EMBO J 17:5001–5014
Suman S, Das TP, Sirimulla S, Alatassi H, Ankem MK, Damodaran C (2016) Withaferin-A suppress AKT induced tumor growth in colorectal cancer cells. Oncotarget 7:13854–13864
Tan C, Tasaka H, Yu KP, Murphy ML, Karnofsky DA (1967) Daunomycin, an antitumor antibiotic, in the treatment of neoplastic disease. Clin Eval Spec Ref Childhood Leukemia Cancer 20:333–353

Tanenbaum ME, Macurek L, Janssen A, Geers EF, Alvarez-Fernandez M, Medema RH (2009) Kif15 cooperates with eg5 to promote bipolar spindle assembly. Curr Biol 19:1703–1711
Thaiparambil JT, Bender L, Ganesh T, Kline E, Patel P, Liu Y, Tighiouart M, Vertino PM, Harvey RD, Garcia A, Marcus AI (2011) Withaferin A inhibits breast cancer invasion and metastasis at sub-cytotoxic doses by inducing vimentin disassembly and serine 56 phosphorylation. Int J Cancer 129:2744–2755
Um HJ, Min KJ, Kim DE, Kwon TK (2012) Withaferin A inhibits JAK/STAT3 signaling and induces apoptosis of human renal carcinoma Caki cells. Biochem Biophys Res Commun 427:24–29
Vogt M, Haggblom C, Yeargin J, Christiansen-Weber T, Haas M (1998) Independent induction of senescence by p16INK4a and p21CIP1 in spontaneously immortalized human fibroblasts. Cell Growth Differ 9:139–146
Vyas AR, Singh SV (2014) Molecular targets and mechanisms of cancer prevention and treatment by withaferin a, a naturally occurring steroidal lactone. AAPS J 16:1–10
Wadhwa R, Kaul SC, Mitsui Y, Sugimoto Y (1993) Differential subcellular distribution of mortalin in mortal and immortal mouse and human fibroblasts. Exp Cell Res 207:442–448
Wadhwa R, Sugihara T, Yoshida A, Nomura H, Reddel RR, Simpson R, Maruta H, Kaul SC (2000) Selective toxicity of MKT-077 to cancer cells is mediated by its binding to the hsp70 family protein mot-2 and reactivation of p53 function. Cancer Res 60:6818–6821
Wadhwa R, Yaguchi T, Hasan MK, Mitsui Y, Reddel RR, Kaul SC (2002) Hsp70 family member, mot-2/mthsp70/GRP75, binds to the cytoplasmic sequestration domain of the p53 protein. Exp Cell Res 274:246–253
Wadhwa R, Takano S, Kaur K, Deocaris CC, Pereira-Smith OM, Reddel RR, Kaul SC (2006) Upregulation of mortalin/mthsp70/Grp75 contributes to human carcinogenesis. Int J Cancer 118:2973–2980
Wadhwa R, Singh R, Gao R, Shah N, Widodo N, Nakamoto T, Ishida Y, Terao K, Kaul SC (2013) Water extract of Ashwagandha leaves has anticancer activity: identification of an active component and its mechanism of action. PLoS One 8:e77189
Wadhwa R, Konar A, Kaul SC (2016) Nootropic potential of Ashwagandha leaves: beyond traditional root extracts. Neurochem Int 95:109–118
Widodo N, Kaur K, Shrestha BG, Takagi Y, Ishii T, Wadhwa R, Kaul SC (2007) Selective killing of cancer cells by leaf extract of Ashwagandha: identification of a tumor-inhibitory factor and the first molecular insights to its effect. Clin Cancer Res 13:2298–2306
Widodo N, Takagi Y, Shrestha BG, Ishii T, Kaul SC, Wadhwa R (2008) Selective killing of cancer cells by leaf extract of Ashwagandha: components, activity and pathway analyses. Cancer Lett 262:37–47
Widodo N, Priyandoko D, Shah N, Wadhwa R, Kaul SC (2010) Selective killing of cancer cells by Ashwagandha leaf extract and its component withanone involves ROS signaling. PLoS One 5:e13536
Xu HJ, Zhou Y, Ji W, Perng GS, Kruzelock R, Kong CT, Bast RC, Mills GB, Li J, Hu SX (1997) Reexpression of the retinoblastoma protein in tumor cells induces senescence and telomerase inhibition. Oncogene 15:2589–2596
Yang H, Wang Y, Cheryan VT, Wu W, Cui CQ, Polin LA, Pass HI, Dou QP, Rishi AK, Wali A (2012) Withaferin A inhibits the proteasome activity in mesothelioma in vitro and in vivo. PLoS One 7:e41214
Yco LP, Mocz G, Opoku-Ansah J, Bachmann AS (2014) Withaferin A inhibits STAT3 and induces tumor cell death in Neuroblastoma and multiple myeloma. Biochem Insights 7:1–13
Yoneyama T, Arai MA, Sadhu SK, Ahmed F, Ishibashi M (2015) Hedgehog inhibitors from Withania somnifera. Bioorg Med Chem Lett 25:3541–3544
Yu Y, Hamza A, Zhang T, Gu M, Zou P, Newman B, Li Y, Gunatilaka AA, Zhan CG, Sun D (2010) Withaferin A targets heat shock protein 90 in pancreatic cancer cells. Biochem Pharmacol 79:542–551

Yu H, Lee H, Herrmann A, Buettner R, Jove R (2014) Revisiting STAT3 signalling in cancer: new and unexpected biological functions. Nat Rev Cancer 14:736–746

Zhang X, Timmermann B, Samadi AK, Cohen MS (2012) Withaferin a induces proteasome-dependent degradation of breast cancer susceptibility gene 1 and heat shock factor 1 proteins in breast cancer cells. ISRN Biochem 2012:707586

Zhu Z, Luo Z, Li Y, Ni C, Li H, Zhu M (2009) Human inhibitor of growth 1 inhibits hepatoma cell growth and influences p53 stability in a variant-dependent manner. Hepatology 49:504–512

Chapter 8
Scientific Evidence for Anticancer Effects of *Withania somnifera* and Its Primary Bioactive Component Withaferin A

Anuradha Sehrawat, Suman K. Samanta, Su-Hyeong Kim, Eun-Ryeong Hahm, and Shivendra V. Singh

Abstract *Withania somnifera* (WS) plant has been used for centuries to cure or treat various disorders in the Ayurvedic medicine. Research over the years has indicated that withanolides are the primary bioactive constituents in WS. Scientific evidence for anticancer effects of WS root extract (WRE) is quite strong, and is derived from both *in vitro* cellular experiments and *in vivo* studies in rodent models of cancer. This article reviews scientific evidence supporting anticancer effects of WRE and its primary withanolide (withaferin A). The primary focus of the present article is on: (a) phytochemistry of WS, (b) withanolide biosynthesis, (c) pharmacokinetics, (d) *in vivo* evidence for anticancer activity of WRE and its primary bioactive component withaferin A (WA), and (e) effect of WA and WRE on cancer stem cell population and/or epithelial-mesenchymal transition. Unpublished results from our own laboratory are presented to demonstrate that WA is the most likely primary anticancer agent in WRE standardized for WA content (sWRE). The mechanisms underlying anticancer effects of WRE and WA have been reviewed extensively by us and others, and therefore are not elaborated in this article.

Keywords *Withania somnifera* • Withaferin A • Anticancer • Phytochemistry • Pharmacokinetics • Clinical trial

---

A. Sehrawat • S.K. Samanta • S.-H. Kim • E.-R. Hahm • S.V. Singh (✉)
Department of Pharmacology and Chemical Biology, University of Pittsburgh Cancer Institute, University of Pittsburgh School of Medicine,
2.32A Hillman Cancer Center, 5117 Centre Avenue, Pittsburgh, PA 15213, USA
e-mail: singhs@upmc.edu

S.C. Kaul, R. Wadhwa (eds.), *Science of Ashwagandha: Preventive and Therapeutic Potentials*, DOI 10.1007/978-3-319-59192-6_8

8.1 Introduction

Withania somnifera (also known as Ashwagandha, Indian winter cherry or Indian Ginseng), a plant belonging to the Solanaceae family, has been used in the Ayurvedic medicine for centuries. In Ayurvedic medicine, *Withania somnifera* (WS) is characterized as a Sattvic Kapha Rasayana (tonic). WS is a key ingredient in hundreds of Ayurvedic formulations. Some of the suggested medicinal benefits of WS include anti-aging effect, revitalization of the body, and mental comfort (Changhadi 1938; Mahima et al. 2012; Sharma 1999). WS grows in many parts of the world, including India, Africa, and Europe (Sharma et al. 2011; Uddin et al. 2012). It is generally available in the form of a fine powder that can be mixed and used with water, ghee (clarified butter) or honey (Singh et al. 2011). The root of WS is believed to be a tonic, diuretic, and stimulant. The leaves of WS are bitter and are recommended for alleviation of fever and painful swelling. The seeds are anthelmintic and combined with rock salt to remove white spots from the cornea as well as to increase sperm count and improve testicular growth (Dar et al. 2015; Singh et al. 2011). WS has been the subject of considerable scientific research, including clinical trials, supporting the use of WS for the treatment of anxiety, cognitive and neurological disorders, senile dementia, and Alzheimer's and Parkinson's diseases (Dar et al. 2015; Tiwari et al. 2014; Uddin et al. 2012; Winters 2006). It is the principal component in many commercially available poly-herbal preparations (Cecchini et al. 2014; Sangwan et al. 2004). Several preclinical studies have been conducted to evaluate the effectiveness of WS for the prevention and treatment of different malignancies. Extracts, derived from the WS root (WRE) or the whole plant, have been shown to possess anticancer activity through diverse yet converging pathways (Rai et al. 2016). Constituents of WS (withaferin A, withanone, etc.) have been shown to suppress proliferation of cancer cells without showing any toxic effects on normal cells (Samadi 2015; Vyas and Singh 2014). WRE significantly reduces lipid peroxidation and increase activities of anti-oxidative enzymes such as superoxide dismutase (SOD) and catalase, thus providing evidence for its anti-oxidant activity (Dhuley 1998; Panda and Kar 1997). Furthermore, the radio-sensitization and chemo-sensitization effects of WS/WA represent other attractive anticancer traits of this plant (Dar et al. 2015). This article reviews experimental evidence for antitumor efficacy of WS and its chemical constituent WA.

8.2 Phytochemistry of *Withania somnifera* (WS)

A variety of chemicals are present in WS, including alkaloids (withanine, somniferine, somninine, somniferinine, withananine, pseudowithanine, tropine, etc.), steroidal lactones (WA, withanolides A-Y, withasomniferin-A, withasomidienone, withasomniferols A-C, withanone, etc.), flavonoids (6,8-dihydroxykaempferol 3-*O*-rutinoside, quercetin and its 3-*O*-rutinoside and 3-*O*-rutinoside-7-*O*-glucoside),

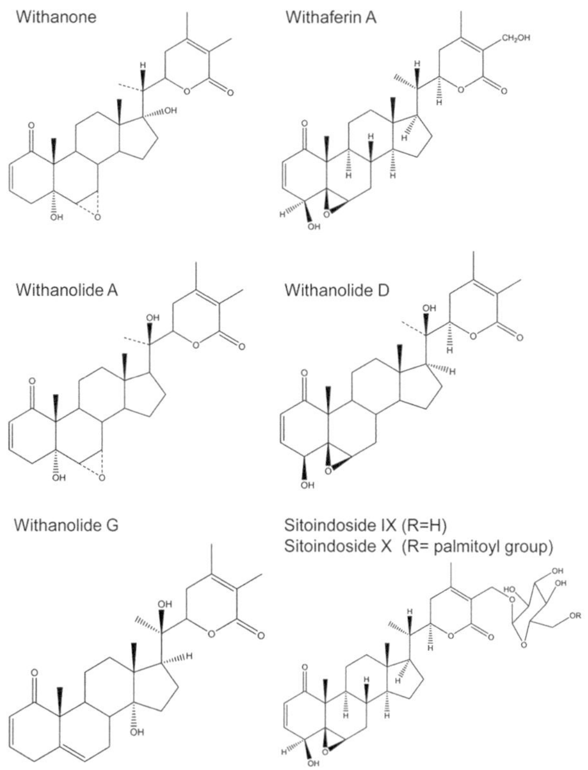

Fig. 8.1 Chemical structures of some chemicals present in WS

saponins, nitrogen containing compounds (withanol, somnisol, somnitol) and other chemicals (cuscohygrine, anahygrine, pseudotropine, anaferine) (Dar et al. 2015; John 2014; Kandil et al. 1994; Mirjalili et al. 2009; Misra et al. 2008). Two acyl steryl glucoside (sitoindoside VII and sitoindoside VIII) and two glycowithanoloids (sitoindoside IX and sitoindoside X) have also been isolated from the root of WS (Matsuda et al. 2001). Other chemical constituents like resin, fat, sugar, phytosterol, ipuranol and fatty acids are also present in WS (Umadevi et al. 2012). The chemical structures of some withanolides in WRE are exemplified in Fig. 8.1.

8.3 Withanolide Biosynthesis

Withanolides are a group of molecules having C-28 steroidal lactone backbone in which C-26 and C-22 or C-26 and C-23 are oxidized to form δ- or γ-lactone ring, respectively (Mirjalili et al. 2009; Misico et al. 2011). The withanolides are generally poly-oxygenated and believed to be produced *via* enzyme system capable of catalyzing oxidation of all carbon atoms in a steroid nucleus (Dhar et al. 2015).

Studies suggest that precursor molecules for the withanolide are the isoprenoids, that are synthesized via classical cytosolic mevalonic acid (MVA) and plastid localized 2-*C*-methyl-*D*-erythritol-4-phosphate (MEP) pathways leading to biosynthesis of 24-methylene cholesterol. This central molecule, through various biochemical reactions (desaturation, hydroxylation, glycosylation, formation of additional rings and addition of the side chains in the steroid backbone), leads to production of various withanolides (Dhar et al. 2015; Gupta et al. 2013a; Mirjalili et al. 2009; Misico et al. 2011). Withanolides are distributed in distinct amount and ratios in different tissues, developmental stages and chemotypes of the plant (Dhar et al. 2015; Gupta et al. 2013a; Sangwan et al. 2008). WA and withanone are the major metabolites in the leaves, whereas withanolide A (WLA) and withanolide D are principal metabolites found in the root (Gupta et al. 2015; Ray and Jha 2001). It is proposed that withanolides are first synthesized in the leaves followed by their transport to the root (Ray and Jha 2001; Singh et al. 2015). Both shoots and roots of WS contain WA and withanolide D, but only withanolide D could be detected in the excised non-transformed as well as transformed root cultures (Ray et al. 1996). This suggests that the *de novo* synthesis of withanolide D probably occurs in the root from primary isoprenogenic precursors. Tissue specific site of synthesis and accumulation of withanolides in WS has also been studied (Bharti et al. 2011; Chaurasiya et al. 2009; Gupta et al. 2011). The expressed sequence tag (EST) and transcriptome databases generated from the leaf and root of WS revealed presence of withanolide biosynthetic genes (Gupta et al. 2013b; Senthil et al. 2010; Senthil et al. 2015). A useful protocol to enhance the withanolide production by a diversity of biotic and abiotic elicitors has been developed in hairy root culture of WS (Sivanandhan et al. 2016). There have been several endeavors to clone and characterize withanolide biosynthetic pathway genes from WS (Akhtar et al. 2013; Dhar et al. 2014; Gupta et al. 2011; Rana et al. 2013). Differential expression of biosynthetic genes in different chemotypes, tissues and in response to elicitor suggested that roots and leaves might possess independent systems for withanolide synthesis (Singh et al. 2015). In addition, various *in vitro* studies with cell suspension, and shoot and root culture and in somaclonal variants have opened new avenues for pathway engineering or withanolide production (Dhar et al. 2015).

8.4 Pharmacokinetics

An understanding of the pharmacokinetic behavior (e.g., absorption, clearance, oral bioavailability, and metabolism) is necessary for drug development. Levels of WA and withanolide A (WLA) were measured in mice plasma after a single oral administration of aqueous extract of WS root at 1 g/kg suspended in 0.5% carboxymethyl cellulose sodium aqueous solution (Patil et al. 2013). The maximum plasma concentrations (C_{max}) for WA and WLA were about 17 ng/mL (~ 35 nM) and 27 ng/mL (~56.5 nM), respectively (Patil et al. 2013). The $T_{1/2}$ (elimination half-life) was about 60 min for WA and 45 min for WLA (Patil et al. 2013).

Following a single intraperitoneal (i.p.) administration of WA (4 mg/kg body weight), a mean C_{max} of 1.8 μM was achieved in mice plasma with a half-life of about 1.4 h (Thaiparambil et al. 2011). This study also indicated that WA is cleared rapidly (0.151 L/hour) as this agent was not detectable in plasma of any mice at the 24-h time point (Thaiparambil et al. 2011). A very recent study designed to determine the tissue distribution of WA in mice after a single i.p. administration (50 mg/kg body) revealed its bioavailability in the liver and intestine but not in the brain (Gambhir et al. 2015). Single oral administration of 500 mg/kg WS to healthy buffalo calves resulted in a mean peak plasma concentration of about 248 μg/mL after 0.75 h (Dahikar et al. 2012). This study also indicated that the mean therapeutic concentration of WS was maintained from 10 min to 3 h (Dahikar et al. 2012). However, the pharmacokinetic behavior of WA in humans is yet to be elucidated.

8.5 *In vivo* Evidence for Anticancer Effects of *Withania somnifera* (WS) Root Extract (WRE) and Withaferin A (WA)

8.5.1 Inhibition of Transplanted Tumor Growth by WS

Devi and colleagues (Devi et al. 1992) were the first to demonstrate *in vivo* growth inhibition of mouse Sarcoma 180 transplanted in BALB/c mice. Administration of the alcoholic extract of WRE (200 to 1000 mg/kg body weight), i.p., for 15 days beginning 24 h after intradermal inoculation of Sarcoma 180 cells resulted in complete tumor regression at the 400 mg/kg and higher doses (Devi et al. 1992). Some mortality was observed at the highest dose of 1000 mg/kg (Devi et al. 1992). *In vivo* anticancer efficacy of WRE against Sarcoma 180 was further established in a follow-up study from the same group of investigators (Devi et al. 1993). WRE administration at 20 mg/mouse dose significantly inhibited pulmonary colonization of B16F-10 melanoma in mice (Leyon and Kuttan 2004). WRE administration also increased survival of mice (Leyon and Kuttan 2004). *In vivo* tumor growth inhibitory effect of ethanol extract of WRE Dunal (200 mg/kg), reflected by a decrease in number of cancer cells, packed cell volume, and tumor weight, was also reported (Christina et al. 2004). The life span of mice treated with WRE Dunal was increased by about 28% when compared to control mice (Christina et al. 2004). Effect of subcutaneous administration of ether extract of WS leaves (termed as i-Extract) on *in vivo* growth of highly aggressive HT1080 fibrosarcoma subcutaneously implanted in female nude mice was determined with three different protocols, including: (a) treatment starting at the time of tumor cell injection; (b) treatment before tumor appearance; and (c) treatment after tumor appearance (Widodo et al. 2007). Tumor growth suppression was observed with all three protocols (Widodo et al. 2007). A mixture of the WRE and the leaf extract of WS in a 1:1 ratio, given i.p. (100–150 mg/kg) or orally (350 mg/kg), inhibited *in vivo* growth of Ehrlich ascitic and

Sarcoma 180 tumors by 41–61% (Malik et al. 2009). *In vivo* anti-metastatic effect of WRE standardized for WA content (termed as sWRE) was investigated using murine 4 T1 mammary carcinoma cells and human basal-like MDA-MB-231 breast cancer cells (Yang et al. 2013). Inhibition of primary tumor growth and pulmonary metastasis were observed in both models after oral administration (3 times/week) of 4 mg sWRE/kg and 8 mg sWRE/kg (Yang et al. 2013). Importantly, both 4 and 8 mg/kg doses administered orally for 35 days were well-tolerated by the female Balb/c mice (Yang et al. 2013). In the MDA-MB-231 model, sWRE (oral treatment) and WA (i.p. treatment) exhibited comparable efficacy suggesting that WA might be the primary anticancer agent at least in sWRE. Data from our laboratory shown later in this article further supports this possibility. Dietary administration of human grade WRE to laying hen resulted in a decrease in incidence and progression of ovarian cancer (Barua et al. 2013). Water extract of WS leaves (termed as ASH-WEX) (140 mg/kg or 4 mL ASH-WEX/kg body weight) administered orally on 21 consecutive days inhibited growth of C6 glioma cells in rats reflected by a 43.6% reduction in tumor volume (Kataria et al. 2015). In conclusion, WRE administration seems to inhibit *in vivo* growth of transplanted tumor in rodents. Because different doses and schedules of WRE administration were used in these studies, it is difficult to comment on treatment conditions optimal for humans. Another potential limitation of these studies relates to lack of standardization for withanolide content (e.g., WA), which can vary greatly in different WRE preparations. Hopefully, future studies will use only well-characterized (standardized for WA or any other withanolide) WRE preparations for antitumor evaluations. In fact, a very recent study showed *in vivo* activity of a standardized WRE preparation (AshwaMAX containing 4.3% WA and 8.4% total withanolides) against patient-derived glioblastoma multiforme cell line GBM39 (P4–9) injected into the right parietal region of the brain of female nude mice (Chang et al. 2016). AshwaMAX was given by gavage every other day at 40 mg/kg/day dose (Chang et al. 2016).

8.5.2 Cancer Chemoprevention by WS

Cancer chemoprevention by WS/WRE was initially studied using chemically induced rodent models. Prevention of 20-methylcholanthrene-induced sarcoma in mice was achieved with i.p. administration of 20 mg WRE/mouse (Davis and Kuttan 2000a). Chemoprevention of 20-methylcholanthrene-induced fibrosarcoma was also observed after oral administration of 400 mg WRE/kg body weight (Prakash et al. 2001). The overall incidence of skin papilloma induced by dimethyl benzanthracene (DMBA) and croton oil was reduced by 50% upon i.p. administration of 20 mg WRE/mouse prior to and after DMBA administration (Davis and Kuttan 2001). Significant increase in antioxidants, including reduced glutathione, glutathione *S*-transferase, glutathione peroxidase, and catalases were observed in the skin and liver of WRE-treated mice when compared with the controls (Davis and Kuttan 2001). Oral administration of 400 mg WRE/kg body weight (3 times per week on alternate days) 1 week before carcinogen challenge and 24 weeks thereafter inhibited incidence and multiplicity of squamous cell carcinoma of the skin induced by DMBA (Prakash et al. 2002). Moreover, WRE administration was able to restore imbalances in antioxidants (e.g., reduced glutathione, superoxide dismutase, catalase, etc.) and oxidant (malondialdehyde) levels (Prakash et al. 2002). Dietary administration of WRE (2.5 and 5%, w/w) to mice inhibited benzo(a)pyrene-induced forestomach papilloma incidence by 60% and 13%, respectively, and DMBA-induced skin papilloma incidence (Padmavathi et al. 2005). Skin tumor multiplicity was also decreased upon feeding of mice with WRE-supplemented diets (Padmavathi et al. 2005). Muralikrishnan et al. (2010) provided evidence for constraint of azoxymethane-induced colon cancer and immune dysfunction in mice by oral administration of 400 mg WS extract/kg once every week for 4 weeks. Chemopreventive effect of WRE against estrogen receptor-positive breast cancer was examined using *N*-methyl-*N*-nitrosourea-rat model (Khazal et al. 2013). WRE (150 mg/kg body weight/day) was administered to rats by daily gavage starting 1 week prior to the carcinogen injection and continuing for the duration of the study (155 days) (Khazal et al. 2013). Even though tumor multiplicity and burden (wet weight) were modestly lower (21–23%) in the WRE treatment group compared with the rats of the control group, the differences were not statistically significant (Khazal et al. 2013). Interestingly, the labeling indices for proliferation markers Ki-67 and PCNA were significantly lower (42% and 38%, respectively) in the WRE treatment group compared with control (Khazal et al. 2013). One potential limitation of the study was that the WRE was not standardized for withanolide content (Khazal et al. 2013). Dietary administration 750 mg WRE/kg diet for 10 months produced similar effects in estrogen receptor-negative breast cancer in MMTV-*neu* transgenic mouse model (i.e., a trend for a decrease in tumor multiplicity and wet tumor weight that was statistically not significant but a significant decrease in tumor cell proliferation markers) (Khazal et al. 2014).

8.5.3 *In vivo Evidence for Anticancer Efficacy of WA*

In vitro anticancer effect of WA was first shown in the late sixties (Shohat et al. 1967), and a few years later, the same group demonstrated its *in vivo* efficacy against Ehrlich ascites tumor model (Shohat and Joshua 1971). The Ehrlich model was used to further characterize the *in vivo* activity of WA by i.p. injection at 10–60 mg/kg, and statistically significant and dose-dependent tumor growth inhibition was reported (Devi et al. 1995). The *in vivo* anticancer activity of WA was subsequently extended to other solid tumor xenograft models. The first evidence of solid tumor xenograft growth inhibition by WA was obtained using PC-3 human prostate cancer cells after daily i.p. injections with 4 or 8 mg/kg body weight (Yang et al. 2007). Tumor growth inhibition of 54 to 70% was reported after 24 days of WA treatment (Yang et al. 2007). Notably, only 7 days of treatment resulted in complete tumor regression in one mouse of the 4 mg WA/kg group (Yang et al. 2007). The same

PC-3 prostate cancer xenograft model was used to demonstrate tumor growth inhibition following intratumoral injections at 5 mg WA/kg body weight (Srinivasan et al. 2007). We demonstrated that five times per week i.p. injection of 4 mg WA/kg resulted in >40% inhibition of orthotopic MDA-MB-231 xenograft growth (Stan et al. 2008). While these examples serve to highlight early studies demonstrating *in vivo* efficacy of WA against solid tumors, growth inhibition has also been reported in xenograft models of Panc-1 pancreatic tumor (Yu et al. 2010), leiomyosarcoma (SKLMS1) and fibrosarcoma (HT1080) (Lahat et al. 2010), medullary thyroid carcinoma (DRO-81-1) (Samadi et al. 2010), CaSki cervical cancer cells (Munagala et al. 2011), glioma (Santagata et al. 2012), 92.1 uveal melanoma cells (Samadi et al. 2012), AB12 murine mesothelioma cells (Yang et al. 2012), MDA-MB-231 basal-like breast cancer (Nagalingam et al. 2014), HCT116 human colon cancer (Choi and Kim 2015), and Akt overexpressing HCT116 colon cancer cells (Suman et al. 2016). WA administration was also effective in suppressing B-cell lymphoma in a syngeneic-graft model (McKenna et al. 2015). Notably, i.p. route was employed in most of these *in vivo* studies. In future studies, it is important to test anticancer efficacy of WA after oral administration. Short plasma half-life of WA also suggests that frequent administration (twice or thrice daily) may produce even greater antitumor response than that previously observed with single daily treatment. Nevertheless, WA shows *in vivo* efficacy against a variety of human and rodent cancer cells.

8.5.4 Evidence for Cancer Chemoprevention by WA

A handful of studies have assessed cancer chemoprevention by WA. Cancer chemopreventive efficacy of WA has been established in DMBA-induced oral cancer model in golden Syrian hamsters (Manoharan et al. 2009). Oral treatment with 20 mg WA/kg body weight for 14 weeks resulted in complete prevention of oral tumorigenesis (Manoharan et al. 2009). Our laboratory was the first to demonstrate mammary cancer prevention by WA in a clinically relevant transgenic mouse model (MMTV-*neu*) (Hahm et al. 2013). Unlike data with dietary WS administration (Khazal et al. 2014), WA administration by i.p. route at a modest regimen of 100 μg/mouse/day (three times per week) resulted in a statistically significant decrease in tumor burden (both macroscopic and microscopic) compared with control MMTV-*neu* mice but the overall incidence was not affected (Hahm et al. 2013). A reduction in tumor incidence may be possible with a more intense dosing regimen (i.e., higher dose and/or daily administration), but further work is needed to explore this possibility. Nevertheless, biomarkers of mammary cancer prevention by WA in MMTV-*neu* mice included inhibition of energy pathways including complex III of the mitochondrial electron transport chain, glycolysis and tricarboxylic acid pathway (Hahm et al. 2013). Breast cancer prevention by WA in this model was also associated with apoptosis induction (Hahm et al. 2013). A more recent study demonstrated prevention of DMBA-initiated and 12-*O*-tetradecanoylphorbol-13-acetate-promoted skin carcinogenesis in DBA/2 mice with topical application of WA (Li et al. 2015a). Further research is needed to establish chemopreventive efficacy of WA for other cancer types.

8.5.5 Anticancer Activity of WA in Combination with Chemotherapy Drugs

Anticancer efficacy of WA in combination with chemotherapy drugs has also been investigated (Cohen et al. 2012; Kakar et al. 2012; Lee et al. 2009; Yang et al. 2011a, b). These studies were undertaken to identify efficacious combinations for minimizing side effects of chemotherapy drugs. Examples include WA in combination with TRAIL, sorafenib, cisplatin, and doxorubicin. For example, WA and sorafenib combination exhibited synergistic efficacy in thyroid cancer cells (Cohen et al. 2012). Sensitization of Caki renal cancer cells to TRAIL-induced apoptosis by WA was also shown (Lee et al. 2009). Treatment of ovarian cancer cell lines with WA and doxorubicin combination exhibited time- and dose-dependent synergistic effect on cell proliferation inhibition and apoptosis induction (Fong et al. 2012). This study also provided *in vivo* evidence for the superiority of the WA/doxorubicin combination (Fong et al. 2012). Purified extract of WS containing 1.5% withanolides protected against doxorubicin-induced cardiotoxicity in rats (Hamza et al. 2008). WA-mediated sensitization of ovarian cancer cells to cisplatin-induced apoptosis *in vitro* has also been reported (Kakar et al. 2012). WA augmented oxaliplatin-mediated growth inhibition in pancreatic cancer cells (Li et al. 2015b). This chemosensitization was attributable to mitochondrial dysfunction and inactivation of the PI3K/AKT pathway (Li et al. 2015b). More importantly, WA/oxaliplatin combination therapy exhibited the strongest antitumor effect compared with single agents *in vivo* without additional toxicity (Li et al. 2015b). However, most of these studies have used *in vitro* cellular models to study antitumor effect of the combination. *In vivo* data establishing anticancer synergy/additivity is needed to warrant clinical investigations of WA in combination with other agents like sorafenib.

8.6 Inhibition of Cancer Stem-Like Cells by WA

The cancer stem-like cell hypothesis implicates this cell population in tumor initiation and progression as well as in resistance to chemotherapy (Akbari-Birgani et al. 2016). Our laboratory was the first to provide *in vitro* as well as *in vivo* evidence for inhibition of breast cancer stem cell (bCSC) by WA (Kim and Singh 2014). Specifically, bCSC markers such as mammosphere frequency, aldehyde dehydrogenase 1 (ALDH1) activity, and $CD44^{high}/CD24^{low}$/epithelial-specific antigen-positive (ESA^{+}) fraction were decreased significantly *in vitro* in the presence of plasma achievable concentrations of WA (Kim and Singh 2014). More importantly, the WA-mediated prevention of breast cancer development in MMTV-*neu* mice *in vivo* was accompanied by inhibition of mammosphere number and ALDH1 activity (Kim and Singh 2014). An induced CSC-like model in MCF-10A cells also showed inhibition of stem cell activity by WA (Nishi et al. 2014). Since these two initial reports showing inhibition of bCSC's, it was shown that WA alone as well as in combination with cisplatin could suppress growth and metastasis of ovarian cancer cells by targeting cancer stem-like cells (Kakar et al. 2014). *In vitro* inhibition of neurosphere formation in glioblastoma multiforme was also shown (Chang et al. 2016). On the other hand, ethanolic extract of WS failed to inhibit side population suggesting lack of any activity against cancer stem-like cells (Maliyakkal et al. 2015). Further research using *in vivo* models is necessary to determine whether WA and/or WRE affect cancer stem-like cells in other types of malignancies.

8.7 Comparative Efficacy of WRE and WA in Breast Cancer

We compared the growth-inhibitory efficacy of sWRE and WA against breast cancer cells using basal-like MDA-MB-231 (Fig. 8.2a) and SUM159 (Fig. 8.2b) cells. WA and sWRE exhibited nearly comparable activity in both cell lines (Fig. 8.2a, b). Recently, we

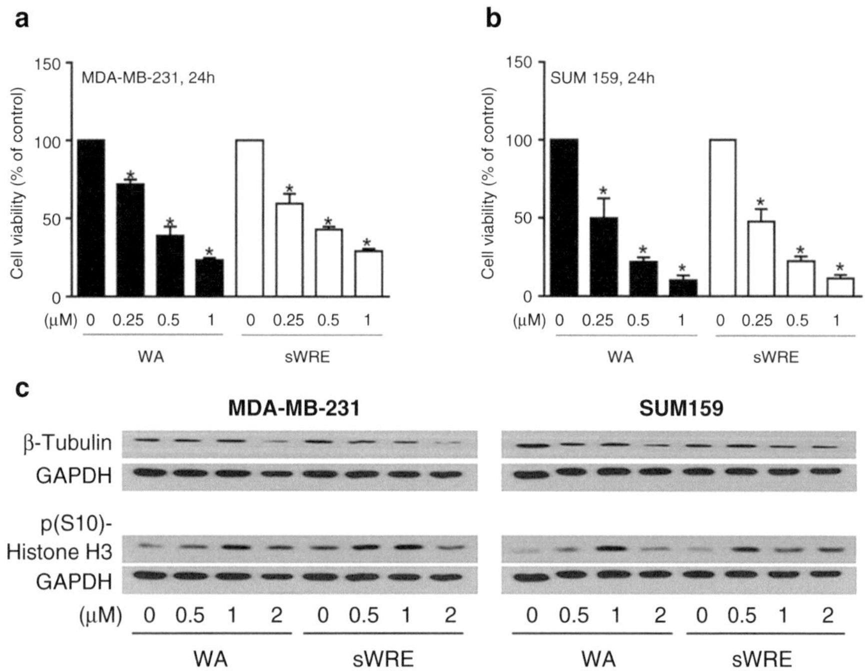

Fig. 8.2 Effect of WA and sWRE on viability of (**a**) MDA-MB-231 and (**b**) SUM159 cells (24 h treatment). Cell viability was determined by trypan blue dye exclusion assay. Results shown are mean ± SD (n = 3). *Significantly different ($P < 0.05$) compared with vehicle-treated control by one-way ANOVA followed by Dunnett's test. The results were consistent in replicate experiments. (**c**) Western blotting for β-Tubulin and p-(S10)-Histone H3 using lysates from MDA-MB-231 and SUM159 cells after 24 h treatment with vehicle (control), WA or sWRE

identified β-tubulin as a novel target of WA-mediated growth arrest in human breast cancer cells (Antony et al. 2014). We observed that in human breast cancer cells growth arrest by WA was associated with downregulation and covalent binding at Cysteine 303 of β-tubulin (Antony et al. 2014). Both WA and sWRE were able to downregulate β-tubulin in these cells (Fig. 8.2c). WA and sWRE treatments also increased levels of S10 phosphorylated histone H3, a marker of mitotic arrest (Fig. 8.2c). Similar to WA, treatment of breast cancer cells with sWRE resulted in a dose-dependent and statistically significant decrease in first- and second-generation mammosphere frequency in both MCF-7 and SUM159 cell lines (Fig. 8.3a, b). Inhibitory effect of sWRE on bCSC fraction was confirmed by flow cytometric analysis of ALDH1 activity and $CD44^{high}/CD24^{low}/ESA^{+}$ fraction (Fig. 8.3c, d). The ALDH1 activity was decreased significantly in the presence of 50 and 100 nM of sWRE in both cell lines when compared with control (Fig. 8.3c). In comparison with vehicle-treated control, the $CD44^{high}/CD24^{low}/ESA^{+}$ fraction was significantly lower in the sWRE-treated MCF-7 and SUM159 cultures (Fig. 8.3d). These results suggested that WA might be the primary anticancer component in sWRE. However, further studies are needed to establish the *in vivo* efficacy of sWRE for elimination of cancer stem-like cells.

8.8 Inhibition of Epithelial to Mesenchymal Transition (EMT)

The EMT is a process by which epithelial cells acquire spindle-shaped morphology leading to reduced cell-cell contact but increased motility (Thiery 2002). EMT is characterized by downregulation of epithelial adherens junction proteins (e.g., E-cadherin) and induction of mesenchymal markers such as vimentin (Thiery 2002). Accumulating evidence indicates that EMT promotes CSC properties such as the ability to self-renew and initiate tumors (Hollier et al. 2009). Because WA inhibited CSC self-renewal, the EMT modifying effects of this agent were also studied. We were the first to demonstrate that both sWRE and WA inhibited EMT (Lee et al. 2015; Yang et al. 2013). sWRE inhibited breast cancer cell motility and invasion at 1 μM dose that was associated with disruption of the vimentin morphology in cell lines (Yang et al. 2013). The TGF-β-induced EMT in MCF-10A cells was prevented in the presence of sWRE (Yang et al. 2013). A follow-up study from our laboratory using WA confirmed its anti-EMT effect in breast cancer cells (Lee et al. 2015). Specifically, experimental EMT induced by exposure of MCF-10A cells to TNF-α and TGF-β combination was partially attenuated in the presence of WA, whereas its close naturally-occurring structural analogs withanone or withanolide A (WLA) (Fig. 8.1) were practically inactive (Lee et al. 2015). Cell migration inhibition by WA was also partially attenuated by TNF-α and TGF-β treatments (Lee et al. 2015). More importantly, vimentin protein was significantly downregulated in the MDA-MB-231 xenografts as well as in the tumors from the WA-treated

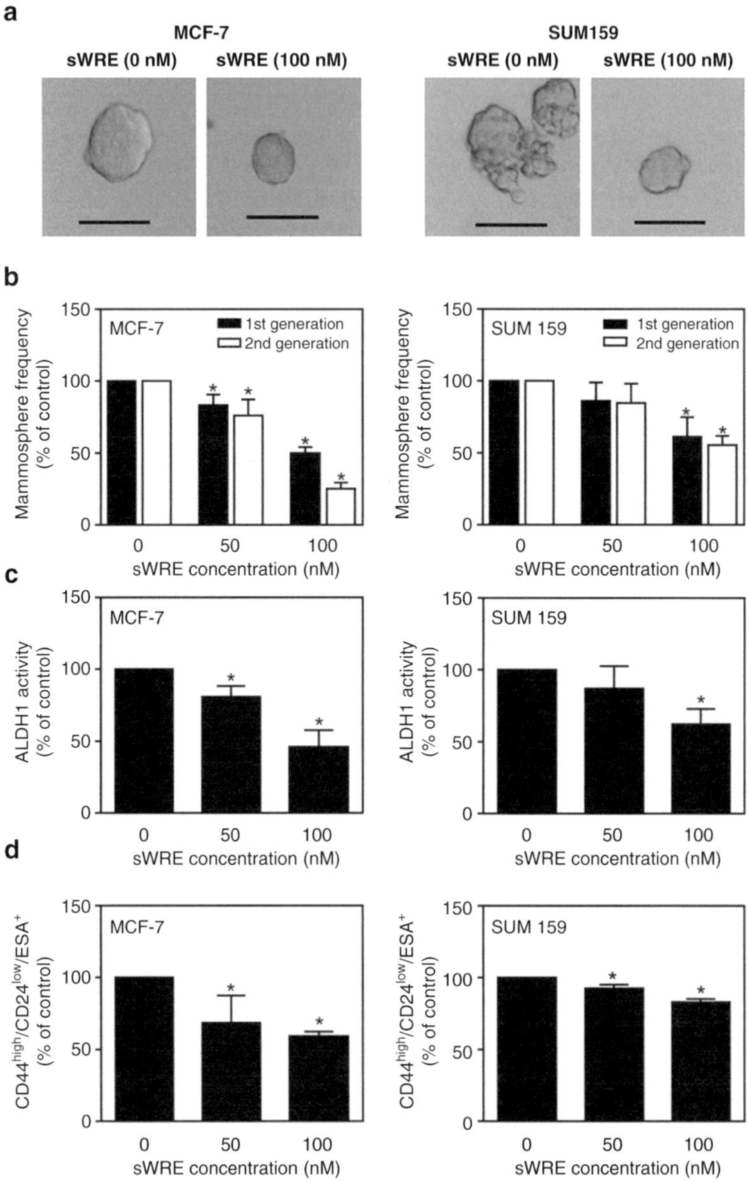

Fig. 8.3 sWRE inhibits bCSC population. (**a**) Representative images of first generation of mammospheres after 5 days of treatment with vehicle (control) or 100 nM sWRE (100× magnification; scale bar = 100 μm). (**b**) Percentage of first (after 5 days of cell seeding) and second generation (after 7 days of cell seeding) mammospheres relative to corresponding vehicle-treated control. (**c**) Percentage of ALDH1 activity (relative to corresponding control) in MCF-7 and SUM159 cells after 72 h treatment with vehicle (control) or 50 or 100 nM of sWRE. (**d**) Bar graphs showing quantitation of CD44 ^{high}/CD24^{low}/ESA^{+} fraction in MCF-7 and SUM159 cells relative to corresponding vehicle-treated control. Results shown are mean ± SD (n = 3). *Significantly different ($P < 0.05$) compared with control by one-way ANOVA with Dunnett's adjustment. Comparable results were observed in two independent experiments. Representative data from one such experiment are shown

MMTV-*neu* mice when compared with controls (Lee et al. 2015) providing first evidence of WA-mediated *in vivo* inhibition of biochemical features of EMT (Lee et al. 2015). These results support anti-metastatic effect of WA reviewed elsewhere (Vyas and Singh 2014).

8.9 Immunomodulatory Effects

The immunomodulatory effects of WS have been described in a number of animal studies with evidence for both immunostimulatory and immunosuppressive properties (Dar et al. 2016; Kuttan 1996; Tiwari et al. 2014; Ziauddin et al. 1996). WS extract showed profound effect on hematopoietic system by acting as an immunoregulator and chemoprotective agent (Tiwari et al. 2014). Extract of WS increased cell-mediated immunity in normal mice by enhancing the proliferation of lymphocytes, bone marrow cells and thymocytes in responses to mitogens (Davis and Kuttan 2002). Davis and Kuttan (2000b) showed that WRE inhibited delayed type hypersensitivity reaction (Mantoux test) and enhanced the phagocytic activity of peritoneal macrophages in mice. WRE enhanced the levels of interferon gamma (IFN-γ), interleukin-2 (IL-2), and granulocyte macrophages colony stimulating factor (GM-CSF) in mice (Davis and Kuttan 1999). Immune enhancement with WS extract was observed in mice with myelosuppression induced by cyclophosphamide (Davis and Kuttan 1998; Davis and Kuttan 1999; Tiwari et al. 2014). Treatment with the extract counteracted the cyclophosphamide, azathioprin or prednisolone-induced immunosuppression by causing a significant increase in hemagglutinin antibody and hemolytic antibody response towards human erythrocytes (Ziauddin et al. 1996). Administration of WS could also enhance total WBC count, bone marrow cellularity as well as α-esterase positive cells in Swiss albino mice treated with cyclophosphamide (Davis and Kuttan 1998). BALB/c mice treated with WS extract exhibited inhibition of ochratoxin A-induced suppression of chemotactic activity and productions of IL-1 and TNF-α by macrophages (Mahima et al. 2012). WS extract could reduce the ovalbumin-induced paw edema in mice, almost similar to that of standard drug disodium chromoglycate (Agarwal et al. 1999). Ethanolic extract of WS enhanced both humoral and cell-mediated immune responses in BALB/c mice under immune suppressed conditions. WRE has been reported to induce helper T lymphocyte (Th1) polarized cell-mediated immune response in BALB/c mice (Malik et al. 2007). Oral administration of aqueous fraction of WRE caused significant increase in the stress-induced depleted T-cell population and increased the expression of Th1 cytokines in chronically stressed mice (Khan et al. 2006). In support of these findings, WS was reported to possess immune potentiating and myelo-protective effect (John 2014). Standardized herbal formulation of WS (WSF) repressed pStat-3, with a selective stimulation of Th1 immunity as demonstrated by enhanced secretion of IFN-γ and IL-2 (Malik et al. 2009). WSF was also harmless when given orally up to 1500 mg/kg to rats for 6 months with augmentation of proliferation of CD4(+)/CD8(+) and NK cells along with an amplified expression of CD40/CD40L/CD80 (Malik et al. 2009). WS treatment lowered the levels of TNF-α (Davis and Kuttan 1999). Administration of WS could reduce leucopenia, enhance bone marrow cellularity and normalize the ratio of normochromatic to polychromatic erythrocytes in mice treated with sub-lethal dose of gamma radiation (Kuttan 1996). Extract of WS showed immunomodulatory effects on azoxymethane-induced colon cancer in Swiss albino mice when treated with the plant extract at 400 mg/kg body weight once a week for 4 weeks (Muralikrishnan et al. 2010). WA and Withanolide E (WE) exhibited specific immunosuppressive effect on human B and T lymphocytes and on mice thymocytes. WE showed specific effect on T lymphocytes whereas WA influenced both B and T lymphocytes (John 2014).

8.10 Clinical Trials

Cancer preventive or therapeutic effects of WRE or WA in humans are yet to be determined. However, a few studies have studied the effect of WS on the immune system, central nervous system, and general growth promotion. Many of these studies were inconclusive due to small sample size, inadequate treatment methods or the short trial duration. In an open-label, prospective, nonrandomized comparative trial on 100 patients with breast cancer of all stages, it was noted that WRE (2 g every 8 h) has the potential to improve quality-of-life with alleviation of cancer-related fatigue (Biswal et al. 2013). In this study, the quality-of-life and fatigue scores were evaluated before, during, and on the last cycle of chemotherapy (Biswal et al. 2013). Another Phase I study on 25 individuals (daily administration of ~6 gm of WS extract for 5 days) aimed to examine the effects on stress, inflammation and immune cell activation (ClinicalTrials.gov Identifier: NCT00817752), showed that WS increased the number and activation of CD4+ T cells (Mikolai et al. 2009). WS has been studied for its effect on anxiety and stress. Sixty-four participants were randomly divided into either placebo group or WS treatment arm. WS arm subjects received one capsule containing 300 mg of high-concentration full-spectrum extract from the root of the WS twice a day for a period of 60 days (Chandrasekhar et al. 2012). Significant differences were found for all outcome measures in WS treated group, including scores on the perceived stress scale, the general health questionnaire, and levels of cortisol in the bloodstream relative to the placebo group (Chandrasekhar et al. 2012). However, validity of the study was compromised by a small sample size (Chandrasekhar et al. 2012). In two independent double-blind intervention studies in India with healthy volunteers (age $\geq$ 55 years), consumption of natural care tea fortified with five herbs including WS significantly improved the NK cell activity of the volunteers in comparison with a population consuming regular tea (Bhat et al. 2010). These limited controlled clinical trials reported no serious adverse events. However, clinical trials are still needed to evaluate cancer preventive or therapeutic effects of WRE and WA.

8.11 Toxicology

WS is considered a safe herb as it has been used for centuries in the Ayurvedic medicine without any toxicity. Safety of WS formulation was studied in eighteen apparently healthy volunteers (12 males and 6 females, age: 18–30 years, and BMI: 19–30). This volunteer study demonstrated that WS, when given in the form of aqueous extract in capsules with gradual escalating doses from 750 to 1250 mg/day, was well-tolerated (Raut et al. 2012). The formulation was found to be safe based on hematological and biochemical organ function tests and also demonstrated muscle strengthening, lipid lowering, and improved quality of sleep. Only one volunteer showed increased appetite, libido, and hallucinogenic effects with vertigo at the lowest dose and was withdrawn from the study (Raut et al. 2012). Malhotra et al. (1965a) tested acute toxicity of 2% suspension of ashwagandholine (total alkaloids from the WRE) prepared in ten-percent glycol using two percent gum acacia as a suspending agent in rats and mice. The acute LD_{50} value was found to be 465 mg/kg (332–651 mg/kg) in rats and 432 mg/kg (299–626 mg/kg) in mice (Malhotra et al. 1965a; Mishra et al. 2000). An effect on spontaneous motor activity was observed at higher doses (Malhotra et al. 1965a, b). In a chronic dosing study, repeated injections of WS extract for 30 days to rats did not produce any mortality (Sharada et al. 1993; Tiwari et al. 2014). In a prenatal developmental toxicity evaluation of WS, root extract at the dose levels of 500, 1000, and 2000 mg/kg/day, given orally to pregnant rats during the period of major organogenesis and histogenesis (days 5 to 19 of gestation), no evidence of maternal or developmental fetal toxicity was observed (Prabu and Panchapakesan 2015). No changes in body weight of parental females, number of corpora lutea, implantations, viable fetuses, and external, skeletal or visceral malformations were observed (Prabu and Panchapakesan 2015). Aphale et al. (1998) investigated the subacute toxicity with WS in rats by measuring various safety parameters. WS administered for a period of 90 days did not show any toxicity (Aphale et al. 1998).

8.12 Concluding Remarks

Preclinical studies in the past have mainly focused on the mechanism by which WA induces apoptosis in cancer cells (reviewed in Vyas and Singh 2014). In contrast, the mechanisms underlying inhibition of EMT and cancer stem-like traits by WRE/WA are still poorly understood. Animal studies have provided *in vivo* evidence for anticancer activity of WRE and WA against some cancer types, which should now be extended to other cancer types because of promising *in vitro* data in cellular cancer models. Cancer chemopreventive effect of WA/WRE has only been studied in a few types of cancer including breast and oral cancers. It is highly likely that these agents are able to prevent other types of cancers, but this speculation requires experimental validation using chemically induced or transgenic models representing different cancers. Clinical development of WRE or WA for cancer prevention and/ or treatment should also be a top priority. Clinical translation of WA is expected to be slow because of regulatory issues. On the other hand, clinical trials with standardized WRE are feasible because of comparable anticancer effects of WA and WRE at least in breast cancer cells (present study).

Acknowledgements The work cited from the corresponding author's laboratory was supported by the US PHS grant CA142604 awarded by the National Cancer Institute.

References

Agarwal R, Diwanay S, Patki P, Patwardhan B (1999) Studies on immunomodulatory activity of *Withania somnifera* (Ashwagandha) extracts in experimental immune inflammation. J Ethnopharmacol 67:27–35

Akbari-Birgani S, Paranjothy T, Zuse A, Janikowski T, Cieślar-Pobuda A, Likus W, Urasińska E, Schweizer F, Ghavami S, Klonisch T, Łos MJ (2016) Cancer stem cells, cancer-initiating cells and methods for their detection. Drug Discov Today 21:836–842

Akhtar N, Gupta P, Sangwan NS, Sangwan RS, Trivedi PK (2013) Cloning and functional characterization of 3-hydroxy-3-methylglutaryl coenzyme A reductase gene from *Withania somnifera*: an important medicinal plant. Protoplasma 250:613–622

Antony ML, Lee J, Hahm ER, Kim SH, Marcus AI, Kumari V, Ji X, Yang Z, Vowell CL, Wipf P, Uechi GT, Yates NA, Romero G, Sarkar SN, Singh SV (2014) Growth arrest by the antitumor steroidal lactone withaferin A in human breast cancer cells is associated with down-regulation and covalent binding at cysteine 303 of β-tubulin. J Biol Chem 289:1852–1865

Aphale AA, Chhibba AD, Kumbhakarna NR, Mateenuddin M, Dahat SH (1998) Subacute toxicity study of the combination of ginseng (*Panax ginseng*) and Ashwagandha (*Withania somnifera*) in rats: a safety assessment. Indian J Physiol Pharmacol 42:299-302

Barua A, Bradaric MJ, Bitterman P, Abramowicz JS, Sharma S, Basu S, Lopez H, Bahr JM (2013) Dietary supplementation of *Ashwagandha* (*Withania somnifera*, Dunal) enhances NK cell function in ovarian tumors in the laying hen model of spontaneous ovarian cancer. Am J Reprod Immunol 70:538–550

Bharti SK, Bhatia A, Tewari SK, Sidhu OP, Roy R (2011) Application of HR-MAS NMR spectroscopy for studying chemotype variations of *Withania somnifera* (L.) Dunal. Magn Reson Chem 49:659–667

Bhat J, Damle A, Vaishnav PP, Albers R, Joshi M, Banerjee G (2010) *In vivo* enhancement of natural killer cell activity through tea fortified with Ayurvedic herbs. Phytother Res 24:129–135

Biswal BM, Sulaiman SA, Ismail HC, Zakaria H, Musa KI (2013) Effect of *Withania somnifera* (Ashwagandha) on the development of chemotherapy-induced fatigue and quality of life in breast cancer patients. Integr Cancer Ther 12:312–322

Cecchini S, Paciolla M, Caputo AR, Bavoso A (2014) Antioxidant potential of the polyherbal formulation "ImmuPlus": a nutritional supplement for horses. Vet Med Int 2014:434239

Chandrasekhar K, Kapoor J, Anishetty S (2012) A prospective, randomized double-blind, placebo-controlled study of safety and efficacy of a high-concentration full-spectrum extract of Ashwagandha root in reducing stress and anxiety in adults. Indian J Psychol Med 34:255–262

Chang E, Pohling C, Natarajan A, Witney TH, Kaur J, Xu L, Gowrishankar G, D'Souza AL, Murty S, Schick S, Chen L, Wu N, Khaw P, Mischel P, Abbasi T, Usmani S, Mallick P, Gambhir SS (2016) AshwaMAX and Withaferin A inhibits gliomas in cellular and murine orthotopic models. J Neurooncol 126:253–264

Changhadi GS (1938) Ashwagandharishta-Ras Tantra Saar and Sidhyaprayog Sangrah. Krishna-Gopal Ayurveda Bhawan (Dharmarth Trust), Nagpur, pp 743–744

Chaurasiya ND, Sangwan RS, Misra LN, Tuli R, Sangwan NS (2009) Metabolic clustering of a core collection of Indian ginseng *Withania somnifera* Dunal through DNA, isoenzyme, polypeptide and withanolide profile diversity. Fitoterapia 80:496–505

Choi BY, Kim BW (2015) Withaferin-A inhibits colon cancer cell growth by blocking STAT3 transcriptional activity. J Cancer Prev 20:185–192

Christina AJ, Joseph DG, Packialakshmi M, Kothai R, Robert SJ, Chidambaranathan N, Ramasamy M (2004) Anticarcinogenic activity of *Withania somnifera* Dunal against Dalton's ascitic lymphoma. J Ethnopharmacol 93:359–361

Cohen SM, Mukerji R, Timmermann BN, Samadi AK, Cohen MS (2012) A novel combination of withaferin A and sorafenib shows synergistic efficacy against both papillary and anaplastic thyroid cancers. Am J Surg 204:895–901

Dahikar PR, Kumar N, Sahni YP (2012) Pharmacokinetics of *Withania somnifera* (Ashwagandha) in healthy buffalo calves. Buffalo Bull 31:219–223

Dar NJ, Hamid A, Ahmad M (2015) Pharmacologic overview of *Withania somnifera*, the Indian Ginseng. Cell Mol Life Sci 72:4445–4460

Dar PA, Singh LR, Kamal MA, Dar TA (2016) Unique medicinal properties of *Withania somnifera*: phytochemical constituents and protein component. Curr Pharm Des 22:535–540

Davis L, Kuttan G (1998) Suppressive effect of cyclophosphamide-induced toxicity by *Withania somnifera* extract in mice. J Ethnopharmacol 62:209–214

Davis L, Kuttan G (1999) Effect of *Withania somnifera* on cytokine production in nol and cyclophosphamide treated mice. Immunopharmacol Immunotoxicol 21:695–703

Davis L, Kuttan G (2000a) Effect of *Withania somnifera* on 20-methylcholanthrene induced fibrosarcoma. J Exp Clin Cancer Res 19:165–167

Davis L, Kuttan G (2000b) Immunomodulatory activity of *Withania somnifera*. J Ethnopharmacol 71:193–200

Davis L, Kuttan G (2001) Effect of *Withania somnifera* on DMBA induced carcinogenesis. J Ethnopharmacol 75:165–168

Davis L, Kuttan G (2002) Effect of *Withania somnifera* on cell mediated immune responses in mice. J Exp Clin Cancer Res 21:585–590

Devi PU, Sharada AC, Solomon FE, Kamath MS (1992) In vivo growth inhibitory effect of *Withania somnifera* (Ashwagandha) on a transplantable mouse tumor, Sarcoma 180. Indian J Exp Biol 30:169–172

Devi PU, Sharada AC, Solomon FE (1993) Antitumor and radiosensitizing effects of *Withania somnifera* (Ashwagandha) on a transplantable mouse tumor, Sarcoma-180. Indian J Exp Biol 31:607–611

Devi PU, Sharada AC, Solomon FE (1995) In vivo growth inhibitory and radiosensitizing effects of withaferin A on mouse Ehrlich ascites carcinoma. Cancer Lett 95:189–193

Dhar N, Rana S, Razdan S, Bhat WW, Hussain A, Dhar RS, Vaishnavi S, Hamid A, Vishwakarma R, Lattoo SK (2014) Cloning and functional characterization of three branch point oxidosqualene cyclases from *Withania somnifera* (L.) Dunal. J Biol Chem 289:17249–17267

Dhar N, Razdan S, Rana S, Bhat WW, Vishwakarma R, Lattoo SK (2015) A decade of molecular understanding of withanolide biosynthesis and in vitro studies in *Withania somnifera* (L.) Dunal: prospects and perspectives for pathway engineering. Front Plant Sci 6:1031

Dhuley JN (1998) Effect of ashwagandha on lipid peroxidation in stress-induced animals. J Ethnopharmacol 60:173–178

Fong MY, Jin S, Rane M, Singh RK, Gupta R, Kakar SS (2012) Withaferin A synergizes the therapeutic effect of doxorubicin through ROS-mediated autophagy in ovarian cancer. PLoS One 7:e42265

Gambhir L, Checker R, Sharma D, Thoh M, Patil A, Degani M, Gota V, Sandur SK (2015) Thiol dependent NF-κB suppression and inhibition of T-cell mediated adaptive immune responses by a naturally occurring steroidal lactone Withaferin A. Toxicol Appl Pharmacol 289:297–312

Gupta P, Akhtar N, Tewari SK, Trivedi P (2011) Differential expression of farnesyl diphosphate synthase gene from *Withania somnifera* in different chemotypes and in response to elicitor. Plant Growth Regul 65:93–100

Gupta P, Agarwal AV, Akhtar N, Sangwan RS, Singh SP, Trivedi PK (2013a) Cloning and characterization of 2-C-methyl-D-erythritol-4-phosphate pathway genes for isoprenoid biosynthesis from Indian ginseng, *Withania somnifera*. Protoplasma 250:285–295

Gupta P, Goel R, Pathak S, Srivastava A, Singh SP, Sangwan RS, Asif MH, Trivedi PK (2013b) De novo assembly, functional annotation and comparative analysis of *Withania somnifera* leaf and root transcriptomes to identify putative genes involved in the withanolides biosynthesis. PLoS One 8:e62714

Gupta P, Goel R, Agarwal AV, Asif MH, Sangwan NS, Sangwan RS, Trivedi (2015) Comparative transcriptome analysis of different chemotypes elucidates withanolide biosynthesis pathway from medicinal plant *Withania somnifera*. Sci Rep 5: 18611

Hahm ER, Lee J, Kim SH, Sehrawat A, Arlotti JA, Shiva SS, Bhargava R, Singh SV (2013) Metabolic alterations in mammary cancer prevention by withaferin A in a clinically relevant mouse model. J Natl Cancer Inst 105:1111–1122

Hamza A, Amin A, Daoud S (2008) The protective effect of a purified extract of *Withania somnifera* against doxorubicin-induced cardiac toxicity in rats. Cell Biol Toxicol 24:63–73

Hollier BG, Evans K, Mani SA (2009) The epithelial-to-mesenchymal transition and cancer stem cells: a coalition against cancer therapies. J Mammary Gland Biol Neoplasia 14:29–43

John J (2014) Therapeutic potential of *Withania somnifera*: a report on phytopharmacological properties. Int J Pharm Sci Res 5:2131–2148

Kakar SS, Jala VR, Fong MY (2012) Synergistic cytotoxic action of cisplatin and withaferin A on ovarian cancer cell lines. Biochem Biophys Res Commun 423:819–825

Kakar SS, Ratajczak MZ, Powell KS, Moghadamfalahi M, Miller DM, Batra SK, Singh SK (2014) Withaferin A alone and in combination with cisplatin suppresses growth and metastasis of ovarian cancer by targeting putative cancer stem cells. PLoS One 9:e107596

Kandil FE, El-Sayed NH, Abou-Douh AM, Mabry TJ (1994) Flavonol glycosides and phenolics from *Withania somnifera*. Phytochemistry 37:1215-1216

Kataria H, Kumar S, Chaudhary H, Kaur G (2015) *Withania somnifera* suppresses tumor growth of intracranial allograft of glioma cells. Mol Neurobiol 53:4143–4158

Khan B, Ahmad SF, Bani S, Kaul A, Suri KA, Satti NK, Athar M, Qazi GN (2006) Augmentation and proliferation of T lymphocytes and Th-1 cytokines by *Withania somnifera* in stressed mice. Int Immunopharmacol 6:1394–1403

Khazal KF, Samuel T, Hill DL et al (2013) Effect of an extract of *Withania somnifera* root on estrogen receptor-positive mammary carcinomas. Anticancer Res 33:1519–1523

Khazal KF, Hill DL, Grubbs CJ (2014) Effect of *Withania somnifera* root extract on spontaneous estrogen receptor-negative mammary cancer in MMTV/*Neu* mice. Anticancer Res 34:6327–6332

Kim SH, Singh SV (2014) Mammary cancer chemoprevention by withaferin A is accompanied by *in vivo* suppression of self-renewal of cancer stem cells. Cancer Prev Res (Phila) 7:738–747

Kuttan G (1996) Use of *Withania somnifera* Dunal as an adjuvant during radiation therapy. Indian J Exp Biol 34:854–856

Lahat G, Zhu QS, Huang KL, Wang S, Bolshakov S, Liu J, Torres K, Langley RR, Lazar AJ, Hung MC, Lev D (2010) Vimentin is a novel anti-cancer therapeutic target; insights from *in vitro* and *in vivo* mice xenograft studies. PLoS One 5:e10105

Lee TJ, Um HJ, Min do S, Park JW, Choi KS, Kwon TK (2009) Withaferin A sensitizes TRAIL-induced apoptosis through reactive oxygen species-mediated up-regulation of death receptor 5 and down-regulation of c-FLIP. Free Radic Biol Med 46:1639–1649

Lee J, Hahm ER, Marcus AI, Singh SV (2015) Withaferin A inhibits experimental epithelial-mesenchymal transition in MCF-10A cells and suppresses vimentin protein level in vivo in breast tumors. Mol Carcinog 54:417–429

Leyon PV, Kuttan G (2004) Effect of *Withania somnifera* on B16F-10 melanoma induced metastasis in mice. Phytother Res 18:118–122

Li W, Zhang C, Du H, Huang V, Sun B, Harris JP, Richardson Q, Shen X, Jin R, Li G, Kevil CG, Gu X, Shi R, Zhao Y (2015a) Withaferin A suppresses the up-regulation of acetyl-coA carboxylase 1 and skin tumor formation in a skin carcinogenesis mouse model. Mol Carcinog 55:1739–1746

Li X, Zhu F, Jiang J, Sun C, Wang X, Shen M, Tian R, Shi C, Xu M, Peng F, Guo X, Wang M, Qin R (2015b) Synergistic antitumor activity of withaferin A combined with oxaliplatin triggers reactive oxygen species-mediated inactivation of the PI3K/AKT pathway in human pancreatic cancer cells. Cancer Lett 357:219–230

Mahima RA, Deb R, Latheef SK, Abdul Samad H, Tiwari R, Verma AK, Kumar A, Dhama K (2012) Immunomodulatory and therapeutic potentials of herbal, traditional/indigenous and ethnoveterinary medicines. Pak J Biol Sci 15:754–774

Malhotra CL, Mehta VL, Das PK, Dhalla NS (1965a) Studies on Withania-Ashwagandha, Kaul (Part-V): the effect of total alkaloids (Ashwagandholine) on the central nervous system. Indian J Physiol Pharmacol 9:127–136

Malhotra CL, Mehta VL, Prasad K, Das PK (1965b) Studies on Withania Ashwagandha, Kaul (Part IV) the effect of total alkaloids on the smooth muscles. Indian J Physiol Pharmacol 9:9–15

Malik F, Singh J, Khajuria A, Suri KA, Satti NK, Singh S, Kaul MK, Kumar A, Bhatia A, Qazi GN (2007) A standardized root extract of *Withania somnifera* and its major constituent withanolide-A elicit humoral and cell-mediated immune responses by up regulation of Th1-dominant polarization in BALB/c mice. Life Sci 80:1525–1538

Malik F, Kumar A, Bhushan S, Mondhe DM, Pal HC, Sharma R, Khajuria A, Singh S, Singh G, Saxena AK, Suri KA, Qazi GN, Singh J (2009) Immune modulation and apoptosis induction: two sides of antitumoural activity of a standardised herbal formulation of *Withania somnifera*. Eur J Cancer 45:1494–1509

Maliyakkal N, Appadath Beeran A, Balaji SA, Udupa N, Ranganath Pai S, Rangarajan A (2015) Effects of *Withania somnifera* and *Tinospora cordifolia* extracts on the side population phenotype of human epithelial cancer cells: toward targeting multidrug resistance in cancer. Integr Cancer Ther 14:156–171

Manoharan S, Panjamurthy K, Menon VP, Balakrishnan S, Alias LM (2009) Protective effect of Withaferin-A on tumour formation in 7,12-dimethylbenz[a]anthracene induced oral carcinogenesis in hamsters. Indian J Exp Biol 47:16–23

Matsuda H, Murakami T, Kishi A, Yoshikawa M (2001) Structures of withanosides I, II, III, IV, V, VI, and VII, new withanolide glycosides, from the roots of Indian *Withania somnifera* DUNAL. and inhibitory activity for tachyphylaxis to clonidine in isolated guinea-pig ileum. Bioorg Med Chem 9:1499–1507

McKenna MK, Gachuki BW, Alhakeem SS, Oben KN, Rangnekar VM, Gupta RC, Bondada S (2015) Anti-cancer activity of withaferin A in B-cell lymphoma. Cancer Biol Ther 16:1088–1098

Mikolai J, Erlandsen A, Murison A, Brown KA, Gregory WL, Raman-Caplan P, Zwickey HL (2009) In vivo effects of Ashwagandha (*Withania somnifera*) extract on the activation of lymphocytes. J Altern Complement Med 15:423–430

Mirjalili MH, Moyano E, Bonfill M, Cusido RM, Palazón J (2009) Steroidal lactones from *Withania somnifera*, an ancient plant for novel medicine. Molecules 14:2373–2393

Mishra LC, Singh BB, Dagenais S (2000) Scientific basis for the therapeutic use of *Withania somnifera* (Ashwagandha): a review. Altern Med Rev 5:334–346

Misico RI, Nicotra VE, Oberti JC, Barboza G, Gil RR, Burton G (2011) Withanolides and related steroids. Prog Chem Org Nat Prod 94:127–229

Misra L, Mishra P, Pandey A, Sangwan RS, Sangwan NS, Tuli R (2008) Withanolides from *Withania somnifera* roots. Phytochemistry 69:1000–1004

Munagala R, Kausar H, Munjal C, Gupta RC (2011) Withaferin A induces p53-dependent apoptosis by repression of HPV oncogenes and upregulation of tumor suppressor proteins in human cervical cancer cells. Carcinogenesis 32:1697–1705

Muralikrishnan G, Dinda AK, Shakeel F (2010) Immunomodulatory effects of *Withania somnifera* on azoxymethane induced experimental colon cancer in mice. Immunol Invest 39:688–698

Nagalingam A, Kuppusamy P, Singh SV, Sharma D, Saxena NK (2014) Mechanistic elucidation of the antitumor properties of withaferin A in breast cancer. Cancer Res 74:2617–2629

Nishi M, Akutsu H, Kudoh A, Kimura H, Yamamoto N, Umezawa A, Lee SW, Ryo A (2014) Induced cancer stem-like cells as a model for biological screening and discovery of agents targeting phenotypic traits of cancer stem cell. Oncotarget 5:8665–8680

Padmavathi B, Rath PC, Rao AR, Singh RP (2005) Roots of *Withania somnifera* inhibit forestomach and skin carcinogenesis in mice. Evid Based Complement Alternat Med 2:99–105

Panda S, Kar A (1997) Evidence for free radical scavenging activity of Ashwagandha root powder in mice. Indian J Physiol Pharmacol 41:424–426

Patil D, Gautam M, Mishra S, Karupothula S, Gairola S, Jadhav S, Pawar S, Patwardhan B (2013) Determination of withaferin A and withanolide A in mice plasma using high-performance liquid chromatography-tandem mass spectrometry: application to pharmacokinetics after oral administration of *Withania somnifera* aqueous extract. J Pharm Biomed Anal 80:203–212

Prabu PC, Panchapakesan S (2015) Prenatal developmental toxicity evaluation of *Withania somnifera* root extract in Wistar rats. Drug Chem Toxicol 38:50–56

Prakash J, Gupta SK, Kochupillai V, Singh N, Gupta YK, Joshi S (2001) Chemopreventive activity of *Withania somnifera* in experimentally induced fibrosarcoma tumours in Swiss albino mice. Phytother Res 15:240–244

Prakash J, Gupta SK, Dinda AK (2002) *Withania somnifera* root extract prevents DMBA-induced squamous cell carcinoma of skin in Swiss albino mice. Nutr Cancer 42:91–97

Rai M, Jogee PS, Agarkar G, dos Santos CA (2016) Anticancer activities of *Withania somnifera*: current research, formulations, and future perspectives. Pharm Biol 54:189–197

Rana S, Lattoo SK, Dhar N, Razdan S, Bhat WW, Dhar RS, Vishwakarma R (2013) NADPH-cytochrome P450 reductase: molecular cloning and functional characterization of two paralogs from *Withania somnifera* (L.) Dunal. PLoS One 8:e57068

Raut AA, Rege NN, Tadvi FM, Solanki PV, Kene KR, Shirolkar SG, Pandey SN, Vaidya RA, Vaidya AB (2012) Exploratory study to evaluate tolerability, safety, and activity of *Ashwagandha* (*Withania somnifera*) in healthy volunteers. J Ayurveda Integr Med 3:111–114

Ray S, Jha S (2001) Production of withaferin A in shoot cultures of *Withania somnifera*. Planta Med 67:432–436

Ray S, Ghosh B, Sen S, Jha S (1996) Withanolide production by root cultures of *Withania somnifera* transformed with *Agrobacterium rhizogenes*. Planta Med 62:571–573

Samadi AK (2015) Potential anticancer properties and mechanisms of action of withanolides. Enzyme 37:73–94

Samadi AK, Mukerji R, Shah A, Timmermann BN, Cohen MS (2010) A novel RET inhibitor with potent efficacy against medullary thyroid cancer in vivo. Surgery 148:1228–1236

Samadi AK, Cohen SM, Mukerji R, Chaguturu V, Zhang X, Timmermann BN, Cohen MS, Person EA (2012) Natural withanolide withaferin A induces apoptosis in uveal melanoma cells by suppression of Akt and c-MET activation. Tumour Biol 33:1179–1189

Sangwan RS, Chaurasiya ND, Misra LN, Lal P, Uniyal GC, Sharma R, Sangwan NS, Suri KA, Qazi GN, Tuli R (2004) Phytochemical variability in commercial herbal products and preparations of *Withania somnifera* (Ashwagandha). Curr Sci 86:461–465

Sangwan RS, Das Chaurasiya N, Lal P, Misra L, Tuli R, Sangwan NS (2008) Withanolide A is inherently de novo biosynthesized in roots of the medicinal plant Ashwagandha (*Withania somnifera*). Physiol Plant 133:278–287

Santagata S, Xu YM, Wijeratne EM, Kontnik R, Rooney C, Perley CC, Kwon H, Clardy J, Kesari S, Whitesell L, Lindquist S, Gunatilaka AA (2012) Using the heat-shock response to discover anticancer compounds that target protein homeostasis. ACS Chem Biol 7:340–349

Senthil K, Wasnik NG, Kim YJ, Yang DC (2010) Generation and analysis of expressed sequence tags from leaf and root of *Withania somnifera* (Ashwgandha). Mol Biol Rep 37:893–902

Senthil K, Jayakodi M, Thirugnanasambantham P, Lee SC, Duraisamy P, Purushotham PM, Rajasekaran K, Nancy Charles S, Mariam Roy I, Nagappan AK, Kim GS, Lee YS, Natesan S, Min TS, Yang TJ (2015) Transcriptome analysis reveals *in vitro* cultured *Withania somnifera* leaf and root tissues as a promising source for targeted withanolide biosynthesis. BMC Genomics 16:14

Sharada AC, Solomon FE, Devi PU (1993) Toxicity of *Withania somnifera* root extract in rats and mice. Int J Pharmacogn 31(3):205–212

Sharma PV (1999) Ashwagandha, Dravyaguna Vijnana, Chaukhambha Viashwabharti. Varanasi, pp 763–765

Sharma V, Sharma S, Pracheta PR (2011) *Withania somnifera*: a rejuvenating Ayurvedic medicinal herb for the treatment of various human ailments. Int J PharmTech Res 3:187–192

Shohat B, Joshua H (1971) Effect of withaferin A on ehrlich ascites tumor cells. II. Target tumor cell destruction *in vivo* by immune activation. Int J Cancer 8:487–496

Shohat B, Gitter S, Abraham A, Lavie D (1967) Antitumor activity of withaferin A (NSC-101088). Cancer Chemother Rep 51:271–276

Singh N, Bhalla M, de Jager P, Gilca M (2011) An overview on Ashwagandha: a Rasayana (rejuvenator) of Ayurveda. Afr J Tradit Complement Altern Med 8(Suppl):208–213

Singh P, Guleri R, Singh V, Kaur G, Kataria H, Singh B, Kaur G, Kaul SC, Wadhwa R, Pati PK (2015) Biotechnological interventions in *Withania somnifera* (L.) Dunal. Biotechnol Genet Eng Rev 31:1–20

Sivanandhan G, Selvaraj N, Ganapathi A, Manickavasagam M (2016) Elicitation approaches for withanolide production in hairy root culture of *Withania somnifera* (L.) Dunal. Methods Mol Biol 1405:1–18

Srinivasan S, Ranga RS, Burikhanov R, Han SS, Chendil D (2007) Par-4-dependent apoptosis by the dietary compound withaferin A in prostate cancer cells. Cancer Res 67:246–253

Stan SD, Hahm ER, Warin R, Singh SV (2008) Withaferin A causes FOXO3a- and Bim-dependent apoptosis and inhibits growth of human breast cancer cells *in vivo*. Cancer Res 68:7661–7669

Suman S, Das TP, Sirimulla S, Alatassi H, Ankem MK, Damodaran C (2016) Withaferin-A suppress AKT induced tumor growth in colorectal cancer cells. Oncotarget 7:13854–13864

Thaiparambil JT, Bender L, Ganesh T, Kline E, Patel P, Liu Y, Tighiouart M, Vertino PM, Harvey RD, Garcia A, Marcus AI (2011) Withaferin A inhibits breast cancer invasion and metastasis at sub-cytotoxic doses by inducing vimentin disassembly and serine 56 phosphorylation. Int J Cancer 129:2744–2755

Thiery JP (2002) Epithelial-mesenchymal transitions in tumour progression. Nat Rev Cancer 2:442–454

Tiwari R, Chakraborty S, Saminathan M, Dharma K, Singh SV (2014) Ashwagandha (*Withania somnifera*): role in safeguarding health, immunomodulatory effects, combating infections and therapeutic applications: a review. J Biol Sci 14:77–94

Uddin Q, Samiulla L, Singh VK, Jamil SS (2012) Phytochemical and pharmacological profile of *Withania somnifera* Dunal: a review. J Appl Pharm Sci 2:170–175

Umadevi M, Rajeswari R, Sharmila Rahale C, Selvavenkadesh S, Pushpa R, Sampath Kumar KP, Bhowmik D (2012) Traditional and medicinal uses of *Withania somnifera*. Pharm Innov 1:102–110

Vyas AR, Singh SV (2014) Molecular targets and mechanisms of cancer prevention and treatment by withaferin A, a naturally occurring steroidal lactone. AAPS J 16:1–10

Widodo N, Kaur K, Shrestha BG, Takagi Y, Ishii T, Wadhwa R, Kaul SC (2007) Selective killing of cancer cells by leaf extract of Ashwagandha: identification of a tumor-inhibitory factor and the first molecular insights to its effect. Clin Cancer Res 13:2298–2306

Winters M (2006) Ancient medicine, modern use: *Withania somnifera* and its potential role in integrative oncology. Altern Med Rev 11:269–277

Yang H, Shi G, Dou QP (2007) The tumor proteasome is a primary target for the natural anticancer compound Withaferin A isolated from "Indian winter cherry". Mol Pharmacol 71:426–437
Yang ES, Choi MJ, Kim JH, Choi KS, Kwon TK (2011a) Combination of withaferin A and X-ray irradiation enhances apoptosis in U937 cells. Toxicol In Vitro 25:1803–1810
Yang ES, Choi MJ, Kim JH, Choi KS, Kwon TK (2011b) Withaferin A enhances radiation-induced apoptosis in Caki cells through induction of reactive oxygen species, Bcl-2 downregulation and Akt inhibition. Chem Biol Interact 190:9–15
Yang H, Wang Y, Cheryan VT, Wu W, Cui CQ, Polin LA, Pass HI, Dou QP, Rishi AK, Wali A (2012) Withaferin A inhibits the proteasome activity in mesothelioma in vitro and in vivo. PLoS One 7:e41214
Yang Z, Garcia A, Xu S, Powell DR, Vertino PM, Singh S, Marcus AI (2013) *Withania somnifera* root extract inhibits mammary cancer metastasis and epithelial to mesenchymal transition. PLoS One 8:e75069
Yu Y, Hamza A, Zhang T, Gu M, Zou P, Newman B, Li Y, Gunatilaka AA, Zhan CG, Sun D (2010) Withaferin A targets heat shock protein 90 in pancreatic cancer cells. Biochem Pharmacol 79:542–551
Ziauddin M, Phansalkar N, Patki P, Diwanay S, Patwardhan B (1996) Studies on the immunomodulatory effects of Ashwagandha. J Ethnopharmacol 50:69–76

Chapter 9
Withaferin-A as a Potential Candidate for Cancer Therapy: Experimental Evidence of Its Effects on Telomerase Plus and Minus Cancer Cells

Zeenia Kaul, Yue Yu, Sunil C. Kaul, and Renu Wadhwa

Abstract Telomeres (specialized ends of eukaryotic chromosomes) shorten with each round of division in normal human cells. In contrast, cancer cells maintain their telomere length and have been defined as their most consistent attribute. It is achieved by activation of telomere maintenance mechanisms that may either involve up regulation of a specialized reverse transcriptase, telomerase (Telomerase positive, TEP) or recombination based Alternative Lengthening of Telomeres (ALT). Telomere shortening is considered as a tumor suppressor mechanism and hence the drugs to activate this mechanism are deemed anticancer agents. In this chapter we describe the basic mechanisms of telomere length maintenance and their targeting in anticancer therapy. Use of Withaferin-A as a potential anticancer drug for TEP and ALT cells is discussed. Whereas ALT cells escape the deleterious effect of telomerase inhibitors in cancer therapy, Withaferin-A kills both TEP and ALT cells effectively. Thus, Withaferin-A emerges as a promising potential candidate for cancer therapy.

Keywords Withania • Withaferin-A • Telomerase • ALT • Cytotoxic • Cancer therapy

---

Z. Kaul
Department of Molecular Virology, Immunology & Medical Genetics, The Ohio State University, Columbus, OH 43210, USA

Y. Yu • S.C. Kaul • R. Wadhwa (✉)
DBT-AIST International Laboratory for Advanced Biomedicine (DAILAB), Biomedical Research Institute, National Institute of Advanced Industrial Science & Technology (AIST), Central 5-41, 1-1-1 Higashi, Tsukuba, Ibaraki 305-8565, Japan
e-mail: renu-wadhwa@aist.go.jp

S.C. Kaul, R. Wadhwa (eds.), *Science of Ashwagandha: Preventive and Therapeutic Potentials*, DOI 10.1007/978-3-319-59192-6_9

9.1 Introduction

Cancer is a complex disease of which initiation, growth and progression to aggressive stages including invasion and metastasis to other tissues involves multiple factors. It is largely defined as a disease of proliferation in which normal cells (divide only a limited number of times in culture) breach controls on cell division and set on to the path of immortalization that can be established into permanent cell lines. It has been established that immortalization, one of the most important aspects of the cancer cell phenotype, results from multiple genetic and environmental changes in multi-step processes. The role of telomeres in normal cell proliferation controls and their breaching in cancer cells has been firmly established in last three decades. In contrast to telomere shortening with every cell division in normal cells, cancer cells evolve mechanisms to maintain their telomere length. This chapter provides a simple outline on structural and functional aspects of telomeres in normal cell proliferation controls and their maintenance in cancer cells by activation of either telomerase or alternative mechanisms. Cytotoxic activity of Withaferin-A on telomerase plus and minus cancer cells is discussed. Also, the experimental evidence showing the mechanism how Withaferin-A kills both telomerase plus and minus cancer cells is discussed.

Telomeres are specialized structures at the ends of chromosomes composed of repetitive 5′ TTAGGG 3′ DNA repeat sequences in all vertebrate species (Blackburn 1991). Human telomeres contain 5–15 kb of repetitive DNA. In the early 1970s, Olovnikov (1971) and Watson (1972) independently proposed that chromosome ends shorten with each round of cell division due to the inability of DNA polymerases to replicate the ends of linear DNA molecules. Olovnikov further hypothesized that the finite doubling capacity of normal mammalian cells is due to the loss of telomeric DNA with each cell division and eventual deletion of essential sequences and proteins (Fig. 9.1).

Harley et al. first demonstrated that the amount and length of telomeric DNA decreases as a function of serial passage *in vitro* (Harley et al. 1990). There is much experimental data to support the role of telomere shortening in senescence *in vitro* (telomere shortening during serial passaging of normal cells), and *in vivo* (telomere length decrease with age) (Wadhwa et al. 2016). The minimum rate of telomere shortening in human cells *in vitro* is about 30–50 base pairs with each cell division (Levy et al. 1992). Telomere shortening has been established as a reliable marker of replicative senescence, and is tightly associated with (i) upregulation of tumor suppressor activities, (ii) accumulation of DNA damage response proteins at telomeres, (iii) increase in intrinsic stress including mitochondrial dysfunction and accumulation of reactive oxygen species (ROS), and (iv) increase in secreted growth factors, matrix remodeling enzymes, and inflammatory cytokines that contribute to age-related pathologies including cancers (Fig. 9.2)(Shay 1997; Colgin and Reddel 2004; von Zglinicki et al. 2001; Passos et al. 2007; Singhapol et al. 2013; Sikora et al. 2014; Maciel-Baron et al. 2016). However, telomere-dependent replicative senescence is not solely determined by telomere length but also by telomere binding

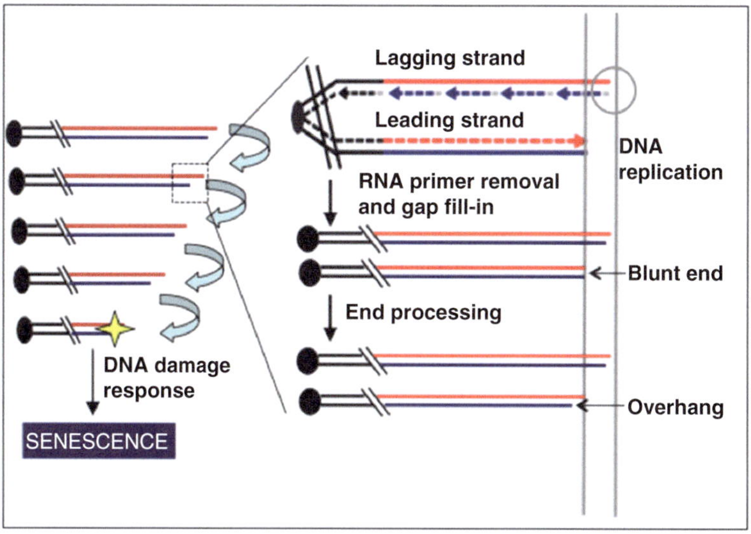

Fig. 9.1 The end replication problem, telomere shortening and senescence. During lagging-strand DNA replication, RNA primers (*grey*) are removed and the resulting gaps are filled in, but the gap resulting from removal of the terminal RNA primer (*circled*) cannot be filled in. This is referred to as the end-replication problem. In addition, the leading-strand replicated telomere is processed by removal of terminal C-strand DNA to form a 3′-overhang. The result of both of these processes is that the template for synthesis of telomeric DNA in the next S-phase of the cell cycle is shorter, and consequently there is progressive telomere shortening with each cell division. It has been hypothesized that telomeric shortening eventually causes a DNA damage response that triggers the onset of senescence

proteins and telomere structure that play an important role (Passos et al. 2007; Kaul et al. 2012; von Figura et al. 2009) in triggering senescence.

9.2 Telomere Binding Proteins

A unique complex containing six proteins, telomeric repeat binding factors 1 and 2 (TRF1 and TRF2), protection of telomeres 1 (POT1), repressor activator protein 1 (Rap1), TIN2-POT1-interacting protein 1 (TPP1), and TRF-interacting nuclear protein 2 (TIN2), referred to as "shelterin" (Palm and de Lange 2008) binds to single-stranded and double-stranded telomeric DNA and protects chromosome ends (Martinez and Blasco 2010) (Fig. 9.3a). Shelterin has a highly specific binding affinity for telomeric DNA, due many specific oligonucleotide binding (OB) folds within the complex. It has been proposed that the shortened telomere may affect the binding as well as the expression of the shelterin proteins leading to an unprotected or exposed telomere. TRF1 and TRF2 are the two most specific double-stranded telomeric DNA binding proteins. Other proteins including POT1, RAD3, TIN2 and

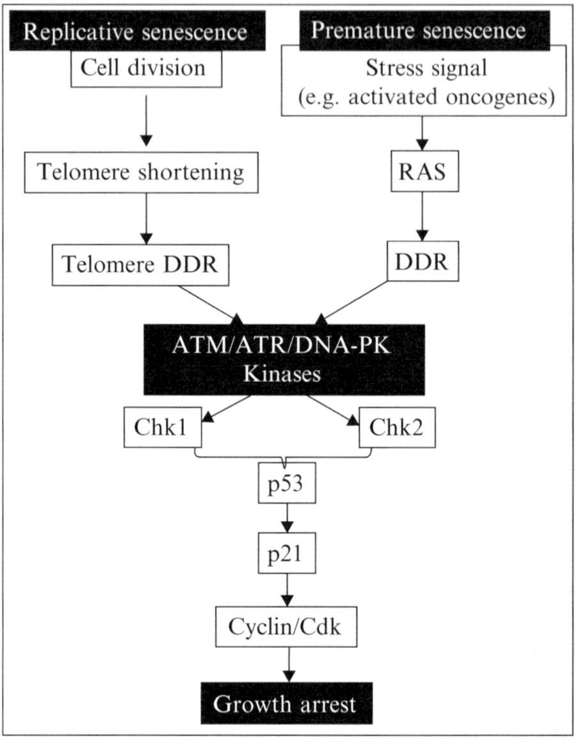

Fig. 9.2 Diagram of relationship between replicative and premature senescence. Replicative senescence is associated with telomere shortening ultimately resulting in increased expression of the cyclin dependent kinase (Cdk) inhibitor, p21. Premature senescence elicited by DNA damaging agents, activated oncogenes etc. leads to activation of the same Cdk inhibitor. Both replicative senescence and premature involve a DNA damage response (DDR) with the recruitment of ATM, ATR and DNA-PK kinases and their substrates including Chk1 and Chk2. ATM-dependent activation of p53 leads to irreversible cell cycle arrest through the activation of p21

TPP1 and Rap1 localize to the telomere, mainly via specific interactions with TRF1 or TRF2 (de Lange 2005; Martinez and Blasco 2011; Patel et al. 2015).

9.3 Telomere Dysfunction

Unprotected or critically short telomeres are readily detected as DNA breaks and trigger DNA damage signaling (Giardini et al. 2014; Hartwig and Collares 2013). A cellular state indistinguishable from replicative senescence may be induced in normal human cells, regardless of their telomere length, by disrupting the function of TRF2, POT1 or TPP1 (van Steensel et al. 1998). Such state was well characterized by DNA damage foci on telomeres as detected by the presence of phosphorylated histone H2AX (γ-H2AX), tumor suppressor p53-binding protein 1 (53BP1) on telomeres. The latter have been referred to as telomere dysfunction induced foci (TIFs) (Mender and Shay 2015; Takai et al. 2003). Similar foci occur in much smaller numbers in nuclei of cells undergoing replicative senescence, and telomeric chromatin can be immunoprecipitated with antibodies against γ-H2AX in senescent cells (d'Adda di Fagagna et al. 2003). The causal relationship between DNA damage response signaling and senescence is also demonstrated by the observation that in

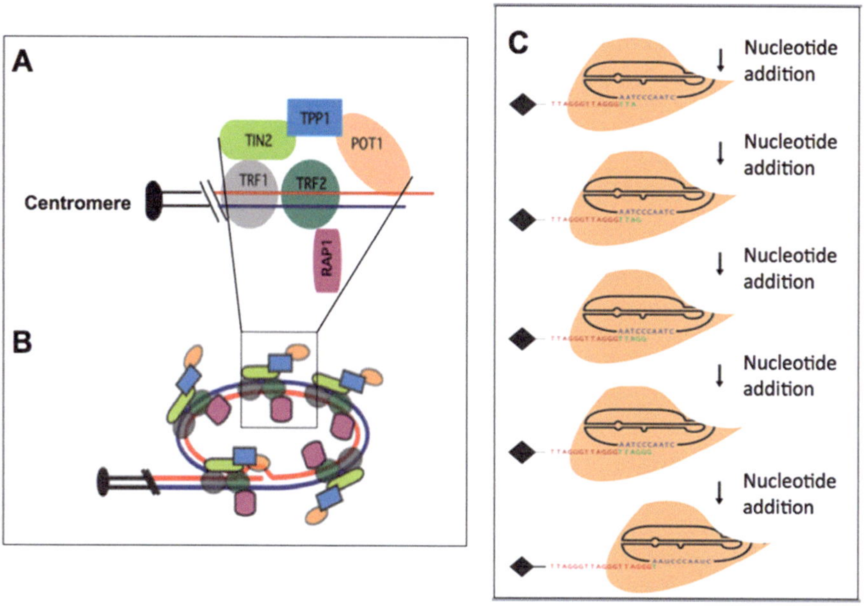

Fig. 9.3 Shelterin complex and schematic presentation of telomerase action. (**a**) The telomere-specific shelterin complex contains two double-stranded telomeric DNA binding proteins, telomeric repeat binding factor 1 (TRF1) and 2 (TRF2), as well as a single-strand G-rich telomeric DNA binding protein, protection of telomeres 1 (POT1). TIN2 bridges TRF1 and TRF2 and while the POT1- and TIN2-interacting protein (TPP1) connects POT1 to TRF1 and TRF2 via TIN2. Repressor and activator protein 1 homologue (RAP1) binds directly to TRF2 and does not interact with the other shelterin proteins. (**b**) Shelterin complex is shown bound along the T-loop. (**c**) The 3′ end of the G-rich strand acts as a substrate for telomerase binding via base pairing to the template within the hTR subunit. Telomerase then catalyses the addition of nucleotides corresponding to the telomeric repeat sequence (nucleotide addition) and after addition of six nucleotides, the enzyme translocates to the telomere end. Continued nucleotide addition and translocation accounts for the processivity of telomerase

both the experimentally induced and replicative senescence the p53-mediated cell cycle arrest is ATM- and/or ATR-dependent (Karlseder et al. 1999; Schmitt et al. 2007). Advances in understanding mammalian telomere dysfunction have been obtained by expressing a TRF2 mutant proteins that resulted in gross chromosome uncapping and induction of DNA damage response, activation of p53 and pRb resulting in senescence (Karlseder et al. 1999; de Lange 2005; van Steensel et al. 1998).

9.4 Immortalization

Normal cells that inactivate one or more tumor suppressor and DNA damage signaling mechanisms can be established as permanent cell lines in culture and are referred to as immortal. These cells have been characterized to possess either of the two telomere maintaining mechanisms (TMM); activation of telomerase enzyme (Blackburn 2001) or Alternative Lengthening of Telomeres (ALT) (Bryan et al. 1995; Conomos et al. 2013).

9.5 Telomerase Enzyme

Telomeres of telomerase positive human immortal and cancer cell lines are relatively homogeneous in length (<10 kb) that is maintained during their serial passaging in culture (Counter et al. 1998; Hastie et al. 1990). In contrast, normal somatic cells possess low or undetectable level of telomerase activity and undergo progressive telomere shortening during continued proliferation.

There is a very close temporal correlation between the immortalization event and the onset of telomerase activity when cells are escaping from crisis (Bryan et al. 1997a). The enzyme uses its own RNA subunit as a template for the *de novo* addition of telomeric repeats to the 3′ terminus of eukaryotic chromosomes (Fig. 9.3b). Telomerase is a specialized reverse transcriptase that is responsible for telomere length maintenance in human cancers. It is a multi-subunit enzyme comprises of, at least, three components: TERT (telomerase reverse transcriptase protein), TR (telomerase RNA subunit), and dyskerin (RNA-binding protein) (Cohen et al. 2007) (Fig. 9.4a). Mass analyses of the complex (650–670-kDa) of and of the separate components, hTR (153-kDa), hTERT (127-kDa) and dyskerin (57-kDa) suggested that the active complex exists as a dimer.

Human telomerase RNA (hTR; coded by chromosome 3q26.438 locus) sequence varies widely from species to species. Furthermore, the template sequence makes up only a small portion of the RNA suggesting that in addition to being an essential template for telomerase reverse transcriptase activity, hTR plays an important structural role within the complex. hTR is defined by a hairpin-hinge-hairpin-tail secondary structure containing a box H motif (ANANNA – where N represents any nucleotide) within the hinge and a box ACA motif (ACA) within the tail. This region of hTR mediates RNA binding to dyskerin and possibly other members of the H/ACA protein family-NHP2, NOP10 and GAR1- in RNA-protein and indirect protein-protein interactions (Mengual Gomez et al. 2014).

Human Telomerase Reverse Transcriptase (hTERT) (127-kDa protein coded by 5p15.33 locus) is a highly specialized and unusual reverse transcriptase (RT), which shares partial sequence and structural homology with known retroviral transcriptases (Cohen et al. 2007; Cong and Shay 2008; Cong et al. 2002). Overexpression of hTR and/or hTERT has demonstrated that both components result in increased telomerase activity leading to telomere lengthening, consistent with the conclusion that both hTR and hTERT levels are limiting for telomerase activity (Cristofari et al. 2007; Bodnar et al. 1998; Cao et al. 2008).

Dyskerin (514 amino acid nuclear protein coded by chromosome Xq28) together with NHP2, NOP10 and specific guide RNAs, form RNP complexes (Vulliamy et al. 2008). It was first identified as the gene mutated in Dyskeratosis Congenita

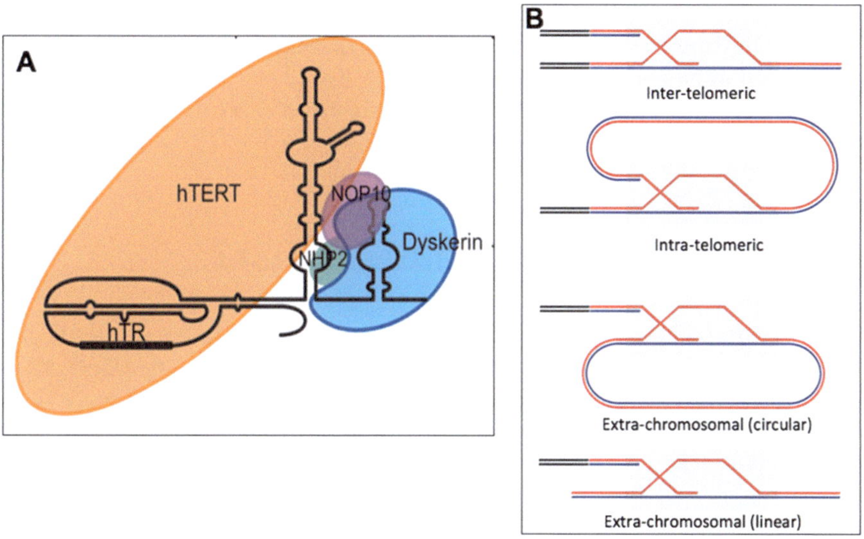

Fig. 9.4 Telomere lengthening processes. (**a**) Schematic diagram of telomerase enzyme complex that adds TTAGGG repeats to chromosome ends is shown. (**b**) Telomeric recombination in ALT cells. According to the HR-dependent replication model, it is proposed that the copy template used for alternative lengthening of telomeres (ALT)-mediated telomere lengthening may be (another telomere, the t-loop of the same telomere, circular extra chromosomal telomeric DNA, or linear extra chromosomal telomeric DNA (as shown from top to bottom)

(DC) patients and causes triad of mucocutaneous abnormalities and bone marrow failure (Heiss et al. 1998). Cells from DC patients have short telomeres and increased susceptibility to cancer. The telomere dysfunction in these cells either arises as a result of a failure to maintain telomere length or an inability to protect the telomere. This relationship highlights the importance of dyskerin for telomere maintenance.

9.6 Telomerase Inhibitors

Telomerase activity is detectable in a large majority of human tumor-derived cell lines and in primary tumor samples. The suppression of telomerase activity in cancer and tumor cells causes telomere shortening that is often associated with onset of apoptosis or inhibition of cell proliferation. Hence, the development of anti-telomerase drugs for cancer therapy has been initiated (White et al. 2001). However, there are several concerns on the efficacy of telomerase inhibitors as anticancer drugs (i) anti-telomerase drugs are expected not to kill cells immediately, but would cause steady telomere shortening until a critical point, (ii) germline cells also use telomerase to prevent telomere shortening, (iii) the very low abundance of endogenous telomerase enzyme even in cancer cells makes it very difficult to purify human telomerase in sufficient quantities to screen for inhibitors, (iv) no high resolution structural information is available for intact active telomerase enzyme, complicating structure-based design strategies, (v) telomerase is a polymerase, and inhibitors would need to be selective for its inhibition relative to other cellular polymerases and (vi) these will be ineffective for cancer cells that possess ALT mechanism (as discussed below) for telomere lengthening.

9.7 Alternative Lengthening of Telomeres (ALT)

Some *in vitro*-immortalized and tumor-derived cell lines are capable of maintaining their telomere length over very large numbers of population doublings in the absence of telomerase activity that suggested the existence of one or more telomerase-independent TMMs referred to collectively as Alternative Lengthening of Telomeres (ALT) (Bryan et al. 1995). Tumors of mesenchymal origin including osteosarcomas and soft tissue sarcomas commonly use ALT to maintain their telomere length (Henson and Reddel 2010; McDonald et al. 2010). Telomerase-independent telomere length maintenance was first as a recombination-mediated mechanism that can either be inter- or intra-molecular (Dunham et al. 2000; Cesare and Reddel 2010) (Fig. 9.4b).

9.8 Characteristic Features of ALT Cancer Cells

(i) ALT cancer cells have heterogeneous telomere length ranging from <3 kb to >50 kb, with an average length of ~20 kb (Bryan et al. 1997b; Conomos et al. 2013).
(ii) ALT cells undergo post-replicative telomeric exchanges at a much higher frequency than non-ALT (normal or telomerase-positive) cells (Londono-Vallejo 2008).
(iii) ALT cells also contain a unique subset of promyelocytic leukaemia (PML) nuclear bodies known as ALT-associated PML bodies (APBs) that contain telomeric repeat DNA and telomere binding proteins in addition to the HR proteins normally associated with PML bodies (Yeager et al. 1999). APBs are nuclear structures and can be visualized by immunostaining of interphase nuclei with anti-PML antibodies and telomeric DNA or telomere binding proteins. ALT cells possess large APBs (Yeager et al. 1999). The role of APBs in the ALT mechanism is not fully understood; it is not known whether APBs act as repositories for the by-products of ALT process or are the sites of telomeric recombination. Live cell imaging experiments have demonstrated that telomeres move to and from APBs suggesting that APBs may be directly involved in the ALT mechanism (Cesare and Reddel 2010).
(iv) ALT cells possess extrachromosomal telomere repeats (ECTR) including ds telomeric circles (T-circles), linear single-stranded DNA and partially double-

stranded telomeric circles named C- or G-circles (Cesare and Reddel 2010). C-circles are self-priming telomeric C-strand templates for rolling circle amplification. Recent findings have established a quantitative relationship between the amount of ALT activity and the levels of C-circles within a cell (Henson and Reddel 2010).

Genetic or epigenetic alterations that underline the ALT mechanisms of telomere maintenance remain poorly defined. Various proteins with diverse activities (tumor suppression, DNA damage, homologous recombination, G2/M checkpoint and genetic stability controls) have been suggested to play role in this process. The MRE11-NBS1-RAD50 complex is the first protein complex to be identified as necessary for ALT mediated telomere elongation. It plays several roles including (i) detection of double-strand breaks (DSB) and initiation of DSB signaling, (ii) recruitment of DNA repair proteins to break foci and (iii) stabilization of telomere structure (Cesare and Reddel 2010).

9.9 Effect of Withaferin-A on Telomere Plus and Minus Cancer Cells

Withaferin-A (Wi-A) is a steroidal lactone found in the medicinal plant, *Withania somnifera* (Ashwagandha). Several independent studies have shown that it possesses anticancer activity that has been attributable to its multimodule action including induction of (i) DNA damage signaling as indicated by an increase in the levels of p-Chk1 and p-Chk2 in Wi-A treated cells, (ii) cell cycle arrest by downregulation of cyclin B1, Cdk2, cyclin A and p-Cdc2, (iv) wild type p53 signaling and accumulation of $p21^{WAF1}$ and (v) disruption of cytoskeleton elements including actin, vimentin and intermediate filaments suggesting that it is a potential natural anticancer drug (Suman et al. 2016; Lv and Wang 2015; Munagala et al. 2011; Petroni et al. 2015). We investigated the effect of Wi-A on isogenic telomerase-plus (TEP) and -minus (ALT) cancer cells and found that it causes a stronger cytotoxicity to ALT cells. We discuss here experimental and computational evidence that Wi-A inhibits the ALT phenotype, induces telomere dysfunction, and these effects are caused by Myc-Mad mediated transcriptional suppression of NBS-1, a MRN complex protein, an essential component of ALT mechanism.

9.10 Wi-A Causes Stronger Cytotoxicity to ALT Cancer Cells by Telomere Dysfunction and Inhibition of ALT Phenotype

We compared the effect of Wi-A on telomerase plus (TEP) and telomerase negative (ALT) isogenic cancer cells and found that it caused stronger cytotoxicity to ALT cells (Fig. 9.5). We investigated the molecular mechanism of its differential effect in response to low dose of Wi-A on TEP and ALT cells. Cell cycle analyses of Wi-A treated and control cells showed that the low doses of Wi-A were enough to cause G_2/M phase arrest and apoptosis in ALT, but not TEP, cells (Fig. 9.5). In order to investigate if Wi-A treatment affected telomere length, we performed qPCR-based telomere length assays in control and Wi-A treated TEP and ALT cells. Both cells did not show any effect of Wi-A on telomere length (Yu et al. 2017) suggesting that it may alter telomere structure or cause telomere damage/de-protection.

Quantitative assay for telomere dysfunction in control and Wi-A treated cells indeed showed significant increase in telomere damage-induced foci (~10% to 25%) in ALT cells specifically (Yu et al. 2017). Wi-A treated TEP and ALT cells were also examined for a well-established marker for ALT, APBs (ALT-associated Promyelocytic Leukemia (PML) nuclear bodies). APBs were specifically detected in ALT cells and showed 20–40% reduction in Wi-A treated cells (Fig. 9.6). On the

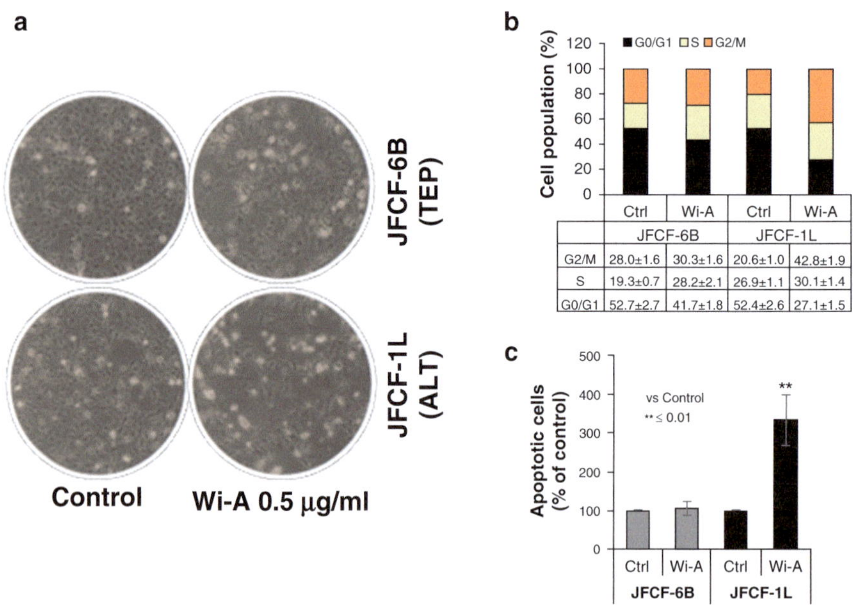

Fig. 9.5 Withaferin-A showed stronger cytotoxicity to ALT cells. Cytotoxic response of isogenic telomerase plus (TEP; JFCF-6B) and telomerase minus (ALT; JFCF-1 L) cells to Withaferin-A. ALT cells were more sensitive (**a**), showed G2/M arrest (**b**) and apoptosis (**c**) (Yu et al. 2017)

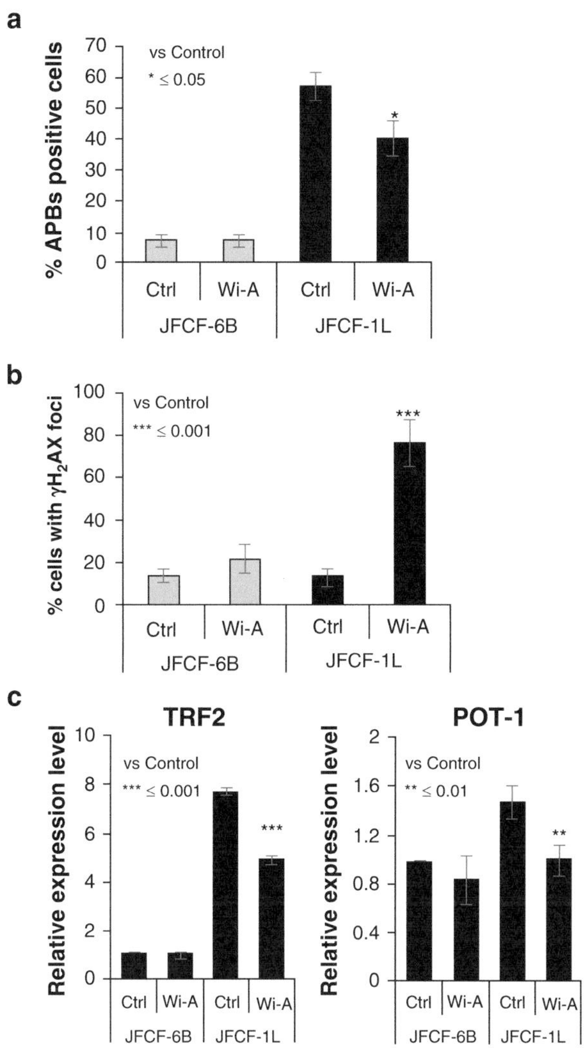

Fig. 9.6 Withaferin-A caused inactivation of ALT mechanism. ALT cells showed decrease in ALT-specific APBs (**a**), increase in γH2AX (**b**) and decrease in TRF2 and POT-1 proteins that are crucial for ALT phenotype (**c**) (Yu et al. 2017)

other hand, telomerase activity was not affected in control as well as Wi-A treated TEP/ALT cells (Yu et al. 2017). The induction of DNA breaks was also examined. Whereas immunostaining did not reveal any difference in γH2AX in control and Wi-A (0.25 μg/ml) treated TEP cells; JFCF-1 L cells showed a large increase in the number of γH2AX positive cells (Fig. 9.6) (Yu et al. 2017). Furthermore, Wi-A treated ALT cells showed lower level of telomere binding proteins, TRF2 and POT1 (shown to be important for ALT) as compared to the TEP cells (Fig. 9.6). These data

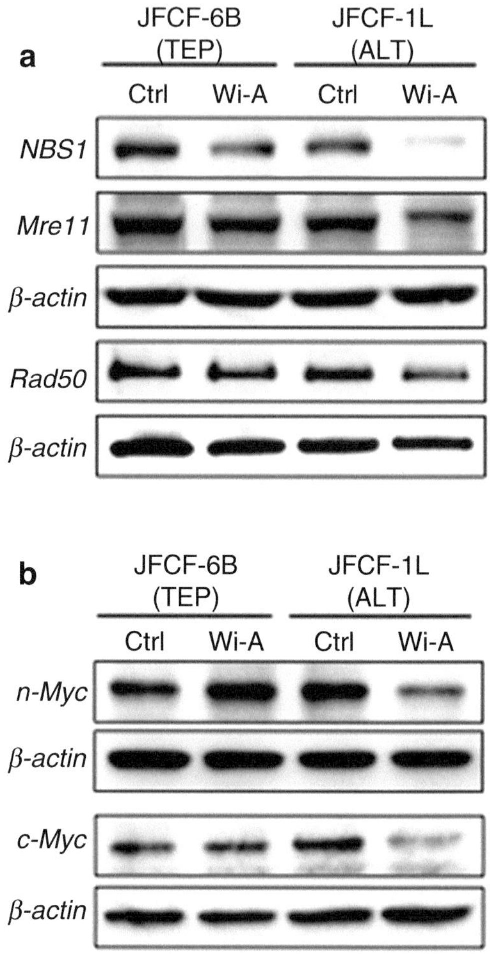

Fig. 9.7 Withaferin-A caused downregulation of MRN complex proteins and Myc, transcriptional regulator of NBS1. Withaferin-A treated ALT cells showed decrease in NBS1, Mre11 and Rad50 (**a**). Transcription factors n-Myc and c-Myc showed decrease in Withaferin-A treated ALT cells (**b**) (Yu et al. 2017)

suggested that stronger cytotoxicity of Wi-A to ALT cells is due to, at least in part, its disruptive effect on telomere protecting proteins.

We examined MRN complex proteins (essential for the ALT mechanism and APB assembly) in control and Wi-A treated TEP and ALT cells. As shown in Fig. 9.7, a sharp decline in NBS1 and MRE11 expression, and a smaller reduction of RAD50 in Wi-A treated ALT cells; TEP cells showed only a small/negligible change (Fig. 9.7). Transcriptional analysis revealed reduction in NBS1, and not in MRE11 and RAD50, in ALT cells suggesting that Wi-A may target NBS1 in ALT cells (Yu et al. 2017). Since MRN complex is regulated by n-Myc and c-Myc at the transcription level, we next examined their status and found that Wi-A led to ~40% decrease in n-Myc and ~55% decrease in c-Myc in ALT cells (Fig. 9.7). Bioinformatics, molecular docking and experimental analyses of targeting potential

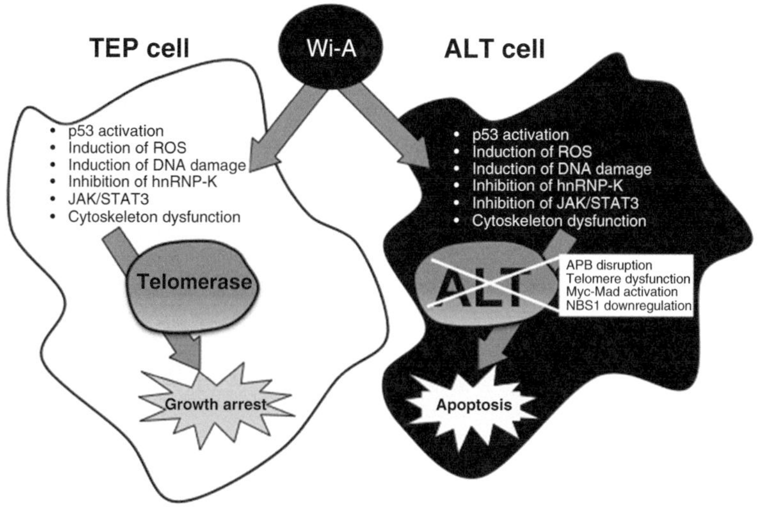

Fig. 9.8 A model showing the effect of Withaferin-A (Wi-A) on TEP and ALT cells. Wi-A targets several cancer promoting proteins and mechanism in these cells (shown in *grey*). In addition, several ALT cell specific mechanisms (APB bodies, down-regulation of NBS1 through Myc-Mad activation and induction of telomere dysfunction) are affected causing apoptosis of ALT cancer cells (Yu et al. 2017)

of Wi-A on Myc-Max and Mad-Max complexes revealed that Wi-A causes a transcriptional suppression of myc by its strong binding to Mad-Max.

Considerable efforts have been put on the development of telomerase inhibitors due to high incidence of telomerase activation as compared to ALT in cancer cells. *In vitro* studies have demonstrated that telomerase and ALT mechanism can co-exist within individual cancer cells. Inhibition of TEP could activate ALT or *vice versa* (Hu et al. 2012; Shay et al. 2012) raising the need for dual action inhibitors. We report Wi-A is a good candidate for overall cancer treatment. It caused strong cytotoxicity to ALT cells by inactivation of ALT mechanisms. Although no effect on telomere length and telomerase activity of TEP cells, they undergo growth arrest or death by telomerase independent mechanisms as reported previously (Um et al. 2012; Thaiparambil et al. 2011; Lee et al. 2015; Gao et al. 2014; Chi et al. 2010; Bargagna-Mohan et al. 2013; Widodo et al. 2007, 2008, 2010; Cui et al. 2014) (Fig. 9.8). Wi-A thus could be a good candidate natural compound for cancer therapy.

References

Bargagna-Mohan P, Deokule SP, Thompson K, Wizeman J, Srinivasan C, Vooturi S, Kompella UB, Mohan R (2013) Withaferin A effectively targets soluble vimentin in the glaucoma filtration surgical model of fibrosis. PLoS One 8:e63881

Blackburn EH (1991) Structure and function of telomeres. Nature 350:569–573

Blackburn EH (2001) Switching and signaling at the telomere. Cell 106:661–673
Bodnar AG, Ouellette M, Frolkis M, Holt SE, Chiu CP, Morin GB, Harley CB, Shay JW, Lichtsteiner S, Wright WE (1998) Extension of life-span by introduction of telomerase into normal human cells. Science 279:349–352
Bryan TM, Englezou A, Gupta J, Bacchetti S, Reddel RR (1995) Telomere elongation in immortal human cells without detectable telomerase activity. EMBO J 14:4240–4248
Bryan TM, Englezou A, Dalla-Pozza L, Dunham MA, Reddel RR (1997a) Evidence for an alternative mechanism for maintaining telomere length in human tumors and tumor-derived cell lines. Nat Med 3:1271–1274
Bryan TM, Marusic L, Bacchetti S, Namba M, Reddel RR (1997b) The telomere lengthening mechanism in telomerase-negative immortal human cells does not involve the telomerase RNA subunit. Hum Mol Genet 6:921–926
Cao Y, Huschtscha LI, Nouwens AS, Pickett HA, Neumann AA, Chang AC, Toouli CD, Bryan TM, Reddel RR (2008) Amplification of telomerase reverse transcriptase gene in human mammary epithelial cells with limiting telomerase RNA expression levels. Cancer Res 68:3115–3123
Cesare AJ, Reddel RR (2010) Alternative lengthening of telomeres: models, mechanisms and implications. Nat Rev Genet 11:319–330
Chi F, Jong TD, Wang L, Ouyang Y, Wu C, Li W, Huang SH (2010) Vimentin-mediated signalling is required for IbeA+ E. coli K1 invasion of human brain microvascular endothelial cells. Biochem J 427:79–90
Cohen SB, Graham ME, Lovrecz GO, Bache N, Robinson PJ, Reddel RR (2007) Protein composition of catalytically active human telomerase from immortal cells. Science 315:1850–1853
Colgin L, Reddel R (2004) Telomere biology: a new player in the end zone. Curr Biol 14:R901–R902
Cong Y, Shay JW (2008) Actions of human telomerase beyond telomeres. Cell Res 18:725–732
Cong YS, Wright WE, Shay JW (2002) Human telomerase and its regulation. Microbiol Mol Biol Rev 66:407–425
Conomos D, Pickett HA, Reddel RR (2013) Alternative lengthening of telomeres: remodeling the telomere architecture. Front Oncol 3:27
Counter CM, Hahn WC, Wei W, Caddle SD, Beijersbergen RL, Lansdorp PM, Sedivy JM, Weinberg RA (1998) Dissociation among in vitro telomerase activity, telomere maintenance, and cellular immortalization. Proc Natl Acad Sci U S A 95:14723–14728
Cristofari G, Sikora K, Lingner J (2007) Telomerase unplugged. ACS Chem Biol 2:155–158
Cui ZG, Piao JL, Rehman MU, Ogawa R, Li P, Zhao QL, Kondo T, Inadera H (2014) Molecular mechanisms of hyperthermia-induced apoptosis enhanced by withaferin A. Eur J Pharmacol 723:99–107
d'Adda di Fagagna F, Reaper PM, Clay-Farrace L, Fiegler H, Carr P, Von Zglinicki T, Saretzki G, Carter NP, Jackson SP (2003) A DNA damage checkpoint response in telomere-initiated senescence. Nature 426:194–198
de Lange T (2005) Shelterin: the protein complex that shapes and safeguards human telomeres. Genes Dev 19:2100–2110
Dunham MA, Neumann AA, Fasching CL, Reddel RR (2000) Telomere maintenance by recombination in human cells. Nat Genet 26:447–450
Gao R, Shah N, Lee JS, Katiyar SP, Li L, Oh E, Sundar D, Yun CO, Wadhwa R, Kaul SC (2014) Withanone-rich combination of Ashwagandha withanolides restricts metastasis and angiogenesis through hnRNP-K. Mol Cancer Ther 13:2930–2940
Giardini MA, Segatto M, da Silva MS, Nunes VS, Cano MI (2014) Telomere and telomerase biology. Prog Mol Biol Transl Sci 125:1–40
Harley CB, Futcher AB, Greider CW (1990) Telomeres shorten during ageing of human fibroblasts. Nature 345:458–460
Hartwig FP, Collares T (2013) Telomere dysfunction and tumor suppression responses in dyskeratosis congenita: balancing cancer and tissue renewal impairment. Ageing Res Rev 12:642–652
Hastie ND, Dempster M, Dunlop MG, Thompson AM, Green DK, Allshire RC (1990) Telomere reduction in human colorectal carcinoma and with ageing. Nature 346:866–868

Heiss NS, Knight SW, Vulliamy TJ, Klauck SM, Wiemann S, Mason PJ, Poustka A, Dokal I (1998) X-linked dyskeratosis congenita is caused by mutations in a highly conserved gene with putative nucleolar functions. Nat Genet 19:32–38
Henson JD, Reddel RR (2010) Assaying and investigating Alternative Lengthening of Telomeres activity in human cells and cancers. FEBS Lett 584:3800–3811
Hu J, Hwang SS, Liesa M, Gan B, Sahin E, Jaskelioff M, Ding Z, Ying H, Boutin AT, Zhang H, Johnson S, Ivanova E, Kost-Alimova M, Protopopov A, Wang YA, Shirihai OS, Chin L, DePinho RA (2012) Antitelomerase therapy provokes ALT and mitochondrial adaptive mechanisms in cancer. Cell 148:651–663
Karlseder J, Broccoli D, Dai Y, Hardy S, de Lange T (1999) p53- and ATM-dependent apoptosis induced by telomeres lacking TRF2. Science 283:1321–1325
Kaul Z, Cesare AJ, Huschtscha LI, Neumann AA, Reddel RR (2012) Five dysfunctional telomeres predict onset of senescence in human cells. EMBO Rep 13:52–59
Lee J, Hahm ER, Marcus AI, Singh SV (2015) Withaferin A inhibits experimental epithelial-mesenchymal transition in MCF-10A cells and suppresses vimentin protein level in vivo in breast tumors. Mol Carcinog 54:417–429
Levy MZ, Allsopp RC, Futcher AB, Greider CW, Harley CB (1992) Telomere end-replication problem and cell aging. J Mol Biol 225:951–960
Londono-Vallejo JA (2008) Telomere instability and cancer. Biochimie 90:73–82
Lv TZ, Wang GS (2015) Antiproliferation potential of withaferin A on human osteosarcoma cells via the inhibition of G2/M checkpoint proteins. Exp Ther Med 10:323–329
Maciel-Baron LA, Morales-Rosales SL, Aquino-Cruz AA, Triana-Martinez F, Galvan-Arzate S, Luna-Lopez A, Gonzalez-Puertos VY, Lopez-Diazguerrero NE, Torres C, Konigsberg M (2016) Senescence associated secretory phenotype profile from primary lung mice fibroblasts depends on the senescence induction stimuli. Age (Dordr) 38:26
Martinez P, Blasco MA (2010) Role of shelterin in cancer and aging. Aging Cell 9:653–666
Martinez P, Blasco MA (2011) Telomeric and extra-telomeric roles for telomerase and the telomere-binding proteins. Nat Rev Cancer 11:161–176
McDonald KL, McDonnell J, Muntoni A, Henson JD, Hegi ME, von Deimling A, Wheeler HR, Cook RJ, Biggs MT, Little NS, Robinson BG, Reddel RR, Royds JA (2010) Presence of alternative lengthening of telomeres mechanism in patients with glioblastoma identifies a less aggressive tumor type with longer survival. J Neuropathol Exp Neurol 69:729–736
Mender I, Shay JW (2015) Telomere Dysfunction Induced Foci (TIF) Analysis. Bio Protoc 5:pii: e1656
Mengual Gomez DL, Armando RG, Farina HG, Gomez DE (2014) Telomerase and telomere: their structure and dynamics in health and disease. Medicina (B Aires) 74:69–76
Munagala R, Kausar H, Munjal C, Gupta RC (2011) Withaferin A induces p53-dependent apoptosis by repression of HPV oncogenes and upregulation of tumor suppressor proteins in human cervical cancer cells. Carcinogenesis 32:1697–1705
Olovnikov AM (1971) Principle of marginotomy in template synthesis of polynucleotides. Dokl Akad Nauk SSSR 201:1496–1499
Palm W, de Lange T (2008) How shelterin protects mammalian telomeres. Annu Rev Genet 42:301–334
Passos JF, Saretzki G, Ahmed S, Nelson G, Richter T, Peters H, Wappler I, Birket MJ, Harold G, Schaeuble K, Birch-Machin MA, Kirkwood TB, von Zglinicki T (2007) Mitochondrial dysfunction accounts for the stochastic heterogeneity in telomere-dependent senescence. PLoS Biol 5:e110
Patel TN, Vasan R, Gupta D, Patel J, Trivedi M (2015) Shelterin proteins and cancer. Asian Pac J Cancer Prev 16:3085–3090
Petroni M, Sardina F, Heil C, Sahun-Roncero M, Colicchia V, Veschi V, Albini S, Fruci D, Ricci B, Soriani A, Di Marcotullio L, Screpanti I, Gulino A, Giannini G (2015) The MRN complex is transcriptionally regulated by MYCN during neural cell proliferation to control replication stress. Cell Death Differ 23:197–206

Schmitt E, Paquet C, Beauchemin M, Bertrand R (2007) DNA-damage response network at the crossroads of cell-cycle checkpoints, cellular senescence and apoptosis. J Zhejiang Univ Sci B 8:377–397

Shay JW (1997) Telomerase in human development and cancer. J Cell Physiol 173:266–270

Shay JW, Reddel RR, Wright WE (2012) Cancer and telomeres-an ALTernative to telomerase. Science 336:1388–1390

Sikora E, Bielak-Zmijewska A, Mosieniak G (2014) Cellular senescence in ageing, age-related disease and longevity. Curr Vasc Pharmacol 12:698–706

Singhapol C, Pal D, Czapiewski R, Porika M, Nelson G, Saretzki GC (2013) Mitochondrial telomerase protects cancer cells from nuclear DNA damage and apoptosis. PLoS One 8:e52989

Suman S, Das TP, Sirimulla S, Alatassi H, Ankem MK, Damodaran C (2016) Withaferin-A suppress AKT induced tumor growth in colorectal cancer cells. Oncotarget 7:13854–13864

Takai H, Smogorzewska A, de Lange T (2003) DNA damage foci at dysfunctional telomeres. Curr Biol 13:1549–1556

Thaiparambil JT, Bender L, Ganesh T, Kline E, Patel P, Liu Y, Tighiouart M, Vertino PM, Harvey RD, Garcia A, Marcus AI (2011) Withaferin A inhibits breast cancer invasion and metastasis at sub-cytotoxic doses by inducing vimentin disassembly and serine 56 phosphorylation. Int J Cancer 129:2744–2755

Um HJ, Min KJ, Kim DE, Kwon TK (2012) Withaferin A inhibits JAK/STAT3 signaling and induces apoptosis of human renal carcinoma Caki cells. Biochem Biophys Res Commun 427:24–29

van Steensel B, Smogorzewska A, de Lange T (1998) TRF2 protects human telomeres from end-to-end fusions. Cell 92:401–413

von Figura G, Hartmann D, Song Z, Rudolph KL (2009) Role of telomere dysfunction in aging and its detection by biomarkers. J Mol Med (Berl) 87:1165–1171

Vulliamy T, Beswick R, Kirwan M, Marrone A, Digweed M, Walne A, Dokal I (2008) Mutations in the telomerase component NHP2 cause the premature ageing syndrome dyskeratosis congenita. Proc Natl Acad Sci U S A 105:8073–8078

Wadhwa R, Kaul Z, Kaul SC (2016) Cell cycle checkpoints and senescence. In: Suresh ISR, Hayflick L (eds) Cellular ageing and replicative senescence. Springer, Cham, pp 145–166

Watson JD (1972) Origin of concatemeric T7 DNA. Nat New Biol 239:197–201

White LK, Wright WE, Shay JW (2001) Telomerase inhibitors. Trends Biotechnol 19:114–120

Widodo N, Kaur K, Shrestha BG, Takagi Y, Ishii T, Wadhwa R, Kaul SC (2007) Selective killing of cancer cells by leaf extract of Ashwagandha: identification of a tumor-inhibitory factor and the first molecular insights to its effect. Clin Cancer Res 13:2298–2306

Widodo N, Takagi Y, Shrestha BG, Ishii T, Kaul SC, Wadhwa R (2008) Selective killing of cancer cells by leaf extract of Ashwagandha: components, activity and pathway analyses. Cancer Lett 262:37–47

Widodo N, Priyandoko D, Shah N, Wadhwa R, Kaul SC (2010) Selective killing of cancer cells by Ashwagandha leaf extract and its component Withanone involves ROS signaling. PLoS One 5(10):e13536

Yeager TR, Neumann AA, Englezou A, Huschtscha LI, Noble JR, Reddel RR (1999) Telomerase-negative immortalized human cells contain a novel type of promyelocytic leukemia (PML) body. Cancer Res 59:4175–4179

von Zglinicki T, Burkle A, Kirkwood TB (2001) Stress, DNA damage and ageing - an integrative approach. Exp Gerontol 36:1049–1062

Yu Y, Katiyar SP, Sundar D, Kaul Z, Miyako E, Zhang Z, Kaul SC, Reddel RR, Wadhwa R (2017) Withaferin-A kills cancer cells with and without telomerase: chemical, computational and experimental evidences. Cell Death Dis 8:e2755

Chapter 10
Withania somnifera Has Potential to Treat Cancer

Babli Halder and Suman S. Thakur

Abstract *Withania somnifera* is universally known as Ashwagandha in Ayurveda. Different parts of the *Withania* are used in medicinal and clinical applications. It possesses a lot of therapeutic action. Among them anticancer activity of *Withania* is the most interesting subject to research with. Recently in various studies *Withania somnifera* extract, different metabolites derived from *Withania* are found to show potent anticancer efficacy against various cancers including pancreatic, breast, lung, cervical, skin etc. The mostly studied Withaferin A (WA), a withanolide purified from *Withania somnifera* is an important bioactive molecule. WA showed promising inhibitory activity against various cancer cells *in vitro* and *in vivo*. WA induces apoptosis in cancer cells by stimulating pro-apoptotic proteins and inhibiting anti-apoptotic proteins. The mechanisms of action of WA in cancer cells are extensively studied all over the world. *Withania somnifera* could be use full medicinal herb as a potent chemotherapeutic agent in diverse clinical circumstances.

Keywords *Withania somnifera* • Cancer • Withafrin A • Withanolide • Traditional medicine • Chemotherapeutic agent

10.1 Introduction

Aswagandha (Withania somnifera) is a very well known medicinal plant and it has a number of ethno pharmacological actions. Various parts of different species of *Withania* having diverse activities (Naz et al. 2009). *Withania somnifera* has a wide range of therapeutic potentials like anti-arthritic, anti-cancer, anti-inflammatory, anti-aging, immunoregulatory, chemo protective, cardio protective, and recovery from neurodegenerative disorders. With the developing realization of its benefits several *in vitro* and *in vivo* preclinical studies are ongoing to validate its therapeutic

B. Halder • S.S. Thakur (✉)
Proteomics and Cell Signaling, Lab E409, Centre for Cellular & Molecular Biology, Habsiguda, Uppal Road, Hyderabad 500 007, Telangana, India
e-mail: sst@ccmb.res.in

S.C. Kaul, R. Wadhwa (eds.), *Science of Ashwagandha: Preventive and Therapeutic Potentials*, DOI 10.1007/978-3-319-59192-6_10

potentials (Singh et al. 2015). Diverse compounds of *Withania* species like withaferin A derived from the leaves, is well recognized to posses anti-cancer properties (Jayaprakasam et al. 2003). Recently quite a lot of studies have been conducted to validate its anti-cancerous effect in *in vitro* and *in vivo* preclinical experiments. In various studies the mechanism of action of this phytochemicals has been dissect out elaborately. *Withania* could modulate different signaling pathways in cancer cell lines (Gambhir et al. 2015; Park et al. 2015; Nishikawa et al. 2015; Sarkar et al. 2014; Roy et al. 2013; Grover et al. 2012; Munagala et al. 2011; Yu et al. 2010). Withaferin A (WA), is a steroidal lactone isolated from medicinal plant *Withania somnifera*, was attenuated tumor development by inducing oxidative stress by increasing the basal reactive oxygen species levels and suppressing NF-κB pathway (Gambhir et al. 2015). Compound derived from Methanolic (MeOH) extract of withaferin A showed cytotoxic effect against human pancreatic (PANC-1), prostate (DU145) and breast (MCF7) cancer cell line by inhibiting hedgehog (Hh) signal (Yoneyama et al. 2015). The smallest inhibitor of apoptosis known as Survivin is up regulated in case of almost cancer cells which plays both the roles in proliferation and survival of the cancer cells. It inhibits caspases and facilitates mitosis by becoming a part of chromosomal passenger complex through its BIR5 domain. Thus it causes inhibition of apoptosis. Withanone, extracted from roots of *Withania somnifera* showed binding affinity with BIR5 domain of survivin and which causes inhibitory action against caspases and as a result induction of apoptosis. Binding of withanone at BIR5 domain of survivin might also inhibited to be a part of chromosomal passenger complex and resulted hindrance of the mitotic process within the cancer cell (Wadegaonkar and Wadegaonkar 2013). Molecular docking studies provided evidence that WA is a promising VEGF inhibitor, which has important role in the angiogenesis in cancer cells (Saha et al. 2013). HSP 90 (Heat shock protein 90) is a molecular chaperone is known to responsible for quality control in cell environment and protein folding. Other than this HSP 90 is involved in the maturation and stabilization of an extensive array of oncogenic client proteins, which are decisive for oncogenesis and malignant development. Computational docking study showed that WA inhibits the interaction between chaperone (Hsp90) to its co-chaperone (Cdc37) by disrupting the stability of attachment of Hsp90 to Cdc37. Thus WA could be a potential anticancer drug as determined by its significant Hsp90-client modulating capability (Grover et al. 2011).

10.2 Effect of *Withania* on Breast Cancer

WA is very important phytochemical to treat various kind of cancer. Among them breast cancer treatment by WA is recently progressed a lot. Szarc vel Szic K et al. in 2014 has shown that WA decreased invasion of the breast cancer cell MDA-MB-231. WA treatment reduced gene expression of extracellular matrix-degrading proteases (uPA, PLAT, ADAM8), cell adhesion molecules like integrins, laminins and along with this increased expression of the breast cancer metastasis suppressor gene

(BRMS1). Preventive effect of *Withania somnifera* root extract was assessed in mammary cancer bearing female transgenic MMTV/*Neu* mice. A 33% reduction of mammary carcinoma was found in treated group than those of the control group. Moreover reduction of the Ki67, a cellular marker of cell proliferation, proliferating cell nuclear antigen marker (PCNA) as well as chemokine CCL2 was also observed in mammary carcinoma of treated mice in comparison to the control mice (Khazal et al. 2014).

In another study, it has been shown that WA inhibited growth of cultured and xenografted human breast cancer cells in nude mice. WA also prevented breast cancer development and pulmonary metastasis in transgenic mice (Lee et al. 2013a, b). Treatment of human breast cancer cells, MDA-MB-231 and MCF-7 to pharmacological concentrations of WA resulted in cleavage of Notch2 as well as Notch4. The levels of both transmembrane and cleaved Notch1 were decreased towards the treatment of WA (Lee et al. 2012). WA also found to inhibit level of XIAP, cIAP-2, and Survivin in human breast cancer cells MDA-MB-231 and MCF-7 thereby induced apoptosis. In *in vivo* MDA-MB-231 xenografted model WA reduced Survivin protein level, which indicated the significant role in the induction of apoptosis (Hahm and Singh 2012). WA showed anti-metastatic activity in human breast cancer cells by mediating through its action on vimentin and vimentin ser56 phosphorylation (Thaiparambil et al. 2011).

Important role of reactive oxygen species (ROS) in WA mediated apoptosis induction in the human breast cancer cells, MCF-7 and MDA-MB-231 was experimentally demonstrated by Hahm et al. in 2011. WA treatment resulted in generation of ROS in both of the human breast cancer cells. The authors also evidenced production of ROS associated with the mitochondrial dysfunction. Expression level of pro-apoptotic protein Bax and Bak were also been increased in WA treated cancer cells.

Widodo and his group (2010) investigated the mechanism of action of the leaf extract of *Withania somnifera*, Ashwagandha (i-Extract) and its purified component Withanone in breast cancer cells MCF-7. They have found selective cancer cell death by induction of ROS (reactive oxygen species) signaling cascade in treated cells. A decrease in the mitochondrial membrane potential was found in treated MCF-7 cells. MCF-7 cells were arrested at G2 phase after treatment with i-Extract and Withanone. i-Extarct, Withanone, Withaferin A were also seen to induce DNA damage in MCF-7 cells by stimulation and phosphorylation of γH2AX, which is a marker of DNA double strand break. An increment of the CDKN1A-p21 level was also observed in treated MCF-7 cells in comparison to untreated cancer cells. Moreover in their bioinformatics study, it was found that the association of p53, apoptosis and insulin/IGF signaling cascade is connected with the ROS signaling.

10.3 Effect of *Withania* on Gastric, Colon and Colorectal Cancer

Withaferin A (WA), purified from *Withania somnifera*, has well known anti-inflammatory and anti-tumor effects. Kim et al. studied that WA could be a protective and therapeutic agent for gastric cancer (Kim et al. 2015). Choi and Kim's study

(2015) suggested that withaferin-A derived from leaf extract of *W. somnifera* deceased interleukin-6-induced activation of STAT3 thus progression of colon cancer. They have generated HCT116 colon cancer cells xenograft tumors in nude mice. Treated mice with Withaferin-A have shown decreased weight and less volume tumor than that of the untreated control mice. It has been also shown that immunomodulatory function of *Withaferin somnifera* might be helpful to treat the colon cancer induced by the carcinogen Azoxymethane *in vivo* in Swiss albino mice (Muralikrishnan et al. 2010a, b). In an *in vivo* model WA showed promising protective agent against colon cancer cells. Azoxymethane induced colon cancer in Swiss albino mice. The comparison has been made on the enzymes involved in tricarboxylic acid (TCA) cycle between WA treated and untreated colon cancer bearing mice. After the stipulated period decreased activities of TCA cycle key enzymes such as isocitrate dehydrogenase (ICDH), succinate dehydrogenase (SDH), malate dehydrogenase (MDH), and alpha-keto glutarate dehydrogenase (alpha-KGDH) were found in colon cancer bearing animals. On the other hand it has been seen that WA normalized the level of the TCA key enzymes, ICDH, SDH, MDH, and alpha-KGDH in colon cancer bearing mice. These results together proposed that the WA has a potential anti cancer efficacy against colon cancer (Muralikrishnan et al. 2010a, b). In colon cancer development Notch signaling has a critical task. WA inhibited Notch-1 signaling and reduced prosurvival pathways, such as Akt/NF-kappaB/Bcl-2, in three human colon cancer cell lines (HCT-116, SW-480, and SW-620). WA induced apoptosis by down regulating mammalian target of rapamycin signaling components, pS6K and p4E-BP1, and triggered c-Jun-NH(2)-kinase. These observations confer emphasis to the anticancer potential of WA, which could be further developed as a targeted chemotherapy in case of colon cancer (Koduru et al. 2010). WA arrested cell cycle at G2/M in colorectal cancer cell lines HCT116 and SW480. It showed induction of apoptosis in a dose dependent manner. WA caused mitotic delay by blocking Spindle assembly checkpoint (SAC) function. Apoptosis induction by WA is associated with proteasomal degradation of Mad2 and Cdc20, an important constituent of the Spindle Checkpoint Complex (Das et al. 2014).

10.4 Effect of *Withania* on Renal Cancer

WA induced apoptosis in human renal cancer cells, Caki by down-regulation of STAT3 and its associated Janus-activated kinase 2 (JAK2) activity, and other genes like Bcl-xL, Bcl-2, cyclin D1 and surviving (Um et al. 2012). WA treatment of human renal Caki cells stimulated a number of ER stress markers, such as phosphorylation of eukaryotic initiation factor-2α (eIF-2 α), ER stress-specific XBP1 splicing, and increment of glucose-regulated protein (GRP)-78. Additionally, WA responsible for up regulation of CAAT/enhancer-binding protein-homologous protein (CHOP), which suggested the generation of ER stress. These studies provide proof that WA incubation is important to induce apoptosis in human renal cancer cells (Choi et al. 2011). In another study WA has shown potent protective effect against human renal cancer Caki cells. In this study it has been observed that combinatorial effect of WA and radiation on induction of apoptosis in Caki cells. This treatment increased cleavage of caspases 3 and PARP and arrest of cell cycle at sub-G1 phase. WA improved radiation stimulated apoptosis in renal cancer cells by means of ROS generation and down regulation of Bcl2 and dephosphorylation of Akt (Yang et al. 2011a, b).

10.5 Effect of *Withania* on Skin Cancer, Osteosarcoma and Mesothelioma

In our study the cytotoxic effect of *Withania somnifera* root extract on human malignant melanoma A375 cells was demonstrated. The crude extract of *Withania* showed time and dose dependant cytotoxicity against A375 cells as observed by MTT assay. MTT assay showed a statistically significant reduction of cell viability with an IC50 value of 350 μg/ml for 24 h of treatment, 250 μg/ml for 48 h of treatment and 200 μg/ml for 72 h of treatment with the root extract (Fig. 10.1). Phase contrast micrography of A375 cells after treatment with IC50 doses of Withania extract showed spherical appearance as compared to the untreated polygonal A375 cells. Fluorescence microscopic observation with DAPI staining showed nuclear blebbing and apoptotic body formation after 48 h and 72 h of treatment respectively. Further a prominent DNA ladder formation was observed in 48 h and 72 h *Withania* extract treated A375 cells. Thus all the observations from the *Withania* root extract treated A375 cells indicate an initiation of apoptotic cell death (Halder et al. 2015).

Withaferin A (WA) was inhibited carcinogen-induced up-regulation of acetyl-CoA carboxylase 1 (ACC1), which was additionally established in a skin cell transformation model. WA could able to suppress ACC1 level in human melanoma cells,

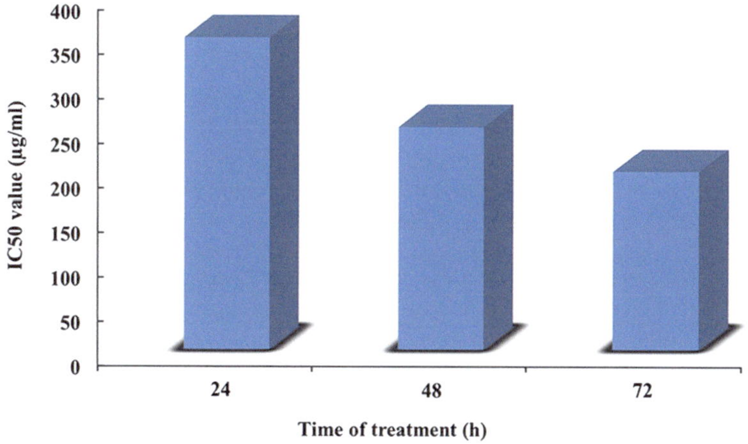

Fig. 10.1 IC_{50} value at different time point of treatment with *Withania somnifera* water extract in human malignant melanoma A375 cells

which suggests that ACC1 could be potential target of WA in clinical studies (Li et al. 2015a, b). Effect of WA has been studied on skin epidermal JB6 P+ cells, which is a famous model for tumor promotion. It has been found that WA inhibited the tumor promoter 12-O-tetradecanoylphorbol 13-acetate (TPA)-induced cell transformation and cell proliferation. Inactivation of isocitrate dehydrogenase 1 (IDH1) by TPA has been observed to be reversed by WA. Mitochondrial functions were suppressed by TPA treatment which was supported by reduced mitochondrial membrane potential, complex I activity and mitochondrial respiration. But WA treatment could inhibit all theses suppressions. Moreover it was examined WA blocked reduction of the level of α-ketoglutarate, a product of IDH1 upon TPA treatment. Tumor promoter treatment caused the elevation of lactate level which is a glycolysis marker but WA inhibited its rise by this it showed its chemopreventive activity in JB6 P+ cells (Li and Zhao 2013). WA has shown to inhibit human malignant melanoma cells by induction of apoptosis in these cell lines as evident from down regulation of Bcl-2, by increment of Bax translocation to mitochondria, generation of Reactive oxygen species (ROS), change of mitochondrial membrane potential, release of cytochrome c in cytosol, activation of caspases 9 and caspases 3 and DNA fragmentation (Mayola et al. 2011). Withaferin A (WA) showed antiproliferative activity by inhibiting cell cycle at the G2/M phase in association with the reduced expression of cyclin B1, cyclin A, Cdk2 and p-Cdc2 (Tyr15) and an increase in the levels of p-Chk1 (Ser345) and p-Chk2 (Thr68) in two types of human osteosarcoma cells, MG-63 and U2OS (Lv and Wang 2015). Wadhwa et al. (2013) have observed that the water extract of Ashwagandha, *Withania somnifera* leaves (ASH-WEX) and TEG (triethylene glycol) derived from ASH-WEX showed potent cytotoxic effect on human osteosarcoma U2OS cells. This treatment revealed an increment of the level of p53 and its downstream p21 protein. Phosphorylated p53 level was also increased due to treatment when compare to the untreated cancer cells. An arrest of G1 phase of cell cycle was observed in ASH-WEX and TEG treated cancer cells. Additionally this treatment caused hypophosphorylation of pRB thus activating its tumor suppressing function. Further a decrease of MMP-3 (matrix Metalloproteinases-3) and MMP-9 in treated U2OS cells was observed, thereby suggesting an antimetastatic role of ASH-WEX and TEG.

WA has potent suppressive effects on the malignant pleural mesothelioma (MPM) growth. Both murine and patient-derived MPM cells growth inhibited by WA treatment. It induces apoptosis evidenced by activation of pro-apoptotic p38 stress activated protein kinase (SAPK) and caspase-3, increased levels of pro-apoptotic Bax protein and cleavage of poly-(ADP-ribose)-polymerase (PARP). Gene-array based analyses also revealed that WA suppressed cell growth and metastasis-promoting genes like c-myc. Expression of the cell cycle and apoptosis regulatory protein (CARP)-1/CCAR1 was also stimulated towards the treatment of WA thus it acts as a promising anti MPM agent (Yang et al. 2012). The Aurora A kinase is a Ser/Thr kinase and its functional copartner TPX2 are the potential candidates for promotion of various cancers. Computational and experimental studies revealed that withanone derived from *Withania* could able to dissociate TPX2-Aurora after binding with this complex thus it might be acted as a potential anticancer drug (Grover et al. 2012).

10.6 Effect of *Withania* on Glioma and Neuroblastoma

An extract of *Withania somnifera* called AshwaMAX is effective on orthotopic Glioblastoma multiforme (GBM) in murine model (Chang et al. 2016). Water extract of *Withania somnifera* blocked the G2/M stage and activated many pro-apoptotic cascades which led to suppression of cyclin D1, bcl-xl, and p-Akt,. It further decreased the expression of polysialylated form of neural cell adhesion molecule (PSA-NCAM) and action of matrix metalloproteinases in glioma cells. In *in vivo* experiment it has been observed that Aswagandha water extract reduced the intracranial tumor volumes and suppressed the tumor-promoting proteins p-nuclear factor kappa B (NF-κB), p-Akt, vascular endothelial growth factor (VEGF), heat shock protein 70 (HSP70), PSA-NCAM, and cyclin D1 in the rat model of orthotopic glioma allograft (Kataria et al. 2015).

Alcoholic extract of *Withania somnifera* leaves (i-Extact) and its different components (Withaferin A, Withanone, Withanolide A) showed an effective dose dependent antiproliferative activity against rat glioma cell line, C6 and human glioma cell line, YKG1. High concentrations of i-extarct and its constituents induced apoptosis but low concentrations resulted in the growth arrest. Cell cycle arrest was found at S phase in the treated C6 cells. Additionally i-Extract and its components were able to change the morphology of glioma cells. Moreover it was observed that there was augmentation of GFAP (glial fibrillary acidic protein) level, decrease in cell motility and enhancement of neuronal cell adhesion molecules (NCAM) in the treated glioma cells (Shah et al. 2009).

In neuroblastoma cells (IMR-32) Ashwagandha water extract (ASH-WEX) showed anti-proliferative and anti migratory activities. ASH-WEX down regulated MMP2 and MMP9 activities. ASH-WEX treatment resulted in cell cycle arrest at G0/G1 phase and augmented early apoptotic population. Down regulation of cell cycle marker Cyclin D1, anti-apoptotic marker bcl-xl and Akt-P provide evidence that ASH-WEX might prove to be a potential therapeutic intervention in case of neuroblastoma related malignancies (Kataria et al. 2013).

10.7 Effect of *Withania* on Pancreatic Cancer

Li et al. have shown that the WA extracted from *Withania* has shown synergistic effect with Oxaliplatin on human pancreatic cells *in vitro* and *in vivo*. Combinatorial treatment oxaliplatin and WA showed significant cytotoxicity in human pancreatic cells by means of induction of apoptosis and reduction of PI3K/AKT thus it could be an effective strategy for the treatment of pancreatic cancer (Li et al. 2015a, b). In pancreatic cancer cells β-catenin plays an essential role in oncogenesis. Pancreatic cancer is very aggressive and weakly prognosed. Withanolide-D (witha-D) extracted from *Withania* has shown to increase apoptosis in pancreatic ductal adenocarcinoma cells by arresting cell cycle dynamics at cell-the G2/M phase. Witha-D down regulated Akt, which in turn failed to promote Gsk3β deactivation. As a result active Gsk3β aided β-catenin demolition in pancreatic carcinoma cells. This observation proposed a new potential molecule in the treatment of pancreatic adenocarcinoma (Sarkar et al. 2014). Different *in vitro* and *in vivo* studies shown that WA exhibited potent anti-proliferative effect against human pancreatic cancer cells Panc-1, MiaPaCa2 and BxPc3. WA was able to induce significant apoptosis in Panc-1 cell line. WA was found to bind with Hsp90, reduce Hsp90 chaperone function through an ATP-independent mechanism. Thus resulted in Hsp90 client protein degradation, and exerted anticancer potential against pancreatic cancer (Yu et al. 2010).

10.8 Effect of *Withania* on Cervical Cancer

Hyperthermia in combination of WA showed significant increase of apoptosis in human cervical cancer HeLa cells by augmenting pro-apoptotic Bcl-2-family proteins tBid and Noxa, and down regulating antiapoptotic Bcl-2 and Mcl-1. Combined treatment with hyperthermia and WA induced significant increases in JNK phosphorylation (p-JNK), and decreases in the phosphorylation of ERK (p-ERK) compared with either treatment alone (Cui et al. 2014). WA treatment down regulated the level of TGF β and suppressed expression and activity of matrix metalloproteinase (MMP) 9 in human cervical cancer cells Caski cell line (Lee et al. 2013a, b). Human papilloma virus (HPV) expressing E6 and E7 oncoproteins is the major cause of cervical cancer. These oncoproteins actually inactivate tumor suppressor proteins p53 and pRb, respectively. So suppression of HPV oncoproteins would result in activation of tumor repressor pathways leads to apoptosis in cancer cells. WA showed inhibitory efficacy against several different cancers. Here WA has shown protective effect in both *in vitro* and *in vivo* models of human cervical cancer cells, CaSki. WA significantly inhibited proliferation of the cervical cancer cells associated with down regulation of expression of HPV E6 and E7 oncoproteins, induction of p53 up regulation, augmentation of p21(cip1/waf1) and its interaction with proliferating cell nuclear antigen (PCNA). Moreover it caused G(2)/M cell cycle arrest and modulation of cyclin B1, p34(cdc2) and PCNA levels. Decreased the levels of STAT3 and its phosphorylation at Tyr(705) and Ser(727) and alter expression levels of p53-mediated apoptotic markers-Bcl2, Bax, caspase-3 and cleaved PARP after treatment with WA are responsible for the induction of apoptosis in human cervical cancer. In vivo study also revealed that the WA treatment resulted in a significant reduction of tumor volume in athymic nude mice injected with CaSki cells (Munagala et al. 2011).

10.9 Effect of *Withania* on Prostate Cancer

Nishikawa et al. (2015) reported that Withaferin A (WA) induces cell death by apoptosis or necrosis in various kinds of tumor cells. WA could induce cell death selectively in two cancer cell lines, androgen-insensitive PC-3 and DU-145 prostate adenocarcinoma cells, but the cytotoxicity was fewer severe in androgen-sensitive LNCaP prostate adenocarcinoma cells and normal human fibroblasts (TIG-1 and KD). Abnormal cell cycle dynamics is one of the major causes of prostate cancer. So most of the chemotherapeutic drugs mainly target the cell cycle regulatory molecules. Treatment with WA in both the prostate cancer cell lines (PC-3 and DU-145) inhibited cell cycle at G2/M phase. It is accompanied by up-regulation of phosphorylated Wee-1, phosphorylated histone H3, p21, and Aurora B while down regulation of cyclins (A2, B1, and E2). A reduction in phosphorylated Cdc2 (Tyr15) was also found. Moreover WA treatment also reduced levels of phospho Chk1 (Ser345) and Chk2 (Thr68) in prostate cancer cells (Roy et al. 2013). Inhibition of purified proteasome by WA has been determined by Yang et al. (2007). They have observed reduction of chymotrypsin like activity of 26S proteasome in human prostate cancer cell lines, PC3 and LNCaP and in solid tumors. This resulted in induction of apoptosis, inhibition of angiogenesis and *in vivo* suppression of tumor growth. WA induced accumulation of proteasome substrate Bax, IκBα, and p27 and also ubiquitinated proteins in androgen receptor (AR) independent human prostate cancer PC3 cells. Proteasomal degradation was associated with the activation of caspase-3 and cleavage of PARP in PC3 cells. It has also been shown that WA treatment reduced AR level in AR dependent LNCaP cells. In their in vivo xenograft model with PC3 cell line, they found that treatment with WA in tumor bearing mice showed a significant regression of tumor growth. Treatment with WA resulted in the reduction of chymotrypsin activity of proteasome and accumulation of proteasome substrate. Apoptotic morphology and TUNEL (Terminal deoxynucleotidyl transferase dUTP nick end labeling) positive apoptotic cells were also observed in WA treated tumor tissue *in vivo*.

10.10 Effect of *Withania* on Myeloid Cells

In tumor progression myeloid cells play a vital role. Myeloid-derived suppressor cells (MDSC) and tumor-associated macrophages (TAMs) are the most common tumor-infiltrating cells. MDSC and TAMS cells support tumor expansion by their intrinsic immune suppressive action and that is enhanced by their cross talk. It has been found that the root extract of the plant *Withania somnifera* reduce tumor development. WA treatment reduced production of IL-10 in MDSC and secretion of IL-6 and TNFα, cytokines that increase MDSC accumulation and function from macrophages (Sinha and Ostrand-Rosenberg 2013).

10.11 Effect of *Withania* on Lymphoma

Activity of heat shock protein (Hsp) 90 inhibited by WA as revealed by a quick increase in Hsp70 expression levels in lymphomas. It has been found that WA robustly inhibited the survival of numerous human and murine B cell lymphoma cells. Moreover, in vivo experiments with syngeneic-graft lymphoma cells proposed that WA inhibits the augmentation of tumor without affecting other proliferative tissues. From this study it could be said that the anti-cancer effects of WA in lymphomas are because of its capability to inhibit Hsp90 function and successive decrease of decisive kinases and cell cycle controllers (McKenna et al. 2015). WA treatment enhances ionizing radiation (IR) induced apoptosis in human lymphoma U937 cells in associated with the PARP cleavage, caspase-3 activation and down regulation of anti-apoptotic protein Bcl-2 which proposed that WA might be acted as a potential radiosensitizer. From the experimental evidences it was revealed that generation of reactive oxygen species (ROS), down regulation of Bcl-2 and MAPKs pathway activation are importantly involved in the apoptosis stimulated by WA and radiation (Yang et al. 2011a, b). Ceramide has diverse biological function and it is a fundamental second messenger. Its main function to induce apoptosis and repress the cell growth. In leukemic cells, deregulation in ceramide metabolism causes its chemoresistance. Thus in leukemia chemotherapeutic regiment, management in ceramide production by physiological and pharmacological modulators provides a strategy for the prevention. Withanolide D derived from *Withania somnifera* enhanced the ceramide accumulation by activating N-SMase 2. WithaD inhibited JNK and p38MAPK and stimulated apoptosis in both in myeloid (K562) and lymphoid (MOLT-4) cells and also with primary cells derived from leukemia patients. All together, it might be suggested that WithaD could consider as a potential chemotherapeutic agent in drug development strategy (Mondal et al. 2010).

10.12 Effect of *Withania* on Liver and Lung Cancer

In HepG2 cell line, crude extract of *Withania* showed potential anticancer activity. Dichloromethane fraction of *Withania* showed significant amount of apoptotic cells W. *somnifera* may have the potential to serve as a template for future anticancer drug development (Abutaha 2015). However, further investigation is required to identify the active compound/s. WA exhibited anti-angiogenic, anti-metastatic and anti-neoplastic activity in human metastatic liver cancer cell line, SK-Hep1 (Lee et al. 2013a, b). WA showed an effective protective action against human lung cancer cell line NCI-H460 (Choudhary et al. 2010).

10.13 Effect of *Withania* on Head and Neck Cancer

WA has been shown to inhibit the proliferation, metastasis, invasion and angiogenesis of head and neck cancer cells, AMC-HN4. WA induced apoptosis, by the increment of sub-G1 cell population and the cleavage of poly (ADP-ribose) polymerase (PARP) (Park et al. 2015).

10.14 Effect of *Withania* on Ovarian Cancer

In case of ovarian cancer chemotherapy is an effective way to treat after cytoreductive surgery. But relapse of the tumor occurs due to chemo-resistance of cancer stem cells. Treatment with WA alone or in combination of cisplatin might reduce significant tumor growth of nude mice bearing orthotopic ovarian tumors, which was generated by injecting human ovarian epithelial cancer cell line (A2780). This combination significantly eliminated cells, which are expressing cancer stem cells markers like -CD44, CD24, CD34, CD117 and Oct4 and down regulated Notch1, Hes1 and Hey1 genes expression. So it might be said that WA alone or in combination with cisplatin could be an efficient therapy for ovarian cancer (Kakar et al. 2014).

10.15 Conclusion

In order to conclude it could be said that extract from *Withania somnifera* or its derivative like WA exhibited significant anti proliferative activity in different human cancer cells. WA shows chemoprevention by facilitating suppression of tumor promoting factors and induction of apoptosis inducing factors in cancer. Chemotherapeutic activity has been proven in *in vitro* as well as *in vivo* model so it serves as a template for prospect anticancer drug development. Nevertheless, additional research is essential to classify the active compound/s and its bioactivity.

Acknowledgements Authors are thankful to the Department of Biotechnology (Grant Number: GAP0405) and Council of Scientific and Industrial Research (Grant Number BSC0208) and AB SCIEX, India for funding and help. We are grateful to all the lab members for fruitful discussion.

References

Abutaha N (2015) *In vitro* antiproliferative activity of partially purified *Withania somnifera* fruit extract on different cancer cell lines. J BUON 20:625–630

Chang E, Pohling C, Natarajan A, Witney TH, Kaur J, Xu L, Gowrishankar G, D'Souza AL, Murty S, Schick S, Chen L, Wu N, Khaw P, Mischel P, Abbasi T, Usmani S, Mallick P, Gambhir SS (2016) AshwaMAX and Withaferin A inhibits gliomas in cellular and murine orthotopic models. J Neuro-Oncol 126:253–264

Choi BY, Kim BW (2015) Withaferin-A inhibits colon cancer cell growth by blocking STAT3 transcriptional activity. J Cancer Prev 20:185–192

Choi MJ, Park EJ, Min KJ, Park JW, Kwon TK (2011) Endoplasmic reticulum stress mediates withaferin A-induced apoptosis in human renal carcinoma cells. Toxicol In Vitro 25:692–698

Choudhary MI, Hussain S, Yousuf S, Dar A, Mudassar A-u-R (2010) Chlorinated and diepoxy withanolides from *Withania somnifera* and their cytotoxic effects against human lung cancer cell line. Phytochemistry 71:2205220–2205229

Cui ZG, Piao JL, Rehman MU, Ogawa R, Li P, Zhao QL, Kondo T, Inadera H (2014) Molecular mechanisms of hyperthermia-induced apoptosis enhanced by withaferin A. Eur J Pharmacol 723:99–107

Das T, Roy KS, Chakrabarti T, Mukhopadhyay S, Roychoudhury S (2014) Withaferin A modulates the Spindle assembly checkpoint by degradation of Mad2-Cdc20 complex in colorectal cancer cell lines. Biochem Pharmacol 91:31–39

Gambhir L, Checker R, Sharma D, Thoh M, Patil A, Degani M, Gota V, Sandur SK (2015) Thiol dependent NF-κB suppression and inhibition of T-cell mediated adaptive immune responses by a naturally occurring steroidal lactone Withaferin A. Toxicol Appl Pharmacol 289:297–312

Grover A, Shandilya A, Agrawal V, Pratik P, Bhasme D, Bisaria VS, Sundar D (2011) Hsp90/Cdc37 chaperone/co-chaperone complex, a novel junction anticancer target elucidated by the mode of action of herbal drug Withaferin A. BMC Bioinf 12(Suppl 1):30

Grover A, Singh R, Shandilya A, Priyandoko D, Agrawal V, Bisaria VS, Wadhwa R, Kaul SC, Sundar D (2012) Ashwagandha derived withanone targets TPX2-Aurora A complex: computational and experimental evidence to its anticancer activity. PLoS One 7:e30890

Hahm ER, Singh SV (2012) Withaferin A-induced apoptosis in human breast cancer cells is associated with suppression of inhibitor of apoptosis family protein expression. Cancer Lett 334:101–108

Hahm ER, Moura MB, Kelley EE, Van Houten B, Shiva S, Singh SV (2011) Withaferin A-induced apoptosis in human breast cancer cells is mediated by reactive oxygen species. PLoS One 6:e23354

Halder B, Singh S, Thakur SS (2015) *Withania somnifera* root extract has potent cytotoxic effect against human malignant melanoma cells. PLoS One 10:e0137498

Jayaprakasam B, Zhang Y, Seeram NP, Nair MG (2003) Growth inhibition of human tumor cell lines by withanolides from Withania somnifera leaves. Life Sci 74:125–132

Kakar SS, Ratajczak MZ, Powell KS, Moghadamfalahi M, Miller DM, Batra SK, Singh SK (2014) Withaferin alone and in combination with cisplatin suppresses growth and metastasis of ovarian cancer by targeting putative cancer stem cells. PLoS One 9:e107596

Kataria H, Wadhwa R, Kaul SC, Kaur G (2013) Withania somnifera water extract as a potential candidate for differentiation based therapy of human neuroblastomas. PLoS One 8:e55316

Kataria H, Kumar S, Chaudhary H, Kaur G (2015) Withania somnifera suppresses tumor growth of intracranial allograft of glioma cells. Mol Neurobiol 53:1–16

Khazal KF, Hill DL, Grubbs CJ (2014) Effect of *Withania somnifera* root extract on spontaneous estrogen receptor-negative mammary cancer in MMTV/Neu mice. Anticancer Res 34:6327–6332

Kim JE, Lee JY, Kang MJ, Jeong YJ, Choi JA, SM O, Lee KB, Park JH (2015) Withaferin A inhibits Helicobacter pylori-induced production of IL-1β in dendritic cells by regulating NF-κB and NLRP3 inflammasome activation. Immune Netw 15:269–277

Koduru S, Kumar R, Srinivasan S, Evers MB, Damodaran C (2010) Notch-1 inhibition by Withaferin-A: a therapeutic target against colon carcinogenesis. Mol Cancer Ther 9:202–210

Lee J, Sehrawat A, Singh SV (2012) Withaferin A causes activation of Notch2 and Notch4 in human breast cancer cells. Breast Cancer Res Treat 136:45–56

Lee DH, Lim IH, Sung EG, Kim JY, Song IH, Park YK, Lee TJ (2013a) Withaferin A inhibits matrix metalloproteinase-9 activity by suppressing the Akt signaling pathway. Oncol Rep 30:933–938

Lee J, Hahm ER, Marcus AI, Singh SV (2013b) Withaferin A inhibits experimental epithelial-mesenchymal transition in MCF-10A cells and suppresses vimentin protein level in vivo in breast tumors. Mol Carcinog 54:417–429

Li W, Zhao Y (2013) Withaferin A suppresses tumor promoter 12-O-tetradecanoylphorbol 13-acetate-induced decreases in isocitrate dehydrogenase 1 activity and mitochondrial function in skin epidermal JB6 cells. Cancer Sci 104:143–148

Li W, Zhang C, Du H, Huang V, Sun B, Harris JP, Richardson Q, Shen X, Jin R, Li G, Kevil CG, Gu X, Shi R, Zhao Y (2015a) Withaferin A suppresses the up-regulation of acetyl-coA carboxylase 1 and skin tumor formation in a skin carcinogenesis mouse model. Mol Carcinog 55:1739–1746

Li X, Zhu F, Jiang J, Sun C, Wang X, Shen M, Tian R, Shi C, Xu M, Peng F, Guo X, Wang M, Qin R (2015b) Synergistic antitumor activity of withaferin A combined with oxaliplatin triggers reactive oxygen species-mediated inactivation of the PI3K/AKT pathway in human pancreatic cancer cells. Cancer Lett 357:219–230

Lv TZ, Wang GS (2015) Antiproliferation potential of withaferin A on human osteosarcoma cells via the inhibition of G2/M checkpoint proteins. Exp Ther Med 10:323–329

Mayola E, Gallerne C, Esposti DD, Martel C, Pervaiz S, Larue L, Debuire B, Lemoine A, Brenner C, Lemaire C (2011) Withaferin A induces apoptosis in human melanoma cells through generation of reactive oxygen species and down-regulation of Bcl-2. Apoptosis 16:1014–1027

McKenna MK, Gachuki BW, Alhakeem SS, Oben KN, Rangnekar VM, Gupta RC, Bondada S (2015) Anti-cancer activity of withaferin A in B-cell lymphoma. Cancer Biol Ther 16:1088–1098

Mondal S, Mandal C, Sangwan R, Chandra S, Mandal C (2010) Withanolide D induces apoptosis in leukemia by targeting the activation of neutral sphingomyelinase-ceramide cascade mediated by synergistic activation of c-Jun N-terminal kinase and p38 mitogen-activated protein kinase. Mol Cancer 9:239

Munagala R, Kausar H, Munjal C, Gupta RC (2011) Withaferin A induces p53-dependent apoptosis by repression of HPV oncogenes and upregulation of tumor suppressor proteins in human cervical cancer cells. Carcinogenesis 32:1697–1705

Muralikrishnan G, Amanullah S, Basha MI, Dinda AK, Shakeel F (2010a) Modulating effect of *Withania somnifera* on TCA cycle enzymes and electron transport chain in azoxymethane-induced colon cancer in mice. Immunopharmacol Immunotoxicol 32:523–527

Muralikrishnan G, Dinda AK, Shakeel F (2010b) Immunomodulatory effects of *Withania somnifera* on azoxymethane induced experimental coloncancer in mice. Immunol Investig 39:688–698

Naz S, Masud T, Nawaz MA (2009) Characterization of milk coagulating properties from the extract of *Withania coagulans*. Int J Dairy Technol 62:315–320

Nishikawa Y, Okuzaki D, Fukushima K, Mukai S, Ohno S, Ozaki Y, Yabuta N, Nojima H (2015) Withaferin A induces cell death selectively in androgen-independent prostate cancer cells but not in normal fibroblast cells. PLoS One 10:e0134137

Park JW, Min KJ, Kim DE, Kwon TK (2015) Withaferin A induces apoptosis through the generation of thiol oxidation in human head and neck cancer cells. Int J Mol Med 35:247–252

Roy RV, Suman S, Das TP, Luevano JE, Damodaran C (2013) Withaferin A, a steroidal lactone from *Withania somnifera*, induces mitotic catastrophe and growth arrest in prostate cancer cells. J Nat Prod 76:1909–1915

Saha S, Islam MK, Shilpi JA, Hasan S (2013) Inhibition of VEGF: a novel mechanism to control angiogenesis by Withania *somnifera's* key metabolite Withaferin A. In Silico Pharmacol 1:11

Sarkar S, Mandal C, Sangwan R, Mandal C (2014) Coupling G2/M arrest to the Wnt/β-catenin pathway restrains pancreatic adenocarcinoma. Endocr Relat Cancer 21:113–125

Shah N, Kataria H, Kaul SC, Ishii T, Kaur G, Wadhwa R (2009) Effect of the alcoholic extract of Ashwagandha leaves and its components on proliferation, migration, and differentiation of glioblastoma cells: combinational approach for enhanced differentiation. Cancer Sci 100:1740–1747

Singh P, Guleri R, Singh V, Kaur G, Kataria H, Singh B, Kaur G, Kaul SC, Wadhwa R, Pati PK (2015) Biotechnological interventions in *Withania somnifera* (L.) Dunal. Genet Eng Rev 19:1–20

Sinha P, Ostrand-Rosenberg S (2013) Myeloid-derived suppressor cell function is reduced by Withaferin A, a potent and abundant component of *Withania somnifera* root extract. Cancer Immunol Immunother 62:1663–1673

Szarc vel Szic K, Op de Beeck K, Ratman D, Wouters A, Beck IM, Declerck K, Heyninck K, Fransen E, Bracke M, De Bosscher K, Lardon F, Van Camp G, Vanden Berghe W (2014) Pharmacological levels of Withaferin A (*Withania somnifera*) trigger clinically relevant anticancer effects specific to triple negative breast cancer cells. PLoS One 9:e87850

Thaiparambil JT, Bender L, Ganesh T, Kline E, Patel P, Liu Y, Tighiouart M, Vertino PM, Harvey RD, Garcia A, Marcus AI (2011) Withaferin A inhibits breast cancer invasion and metastasis at sub-cytotoxic doses by inducing vimentin disassembly and serine 56 phosphorylation. Int J Cancer 129:2744–2755

Um HJ, Min KJ, Kim DE, Kwon TK (2012) Withaferin A inhibits JAK/STAT3 signaling and induces apoptosis of human renal carcinoma Caki cells. Biochem Biophys Res Commun 427:24–29

Wadegaonkar VP, Wadegaonkar PA (2013) Withanone as an inhibitor of survivin: a potential drug candidate for cancer therapy. J Biotechnol 168:229–233

Wadhwa R, Singh R, Gao R, Shah N, Widodo N, Nakamoto T, Ishida Y, Terao K, Kaul SC (2013) Water extract of Ashwagandha Leaves has anticancer activity: identification of an active component and its mechanism of action. PLoS One 8:e77189

Widodo N, Priyandoko D, Shah N, Wadhwa R, Kaul SC (2010) Selective killing of cancer cells by Ashwagandha Leaf extract and its component Withanone involves ROS signaling. PLoS One 5:e13536

Yang H, Shi G, Dou QP (2007) The tumor proteasome is a primary target for the natural anticancer compound Withaferin A isolated from "Indian winter cherry". Mol Pharmacol 71:426–437

Yang ES, Choi MJ, Kim JH, Choi KS, Kwon TK (2011a) Combination of withaferin A and X-ray irradiation enhances apoptosis in U937 cells. Toxicol In Vitro 25:1803–1810

Yang ES, Choi MJ, Kim JH, Choi KS, Kwon TK (2011b) Withaferin A enhances radiation-induced apoptosis in Caki cells through induction of reactive oxygen species, Bcl-2 downregulation and Akt inhibition. Chem Biol Interact 190:9–15

Yang H, Wang Y, Cheryan VT, Wu W, Cui CQ, Polin LA, Pass HI, Dou QP, Rishi AK, Wali A (2012) Withaferin A inhibits the proteasome activity in mesothelioma *in vitro* and *in vivo*. PLoS One 7:e41214

Yoneyama T, Arai MA, Sadhu SK, Ahmed F, Ishibashi M (2015) Hedgehog inhibitors from *Withania somnifera*. Bioorg Med Chem Lett 25:3541–3544

Yu Y, Hamza A, Zhang T, Gu M, Zou P, Newman B, Li Y, Gunatilaka AA, Zhan CG, Sun D (2010) Withaferin A targets heat shock protein 90 in pancreatic cancer cells. Biochem Pharmacol 79:542–551

Chapter 11
The Effects of Withaferin A on Normal and Malignant Immune Cells

Sara S. Alhakeem, Mary K. McKenna, Karine Z. Oben, Beth W. Gachuki, Sunil K. Noothi, Natarajan Muthusamy, Vivek M. Rangnekar, Ramesh C. Gupta, and Subbarao Bondada

Abstract Ashwagandha, an ayurvedic medicinal plant, has been found to be a major contributor in treating medical ailments. Ayurveda is one of the world's oldest traditional medical systems originated in India through various indigenous theories, beliefs and experiences. Often translated as science of "ayus" (life) and "ved" (knowledge), it emphasizes the use of herbs and their derivatives for prevention of disease, rejuvenation of body systems, and improving the quality of life. In this chapter, we will discuss some of the immuno-modulatory effects of withaferin A (WFA) as well as its growth-inhibitory effects on B cell lymphoma, leukemia and myelodysplastic syndrome.

Keywords *Withania somnifera* • Withaferin A • Immuno-modulatory effects • B cell lymphoma • Growth-inhibitory effects

S.S. Alhakeem • M.K. McKenna • K.Z. Oben • B.W. Gachuki • S.K. Noothi • S. Bondada (✉)
Department of Microbiology, Immunology and Molecular Genetics, Markey Cancer Center, University of Kentucky, Lexington, KY 40536, USA
e-mail: bondada@email.uky.edu

N. Muthusamy
Department of Internal Medicine, Ohio State University, Columbus, OH 43210, USA

V.M. Rangnekar
Department of Radiation Medicine, Markey Cancer Center, University of Kentucky, Lexington, KY, USA

R.C. Gupta
Department of Pharmacology and Toxicology, and James Graham Brown Cancer Center, University of Louisville, Louisville, KY 40202, USA

S.C. Kaul, R. Wadhwa (eds.), *Science of Ashwagandha: Preventive and Therapeutic Potentials*, DOI 10.1007/978-3-319-59192-6_11

11.1 Introduction

Ashwagandha, an ayurvedic medicinal plant, has been found to be a major contributor in treating medical ailments. Ayurveda is one of the world's oldest traditional medical systems that originated in India through various indigenous theories, beliefs and experiences. Often translated as science of "ayus" (life) and "ved" (knowledge), it emphasizes the use of herbs and their derivatives for prevention of disease, rejuvenation of body systems, and improving the quality of life (Mishra et al. 2000). Ashwagandha (*Withania somnifera (WS)* or Indian Ginseng) is a small evergreen woody shrub found in Africa, the Mediterranean, and India. It has been used as an adaptogen, liver tonic, rejuvenating herbal drug, anti-inflammatory, and immunostimulatory agent (Mishra et al. 2000). In addition, despite the limited laboratory evidence, more recently, its use has expanded to treat specific conditions including anxiety, asthma, bronchitis, emaciation, ulcers, insomnia, inflammation, cognitive and neurological disorders, and senile dementia (Widodo et al. 2009; Malik et al. 2007). Withanolides and alkaloids are two of the major classes of compounds characterized from *W. somnifera* (Mishra et al. 2000). Withaferin A (WFA) represented the first natural lactone of the withanolide series isolated (Mishra et al. 2000; Malik et al. 2007). WFA, a steroidal lactone, was demonstrated to have growth inhibitory effects for cancers of breast, cervix, lung, prostate, pancreas, and oral cancers as well as anti-angiogenic effects (Vanden Berghe et al. 2012). It has been reported in various studies to possess both immune-activating and immuno-suppressive properties. In this chapter, we will discuss some of the immuno-modulatory effects of WFA as well as its growth-inhibitory effects on B cell lymphoma, leukemia and myelodysplastic syndrome.

11.2 Immuno-Modulatory Activity of Ashwagandha

Ashwagandha or *WS* is a well-known herb that has been shown to exhibit many health benefits. Steroidal alkaloids and lactones (withanolides) are the active constituents from the roots and leaves of this plant (Mishra et al. 2000). The withanolides have been implicated in the wide ranging therapeutic activities of *WS* in promoting good physical and mental health, diuretic, general tonic and rejuvenator, stress reliever, anti-aging, anti-oxidant, and anti-inflammatory functions in addition to anticancer activities (Mishra et al. 2000). It is important to understand the mechanisms by which some of the chemical components in *WS* function, because of their wide range of activities. In this section we review both anti-inflammatory and immunostimulatory functions of the ashwagandha plant extracts and purified WFA. Early studies have shown that preparations of *WS* extract caused the activation of peritoneal macrophages and phagocytosis. Administration of *WS* extract to normal Balb/c mice enhanced cytokine production such as IFN-γ and IL-2, which indicated its immune-potentiating and myeloprotective effects (Davis and Kuttan 1999, 2000).

In addition, *WS* extract had a protective effect against cyclophosphamide-induced myelosuppression in a mouse model (Agarwal et al. 1999). A study by Iuvone et al. found that a WS root extract caused a significant increase in nitric oxide (NO) production by J774 macrophages (Iuvone et al. 2003). NO is normally synthesized by oxidation of a terminal guanidine nitrogen of L-arginine, a reaction catalyzed by the NO synthase (NOS) (Moncada et al. 1991). An isoform of NOS called inducible NOS (iNOS) is typically synthesized in response to pro-inflammatory mediators by a number of immune cells and its expression has been shown to be beneficial in host defense and modulating the immune response (Bogdan 2001). For example, iNOS-derived production of NO by macrophages inhibits the growth of many pathogens, including bacteria, fungi, viruses, and parasites (Bogdan 2001). Therefore, the increased NO production by macrophages in response to *WS* root and plant extracts could account for some of the immune-stimulant properties (Iuvone et al. 2003).

A few other studies have reported immune-enhancing properties of *WS* extracts including those that contain WFA. In evaluating the potential immune-prophylactic efficacies of these extracts against an infective pathogen, Kushwaha et al. (2012) demonstrated that oral administration of crude root extracts as well as purified WFA offered protection for susceptible rodents (*Mastynys coucha*) against infection with the filarial parasite *B. malayi*. Additionally, treatment with *WS* extracts led to defective embryogenesis in females of the parasitic worm (Kushwaha et al. 2012). This suggested that *WS* extracts might be used as immuno-prophylactic agents (Kushwaha et al. 2012). The root extracts of *WS* and pure WFA were investigated recently in another model of parasite infection, which involved *Leishmania donovani* infection of hamsters (Tripathi et al. 2014). In this study, naïve animals were fed orally a number of different *WS* extracts or WFA 5 days prior to challenge with Leishmania parasites (Tripathi et al. 2014). There was a significant increase in the mRNA expression of iNOS, interferon-γ (IFN-γ), interleukin (IL)-12 and tumor necrosis factor-α (TNF-α) but a decrease in IL-4, IL-10 and transforming growth factor (TGF)-β in mononuclear cells upon treatment of animals with WS extract or WFA (Tripathi et al. 2014). Also enhanced were Leishmania-specific lymphocyte proliferation as well as reactive oxygen species (ROS), NO and anti-leishmanial IgG2 levels, resulting in a better control of the parasites (Tripathi et al. 2014).

The effects of root extracts of *WS* on humoral immune responses were investigated using a sheep red blood cell (SRBC) immunization study (Malik et al. 2007). SRBC is a T-cell dependent antigen that is often used to test immunotoxicity of potentially toxic chemicals in mice (Ladics 2007). The quantification of specific IgM antibody-forming cell response is often used to measure the immunotoxicity of the test substances (Ladics 2007). Mice were administered a chemically standardized aqueous alcoholic (1:1) root extract of *WS* orally for 15 days and immunized with SRBC intraperitoneally on day 9 of drug treatment (Malik et al. 2007). The *WS* extract enhanced cell mediated as well as humoral immunity as reflected by an increase in SRBC-specific IgM and IgG titers with an additional increase in proliferation and differentiation of both T and B lymphocytes (Malik et al. 2007). Furthermore, authors found that the extract selectively, induced type 1 immunity since there was an increase in expression of T helper cell (Th)1 cytokines including IFN-γ and IL-2 while there was a decline in the Th2 cytokine, IL-4 (Malik et al. 2007). Interestingly, they found that withanolide-A was mostly responsible for the Th1 skewing effect observed (Malik et al. 2007). Finally, in agreement with previous studies mentioned, *WS* root extract strongly activated macrophage functions indicated by enhanced secretion of nitrite, IL-12 and TNF-α (Malik et al. 2007). Together, these studies have demonstrated the potential immuno-stimulatory effects of *WS* extracts and WFA and suggest its potential applications against intracellular pathogen infections and in the management of diseases involving immunosuppression (Malik et al. 2007; Kushwaha et al. 2012; Tripathi et al. 2014).

Despite some literature about the immuno-stimulatory effects, there have also been several studies that indicated an immuno-suppressive role for WS extracts. Sohat et al. found that WFA and withanolide E, two steroidal lactones, had immuno-suppressive effects on proliferation of human B and T lymphocytes as well as of murine thymocytes (Shohat et al. 1978). In this study, WFA inhibited the formation of rosettes with SRBC alone (E-rosettes) and SRBC coated with antibody and complement (EAC-rosettes) by normal human T and B lymphocytes, suggesting an interference with some of the surface receptors involved in rosette formation. The effect of WFA on the functional activity of normal human T lymphocytes was tested using a local xenogeneic graft versus host reaction (GVHR). Administration of low doses of WFA for three consecutive days resulted in an inhibition of the graft versus host reaction (Shohat et al. 1978).

The results from all the studies reviewed raise the question of the actual role of *WS* extracts on the immune system. It is worth mentioning that most studies showing immune-stimulatory effects use a crude root extract from the plant. On the other hand, many of the recent studies demonstrating an anti-cancer or immuno-suppression effects of ashwagandha use purified forms of some of the major components found in *WS* extracts including WFA. Also many of the studies demonstrating immune-enhancing effects of WFA or *WS* extracts used oral administration whereas many but not all of the recent studies demonstrating immune suppression use intraperitoneal route to administer WFA. This might account for some of the contradictory results found in literature. Also the actual chemical composition of various extracts may be slightly different leading to variable results in different systems. Since the majority of the anti-cancer studies use a purified form of WFA, we became interested in studying the immuno-modulatory effects of WFA. In the next section, we discuss some of our results that demonstrate immuno-suppressive effects of purified WFA.

11.2.1 WFA Inhibits Both T and B Cell Proliferation in Response to Stimulation

In spite of extensive literature on the anti-tumor effects of WFA, little is known about how it affects the normal cells. Hence the immuno-modulatory properties of WFA were investigated by testing its effects on the proliferation of purified murine

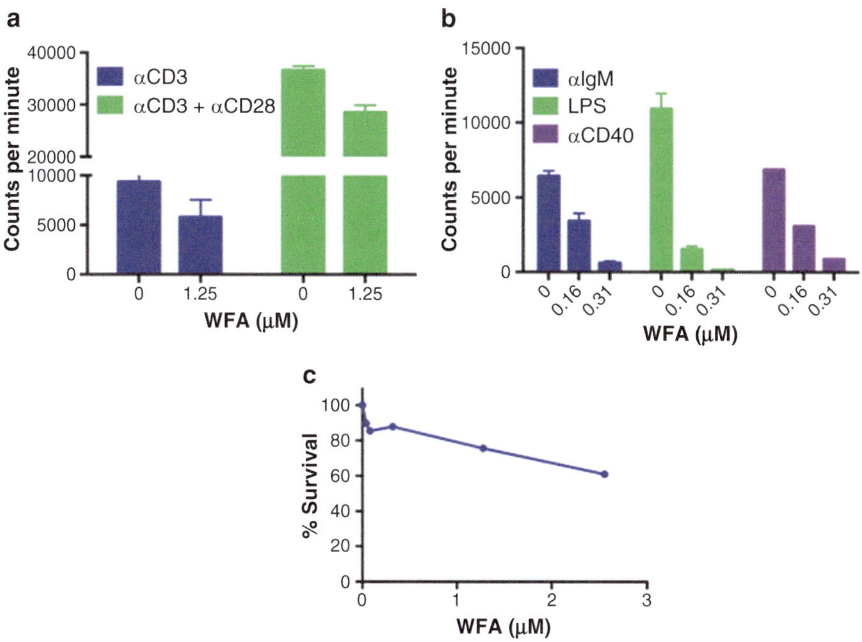

Fig. 11.1 WFA inhibits both T and B cell proliferation in response to various stimuli. (**a**) Using tritiated thymidine incorporation assay, T cell response was measured in response to T cell receptor (CD3) ligation using an anti-CD3 antibody with or without anti-CD28. (**b**) B cell mitogenic response was evaluated in response to LPS, anti-IgM or by anti-CD40 to respectively mimic T-independent and T-dependent B cell activation. All samples were evaluated in triplicates. Bars indicate standard deviation. (**c**) Cytotoxic effects of WA to lymphocytes were evaluated by trypan blue staining

T and B cells. Only purified WFA preparations were used for these studies. As surrogate for antigen, purified normal T cells were stimulated with anti-CD3 that ligates the T cell receptor (TCR) complex or antibodies to CD3 and CD28, a key costimulatory molecule on T cells that enhances TCR signals. Low concentrations of WFA induced a dose-dependent inhibition of T-cell proliferation in response to both the stimuli tested (Fig. 11.1a). In addition, low doses of WFA induced striking inhibitory effects on the proliferation responses of purified B cells stimulated via a T-cell dependent (anti-CD40) or T-cell independent (lipopolysaccharide or LPS) mechanisms (Fig. 11.1b). On average EC_{50} values for WFA mediated suppression of B cell responses (~0.25 μM) were less than those for T cells (~1.5 μM). The susceptibility of B cell responses induced by B cell receptor ligation with anti-IgM to WFA-mediated suppression was also similar to the effect of WFA on LPS and anti-CD40 induced B cell expansion. Interestingly, WFA treatment induced only a small decrease in the survival of unstimulated lymphocytes especially at the submicromolar concentrations that inhibited T and B lymphocyte proliferation (Fig. 11.1c). These results indicate that the observed immunosuppression was not due to the cytotoxic effects of WFA on normal unstimulated lymphocytes. The immuno-suppressive effects of WFA were shown by another study, where WFA inhibited both Con-A- induced lymphocyte proliferation and anti-CD3/CD28 mAb-induced proliferation of $CD4^+$ T cells and $CD8^+$ T cells *in vitro* (Gambhir et al. 2015). This study also found that pre-treatment of cells with different concentrations of WFA significantly suppressed LPS-induced proliferation of B cells *in vitro* (Gambhir et al. 2015).

11.2.2 WFA Inhibits Both T and B Cell Differentiation Responses

The effects of WFA on the differentiation responses of normal T and B cells were studied using the enzyme-linked immunosorbent assay (ELISA) to measure secreted antibodies and cytokines. Purified B cells were stimulated with T-dependent (anti-CD40 + IL-4 + IL-5) or T-independent stimuli (LPS) in the presence or absence of WFA and both IgM and IgG secretion were measured by ELISA. The results indicated that IgM and IgG secretion response induced by LPS stimulation was more sensitive to the effects of WFA than the antibody secretion response induced by the T-dependent (TD) B cell activation (EC_{50} of ~0.1 μM for LPS versus ~0.5 μM for TD response) (Fig. 11.2a). This is consistent with the greater resistance of T-dependent (anti-CD40 + IL-4 + IL-5) than LPS induced B cell proliferation (Fig. 11.1b versus Fig. 11.2a). Nonetheless, in both cases WFA was able to significantly reduce IgM and IgG secretion by stimulated B cells.

IL-2 was used as a measure of overall T cell activation. IL-4 and IL-5 were used as cytokines characteristic of Th2 type of T-helper cells, whereas IFN-γ was used as a marker for Th1 cells. IL-17A was employed to assess inflammatory T-helper cell subset, Th17, whereas IL-10 was quantified to determine the effects of WFA on anti-inflammatory cytokines. Highly purified $CD4^+$ T cells stimulated with anti-CD3 + anti-CD28 mAb secreted significantly higher levels of all the cytokines (IL-2, IL-4, IL-5, IL-10, IL-17A and IFN-γ) measured as compared to unstimulated cells. Treatment with WFA inhibited anti-CD3/CD28 induced secretion of cytokines (Fig. 11.2b). The cytokine secretion response was more sensitive than the proliferation response of T cells (Fig. 11.1a versus Fig. 11.2b) with an EC_{50} of 0.1–0.3 μM for cytokine secretion as against an EC_{50} of 1.5 μM for the T-cell proliferation response. Although the inhibitory effects of WFA were not specific to Th1, Th2, or Th17 responses, secretion of IL-2, IL-4 and γ-IFN was more sensitive than that of IL-5, IL-10 and IL-17A to the WFA induced inhibition (Fig. 11.2b).

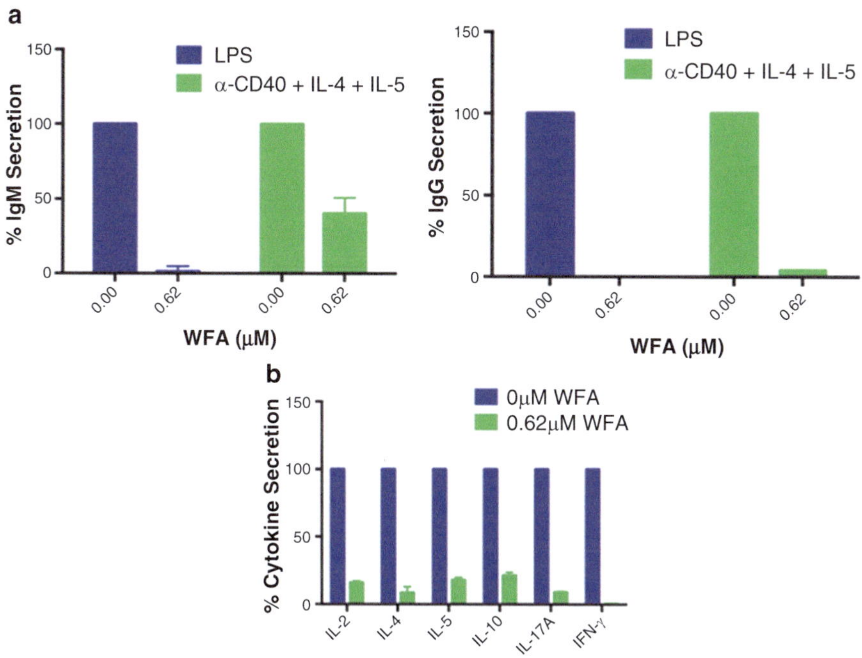

Fig. 11.2 WFA inhibits both T and B cell differentiation responses. (**a**) Differentiation responses of B cells indicated by IgM (*left*) and IgG (*right*) production were measured by ELISA. (**b**) Differentiation responses of T cells as represented by secretion of IL-10, IL-17A, IFN-ϒ, IL-2, IL-4 and IL-5 were measured by ELISA. The unstimulated differentiation response was set to 100%. All samples were evaluated in triplicates. Bars indicate standard deviation

11.3 Mechanisms Underlying Immunomodulatory Activity of WFA

The mechanism of action of WFA has been studied in multiple systems including cancer and inflammation. This agent inhibits inducible nitric oxide synthase and nitric oxide, which are known to have beneficial immuno-modulatory effects in acute and chronic inflammatory disorders, by downregulation of NF-κB in macrophage type cells (RAW264.7) (Oh et al. 2008). It should be noted that *WS* root extract had the opposite effect, i.e. enhanced NOS in the J774 macrophage cells, which could be due to other components in the extracts that opposed the inhibitory effects of WFA or due to the use of different cell lines (Iuvone et al. 2003). Also, WFA downregulated cell adhesion molecules in pulmonary epithelial cells, at least in part, by blocking Akt and down-regulating NF-κB activity (Bogdan 2001). Other studies including a report by Kaileh M et al., using L929sA mouse fibroblast cells, showed that WFA inhibits NF-κB activation by directly suppressing IKKβ activity, which in turn blocked IκBα phosphorylation thereby preventing its ubiquitination and degradation (Kaileh et al. 2007). The NF-κB inhibiting activity of WFA has been studied in several different tumor cell types (McKenna et al. 2015; Vanden Berghe et al. 2012). These reports and others clearly indicate that NF-κB pathway, which regulates inflammation, apoptosis and tumorigenesis is one of the main targets that mediates the biological effects of WFA. Although there are few studies on the immuno-modulatory effects of WFA, even fewer reports discuss the possible mechanisms leading to those effects. In a recent study investigating the immuno-modulatory effects of WFA, authors found that modulation of cellular thiols plays a central role in the observed immuno-suppressive effects of WFA in lymphocytes (Gambhir et al. 2015). WFA induced oxidative stress by increasing the basal reactive oxygen species. This study demonstrated the ability of WFA to directly interact with free thiols. WFA inhibited NF-κB nuclear translocation in lymphocytes and prevented the direct binding of nuclear NF-κB to its consensus sequence. Therefore, in agreement with the effects of WFA in other systems, direct targeting of NF-κB by WFA may also be one of the molecular mechanisms responsible for the observed inhibition of T cell mediated immune responses (Gambhir et al. 2015). The mechanisms by which *WS* extracts of WFA enhance immune responses are not understood at the present time.

11.4 Effect of WFA on B Cell Lymphomas

Non-Hodgkin Lymphoma (NHL) represents over 4% of all cancer cases in the United States with a majority being B cell in origin (Howlader et al. 2013). Diffuse Large B cell lymphoma (DLBCL) is one of the most common and aggressive NHLs that contributes to 30% of newly diagnosed cases of B-cell lymphoma (Young and Staudt 2013). Current therapy for DLBCL includes the R-CHOP regimen (rituximab, cyclophosphamide, doxorubicin, vincristine, prednisone) that may lead to the development of resistance with a ~ 40% patient mortality, which therefore warrants new treatment options (Young and Staudt 2013; Shaffer et al. 2012). DLBCL have been classified as germinal center type (GC) and activated B cell lymphoma (ABC) (Shaffer et al. 2012). Our investigations of the effects of WFA treatment on B cell lymphoma demonstrated its potent anti-proliferative activity against a variety of B cell lymphoma lines *in vitro* and effective tumor reducing capability *in vivo* (McKenna et al. 2015). Interestingly, WFA was effective against both GC and ABC types of DLBL. WFA also inhibited B cell lymphoma growth in a syngeneic model in which luciferase expressing A20 lymphoma cells were injected subcutaneously into Balb/c mice. Tumor growth was measured through bioluminescence imaging. WFA treatment administered intraperitoneally every other day significantly reduced tumor growth compared to the vehicle treatment (McKenna et al. 2015). Despite its anti-proliferative effects on injected B-lymphoma cells, WFA did not affect endogenous epithelial cell proliferation as shown by near normal colon histology in WFA treated animals.

Previously we had shown that B-cell receptor (BCR) signaling is important for B-lymphoma survival and growth *in vitro* (Gururajan et al. 2007; Gururajan et al.

2006). BCR signaling has also been shown to be important in patients by identification of mutations in the BCR signaling pathway (Shaffer et al. 2012). BCR signaling involves activation of proximal protein tyrosine kinases (PTK), Lyn, Syk and Btk, all of which are constitutively activated in B-cell lymphomas. Interestingly WFA reduced activation of all these three PTKs rapidly after addition to B-lymphoma cultures. Reduction of the multiple PTK activities could be due to the down regulation of the activity of heat shock protein 90 (HSP90). Consistent with a reduction in HSP90 activity, WFA rapidly induced HSP70, which compensates partially for HSP90. Thus WFA appears to affect B-lymphoma growth in a novel manner by affecting BCR signaling.

In our lymphoma models (SUDHL-6 and LY-10 cell lines) WFA reduced activation of IKKα/β, a kinase required for degradation of the inhibitor of NF-κB (IκB) and nuclear translocation of NF-κB. BCR signaling is also known to lead to NF-κB activation. Hence the reduction in NF-κB activity in WFA treated B lymphoma cells could be due to a reduction in BCR signaling, though a role for other mechanisms cannot be ruled out. Other studies reported on the direct inhibitory effects of WFA on NF-κB activity suggesting multiple mechanisms of action (McKenna et al. 2015). Analysis utilizing in silico molecular docking by Grover et al. revealed that WFA interacts with the NEMO subunit (IKKγ) of the IKK complex and prevents complete kinase complex formation and downstream IκB phosphorylation required for NF-κB activity (Grover et al. 2010). In human lung epithelial cells WFA inhibited TNFα-induced activity of NF-κB by preventing DNA binding (Oh and Kwon 2009). Jackson et al. (2015) suggest that WFA blocks TNFα- induced NF-κB activity by preventing the ubiquitin dependent NEMO reorganization required for activation of the IKK complex. Similar to studies in pancreatic cancer, WFA was found to inhibit the activity of heat shock protein-90 (HSP90) resulting in the degradation of key target proteins that are found to be involved in cell survival (Akt), cell cycle progression (Cyclin B, CDK1, CDK4), and specifically B cell receptor signaling (Lyn, Syk, Btk) (Caplan et al. 2007; McKenna et al. 2015). While it has not been confirmed in B cell lymphoma, a study suggests that WFA inhibits HSP90 by inhibiting HSP90 complex with the co-chaperone Cdc37 and preventing its action on target proteins (Yu et al. 2010). Client proteins of the HSP90-Cdc37 complex include kinases such as Akt, Lyn, Syk, CDKs, and the IKK complex (Vanden Berghe et al. 2012; Taipale et al. 2012). We hypothesize that WFA is targeting the function of HSP90-Cdc37 leading to a down regulation of client proteins that are important for BCR signaling in DLBCL and, therefore, results in inhibition of NF-κB and cell death of B-cell lymphoma (Fig. 11.3). The most aggressive DLBCLs are associated with constitutive BCR signaling and NF-κB activity. Hence treatment with WFA, which interferes with both BCR signaling and NF-κB activation has potential to be a novel therapy for DLBCLs (Davis et al. 2001).

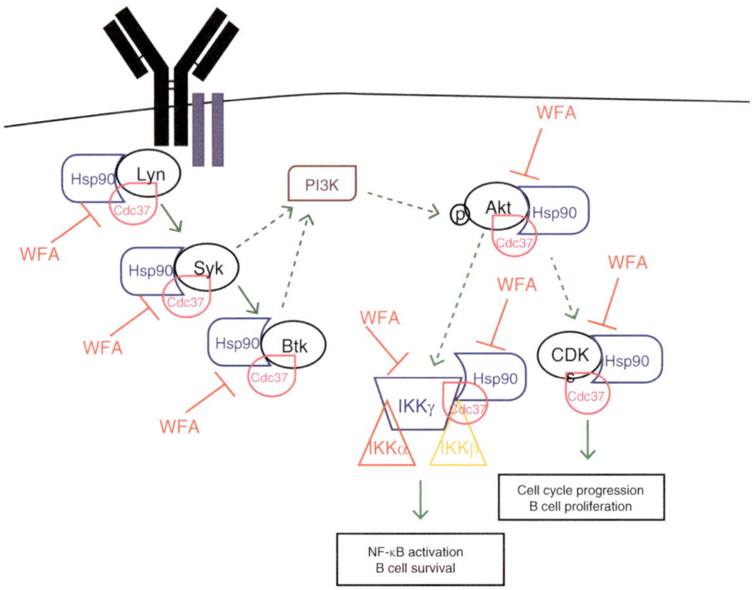

Fig. 11.3 Proposed mechanism of action of WFA on B cell lymphoma tumors. WFA targets the Cdc37-HSP90 complex preventing the activity on client proteins such as proximal protein tyrosine kinases involved in BCR signaling, prosurvival and growth supportive protein kinases, as well as components of the IKK complex. Loss of HSP90 activity in these cells leads to client degradation and eventual cell death

11.5 Effect of WFA on B-Cell Chronic Lymphocytic Leukemia

NHL includes a heterogeneous group of B cell malignancies that can be grouped into aggressive and fast growing types (i.e. DLBCL, Mantle Cell Lymphoma and Burkitt's Lymphoma) as well as slow growing or indolent types such as Chronic Lymphocytic Leukemia (CLL) and follicular lymphoma (Howlader et al. 2016). CLL is the most common adult leukemia in the western world. Unlike many DLBCL that are characterized by uncontrolled proliferation, leukemic cells are defined by abnormal lymphocytes produced in the bone marrow that have defects in apoptosis, or other cell death mechanisms, leading to the accumulation of cells in the peripheral blood and secondary lymphoid structures (Chiorazzi et al. 2005). CLL are thought to derive from a unique subset of B cells called B-1 cells. We used a well-studied mouse model of CLL to determine if WFA can be effective against this indolent type of B cell malignancy. The model is a Eμ-Tcl1 transgenic mouse that overexpresses Tcl1, an oncogene derived from a human T-cell leukemia, in a B cell specific manner using Ig-VH promoter and Eμ-enhancer (Bichi et al. 2002). These mice reproducibly develop CLL between 10–15 months of age. Our in vitro studies showed that survival of CLL cells derived from the Eμ-Tcl1 transgenic

Table 11.1 Effective inhibitory concentrations of WFA for murine B cell malignancies. Values represent an average of at least two experiments including triplicate samples determined from a linear line of best fit on a logarithmic scale

	NHL Subtype	EC_{50} (μM)
A20-Luc Lymphoma	GCB-DLBCL	5.35
BKS-2 Lymphoma	Immature Lymphoma	4.41
Eμ-Tcl1 B-CLL	Primary Leukemia (CLL)	0.49

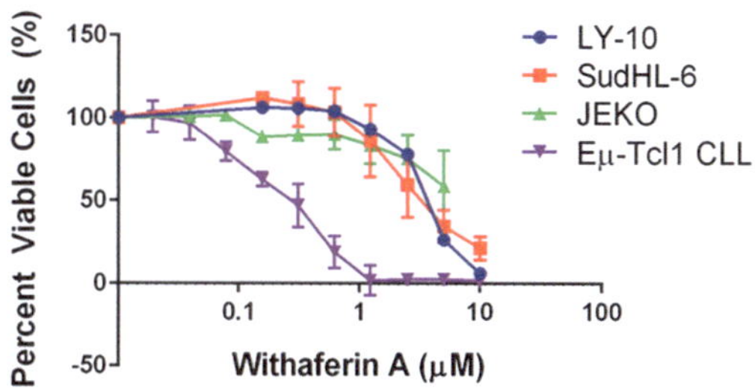

Fig. 11.4 Effect of WFA on human B-lymphoma lines compared to B-CLL. Human B cell lymphoma cells and murine CLL cells were treated with different concentrations of Withaferin A for 48 hr. and then proliferation and viability were measured by MTT assay. Data points indicate percentage of viable cells of triplicate cultures and an average of at least two experiments. The curves are plotted on a log scale with 0.01 μM drug concentration representing no drug added. Bars indicate mean ± standard error

mouse model was profoundly blocked by WFA and that these CLL cells are more sensitive to WFA treatment than other mouse or human lymphoma cells (Table 11.1, Fig. 11.4). This is an interesting discovery as many of the previous studies of anticancer effects of WFA utilized models that involve extensive proliferation of cancer cells *in vitro* and *in vivo*, whereas CLL cells are known to be quiescent in cultures *in vitro* and in peripheral blood, *in vivo*. Also the quiescent CLL cells are more susceptible to the inhibitory effects of WFA than normal lymphocytes, which are also quiescent *in vitro* and *in vivo* (compare Fig. 11.2c). One possible explanation is that CLL cells have a higher constitutive level of BCR signaling compared to unstimulated normal B lymphocytes and WFA treatment results in degradation of Lyn, Btk, and Syk kinases that are required for the survival of CLL cells (Stevenson et al. 2011). Studies are being performed to confirm this hypothesis. Mandal et al. investigated the effects WFA on leukemic cells from both myeloid and lymphoid origin. These studies showed that WFA has the capacity to induce apoptosis of primary leukemic cells with no toxicity to normal human lymphocytes (Mandal et al. 2008). Interestingly, Mandal et al. found that WFA was slightly more toxic to an acute B-lymphoblastic leukemia cell line (REH) compared to T-cell (Jukrat, MOLT-4) and myeloid leukemia (K562) cell lines [Ref 35, Fig. 11.1b]. Data also suggested that WFA induced activation of caspase 9, 3 and PARP cleavage resulting in apoptosis (Mandal et al. 2008). In primary murine CLL cells caspase 3

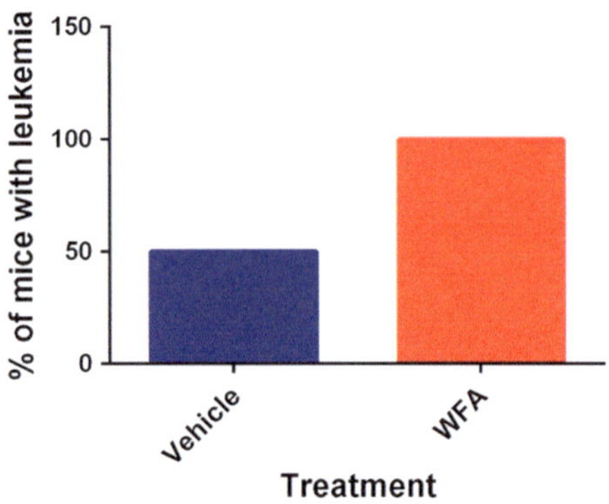

Fig. 11.5 Effect of WFA on leukemic growth. Percent of leukemic engraftment in animals injected with primary Eμ-Tcl1 CLL cells treated with WFA or vehicle. Animals treated with WFA had overall greater engraftment after 29 days of growth compared to vehicle. Each bar represents percent engraftment with six animals per treatment group

cleavage was increased after WFA treatment suggesting apoptosis as a major mechanism of the inhibitory effect of WFA (unpublished data).

To determine the effect of WFA on CLL *in vivo*, an adoptive transfer model was used, which consisted in injection of EμTcl1 CLL spleen cells intravenously (IV) into syngeneic recipients and monitoring CLL in peripheral blood by flow cytometry. The recipient mice with CLL burden exhibit splenomegaly and accumulation of leukemic cells in the blood, which is also the phenotype of Eμ-Tcl1 mice that develop CLL de novo. Surprisingly, WFA treatment did not decrease CLL burden in blood at all times after CLL injection. Furthermore, splenomegaly was also not attenuated by WFA (unpublished data). Interestingly even the growth of A20 B-lymphoma cells was not blocked by WFA when lymphoma cells were injected intravenously to induce systemic growth of tumor cells. On the other hand, WFA blocked A20 B-lymphoma growth when injected subcutaneously that results in solid tumors (McKenna et al. 2015). These results suggest that WFA may be targeting other factors *in vivo* to prevent the growth of solid tumors. The role of WFA and other Ashwagandha components on angiogenesis has been well studied. Results indicated that WFA has potent anti-angiogenic properties *in vitro* and *in vivo* at doses lower than required to observe anti-tumor effects, primarily by targeting vimentin (Mohan et al. 2004; Bargagna-Mohan et al. 2007). In addition, in silico studies also suggest that WFA may bind directly to vascular endothelial growth factor (VEGF), a key regulator of angiogenesis, and inhibits its ability to promote vasculogenesis (Saha et al. 2013). These findings may provide an explanation as to why WFA does not inhibit systemic tumors such as IV injected lymphoma cells as well as CLL cells *in vivo*, as they do not require angiogenesis but further studies must be performed to confirm this mechanism. Unexpectedly WFA treatment enhanced the proportion of recipients that were successfully engrafted with transferred CLL cells (Fig. 11.5). This result suggests that an immunomodulatory effect of WFA may be playing a role to prevent any immune response towards leukemic growth in a hematologic cancer model. Further studies examining the function of immune cells in this model are required to understand the ability of WFA to promote growth of systemically injected hematological tumor cells.

11.6 Anticancer Mechanisms of Withaferin A Are Pleiotropic

Several studies over the years have demonstrated the anti-cancer potential of Withaferin A in a number of cancer models. However, the mechanisms by which WFA exhibits its effects are still not completely understood. This is mostly due to the fact that the anti-cancer effects of WFA are pleiotropic. WFA has been shown to inhibit proliferation, angiogenesis and metastasis, and to induce apoptosis and autophagy (Lee and Choi 2016). All of these effects culminate into one overall outcome -decrease in the numbers of cancer cells. WFA can alter the expression and activity of several proteins ranging from cytoskeletal proteins, cell cycle proteins, proteasome proteins, endoplasmic reticulum (ER) proteins, heat shock proteins, and transcription factors (Samadi 2015; Lee and Choi 2016). The pathways and proteins targeted vary greatly in different systems. WFA induced G2/M cell cycle arrest and apoptosis, and inhibited HSP90 activity and NF-κB nuclear translocation in B lymphoma and in several solid tumor models (McKenna et al. 2015). Studies from our laboratory in a human myelodysplastic syndrome cell line (MDS-L) model revealed that in addition to a G2/M cell cycle arrest, WFA also induced an S phase arrest in this system. Moreover, no apparent inhibitory effect of WFA on NF-κB nuclear translocation was observed in the MDS-L cells despite significant growth inhibition (unpublished data). Preliminary results in this MDS-L model suggest that WFA may induce JNK, which has been previously shown to be involved in induction of apoptosis in several but not all cell types. This would suggest that elucidation of the anti-cancer mechanisms of WFA in individual models is going to be crucial for its further development as an anti-cancer drug. Most studies have focused on the overall effect of WFA in cancer cell growth via cell survival or cell cycle mechanisms. However, a few studies have demonstrated direct targets of WFA by exploiting the thiol reactivity of the steroid lactone moiety of WFA. This led to identification of a few direct targets of WFA, which include vimentin, VEGF, p65 of NF-κB, IKK-γ and HSP90. However, their generality for more than one system has not yet been established. Hence better characterization of the proximal or direct targets of WFA may provide better answers about its mechanism of action, which could pave the way for clinical studies.

11.7 Future Perspectives

Crude extracts from the plant *WS* have been widely used in Indian medicine despite limited laboratory evidence (Mishra et al. 2000). Based on the studies reviewed in this chapter, some of these benefits could be explained by the observed immunostimulatory effects of *WS* extracts by enhancing immune responses. In particular *WS* extracts increased pro-inflammatory responses that are protective against intracellular pathogens and in the management of diseases associated with immunosuppression (Tripathi et al. 2014; Kushwaha et al. 2012; Iuvone et al. 2003). WFA, a key cytotoxic component of *WS* extracts has recently become an attractive natural product under investigation in various preclinical studies for treatment of different cancer types. Our studies have extended the anti-cancer effects of WFA to hematological cancers, such as B-lymphoma (McKenna et al. 2015) and myeloid dysplastic syndrome (unpublished data). The thiol-reactive properties of WFA, could possibly explain its pleiotropic effects. Future studies have to better define the proximal targets of WFA so that it can be targeted to specific cancers without inhibiting the immune system. Also further studies of the effects of WFA on immune system in cancer models are required in order to advance this plant therapeutic to clinical trials, so that any adverse effects can be minimized. The beneficial effects of WFA have been mostly associated with inhibition of NF-κB signaling, which is necessary for inflammatory responses. Many acute and chronic inflammatory diseases such as arthritis, atherosclerosis, and Crohn's disease as well as autoimmune diseases such as lupus are associated with inflammation, which requires NF-κB activation as one of the first steps. The ability of WFA to block NF-κB activation makes it a potential therapeutic for many of these conditions too.

References

Agarwal R, Diwanay S, Patki P, Patwardhan B (1999) Studies on immunomodulatory activity of Withania somnifera (Ashwagandha) extracts in experimental immune inflammation. J Ethnopharmacol 67:27–35

Bargagna-Mohan P, Hamza A, Kim YE, Khuan Abby Ho Y, Mor-Vaknin N, Wendschlag N, Liu J, Evans RM, Markovitz DM, Zhan CG, Kim KB, Mohan R (2007) The tumor inhibitor and antiangiogenic agent withaferin a targets the intermediate filament protein vimentin. Chem Biol 14:623–634

Bichi R, Shinton SA, Martin ES, Koval A, Calin GA, Cesari R, Russo G, Hardy RR, Croce CM (2002) Human chronic lymphocytic leukemia modeled in mouse by targeted TCL1 expression. Proc Natl Acad Sci U S A 99:6955–6960

Bogdan C (2001) Nitric oxide and the immune response. Nat Immunol 2(10):907–916

Caplan AJ, Mandal AK, Theodoraki MA (2007) Molecular chaperones and protein kinase quality control. Trends Cell Biol 17:87–92

Chiorazzi N, Rai KR, Ferrarini M (2005) Chronic lymphocytic leukemia. N Engl J Med 352:804–815

Davis L, Kuttan G (1999) Effect of Withania somnifera on cytokine production in normal and cyclophosphamide treated mice. Immunopharmacol Immunotoxicol 21:695–703

Davis L, Kuttan G (2000) Immunomodulatory activity of Withania somnifera. J Ethnopharmacol 71:193–200
Davis RE, Brown KD, Siebenlist U, Staudt LM (2001) Constitutive nuclear factor kappaB activity is required for survival of activated B cell-like diffuse large B cell lymphoma cells. J Exp Med 194:1861–1874
Gambhir L, Checker R, Sharma D, Thoh M, Patil A, Degani M, Gota V, Sandur SK (2015) Thiol dependent NF-kappaB suppression and inhibition of T-cell mediated adaptive immune responses by a naturally occurring steroidal lactone Withaferin A. Toxicol Appl Pharmacol 289:297–312
Grover A, Shandilya A, Punetha A, Bisaria VS, Sundar D (2010) Inhibition of the NEMO/IKKbeta association complex formation, a novel mechanism associated with the NF-kappaB activation suppression by Withania somnifera's key metabolite withaferin A. BMC Genomics 11(Suppl 4):S25
Gururajan M, Jennings CD, Bondada S (2006) Cutting edge: constitutive B cell receptor signaling is critical for basal growth of B lymphoma. J Immunol 176:5715–5719
Gururajan M, Dasu T, Shahidain S, Jennings CD, Robertson DA, Rangnekar VM, Bondada S (2007) Spleen tyrosine kinase (Syk), a novel target of curcumin, is required for B lymphoma growth. J Immunol 178:111–121
Howlader NNA, Krapcho M, Garshell J, Miller D, Altekruse SF, Kosary CL, Yu M, Ruhl J, Tatalovich Z, Mariotto A, Lewis DR, Chen HS, Feuer EJ, Cronin KA (2013) SEER cancer statistics review, 1975–2011. National Cancer Institute Bethesda, Maryland (based on November 2013 SEER data submission)
Howlader N, Morton LM, Feuer EJ, Besson C, Engels EA (2016) Contributions of subtypes of non-Hodgkin lymphoma to mortality trends. Cancer Epidemiol Biomark Prev 25:174–179
Iuvone T, Esposito G, Capasso F, Izzo AA (2003) Induction of nitric oxide synthase expression by Withania somnifera in macrophages. Life Sci 72:1617–1625
Jackson SS, Oberley C, Hooper CP, Grindle K, Wuerzberger-Davis S, Wolff J, McCool K, Rui L, Miyamoto S (2015) Withaferin A disrupts ubiquitin-based NEMO reorganization induced by canonical NF-kappaB signaling. Exp Cell Res 331:58–72
Kaileh M, Vanden Berghe W, Heyerick A, Horion J, Piette J, Libert C, De Keukeleire D, Essawi T, Haegeman G (2007) Withaferin a strongly elicits IkappaB kinase beta hyperphosphorylation concomitant with potent inhibition of its kinase activity. J Biol Chem 282:4253–4264
Kushwaha S, Soni VK, Singh PK, Bano N, Kumar A, Sangwan RS, Misra-Bhattacharya S (2012) Withania somnifera Chemotypes NMITLI 101R, NMITLI 118R, NMITLI 128R and withaferin A protect Mastomys coucha from Brugia malayi infection. Parasite Immunol 34:199–209
Ladics GS (2007) Primary immune response to sheep red blood cells (SRBC) as the conventional T-cell dependent antibody response (TDAR) test. J Immunotoxicol 4:149–152
Lee IC, Choi BY (2016) Withaferin-A--A natural anticancer agent with Pleitropic mechanisms of action. Int J Mol Sci 17:290
Malik F, Singh J, Khajuria A, Suri KA, Satti NK, Singh S, Kaul MK, Kumar A, Bhatia A, Qazi GN (2007) A standardized root extract of Withania somnifera and its major constituent withanolide-A elicit humoral and cell-mediated immune responses by up regulation of Th1-dominant polarization in BALB/c mice. Life Sci 80:1525–1538
Mandal C, Dutta A, Mallick A, Chandra S, Misra L, Sangwan RS, Mandal C (2008) Withaferin A induces apoptosis by activating p38 mitogen-activated protein kinase signaling cascade in leukemic cells of lymphoid and myeloid origin through mitochondrial death cascade. Apoptosis 13:1450–1464
McKenna MK, Gachuki BW, Alhakeem SS, Oben KN, Rangnekar VM, Gupta RC, Bondada S (2015) Anti-cancer activity of withaferin A in B-cell lymphoma. Cancer Biol Ther 16:1088–1098
Mishra LC, Singh BB, Dagenais S (2000) Scientific basis for the therapeutic use of Withania somnifera (ashwagandha): a review. Altern Med Rev 5:334–346

Mohan R, Hammers HJ, Bargagna-Mohan P, Zhan XH, Herbstritt CJ, Ruiz A, Zhang L, Hanson AD, Conner BP, Rougas J, Pribluda VS (2004) Withaferin A is a potent inhibitor of angiogenesis. Angiogenesis 7:115–122
Moncada S, Palmer RM, Higgs EA (1991) Nitric oxide: physiology, pathophysiology, and pharmacology. Pharmacol Rev 43:109–142
Oh JH, Kwon TK (2009) Withaferin A inhibits tumor necrosis factor-alpha-induced expression of cell adhesion molecules by inactivation of Akt and NF-kappaB in human pulmonary epithelial cells. Int Immunopharmacol 9:614–619
Oh JH, Lee TJ, Park JW, Kwon TK (2008) Withaferin A inhibits iNOS expression and nitric oxide production by Akt inactivation and down-regulating LPS-induced activity of NF-kappaB in RAW 264.7 cells. Eur J Pharmacol 599:11–17
Saha S, Islam MK, Shilpi JA, Hasan S (2013) Inhibition of VEGF: a novel mechanism to control angiogenesis by Withania somnifera's key metabolite Withaferin A. In Silico Pharmacol 1:11
Samadi AK (2015) Potential anticancer properties and mechanisms of action of Withanolides. The Enzymes 37:73–94
Shaffer AL 3rd, Young RM, Staudt LM (2012) Pathogenesis of human B cell lymphomas. Annu Rev Immunol 30:565–610
Shohat B, Kirson I, Lavie D (1978) Immunosuppressive activity of two plant steroidal lactones withaferin A and withanolide E. Biomedicine 28:18–24
Stevenson FK, Krysov S, Davies AJ, Steele AJ, Packham G (2011) B-cell receptor signaling in chronic lymphocytic leukemia. Blood 118:4313–4320
Taipale M, Krykbaeva I, Koeva M, Kayatekin C, Westover KD, Karras GI, Lindquist S (2012) Quantitative analysis of HSP90-client interactions reveals principles of substrate recognition. Cell 150:987–1001
Tripathi CD, Gupta R, Kushawaha PK, Mandal C, Misra Bhattacharya S, Dube A (2014) Efficacy of Withania somnifera chemotypes NMITLI – 101R, 118R and Withaferin A against experimental visceral leishmaniasis. Parasite Immunol 36:253–265
Vanden Berghe W, Sabbe L, Kaileh M, Haegeman G, Heyninck K (2012) Molecular insight in the multifunctional activities of Withaferin A. Biochem Pharmacol 84:1282–1291
Widodo N, Shah N, Priyandoko D, Ishii T, Kaul SC, Wadhwa R (2009) Deceleration of senescence in normal human fibroblasts by withanone extracted from ashwagandha leaves. J Gerontol A Biol Sci Med Sci 64:1031–1038
Young RM, Staudt LM (2013) Targeting pathological B cell receptor signalling in lymphoid malignancies. Nat Rev Drug Discov 12:229–243
Yu Y, Hamza A, Zhang T, Gu M, Zou P, Newman B, Li Y, Gunatilaka AA, Zhan CG, Sun D (2010) Withaferin A targets heat shock protein 90 in pancreatic cancer cells. Biochem Pharmacol 79:542–551

Chapter 12
Ashwagandha for Cancer Metastasis: Bioactives and Basics of Their Function

Sunil C. Kaul, Priyanshu Bhargava, and Renu Wadhwa

Abstract Cancer metastasis is an extremely complex multistep process that is regulated by multiple module interactions of genetic, environmental factors. It leads to movement of cancer cells from the primary tumor site to distant sites in the body through blood or lymph circulation. Tightly regulated by migratory, invasive characteristics of the tumor cells as well as micro-environment, it poses major hurdles to the cancer therapy. In this chapter, we sketch a simple picture of ayurvedic description of cancer and metastasis, provide a review on the metastatic targets and discuss findings on Ashwagandha bioactives for therapy of metastatic cancers.

Keywords Ashwagandha • Bioactives • Prevention • Cancer • Metastasis

12.1 Ayurvedic Description of Cancer and Metastasis

Ayurveda (Ayur – life; Veda – knowledge) is the world's oldest existing literature (1200 years old) recorded in Sanskrit, an ancient language of India. It describes experiences and teachings on (i) spiritual practices such as, meditation & yoga, and (ii) use of herbs to cure diseases and sustain health in the form of home remedies practiced in Indian continent through thousands of years. Ayurvedic philosophy brings together a series of conceptual systems, such as balance and disorder between mind, body and soul. Three fundamental energy complexes (called doshas) – Vata, Pitta and Kapha, describe the human kinds in terms of their functional nature and hence the disease types. Imbalance in energies influences the physiology in several ways. Vata defines the mobile, energetic and non-material aspects. Vata type

S.C. Kaul • R. Wadhwa (✉)
DBT-AIST International Laboratory for Advanced Biomedicine (DAILAB), Biomedical Research Institute, National Institute of Advanced Industrial Science & Technology (AIST), Central 5-41, 1-1-1 Higashi, Tsukuba, Ibaraki 305-8565, Japan
e-mail: renu-wadhwa@aist.go.jp

P. Bhargava
Graduate School of Life & Environmental Sciences, University of Tsukuba, Ibaraki 305 8575, Japan

S.C. Kaul, R. Wadhwa (eds.), *Science of Ashwagandha: Preventive and Therapeutic Potentials*, DOI 10.1007/978-3-319-59192-6_12

individuals often show malfunction of digestive system, arthritis and anxiety. Pitta describes transformative or intelligence aspects; individuals of this type tends to have inflammations, infections and ulcers, and Kapha (defines the structural & physical aspects) types tend to be overweight, diabetic and shown congestive disorders. Ayurvedic medicine aims to cure or prevent disease by increasing energy and well-being, balancing mind, body and spirit and decreasing stress by use of herbs, massage, yoga and breathing & relaxation techniques. According to Ayurveda, cancer is state of high disorder and imbalance between mind, body and spirit caused by physical, emotional and spiritual factors. According to doshas, it is mainly categorized into three types, (i) **Vatta Pradhan Arbuda** – Cancer caused in this category due to imbalance in vatta dosha. The tumors are well vascularized, soft and black in color, (ii) **Pitta Pradhan Arbuda** – Cancer caused in this category is due to agni factor turning vicious. Tumor is yellowish or red, exudes large amount of warm blood and feels severe burning sensation, and (iii) **Kapha Pradhan Arbuda** – Cancer caused in this category is described to be due to vitiation of kapha dosha. Tumor is solid and hard. Although mostly painless it exudes white thick fluid often causing itching.

Cancer metastasis is a complex and multi-step process of dissociation of cancer cells from the primary tumor site followed by migration, invasion, adhesion and proliferation at a distant site within the body through bloodstream or the lymph system. It involves multiple factors that determine this process that requires (i) the acquisition of the migratory characteristics by the tumor cells by which they leave the primary site, (ii) capacity to proteolytically digest surrounding connective tissues, (iii) characteristics to enter the lymphatic or blood vessels by which they travel to distant site of the body and (iv) ability to proliferate at the new site to establish into tumor. All of these processes are highly influenced by origin, type and stage of the cancer, and in turn determine the type and outcome of the treatment. Matrix metalloproteinases (MMPs) that regulate extracellular matrix play an important role in tumor metastasis and angiogenesis. They are regulated by cytokines, growth factors via interlinked signaling pathways. The choice of cancer treatment (either surgery, chemotherapy, radiation therapy, hormone therapy, laser-immunotherapy, or a combination) depends on the type of primary cancer, size and location of the metastasis. Whereas the most direct treatment of any kind of cancer is surgical removal of the tumor, the others include achieving reduction in tumor growth by synthetic or herbal drugs either consumed orally or applied over the tumor as a paste. These either act directly on the cancer or support a deeper healing by supporting the immune system. Most of the times, the combinational chemotherapy that involves combination of drugs to kill cancer cells and boost the immune system are used. However, the treatment is complicated by metastasis, a multi-step, complex process of cancer cell movement from their primary to a secondary site within the body through the bloodstream. Metastasis also accounts for tumor relapse in 90% of the cancers accounting for failure of therapy and high mortality. Current options to cure metastatic cancers are very limited and there is a substantial need to understand the phenomenon of metastasis, uncover the factors that influence cellular migration and adhesion characteristics, and formulate new strategies and reagents for safe and effective cancer treatment.

Stepwise metastasis progression has been well characterized in several types of pre-malignant lesions, such as dysplasia and hyperplasia, which can be detected in various organs before localized as fully malignant invasive tumors. The process of invasion in which tumor cells acquire the ability to migrate and penetrate the surrounding tissues, pass through the basement membrane and extracellular matrix, and penetrate the lymphatic or vascular circulation. In this process, the recruitment of new blood vessels (called angiogenesis) that provide route for intravasation of cancer cells from the primary tumor sites to circulation and distant tissues is an important player. Angiogenesis is controlled by multiple endogenous and exogenous chemical signals from tumor cells and microenvironment in a phase of rapid growth. Inhibition of these signals has been shown to lead to necrosis or apoptosis and has hence been proposed as targets for tumor therapy. Angiogenesis involves four main intertwined steps – (i) tissues basement membrane is injured locally, (ii) endothelial cells are activated by angiogenic factors like vascular endothelial growth factor (VEGF), basic fibroblast growth factor (bFGF), angiogenin, transforming growth factor (TGF) – α, β, tumor necrosis factor (TNF-α), platelet-derived endothelial growth factor etc. (Folkman 1995a) that arise from rapidly dividing tumor cells and transforming tumor microenvironment, (iii) endothelial cells proliferate, stabilize and undergo cell-cell and cell-microenvironment interactions leading to activation of tumor promoting and transforming signaling and (iv) angiogenic factors promote the process of blood vessel formation and extension, and provide route for motile cells to invade to distant sites. Since these signaling are connected by positive feed-backs, the tumors grow progressively in an uncontrolled manner and are difficult to treat (Folkman 1995b). Rapidly growing mass of tumor cells often experience lack of nutrients and hypoxia environment responsible for induction of hypoxia signaling that in turn provides positive feedback to proliferative and metastatic features of the tumors. Upregulation of hypoxia-inducible factor 1- alpha (HIF-1α) and secretion of vascular endothelial growth factor (VEGF), matrix metalloproteinase (MMPs) and increased levels of adhesion factor α/β-integrins have been well established as hallmarks of metastasis. During this process, metastatic cells invade and changes in their plasticity based on morphological or phenotypical conversion includes Epithelial to Mesenchymal Transition (EMT), Collective to Amoeboid Transition (CAT) and Mesenchymal to Amoeboid Transition (MAT) (Nieto 2009).

12.2 Epithelial to Mesenchymal Transition (EMT)

EMT is a highly conserved process in embryogenesis. MET (mesenchymal to epithelial transition) is a critical developmental process including gastrulation, neural crest formation and myogenesis. These are characterized by change in the phenotype of cells. Whereas epithelial cells are tightly connected to each other by junctions (tight, gap and adherens junctions) and have an apico-basal polarity determined by cytoskeleton, the mesenchymal cells are spindle shaped, interact with each other only through focal points, lack polarization. Epithelial cells express high levels of E-cadherin; mesenchymal cells express N-cadherin, fibronectin and vimentin. Based on the biological context, EMT has been categorized into 3 types – developmental (Type I), fibrosis and wound healing (Type II) and cancer (Type III) (Biganzoli et al. 2016; Kalluri and Weinberg 2009; Nieto 2009). During the process of epithelial-mesenchymal transition (EMT), the epithelial cells lose their polarity and cell-cell adhesion (epithelial characteristics), and acquire migratory and invasive properties (mesenchymal characteristics) by multiple pathways, several of these still remain unclear. However, alterations in protein profiles and cytoskeleton characteristics are most often considered as reliable biomarkers of EMT. Cell surface proteins, E-cadherin (biomarker of epithelial cells) and integrins, are replaced by mesenchymal markers (N-cadherin, vimentin or fibronectin). These processes are determined by several regulatory proteins including transcriptional factors, snail, slug and twist (Thiery et al. 2009). Not only that these proteins are enriched in cancer cells, their ectopic expression has been shown to induce EMT, downregulation of E-cadherin and increase in tumor metastasis (Harris and Tepass 2010). Of note, TGF-β acts as a tumor suppressor factor at early stages of cancer and promotes its malignant characteristics at the later stages (Zavadil and Bottinger 2005). p53 has been shown to inhibit EMT by transcriptional repression of SNAIL and non-coding regulation mediated by miRNAs (Kim et al. 2011). EMT has also been well correlated with the drug resistance feature of cancer cells. Tamoxifen resistant breast cancer cells showed aggressive and invasive behavior that could be reversed upon the inhibition of the autocrine epidermal growth factor receptor (EGFR) pathway, a key determinant of EMT (Xu et al. 2010).

12.3 Collective to Amoeboid Transition (CAT)

The individual cell invasion known as collective to amoeboid transition involves detachment of individual cells from the cell clusters especially in melanoma (Hegerfeldt et al. 2002). These cells show reduced cell-epithelial cell membrane (ECM) interaction as compared to EMT, which enables them to cross the ECM barriers easily and independently. The major marker studied in this kind of transition is integrin β-1 (Hegerfeldt et al. 2002).

12.4 Mesenchymal to Amoeboid Transition (MAT)

It was detected especially in breast cancer, melanoma and fibrosarcoma and is regulated by proteins such as EphA2kinase. Re-expression of EphA2 in mesenchymal melanoma cells triggered nonproteolytic invasive signaling causing amoeboid motility in the cells (Parri et al. 2009). $p27^{kip1}$ and p53 tumor suppressor proteins have been shown to be involved in this transition; their loss of function enhanced the aggressiveness of cancer cells (Berton et al. 2009).

The organs that are mostly affected by metastasis are lung, liver, brain and bones; lung and liver are the top metastatic sites for maximum type of cancers. The lungs are the first filter for the tumor cells because blood circulation in malignancies flow directly through the lungs via venous drainage (Abbruzzese et al. 1995). Tumor types, which prominently metastasized liver, are melanoma, lymphomas and rarely sarcomas. Patients with lung, breast, melanoma, renal and colorectal cancer are known to have higher incidence of brain metastasis (Schouten et al. 2002). Bone metastasis is commonly seen in prostate, breast and lung cancer and it leads to severe morbidities including severe pain in bone, fractures, hypercalcemia and paralysis (Mundy 2002).

The microenvironment of tumor plays an important role in metastasis. It comprises various kind of cells including endothelial cells, heterogenous cancer cells, lymph, blood vessels, pericytes, stromal fibroblast and bone marrow derived cells like macrophages, neutrophils and mesenchymal stem cells (Joyce and Pollard 2009). Tumor microenvironment provides growth factors and cytokines e.g. TGF-β, hepatocyte growth factor (HGF), fibroblast growth factor (FGF) with ECM arrangement to facilitate tumor growth and metastasis. Besides these, increased level of MMPs, chemokines and cytokines provide enhanced proliferative, migratory and angiogenic signaling to the growing tumor cells (Strzyz 2016; Coussens and Werb 2002). The tight connection in metastasis and inflammation has been demonstrated by high content of inflammatory cells, particularly tumor-associated macrophages, in invasive fronts of advanced carcinoma (Condeelis and Pollard 2006). Since cancer metastasis is regulated by complex genetic alterations, cell-cell and cell-environment interactions, it requires multi-module treatment.

Of the 121 prescription anticancer drugs (often called chemotherapeutic or chemo-preventive agents) in use today, 90 have been derived from plants. Between 1981 and 2002, 48 out of 65 approved drugs were derived from natural products and many of these phytochemicals from Ayurvedic herbs that, in contrast to the modern medicine (a single chemical entity), contain multiple active components working synergistically to produce therapeutic benefits and lower the adverse effects. Several herbs possessing cytotoxicity to cancer cells have been identified. While description of all of these is out of the scope of this chapter, we describe here the effects of Ashwagandha on cancer metastasis.

12.5 Bioactives of Ashwagandha

Ashwagandha (*Withania sominifera*) has been used as traditional medicine from decades for the treatment of various ailments. The various parts of Ashwagandha and its constituents are effective in prevention and treatment of different cancers. The active ingredients are alkaloids, (Isopelletierine, Anaferine, Cuscohygrine,

Table 12.1 A most common chemical profile of Ashwagandha

Chemical composition in Ashwagandha		Weight in mg/g			
Type	Name	Roots	Leaves	Fruits	References
Steroidal lactones	Withanone	5.54+/−0.4	18.42+/−0.8	–	Chatterjee et al. (2010)
	27-deoxywithanone	3.94+/−0.4	1.63+/−0.2	–	
	27-hydroxywithanone	0.50+/−0.1	0.50+/−0.1	–	
	5,6-epoxy Withaferin-A	0.92+/−0.4	22.31+/−1	–	
	17-hydroxy-27-deoxy-Withaferin-A	0.66+/−0.2	3.61+/−0.5	–	
	Withanolide A	3.88+/−0.7	2.11+/−0.5	–	
	27-hydroxy Withanolide B	0.55+/−0.2	2.78+/−0.5	–	
	Withanoside IV	0.44+/−0.1	1.60+/−0.2	–	
	Withanoside VI	3.74+/−0.2	1.90+/−0.2	–	
	12-deoxywithastromonolide	1.90+/−0.5	2.15+/−0.5	–	
	Physagulin	–	3.46 ± 0.4	–	
Phenolic acid	Kaempferol	–	–	0.06	Alam et al. (2011)
	Naringenin	–	–	0.5	
	(+)-Catechin	12.82	28.38	19.48	
	Gallic acid	–	0.18	–	
	Syringic acid	–	0.3	–	
	p-coumaric acid	–	0.8	–	
	Vanillic acid	–	0.15	–	
	Benzoic acid	–	0.8	–	
	Trigonelline	–	1.33+/−0.3	–	Chatterjee et al. (2010)
	Palmitic acid	1.18+/−0.2	3.55+/−0.5	–	
	Oleic acid	0.39+/−0.1	0.71+/−0.1	–	
	Linoleic acid	1.31+/−0.2	1.52+/−0.2	–	
	Linolenic acid	0.15+/−0.1	4.38+/−0.5	–	

Anahygrine, etc.), Steroidal lactones (Withanolides, Withaferins) and Saponins (Mirjalili et al. 2009). Sitoindosides and Acylsterylglucosides have also been to possess anti-stress properties. Active principles of Ashwagandha have also been shown to possess significant anti-stress activity against acute models of experimental stress (Madina et al. 2007). Some clinical studies have suggested that Ashwagandha could be useful as anti-tumor, immunomodulator agent in sarcoma, brain cancers, uterine tumors, fibroids and endodermal carcinomas. The active ingredients of Ashwagandha are its secondary metabolites including alkaloids, (isopelletierine, anaferine, cuscohygrine, anahygrine, etc.), steroidal lactones (withanolides, withaferins, withanone) and saponins (Mirjalili et al. 2009). A most common chemical profile of Ashwagandha root and leaf is described in Table 12.1.

12.6 Anti-metastasis and Anti-angiogenic Activity of Ashwagandha – Molecular Targets

Several studies have described anti-metastasis and anti-angiogenic activities in Ashwagandha extracts and their purified components. (Yang et al. 2012) reported that Withaferin-A (Wi-A) inhibits the proteasome activity in mesothelioma. Regression of murine MPM (malignant pleural mesothelioma) tumor growth by Wi-A treatments caused a complete regression of tumor growth in one Wi-A – treated animal, suggested that Wi-A can facilitate its direct targeting and inhibition of tumor cellular proteasome, resulting in elevated apoptosis induction and tumor growth inhibition. (Saha et al. 2013) reported that Wi-A is a potent anti-VEGF agent as ascertained by its potential interaction with VEGF. It was proposed as an anti-angiogenic drug. Gao et al. (2014), Hahm et al. (2013), Wadhwa et al. (2013) have reported metabolic alterations in mammary cancer prevention by Wi-A. It resulted in significant decrease in tumor mass and the incidence of pulmonary metastasis. These effects of Wi-A were associated with increased apoptosis, inhibition of complex III activity and reduced levels of glycolysis intermediates. This report clearly indicated that Wi-A inhibited tumor burden but not tumor incidence. The possible interpretation is that inhibition of mammary cancer incidence require daily dose of Wi-A rather than intermittent schedule performed in this study. The another possible reason is notch signaling which is often hyperactive in human breast cancer (Lee et al. 2012) impedes prevention efficacy of Wi-A. In context of *in vivo* results, higher levels of 8-hydroxy-2′-deoxyguanosine (8-OhdG) found in tumor from mice group treated with Wi-A as compared to control suggested that Wi-A caused Reactive Oxygen Species (ROS)-dependent apoptosis in tumor cells. Wi-A was also reported to trigger clinically relevant anticancer effects specific to triple negative breast cancer cells. In this study, a comparative pathway based transcriptome analysis was performed in epithelial like MCF7 and triple negative mesenchymal MDA-MB-231 breast cancer cells. Wi-A showed strong decrease invasion in single cell collagen invasion assay in MDA-MB-231 as compared to MCF7 cells. Furthermore, transcriptional downregulation of uPA (extracellular matrix-degrading protease), PLAT, ADAM8, integrins, laminins, TNFSF12, interlukin 6, ANGPTL2, CSF1R was detected in therapy-resistant triple negative breast cancer cells (Szarc vel Szic et al. 2014). Wi-A induced suppression of Vimentin, MMP9 and Akt causing EMT inhibition has been reported in breast and ovarian cancer cells (Lee et al. 2013; Yang et al. 2013; Lee et al. 2015). Wi-A alone and its combination with cisplatin caused strong suppression of growth and metastasis of ovarian tumors (Yang et al. 2012; Kakar et al. 2014; Hahm et al. 2013). The current understanding on the anti-metastatic role of Wi-A involves the following targets:

12.6.1 Vimentin

Yokota et al. (2006) first discovered that Wi-A caused degradation of a 56-kDa protein in HUVEC cells. The protein was identified as vimentin, an intermediate filament protein that regulates wound healing, angiogenesis, cancer growth and metastasis (van Beijnum et al. 2006; Eckes et al. 2000). Docking studies showed that Wi-A docks to the specific amino acid residues, Gln234, Cys328 and Asp331, of vimentin and this bonding does not block aggregation of vimentin into a tetramer. However, it alters its binding efficacy and induces fragmentation or depolymerization (Thaiparambil et al. 2011; Bargagna-Mohan et al. 2007). Wi-A also phosphorylates serine 56 residue of vimentin, a site phosphorylated prior to disassembly. The C3 carbon on the A-ring of Wi-A is critical for phosphorylation (Thaiparambil et al. 2011). The reduction in vimentin expression is concentration and time dependent process and can occur at low (nanomolar) concentrations (Lahat et al. 2010). Wi-A may also reversed reduce TGF-β induced increase in vimentin expression (Yang et al. 2013). It was shown that Wi-A directly binds to vimentin and induces its degradation. Besides, the other filament proteins, such as KIFs (keratin heteropolymer Ifs), PFs (Peripherin filaments), NIFs (neurofilament triplet protein) and VIFs (vimentin inhibitory filaments) were affected in a similar manner. The study showed that there is disruption of microtubules and microfilament formation in the cell's cytoskeleton and increase in actin stress fibers in Wi-A treated cells (Grin et al. 2012).

12.6.2 NF-kB

Several genes that regulate cellular proliferation, carcinogenesis, metastasis and inflammation are regulated by activation of nuclear factor-kappaB (NF-kappaB). NF-kB is kept inactive by its inhibitor IkB that forms a complex, keeping NF-kB in the cytoplasm and prevents NF-kB activation. IkB is phosphorylated by IKK (IkB Kinase) to release and activate NF-kB (Karin and Delhase 2000). IKK is a complex of two subunits, IKKα and IKKβ and a regulatory subunit called NEMO. Withanolide A has been shown to directly interact with the NEMO in its binding pocket at Glu89, Glu99, Phe92, Leu 93, Phe 97 and Ala 100. Such interaction was shown to interfere with NEMO:IKKβ interaction (Grover et al. 2010b). It has also been shown that Wi-A stops degradation of IkBα via blocking IKKβ and MEK1/Erk pathway and causes sustained inhibition of transcriptional activation function of NF-kB (Kaileh et al. 2007). Lactones have been reported to directly bind to cys-179 of IKKβ kinase and inhibit its complex formation (Na and Surh 2006). Resorcylic acid lactones were shown to target mitogen-activated protein kinase pathways at multiple levels including inactivation of receptors, protein kinases and effector proteins (Schirmer et al. 2006). Furthermore, co-treatment of Withaferin-A and DTT (dithiothreitol) reversed the inhibitory effects of Withaferin-A on TNF-induced NFkB activation (Kaileh et al. 2007).

12.6.3 20S Proteasome

Proteasomes are large multicatalytic proteinase complexes present in cytosol and nucleus of the eukaryotic cells. The ubiquitin-proteasome system is main site for non-lysosomal degradation of most intracellular proteins and hence it plays an important role in tightly regulated cellular processes e.g. cell cycle progression, proliferation, differentiation, angiogenesis, apoptosis and homeostasis (Yang et al. 2012; Chen and Dou 2010). In this process, proteins are first modified by polyubiquitin chains and then degraded in small peptides by the proteasome; ubiquitin is released and recycled. Several cell cycle-, inflammation-, carcinogenesis-, metastasis- and angiogenesis-regulatory transcriptional factors such as NFkB, p53, c-Jun, c-Myc, c-Fos, HIF-1-α are known to be regulated by proteasome (Mani and Gelmann 2005; Godwin et al. 2013; Zhang and Peng 2007; Helbig et al. 2003). Furthermore, alterations in proteasomal activity have been documented to be directly associated with the pathological state of these disorders. Wi-A has been shown to target catalytic β subunit of the 20S-proteosome specifically resulting in decreased chymotryptic activity and increase in pro-apoptotic proteins e.g. p21, Bax, IkBα (Grover et al. 2010a; Yang et al. 2012). Molecular docking studies revealed that Wi-A inhibits activity of 20S proteasomes by blocking the nucleophilic function of N-terminal Thr1 (Grover et al. 2010a). (Grin et al. 2012) demonstrated that Wi-A caused aggregation of intermediate filament protein that impaired the function of the proteasome and caused accumulation of its target proteins, Bax, IkB-α and $p27^{kip1}$, involved in apoptotic cell death (Grin et al. 2012).

12.6.4 Survivin

Survivin is an apoptosis and mitotic signaling regulatory protein, also involved in angiogenesis and chemo-resistance of cancer cells. It is the smallest member of the inhibitor of apoptosis protein family and is enriched in majority of tumors (Groner and Weiss 2014). High level of expression of surviving is associated with advanced aggressive tumor stages and metastasis disease (Werner et al. 2016). It has been documented that surviving inhibits caspases and apoptosis of cancer cells. Molecular docking studies showed that Withanone binds to surviving at its BIR5 domain and interferes with its inhibitory action on caspases (Wadegaonkar and Wadegaonkar 2013). Survivin expression is regulated by transcriptional and post-transcriptional mechanisms and is associated with growth factor signaling including epidermal growth factor receptor (EGFR) (Asanuma et al. 2005), EGFR2 (Zhu et al. 2010) and insulin-like growth factor (IGF-1) (Vaira et al. 2007). Based on the clinical studies, it has been reported that EGFR induces surviving expression in breast cancer patients and contributes to their resistance to etoposide-induced apoptosis (Asanuma et al. 2005). It has been established that surviving is phosphorylated by cyclin dependent kinase and the phosphorylated surviving translocates to nucleus wherein it regulates p53, Wnt, hypoxia, TGF, and Notch signaling pathways. Activation of the Wnt signaling pathway leads to accumulation of β-catenin in the cytoplasm. B-catenin translocates to the nucleus and upregulates surviving by forming an active transcription factor complex with TCF enhancer factor. HIF-1α and STAT3 have been also been shown to bind to the surviving promoter and function as its transcriptional activators. On the other hand, TGF-β acts as a negative regulator of surviving through Smads 2, 3, and 4 proteins (Chen et al. 2016). The potential mechanism by which surviving mediates metastasis and other cellular events involving local invasion, intravasation into the circulation, survival and transportation in the circulation, extravasation from the bloodstream and growth in the metastatic site still needs to be more clear (Chen et al. 2016; Gopalkrishnan et al. 2001).

12.6.5 Mortalin

Mortalin/GPR75/heat sock protein (HSP70)/cytosolic heat shock protein is a ubiquitous mitochondrial chaperone and plays an important role in human carcinogenesis by increasing cancer cell proliferation, protection of cancer cells against apoptosis (Saxena et al. 2013; Lu et al. 2011a; Wadhwa et al. 2006). It interacts with tumor suppressor protein p53 and inactivates its functions including transcriptional activation and control of centrosome duplication leading to uncontrolled proliferation, a hallmark of cancer cell (Wadhwa et al. 1998; Lu et al. 2011a, b). Most recently, it was shown to activate telomerase and hnRNP-K proteins and contribute to malignant phenotype of cancer cells (Gao et al. 2013). In agreement with these reports, knockdown of mortalin in cancer cells was shown to activate p53 function and cause their growth arrest or apoptosis (Nigam et al. 2015; Lu et al. 2011b). Some studies have shown correlation of mortalin expression level with metastatic potential and tumor recurrence in case of hepatocellular carcinoma, suggesting clinical application of mortalin as a chemotherapeutic drug target (Lu et al. 2011a). Chen et al also showed that the genetically isogenic cell lines with variable metastatic potentials possess tight correlation with the level of expression of mortalin, suggesting its role in metastatic hepatocellular carcinoma (HCC) (Chen et al. 2011). Furthermore, mortalin-compromised cells showed inhibition of EMT suggesting mortalin as a therapeutic target for HCC metastasis. Ashwagandha withanolide, Withanone was shown to binds with mortalin, like MKT-077 (mortalin inhibitor) at the Phe 272 and Asn 139 (lactone group in withanone) (Grover et al. 2012a). It resulted in abrogation of mortalin-p53 complexes and reactivation of p53 proteins leading to growth arrest of cancer cells (Wadhwa et al. 2016).

12.6.6 Aurora A

Aurora A, centrosome associated a protein member of the serine/threonine kinase family regulating centrosome function, spindle assembly, chromosome segregation and cytokinesis (Bischoff and Plowman 1999). It has gained attention in recent years because of its increase expression levels in several types of tumors. However, the mechanism by which Aurora A mediates its invasive effects still remain elusive. TPX2, a microtubule-associated protein and downstream of Ran-GTP, plays role in spindle formation. TPX-2 is highly expressed in colon cancer tissues and cell lines via PI3K/Akt signaling pathway (Wei et al. 2013). Binding studies clearly shown that the NH_2 terminus of TPX-2 can directly interact with the COOH terminal catalytic domain of Aurora A and make a complex (Kufer et al. 2003). Withanone has showed direct semi-flexible docking via hydrogen bonding to His 280 with Aurora A, which is a residue required binding for TPX2 (Grover et al. 2012b). So these two proteins cannot bind and influence the genome resulting reduction in invasiveness.

12.6.7 Hsp90

Heat Shock protein 90 is an abundant molecular chaperon in cancer cells that maintain the function and stability of numerous client proteins (Neckers 2002). Hsp90 clients are frequently mutated or activated in cancer and their interaction is regulated by the adenine nucleotide binding status of Hsp90. Hsp90 inhibitors inhibit cancer cell proliferation and function by competing with ATP binding, freezing chaperon cycle, decrease the affinity of Hsp90 for client proteins and finally leads to proteasome-mediated client protein degradation (Grenert et al. 1997). Moreover, the level of cell surface Hsp90 has been shown to increase and correlate with carcinogenesis and metastasis, respectively (Becker et al. 2004). Withaferin-A has been shown to inhibit Hsp90 signaling in pancreatic cancer cells (Yu et al. 2010). Recently, it was shown that Withaferin-A docks to the active site of Hsp90 resulting in structural disruption of the site where Cdc37 binds to Hsp90 causing inhibition of complex formation (Narayan et al. 2015).

12.6.8 VEGF and hnRNP-K

Vascular Endothelial Growth Factor (VEGF) is a potent and essential angiogenic factor requires for vascular endothelial cells. It is also known as vascular permeability factor, produced by many cell types including cancer cells, macrophages, platelets, keratinocytes (Prager et al. 2010). It works as survival, growth and metastasis factor (Pidgeon et al. 2001). hnRNP-K is a RNA binding multifunctional protein that regulates ERK1/2, MMPs and VEGF resulting migration, invasion and ascites

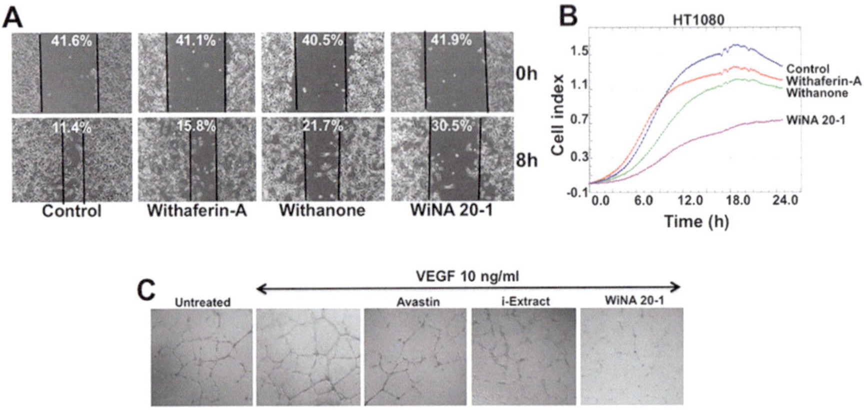

Fig. 12.1 Withaferin-A, Withanone and their combination WiNA 20-1 caused inhibition of cancer cell migration (**a** and **b**) and tube formation in HUVEC cells (**c**) (Adopted from Gao et al. 2014)

formation (Gao et al. 2013). Withaferin-A was shown to target hnRNP-K and VEGF and inhibits metastasis signaling (Gao et al. 2014). A combination of Withanone and Withaferin-A in a ratio of 20:1 (WiNA 20-1) was shown to be selectively toxic to cancer cells. Migration, invasion & metastasis ability of fibrosarcoma (HT1080) was significantly reduced after WiNA 20-1 treatment. The data was supported by quantitative assays for cell invasion and migration (Fig. 12.1a). Furthermore, by tube formation assay using human umbilical vein endothelial cells (HUVEC) treated with WiNA 20-1, it was found that VEGF-induced tube formation was strongly inhibited by the combination (Gao et al. 2014) (Fig. 12.1c) and was comparable to Avastin (50 μg/ml), an established inhibitor of angiogenesis. *In vivo* validation of anti-metastasis activity by intraperitoneal injections of either Withanone (1 mg/kg), Withaferin-A (0.5 mg/kg), or their combination (WiNA) in 100 μL of 0.5% DMSO in HT1080 subcutaneous xenograft and lung metastasis nude mice models also support anticancer and anti-metastasis activity by downregulation of hnRNP-K in treated tumors as compared to control tumors (Gao et al. 2014) (Fig. 12.2a). Whereas there were no significant differences found in body weight of control, Withaferin-A, Withanone and WiNA-injected mice. WiNA-injected mice showed strong suppression of subcutaneous HT1080 tumor-xenografts. In the lung metastasis model, big lung tumors were detected only in control mice suggesting that WiNA has significant anti-cancer and anti-metastasis activities, *in vivo*. Bioinformatics, molecular docking and experimental studies to dissect the molecular mechanism of anti-migratory activity of Withanone, Withaferin-A and WiNA 20-1 revealed targeting of mortalin and hnRNP-K, established players of metastasis and angiogenesis (Gao et al. 2014). Molecular docking of the crystal structure of KH3 domain of hnRNP-K protein with DNA (DNA-Protein complex) revealed interaction of hnRNP-K with Withaferin-A/ Withanone at Lys22, Asp23, Ala25, Ile29, Lys31, Arg40, Lys47, Ile48, Asp50, Tyr84 and Ser85 of KH3_hnRNP-K protein (Gao et al. 2014) (Fig. 12.3c). Furthermore, Withanone was also shown to

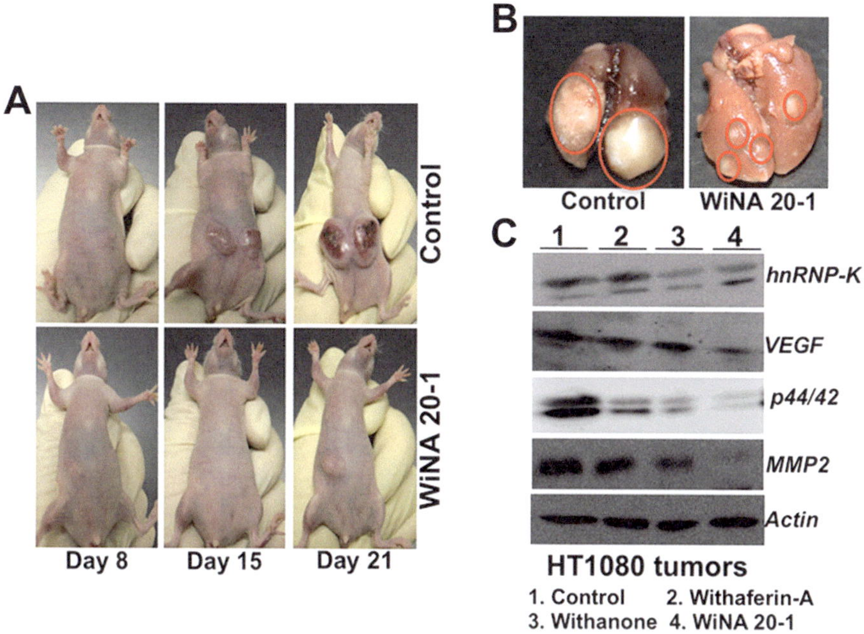

Fig. 12.2 Combination of Withaferin-A and Withanone (WiNA 20-1) caused inhibition of tumor growth and metastasis in nude mice subcutaneous xenografts (**a**) and lung metastasis (**b**) models. Downregulation of hnRNP-K, VEGF, P44/42 and MMP2 was observed in small tumors from WiNA treated mice as compared to the big tumors from control mice (**c**) (Adopted from Gao et al. 2014)

target TPX2 oncogene, a prime regulator of Aurora A kinase that plays a critical role during mitosis and cytokinesis (Grover et al. 2012b).

Ashwagandha leaves (water extract) have its active component identified as triethylene glycol (TEG), which activates tumor suppressor proteins p53, pRB, and down-regulates MMP-3 and MMP-9 (regulators of metastasis) (Wadhwa et al. 2013). TEG is a member of dihydroxy alcohol family and is hydrophilic nature makes it more available for water molecules to interact. It has been well established that at low pressure TEG as a low toxicity mild disinfectant towards a variety of airborne, solution and surface bound microbes including bacteria and viruses (Rudnick et al. 2009). An old study on dogs having inoperable malignant and metastatic tonsillar epithelioma and reticulum cell sarcoma reported the regression of tumors in response to the treatment with TEG without any toxicity (Owen 1962). Triethylene tetramine (TETA) was reported as a novel ligand for γ-quadruplex that has many kinds of biological activities, including telomerase inhibition and induction of senescence in tumor cells. Low doses of TETA had limited ability to inhibit the growth of tumor cells in short-term culture, but it could significantly enhance anti-tumor activity of traditional anti-tumor drugs *in vitro* and *in vivo* (Liu et al. 2008). We found that water extract of Ashwagandha leaves (WEX) that showed good anticancer activity *in vitro* (Wadhwa et al. 2013) also was effective in tumor

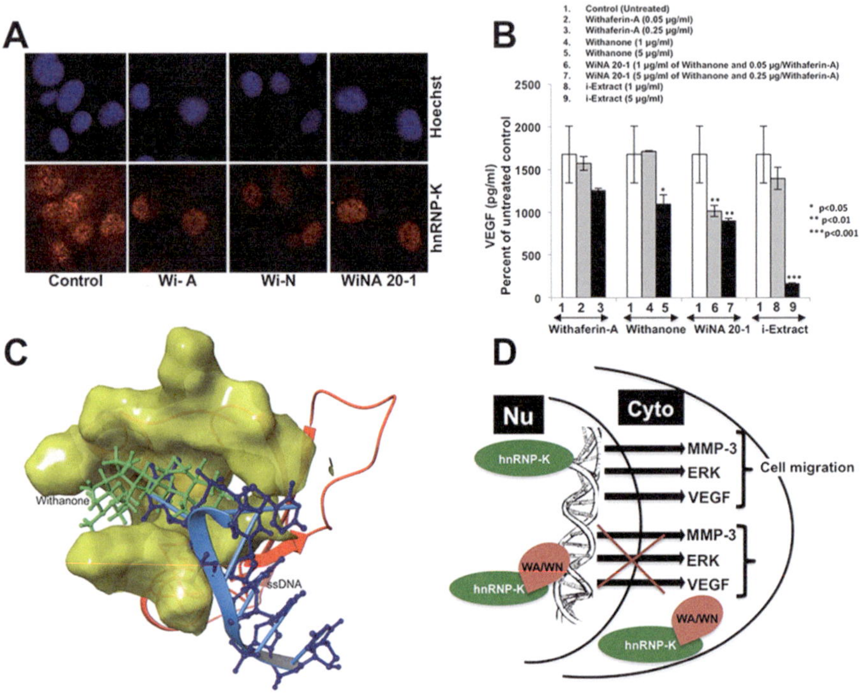

Fig. 12.3 Withaferin-A, Withanone and their combination WiNA 20-1 targets hnRNP-K (**a** and **c**), downregulates VEGF (**b**) and inhibits cancer cell migration (**d**) (Adopted from Gao et al. 2014)

suppression in nude mice models. Mice were fed with 500 mg/kg body weight of WEX on every alternate day for 60 days. No toxicity in terms of body weight and physical activity was observed. In addition to the strong tumor suppression in subcutaneous tumor model, ASH-WAX and its active component TEG both showed significant reduction in number and volume of lung tumors as compared to control. Both oral feeding and intraperitoneal injections of TEG were effective to significantly reduce the tumor volume and lung metastasis. Anti-metastatic activities of TEG and ASH-WEX were supported by molecular analyses that revealed remarkable decrease in matrix metalloproteases and activation of tumor suppressor protein pRB (Fig. 12.4).

In lights of the findings that alcoholic and water extracts of Ashwagandha leaves possess anticancer bioactive compounds, we aimed to enrich the water extracts with active withanolides by recruiting various isoforms of cyclodextrin (CD). The latter are natural derivatives of starch or polymer of glucose that possess circular structure and are widely used in food, pharmaceutical, agriculture and drug delivery. They possess unique structure (hydrophobic inside and a hydrophilic outside) that enhances the solubility and bioavailability of compounds. γCD, consists of 8 glucose monomers arranged in the form of a cyclic ring, is widely accepted as food constituent and has been reported to enhance the bioavailability of hydrophobic

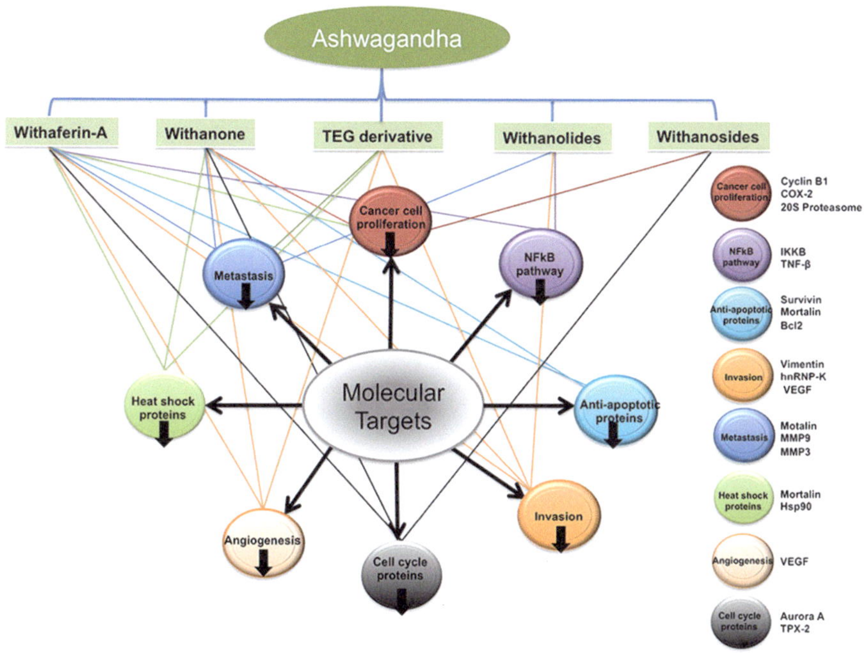

Fig. 12.4 Cellular targets of Ashwagandha-derived bioactives for cancer cell metastasis

ingredients such as coenzyme Q10 (Terao et al. 2006; Miyoshi et al. 2011). We found that CD-derived extracts of Ashwagandha leaves contained high level of Withanone and Withaferin-A. By cell-based assays, we found that the CD extracts of Ashwagandha leaves have enhanced cancer cell cytotoxicity as compared to the conventional water extract (Wadhwa et al. 2013). HPLC analysis of Withanone and Withaferin-A in γCD residual precipitate revealed that they contained 17-fold higher Withanone than Withaferin-A and as similar to our WiNA combination (Gao et al. 2014). Taken together, it is strongly suggestive that Ashwagandha possess multiple bioactives useful for the treatment of cancer cell metastasis. These natural bioactives work through multiple pathways and offer high rate of successful treatment. Furthermore, NEW (natural Efficient and Welfare) technologies such as use of cyclodextrins can be useful for further enhancing the anticancer potential.

References

Abbruzzese JL, Abbruzzese MC, Lenzi R, Hess KR, Raber MN (1995) Analysis of a diagnostic strategy for patients with suspected tumors of unknown origin. J Clin Oncol 13:2094–2103

Alam N, Hossain M, Khalil MI, Moniruzzaman M, Sulaiman SA, Gan SH (2011) High catechin concentrations detected in Withania somnifera (ashwagandha) by high performance liquid chromatography analysis. BMC Complement Altern Med 11:65

Asanuma H, Torigoe T, Kamiguchi K, Hirohashi Y, Ohmura T, Hirata K, Sato M, Sato N (2005) Survivin expression is regulated by coexpression of human epidermal growth factor receptor 2 and epidermal growth factor receptor via phosphatidylinositol 3-kinase/AKT signaling pathway in breast cancer cells. Cancer Res 65:11018–11025

Bargagna-Mohan P, Hamza A, Kim YE, Khuan Abby Ho Y, Mor-Vaknin N, Wendschlag N, Liu J, Evans RM, Markovitz DM, Zhan CG, Kim KB, Mohan R (2007) The tumor inhibitor and antiangiogenic agent withaferin A targets the intermediate filament protein vimentin. Chem Biol 14:623–634

Becker B, Multhoff G, Farkas B, Wild PJ, Landthaler M, Stolz W, Vogt T (2004) Induction of Hsp90 protein expression in malignant melanomas and melanoma metastases. Exp Dermatol 13:27–32

Berton S, Belletti B, Wolf K, Canzonieri V, Lovat F, Vecchione A, Colombatti A, Friedl P, Baldassarre G (2009) The tumor suppressor functions of p27(kip1) include control of the mesenchymal/amoeboid transition. Mol Cell Biol 29:5031–5045

Biganzoli L, Aapro M, Loibl S, Wildiers H, Brain E (2016) Taxanes in the treatment of breast cancer: Have we better defined their role in older patients? A position paper from a SIOG Task Force. Cancer Treat Rev 43:19–26

Bischoff JR, Plowman GD (1999) The Aurora/Ipl1p kinase family: regulators of chromosome segregation and cytokinesis. Trends Cell Biol 9:454–459

Chatterjee S, Srivastava S, Khalid A, Singh N, Sangwan RS, Sidhu OP, Roy R, Khetrapal CL, Tuli R (2010) Comprehensive metabolic fingerprinting of Withania somnifera leaf and root extracts. Phytochemistry 71:1085–1094

Chen D, Dou QP (2010) The Ubiquitin-Proteasome System as a Prospective Molecular Target for Cancer Treatment and Prevention. Curr Protein Pept Sc 11:459–470

Chen X, Xu B, Li H, Yang L, Zuo J, Liu W, Liu C (2011) Expression of mortalin detected in human liver cancer by tissue microarrays. Anat Rec (Hoboken) 294:1344–1351

Chen X, Duan N, Zhang C, Zhang W (2016) Survivin and tumorigenesis: Molecular mechanisms and therapeutic strategies. J Cancer 7:314–323

Condeelis J, Pollard JW (2006) Macrophages: obligate partners for tumor cell migration, invasion, and metastasis. Cell 124:263–266

Coussens LM, Werb Z (2002) Inflammation and cancer. Nature 420:860–867

Eckes B, Colucci-Guyon E, Smola H, Nodder S, Babinet C, Krieg T, Martin P (2000) Impaired wound healing in embryonic and adult mice lacking vimentin. J Cell Sci 113:2455–2462

Folkman J (1995a) Angiogenesis in cancer, vascular, rheumatoid and other disease. Nat Med 1:27–31

Folkman J (1995b) Seminars in medicine of the Beth Israel Hospital, Boston – Clinical applications of research on angiogenesis. New Engl J Med 333:1757–1763

Gao R, Yu Y, Inoue A, Widodo N, Kaul SC, Wadhwa R (2013) Heterogeneous nuclear ribonucleoprotein K (hnRNP-K) promotes tumor metastasis by induction of genes involved in extracellular matrix, cell movement, and angiogenesis. J Biol Chem 288:15046–15056

Gao R, Shah N, Lee JS, Katiyar SP, Li L, Oh E, Sundar D, Yun CO, Kaul SC, Wadhwa R (2014) Withanone rich combination of Ashwagandha withanolides restricts metastasis and angiogenesis through hnRNP-K. Mol Cancer Ther 13:2930–2940

Godwin P, Baird AM, Heavey S, Barr MP, O'Byrne KJ, Gately K (2013) Targeting nuclear factor-kappa B to overcome resistance to chemotherapy. Front Oncol 3:120

Gopalkrishnan RV, Kang DC, Fisher PB (2001) Molecular markers and determinants of prostate cancer metastasis. J Cell Physiol 189:245–256

Grenert JP, Sullivan WP, Fadden P, Haystead TAJ, Clark J, Mimnaugh E, Krutzsch H, Ochel HJ, Schulte TW, Sausville E, Neckers LM, Toft DO (1997) The amino-terminal domain of heat shock protein 90 (hsp90) that binds geldanamycin is an ATP/ADP switch domain that regulates hsp90 conformation. J Biol Chem 272:23843–23850

Grin B, Mahammad S, Wedig T, Cleland MM, Tsai L, Herrmann H, Goldman RD (2012) Withaferin a alters intermediate filament organization, cell shape and behavior. PLoS One 7:e39065

Groner B, Weiss A (2014) Targeting survivin in cancer: novel drug development approaches. BioDrugs 28:27–39

Grover A, Shandilya A, Bisaria VS, Sundar D (2010a) Probing the anticancer mechanism of prospective herbal drug Withaferin A on mammals: a case study on human and bovine proteasomes. BMC Genomics 11 Suppl 4: S15

Grover A, Shandilya A, Punetha A, Bisaria VS, Sundar D (2010b) Inhibition of the NEMO/IKKbeta association complex formation, a novel mechanism associated with the NF-kappaB activation suppression by Withania somnifera's key metabolite withaferin A. BMC Genomics 11 Suppl 4: S25

Grover A, Priyandoko D, Gao R, Shandilya A, Widodo N, Bisaria VS, Kaul SC, Wadhwa R, Sundar D (2012a) Withanone binds to mortalin and abrogates mortalin-p53 complex: computational and experimental evidence. Int J Biochem Cell Biol 44:496–504

Grover A, Singh R, Shandilya A, Priyandoko D, Agrawal V, Bisaria VS, Wadhwa R, Kaul SC, Sundar D (2012b) Ashwagandha derived withanone targets TPX2-Aurora A complex: computational and experimental evidence to its anticancer activity. PLoS One 7:e30890

Hahm ER, Lee J, Kim SH, Sehrawat A, Arlotti JA, Shiva SS, Bhargava R, Singh SV (2013) Metabolic alterations in mammary cancer prevention by withaferin A in a clinically relevant mouse model. J Natl Cancer Inst 105:1111–1122

Harris TJ, Tepass U (2010) Adherens junctions: from molecules to morphogenesis. Nat Rev Mol Cell Biol 11:502–514

Hegerfeldt Y, Tusch M, Brocker EB, Friedl P (2002) Collective cell movement in primary melanoma explants: plasticity of cell-cell interaction, beta1-integrin function, and migration strategies. Cancer Res 62:2125–2130

Helbig G, Christopherson KW 2nd, Bhat-Nakshatri P, Kumar S, Kishimoto H, Miller KD, Broxmeyer HE, Nakshatri H (2003) NF-kappaB promotes breast cancer cell migration and metastasis by inducing the expression of the chemokine receptor CXCR4. J Biol Chem 278:21631–21638

Joyce JA, Pollard JW (2009) Microenvironmental regulation of metastasis. Nat Rev Cancer 9:239–252

Kaileh M, Vanden Berghe W, Heyerick A, Horion J, Piette J, Libert C, De Keukeleire D, Essawi T, Haegeman G (2007) Withaferin a strongly elicits IkappaB kinase beta hyperphosphorylation concomitant with potent inhibition of its kinase activity. J Biol Chem 282:4253–4264

Kakar SS, Ratajczak MZ, Powell KS, Moghadamfalahi M, Miller DM, Batra SK, Singh SK (2014) Withaferin a alone and in combination with cisplatin suppresses growth and metastasis of ovarian cancer by targeting putative cancer stem cells. PLoS One 9:e107596

Kalluri R, Weinberg RA (2009) The basics of epithelial-mesenchymal transition. J Clin Invest 119:1420–1428

Karin M, Delhase M (2000) The I kappa B kinase (IKK) and NF-kappa B: key elements of proinflammatory signalling. Semin Immunol 12:85–98

Kim NH, Kim HS, Li XY, Lee I, Choi HS, Kang SE, Cha SY, Ryu JK, Yoon D, Fearon ER, Rowe RG, Lee S, Maher CA, Weiss SJ, Yook JI (2011) A p53/miRNA-34 axis regulates Snail1-dependent cancer cell epithelial-mesenchymal transition. J Cell Biol 195:417–433

Kufer TA, Sillje HHW, Koerner R, Gruss OJ, Meraldi P, Nigg EA (2003) Human TPX2 is required for targeting Aurora-A kinase to the spindle. Cell Motil Cytoskel 54:179–179

Lahat G, Zhu QS, Huang KL, Wang SZ, Bolshakov S, Liu J, Torres K, Langley RR, Lazar AJ, Hung MC, Lev D (2010) Vimentin Is a Novel Anti-Cancer Therapeutic Target; Insights from In Vitro and In Vivo Mice Xenograft Studies. PLoS One 5:e10105

Lee J, Sehrawat A, Singh SV (2012) Withaferin A causes activation of Notch2 and Notch4 in human breast cancer cells. Breast Cancer Res Treat 136:45–56

Lee DH, Lim IH, Sung EG, Kim JY, Song IH, Park YK, Lee TJ (2013) Withaferin A inhibits matrix metalloproteinase-9 activity by suppressing the Akt signaling pathway. Oncol Rep 30:933–938

Lee J, Hahm ER, Marcus AI, Singh SV (2015) Withaferin A inhibits experimental epithelial-mesenchymal transition in MCF-10A cells and suppresses vimentin protein level in vivo in breast tumors. Mol Carcinog 54:417–429

Liu J, Guo L, Yin F, Zheng X, Chen G, Wang Y (2008) Characterization and antitumor activity of triethylene tetramine, a novel telomerase inhibitor. Biomed Pharmacother 62:480–485

Lu WJ, Lee NP, Kaul SC, Lan F, Poon RT, Wadhwa R, Luk JM (2011a) Induction of mutant p53-dependent apoptosis in human hepatocellular carcinoma by targeting stress protein mortalin. Int J Cancer 129:1806–1814

Lu WJ, Lee NP, Kaul SC, Lan F, Poon RTP, Wadhwa R, Luk JM (2011b) Mortalin-p53 interaction in cancer cells is stress dependent and constitutes a selective target for cancer therapy. Cell Death Different 18:1046–1056

Madina BR, Sharma LK, Chaturvedi P, Sangwan RS, Tuli R (2007) Purification and physico-kinetic characterization of 3beta-hydroxy specific sterol glucosyltransferase from Withania somnifera (L) and its stress response. Biochim Biophys Acta 1774:392–402

Mani A, Gelmann EP (2005) The ubiquitin-proteasome pathway and its role in cancer. J Clin Oncol 23:4776–4789

Mirjalili MH, Moyano E, Bonfill M, Cusido RM, Palazon J (2009) Steroidal Lactones from Withania somnifera, an Ancient Plant for Novel Medicine. Molecules 14:2373–2393

Miyoshi N, Wakao Y, Tomono S, Tatemichi M, Yano T, Ohshima H (2011) The enhancement of the oral bioavailability of g-tocotrienol in mice by g-cyclodextrin inclsion. J Nutr Biochem 22:1121–1126

Mundy GR (2002) Metastasis to bone: causes, consequences and therapeutic opportunities. Nat Rev Cancer 2:584–593

Na HK, Surh YJ (2006) Transcriptional regulation via cysteine thiol modification: a novel molecular strategy for chemoprevention and cytoprotection. Mol Carcinog 45:368–380

Narayan M, Zhang J, Braswell K, Gibson C, Zitnyar A, Lee DC, Varghese-Gupta S, Jinwal UK (2015) Withaferin A regulates LRRK2 levels by interfering with the Hsp90-Cdc37 chaperone complex. Curr Aging Sci 8:259–265

Neckers L (2002) Hsp90 inhibitors as novel cancer chemotherapeutic agents. Trends Mol Med 8:S55–S61

Nieto MA (2009) Epithelial-Mesenchymal transitions in development and disease: old views and new perspectives. Int J Dev Biol 53:1541–1547

Nigam N, Grover A, Goyal S, Katiyar SP, Bhargava P, Wang PC, Sundar D, Kaul SC, Wadhwa R (2015) Targeting mortalin by embelin causes activation of tumor suppressor p53 and deactivation of metastatic signaling in human breast cancer cells. PLoS One 10:e0138392

Owen LN (1962) Treatment of 30 cases of spontaneous tumours in dogs with triethylene glycol diglycidyl ether. Br J Cancer 16:441–452

Parri M, Taddei ML, Bianchini F, Calorini L, Chiarugi P (2009) EphA2 reexpression prompts invasion of melanoma cells shifting from mesenchymal to amoeboid-like motility style. Cancer Res 69:2072–2081

Pidgeon GP, Barr MP, Harmey JH, Foley DA, Bouchier-Hayes DJ (2001) Vascular endothelial growth factor (VEGF) upregulates BCL-2 and inhibits apoptosis in human and murine mammary adenocarcinoma cells. Br J Cancer 85:273–278

Prager GW, Lackner EM, Krauth MT, Unseld M, Poettler M, Laffer S, Cerny-Reiterer S, Lamm W, Kornek GV, Binder BR, Zielinski CC, Valent P (2010) Targeting of VEGF-dependent transendothelial migration of cancer cells by bevacizumab. Mol Oncol 4:150–160

Rudnick SN, McDevitt JJ, First MW, Spengler JD (2009) Inactivating influenza viruses on surfaces using hydrogen peroxide or triethylene glycol at low vapor concentrations. Am J Infect Control 37:813–819

Saha S, Islam MK, Shilpi JA, Hasan S (2013) Inhibition of VEGF: a novel mechanism to control angiogenesis by Withania somnifera's key metabolite Withaferin A. In Silico Pharmacol 1:11

Saxena N, Katiyar SP, Liu Y, Grover A, Gao R, Sundar D, Kaul SC, Wadhwa R (2013) Molecular interactions of Bcl-2 and Bcl-xL with mortalin: identification and functional characterization. Biosci Rep 33

Schirmer A, Kennedy J, Murli S, Reid R, Santi DV (2006) Targeted covalent inactivation of protein kinases by resorcylic acid lactone polyketides. Proc Natl Acad Sci USA 103:4234–4239

Schouten LJ, Rutten J, Huveneers HA, Twijnstra A (2002) Incidence of brain metastases in a cohort of patients with carcinoma of the breast, colon, kidney, and lung and melanoma. Cancer 94:2698–2705

Strzyz P (2016) Cancer biology: TGFbeta and EMT as double agents. Nat Rev Mol Cell Biol 17:202–203

Szarc vel Szic K, Op de Beeck K, Ratman D, Wouters A, Beck IM, Declerck K, Heyninck K, Fransen E, Bracke M, De Bosscher K, Lardon F, Van Camp G, Vanden Berghe W (2014) Pharmacological levels of Withaferin A (Withania somnifera) trigger clinically relevant anticancer effects specific to triple negative breast cancer cells. PLoS One 9:e87850

Terao K, Nakata D, Fukumi H, Schmid G, Arima H, Hirayama F, Uekama K (2006) Enhancement of oral bioavailability of coenzyme Q10 by complexation with γ-cyclodextrin in healthy adults. Nutr Res 26:503–508

Thaiparambil JT, Bender L, Ganesh T, Kline E, Patel P, Liu Y, Tighiouart M, Vertino PM, Harvey RD, Garcia A, Marcus AI (2011) Withaferin A inhibits breast cancer invasion and metastasis at sub-cytotoxic doses by inducing vimentin disassembly and serine 56 phosphorylation. Intl J Cancer 129:2744–2755

Thiery JP, Acloque H, Huang RY, Nieto MA (2009) Epithelial-mesenchymal transitions in development and disease. Cell 139:871–890

Vaira V, Lee CW, Goel HL, Bosari S, Languino LR, Altieri DC (2007) Regulation of survivin expression by IGF-1/mTOR signaling. Oncogene 26:2678–2684

van Beijnum JR, Dings RP, van der Linden E, Zwaans BM, Ramaekers FC, Mayo KH, Griffioen AW (2006) Gene expression of tumor angiogenesis dissected: specific targeting of colon cancer angiogenic vasculature. Blood 108:2339–2348

Wadegaonkar VP, Wadegaonkar PA (2013) Withanone as an inhibitor of survivin: A potential drug candidate for cancer therapy. J Biotechnol 168:229–233

Wadhwa R, Takano S, Robert M, Yoshida A, Nomura H, Reddel RR, Mitsui Y, Kaul SC (1998) Inactivation of tumor suppressor p53 by mot-2, a hsp70 family member. J Biol Chem 273:29586–29591

Wadhwa R, Takano S, Kaur K, Deocaris CC, Pereira-Smith OM, Reddel RR, Kaul SC (2006) Upregulation of mortalin/mthsp70/Grp75 contributes to human carcinogenesis. Intl J Cancer 118:2973–2980

Wadhwa R, Singh R, Gao R, Shah N, Widodo N, Nakamoto T, Ishida Y, Terao K, Kaul SC (2013) Water extract of Ashwagandha leaves has anticancer activity: identification of an active component and its mechanism of action. PLoS One 8:e77189

Wadhwa R, Nigam N, Bhargava P, Dhanjal JK, Goyal S, Grover A, Sundar D, Ishida Y, Terao K, Kaul SC (2016) Molecular characterization and enhancement of anticancer activity of caffeic acid phenethyl ester by gamma cyclodextrin. J Cancer 7:1755–1771

Wei P, Zhang N, Xu Y, Li XX, Shi DB, Wang YW, Li DW, Cai SJ (2013) TPX2 is a novel prognostic marker for the growth and metastasis of colon cancer. J Transl Med 11:313

Werner TA, Tamkan-Olcek Y, Dizdar L, Riemer JC, Wolf A, Cupisti K, Verde PE, Knoefel WT, Krieg A (2016) Survivin and XIAP: two valuable biomarkers in medullary thyroid carcinoma. Br J Cancer 114:427–434

Xu J, Chen Y, Olopade OI (2010) MYC and Breast Cancer. Genes Cancer 1:629–640

Yang H, Wang Y, Cheryan VT, Wu W, Cui CQ, Polin LA, Pass HI, Dou QP, Rishi AK, Wali A (2012) Withaferin A inhibits the proteasome activity in mesothelioma in vitro and in vivo. PLoS One 7:e41214

Yang Z, Garcia A, Xu S, Powell DR, Vertino PM, Singh S, Marcus AI (2013) Withania somnifera root extract inhibits mammary cancer metastasis and epithelial to mesenchymal transition. PLoS One 8:e75069

Yokota Y, Bargagna-Mohan P, Ravindranath PP, Kim KB, Mohan R (2006) Development of withaferin A analogs as probes of angiogenesis. Bioorg Med Chem Lett 16:2603–2607

Yu Y, Hamza A, Zhang T, Gu M, Zou P, Newman B, Li Y, Gunatilaka AA, Zhan CG, Sun D (2010) Withaferin A targets heat shock protein 90 in pancreatic cancer cells. Biochem Pharmacol 79:542–551

Zavadil J, Bottinger EP (2005) TGF-beta and epithelial-to-mesenchymal transitions. Oncogene 24:5764–5774

Zhang J, Peng B (2007) In vitro angiogenesis and expression of nuclear factor kappaB and VEGF in high and low metastasis cell lines of salivary gland Adenoid Cystic Carcinoma. BMC Cancer 7:95

Zhu H, Zhang G, Wang Y, Xu N, He S, Zhang W, Chen M, Liu M, Quan L, Bai J, Xu N (2010) Inhibition of ErbB2 by Herceptin reduces survivin expression via the ErbB2-beta-catenin/TCF4-survivin pathway in ErbB2-overexpressed breast cancer cells. Cancer Sci 101:1156–1162

Chapter 13
Computational Methods to Understand the Anticancer Mechanism of Withanolides

Shashank P. Katiyar, Vidhi Malik, Anjani Kumari, and Durai Sundar

Abstract With many excellent biological properties, withanolides have shown to be promising against few of the most widely spread and incurable diseases such as cancer and Alzheimer's disease. To further improve the potency and selectivity of withanolides, it becomes absolutely necessary to understand their action mechanism. Though many experimental studies have revealed the biological targets of withanolides, still their mode of action mechanism at molecular level needs better understanding. Computational methods are especially reliable and fast to understand the mechanism at molecular level, facilitating the study of dynamics of molecules that would not be possible otherwise. In this chapter, the computational methods to identify the biological targets and the study of mechanism of withanolides are discussed.

Keywords Withanolides • Anticancer • Computational methods • Targets • Mechanism

13.1 Introduction

Withanolides, naturally occurring steroidal alkaloids and lactones, are major chemical ingredient extracted from various parts of Ashwagandha. Pharmacological studies on withanolides have revealed their effectiveness in treatment of various kinds of cancers like breast cancer, prostate cancer, blood cancer, lung cancer, skin cancer, colon cancer, renal cancer and pancreatic cancer (Kumari 2013). Withaferin-A and withanone are two major and most studied components of Ashwagandha extract. Research on activity of withaferin-A and withanone show that these molecules have anti-migratory properties and cause downregulation of migration-promoting

S.P. Katiyar • V. Malik • A. Kumari • D. Sundar (✉)
DAILAB, Department of Biochemical Engineering & Biotechnology, DBT-AIST International Laboratory for Advanced Biomedicine (DAILAB), Indian Institute of Technology (IIT) Delhi, Hauz Khas, New Delhi 110016, India
e-mail: sundar@dbeb.iitd.ac.in

S.C. Kaul, R. Wadhwa (eds.), *Science of Ashwagandha: Preventive and Therapeutic Potentials*, DOI 10.1007/978-3-319-59192-6_13

protein, (heterogeneous nuclear ribonucleoprotein K (hnRNP-K) (Gao et al. 2014). Moreover, study of the effect of these two withanolides in different cancer cell types shows selective killing of cancer cells and less toxicity towards normal cells (Grover et al. 2010, 2012; Saxena et al. 2013; Gao et al. 2014). This differential response is instigated due to differential binding efficiencies with target molecules in cancer and normal cells (Vaishnavi et al. 2012). The reasons for differential binding and selective killing of cancer cells are not completely understood and implementation of computational techniques provide better understanding of interactions of withanolides with their biological targets at molecular level. In this chapter, the computational methods and techniques are discussed that can be employed to select the potential molecular targets related to diseases and can provide insights into the action mechanisms of withanolides.

13.1.1 Importance of Understanding the Molecular Interactions Between Withanolides and the Biological Targets

Despite major advances in experimental techniques, suggesting the role of withanolides against cancer and other diseases, the nature of their interactions with biological targets at molecular level remains unclear. It has been seen that a minor change in the chemical structure of a withanolide can completely alter its functionality, for example an addition of β-methoxy group at C3 position of withaferin-A diminishes its activity (Huang et al. 2015). *In silico* methods can reveal the molecular action mechanism of withanolides and precisely predict the reasons behind the structure dependent differences in activity. Such a study will provide insights regarding the descriptors of withanolides. Descriptors of the withanolides can be utilized for various purposes such as quantification of their activity relationship with their structures, searching of other molecules with similar descriptors, modification of withanolides to improve the activity and to make them more selective. Additionally, *in silico* methods not only does help to understand the molecular mechanism of action, but also save much of the resources, money and time by predicting possible targets of withanolides before proceeding with wet laboratory experiments. Hence, it becomes very important to study the interactions of withanolides with their biological targets.

The next key issue to know about is how *in silico* methods could be applied to study the action mechanism of withanolides. Understanding the action mechanism of a small molecule requires two important things; first is the structural information of the target biological molecule or at least of the surroundings of the active site and second is the possible binding conformations of the small molecules. Structure of a biological molecule such as protein can either be obtained experimentally or modeled computationally. Conformations of withanolides can be generated using *in silico* methods depending on the chemical nature of the surrounding environment.

Once, the information about the structures and conformations of target and small molecules are known, *in silico* methods such as molecular docking and molecular dynamics simulations (MDS) are used to predict the bound complexes between them. The following sections describe the details of identification of biological targets and structure of withanolides for studying their interactions.

13.2 Identification of Biological Molecular Targets

Discovery of a new drug molecule and approval for its release in the market takes almost 15 years with the investment of up to 800 million dollars (Dai and Zhao 2015). Though activities of natural products, such as withanolides are mostly known, their specific targets are usually not known. Also, many times, instead of identifying a new molecule from scratch, starting with the derivatives of natural products is a better and more efficient way. In any case, the first step of identification of potential target molecule related to a disease is the most important and critical step in the field of drug discovery. Before the release of human genome sequence, target molecules were selected on the basis of experimental findings by research groups for a particular disease (Wang et al. 2004). As a result, most of the marketed drug molecules target small number of protein classes, such as, G-protein coupled receptors, ion channels, kinases, proteases and nuclear receptors (Wang et al. 2004). With the release of complete human genome sequence and advent of techniques for quantifying gene expression level, large pool of potential target molecules has become available (Hughes et al. 2011). Comparison between the expression level information (mRNA or Protein) in diseased and normal state and its effect on the biological pathways have a tendency to accelerate diverse range of potential target identification process for drug development (Tamames et al. 2002). Some of the computational techniques that can be employed for selection of target molecules of withanolides are discussed here.

13.2.1 Literature Survey

Traditional approach to identify a biological target molecule for drug discovery was to explore reported experimental findings related to a particular disease. Dysregulated signaling pathways in diseased state were usually targeted by selection of critical protein involved in that pathway. As in the case of cancer, proteins involved in pathways like DNA repair, apoptosis, cell cycle, stress response and proteasomal degradation can be targeted for drug discovery. Most commonly targeted proteins belong to the classes of receptors involved in signaling pathways, kinases, proteases, etc. (Wang et al. 2004). Experimental results indicating difference in expression, localization and phosphorylation state of proteins in normal and diseased state are explored for target identification. For example, mortalin, a heat shock protein, is

Table 13.1 Popular microarray data and next generation sequencing (NGS) read databases

Data type	Database	Availability
Gene expression Microarray data	Array Express	http://www.ebi.ac.uk/arrayexpress/
Gene expression Microarray data	Gene expression omnibus (GEO)	http://www.ncbi.nlm.nih.gov.geo/
Gene expression Microarray data	GeneX	http://www.ncgr.org/genex
Gene expression Microarray data	Stanford Microarray data (SMD)	http://genomewww5.stanford.edu/microarray/smd/
Raw sequence data from NGS platforms	Sequence read archive (SRA)	http://www.ncbi.nlm.nih.gov/sra

found to be overexpressed in cancer cells, which interacts with tumor suppressor protein p53 and leads to cytoplasmic retention and functional inactivation of p53 (Kaul et al. 2005). Therefore, functional activation of p53 by abrogating mortalin-p53 interaction makes mortalin-p53 complex a biological target for drug development. Withaferin-A has been reported to abrupt the interactions between mortalin and p53 protein (Grover et al. 2012).

13.2.2 Expression Data and Biological Pathways Analysis

With the advent of techniques for measuring gene expression like microarray and RNASeq, it has become possible to check the expression of whole transcriptome of particular cell type at high speed and low cost. Comparison of gene expression between (a) diseased and normal cells and (b) drug-treated cell and untreated cell, has a potential to provide insight into the biology of disease and mechanism of action (MOAs) of drug molecule respectively (Dai and Zhao 2015). Wealth of expression data from microarray and RNASeq platforms for different diseased state under varying sets of conditions are available in public databases (some are listed in Table 13.1). Analysis and comparison of gene expression data of diseased and normal cells from these databases can give a list of genes showing contrary behavior and have a potential to provide better insight in dysregulated pathways in diseased state. Target molecule from this list can be selected by performing pathway analysis to identify the genes involved in dysregulated pathway (Young et al. 2014).

Different pipelines are reported in literature to carry out analysis of high throughput sequencing data. One of the NGS analysis pipeline developed by Trapnell et al. (2009), is described here - the raw sequencing reads are first preprocessed via trimming of adapter and low quality reads. Preprocessed reads are then aligned onto the human reference genome using *TopHat* (spliced read aligner) (Trapnell et al. 2009). *Spliced read aligners* work in the following two steps:

Step 1 In the first step, all reads are mapped to the genome using *Bowtie*, an unspliced read aligner. It maps all non-junction reads to the genome using *Burrows-Wheeler transform* method. This *Burrows-Wheeler* indexing approach makes it ultrafast and memory efficient program for alignment of short reads when exact reference genome is available (Langmead et al. 2009).

Step 2 In the second step, unmapped reads are split into smaller segments that are then independently aligned to reference genome. These mapped reads are then extended to find possible splice sites to determine spliced alignment of initially unmapped read (Trapnell et al. 2009).

Above steps help to generate alignment files that contain the information on mapped reads. Mapping of reads is followed by transcriptome reconstruction by using *Cufflinks* that works by generating overlap graph and subsequent parsing of the graph into transcripts by covering minimum path. The set of transcripts obtained are merged based on reference transcripts available and the redundant transcripts are subsequently removed by using *Cuffmerge*. Differential expression of these transcripts in different samples can be checked by using *Cuffdiff*, part of *cufflink suite* (Trapnell et al. 2012). The genes showing differential response can be linked to the dysregulated pathways, which can be targeted by drug molecules (Young et al. 2014). As an example, we describe here this approach that was used to compare the post-irradiation transcriptome of radiation sensitive (LNCaP) and radiation resistant (PC-3) cells (Young et al. 2014). 399 genes were found to be differentially expressed in radio-resistant PC-3 cell lines at 6 hour while only 89 genes were showing differential regulation in LNCaP cells. Ingenuity Pathway Analysis (IPA) was performed to link the differential response observed after radiotherapy to dysregulated pathways. IPA analysis revealed two pathways- DNA repair and DNA replication pathway, showing contrary behavior in radiosensitive and radio-resistant cell line after 24 h. Analysis of both pathways provide list of genes showing contrary behavior in both cell lines. BRCA1, RAD51 and FANCG genes of DNA repair pathways were differentially regulated in PC-3 and LNCaP cell lines. Three genes of DNA replication pathway namely, ORC1, CDC6 and MCM showed similar behavior. Hence, comparison of gene expression data between different samples followed by pathway analysis can provide a small list of genes that may serve as potential target molecules of withanolides.

Similarly microarray expression data analysis can provide a list of potential target molecules. General outline of steps involved in microarray and RNASeq data analysis is shown in Fig. 13.1. Main steps involved in microarray data analysis are

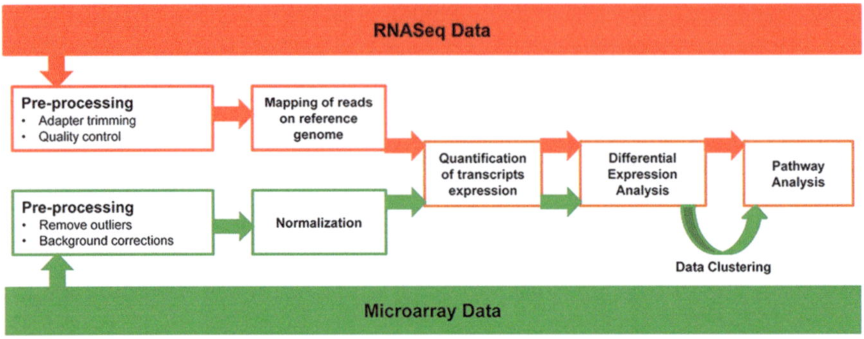

Fig. 13.1 Graphical depiction of the methodology for transcriptome analysis of particular cell line

filtering, normalization, differential expression analysis, clustering, and functional analysis. Various open source software packages including R based *Bioconductor* packages are available for complete analysis of microarray data (Saeed et al. 2003; Tong et al. 2003; Smyth 2005; Subramanian et al. 2005; Kallio et al. 2011;). Some of the R based packages were used in a study to check dysregulated pathways in bladder carcinoma (Xiao and Yiqing 2013). Gene expression profile for bladder carcinoma and normal cells was obtained from GEO database to examine differential gene expression between two datasets. Initially data was normalized using RMA (Robust multichip averaging) package, followed by differential expression analysis using Limma package. 6562 genes were found to be differentially expressed in bladder carcinoma cells. These genes were clustered based on Gene Ontology (GO) using DAVID (Database for Annotation, Visualization and Integrated Discovery). Differentially expressed genes were linked to pathways to identify dysregulated pathways in bladder carcinoma cells via KEGG pathway enrichment analysis. Eight dysregulated pathways were identified using this approach. Analysis of role of differentially expressed genes in these eight dysregulated pathways can provide a clue for potential target molecule. In this way, expression data analysis can provide a list of target molecules of withanolides. Once target molecules are identified, interaction of withanolides with these molecules can be studied using molecular docking and molecular dynamics simulation.

13.3 Molecular Mechanisms of Withanolides

The previous section described the methods for identifying targets of withanolides. Once, a biological target has been identified, there is a necessity to identify its three dimensional structure. A three dimensional structure of a biological molecule such as a protein provides information about the shape and chemical nature of binding pockets. A binding pocket of a protein can either be an active site or an allosteric site and a probable target site for withanolides. Binding of withanolides with a protein at a binding pocket can bring significant conformational changes and alter the functionality of that protein. Alteration in the functionality includes decrease, increase or loss of the functions, which depends on the type of conformational changes brought by withanolides. Withanolides can either bind at the active sites and compete with natural substrate, acting as a competitive inhibitor or bind at an allosteric site. Furthermore, withanolides may also bind at such a binding pocket that is not part of any active site but rather is in contact with other biological molecule such as DNA. In this section, few case studies are discussed as an example, to describe the methodology followed, to reveal the mechanism and mode of interactions of withanolides.

13.3.1 Identification of Molecular Structure of Identified Targets

The most critical step in the field of drug discovery is identification of target molecule. Once target molecule is selected, knowledge of its structure is required to study its interaction with withanolides. Moreover, structure of protein provides knowledge about the environment, surface area and shape of its active site and other binding pockets. Information about the residues present in binding pocket is necessary to predict the binding pose and energy of withanolides. The structure of the target molecule may be obtained from databases if experimentally determined structure is available. Otherwise, comparative modeling based on an experimentally determined structure of a related protein, can provide three-dimensional structure of target molecule.

13.3.1.1 Experimentally-Solved Molecular Structure

With the advent of biophysical techniques like X-ray crystallography and NMR spectroscopy, large numbers of protein and protein-ligand complex structures are now available in databases. Protein data bank (PDB: http://www.rcsb.org/pdb) is the most widely used biomolecule structures database that currently houses around 1,10,777 experimentally determined protein structures. The best way to start a docking experiment is to retrieve target molecule structure from this database. If target protein is available in complex with ligand, then this will be the ideal structure for carrying out interaction studies with withanolides.

13.3.1.2 Computational Prediction of Molecular Structure

Computational methods can be used to predict three-dimensional structure of protein if experimentally determined structure is not available. Homology modeling is a technique used to predict tertiary structure of protein based on the fact that proteins with similar sequences have similar structures and the protein structure is evolutionarily more conserved than its sequence. Therefore, in homology modeling, protein structure is predicted by using experimentally determined structure of homologous protein as a template. Determining the complete structure of protein computationally is a multistep process that involves template identification, multiple sequence alignment, backbone generation, loop modeling, side-chain modeling, model optimization and verification. For example, in a study carried out to inhibit formation of mortalin-p53 complex using withanone as inhibitor, mortalin structure was computationally modeled using multiple template comparative homology modeling (Grover et al. 2012). Six X-ray crystal structures were obtained from PDB showing significant similarity to human mortalin (PDB Ids: 1BA0, 2E88, 3FE1, 3GL1, 2KHO, 2V7Y). MODELLER was used to predict the structure of mortalin using these structures as templates. Five modeled structures were reported, out of which the structure with the least DOPE score was selected for the study (Grover et al. 2012).

13.3.2 Identification of Biologically Active Conformations of Withanolides

Structure of withanolides can be obtained either from databases like PubChem (http://pubchem.ncbi.nlm.nih.gov/) or can be drawn manually using software like Marvin sketch. If the structure of withanolide is retrieved from databases, then it is important to first pre-process it to get the clean structure with correct molecular mechanics-parameters and atom types. Various software are available for ligand preparation and optimization. Most software follow several steps during the ligand optimization; a) ligand structures are deposited in databases in the form of salt, so at first instance, all these counter-ions are removed from the parent molecule. b) After desalting, ligand is protonated, charged groups are neutralized followed by ionized and finally tautomeric states of ligand are generated. Ligand optimization program also generates stereoisomers of the ligand. These steps will generate many structures depending on the number of chiral carbon present in the structure. All generated structures are minimized to relax the structures in three-dimensional space. It is recommended to check the formal charge and bond order of nitrogen manually before proceeding to docking studies (Chen and Foloppe 2010). These steps generate active conformations of withanolides that are suitable for carrying out interaction studies with target molecules.

13.3.3 Interactions Between Biological Targets and Withanolides

Once, the structural information of target molecule and conformational information of withanolides are known, computational methods such as molecular docking and molecular simulations are used to explore the interactions between withanolides and biological targets that reveal their mechanism of action. Along with the structural information, spatial information about the site of action such as active site, allosteric site or binding interface will also be required. The identification of such sites requires some knowledge about the function of biological molecule and the relation of that function with its structure. For example, tumor suppressor protein p53 is a multifunction transcription factor, which not only interacts with DNA but also interacts with many other cellular proteins. Interactions of p53 with other biological molecules significantly change in cancer cells because of various mutations on p53 surfaces. Mortalin is another multifunction protein that interacts with mutant form of p53 and inhibits apoptosis of cancer cells. This functional information is related to structural information of interactions between mutant p53 and mortalin, which is the "latch region" of mortalin protein across residues 439 to 597 (Vaishnavi et al. 2012). In spite of targeting multifunctional proteins, specific function-dependent structural information was required to target mortalin-mutant p53 protein complex. In another study, alcoholic extract of Ashwagandha leaves was found to down-regulate migration-promoting proteins hnRNP-K, VEGF, and metalloproteases while the exact components and biological targets were unclear from the experiments. hnRNP-K is a RNA binding protein and computational analysis revealed that two major components of the extract, withanone and withaferin-A, inhibit function of hnRNP-K by blocking its ssDNA/RNA binding site (Gao et al. 2014).

Experimental information about the binding site of withanolides may not always be available. In the absence of experimental information, computational methods such as protein cavity analysis and protein-protein binding interface analysis are used to predict the binding site of withanolides (Saxena et al. 2013). Also, there are cases when targeting active site of a protein may produce side effects such as targeting ATP binding site of kinase enzymes. In such cases, identifying an alternate site of action that is specific for a particular kinase becomes important. Withanone is reported to target Aurora-A kinase, a Ser/Thr kinase. The function of Aurora-A kinase is regulated by its interactions with another protein, TPX2. Withanone aims for Aurora A-TPX2 interaction site, leaving kinase active site untouched (Grover et al. 2012).

13.3.3.1 Study of Static Conformations Between Withanolides and the Target Molecules

Information about the possible binding conformations of withanolides, structure of target molecule and binding site is collectively used to calculate the structure of withanolide-bound biological molecule using various available docking algorithms such as genetic algorithms. A docking algorithm depends on a search algorithm to predict a binding pose of withanolide at the binding site and a scoring function to score the binding pose (Halperin et al. 2002). Various software packages are available to automate the docking processes; AutoDock is the most widely used open access docking software (Goodsell et al. 1996; Cecchini et al. 2004).

In a study, withaferin-A and withanone were reported to target the KH3 domain of hnRNP-K and inhibit its ssDNA/RNA binding function. Here, AutoDock was used to dock both withaferin-A and withanone at the KH3 domain of hnRNP-K protein, which is the site of ssDNA/RNA binding. ssDNA binds with the Lys22, Asp23, Ala25, Ile29, Lys31, Arg40, Lys47, Ile48, Asp50, Tyr84, and Ser85 of KH3 domain of hnRNP-K protein (Fig. 13.2a); this binding site information was extracted using an experimentally-solved hnRNP-K and ssDNA complex structure (PDBID: 1ZZI). Docking of withaferin-A and withanone specifically at ssDNA/RNA-binding site generated their bound conformations with hnRNP-K, scoring binding energies of -8.9 kcal/mol and -9.19 kcal/mol respectively. Both withaferin-A and withanone were in contact with residues Gly26, Ser27, Gly30, Lys31, Gln34, Gln83, Ser85, and Lys87 of hnRNP-K via hydrogen bonds (H-bonds) and hydrophobic interactions (Fig. 13.2b, c). Docking scores and binding poses alone might be misleading if one does not check their biological significance. The residues interacting with withaferin-A and withanone were all ssDNA binding site residues; also, biological significance of binding of withaferin-A and withanone was assessed by the fact that the second nucleotide Thymine (DT) of ssDNA also interacts with residues Tyr84 and Ser85. Binding of withaferin-A and withanone with Tyr84 and Ser85 residues can hinder binding of hnRNP-K with ssDNA/RNA completely or partially because of the overlap with the binding of first few nucleotides that play key role in its transcription activation/deactivation function (Fig. 13.2d, e).

13.3.3.2 Study of Dynamics Between Withanolides and the Target Molecules

Experimental techniques such as X-ray crystallography, nuclear magnetic resonance (NMR) and electron microscopy allow determination of the molecular structures. These solved structures are snapshots of the most stable spatial conformation in a particular environment. Similarly, molecular complexes obtained from molecular modeling and docking studies are the most probable snapshots. Though these static snapshots provide quite detailed spatial description of molecular structures and binding between two molecules, it does not explain any dynamic behavior of the molecules and their interacting molecules in temporal context. Studying the

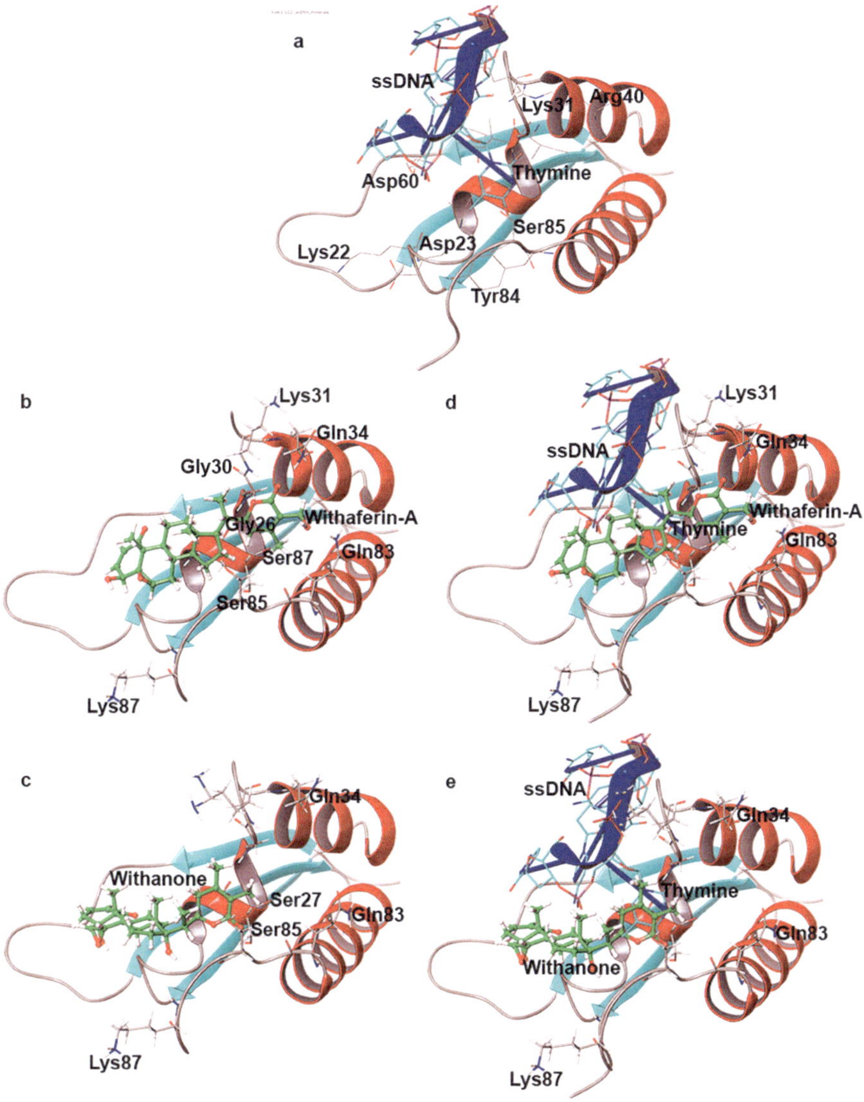

Fig. 13.2 Interactions of withaferin-A and withanone at the ssDNA binding site of KH3-hnRNP-K protein. (**a**) Interactions of ssDNA with KH3 domain of hnRNP-K protein. Thymine binds with KH3 domain where residue Ser85 surrounds it. (**b**) Withaferin-A binding at ssDNA binding site of KH3 domain. (**c**) Binding of withanone at the KH3 domain that occupies the same space where ssDNA binds. (**d**) Superimposition of ssDNA and withaferin-A at KH3 domain. (**e**) Superimposition of ssDNA and withanone at the KH3 domain. Superimposition of ssDNA with both withaferin-A and withanone clearly reveals that binding of any of the ligands hinders the binding of thymine nucleotide of ssDNA

dynamics of molecules is very important for many reasons, in fact more important than studying just a static snapshot. Computer simulations are used to study the dynamics of molecules and loosely known as molecular dynamics simulations (MDS). Computer simulations make possible testing and improving theoretical models using a more realistic representation of nature and gives freedom to cross the barriers to test theories that would have been impossible otherwise. MDS also offer new insights into mechanisms and processes, which are not directly accessible through experiments (Karplus and McCammon 2002).

After withanolides are found to interact with a protein molecule by molecular docking studies, next task is to study the stability of binding pose and biologically significant interactions. Earlier in this chapter, binding of withaferin-A and withanone with KH3 domain of hnRNP-K was discussed. Stability of withaferin-A within the ssDNA binding site was reported from a study of 60ns long MDS. During the MDS, binding small molecules are expected to retain important interactions with the protein residues in order to be considered a good lead molecule. If molecular docking studies are indicating the interactions of a small molecule with protein at the critical site, but during the MDS study if molecule fails to retain its position and interaction, that molecule might not be usable for further studies. Withaferin-A showed a highly stable behavior at the KH3 domain of hnRNP-K during 60ns MDS. The stability of withaferin-A can be assessed on the basis of its movements within the pocket and root mean square deviations (RMSD) measurement. Withaferin-A had very low RMSD during whole 60ns simulation and retained all of its important interactions with KH3 domain of hnRNP-K (Fig. 13.3). Withanone was a more complex case to analyze. Molecular docking studies actually presented two close binding poses of withanone: one with -9.1 kcal/mol and other with -8.92 kcal/mol. Both the poses were at the ssDNA-binding site, thereby making it impossible to select one without further information. MDS studies showed that

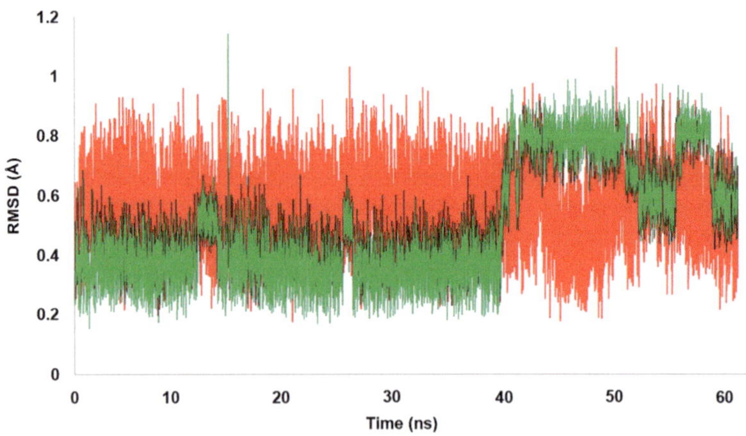

Fig. 13.3 RMSD deviations of withaferin-A (*red*) and withanone (*green*) during 60 ns molecular dynamics simulations. RMSD of both the trajectories suggest highly stable behavior of withaferin-A and withanone at the ssDNA binding site of hnRNP-K protein

-9.1 kcal/mol scoring conformation of withanone retained more of the critical residues during the simulation studies, including those that were involved with the binding with second thymine of ssDNA. RMSD of withanone, bound to KH3 domain suggested the highly stable state at the ssDNA-binding site (Fig. 13.3). Hence, the conformation of withanone with higher score was proposed as the active conformation. These molecular docking and dynamics study have stabilized the hnRNPK-RNA binding inhibiting role of withaferin-A and withanone and have complemented the experimental findings that alcoholic extracts of Ashwagandha down regulates the migration promoting proteins.

13.4 Experimental Validation of the Computational Results

Experimental information and *in silico* results complement each other and make better contributions together. *In silico* studies often requires initial inputs from experimental results for target identification, target structure and sequence information, function and structure correlation of target or active and inactive conformations of a molecule. Not only, experimental information is used to accurately generate models and predictions but also often *in silico* predictions require experimental validations because of high possibilities of false positive and negative results introduced at various levels (Fig. 13.4). Inaccuracies might arise because of human errors, machine inaccuracies, and flaws in models or algorithms. Hence, experimental validation of *in silico* results is a critical part to detect any bias and imprecisions.

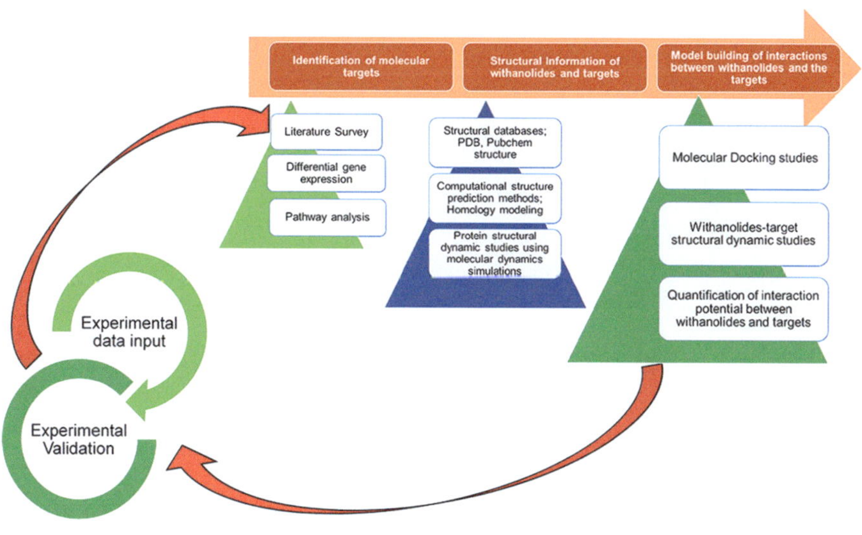

Fig. 13.4 Depiction of the relationship between *in silico* studies and experimental techniques and their dependence on each other

Study of mechanism of withanolides may directly not be possible using the existing experimental techniques or may prove costly, but certainly is required to assist *in silico* studies. Considering the huge number of possible biological targets, searching a target blindly is not considered a good approach, even computationally. Hence, initial identification of possible targets from different experimental studies and expression analysis narrows down the possible targets. In the previous sections, where withanolides were identified to target specifically the ssDNA binding domains of hnRNP-K protein using computational methods, experimental results gave the leads by suggesting hnRNP-K protein. Once, the mechanism of inhibition was known using computational methods, anti-metastatic activity of the combination of withanone:withaferin-A (20:1) was confirmed using *in vivo* studies.

13.5 Challenges and the Future Scope

Computational methods are the most effective approach to study the dynamics of biological molecules and their interactions with other biological molecules. The previous section discussed how computer simulations were used to study the dynamics of withanolides with their biological target molecules. Not only the dynamics were studied, computational methods were also used to predict the binding affinity and the binding poses of withanolides against the biological targets. All the computational studies were carried out either based on the available experimental data or that were required for experimental validations. For most cases, *in silico* results were in well accordance with the experimental results but there have been few instances where the contrary was observed. Withanolides were observed to bind with protein target *in silico* but experimental results did not support the *in silico* findings. For example, withanoside IV and withanoside V were found to bind with the vascular endothelial growth factor receptor -2 (VEGFR-2) at the vascular endothelial growth factor (VEGF) binding site during an *in silico* study. These *in silico* findings suggest that withanoside IV and withanoside V should have interfered with the activation process of VEGFR2, eventually inhibiting the VEGF pathway. MTT assay reflect the number of viable cells by assessing the cell metabolic activity. If withanoside IV and withanoside V were inhibiting the VEGF pathway in cancer cells, the reduction in viable cell numbers was an expected result. But when HELA cells were treated with withanoside IV and withanoside V, the viability of cells was not affected, even at very high concentrations (data not shown).

Membrane permeability and metabolism of withanolides are two other factors that are critical for corroboration of *in silico* and experimental findings. Not all withanolides are equally permeable and may have different metabolic fate. Studying the permeability and metabolism of any small molecule is highly important, but at the same time a very challenging task. In the following section, we discuss the importance and the challenges in studying membrane permeability and metabolism of withanolides.

13.5.1 Permeability of Withanolides

For a molecule to be like a drug, it needs to have good adsorption, distribution, metabolism and excretion (ADME) properties. Permeability assessment is a crucial component in selecting and optimizing the lead compounds with good ADME properties. A common and quick method is to comply with rules-of-thumbs for drug-likeness such as Lipinski's Rules of five (Lipinski et al. 2001). Rule of five suggests that a good drug like compound should have number of hydrogen bond donors between zero to five, less than 10 hydrogen bond acceptors, molecular mass less than 500 Daltons and octane-water partition coefficient *log P* lower than 5. Collectively, ADME properties of a molecule come under pharmacokinetic properties of drug in human body. Most of drug molecules fall under Rules of five, however, natural products have been proved to be exception from few of the rules (Keller et al. 2006).

Current predictive models for permeability assessment are quantitative structure permeability relationship (QSPR) methods. The QSPR approach has widely been applied to predict permeability of various biological membranes such as oral bioavailability (Yoshida and Topliss 2000), intestinal absorption (Suenderhauf et al. 2011), skin permeability (Potts and Guy 1995) and brain permeability (Liu and So 2001). QSPR models are based on descriptors such as molecular weight, polar surface area, partition coefficients and number of hydrogen bonds and their correlation with known rate of permeation (Leung et al. 2012). Because QSPR models depend on available training data set, the main limitation of such an approach is transferability. The accuracy of such models depends highly on the chemical structural similarities between target and training compounds.

Withanolides are natural products and do not necessarily comply with the Rules of five nor can any other QSPR model be applied to predict the permeability of withanolides. One needs to generate QSPR models for withanolides using experimental permeability data of withanolides but number of withanolides is not sufficient enough to generate such models. Alternatively, methods based on direct physical interactions can be used to study the permeability of withanolides. The trans-membrane passive diffusion is a major factor in intestinal absorption, crossing blood brain-barrier and through plasma membrane of a cell. All atomic MDS can be used to study the interaction of withanolides with bi-layered lipid membrane and water solvent environment to gain knowledge of energetics and thermodynamics of passive membrane permeability. However, the main challenge using MDS to study permeability is requirement of high computational power.

Considering the limited number of withanolides, experimental assessment of their membrane permeability followed by QSPR model generation should be the way to study the passive membrane permeability of withanolides. Generation of QSPR model becomes more important when one wants to modify and improve the structure of withanolides. Because the chemical skeleton and space of modified withanolide derivatives will be similar to that of the original withanolide, QSPR models generated using withanolides can safely be applied to their derivatives.

13.5.2 Metabolism of Withanolides

Metabolism of a drug inside body is another factor involved in identification of drug like molecules. The metabolic system is one of the main defense systems against the foreign harmful molecules by transforming them into excretory metabolites. Natural products have been evolved along with humans and tend to escape metabolic excretion. Hence, natural products have inherent advantages over the synthetic molecules and they do not necessarily have to be like a drug like molecule. Though natural products may not always be drug-like, understanding their metabolic fate is important to gain the knowledge of active molecules and their mechanisms (Ntie-Kang et al. 2013). Also, modifications of the natural products to make their derivatives need methods to assess the metabolic fate of derived products (Kirchmair et al. 2015).

Next challenge in learning about the withanolides is to know their metabolic fate, by both experimentally or computationally studying their interactions with enzymes of metabolic systems such as cytochrome P450, dehydrogenases, hydrolases, peroxidases, glutathione S-transferases and many more (Testa et al. 2012). There are abundant experimental methods to explore the metabolic fate of withanolides but are limited by the demand for expertise, cost and time. Meanwhile, newer computational methods and tools are being developed to predict drug metabolism with the advantage of higher throughput, lower cost, less time consumption over the experimental methods. Cytochrome P450 is the most widely studied system using both experimental and *in silico* approaches (Potts and Guy 1995; Zhang et al. 2011; Andrade et al. 2014). However, metabolic fate of withanolides and their derivatives using computational methods is yet to be explored. Such a study will largely improve the understanding of action mechanism of withanolides.

References

Andrade CH, Silva DC, Braga RC (2014) In silico prediction of drug metabolism by P450. Curr Drug Metab 15:514–525

Cecchini M, Kolb P, Majeux N, Caflisch A (2004) Automated docking of highly flexible ligands by genetic algorithms: a critical assessment. J Comput Chem 25:412–422

Chen I-J, Foloppe N (2010) Drug-like bioactive structures and conformational coverage with the LigPrep/ConfGen suite: comparison to programs MOE and catalyst. J Chem Inf Model 50:822–839

Dai Y-F, Zhao X-M (2015) A Survey on the computational approaches to identify drug targets in the postgenomic era. BioMed Res Int 2015:239654

Gao R, Shah N, Lee J-S, Katiyar SP, Li L, Oh E, Sundar D, Yun C-O, Wadhwa R, Kaul SC (2014) Withanone-rich combination of Ashwagandha withanolides restricts metastasis and angiogenesis through hnRNP-K. Mol Cancer Ther 13:2930–2940

Goodsell DS, Morris GM, Olson AJ (1996) Automated docking of flexible ligands: applications of AutoDock. J Mol Recog JMR 9:1–5

Grover A, Priyandoko D, Gao R, Shandilya A, Widodo N, Bisaria VS, Kaul SC, Wadhwa R, Sundar D (2012) Withanone binds to mortalin and abrogates mortalin–p53 complex: Computational and experimental evidence. Int J Biochem Cell Biol 44:496–504

Grover A, Shandilya A, Punetha A, Bisaria VS, Sundar D (2010) Inhibition of the NEMO/IKKβ association complex formation, a novel mechanism associated with the NF-κB activation suppression by Withania somnifera's key metabolite withaferin A. BMC Genom 11(Suppl 4):S25

Halperin I, Ma B, Wolfson H, Nussinov R (2002) Principles of docking: An overview of search algorithms and a guide to scoring functions. Proteins 47:409–443

Huang C, Vaishnavi K, Kalra R, Zhang Z, Sekar K, Kaul SC, Wadhwa R (2015) 3beta-methoxy derivation of Withaferin-a Attenuates its Anticancer potency: Bioinformatics and molecular evidences. Med Aromat Plants 4:1–8

Hughes J, Rees S, Kalindjian S, Philpott K (2011) Principles of early drug discovery. Br J Pharmacol 162:1239–1249

Kallio MA, Tuimala JT, Hupponen T, Klemelä P, Gentile M, Scheinin I, Koski M, Käki J, Korpelainen EI (2011) Chipster: user-friendly analysis software for microarray and other high-throughput data. BMC Genom 12:1

Karplus M, McCammon JA (2002) Molecular dynamics simulations of biomolecules. Nat Struc Biol 9:646–652

Kaul SC, Aida S, Yaguchi T, Kaur K, Wadhwa R (2005) Activation of wild type p53 function by its mortalin-binding, cytoplasmically localizing carboxyl terminus peptides. J Biol Chem 280:39373–39379

Keller TH, Pichota A, Yin Z (2006) A practical view of 'druggability'. Curr Opin Chem Biol 10:357–361

Kirchmair J, Goller AH, Lang D, Kunze J, Testa B, Wilson ID, Glen RC, Schneider G (2015) Predicting drug metabolism: experiment and/or computation? Nat Rev Drug Discov 14:387–404

Kumari M (2013) Withania somnifera: A Wild Plant Effective in Treatment of Various Types of Cancer. J Chem Biol Phys Sci 4:379

Langmead B, Trapnell C, Pop M, Salzberg SL (2009) Ultrafast and memory-efficient alignment of short DNA sequences to the human genome. Genome Biol 10:R25

Leung SS, Mijalkovic J, Borrelli K, Jacobson MP (2012) Testing physical models of passive membrane permeation. J Chem Inf Model 52:1621–1636

Lipinski CA, Lombardo F, Dominy BW, Feeney PJ (2001) Experimental and computational approaches to estimate solubility and permeability in drug discovery and development settings. Adv Drug Deliv Rev 46:3–26

Liu R, So SS (2001) Development of quantitative structure-property relationship models for early ADME evaluation in drug discovery. 1. Aqueous solubility. J Chem Inf Comput Sci 41:1633–1639

Ntie-Kang F, Lifongo LL, Mbah JA, Owono Owono LC, Megnassan E, Mbaze LM, Judson PN, Sippl W, Efange SM (2013) In silico drug metabolism and pharmacokinetic profiles of natural products from medicinal plants in the Congo basin. In Silico Pharmacol 1:12

Potts RO, Guy RH (1995) A predictive algorithm for skin permeability: the effects of molecular size and hydrogen bond activity. Pharm Res 12:1628–1633

Saeed A, Sharov V, White J, Li J, Liang W, Bhagabati N, Braisted J, Klapa M, Currier T, Thiagarajan M (2003) TM4: a free, open-source system for microarray data management and analysis. Biotechniques 34:374

Saxena N, Katiyar SP, Liu Y, Grover A, Gao R, Sundar D, Kaul SC, Wadhwa R (2013) Molecular interactions of Bcl-2 and Bcl-xL with mortalin: identification and functional characterization. Biosci Rep 33:797–806

Smyth GK (2005) Limma: linear models for microarray data. In: Bioinformatics and computational biology solutions using R and bioconductor. Springer, New York, pp 397–420

Subramanian A, Tamayo P, Mootha VK, Mukherjee S, Ebert BL, Gillette MA, Paulovich A, Pomeroy SL, Golub TR, Lander ES (2005) Gene set enrichment analysis: a knowledge-based approach for interpreting genome-wide expression profiles. Proc Natl Acad Sci U S A 102:15545–15550

Suenderhauf C, Hammann F, Maunz A, Helma C, Huwyler J (2011) Combinatorial QSAR modeling of human intestinal absorption. Mol Pharm 8:213–224

Tamames J, Clark D, Herrero J, Jn D, Blaschke C, Fernández JM, Oliveros JC, Valencia A (2002) Bioinformatics methods for the analysis of expression arrays: data clustering and information extraction. J Biotechnol 98:269–283

Testa B, Pedretti A, Vistoli G (2012) Reactions and enzymes in the metabolism of drugs and other xenobiotics. Drug Discov Today 17:549–560

Tong W, Cao X, Harris S, Sun H, Fang H, Fuscoe J, Harris A, Hong H, Xie Q, Perkins R (2003) ArrayTrack—supporting toxicogenomic research at the US Food and Drug Administration National Center for Toxicological Research. Environ Health Perspect 111:1819

Trapnell C, Pachter L, Salzberg SL (2009) TopHat: discovering splice junctions with RNA-Seq. Bioinformatics 25:1105–1111

Trapnell C, Roberts A, Goff L, Pertea G, Kim D, Kelley DR, Pimentel H, Salzberg SL, Rinn JL, Pachter L (2012) Differential gene and transcript expression analysis of RNA-seq experiments with TopHat and Cufflinks. Nat Protoc 7:562–578

Vaishnavi K, Saxena N, Shah N, Singh R, Manjunath K, Uthayakumar M, Kanaujia SP, Kaul SC, Sekar K, Wadhwa R (2012) Differential activities of the two closely related withanolides, Withaferin A and Withanone: bioinformatics and experimental evidences. PLoS One 7:e44419

Wang S, Sim TB, Kim Y-S, Chang Y-T (2004) Tools for target identification and validation. Curr Opin Chem Biol 8:371–377

Xiao J, Yiqing Q (2013) Bioinformatics analysis of the gene expression profile in Bladder carcinoma. Genet Mol Biol 36:287–291

Yoshida F, Topliss JG (2000) QSAR model for drug human oral bioavailability. J Med Chem 43:2575–2585

Young A, Berry R, Holloway AF, Blackburn NB, Dickinson JL, Skala M, Phillips JL, Brettingham-Moore KH (2014) RNA-seq profiling of a radiation resistant and radiation sensitive prostate cancer cell line highlights opposing regulation of DNA repair and targets for radiosensitization. BMC cancer 14:808

Zhang T, Chen Q, Li L, Liu LA, Wei DQ (2011) In silico prediction of cytochrome P450-mediated drug metabolism. Comb Chem High Throughput Screen 14:388–395

Part III
Ashwagandha for Brain

Chapter 14
Ashwagandha for Brain Health: Experimental Evidence for Its Neuroregenerative Activities

Renu Wadhwa, Rajkumar S. Kalra, Anupama Chaudhary, and Sunil C. Kaul

Abstract Rapidly growing population, social and environmental factors are continuously exerting impact on social and psychological human health in multiple ways. In particular, these factors cause cumulative stress affecting cognitive functions and elevate risks of diverse neuronal dysfunction and malignant brain diseases. Maintenance of brain health has hence emerged as a new challenge in increasing old age populations worldwide. Use of herbal drug constituents has emerged as a preferred choice due to enormous undesirable side effects of pharmacological drugs. Ashwagandha, a sub-tropical medicinal plant has been extensively used in Indian traditional home medicine system i.e. called "Ayurveda" for its stress-, hypertension-, aging-, neurological dysfunction- and malignant growth-inhibitory properties. Detailed molecular insights of these activities and mechanism(s) are only beginning to be demonstrated by laboratory studies. In this chapter, we attempt to sketch research evidence of the benefits of Ashwagandha bioactives for brain health.

Keywords Ashwagandha • Withanolides • Neurodegenerative disease • Neuroprotection • Brain health

14.1 Introduction

Human life has witnessed a major transition in its every aspect over the last century. Advancement of healthcare, quality of life and living standards improved significantly. Increasing population with diversely complicated social organization led to a change in social and psychological health. Industrialization of life style with excessive use of chemicals has led to a growing sense of stress and anxiety. Together, these scenarios have posed new challenges to maintain brain health. Risk of such factors

R. Wadhwa • R.S. Kalra • A. Chaudhary • S.C. Kaul (✉)
DBT-AIST International Laboratory for Advanced Biomedicine (DAILAB), Biomedical Research Institute, National Institute of Advanced Industrial Science & Technology (AIST), Central 5-41, 1-1-1 Higashi, Tsukuba, Ibaraki 305-8565, Japan
e-mail: s-kaul@aist.go.jp

S.C. Kaul, R. Wadhwa (eds.), *Science of Ashwagandha: Preventive and Therapeutic Potentials*, DOI 10.1007/978-3-319-59192-6_14

exceeds cumulatively at old age and is associated with high incidence of neurodegenerative diseases, functional abnormalities, disorders, psychological conditions and malignancies. Complex etiology of brain pathologies, their progression and consequent symptomatic dysfunctions are marked by sequential molecular alterations that have not been fully characterized as yet. Therefore, protective/early stage preventive or therapeutic approaches are considered as an alternative strategy and hope. Traditional home medicine systems describe several natural reagents that possess health-supplementing activities. Although these are trusted and have been used through centuries, the science of their effects has not been sufficiently demonstrated.

14.2 Ashwagandha and Its Active Constituents

Ayurveda (Ayur = life, Veda = science or knowledge) is a traditional Indian home medicine system practiced since ancient times. Carried through several generations, it continues to be a living tradition with a history of about 5000 years. It is largely based on the use of herbs to provide a unique balance between physical and psychological fitness (Singh et al. 2011). Ashwagandha (*Withania somnifera*), also called "Queen of Ayurveda", "Winter Cherry" or "Indian ginseng" is a sub-tropical herb, indigenous of Indian sub-continent, and is highly regarded in Ayurveda. Ashwagandha has been recognized for its ability to boost neuronal function, reduce stress and anxiety to promote cerebral calmness. It is a key component of rasāyana (a rejuvenative herbal tonic) that is used to combat stress, anxiety and depression, and hence promoting vitality, longevity and several aspects of brain and immune functions that are supported by laboratory studies (Kulkarni and Dhir 2008; Singh et al. 2008, 2011; Ven Murthy et al. 2010; Ahmad et al. 2005; Baitharu et al. 2013). However, the bioactive components and molecular mechanism(s) of such beneficial effects are only beginning to be resolved. Several studies have referred a reasonable dose of Ashwagandha herb, as safe, non-toxic, and comestible. Based on extensive toxicological studies, it has been categorized in GRASE (Generally Recognized As Safe and Effective) family of herbs. Its major bioactive constituents include steroidal alkaloids and ester lactones, a category of constituents collectively known as withanolides (steroidal lactones comprising ergaostane skeleton). These forms with C9 side chain along with six-membered lactone ring on a C28 steroidal skeletal nucleus. Besides withanoloids, numerous sitoindosides also have been identified.

14.3 Ashwagandha as a Brain Tonic: Experimental Evidences

A large number of experimental studies on rodent models have shown that Ashwagandha extracts improve neurophysiological functions under chronic stress conditions including fatigue, ischemic stroke and dyskinesia (Bhattacharya et al.

2002; Singh et al. 2002; Chaudhary et al. 2003; Naidu et al. 2003). These effects have been attributed to protect against oxidative stress (a common causative factor for these disorders) (Jain et al. 2001; Ahmad et al. 2010; Parihar et al. 2004; Shukla et al. 2011a) and often lead to physiological and behavioral rejuvenation (Kumar et al. 2005; Naidu et al. 2006; Sankar et al. 2007; Bhattacharya et al. 2001) (Fig. 14.1). Cell culture studies have exhibited that such effect is modulated by NRF2, a key antioxidant protein (Reuland et al. 2013). Several experimental oxidative stress models in rodents including exposure to streptozotocin, maneb, paraquat, bromobenzene, nicotinamide streptozotocin and rotenone have endorsed antioxidant and neuroprotective activity of Ashwagandha extracts (Ahmed et al. 2013; Prakash et al. 2013; Vedi et al. 2014; Manjunath and Muralidhara 2013; Shukla et al. 2011a, 2012; Anwer et al. 2012). The effects have been assigned to active constituents including withanolide A (Soman et al. 2012) and glyco-withanolides (Bhattacharya and Muruganandam 2003). Ashwagandha extract was shown to be effective for chronic fatigue (Singh et al. 2002; Kaur et al. 2004) and protected from swimming-induced oxidative stress (Misra et al. 2009) in rodents. Similarly, we have observed that Ashwagandha extracts and withanone, at a lower non-toxic doses diminish oxidative stress and protect glial and neuronal cells (Shah et al. 2015). We

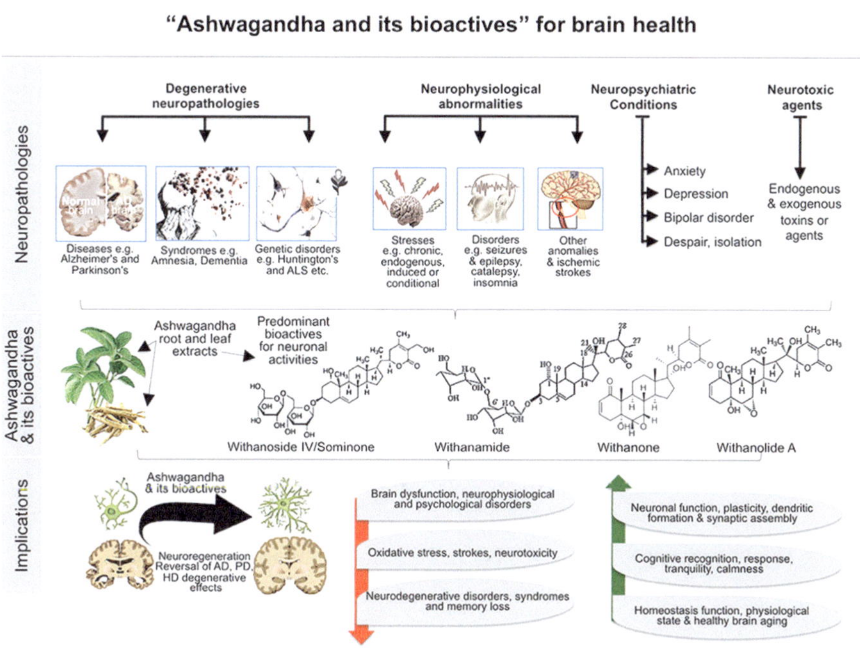

Fig. 14.1 Ashwagandha and its bioactives for brain health. Schematic outline of common neuropathologies involving degenerative, neurophysiological, neuropsychiatric or neurotoxicity related conditions and benefits of Ashwagandha and its bioactive compounds-based approaches towards effective treatment and their implications in maintaining brain function, oxidative stress and memory loss by augmenting neuronal function, cognitive response and reinstating homeostasis

found that an alcoholic extract of Ashwagandha leaves (i-Extract) reduced scopolamine-induced oxidative stress in the brain (Gautam et al. 2015; Konar et al. 2011). Treatment of water extract of Ashwagandha leaves (ASH-WEX) also reduced risk of glutamate-induced oxidative stress and excitotoxicity both in glial and neuronal cell culture-based assays (Kataria et al. 2012).

Several laboratory studies have shown positive effects of Ashwagandha extracts and active phytochemicals on cognitive functions of the brain (Fig. 14.2). Kuboyama and colleagues reported that withanolide A caused neuritic regeneration and synaptic reconstruction in critically damaged neurons. In cultured cortical neurons, withanolide A instigated significant renewal of axons and dendrites and restored pre- and post-synapses, damaged by Amyloid-β (25-35) treatment. It was found to limit damage of the cerebral cortex and core hippocampus synapses, and form active neuronal

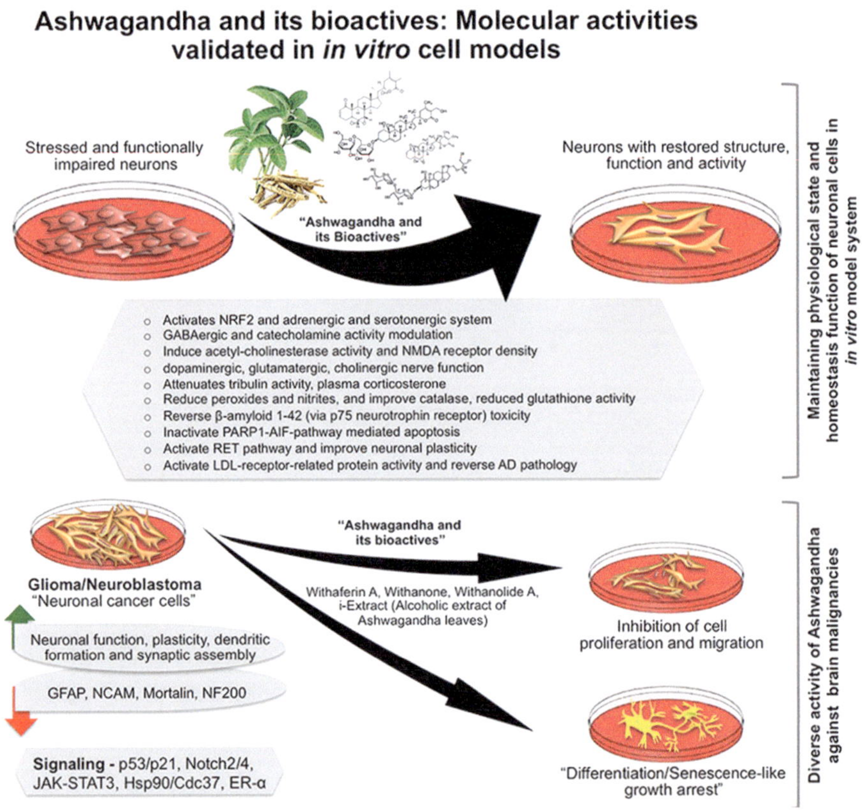

Fig. 14.2 Ashwagandha and its bioactives: molecular activities validated in *in vitro* cell models. Administration of Ashwagandha and its bioactives to stressed neurons restores their structure, function and activity via regulating diverse molecular signaling pathways (*upper half*). Numerous Ashwagandha purified bioactives e.g. Withaferin-A, Withanone, Withanolide A, i-Extract possess activity against brain malignancies (*lower half*) by inhibiting cancer cell growth and by causing differentiation leading to senescence-like growth arrest along with modulation of diverse molecular phenotypes and signaling pathways

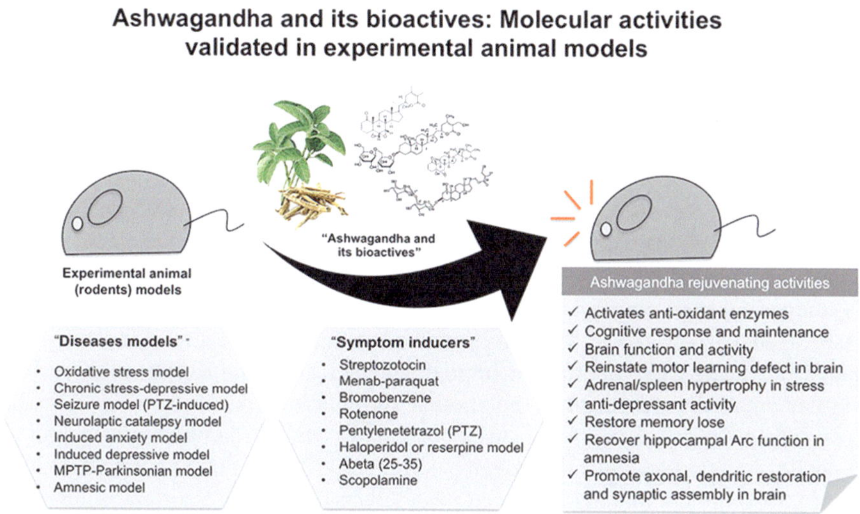

Fig. 14.3 Ashwagandha and its bioactives: diverse molecular activities of Ashwagandha and its bioactives against variety of neural disease/symptoms induced by artificial agents in experimental animal models. Multiple Ashwagandha neuroregenerating activities including augmenting brain function, cognitive response, antioxidant activity, antidepressant and memory restoration validated in rodent models

networks as observed by MAP2, NF-H, synaptophysin, and PSD-95 functional neuronal markers (Kuboyama et al. 2005). Based on these findings, withanolide A was suggested to be a vital compound to boost brain health. Similarly, Ashwagandha extracts and derivatives were found to support brain functions under potential anomalies induced by either drugs/environmental insults or psychological conditions (Sankar et al. 2007; Kumar and Kumar 2009; Kasture et al. 2009; Nair et al. 2007, 2008) (Fig. 14.3). Ashwagandha root extract was shown to restore cognitive brain function, behavioral and the oxidant-antioxidant imbalance in the midbrain of MPTP-intoxicated mice (Sankar et al. 2007). Similarly, it restored 3-Nitropropionic acid, a potent neurotoxin, induced oxidative/nitrosative stress and neurobehavioral and biochemical alterations, as studied in the striatum and cortex regions of rat brain (Kumar and Kumar 2009). In the haloperidol-induced catalepsy mice model, Ashwagandha derivatives containing formulation restored posture anomalies and augmented anti-oxidant potential involving SOD activity (Nair et al. 2007).

14.4 Ashwagandha for Neurological Disorders: Experimental Evidences

Benefits of Ashwagandha have been reported in a number of central nervous system disorders including seizures, epilepsy, catalepsy, schizophrenia and insomnia (Prasad and Malhotra 1968; Rasheed et al. 2013) (Fig. 14.1). Kulkarni and colleagues have provided experimental evidence for the anti-convulsant effect of Ashwagandha root extract against seizure in pentylenetetrazol (PTZ) seizure mouse model (Kulkarni et al. 2008a). The extract was seen to modulate GABAAergic activity. Similar model provided evidence to the anti-epileptic activity (Kulkarni and Dhir 2008; Akula et al. 2009; Soman et al. 2012). It was shown that Ashwagandha, besides its anti-oxidant function, modulated NMDA receptor density, improved motor learning defects by modulating AMPA receptor function in the rat cerebellum (Soman et al. 2013). **Catalepsy**, a neurological disorder that is characterized by seizures with a loss of consciousness and acquiring immobilized rigid state of body, is often associated with hysteria, and other complex cerebellar nervous disorders. Ashwagandha was shown to provide protection against haloperidol or reserpine-induced catalepsy in a neuroleptic animal model (Kumar and Kulkarni 2006). In a similar approach, treatment of polyherbal formulation containing Ashwagandha was shown to reduce neurological abnormalities and oxidative stress in mouse model of haloperidol-induced catalepsy (Nair et al. 2008). **Schizophrenia**, neurological disorder comprising abnormal metabolism/function of neurotransmitters in the brain that involves dopamine, glutamate, GABA, acetylcholine, and serotonin abnormalities (Nandhagopal et al. 2008; Felicio et al. 2014). Ashwagandha has been shown to augment dopaminergic (Ahmad et al. 2005; Prakash et al. 2013), glutamatergic (Kataria et al. 2012), GABAergic (Yin et al. 2013) and cholinergic (Schliebs et al. 1997) nervous functions and to provide neuroprotection. Use of Ashwagandha was also shown to reduce risk of behavioral and extrapyramidal side effects of common antipsychotics e.g. haloperidol towards their use in schizophrenia (Nair et al. 2007). Furthermore, treatment with Ashwagandha has been shown to reduce schizophrenia and alleviate effect of associated neurological anomalies (Agnihotri et al. 2013). **Insomnia**, sleep disorder that comprises inability to sleep, irregular patterns and lack of quality sleep, has been treated with Ashwagandha in traditional home medicine practice. It has been reported to induce physical calmness and relaxation and alleviate insomnia (Russo et al. 2001; Kuboyama et al. 2014; Manjunath and Telles 2005) that is often associated with reduced risk of inevitable neurodegenerative vulnerability (Arvanitakis et al. 2007). Sleep related anomalies are believed to potentiate risks of serious neurodegenerative diseases e.g. Alzheimer, Parkinson's disease, dementia and lateral sclerosis (Uitti and Wszolek 2003; Boeve et al. 2007). Alkaloids derived from Ashwagandha or their combination with diazepam was shown to restore sleep loss in mice models of sleep. They were shown to reduce lipid peroxidation, nitrites levels, improve catalase, and reduce glutathione activities and activate GABAergic signaling (Kumar and Kalonia 2007, 2008). Healthy human volunteers using Ashwagandha reported improvement in quality of sleep (Raut et al. 2012).

Cognitive dysfunctions often lead to a variety of neurophysiological anomalies (Spector et al. 2011) (Fig. 14.3). Consistent with the positive effects in the use of Ashwagandha for cognitive function and behavior, it was found to improve the key symptomic dysfunctions including auditory-verbal memory, reaction time, social cognition, associated with bipolar disorder and other neuropsychiatric conditions (Chengappa et al. 2013; Bhattacharya et al. 2000c; Gupta and Rana 2008; Maity et al. 2011; Bhattacharya et al. 2000b; Shreevathsa et al. 2013; Ramanathan et al. 2011; Muruganandam et al. 2002; Sarris et al. 2013), and improved locomotion and cognitive functions during ageing (Singh et al. 2008; Russo et al. 2001; Pingali et al. 2014). (Schliebs et al. 1997) reported that it reduced neuritic atrophy and synaptic loss and activated procholinergic effects in the rodent brain (Schliebs et al. 1997). (Kuboyama et al. 2005) have shown that withanolide A, a predominant withanolide in Ashwagandha extracts, reverses memory loss by rejuvenation of dendritic axons and spines in mice models. Jain et al. (2001) reported that Ashwagandha attenuates damage to CA2 and CA3 region of hippocampal neurons in rat model of stress. In agreement with these studies, several other reports showed that the rodents pre-treated with Ashwagandha exhibited significant reduction in hypercortisolemia and other stress symptoms (Kaur et al. 2003; Bhattacharya et al. 2000a; Bhattacharya and Muruganandam 2003).

Above described diverse Ashwagandha activities at molecular level, were driven by modulation of GABAAergic activity, AMPA receptor function and by reinstating change in NMDA receptor density, and thus cumulatively augmenting neuroprotective anti-oxidant and anti-inflammatory activities in brain of experimental models (Fig. 14.3). Taken together, Ashwagandha extracts and derivatives improve dopaminergic, glutamatergic, GABAergic and cholinergic nervous functions, and thus mechanistically provide protection from the risk of diverse neurological disorders.

14.5 Ashwagandha for Neuropsychiatric Conditions: Experimental Evidences

Consistent to the fact that Ashwagandha supplementation induces relaxation and calmness in the body, a number of reports have suggested the use of Ashwagandha in treatment of a variety of neuropsychiatric conditions including **anxiety and depression** (Fig. 14.1). Study by Andrade (2009) has shown anxiolytic potential of ethanolic extract of Ashwagandha in patients having ICD-10 anxiety disorders. Consistent to this study, treatment with Ashwagandha-derived bioactive glycowithanolides was found to reduce the levels of tribulin (i.e. an anxiety marker and induced by an anxiogenic agent, pentylenetetrazole) in rat brain (Bhattacharya et al. 2000c) (Fig. 14.3). An oral administration of Ashwagandha attenuated ethanol withdrawal-induced anxiety in rats addicted to chronic ethanol consumption and suggested its use for management of ethanol withdrawal related anxiety (Gupta and Rana 2008; Mohan et al. 2011). Several other positive behavioral and physiological effects of Ashwagandha were also identified (Wscieklica et al. 2016). Anxiolytic effect of Ashwagandha in comparison to diazepam, a standard anti-anxiety drug, along with a number of commercial anti-psychotics drugs including fluoxetine, gabapentin, citalopram, clozapine, resperidone and venlafaxine was also analyzed. The study confirmed anti-anxiety activity of Ashwagandha in a comparative behavioral based study tested on different mazes (Kulkarni et al. 2008b). In a randomized trail on company employees, with moderate to severe anxiety levels, Cooley and colleagues conducted a comparative survey on individuals taking naturopathic care (comprising dietary, relaxation and Ashwagandha supplementation) or standardized psychotherapy intervention (psychotherapy, relaxation and placebo) over a 12 weeks period (Cooley et al. 2009). Although both approaches demonstrated a significant reduction in patient's anxiety, decline in anxiety levels in the Ashwagandha group over the psychotherapy group were reported (Cooley et al. 2009; Chandrasekhar et al. 2012; Dar et al. 2016). Besides anxiety, Ashwagandha was also found to reduce levels of plasma corticosterone and hypertrophy of adrenals and spleen induced by chronic stress in Wister rats (Pramanik et al. 2010). Ashwagandha supplementation was also shown to improve obsessive-compulsive disorder induced anxiety in mice models (Kaurav et al. 2012). Effect of Ashwagandha in comparison to a number of other herbal anxiolytic medicines was analyzed based on pre-clinical evidences that described distinct anxiolytic advantages of Ashwagandha (Sarris et al. 2013). **Anti-depressant** effects of Ashwagandha have been established in a variety of comparative experimental models showing that its anti-depressant activity is comparable with imipramine (Tripathi et al. 1998; Bhattacharya et al. 2000c). It is widely accepted as a mood stabilizer in clinical conditions of psychiatric and alleviates depression (Maity et al. 2011; Chatterjee et al. 2001; Muruganandam et al. 2002; Krishnamurthy and Telles 2007). The effect was shown to be mediated by activation of adrenergic and serotonergic system (Ramanathan et al. 2011). In a mice model of depression, an Ashwagandha containing herbal formulation found to significant inhibit despair, moderate anti-reserpine activity, sedation and relative improvement in chronic fatigue syndrome (Shreevathsa et al. 2013). With improving behavioral, psychological and biochemical activities, Ashwagandha supplementation was suggested to have potent anti-depressant activity.

14.6 Ashwagandha for Protection Against Neurotoxic Agents: Experimental Evidence

Ashwagandha has been shown to reverse cytotoxicity induced by a variety of toxins, including endogenous toxins (neurotoxic peptide fragment and a predominant form of β-amyloid enriched in the brain of patients of Alzheimer's disease or Down's syndrome) and HIV (Human Immunodeficiency Virus) Type 1 in human neuron cells (Kurapati et al. 2013, 2014) (Fig. 14.2). Furthermore, cytoprotective activity of Ashwagandha extracts and purified components against various exogenous neurotoxic agents including cyclosphamide and derivatives (Praveenkumar et al. 1994; Davis and Kuttan 1998; Jena et al. 2003), Kainic acid (Parihar and Hemnani 2003), Streptozotocin (Parihar et al. 2004), 6-hydroxydopamine (Ahmad et al. 2005), methoxy acetic acid (Priyandoko et al. 2011), lead (Kumar et al. 2014), maneb-paraquat (Prakash et al. 2013) and scopolamine (Konar et al. 2011) have been reported (Fig. 14.3). Protection against neurotoxic drugs, hydrogen peroxide and viruses induced neurocognitive disorders has also been reported (Kurapati et al.

2013, 2014; Kumar et al. 2010; Patil et al. 2010; Shukla et al. 2011b; Kuboyama et al. 2014). Our laboratory studies demonstrated that Ashwagandha derivative withanone could protect human cells from methoxy acetic acid-induced toxicity. Withanone was found to protect cells from its toxicity by attenuating the ROS level, DNA and mitochondrial damage, and by inducing NRF2 cell defense and proteasomal degradation pathways (Priyandoko et al. 2011). We also found that the Ashwagandha extracts and withanone at a lower dose, protect glial and neuronal cells from oxidative and glutamate stress (Shah et al. 2015). Moreover, we analyzed neuroprotective response of aqueous Ashwagandha leaf extract on glutamate-induced 'excitotoxicity' and found that its pre-treatment reverted such cytotoxicity, mediated through changes in heat shock protein (Hsp) 70 levels, both in glial and neuronal cancer cells (Kataria et al. 2012) (Fig. 14.4).

Atherosclerosis or steady cholesterol deposition in the arteries is a common cause of ischemia (deprivation of oxygen and key nutrients in the brain) that causes significant brain damage (Ross et al. 2013). Use of Ashwagandha was found to provide cytoprotection from ischemia and reperfusion injuries in animal models (Gupta et al. 2004; Mohanty et al. 2008; Raghavan and Shah 2014). In a rodent

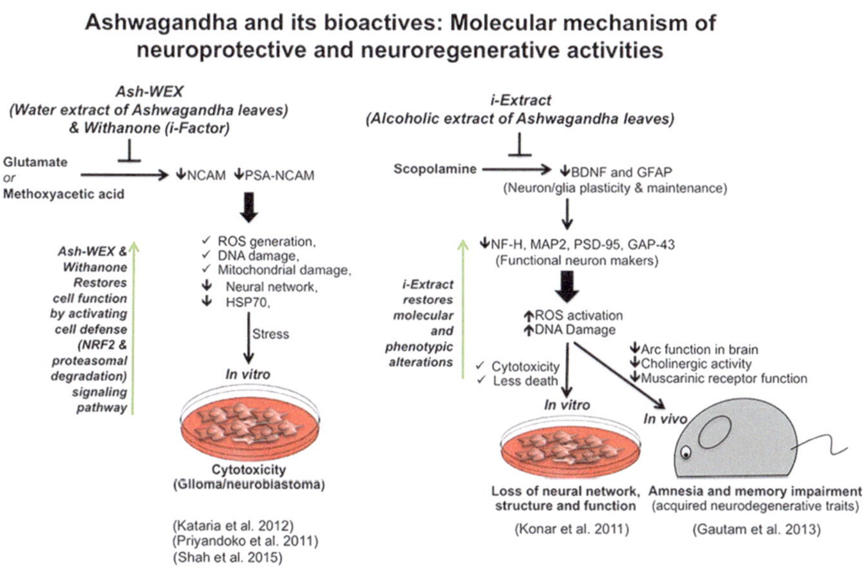

Fig. 14.4 Ashwagandha and its bioactives: molecular mechanism of neuroprotective and neuroregenerative activities. Ash-WEX (Water extract of Ashwagandha leaves) and withanone, a purified bioactive compound protects neurons from neurotoxic agents (glutamate and methoxyacetic acid) essentially by neutralizing oxidative damage, as validated in *in vitro* models (*left*, Kataria et al. 2012; Priyandoko et al. 2011; Shah et al. 2015). i-Extract (Alcoholic extract of Ashwagandha leaves) protects cell from scopolamine-induced oxidative stress and restores neural network, structure, function and memory anomalies validated in *in vitro* and *in vivo* models (*right*, Konar et al. 2011; Gautam et al. 2013)

model of ischemia and reperfusion injury, Ashwagandha administration was shown to reduce stroke/injury effect by augmented levels of endogenous antioxidants i.e., GSH, GSHPx, SOD and CAT (Gupta et al. 2004). Similarly, pre-treatment with Ashwagandha restored oxidant-antioxidant balance and caused anti-apoptotic effects by modulating Bcl-2 and Bax levels (Mohanty et al. 2008) (Fig. 14.2). Moreover, Ashwagandha was found to enhance hemeoxygenase 1 (HO1) levels and suppress the PARP1 via the PARP1-AIF pathway, thereby preventing apoptosis-inducing factor (AIF) nuclear translocation and eventual apoptosis. Herein, Ashwagandha treatment reduced the level of Sema3A, while Wnt, pGSK-3β, and pCRMP2 levels remained unchanged (Raghavan and Shah 2014). These studies suggested a crosstalk of antioxidant-antiapoptic pathways and a possible association of angiogenesis in addition to its antiapoptotic and antioxidant properties.

14.7 Ashwagandha for Treatment of Age-Related Neurodegenerative Disorders

Neurodegeneration is an intricate process comprising of progressive loss of neural structure and function in older age. It leads to neurodegenerative diseases such as Parkinson's, Alzheimer's and Huntington's, categorized based on their distinct origin, and are frequently seen in elderly population. Such progressive neurodegeneration also leads to multiple neurophysiological syndromes such as amnesia, dementia etc.. Worldwide, approximately 18 million people are presently undertreated for Alzheimer's disease and it is expected to escalate to 34 million by 2025. Old-age neurodegenerative diseases impact not only the patient's life expectancy and quality of life, but also severely affect family, social and global economy towards management and treatment of the disease. Their prevention and treatment by natural drugs has hence been highly desired and valued.

14.7.1 Alzheimer's Disease (AD)

AD is a chronic neurodegenerative old age disease causing progressive loss of memory (accounts for 60-70% of dementia cases) and cognitive functions (motion, balance and behavior related anomalies). Herbal formulation containing Ashwagandha and various withanolides were shown to inhibit acetyl-cholinesterase activity, a widely accepted therapeutic regime for AD (Schliebs et al. 1997; Choudhary et al. 2005; Grover et al. 2012). Withanoside IV, a key withanolide in Ashwagandha extracts was found to reinstate memory loss in amyloid-β (25-35)-treated mice and prevent structural and functional damage to neurons

(Kuboyama et al. 2006; Kulkarni and Dhir 2008; Tohda 2008; Mirjalili et al. 2009; Tohda and Joyashiki 2009; Jayaprakasam et al. 2010). Of note, reversal of Alzheimer's disease progression (Sehgal et al. 2012), and attenuating amyloid-β (1-42) induced toxicity in neuronal cells (Kurapati et al. 2013) have been reported suggesting the role of Ashwagandha as a brain heath remedy by supporting its neurophysiological and homeostasis function (Fig. 14.2). Hynd and colleagues in 2004 described the risk of glutamate-mediated excitotoxicity and its role in progression of neurodegenerative effect in AD (Hynd et al. 2004). Sehgal and colleagues found that oral administration of its root extract (predominantly withanolides and withanosides) reversed the behavioral discrepancies, plaque, buildup of oligomers and beta-amyloid peptides (Aβ) in the brain of adult AD transgenic mice (Sehgal et al. 2012). Ashwagandha root extract-treated mice showed clearance of brain Aβ peptides to periphery. Subsequently, elevated levels of low-density lipoprotein receptor-related protein (LRP) and neprilysin (NEP, an Aβ-degrading protease) was observed in mice brain microvessels, suggested an important role of Ashwagandha extract in Aβ clearance and diagnosis as seen in AD models (Sehgal et al. 2012). Another study demonstrated that Ashwagandha derivative withanamides A and C specifically bind to the beta-amyloid (25-35) active motif and prevents amyloid fibril formation in the rat neuronal cells (Jayaprakasam et al. 2010). Above reports suggested that Ashwagandha extract and derivatives have potential to prevent and reverse neurodegenerative effect of AD.

14.7.2 Parkinson's Disease (PD)

PD on the other hand, is a neurodegenerative disease comprising of deterioration of dopamine producing neurons (Fig. 14.3). Muscular rigidity, tremors, and senile motion, speech and postures have been shown to be improved by Ayurvedic formulation consisting of Ashwagandha (Nagashayana et al. 2000). In a rat model of 6-hydroxydopamine-induced Parkinsonism, Ashwagandha treatment was shown to provide neuroprotective effects against neuronal injury caused by oxidative agents and free radicals (Ahmad et al. 2005). In MPTP (1-methyl 4-phenyl 1,2,3,6-tetrahydropyridine)-induced Parkinsonism, Ashwagandha supplementation was shown to improve antioxidant and catecholamines level leading to improve ment in animal behavior and physical anomalies (Sankar et al. 2007; RajaSankar et al. 2009a, b). Similarly, in maneb-paraquat induced PD mice showed neuroprotective effects of Ashwagandha (Prakash et al. 2013). At molecular level, Ashwagandha root extract was shown to enhance DOPAC, DA, and HVA levels and to normalize TBARS expressions in the brain of PD mouse. Moreover, its treatment exhibited improved motor function and higher GSH and GPx levels in the mice brain. However, reduced oxidative stress and improved catecholamines level largely considered as two major mechanism of Ashwagandha activity in PD models.

14.7.3 Amnesia

It is a neurodegenerative disorder, defined by deficit in memory caused by brain damage or psychological trauma, and has also been treated by Ashwagandha. In mice model of amnesia, i-Extract and withanone were shown to improve amnesia symptoms and cognitive response that was attributed to induction of Brain-derived neurotrophic factor (BDNF) and activity-regulated cytoskeleton-associated (ARC) signalings (Konar et al. 2011; Gautam et al. 2013, 2015) (Fig. 14.4). In a rodent model of scopolamine-induced amnesia, decline in central cholinergic neuronal activity resulted in decreased BDNF and glial fibrillary acidic protein (GFAP) levels (in dose and time dependent manner) and blockade of muscarinic receptor activity (Konar et al. 2011). Expression of BDNF, a neurotrophin family growth factor, increased in response to i-Extract and withanone treatment in brain-derived cells. Another study on the effect of i-Extract in recovery of Arc levels in the cerebral cortex and hippocampus in the brain of a scopolamine-induced amnesic mice model, demonstrated that Arc might play a role in i-Extract based recovery from amnesia (Gautam et al. 2013). i-Extract pre-treatment restored F-actin polymerization in scopolamine-induced amnesia or with Arc antisense oligodeoxynucleotides (Gautam et al. 2015). It suggested that withanone and i-Extract both prevent progression of induced amnesia and restores homeostatic function.

14.7.4 Huntington's Disease (HD)

Ashwagandha has also been shown to prevent symptomatic aberrations genetic neurodegenerative disorders including Huntington's disease (HD), a common autosomal dominant genetic neurodegenerative disorder marked by abnormalities in muscle coordination, cognitive and behavioral responses (Hynd et al. 2004). In HD animal models, it was shown to protect neuronal cells against 3-nitropropionic acid that is know to induce neurophysiological anomalies including biochemical, behavioral and mitochondrial dysfunction (Kumar and Kumar 2009; Choudhary et al. 2013). Ashwagandha was shown to mediate such effects via its antioxidant activity, which remained common among all neurodegenerative disease including AD, PD and HD (Shukla et al. 2011b; Singh et al. 2011).

14.7.5 Amyotrophic Lateral Sclerosis (ALS)

It is a neurodegenerative disorder caused by progressive deterioration of myelin sheath that protects brain and spinal cord neurons, was also reported to be benefited by Ashwagandha (Shukla et al. 2011b).

In these studies, in addition to the activation of anti-oxidative defense mechanisms, withanolides were shown to activate muscarinic receptors function (Schliebs et al. 1997; Ali and Meitei 2011), inhibit cholinesterase (Choudhary et al. 2004, 2005) and promote neuronal differentiation, regeneration and plasticity signaling cascades (Kataria et al. 2012; Choudhary et al. 2013) (Fig. 14.4). Furthermore, majority of neurodegenerative diseases are marked with a loss of mitochondrial function and elevated production of toxic peptide/proteins in the brain. Such events gradually lead to a loss of cognitive abilities such as decision-making and memory. We have also found that withanone protects against Methoxyacetic acid, a key industrial metabolite (Priyandoko et al. 2011). Methoxyacetic acid exposure evoked ROS levels, mitochondrial and DNA damage in normal human cells and caused premature senescence. Withanone was found to suppress such effect by instigating cell defense pathways involving NRF2 and proteasomal degradation (Priyandoko et al. 2011). Similarly, DNA damage induced by scopolamine or H_2O_2 to brain derived cells was inhibited by Ashwaganda extract or its purified component, withanone (Konar et al. 2011; Shah et al. 2015).

14.8 Ashwagandha for Treatment of Brain Malignancies

A growing body of evidence has suggested multiple benefits of Ashwagandha in number of brain malignancies. Cytotoxic and tumor-sensitizing properties of Ashwagandha in several brain cancers were reported (Devi 1996; Kuttan 1996). We found that Ashwagandha leaf extract and its derived components possess differentiation activity for glioblastoma and neuroblastoma cells in culture (Shah et al. 2009, 2015) (Fig. 14.5). The cells treated with low doses of i-Extract or purified compounds such as withaferin A, withanone, withanolide A showed differentiation morphology and senescence-like growth arrest. At molecular level, i-Extract and withanone were found to upregulate GFAP, neuronal cell adhesion (NCAM) molecules. Water extract (Ash-WEX) was also shown to modulate expression levels of GFAP, NCAM, NF200 and mortalin (Kataria et al. 2011, 2013) and protected differentiated glioblastoma (C6) and neuroblastoma (IMR-32) cells from glutamate-induced excitotoxicity (Kataria et al. 2012). Use of Ashwagandha extract was also found to induce morphological and physiological changes in GnV-3, an immortalized rat hypothalamic cell line (Kataria et al. 2015a). In *in vivo* glioblastoma model, ASH-WEX reduced the tumor burden and suppressed expression of NF-kappaB, p-Akt, VEGF, HSP70, PSA-NCAM, and cyclin D1, proteins involved in tumor progression, metastases and survival (Kataria et al. 2015b). This was also found to suppress expression of vascular endothelial growth factor (VEGF), PSA-NCAM, HSP70, and cyclin D1 protein, generally upregulated in metastatic glioma cells. Permeability to blood-brain barrier that separate blood from cerebrospinal fluid in brain greatly contributes to efficacy of a drug in several neuropathologies. To test

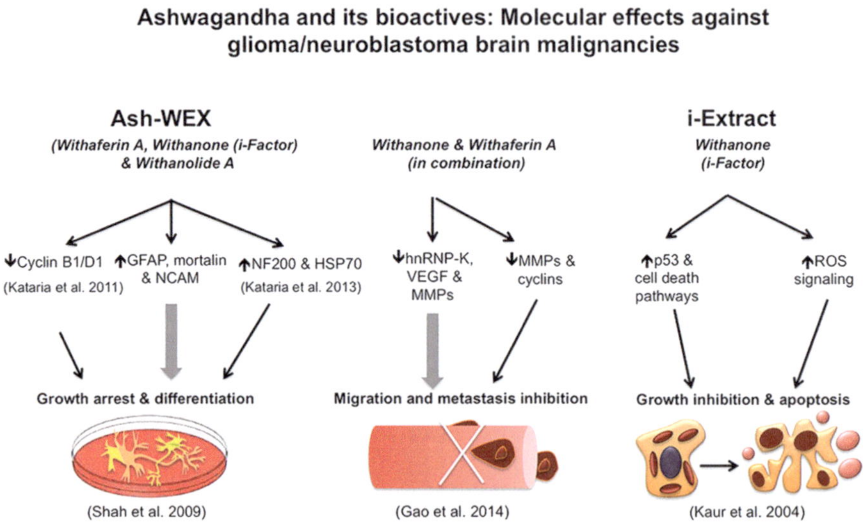

Fig. 14.5 Ashwagandha and its bioactives: molecular effects against glioma/neroblastoma brain malignancies. Ash-WEX and its purified bioactives, Withaferin-A, withanone and withanolide A suppress glioblastoma cell proliferation by modulating molecular expression of markers involved in growth arrest and differentiation (*left*, Shah et al. 2009). Withanone and withaferin A inhibit cancer cell migration and metastasis (middle, Gao et al. 2014); whereas i-Extract and withanone promote cancer cell growth inhibition and apoptosis mediated by p53 and ROS signaling (*right*, Kaur et al. 2004)

this, *in vivo* study has demonstrated the presence of withanamides in the blood and brain of mice administered with Ashwagandha fruit extract (Vareed et al. 2014), emphasizing efficacy of Ashwagandha bioactives in brain pathologies. Withanolides derived from Ashwagandha were shown to inhibit growth and invasion through suppression of nuclear factor-kappa B (NF-kB) and its transcription regulatory activity (Ichikawa et al. 2006). (Lee et al. 2012) suggested that withaferin A and its structural analogue withanone and withanolide A cause anti-migration effect through activation of Notch pathway. Withaferin A was also found to inhibit epithelial-mesenchymal transition process by suppressing rearrangement of vimentin, a type III intermediate filament (IF) protein involved in cell motility and migration (Lee et al. 2013b; Yang et al. 2013; Antony et al. 2014; Grin et al. 2012). It was shown to cause vimentin disassembly and Ser56 phosphorylation that regulates structural arrangement of its filaments in migration and metastases (Thaiparambil et al. 2011) and inhibit MMP-9 activity by repressing Akt signaling (Lee et al. 2013a). A combination of withanone and withaferin-A was also shown to inhibit cancer cell migration and metastasis by inactivation of hnRNP-K protein (Gao et al. 2014).

14.9 Current Perspectives and Next: Managing Risks of Brain Aging and Old Age Neurological Pathologies

Complex etiology of brain pathologies makes their diagnosis and therapeutics challenging. Accumulative cellular stress is recognized as a key trigger in the progression of several neurological anomalies. As discussed in this chapter Ashwagandha bioactives are useful to neutralize accumulative stress in cell, essentially by providing protection against a variety of agents/factors and conditions causing neurological pathologies. Improved brain function and cognitive response in a variety of neuropathological conditions including Alzheimer's and Parkinson's and their associated anomalies e.g. amnesia, dementia has been reported. Several studies have provided evidence to its anticancer activities including induction of differentiation, apoptosis or inhibition of proliferation and metastases by modulation of several molecular signaling pathways.

Rasayana is an Ayurvedic micronutrient remedy that is known to promote healthy aging and longevity. Among its all components, Ashwagandha has consistently been reported as an essential adaptogenic and anti-stress herb that retards brain aging and memory ameliorating effects, and promotes cognitive response, neuroregeneration (Singh et al. 2008; Bhattacharya et al. 2000c). As discussed in this chapter, a large number of studies have provided evidence to the benefits of Ashwagandha phytochemicals to behavioral and psychological brain fitness in preventive and therapeutic ways. Of note, these have not only been reported to protect from the risk of degenerative neuropathologies, but also found to reverse effects of such pathologies at advanced stage in multiple ways. Protection against oxidative stress, activation of neuronal signalings and reconstruction of nerve cell networks by Ashwagandha bioactives can essentially provide a hope for targeted therapeutics. Bioinformatics and computational evidences have beginning to suggest the molecular modes of action of the specific bioactives and are opening a new avenue of exploring Ashwagandha activities. Druggability of predicted protein targets, pharmokinetics for brain pathologies defines another avenue that warrants further investigations and validation of their effects in animal models.

References

Agnihotri AP, Sontakke SD, Thawani VR, Saoji A, Goswami VS (2013) Effects of Withania somnifera in patients of schizophrenia: a randomized, double blind, placebo controlled pilot trial study. Indian J Pharmacol 45:417–418

Ahmad M, Saleem S, Ahmad AS, Ansari MA, Yousuf S, Hoda MN, Islam F (2005) Neuroprotective effects of Withania somnifera on 6-hydroxydopamine induced Parkinsonism in rats. Hum Exp Toxicol 24:137–147

Ahmad MK, Mahdi AA, Shukla KK, Islam N, Rajender S, Madhukar D, Shankhwar SN, Ahmad S (2010) Withania somnifera improves semen quality by regulating reproductive hormone levels and oxidative stress in seminal plasma of infertile males. Fertil Steril 94:989–996

Ahmed ME, Javed H, Khan MM, Vaibhav K, Ahmad A, Khan A, Tabassum R, Islam F, Safhi MM, Islam F (2013) Attenuation of oxidative damage-associated cognitive decline by Withania somnifera in rat model of streptozotocin-induced cognitive impairment. Protoplasma 250:1067–1078

Akula KK, Dhir A, Kulkarni SK (2009) Effect of various antiepileptic drugs in a pentylenetetrazol-induced seizure model in mice. Methods Find Exp Clin Pharmacol 31:423–432

Ali SA, Meitei KV (2011) On the action and mechanism of withaferin-A from Withania somnifera, a novel and potent melanin dispersing agent in frog melanophores. J Recept Signal Transduct Res 31:359–366

Andrade C (2009) Ashwagandha for anxiety disorders. World J Biol Psychiatry 10:686–687

Antony ML, Lee J, Hahm ER, Kim SH, Marcus AI, Kumari V, Ji X, Yang Z, Vowell CL, Wipf P, Uechi GT, Yates NA, Romero G, Sarkar SN, Singh SV (2014) Growth arrest by the antitumor steroidal lactone withaferin A in human breast cancer cells is associated with down-regulation and covalent binding at cysteine 303 of beta-tubulin. J Biol Chem 289:1852–1865

Anwer T, Sharma M, Pillai KK, Khan G (2012) Protective effect of Withania somnifera against oxidative stress and pancreatic beta-cell damage in type 2 diabetic rats. Acta Pol Pharm 69:1095–1101

Arvanitakis Z, Witte RJ, Dickson DW, Tsuboi Y, Uitti RJ, Slowinski J, Hutton ML, Lin SC, Boeve BF, Cheshire WP, Pooley RA, Liss JM, Caviness JN, Strongosky AJ, Wszolek ZK (2007) Clinical-pathologic study of biomarkers in FTDP-17 (PPND family with N279K tau mutation). Parkinsonism Relat Disord 13:230–239

Baitharu I, Jain V, Deep SN, Hota KB, Hota SK, Prasad D, Ilavazhagan G (2013) Withania somnifera root extract ameliorates hypobaric hypoxia induced memory impairment in rats. J Ethnopharmacol 145:431–441

Bhattacharya A, Ghosal S, Bhattacharya SK (2001) Anti-oxidant effect of Withania somnifera glycowithanolides in chronic footshock stress-induced perturbations of oxidative free radical scavenging enzymes and lipid peroxidation in rat frontal cortex and striatum. J Ethnopharmacol 74:1–6

Bhattacharya A, Muruganandam AV, Kumar V, Bhattacharya SK (2002) Effect of poly herbal formulation, EuMil, on neurochemical perturbations induced by chronic stress. Indian J Exp Biol 40:1161–1163

Bhattacharya A, Ramanathan M, Ghosal S, Bhattacharya SK (2000a) Effect of Withania somnifera glycowithanolides on iron-induced hepatotoxicity in rats. Phytother Res 14:568–570

Bhattacharya SK, Bhattacharya A, Chakrabarti A (2000b) Adaptogenic activity of Siotone, a polyherbal formulation of Ayurvedic rasayanas. Indian J Exp Biol 38:119–128

Bhattacharya SK, Bhattacharya A, Sairam K, Ghosal S (2000c) Anxiolytic-antidepressant activity of Withania somnifera glycowithanolides: an experimental study. Phytomedicine 7:463–469

Bhattacharya SK, Muruganandam AV (2003) Adaptogenic activity of Withania somnifera: an experimental study using a rat model of chronic stress. Pharmacol Biochem Behav 75:547–555

Boeve BF, Silber MH, Saper CB, Ferman TJ, Dickson DW, Parisi JE, Benarroch EE, Ahlskog JE, Smith GE, Caselli RC, Tippman-Peikert M, Olson EJ, Lin SC, Young T, Wszolek Z, Schenck CH, Mahowald MW, Castillo PR, Del Tredici K, Braak H (2007) Pathophysiology of REM sleep behaviour disorder and relevance to neurodegenerative disease. Brain 130:2770–2788

Chandrasekhar K, Kapoor J, Anishetty S (2012) A prospective, randomized double-blind, placebo-controlled study of safety and efficacy of a high-concentration full-spectrum extract of ashwagandha root in reducing stress and anxiety in adults. Indian J Psychol Med 34:255–262

Chatterjee S, Rao AT, Das SN, Agrawal SK (2001) Anti-atherogenic Action of "Cardipro" – A Herbal Proprietary Formulation. Anc Sci Life 20:81–85

Chaudhary G, Sharma U, Jagannathan NR, Gupta YK (2003) Evaluation of Withania somnifera in a middle cerebral artery occlusion model of stroke in rats. Clin Exp Pharmacol Physiol 30:399–404

Chengappa KN, Bowie CR, Schlicht PJ, Fleet D, Brar JS, Jindal R (2013) Randomized placebo-controlled adjunctive study of an extract of withania somnifera for cognitive dysfunction in bipolar disorder. J Clin Psychiatry 74:1076–1083

Choudhary MI, Nawaz SA, ul-Haq Z, Lodhi MA, Ghayur MN, Jalil S, Riaz N, Yousuf S, Malik A, Gilani AH, ur-Rahman A (2005) Withanolides, a new class of natural cholinesterase inhibitors with calcium antagonistic properties. Biochem Biophys Res Commun 334:276–287

Choudhary MI, Yousuf S, Nawaz SA, Ahmed S, Atta ur R (2004) Cholinesterase inhibiting withanolides from Withania somnifera. Chem Pharm Bull (Tokyo) 52:1358–1361

Choudhary S, Kumar P, Malik J (2013) Plants and phytochemicals for Huntington's disease. Pharmacogn Rev 7:81–91

Cooley K, Szczurko O, Perri D, Mills EJ, Bernhardt B, Zhou Q, Seely D (2009) Naturopathic care for anxiety: a randomized controlled trial ISRCTN78958974. PLoS One 4:e6628

Dar PA, Singh LR, Kamal MA, Dar TA (2016) Unique Medicinal Properties of Withania Somnifera: Phytochemical Constituents and Protein Component. Curr Pharm Des 22:535–540

Davis L, Kuttan G (1998) Suppressive effect of cyclophosphamide-induced toxicity by Withania somnifera extract in mice. J Ethnopharmacol 62:209–214

Devi PU (1996) Withania somnifera Dunal (Ashwagandha): potential plant source of a promising drug for cancer chemotherapy and radiosensitization. Indian J Exp Biol 34:927–932

Felicio AC, Dinelle K, Agarwal PA, McKenzie J, Heffernan N, Road JD, Appel-Cresswell S, Wszolek ZK, Farrer MJ, Schulzer M, Sossi V, Stoessl AJ (2014) *In vivo* dopaminergic and serotonergic dysfunction in DCTN1 gene mutation carriers. Mov Disord 29:1197–1201

Gao R, Shah N, Lee JS, Katiyar SP, Li L, Oh E, Sundar D, Yun CO, Kaul SC, Wadhwa R (2014) Withanone Rich Combination of Ashwagandha Withanolides Restricts Metastasis And Angiogenesis Through hnRNP-K. Mol Cancer Ther 13:2930–2940

Gautam A, Kaul SC, Thakur MK (2015) Alcoholic Extract of Ashwagandha Leaves Protects Against Amnesia by Regulation of Arc Function. Mol Neurobiol 53:1760–1769

Gautam A, Wadhwa R, Thakur MK (2013) Involvement of hippocampal Arc in amnesia and its recovery by alcoholic extract of Ashwagandha leaves. Neurobiol Learn Mem 106:177–184

Grin B, Mahammad S, Wedig T, Cleland MM, Tsai L, Herrmann H, Goldman RD (2012) Withaferin a alters intermediate filament organization, cell shape and behavior. PLoS One 7:e39065

Grover A, Shandilya A, Agrawal V, Bisaria VS, Sundar D (2012) Computational evidence to inhibition of human acetyl cholinesterase by withanolide a for Alzheimer treatment. J Biomol Struct Dyn 29:651–662

Gupta GL, Rana AC (2008) Effect of Withania somnifera Dunal in ethanol-induced anxiolysis and withdrawal anxiety in rats. Indian J Exp Biol 46:470–475

Gupta SK, Mohanty I, Talwar KK, Dinda A, Joshi S, Bansal P, Saxena A, Arya DS (2004) Cardioprotection from ischemia and reperfusion injury by Withania somnifera: a hemodynamic, biochemical and histopathological assessment. Mol Cell Biochem 260:39–47

Hynd MR, Scott HL, Dodd PR (2004) Glutamate-mediated excitotoxicity and neurodegeneration in Alzheimer's disease. Neurochem Int 45:583–595

Ichikawa H, Takada Y, Shishodia S, Jayaprakasam B, Nair MG, Aggarwal BB (2006) Withanolides potentiate apoptosis, inhibit invasion, and abolish osteoclastogenesis through suppression of nuclear factor-kappaB (NF-kappaB) activation and NF-kappaB-regulated gene expression. Mol Cancer Ther 5:1434–1445

Jain S, Shukla SD, Sharma K, Bhatnagar M (2001) Neuroprotective effects of Withania somnifera Dunn. in hippocampal sub-regions of female albino rat. Phytother Res 15:544–548

Jayaprakasam B, Padmanabhan K, Nair MG (2010) Withanamides in Withania somnifera fruit protect PC-12 cells from beta-amyloid responsible for Alzheimer's disease. Phytother Res 24:859–863

Jena GB, Nemmani KV, Kaul CL, Ramarao P (2003) Protective effect of a polyherbal formulation (Immu-21) against cyclophosphamide-induced mutagenicity in mice. Phytother Res 17:306–310

Kasture S, Vinci S, Ibba F, Puddu A, Marongiu M, Murali B, Pisanu A, Lecca D, Zernig G, Acquas E (2009) Withania somnifera prevents morphine withdrawal-induced decrease in spine density in nucleus accumbens shell of rats: a confocal laser scanning microscopy study. Neurotox Res 16:343–355

Kataria H, Gupta M, Lakhman S, Kaur G (2015a) Withania somnifera aqueous extract facilitates the expression and release of GnRH: *In vitro* and *in vivo* study. Neurochem Int 89:111–119

Kataria H, Kumar S, Chaudhary H, Kaur G (2015b) Withania somnifera Suppresses Tumor Growth of Intracranial Allograft of Glioma Cells. Mol Neurobiol 53:4143–4158

Kataria H, Shah N, Kaul SC, Wadhwa R, Kaur G (2011) Water extract of ashwagandha leaves limits proliferation and migration, and induces differentiation in glioma cells. Evid Based Complement Alternat Med 2011:267614

Kataria H, Wadhwa R, Kaul SC, Kaur G (2012) Water extract from the leaves of Withania somnifera protect RA differentiated C6 and IMR-32 cells against glutamate-induced excitotoxicity. PLoS One 7:e37080

Kataria H, Wadhwa R, Kaul SC, Kaur G (2013) Withania somnifera water extract as a potential candidate for differentiation based therapy of human neuroblastomas. PLoS One 8:e55316

Kaur K, Rani G, Widodo N, Nagpal A, Taira K, Kaul SC, Wadhwa R (2004) Evaluation of the anti-proliferative and anti-oxidative activities of leaf extract from *in vivo* and *in vitro* raised Ashwagandha. Food Chem Toxicol 42:2015–2020

Kaur P, Sharma M, Mathur S, Tiwari M, Divekar HM, Kumar R, Srivastava KK, Chandra R (2003) Effect of 1-oxo-5beta, 6beta-epoxy-witha-2-ene-27-ethoxy-olide isolated from the roots of Withania somnifera on stress indices in Wistar rats. J Altern Complement Med 9:897–907

Kaurav BP, Wanjari MM, Chandekar A, Chauhan NS, Upmanyu N (2012) Influence of Withania somnifera on obsessive compulsive disorder in mice. Asian Pac J Trop Med 5:380–384

Konar A, Shah N, Singh R, Saxena N, Kaul SC, Wadhwa R, Thakur MK (2011) Protective role of Ashwagandha leaf extract and its component withanone on scopolamine-induced changes in the brain and brain-derived cells. PLoS One 6:e27265

Krishnamurthy MN, Telles S (2007) Assessing depression following two ancient Indian interventions: effects of yoga and ayurveda on older adults in a residential home. J Gerontol Nurs 33:17–23

Kuboyama T, Tohda C, Komatsu K (2005) Neuritic regeneration and synaptic reconstruction induced by withanolide A. Br J Pharmacol 144:961–971

Kuboyama T, Tohda C, Komatsu K (2006) Withanoside IV and its active metabolite, sominone, attenuate Abeta(25-35)-induced neurodegeneration. Eur J Neurosci 23:1417–1426

Kuboyama T, Tohda C, Komatsu K (2014) Effects of Ashwagandha (roots of Withania somnifera) on neurodegenerative diseases. Biol Pharm Bull 37:892–897

Kulkarni SK, Akula KK, Dhir A (2008a) Effect of Withania somnifera Dunal root extract against pentylenetetrazol seizure threshold in mice: possible involvement of GABAergic system. Indian J Exp Biol 46:465–469

Kulkarni SK, Dhir A (2008) Withania somnifera: an Indian ginseng. Prog Neuropsychopharmacol Biol Psychiatry 32:1093–1105

Kulkarni SK, Singh K, Bishnoi M (2008b) Comparative behavioural profile of newer antianxiety drugs on different mazes. Indian J Exp Biol 46:633–638

Kumar A, Kalonia H (2007) Protective effect of Withania somnifera Dunal on the behavioral and biochemical alterations in sleep-disturbed mice (Grid over water suspended method). Indian J Exp Biol 45:524–528

Kumar A, Kalonia H (2008) Effect of Withania somnifera on Sleep-Wake Cycle in Sleep-Disturbed Rats: Possible GABAergic Mechanism. Indian J Pharm Sci 70:806–810

Kumar A, Kulkarni SK (2006) Effect of BR-16A (Mentat), a polyherbal formulation on drug-induced catalepsy in mice. Indian J Exp Biol 44:45–48

Kumar P, Kumar A (2009) Possible neuroprotective effect of Withania somnifera root extract against 3-nitropropionic acid-induced behavioral, biochemical, and mitochondrial dysfunction in an animal model of Huntington's disease. J Med Food 12:591–600

Kumar P, Singh R, Nazmi A, Lakhanpal D, Kataria H, Kaur G (2014) Glioprotective Effects of Ashwagandha Leaf Extract against Lead Induced Toxicity. Biomed Res Int 2014:182029

Kumar S, Seal CJ, Howes MJ, Kite GC, Okello EJ (2010) *In vitro* protective effects of Withania somnifera (L.) dunal root extract against hydrogen peroxide and beta-amyloid(1-42)-induced cytotoxicity in differentiated PC12 cells. Phytother Res 24:1567–1574

Kumar V, Murthy KN, Bhamid S, Sudha CG, Ravishankar GA (2005) Genetically modified hairy roots of Withania somnifera Dunal: a potent source of rejuvenating principles. Rejuvenation Res 8:37–45

Kurapati KR, Atluri VS, Samikkannu T, Nair MP (2013) Ashwagandha (Withania somnifera) reverses beta-amyloid1-42 induced toxicity in human neuronal cells: implications in HIV-associated neurocognitive disorders (HAND). PLoS One 8:e77624

Kurapati KR, Samikkannu T, Atluri VS, Kaftanovskaya E, Yndart A, Nair MP (2014) beta-Amyloid1-42, HIV-1Ba-L (clade B) infection and drugs of abuse induced degeneration in human neuronal cells and protective effects of ashwagandha (Withania somnifera) and its constituent Withanolide A. PLoS One 9:e112818

Kuttan G (1996) Use of Withania somnifera Dunal as an adjuvant during radiation therapy. Indian J Exp Biol 34:854–856

Lee DH, Lim IH, Sung EG, Kim JY, Song IH, Park YK, Lee TJ (2013a) Withaferin A inhibits matrix metalloproteinase-9 activity by suppressing the Akt signaling pathway. Oncol Rep 30:933–938

Lee J, Hahm ER, Marcus AI, Singh SV (2013b) Withaferin A inhibits experimental epithelial-mesenchymal transition in MCF-10A cells and suppresses vimentin protein level *in vivo* in breast tumors. Mol Carcinog 54:417–429

Lee J, Sehrawat A, Singh SV (2012) Withaferin A causes activation of Notch2 and Notch4 in human breast cancer cells. Breast Cancer Res Treat 136:45–56

Maity T, Adhikari A, Bhattacharya K, Biswas S, Debnath PK, Maharana CS (2011) A study on evalution of antidepressant effect of imipramine adjunct with Aswagandha and Bramhi. Nepal Med Coll J 13:250–253

Manjunath MJ, Muralidhara (2013) Effect of Withania somnifera supplementation on rotenone-induced oxidative damage in cerebellum and striatum of the male mice brain. Cent Nerv Syst Agents Med Chem 13:43–56

Manjunath NK, Telles S (2005) Influence of Yoga and Ayurveda on self-rated sleep in a geriatric population. Indian J Med Res 121:683–690

Mirjalili MH, Moyano E, Bonfill M, Cusido RM, Palazon J (2009) Steroidal lactones from Withania somnifera, an ancient plant for novel medicine. Molecules 14:2373–2393

Misra DS, Maiti R, Ghosh D (2009) Protection of swimming-induced oxidative stress in some vital organs by the treatment of composite extract of Withania somnifera, Ocimum sanctum and Zingiber officinalis in male rat. Afr J Tradit Complement Altern Med 6:534–543

Mohan L, Rao US, Gopalakrishna HN, Nair V (2011) Evaluation of the Anxiolytic Activity of NR-ANX-C (a Polyherbal Formulation) in Ethanol Withdrawal-Induced Anxiety Behavior in Rats. Evid Based Complement Alternat Med 2011:2011

Mohanty IR, Arya DS, Gupta SK (2008) Withania somnifera provides cardioprotection and attenuates ischemia-reperfusion induced apoptosis. Clin Nutr 27:635–642

Muruganandam AV, Kumar V, Bhattacharya SK (2002) Effect of poly herbal formulation, EuMil, on chronic stress-induced homeostatic perturbations in rats. Indian J Exp Biol 40:1151–1160

Nagashayana N, Sankarankutty P, Nampoothiri MR, Mohan PK, Mohanakumar KP (2000) Association of L-DOPA with recovery following Ayurveda medication in Parkinson's disease. J Neurol Sci 176:124–127

Naidu PS, Singh A, Kulkarni SK (2003) Effect of Withania somnifera root extract on haloperidol-induced orofacial dyskinesia: possible mechanisms of action. J Med Food 6:107–114

Naidu PS, Singh A, Kulkarni SK (2006) Effect of Withania somnifera root extract on reserpine-induced orofacial dyskinesia and cognitive dysfunction. Phytother Res 20:140–146

Nair V, Arjuman A, Dorababu P, Gopalakrishna HN, Chakradhar Rao U, Mohan L (2007) Effect of NR-ANX-C (a polyherbal formulation) on haloperidol induced catalepsy in albino mice. Indian J Med Res 126:480–484

Nair V, Arjuman A, Gopalakrishna HN, Nandini M (2008) Effect of Withania somnifera root extract on haloperidol-induced catalepsy in albino mice. Phytother Res 22:243–246

Nandhagopal R, Mak E, Schulzer M, McKenzie J, McCormick S, Sossi V, Ruth TJ, Strongosky A, Farrer MJ, Wszolek ZK, Stoessl AJ (2008) Progression of dopaminergic dysfunction in a LRRK2 kindred: a multitracer PET study. Neurology 71:1790–1795

Parihar MS, Chaudhary M, Shetty R, Hemnani T (2004) Susceptibility of hippocampus and cerebral cortex to oxidative damage in streptozotocin treated mice: prevention by extracts of Withania somnifera and Aloe vera. J Clin Neurosci 11:397–402

Parihar MS, Hemnani T (2003) Phenolic antioxidants attenuate hippocampal neuronal cell damage against kainic acid induced excitotoxicity. J Biosci 28:121–128

Patil SP, Maki S, Khedkar SA, Rigby AC, Chan C (2010) Withanolide A and asiatic acid modulate multiple targets associated with amyloid-beta precursor protein processing and amyloid-beta protein clearance. J Nat Prod 73:1196–1202

Pingali U, Pilli R, Fatima N (2014) Effect of standardized aqueous extract of Withania somnifera on tests of cognitive and psychomotor performance in healthy human participants. Pharmacognosy Res 6:12–18

Prakash J, Yadav SK, Chouhan S, Singh SP (2013) Neuroprotective role of Withania somnifera root extract in maneb-paraquat induced mouse model of parkinsonism. Neurochem Res 38:972–980

Pramanik SS, Sur TK, Debnath PK, Bhattacharyya D (2010) Effect of Pueraria tuberosa tuber extract on chronic foot shock stress in Wistar rats. Nepal Med Coll J 12:234–238

Prasad S, Malhotra CL (1968) Studies on Withania ashwagandha Kaul. VI. The effect of the alkaloidal fractions (acetone, alcohol and water soluble) on the central nervous system. Indian J Physiol Pharmacol 12:175–181

Praveenkumar V, Kuttan R, Kuttan G (1994) Chemoprotective action of Rasayanas against cyclosphamide toxicity. Tumori 80:306–308

Priyandoko D, Ishii T, Kaul SC, Wadhwa R (2011) Ashwagandha leaf derived withanone protects normal human cells against the toxicity of methoxyacetic acid, a major industrial metabolite. PLoS One 6:e19552

Raghavan A, Shah ZA (2014) Withania somnifera Improves Ischemic Stroke Outcomes by Attenuating PARP1-AIF-Mediated Caspase-Independent Apoptosis. Mol Neurobiol 52:1093–1105

RajaSankar S, Manivasagam T, Sankar V, Prakash S, Muthusamy R, Krishnamurti A, Surendran S (2009a) Withania somnifera root extract improves catecholamines and physiological abnormalities seen in a Parkinson's disease model mouse. J Ethnopharmacol 125:369–373

Rajasankar S, Manivasagam T, Surendran S (2009b) Ashwagandha leaf extract: a potential agent in treating oxidative damage and physiological abnormalities seen in a mouse model of Parkinson's disease. Neurosci Lett 454:11–15

Ramanathan M, Balaji B, Justin A (2011) Behavioural and neurochemical evaluation of Perment an herbal formulation in chronic unpredictable mild stress induced depressive model. Indian J Exp Biol 49:269–275

Rasheed A, Satyanarayana KV, Gulabi PS, Rao MS (2013) Chemical and pharmacological standardization of Ashwagandhadi lehyam: an ayurvedic formulation. J Complement Integr Med 15:10

Raut AA, Rege NN, Tadvi FM, Solanki PV, Kene KR, Shirolkar SG, Pandey SN, Vaidya RA, Vaidya AB (2012) Exploratory study to evaluate tolerability, safety, and activity of Ashwagandha (Withania somnifera) in healthy volunteers. J Ayurveda Integr Med 3:111–114

Reuland DJ, Khademi S, Castle CJ, Irwin DC, McCord JM, Miller BF, Hamilton KL (2013) Upregulation of phase II enzymes through phytochemical activation of Nrf2 protects cardiomyocytes against oxidant stress. Free Radic Biol Med 56:102–111

Ross OA, Soto-Ortolaza AI, Heckman MG, Verbeeck C, Serie DJ, Rayaprolu S, Rich SS, Nalls MA, Singleton A, Guerreiro R, Kinsella E, Wszolek ZK, Brott TG, Brown RD Jr, Worrall BB, Meschia JF (2013) NOTCH3 variants and risk of ischemic stroke. PLoS One 8:e75035

Russo A, Izzo AA, Cardile V, Borrelli F, Vanella A (2001) Indian medicinal plants as antiradicals and DNA cleavage protectors. Phytomedicine 8:125–132

Sankar SR, Manivasagam T, Krishnamurti A, Ramanathan M (2007) The neuroprotective effect of Withania somnifera root extract in MPTP-intoxicated mice: an analysis of behavioral and biochemical variables. Cell Mol Biol Lett 12:473–481

Sarris J, McIntyre E, Camfield DA (2013) Plant-based medicines for anxiety disorders, part 2: a review of clinical studies with supporting preclinical evidence. CNS Drugs 27:301–319

Schliebs R, Liebmann A, Bhattacharya SK, Kumar A, Ghosal S, Bigl V (1997) Systemic administration of defined extracts from Withania somnifera (Indian Ginseng) and Shilajit differentially affects cholinergic but not glutamatergic and GABAergic markers in rat brain. Neurochem Int 30:181–190

Sehgal N, Gupta A, Valli RK, Joshi SD, Mills JT, Hamel E, Khanna P, Jain SC, Thakur SS, Ravindranath V (2012) Withania somnifera reverses Alzheimer's disease pathology by enhancing low-density lipoprotein receptor-related protein in liver. Proc Natl Acad Sci U S A 109:3510–3515

Shah N, Kataria H, Kaul SC, Ishii T, Kaur G, Wadhwa R (2009) Effect of the alcoholic extract of Ashwagandha leaves and its components on proliferation, migration, and differentiation of glioblastoma cells: combinational approach for enhanced differentiation. Cancer Sci 100:1740–1747

Shah N, Singh R, Sarangi U, Saxena N, Chaudhary A, Kaur G, Kaul SC, Wadhwa R (2015) Combinations of Ashwagandha Leaf Extracts Protect Brain-Derived Cells against Oxidative Stress and Induce Differentiation. PLoS One 10:e0120554

Shreevathsa M, Ravishankar B, Dwivedi R (2013) Anti depressant activity of Mamsyadi Kwatha: An Ayurvedic compound formulation. Ayu 34:113–117

Shukla K, Dikshit P, Tyagi MK, Shukla R, Gambhir JK (2012) Ameliorative effect of Withania coagulans on dyslipidemia and oxidative stress in nicotinamide-streptozotocin induced diabetes mellitus. Food Chem Toxicol 50:3595–3599

Shukla KK, Mahdi AA, Mishra V, Rajender S, Sankhwar SN, Patel D, Das M (2011a) Withania somnifera improves semen quality by combating oxidative stress and cell death and improving essential metal concentrations. Reprod Biomed Online 22:421–427

Shukla V, Mishra SK, Pant HC (2011b) Oxidative stress in neurodegeneration. Adv Pharmacol Sci 2011:572634

Singh A, Naidu PS, Gupta S, Kulkarni SK (2002) Effect of natural and synthetic antioxidants in a mouse model of chronic fatigue syndrome. J Med Food 5:211–220

Singh N, Bhalla M, de Jager P, Gilca M (2011) An overview on ashwagandha: a Rasayana (rejuvenator) of Ayurveda. Afr J Tradit Complement Altern Med 8:208–213

Singh RH, Narsimhamurthy K, Singh G (2008) Neuronutrient impact of Ayurvedic Rasayana therapy in brain aging. Biogerontology 9:369–374

Soman S, Anju TR, Jayanarayanan S, Antony S, Paulose CS (2013) Impaired motor learning attributed to altered AMPA receptor function in the cerebellum of rats with temporal lobe epilepsy: ameliorating effects of Withania somnifera and withanolide A. Epilepsy Behav 27:484–491

Soman S, Korah PK, Jayanarayanan S, Mathew J, Paulose CS (2012) Oxidative stress induced NMDA receptor alteration leads to spatial memory deficits in temporal lobe epilepsy: ameliorative effects of Withania somnifera and Withanolide A. Neurochem Res 37:1915–1927

Spector AR, Dugger BN, Wszolek ZK, Uitti RJ, Fredrickson P, Kaplan J, Boeve BF, Dickson DW, Strongosky A, Lin SC (2011) Anatomy of disturbed sleep in pallido-ponto-nigral degeneration. Ann Neurol 69:1014–1025

Thaiparambil JT, Bender L, Ganesh T, Kline E, Patel P, Liu Y, Tighiouart M, Vertino PM, Harvey RD, Garcia A, Marcus AI (2011) Withaferin A inhibits breast cancer invasion and metastasis at sub-cytotoxic doses by inducing vimentin disassembly and serine 56 phosphorylation. Int J Cancer 129:2744–2755

Tohda C (2008) Overcoming several neurodegenerative diseases by traditional medicines: the development of therapeutic medicines and unraveling pathophysiological mechanisms. Yakugaku Zasshi 128:1159–1167

Tohda C, Joyashiki E (2009) Sominone enhances neurite outgrowth and spatial memory mediated by the neurotrophic factor receptor, RET. Br J Pharmacol 157:1427–1440

Tripathi AK, Dey S, Singh RH, Dey PK (1998) Alterations in the sensitivity of 5(th) receptor subtypes following chronic asvagandha treatment in rats. Anc Sci Life 17:169–181

Uitti RJ, Wszolek ZK (2003) Dopamine agonists, sleep disorders, and driving in Parkinson's disease. Adv Neurol 91:343–349

Vareed SK, Bauer AK, Nair KM, Liu Y, Jayaprakasam B, Nair MG (2014) Blood-Brain Barrier Permeability of Bioactive Withanamides Present in Withania somnifera Fruit Extract. Phytother Res 28:1260–1264

Vedi M, Rasool M, Sabina EP (2014) Protective effect of administration of Withania somifera against bromobenzene induced nephrotoxicity and mitochondrial oxidative stress in rats. Ren Fail 36:1095–1103

Ven Murthy MR, Ranjekar PK, Ramassamy C, Deshpande M (2010) Scientific basis for the use of Indian ayurvedic medicinal plants in the treatment of neurodegenerative disorders: ashwagandha. Cent Nerv Syst Agents Med Chem 10:238–246

Wscieklica T, de Barros VM, Le Sueur ML, Pouza KC, Spadari RC, Cespedes IC (2016) Alcohol consumption increases locomotion in an open field and induces Fos-immunoreactivity in reward and approach/withdrawal-related neurocircuitries. Alcohol 50:73–82

Yang Z, Garcia A, Xu S, Powell DR, Vertino PM, Singh S, Marcus AI (2013) Withania somnifera root extract inhibits mammary cancer metastasis and epithelial to mesenchymal transition. PLoS One 8:e75069

Yin H, Cho DH, Park SJ, Han SK (2013) GABA-mimetic actions of Withania somnifera on substantia gelatinosa neurons of the trigeminal subnucleus caudalis in mice. Am J Chin Med 41:1043–1051

Chapter 15
Cellular and Molecular Targets Underpinning Memory Enhancement by Ashwagandha

Arpita Konar and Mahendra K. Thakur

Abstract The search for therapeutic candidates of memory disorders including gene targets and compounds both synthetic and natural has been a prime arena of neurobiology research. Amongst suggested therapeutic compounds, several herbal products with a long history of use in Ayurveda have gained attention in modern medicine. Ashwagandha (*Withania somnifera*) also referred to as "Queen of Ayurveda" is at the zenith of Ayurvedic herbs owing to its tremendous potential to recover memory decline in aging and neurodegenerative pathologies as well as enhance basal memory function of healthy individuals. Despite such promising effects, limited mechanistic evidences have hindered its acceptance in modern medicine. However, technical advances in neuroscience research over the past decade have filled-in some gaps in understanding of molecular and mechanistic biology of Ashwagandha effects. In this chapter, we highlight the studies that have deciphered the cellular and molecular mechanisms of memory enhancing potential of Ashwagandha in various disease models. Cellular targets of Ashwagandha include (i) activation of antioxidant defence system rescuing nerve cells from apoptosis, oxidative stress and DNA damage, (ii) induction of cholinergic system and (iii) up-regulation of memory linked neuroplasticity genes and neuronal arborisation. All of these molecular effects translate into increase in memory. Such multiple-module action has intrigued research to unravel upstream master regulators of Ashwagandha effects on gene expression, cell physiology and behaviour.

Keywords Ashwagandha • Anti-oxidant • Induction of cholinergic system • Enhance memory function • Multiple-module action

A. Konar (✉)
CSIR-Institute of Genomics & Integrative Biology, New Delhi 110025, India
e-mail: arpita.konar@igib.in

M.K. Thakur
Biochemistry and Molecular Biology Laboratory, Brain Research Centre, Department of Zoology, Banaras Hindu University, Varanasi 221005, India
e-mail: mkt_bhu@yahoo.com

S.C. Kaul, R. Wadhwa (eds.), *Science of Ashwagandha: Preventive and Therapeutic Potentials*, DOI 10.1007/978-3-319-59192-6_15

15.1 Memory Disorders

Memory is a higher order cognitive brain function that allows us to remember joyful and horrific past experiences, and connect them to present and future. However, this unique possession, that enriches our lives as individuals, gets lost and destroys our sense of self. Several mental health trajectories including neurodegenerative disorders like Alzheimer's disease, psychiatric illness of schizophrenia and depression, traumatic brain injuries, and brain tumours are associated with drastic memory decline (Fig.15.1). Memory decline or amnesia is generally classified as anterograde when the individual is unable to form and retain new information or retrograde when already encoded memory is not retrieved during recall (Gold 2006; Budson and Price 2005; Bartsch and Deuschl 2010). Depending on the time of information, it can be a short-term or long-term memory loss.

Amongst memory disorders, dementia particularly Alzheimer's disease (AD) stands out for significant loss of both short-term and long-term declarative memory even at the early stages. Anatomical and functional deterioration of brain correlates with impaired memory of patients (Holger 2013). Other forms of dementia including vascular and fronto-temporal involves decline in declarative memory. Psychiatric illness in general results in deficits of declarative and procedural memory as evident from depression, schizophrenia, attention deficit–hyperactivity disorder, and obsessive–compulsive disorder patients. Specific impairment of episodic memory is asso-

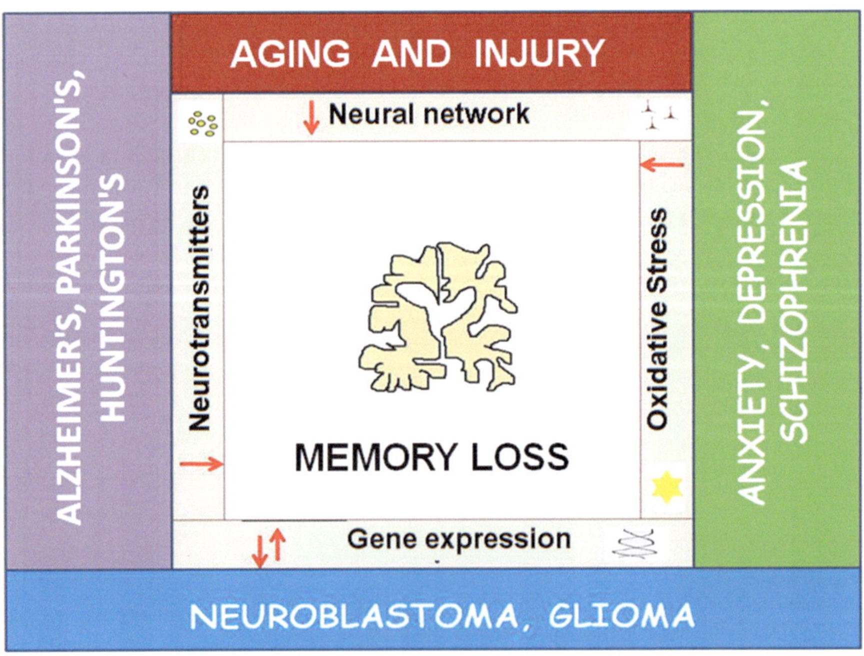

Fig. 15.1 Memory disorders and their causative factors

ciated with mild cognitive impairment, encephalitis, Korsakoff's syndrome, traumatic brain injury, seizure, hypoxic–ischemic injury, cardiopulmonary bypass, side effects of medication, deficiency of vitamin B12, hypoglycemia, anxiety and temporal-lobe surgery, whereas semantic memory loss occurs in traumatic brain injury (McKay and Kopelman 2009).

Besides pathological conditions, normal brain aging is associated with specific impairment of learning and memory. Similar to other organs, brain also undergoes several anatomical and physiological changes including shrinkage in volume, expansion in ventricular system, decrease in number of synapses and synaptic spines, and length of myelinated axons. These changes eventually contribute to altered brain plasticity and network dynamics leading to age-related memory deficits (Morrison and Baxter 2012). Detailed anatomical and electrophysiological studies reveal that aged brain exhibits loss of hippocampal synapses leading to impairment of synaptic plasticity including deficits in LTP and LTD. Such disruption of LTP and LTD impairs the encoding of new memories and enhances the erasure of stored memories, contributing to memory deficits with physiological aging (Kidd 2008; Bano et al. 2011). Although the brain aging leads to delayed recall of information and greater difficulty in learning new information process, all types of memory are not affected equally in this process. Aging basically affects episodic memory related to specific events and experiences. Usually, aging does not exhibit significant decline in semantic memory but aged people fail to remember specific details. Aged people also show difficulty in manipulation, reorganization and integration of contents of working memory (Schulze et al. 2011).

15.2 Ayurvedic Interventions for Memory Enhancement

Most recently, it has been established that memory is not erased permanently. It is temporarily lost with stress and age, and could be recovered. In this regard, elucidation of mechanisms underlying memory loss has been considered central and prioritized for formulation of targeted therapies. The challenges lie in elucidation of signalling pathways underlying the multiple causes of memory deficits. Moreover, individual cases of memory impairment resulting from psychiatric and neurodegenerative disorder have genetic and environmental heterogeneity conferring additional hurdle to the designing of drugs. Several drugs and bio-molecules have been tested to ameliorate amnesia but due to complex etiology and multi-factorial nature and side effects, an ideal recovery agent is lacking till date (Lee and Silva 2009). These limitations necessitated the development of an alternative approach that can exert a holistic impact on memory and other cognitive abilities with minimum side effects and can restore deficits irrespective of the specific causative factor.

An answer to this bewildering problem came from world's oldest traditional medicinal practice of Ayurveda that has a holistic approach of promoting good health through maintenance of body homeostasis. Ayurveda gained prominence in diverse fields including immunology, geriatrics, genetics, neurology and psychiatry.

The ayurvedic home medicines comprise of herbal extracts combined seldom with supplements and those specific for nervous system are classified as Medhya Rasayana referring to the herbal preparation that can improve mental health (Mishra et al. 2000; Winters 2006; Deocaris et al. 2008; Modak et al. 2007). These extracts exert anti-aging and cognition enhancing effects by amelioration of stress and regeneration of damaged neural cells (Ven Murthy et al. 2010). Ashwagandha (*Withania somnifera*), Brahmi (*Bacopa monnieri*), Sankhapuspi (*Convolvulus pluricaulis*), Jatamansi, Jyotishmati, Mandukparni and Vacha are the renowned medhya rasayanas of Ayurvedic literature (Singh et al. 2011).

15.3 Ashwagandha as a Promising Memory Enhancer

Ashwagandha is the most recognized nootropic or natural memory and cognitive enhancer-herb of Ayurveda. It belongs to Solanaceae family and found in continents of Asia and Africa. It is cultivated on a large scale in India predominantly in the states of Madhya Pradesh, Uttar Pradesh, Punjab, Gujarat and Rajasthan. Almost all parts of this shrub are used for various health benefits. The nomenclature Ashwagandha originated from the odor of its root that resembled that of a horse. Its scientific name *Withania somnifera* signifies its well-known sleep inducing activity that might be related to its stress relieving capacity. It is also known asWinter Cherry/ Indian Ginseng/ Queen of Ayurveda. The wide spectrum of phytochemicals and chemotypes of Ashwagandha have been attributed to its diverse implications and health benefits. Laboratory and preclinical studies have demonstrated its antioxidant, anti-inflammatory, anti-stress, anti-diabetic, cardio- and neuroprotective potential. Although diverse in application, Ashwagandha extracts have been considered particularly relevant for treatment of neurological disorders like brain malignancies, neurodegeneration and memory-afflicting injuries (Kuboyama et al. 2005; Kulkarni and Dhir 2008; Singh et al. 2008; Rajasankar et al. 2009).

In both *in vitro* and animal models of memory disorders, Ashwagandha extracts have been proven to be a promising therapeutic candidate natural drug. The extracts show enormous potential to modulate neurotransmitters, rescue oxidative damage, enhance neural plasticity and regenerate damaged neurons and synapses. Although there are a large number of studies, some are worth mentioning in which drastic and convincing effects have been reported. In different transgenic mouse models of neurodegeneration, Ashwagandha extracts ameliorated memory impairment (Kuboyama et al. 2014). APP/PS1 AD transgenic mice showed deficits in memory as evident by specific behavioural paradigms including water maze and radial arm maze test. Prolonged oral administration of methanol extracts of Ashwagandha root recovered memory loss and cleared amyloid beta (Aβ) plaques in cortex and hippocampus, hallmark of AD (Sehgal et al. 2012). Aβ based cellular and animal models of AD have also been instrumental in screening therapeutic agents. The cell culture models showed significant disruption of neural network involving both dendrite and axonal atrophy and premature neurodegeneration. The crude extract and some of its active principles rendered neuro-protection and neuroregeneration against Aβ-induced neurotoxicity. The major components include Withanolide A, Withanoside IV, and Withanoside VI isolated from root extracts that stimulated dendrite and axonal growth and reconstructed damaged synapses in Aβ-induced neuroblastoma cell lines, primary cortical neuronal culture and AD mice brain (Tohda 2008; Kumar et al. 2010; Jayaprakasam et al. 2010; Liu et al. 2012; Kurapati et al. 2013; Jansen et al. 2014). The extracts also revived Aβ-induced loss in dendrite spine density, reduction in length and area, and distortion of morphology. Withanoside IV derivative, sominone also recovered damage in neuronal processing and connectivity and loss in memory acquisition in AD transgenic mice. Sominone also enhanced the basal memory capacities of control animals (Kuboyama et al. 2006; Tohda and Joyashiki 2009). Sankar et al. (2007) reported that in Parkinson's disease animal models, Ashwagandha root extracts also exhibited neuro-protection as manifested by biochemical and behavioural correlates.

Ashwagandha also improved memory in pharmacological and physiological animal models of amnesia. Scopolamine is a potent acetylcholine receptor antagonist that inhibits central cholinergic neuronal activity and impairs memory as evident by several behavioral parameters like water maze and passive avoidance latency test. Scopolamine-induced amnesic mouse model is one of the well-recognized pharmacological animal models of memory dysfunction. In our studies, we found anti-amnesic effect of Ashwagandha leaf extract in scopolamine-challenged mice (Konar et al. 2011). Purified active component, withanone, also restored scopolamine mediated degenerative alterations in brain. Ashwagandha also markedly attenuated memory decline in hypoxic rats and withanolide A contributed primarily to the recovery potential (Baitharu et al. 2013). Although several studies corroborate the nootropic potential of Ashwagandha as mentioned in Ayurveda, clinical trials warrant a comprehensive understanding of cellular and molecular mechanism underlying memory enhancing properties of Ashwagandha. Therefore, the present chapter summarizes the mechanistic evidences revealed till date and the future directions of study, needed for therapeutic application.

15.4 Cellular and Molecular Mechanism of Memory-Enhancing Potential of Ashwagandha in Diverse Model Systems

15.4.1 Activation of Antioxidant Defence System

Brain is vulnerable to oxidative damage owing to high rate of ROS (Reactive Oxygen Species) generation and inadequate antioxidant mechanisms. Consequently, oxidative stress contributes to majority of brain dysfunctions including memory impairment (Mattson 2004). Aging and neurodegenerative diseases such as Parkinson's, Alzheimer and Huntington's associated memory decline results largely from oxidative damage (Di Carlo et al. 2012; Shukla et al. 2011). Ashwagandha extracts exhibit enormous potential of scavenging free radicals and activating antioxidant signalling pathways in various chemically induced and physiological models of memory loss both *in vitro* and *in vivo*.

Scopolamine hydrobromide is a plant alkaloid derivative used for inducing memory loss mimicking aging and dementia, and hence screening of active recovery agents. Till date its amnesic action has been attributed to impaired cholinergic neurotransmission, and it has been characterized as acetylcholine receptor antagonist (Klinkenberg and Blokland 2010). We reported that scopolamine causes significant cytotoxicity, as evident by TUNEL assay, and is associated with dramatic increase in DNA damage and oxidative stress markers in neuronal and glial cell lines. Alcoholic extract of Ashwagandha leaves (i-Extract) and its active principle withanone considerably protected the cells from oxidative damage, indicated by reduced ROS level. It caused significant reduction in the level of scopolamine-induced DNA damage as shown by decreased gamma H2AX (Konar et al. 2011). Rotenone is a neurotoxic compound that mimics Parkinson's associated memory and cognitive impairment. Several studies have correlated the degenerating action of rotenone with oxidative stress. Rotenone when intraperitoneally administered to mice at a dose of 1mg/kg BW for 7 days induced ROS malondialdehyde (MDA), hydroperoxides (HP) and nitric oxide (NO) levels in cerebellum and striatum. It also hindered the antioxidant defence mechanism by reducing glutathione (GSH) levels, inactivating antioxidant enzymes and causing mitochondrial dysfunctions. The oral administration of Withania root extract (400mg/kgBW) for 4 weeks normalized the level of oxidative stress markers and restored GSH level (Manjunath and Murlidhar 2013). Ashwagandha i-Extract at the dose of 100 mg/kgBW for 7 days also restored memory loss by rescuing from oxidative damage in other PD mouse models. It significantly induced superoxide dismutase, catalase and malondialdehyde levels as well as reduced the levels of glutathione and glutathione peroxidase in corpus striatum of PD mouse (Rajasankar et al. 2009).

Amongst physiological models, hypoxia induced memory impairment gains utmost prominence in replicating oxidative damage in memory disorders. The lack of oxygen supply at high altitude causes oxidative stress, degenerates neural cells and affects memory function. This phenomenon known as hypobaric hypoxia is modelled experimentally and Ashwagandha extracts with strong antioxidant potential has been extremely successful in ameliorating stress and restoring memory. Baitharu et al. (2013) reported that crude root extracts decreased the hippocampal level of nitric oxide, corticosterone and ROS whereas withanolide-A increased GSH influencing the free radical scavenging mechanism attenuating overall oxidative stress of hypoxic rats. Withanolide-A also reduced neuronal apoptosis by decreasing activity of caspase-3 in hippocampal cells. Molecular analysis revealed that withanolide-A induced glutathione biosynthesis in neurons by upregulating GCLC level through Nrf2-mediated antioxidant signaling cascade (Baitharu et al. 2014).

In an experimental animal model of epilepsy, Ashwagandha crude root extract as well as active component withanolide-A was neuro-protective and recovered spatial memory loss. Its prime mechanism of action was reduction of oxidative stress, thereby preventing neurodegeneration and inhibition of neurotransmission. Epileptic rodents exhibited significant degeneration of hippocampal neurons accompanied by elevation of oxidative damage as evident by increase in superoxide dismutase and catalase activity. These animals were also deficient in hippocampal dependent spatial memory assessed by radial arm maze paradigm. The administration of total root extract and individual withanolide A substantially attenuated the increase in gene expression and activation of enzymes and almost restored it to the level of control animals. Lipid peroxidation increase in experimental rats was also reduced by the extract. These antioxidant activities of the extract were well reflected in rescue of hippocampal neuronal death and enhancement of memory (Soman et al. 2012).

15.4.2 Modulation of Neurotransmitter Level and Metabolism

The active principles of plant extracts have been reported to activate and inhibit neurotransmitter signalling and thereby modulate neural excitability and transmission. Acetylcholine signalling has been the most attractive target for most of the agents, since it is abundant in brain and implicated in learning and memory (Kumar and Khanum 2012).

The impairment of brain cholinergic neurotransmission owing to degeneration of cholinergic neurons or deficits in receptor signalling cascade is considered crucial for memory decline (Schliebs and Arendt 2011). Cholinergic hypo-function is pivotal for aging and AD related memory deficits and consequently agents stimulating cholinergic system are important for recovery (McKinney and Jacksonville 2005). Withaferin and sitoindosides VII-X isolated from methanolic root extracts enhanced muscarinic receptor binding and acetylcholinesterase (AChE) activity in frontal cortex of male Wistar rats (Mishra et al. 2000). As discussed above, we also demonstrated that the i-Extract attenuates amnesic action of muscarinic antagonist, scopolamine hydrobromide. Gautam et al. (2013) reported that i-Extract normalized the scopolamine mediated increase in AChE activity in mouse cerebral cortex and hippocampus. Ashwagandha root extract and its components withaferin A and withanolide A also modulate the inhibitory GABA receptors of brain (Candelario et al. 2015). Bhattarai et al. (2010) had also reported earlier that the methanolic root extract acts on $GABA_A$ receptor ($GABA_AR$) receptors at the synapses, thus may be crucial for neuroprotection and restoration of lost memory. Ashwagandha extracts also modulated NMDA receptors in epilepsy-induced memory impairment model of rats. Oxidative insult severely affected memory by decreasing the expression and density of NMDA receptors. The root extract of Ashwagandha increased the level of NMDA receptors and recovered memory loss (Soman et al. 2012). In another study by Bhattarai et al. (2013), methanolic extract of Ashwagandha root was reported to act on hippocampal NMDA receptors by activation of the glycine binding site. Hippocampal induction of LTP requires glycinergic activation of NMDA receptors and LTP forms the chief cellular mechanism of memory. Therefore, it was speculated that Ashwagandha by acting as a glycine mimetic and increasing NMDA receptor mediated neurotransmission might induce LTP and enhance memory.

15.4.3 Enhancement of Neural Network

Repair of the damaged neural network including dendrite and axonal regeneration and reconstruction of synaptic connections are the prerequisites for recovery of memory loss. Moreover, enhancement of the neural network in healthy individuals by processes such as neurogenesis and synaptogenesis is implicated in gain of memory function. Interestingly, various extracts of Ashwagandha have exhibited the property of regenerating damaged neural cells in memory disorder models as well as enhancing the existing network in healthy subjects.

It is well known that dementia, particularly AD pathology, entails significant loss of memory interfering with the molecular machinery and neural connections (Holger 2013). The administration of Aβ in neuroblastoma cell lines, primary cortical neurons and mouse brain has been successful in mimicking the network damage in AD including atrophy of axons and decrease in synaptic densities (Kurapati et al. 2013; Jayaprakasam et al. 2010; Liu et al. 2012). The administration of Ashwagandha crude root extract as well as its active principles (Withanolide A, Withanoside IV, Withanoside VI and sominone) increased axonal growth and reformed synapses (Zhao et al. 2002; Kuboyama et al. 2006) in cell culture models. These phytochemicals also demonstrated their regenerating potential in damaged cortical neurons and enhanced spatial and recognition memory in 5XFAD and Aβ25–35 treated mouse model of AD (Tohda 2008). Sominone, the derivative of Withanoside VI, stimulated neurite growth and increased memory even in normal mice (Tohda and Joyashiki 2009).

As already discussed, scopolamine-induced amnesia is another well-recognized model replicating the cellular and molecular changes observed in age related memory decline and neurodegenerative disorders. Scopolamine significantly decreased dendrite length and branching of cultured hippocampal neurons. It also distorted the mushroom shape of dendritic spines, and reduced the overall spine density in mouse hippocampus *in vivo*. Pre-treatment with Ashwagandha leaf extract prevented scopolamine-induced damages to dendrites whereas its post-treatment restored the loss of dendrite growth and morphology both *in vitro* and *in vivo* models. Gene expression analysis revealed that the leaf extract up regulates hippocampal microtubule associated protein (MAP2) whose deficiency is known to affect dendrite elongation, maturation of hippocampal neurons and neuronal migration (Harada et al. 2002).

The alteration in dendrite spine morphology relies primarily on the rearrangement of actin cytoskeleton. The monomeric G-actin and filamentous polymeric, F-actin are present in spines and the degree of actin polymerization is an integral phenomenon of spine formation, maturation and plasticity contributing to memory formation (Cingolani and Goda 2008). Loss of actin-polymerization has been implicated in age-associated memory loss (Johnson et al. 2015). Drug abuse and psychiatric disorders related memory loss has also been linked with the preservation of polymerized actin and thus structural plasticity of neurons (Young et al. 2015).

We observed that scopolamine drastically reduced actin polymerization as depicted by phalloidin staining and also down-regulated the molecular marker Arc (activity- regulated cytoskeletal associated protein). Moreover, intra-hippocampal Arc knockdown also reduced polymerization of actin comparable to that of scopolamine (Gautam et al. 2015). i-Extract increased dendrite spine density in the hippocampus of scopolamine-treated amnesic mice as evident from Golgi staining by up regulating the Arc expression and attenuating the loss of polymerized actin.

15.4.4 Induction of Memory Linked Gene Expression

The etiology of memory loss in aging and age-associated brain disorders is multidimensional, but accumulating evidence suggests that change in gene expression is fundamental. Over the past decade, it has been substantially demonstrated that de novo gene expression in an orchestrated manner in cortex and hippocampus is required for long-term memory formation. Moreover, the traditional techniques of genomics combined with next generation sequencing methods revealed that memory function involves dynamic expression pattern of a wide spectrum of genes (Cavallaro et al. 2002; Igaz et al. 2004). These include transcription factors, immediate early genes, neurotrophic factors, cytoskeletal elements, synaptic proteins, membrane receptors and intracellular kinases. Interestingly, Ashwagandha extracts modulated the expression of almost all the categories in different model systems of memory loss.

In search of genes regulated by neuronal activity, a subset of immediate early genes (IEGs) was discovered that are crucial for synaptic plasticity and memory consolidation. Amongst these IEGs, Arc is acknowledged as the master regulator of protein synthesis dependent synaptic plasticity and declarative as well as procedural memory consolidation. Arc regulates actin cytoskeletal dynamics and AMPA receptor endocytosis integrating neuronal activity with synaptic strength (Bramham et al. 2009). Arc gene ablation significantly impaired late phase of LTP and long term memory (Guzowski et al. 2000). Studies in our laboratory reported that in scopolamine-treated mice, Arc expression gets drastically reduced in the cerebral cortex and hippocampus, and was closely associated with impaired spatial memory. The oral administration of i-Extract recovered memory decline and simultaneously up regulated Arc expression in both the regions of brain. i-Extract also caused increase in Arc level in control animals. Most noticeably, changes in Arc expression in control and scopolamine-treated mice were more conspicuous in hippocampus than cortex (Gautam et al. 2013). These studies have demonstrated that ARC is a cellular target for scopolamine-induced loss of memory and its recovery by i-Extract (Gautam et al. 2013). Arc protein translation in synapses is regulated by brain derived neurotrophic factor (BDNF), another activity evoked gene and a member of the nerve growth factor family that regulates proliferation and differentiation of neurons and maintains their architecture (Yamada and Nabeshima 2003). BDNF is crucial for synaptogenesis, maturation and stabilization, thereby influencing both short- and long-term forms of synaptic plasticity and memory. A growing body of evidence suggests that BDNF is a molecular marker of memory decline resulting from aging, Alzheimer's and psychiatric illness of major depressive disorder (Hubka 2006; Waterhouse and Xu 2009; Bekinschtein et al. 2013; Leal et al. 2013). As anticipated, in our studies, BDNF showed a marked decrease in the brain of scopolamine-treated amnesic mice. i-Extract-treated control and amnesic mice showed up regulation in BDNF mRNA and protein expression.

Ashwagandha was also shown to regulate the expression of various synaptic and cytoskeletal proteins. In sleep-deprived model of memory impairment, significant up regulation of PSA-NCAM, NCAM, and pre-synaptic marker SNAP-25 was observed in hippocampus and frontal cortex of rat brain. These protein levels remained unaltered in ASH-WEX (water extract of Ashwagandha leaves) treated animals, indicating that it rescued stress-induced synaptic rearrangements, thereby maintaining synaptic plasticity. Cytoskeletal elements including GAP-43, NF-H and NeuN marker of neuronal proliferation also showed increase in the hippocampus of 6h sleep-deprived rats. Pre-treatment with ASH-WEX restored gene expression. i-Extract also rescued scopolamine-induced damage to neuronal cells by up-regulation of the expression of neuroplasticity markers, NF-H, MAP2, PSD-95 and GAP-43. An emerging aspect of memory biology is the role of glial cells, which are considered to support neurons. Accumulating experimental evidences have suggested their involvement in modulating neuronal metabolic activity, plasticity and thus memory. In particular, astrocytes actively control synaptic transmission by participating in the formation of neuronal processes, cytoskeletal framework and myelin sheath (Ben Achour and Pascual 2010). Glial fibrillary acid protein (GFAP) is an astrocytic protein, an established marker of astrogenesis and astrocyte activation. Decreased GFAP expression in hippocampus has often been linked to brain injury and chemically induced memory loss (Havekes et al. 2012; Silva et al. 2013). We observed that whereas amnesic drug scopolamine reduced GFAP expression in mouse brain and C6 glioma cell line, i-Extract attenuated the decrease both *in vitro* and *in vivo* (Konar et al. 2011).

15.5 Conclusion

The *in vitro* and animal models of memory disorders have demonstrated Ashwagandha as a promising recovery herb with high translational value. The traditional cellular, biochemical and molecular approaches combined with novel in-silico tools have deciphered its active principles and mechanisms substantiating the application in clinical trials. The standardized root and leaf extracts of Ashwagandha as well as their bioactives have exhibited preventive and therapeutic potential in diverse pharmacological, genetic and physiological models of memory loss. The wide spectrum of neuromodulatory effects of these extracts including antioxidant

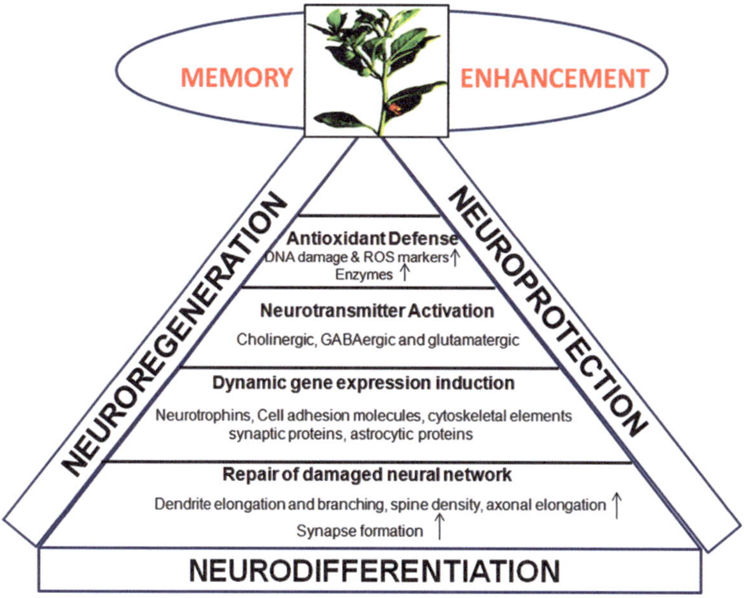

Fig. 15.2 Holistic cellular and molecular targets of Ashwagandha in memory enhancement

defense, activation of neurotransmitter signalling, gene expression changes and regeneration of damaged neural network makes it an appropriate therapeutic candidate (Fig. 15.2). Ashwagandha upregulates the expression of a wide range of memory linked genes coding for neurotrophins, cytoskeletal elements, cell adhesion molecules and synaptic proteins. These observations also intrigues to search for a master regulatory mechanism probably any epigenetic switch attributing to such holistic action of the herb. Taken together, with such convincing evidences for nootropic potential of Ashwagandha it's high time for clinical trials in neurological disorders involving memory deficiencies.

Acknowledgements The work cited in this article from authors' laboratory has been supported by grants from the Department of Science and Technology (SR/SO/HS-54/2008) and Department of Biotechnology (BT/PR3996/MED/97/57/2011), Government of India.

References

Baitharu I, Jain V, Deep SN, Hota KB, Hota SK, Prasad D, Ilavazhagan G (2013) Withania somnifera root extract ameliorates hypobaric hypoxia induced memory impairment in rats. J Ethnopharmacol 145:431–441

Baitharu I, Jain V, Deep SN, Shroff S, Sahu JK, Naik PK, Ilavazhagan G (2014) Withanolide A prevents neurodegeneration by modulating hippocampal glutathione biosynthesis during hypoxia. PLoS One 9:e105311

Bano D, Agostini M, Melino G, Nicotera P (2011) Ageing, neuronal connectivity and brain disorders: an unsolved ripple effect. Mol Neurobiol 43:124–130
Bartsch T, Deuschl G (2010) Transient global amnesia: functional anatomy and clinical implications. Lancet Neurol 9:205–214
Bekinschtein P, Cammarota M, Medina JH (2013) BDNF and memory processing. Neuropharmacol 76:677–683
Ben Achour S, Pascual O (2010) Glia: the many ways to modulate synaptic plasticity. Neurochem Int 57:440–445
Bhattarai JP, Ah Park S, Han SK (2010) The methanolic extract of Withania somnifera ACTS on GABAA receptors in gonadotropin releasing hormone (GnRH) neurons in mice. Phytother Res 24:1147–1150
Bhattarai JP, Park SJ, Han SK (2013) Potentiation of NMDA receptors by Withania somnifera on hippocampal CA1 pyramidal neurons. Am J Chin Med 41:503–513
Bramham CR, Alme MN, Bittins M, Kuipers SD, Nair RR, Pai B, Panja D, Schubert M, Soule J, Tiron A, Wibrand K (2009) The Arc of synaptic memory. Exp Brain Res 200:125–140
Budson AE, Price BH (2005) Memory dysfunction. N Engl J Med 352:692–699
Candelario M, Cuellar E, Reyes-Ruiz JM, Darabedian N, Feimeng Z, Miledi R, Russo-Neustadt A, Limon A (2015) Direct evidence for GABAergic activity of Withania somnifera on mammalian ionotropic GABAA and GABAρ receptors. J Ethnopharmacol 171:264–272
Cavallaro S, D'Agata V, Manickam P, Dufour F, Alkon DL (2002) Memory-specific temporal profiles of gene expression in the hippocampus. Proc Natl Acad Sci U S A99:16279–16284
Cingolani LA, Goda Y (2008) Actin in action: the interplay between the actin cytoskeleton and synaptic efficacy. Nat Rev Neurosci 9:344–356
Deocaris CC, Widodo N, Wadhwa R, Kaul SC (2008) Merger of ayurveda and tissue culture-based functional genomics: inspirations from systems biology. J Transl Med 6:1–8
Di Carlo M, Giacomazza D, Picone P, Nuzzo D, San Biagio PL (2012) Are oxidative stress and mitochondrial dysfunction the key players in the neurodegenerative diseases? Free Radic Res 46:1327–1338
Gautam A, Kaul SC, Thakur MK (2015) Alcoholic extract of Ashwagandha leaves protects against amnesia by regulation of Arc function. Mol Neurobiol 53:1760–1769
Gautam A, Wadhwa R, Thakur MK (2013) Involvement of hippocampal Arc in amnesia and its recovery by alcoholic extract of Ashwagandha leaves. Neurobiol Learn Mem 106:177–184
Gold PE (2006) The many faces of amnesia. Learn Mem 13:506–514
Guzowski JF, Lyford GL, Stevenson GD, Houston FP, McGaugh JL, Worley PF, Barnes CA (2000) Inhibition of activity-dependent arc protein expression in the rat hippocampus impairs the maintenance of long-term potentiation and the consolidation of long-term memory. J Neurosci 20:3993–4001
Harada A, Teng J, Takei Y, Oguchi K, Hirokawa N (2002) MAP2 is required for dendrite elongation, PKA anchoring in dendrites, and proper PKA signal transduction. J Cell Biol 158:541–549
Havekes R, Vecsey CG, Abel T (2012) The impact of sleep deprivation on neuronaland glial signaling pathways important for memory and synaptic plasticity. Cell Signal 24:1251–1260
Holger J (2013) Memory loss in Alzheimer's disease. Dialogues Clin Neurosci 15:445–454
Hubka P (2006) Neural network plasticity, BDNF and behavioral interventions in Alzheimer's disease. Bratisl Lek Listy 107:395–401
Igaz LM, Bekinschtein P, Vianna MM, Izquierdo I, Medina JH (2004) Gene expression during memory formation. Neurotox Res 6:189–204
Jansen RL, Brogan B, Whitworth AJ, Okello EJ (2014) Effects of five Ayurvedic herbs on locomotor behaviour in a Drosophila melanogaster Parkinson's disease model. Phytother Res 28:1789–1795
Jayaprakasam B, Padmanabhan K, Nair MG (2010) Withanamides in Withania somnifera fruit protect PC-12 cells from beta-amyloid responsible for Alzheimer's disease. Phytother Res 24:859–863

Johnson JL, Huang W, Roman G, Costa-Mattioli M (2015) TORC2: a novel target for treating age-associated memory impairment. Sci Rep 5:15193

Kidd PM (2008) Alzheimer's disease, amnestic mild cognitive impairment, and age-associated memory impairment: current understanding and progress toward integrative prevention. Altern Med Rev 13:85–115

Klinkenberg I, Blokland A (2010) The validity of scopolamine as a pharmacological model for cognitive impairment: a review of animal behavioral studies. Neurosci Biobehav Rev 34:1307–1350

Konar A, Shah N, Singh R, Saxena N, Kaul SC, Wadhwa R, Thakur MK (2011) Protective role of Ashwagandha leaf extract and its component withanone on scopolamine-induced changes in the brain and brain-derived cells. PLoS One 6:e27265

Kuboyama T, Tohda C, Komatsu K (2005) Neuritic regeneration and synaptic reconstruction induced by withanolide A. Br J Pharmacol 144:961–971

Kuboyama T, Tohda C, Komatsu K (2006) Withanoside IV and its active metabolite, sominone, attenuate Abeta(25-35)-induced neurodegeneration. Eur J Neurosci 23:1417–1426

Kuboyama T, Tohda C, Komatsu K (2014) Effects of Ashwagandha roots of Withania somnifera on neurodegenerative diseases. Biol Pharm Bull 37:892–897

Kulkarni SK, Dhir A (2008) Withania somnifera: an Indian ginseng. Prog Neuropsychopharmacol Biol Psychiatry 32:1093–1105

Kumar GP, Khanum F (2012) Neuroprotective potential of phytochemicals. Pharmacogn Rev 6:81–90

Kumar S, Seal CJ, Howes MJ, Kite GC, Okello EJ (2010) In vitro protective effects of Withania somnifera (L.) dunal root extract against hydrogen peroxideand beta-amyloid (1-42)-induced cytotoxicity in differentiated PC12 cells. Phytother Res 24:1567–1574

Kurapati KR, Atluri VS, Samikkannu T, Nair MP (2013) Ashwagandha (Withania somnifera) reverses beta-amyloid1-42 induced toxicity in human neuronal cells:implications in HIV-associated neurocognitive disorders (HAND). PLoS One 8:e77624

Leal G, Comprido D, Duarte CB (2013) BDNF-induced local protein synthesis and synaptic plasticity. Neuropharmacology 76:639–656

Lee Y, Silva A (2009) The molecular and cellular biology of enhanced cognition. Nat Rev Neurosci 10:126–140

Liu LF, Durairajan SS, Lu JH, Ko I, Li M (2012) In vitro screening on amyloid precursor protein modulation of plants used in Ayurvedic and traditional Chinese medicine for memory improvement. J Ethnopharmacol 141:754–760

Manjunath MJ, Muralidhara (2013) Effect of Withania somnifera supplementation on rotenone-induced oxidative damage in cerebellum and striatum of the male mice brain. Am J Chin Med 41:503–513

Mattson MP (2004) Pathways towards and away from Alzheimer's disease. Nature 430:631–639

McKay G, Kopelman MD (2009) Psychogenic amnesia: when memory complaints are medically unexplained. Adv Psychiatr Treat 15:152–158

McKinney M, Jacksonville MC (2005) Brain cholinergic vulnerability: relevance to behavior and disease. Biochem Pharmacol 70:1115–1124

Mishra LC, Singh BB, Dagenais S (2000) Scientific basis for the therapeutic use of Withania somnifera (ashwagandha): a review. Altern Med Rev 5:334–346

Modak M, Dixit P, Londhe J, Ghaskadbi S, Paul ADT (2007) Indian herbs and herbal drugs used for the treatment of diabetes. J Clin Biochem Nutr 40:163–173

Morrison JH, Baxter MG (2012) The ageing cortical synapse: hallmarks and implications for cognitive decline. Nat Rev Neurosci 13:240–250

RajaSankar S, Manivasagam T, Sankar V, Prakash S, Muthusamy R, Krishnamurti A, Surendran S (2009) Withania somnifera root extract improves catecholamines and physiological abnormalities seen in a Parkinson's disease model mouse. J Ethnopharmacol 125:369–373

Sankar SR, Manivasagam T, Krisnamurt A, Ramanathan M (2007) The neuroprotective effect of Withania somnifera root extract in MPTP-intoxicated mice: An analysis of behavioural and biochemical variables. Cell Mol Biol Lett 12:473–481
Schliebs R, Arendt T (2011) The cholinergic system in aging and neuronal degeneration. Behav Brain Res 221:555–563
Schulze ET, Geary EK, Susmaras MT, Paliga JT, Maki PM, Little DM (2011) Anatomical correlates of age-related working memory declines. J Aging Res 1–9
Sehgal N, Gupta A, Valli RK, Joshi SD, Mills JT, Hamel E, Khanna P, Jain SC, Thakur SS, Ravindranath V (2012) Withania somnifera reverses Alzheimer's disease pathology by enhancing low-density lipoprotein receptor related protein in liver. Proc Natl Acad Sci USA 109:3510–3515
Shukla KK, Mahdi AA, Mishra V, Rajender S, Sankhwar SN, Patel D, Das M (2011) Withania somnifera improves semen quality by combating oxidative stress and cell death and improving essential metal concentrations. Reprod Biomed Online 22:421–427
Silva AF, Aguiar MS, Carvalho OS, Santana Lde N, Franco EC, Lima RR, Siqueira NV, Feio RA, Faro LR, Gomes-Leal W (2013) Hippocampal neuronal loss, decreased GFAP immunoreactivity and cognitive impairment following experimental intoxication of rats with aluminum citrate. Brain Res 1491:23–33
Singh N, Bhalla M, de Jager P, Gilca M (2011) An overview on ashwagandha: a Rasayana (rejuvenator) of Ayurveda. Afr J Tradit Complement Altern Med 8:208–213
Singh RH, Narsimhamurthy K, Singh G (2008) Neuronutrient impact of Ayurvedic Rasayana therapy in brain aging. Biogerontol 9:369–374
Soman S, Korah PK, Jayanarayanan S, Mathew J, Paulose CS (2012) Oxidative stress induced NMDA receptor alteration leads to spatial memory deficits in temporal lobe epilepsy: ameliorative effects of Withania somnifera and Withanolide A. Neurochem Res 37:1915–1927
Tohda C (2008) Overcoming several neurodegenerative diseases by traditional medicines: the development of therapeutic medicines and unravelling pathophysiological mechanisms. Yakugaku Zasshi 128:1159–1167
Tohda C, Joyashiki E (2009) Sominone enhances neurite outgrowth and spatial memory mediated by the neurotrophic factor receptor, RET. Br J Pharmacol 157:1427–1440
Ven Murthy MR, Ranjekar PK, Ramassamy C, Deshpande M (2010) Scientific basis for the use of Indian ayurvedic medicinal plants in the treatment of neurodegenerative disorders: ashwagandha. Cent Nerv Syst Agents Med Chem 10:238–246
Waterhouse EG, Xu B (2009) New insights into the role of brain-derived neurotrophic factor in synaptic plasticity. Mol Cell Neurosci 42:81–89
Winters M (2006) Ancient medicine, modern use: Withania somnifera and its potential role in integrative oncology. Altern Med Rev 11:269–277
Yamada K, Nabeshima T (2003) Brain-derived neurotrophic factor/TrkB signaling in memory processes. J Pharmacol Sci 91:267–270
Young EJ, Briggs SB, Miller CA (2015) The Actin Cytoskeleton as a Therapeutic Target for the Prevention of Relapse to Methamphetamine Use. CNS Neurol Disord Drug Targets 14:731–737
Zhao J, Nakamura N, Hattori M, Kuboyama T, Tohda C, Komatsu K (2002) Withanolide derivatives from the roots of Withania somnifera and their neurite outgrowth activities. Chem Pharm Bull 50:760–765

Chapter 16
Using Human Neural Stem Cells as a Model to Understand the "Science of Ashwagandha"

Manju Tewari, Hriday S. Pandey, and Pankaj Seth

All that man needs for health and healing has been provided by God in nature, the challenge of science is to find it. Philippus Theophrastus Bombast that of Aureolus

~Paracelsus (1493–1541)

Abstract Ashwagandha (*Withania somnifera*), a traditional Indian herb, has been widely employed in ayurvedic medicine. Various compounds isolated from leaf and root of this plant are used in treating human illness ranging from weakness, anxiety, rheumatic pain, diabetes, infertility, oxidative stress to cancer. In past two decades, scientific evidence for the neuroprotective effect of Ashwagandha further validates its use for treatment of Alzheimer's, Parkinson's, Huntington's disease and spinal cord injury. This chapter discusses the neuroprotective effects of various components of Ashwagandha in neurological disorders, majority of which have been studied using animal models or cell lines. Extensive explorations into mechanistic aspects of Ashwagandha are mandatory to validate the findings obtained from animal models and to confirm their therapeutic potential in human system. In order to explore the therapeutic potential of the drug for treating brain disorders, it is important to investigate the effects of this herbal drug on primary human brain cells. This chapter emphasizes on the potential of using human neural stem cells (NSCs) as an *in vitro* model to study the neuroprotective effects of Ashwagandha and to gain novel insights into the underlying mechanism of action under physiological and pathological conditions. Different sources of neural stem cells have also been described in the chapter with a detailed insight on the method of deriving NSCs.

Keywords Neurological diseases • *Withania somnifera* • Neuroprotective effects • Human fetal neural stem cells • *In vitro* model

---

M. Tewari • H.S. Pandey • P. Seth (✉)
Cellular and Molecular Neuroscience, National Brain Research Centre (NBRC), NH-8, Nainwal Mode, Manesar, Haryana 122051, India
e-mail: pseth@nbrc.ac.in

S.C. Kaul, R. Wadhwa (eds.), *Science of Ashwagandha: Preventive and Therapeutic Potentials*, DOI 10.1007/978-3-319-59192-6_16

16.1 Introduction

Brain disorders and age-associated neurological illness are largely limited to humans and rarely occur spontaneously in rodents. Animal models of diseases are crucial for understanding the fundamental mechanisms associated with the disease and to large extent, these models have improved understanding of the human diseases. Animal studies are also a prerequisite for clinical toxicological assessment of a drug to minimize any adverse effects on humans. The xenograft studies and genetically engineered mouse models (GEM) are currently the most widely used approaches for effectively identifying the sequence of pathological changes that culminate into a disease. Although quite successful in understanding pathological diseases, the animal models have limitations. Often, the cellular and behavioral changes and the resulting phenotypes observed in rodents are very different from that observed in humans under neurological conditions. Species-specific variations between rodents and humans are perhaps the basis for the observed discrepancies. In addition, from the evolutionary point of view, the brain structure of humans is much more complex than rodents and their cognitive abilities are much superior. Rodents are more vulnerable to transgenes or toxins as compared to humans and perhaps for these reasons therapeutic strategies designed using animal disease models often fail to translate to the treatments of human diseases.

Post-mortem brain tissue studies from patients provide highly relevant information but the availability of the tissues is very limited. Moreover, the post-mortem samples represent the end stage of the disease, as samples are collected only when the patient succumbs to the disease. Anatomical studies, cellular and biochemical analysis of the post-mortem tissue identifies the protein alterations and toxins or the mutations associated with the disease. However, an understanding of the sequential cellular and molecular events underlying the progression of the disorder remains elusive in studies performed on autopsy samples. Despite the fact that use of human autopsy brain tissues acquired at various stages help to overcome some problems, however the genetic variability and the occurrence of the different opportunistic infection augment the limitations of its use even in age-matched control studies.

For an enhanced understanding of the pathogenesis of a disease, or to determine the efficacy of tested drugs at cellular and molecular levels, animal studies can be complemented with *in-vitro* human brain cultures before performing clinical trials. This will also help to account for deviations due to species variations to a great extent. In this regard, human stem cell cultures offer flexibility to explore the disease mechanisms in detail at cellular and molecular levels. Stable *in vitro* propagation and differentiation of neural stem cells can offer expanded opportunities to understand the fundamental molecular, cellular and developmental mechanisms and to model neurodegenerative disorders. In this chapter, we highlight recent advances in the field of human neural stem cells, their differentiation into different brain cell types that enhance our current understanding of the disease mechanisms and their potential for use for treating the most debilitating human diseases.

Ashwagandha (*Withania somnifera*; L.dunal) an Indian ginseng has been widely used in traditional ayurvedic medicines to treat an array of human diseases (Ven

Murthy et al. 2010; Singh et al. 2011). In the past two decades, Ashwagandha has been identified to play a protective role in neurodegenerative disorders like Alzheimer's, Huntington's, Parkinson's disease and spinal cord injury (Kulkarni and Dhir, 2008; Kuboyama et al. 2014). This chapter discusses the recent advances in understanding the science of Ashwagandha, and their importance in neurodegenerative disease, with special emphasis on the use of neural stem cell to explore the underlying regulatory mechanisms and potential neuroprotective functions of Ashwagandha.

16.2 Neural Stem Cell: An Introduction

Neural stem cells (NSCs) have been proposed as a powerful tool to dissect out the basic cellular, molecular and developmental processes and to model neurodegenerative diseases (Harris et al. 2016). NSCs are immature cells that are mitotically active, self-renewing and multipotent. Being mitotic in nature, these are a renewable source of cells and they do not require genetic transformation (Casarosa et al. 2014). NSCs can be derived from embryonic, fetal or adult tissues (De Filippis and Binda 2012). Under *in-vitro* conditions, NPCs are capable of forming neurons, astrocytes and oligodendrocytes in presence of specific growth factors (De Filippis and Binda 2012; Louis et al. 2013). *In-vivo,* they differentiate into cell type appropriate to the brain regions where they are grafted (Flax et al. 1998; McMahon et al. 2010; Gupta et al. 2012). Their differentiation properties can be utilized to understand the effect of different toxins, drugs or mutations in specific brain cell types.

Ideal neural stem cells exhibit three cardinal features, namely, (1) Capacity to self-renew for extended period of time, around 5–10 passages, (2) In parallel, the cells also exhibit a great proliferating potential, thus, generating progeny several order of magnitude than the starting population, and (3) Stem cells hold differentiation characteristics or are tripotent, giving rise to neuron, astrocytes and oligodendrocytes, even after prolong passaging and should be transplantable (Gage 2000). Four different types of human neural stem cells (embryonic, adult, fetal and induced pluripotent stem cell derived) have been isolated and characterized till date (Fig. 16.1) and have proved to be quite successful although they have their own benefits and limitations, which will be discussed in the chapter.

16.2.1 Human Embryonic Stem Cells (hESCs)

Embryonic stem (ES) cells are pluripotent cells derived from pre-implantation embryos and have a proven ability to form embryoid bodies, which in turn can be induced to differentiate into all three embryonic germ layers namely, endoderm, mesoderm and ectoderm (Chen and Lai 2015). The three germ layers give rise to specific cells that generate tissues of the fetal and adult organism and are pluripotent. ES cells are derived either from the inner cell mass of the blastocyst or from

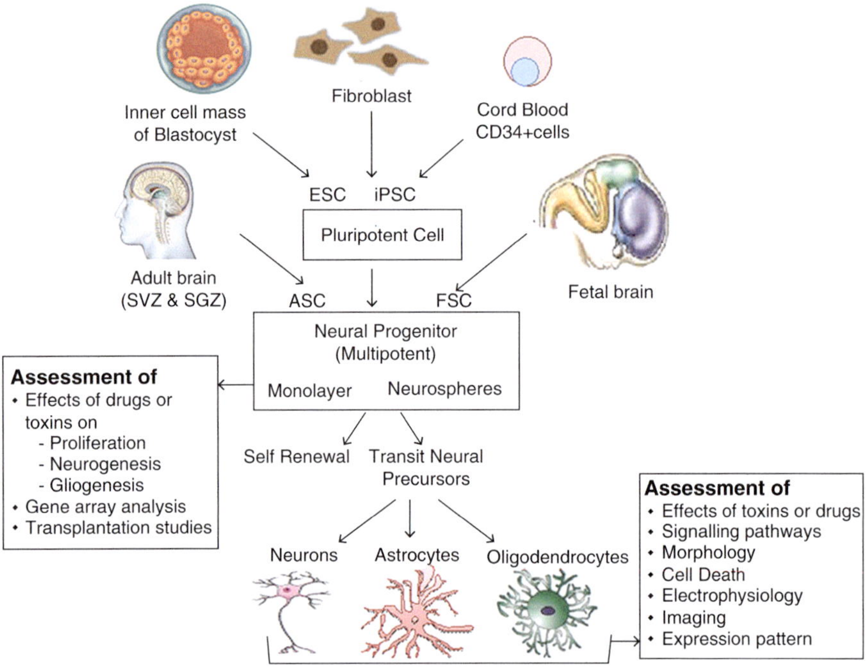

Fig. 16.1 Human neural stem cells (hNSCs) derived from embryonic, fetal and adult tissues and their characteristic properties

primordial germ line (EG cells) (Matsui et al. 1992; Rohwedel et al. 1996; Gepstein 2002). ES cells were first established *in vitro* in 1980s from mouse embryo-derived blastocyst. Later they were successfully derived from primates as well as human embryo (Thomson et al. 1998; Marshall et al. 2001).

Ectodermal layer derived from embryoid bodies, in the presence of specific mitogens can be differentiated into neurons and glia from the intermediate neural precursor cells, which are yet to be characterized (Hoffman and Carpenter 2005; Trounson 2005, 2006; Alsanie et al. 2013). The properties of the hESC-derived progenitors, neurons and glial cells are alike primary brain cells (Carpenter et al. 2001; Pollard et al. 2006).

Since ES cells easily self-renew *in vitro*, they are an apt choice for modeling early developmental studies like embryogenesis as well as developmental and genetic abnormalities which include generation of cell lines with targeted gene alterations and engineered chromosomes (Wobus and Boheler 2005). Pharmacological drug screening, gene delivery system and cell-based transplantation therapy (cell replacement therapies) can also be modelled using these cells (Zhang et al. 2001; Jensen et al. 2009; Zou et al. 2009). Ethical consideration related to procurement of embryonic tissue and the complexity because of tumor formation after transplantation of embryonic neural stem cell critically demands for an alternative approach to get neural stem cells (Wobus 2001; Nikol'skii et al. 2007).

16.2.1.1 Human Adult Neural Stem Cells

In 1965, the pioneering discovery by Altman and Das provided anatomical evidence for the formation of new neural precursor cell throughout life (Altman and Das 1965). Further, the functional integration of new neurons in adult CNS was successfully shown in songbirds thus introducing the concept of adult neurogenesis (Alvarez-Buylla 1992). In the early 1990s, the successful isolation of adult neural stem cells from the mammalian brain revolutionized the field of neuroscience. Adult neural stem cells (aNSCs) reside in specialized neurogenic niches located in the sub-ventricular zone (SVZ) of lateral ventricles and dentate region of the hippocampus that sustain the capacity of self-renewal and multipotency. In the hippocampus, the whole process of adult neurogenesis occurs in the dentate gyrus. However, in the SVZ region, the neuroblasts organize themselves in the chain, forming the rostral migratory stream (RMS) thus connecting the SVZ to the olfactory bulb. In olfactory bulb, the neuroblasts progressively mature into interneurons (GABA-ergic granule neuron) and integrate into the existing neural network (Whitman and Greer 2009; Lim and Alvarez-Buylla 2016). Neurons derived from adult neural stem cells contribute to learning and memory and facilitate brain repair after injury (Sohur et al. 2006; Yu et al. 2014; Horgusluoglu et al. 2016; Jin 2016; Lim and Alvarez-Buylla 2016). Unlike, ES cells, adult neural stem cells (aNSCs) are only capable of expansion and differentiation into neurons and glia up to a limited extent and their long-term effects remain elusive. Undoubtedly, the concept of neural stem cells and adult neurogenesis has tremendously transformed our understanding of the function and plasticity of the brain. Henceforth, there is an increasing excitement about exploiting the intrinsic neurogenic property of postnatal neural stem cells to rescue proper neuronal function after injury or brain diseases.

In vitro, the adult NSCs were first isolated in 1990s from mouse striatum, expanded as large spheres and termed as neurospheres (Reynolds and Weiss 1992). An entire neurosphere can be generated from single neural stem cell and when dissociated can produce new neurospheres (Azari et al. 2010). When given appropriate conditions, aNSCs can further differentiate into neurons, astrocytes and oligodendrocytes. The cultured aNSCs have been used in drug screening and cell replacement therapies (Nam et al. 2015). However, the adult neural stem cells cannot expand indefinitely and undergo senescence after a certain passage in culture conditions. This restrains their uses in clinical practices as compared to embryonic or fetal neural stem cells; however, these cells prove to be an excellent source of cells for CNS repair.

16.2.1.2 Human Fetal Neural Stem Cells

Neural precursor cells can also be isolated from aborted fetal brain tissue of 8–20 week gestation. The isolated fetal neural stem cells can be expanded in cultures in the presence of mitogens epidermal growth factor (EGF) and/or fibroblast growth factor 2 (FGF2). The cultured fetal NSCs are indistinguishable in all characteristic features from those derived from embryonic tissue (Li et al. 2005). Similar to the

NSCs derived from embryonic tissue, the human fetal NSCs in culture either form neurospheres or expand as primary monolayer neural precursor cells. These cells have huge proliferative potential and hence, are easy to expand. Upon removal of mitogens or by replacing EGF and FGF with specific lineage inducing mitogens, the precursor cells can be differentiated to a large number of neurons, astrocytes and oligodendrocytes (Messam et al. 2003; Li et al. 2008; Monaco et al. 2012). Moreover, when used in replacement therapies, they engraft very well and unlike ES cells they do not form teratomas following engraftment (Ishii and Eto 2014). In this chapter, we will briefly summarize the isolation and characterization of human fetal brain-derived NSCs *in vitro*. The application of fetal NSCs as an *in vitro* model to study neurodevelopment, brain diseases and their application in stem cell transplantation to cure neurological disorders will be subsequently discussed.

16.2.2 Induced Pluripotent Stem Cells (iPSCs)

iPSCs are adult pluripotent stem cells that have been genetically reprogrammed from different cell types such as skin fibroblast or blood cells to an intermediate ES cell-like state that has the capability to form every other cell types of the tissue/organism (Wang et al. 2013; Wang et al. 2015). iPSC were first generated in 2006 from mouse fibroblasts. Similar to ES cells, these cells show self-renewal, express all the stem cell markers, form teratomas and give rise to all three germ layers (Hu et al. 2010; Yamanaka 2012). In subsequent years, human iPSCs were successfully generated from various cell types such as blood cells, skin fibroblasts and renal epithelial cells (Lowry et al. 2008; Loh et al. 2009; Yamanaka 2012; Zhou et al. 2012). iPSCs generated from cells of individuals suffering from genetic and other disorders provide powerful means to extensively study the patient-specific genetic alterations (Avior et al. 2016; Csobonyeiova et al. 2016). The next section of this chapter will emphasize on the method of deriving NSCs from human fetal brain tissue.

16.3 *In vitro* Culture Model System of Human Fetal Neural Stem Cells

Many laboratories have provided methods for isolation and characterization of human fetal neural stem cells (Vescovi et al. 1993; Messam et al. 2003). Here, we are discussing an *in vitro* model successfully developed in our laboratory at National Brain Research Centre, India. These cells exhibit the proliferation and multipotent characteristics of neural stem cells. These largely expandable NS cells retain neuronal and glial differentiation potential after prolonged passaging and can be used as an *in-vitro* model to study HIV neuropathogenesis, neuro-glia interactions, neurodevelopmental diseases as well as for transplantation studies.

According to the standardized protocol, the human fetus of 8–20 weeks gestation is obtained from the elective medical termination of first trimester pregnancies with the written consent of the mother. The brain samples are handled as per the protocols approved by the institutional ethical committee in compliance with the recommendations of the Indian Council of Medical Research (ICMR) and DBT-Stem cell ethics committee. Tissue encompassing the sub ventricular zone surrounding the lateral wall of the forebrain ventricle is dissected and collected using 19-gauge needle. The tissue is mechanically dissociated using a pipette by gentle pipetting. Cells are washed once in the neurobasal media and then plated into a 75 cm^2 flask (Nunc) in neurobasal media plus N2 supplement (Invitrogen), neural survival factor (NSF), and bFGF and EGF (25 ng/ml, 20 ng/ml respectively, Sigma-Aldrich or Peprotech). Cultures are maintained for 1–2 weeks, with a half volume of media changed every week. Small neurospheres formed within this period are collected by centrifugation at 700 rpm for 30 seconds. The pellet is resuspended in trypsin (0.025%) and cells are dissociated by repeated pipetting. At higher densities and at earlier passages NS cells can tend to aggregate and spontaneously form neurospheres. This may be reduced substantially through plating the cells onto a tissue culture flask pre-coated with poly-D-lysine (PDL; 30 min treatment of 50 μg/ml PDL at RT) or PDL/laminin (30 min treatment with poly-D-lysine solution 50 μg/ml; wash twice in PBS and then add a 10 μg/ml laminin solution for at least 3 hr at 37^{0}C). Upon replating the newly formed spheres onto a fresh poly-D-lysine (PDL) coated 75 cm^2 flasks, attachment and outgrowth occur within 2–5 days. These cells show typical morphology of neural stem cells. They are subsequently expanded as adherent cultures. Cells undergo rapid expansion and at 70% confluency can be passaged every 3–4 days for more than 25 passages (Mishra et al. 2010). These cells can be cryopreserved at various passages and can be revived later as a renewable source of neural stem cells for experiments.

16.3.1 In vitro *Neuronal and Astrocytic Differentiation from Human Fetal NSCs and Characterization of Derived Neurons and Astrocytes*

NSCs can be differentiated to neuron either by removal of growth factors EGF and bFGF or by replacing these growth factors with BDNF and PDGF (10 ng/ml each; Sigma-Aldrich, MO, USA) (Vescovi et al. 1993). NSCs can also be induced to form astrocytes by changing the stem/progenitor media to minimum essential medium (MEM) supplemented with 10% heat inactivated fetal bovine serum (FBS; Invitrogen, CA, USA). Half of the medium was replaced with fresh culture media on alternate days. The differentiation protocol to achieve fully mature neurons or astrocytes takes around 21 days after which the differentiated cells are characterized using lineage-specific markers (Fig. 16.2) (Tewari et al. 2015). Fetal NSCs retain neuronal and glial differentiation potential even after prolonged passaging and are

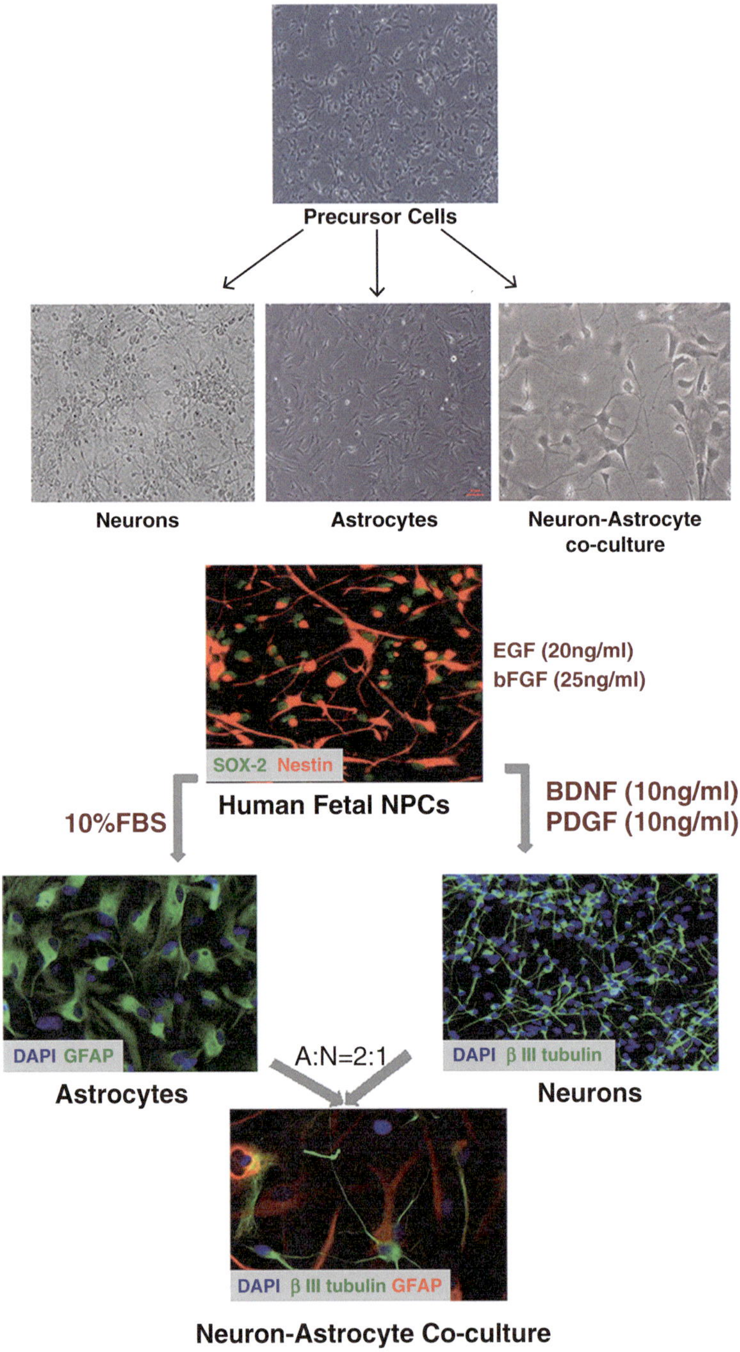

Fig. 16.2 Schematic representation of human fetal brain-derived cell culture model. Primary human fetal brain-derived multipotent neural progenitor cells (NPCs) proliferate *in vitro*. The NPCs can differentiate into neurons, astrocytes as shown in bright field images in *upper panel* and by cell lineage specific markers in *lower panel*

appropriate for experimental studies and transplantation experiments. Moreover, studying astrocytes and neurons at various stages of differentiation can provide the developmental insights of the underlying disease.

We have used human fetal brain derived neural stem cells in rat cortical injury model, mouse retinal degeneration model and rat spinal cord injury model. We found that these cells not only reach to the site of injury but also survive and differentiate into glial and neuronal cells, after transplantation (Unpublished observations). Similar model systems can be expanded to study the modulation of such injuries in presence of Ashwagandha (Table 16.1).

Table 16.1 Transplantation studies of NPCs in animal models of CNS disorders

Disorder	Model	Tested animal species	NSC properties after transplantation	References & Journal
Intracerebral hemorrhage (ICH) stroke	Immortalized cell lines of human NSCs from primary human fetal telencephalon cultures	Mouse	NSCs differentiate into neurons, astrocytes, and oligodendrocytes, and maintained stable high levels of GDNF expression	(Lee et al. 2007) Stem cells
Spinal cord injury (SCI)	Human neural stem cells	Mouse	NSCs differentiate into neurons, oligodendrocytes and promote locomotor recovery	(Cummings et al. 2005) PNAS
	Human neurospheres derived from CNS stem cells	Rat	Proliferate continuously and differentiate into cells of neuronal and glial lineage	(Kelly et al. 2004) PNAS
	C17.2 NSCs mouse	Rat	NSCs secrete neurotrophic factors and promote axonal growth	(Lu et al. 2003) Exp Neurol
Hemi-section model of Spinal cord injury (SCI)	Murine NSCs (clone C17.2)	Rat	NSCs secrete neurotrophic nerve growth factor, brain-derived neuro-trophic factor, and glial cell line-derived neurotrophic factor	(Teng et al. 2002) PNAS
Contusion lesion in spinal cord	Rat fetal NSC	Rat	NPCs differentiate into neurons, astrocytes and oligodendrocytes	(Fujiwara et al. 2004) Neurosci Lett

(continued)

Table 16.1 (continued)

Disorder	Model	Tested animal species	NSC properties after transplantation	References & Journal
Multiple sclerosis (Experimental autoimmune encephalomyelitis, EAE)	Adult neural stem cell cultures	Mouse	Donor NSCs do not appear to be immunopositive for NF, also not largely immunopositive for GFAP. Remain largely nestin positive	(Pluchino et al. 2003) Nature
Elapsing-remitting experimental autoimmune encephalomyelitis (R-EAE)	Sub ventricular-zone-derived syngenic adult NPCs (aNPCs) from 6–8-week-old SJL and C57BL/6 mice	Mouse	Oligodendrocyte progenitors markedly increased	(Pluchino et al. 2005) Nature
Acute experimental allergic encephalomyelitis (EAE) rats	Rat neural precursor cells (NPCs)			(Einstein et al. 2003) Mol and Cell Neurosci
X-irradiation and ethidium bromide injection (EB-X) model kill glial cells and induce demyelination	NPCs derived from the adult human brain	Rat	Neurospheres to glial cells such as astrocytes and oligodendrocytes	(Akiyama et al. 2001) Exp Neurol
Traumatic brain injury	Mouse neural stem cell (NSC) clone C17.2	C57BL/6 mice	NSC into neuron and astrocytes but not oligodendrocytes	(Riess et al. 2002) Neurosurgery
Traumatic brain injury induced using controlled cortical impact (CCI)	MHP (mouse embryonic hippocampus derived	C57BL/6 mice	NSC to NG2 (an oligodendrocyte progenitor cell marker), but no neuron, astrocyte or microglia	(Shear et al. 2004) Brain Res
Experimental intracerebral hemorrhage (ICH)	HB1.F3 embryonic human brains of 15 weeks gestation	Rat	NSCs to neurons and glia	(Jeong et al. 2003) Stroke
Ischemia (middle cerebral artery occlusion, MCAO)	Human NSCs hb1.f3	Rat	NSCs to neurons and astrocytes	(Chu et al. 2004) Brain Res

(continued)

Table 16.1 (continued)

Disorder	Model	Tested animal species	NSC properties after transplantation	References & Journal
1-methyl-4-phenyl-1,2,3,6-tetra-hydropyridine (MPTP) induced Parkinson's disease (PD)	Murine NSCs	Mouse	NSC to dopaminergic neurons	(Ourednik et al. 2002) Nat Biotechnol
	Human fetal NSCs	Monkey	Prevent the MPTP-induced up-regulation of TH+ cells, long term survival of NSCs	(Bjugstad et al. 2005) Cell Transplant
	Human fetal NSCs	Mouse	Long-term (up to 90 days) survival of human stem cell xenograft	(Liker et al. 2003) Brain Res
6-hydroxy-dopamine model of Parkinson's disease (PD)	c17.2 mouse NSCs	Mouse	Neurosphere differentiate to neurons and oligodendrocytes	(Akerud et al. 2001) J Neurosci
	Neonatal mouse SVZ NPCs	Chick embryo	Long-term survival of transplanted fetal neurons, secrete mitogen (Shh)	(Rafuse et al. 2005) Neuroscience
	Human fetal NPCs	Rat	NSCs differentiate into both neurons and astrocytes	(Svendsen et al. 1997) Exp Neurol
	Embryonic rat striatum cells	Rat	NSCs differentiate into astrocytes and TH+ neurons	(Tang et al. 2002) Neurosci Lett
	Embryonic NPC mesencephalic (MS) and cortical (CTx) stem (progenitor) cells	Rat	Less proliferation and more differentiation of MS stem (progenitor) cells to differentiate into mature TH+ neurons.	(Nishino et al. 2000) Exp Neurol
	Adult rat SEZ NPC	Rat	Maintaining a progenitor phenotype	(Richardson et al. 2005) Brain Res
Quinolinic acid (QA) induced Huntington's disease (HD) model	Human fetal CORTEX NSC	Rat	Differentiate into neurons and astrocytes.	(McBride et al. 2004) J Comp Neurol
	Human FETAL neural stem cells	Rat	Secretion of BDNF	(Ryu et al. 2004) Neurobiol Dis

16.4 Ashwagandha: A Traditional Ayurvedic Herb

Ashwagandha (*Withania somnifera*), an Indian ginseng has been used as a safe herbal drug to treat a wide variety of disorders ranging from stress, cancer, diabetes, inflammation to neurological and neurodevelopmental disease (Kuboyama et al. 2014). It also improves intellectual abilities and enhances the memory and life span of individuals. Different chapters of this book will provide detailed insights into the role of this wonder herb as a drug to cure a wide variety of human diseases. In this chapter, we will touch upon the role of Ashwagandha in different neurological disorders that will be described in more details in subsequent chapters of the book. We will also discuss the experimental strategies for using human neural stem cells as an *in vitro* model to study the effect of this wonder drug and to gain insights into the underlying cellular and molecular mechanisms of action of Ashwagandha under physiological and pathological conditions. The various experimental observations signifying the protective effect of *W. somnifera* is summarized in Table 16.2 of this chapter.

16.4.1 Use of Stem Cell to Study the Potential Neuroprotective Effect of Ashwagandha Against Neurodegenerative Diseases

Table 16.2 demonstrates that most of the published studies on protective effects of Ashwagandha are limited to either using the animal models of disease or the *in-vitro* rodent primary cultures or cell lines. Till date, the effects of this herbal drug have not been studied in detail in human cell culture system. Undoubtedly, when tested on human brain cells, it will be invaluable validation of *in vivo* observations in animal systems and will provide promising acceptance of its protective effects. Because of the availability constraints and experimental challenges in the expansion and maintenance of human brain cells, NSCs offer to be a good alternative system. Their ability of self-renewal and multipotency can be used to study useful effects of Ashwagandha on neurodevelopment, neurogenesis, gliogenesis, memory enhancement, disease pathology and for toxicological assessment.

16.4.1.1 Alzheimer's Disease

Alzheimer's disease (AD) is a chronic neurodegenerative disorder accounting for 60 to 80% dementia cases worldwide, mainly in elderly. Neuropathological changes in AD includes abundant amyloid-β deposition, plaque formation and formation of neurofibrillary tangles consisting of hyper-phosphorylated tau protein (Ballatore et al. 2007; Loeffler et al. 2008; Ji et al. 2016). The neurodegenerative changes commonly observed in AD patients are early damage to synapse, which positively

Table 16.2 Effect of derivatives of Ashwagandha in neurodegenerative disorders

Disease	Type of Ws Extract	Components	Model	Outcome	References
Alzheimer's disease (AD)	Semipurified extract of the roots of WS	Withanolides and withanosides	APP/PS1 Alzheimer's disease transgenic mice	Improved	Sehgal et al. (2012)
	Methanol extract of Ashwagandha (roots of Withania somnifera)	Withanolide A, withanoside IV, and withanoside VI	Human neuroblastoma SK-N-SH cells	Neurite outgrowth	Tohda et al. (2000)
	Roots of Withania somnifera	Withanolide A, withanoside IV, withanoside VI	Human neuroblastoma SH-SY5Y	Neurite outgrowth	Zhao et al. (2002)
	Methanol:Chloroform (3:1) extracts of ashwagandha		Human neuronal SK-N-MC cell line	Neuroprotective effect	Kurapati et al. (2013)
	The roots of Withania somnifera	Withanoside IV and sominone (metabolic product of withanalide IV)	Oral administration; Rat cortical neurons	Improved memory deficits, neurite outgrowth	Kuboyama et al. (2006))
Spinal cord injury (SCI)		Withanoside IV, sominone, an aglycone of withanoside IV,	Oral administration in mice	Improved locomotor functions in mice, axonal density and peripheral nervous system myelin level increased.	Tohda (2008)
	Synthetic compound (analogous to sominone)	Denosomin	Oral administration in the mice. Treatment to cultured mice cortical neurons	Improved hind limb motor functions and axonal growth	Teshigawara et al. (2013))
		Withanoside IV	Oral administration in mice	Axonal density and peripheral nervous system (PNS) myelin level increased, ameliorate locomotor functions	Nakayama and Tohda (2007)

(continued)

Table 16.2 (continued)

Disease	Type of Ws Extract	Components	Model	Outcome	References
PD-OHDA	Withania somnifera extract		Oral administration in rats	Neuroprotective, antioxidant, functional recovery	Ahmad et al. (2005)
PD-MPTP	Withania somnifera root extract		Oral administration in mice	Improved motor function treating catecholamines, oxidative damage and physiological abnormalities	RajaSankar et al. (2009b)
	Withania somnifera root extract		Oral administration in mice	Improvement in the mice's behavior and antioxidant status, reduction in the level of lipid peroxidation	Sankar et al. (2007)
	Withania somnifera leaf extract		Oral administration in mice	Treat oxidative damage and physiological abnormalities	Rajasankar et al. (2009a)
PD-paraquat (**PQ**) and maneb (MBPQ)	Withania somnifera root extract		Oral administration in mice	Improvement in the behavioral, anatomical and the biochemical deformities. Neuroprotection	Prakash et al. (2013)
3-Nitro-propionic acid (3-NP) HD model	Withania somnifera root extract		Oral administration in rats	Antioxidant activity	Kumar and Kumar (2009)

correlate with cognitive impairments (Terry et al. 1991; Reddy and Beal 2008). The clinical profile of AD includes memory impairment, apraxia, deficits in visuo-spatial abilities and executive functions affecting day-to-day life in elderly (Molinuevo and Casado-Naranjo 2014; Tang et al. 2016). The effect of root extract of Ashwagandha has been investigated in Alzheimer's disease using different human neuroblastoma cell lines such as SH-SY-5Y and S-K-N-MC. It was observed that Ashwagandha promotes neurite outgrowth, increases the synaptic density and improves memory (Tohda et al. 2000; Zhao et al. 2002; Kurapati et al. 2013). The root extracts have also been demonstrated to be neuroprotective after oral administration in mouse models of Alzheimer's disease (Sehgal et al. 2012). In AD mouse model, predominantly, the astrocyte activation and atrophy affects the neuronal health and functions (Kuboyama et al. 2006). The studies on human neuroblastoma cell lines provide the protective effect of Ashwagandha in neuronal population. To study the effect of the Ashwagandha on astrocytes, glioblastoma cell line such as U373 can be used. Alternatively, studies can be performed using primary astrocyte culture from rodents as they can be isolated from every region of CNS, but cultures should be performed on 2–3 days postnatal as the astrogliogenesis predominantly occurs at this stage. However, with these cultures, there is a possibility of contamination of astrocytes with other cells like microglia, oligodendrocytes, neurons and endothelial cells. Hence, these should be cautiously cultured and tested for astrocyte specific markers such as GFAP, GLAST, S100β, glutamine synthase or ALDH1L1. Considering these limitations, NSCs and its differentiated cell types may provide a better *in vitro* model system to assess the effects of Ashwagandha on amyloid beta-induced disturbances in astrocyte physiology. NSCs derived neuron-glia co-culture system can be used as an alternative to study the protective properties of Ashwagandha, in the disturbed astrocyte-neuron interactions associated with the disease.

16.4.1.2 Parkinson's Disease

Parkinson's disease (PD) is a slowly progressive nervous system disease caused by loss of dopaminergic neurons in substantia nigra (Beitz 2014). In addition to the dopaminergic cell loss, formation of Lewy bodies, mainly consisting aggregated alpha-synuclein (α-Syn) has been suggested to be important in the pathogenesis of the disease (Atik et al. 2016; Benskey et al. 2016; Saito et al. 2016). Clinically, the disease is characterized by motor dysfunction such as muscle rigidity, resting tremors, bradykinesia and postural instability (Marras et al. 2002). Various cell culture systems have been utilized to study the disease process such as the human cell lines HEK293, SHSY-5Y, NT2 (derived from embryonic carcinomas), LUHMES (derived from human mesencephalic cells) and the embryonic primary midbrain dopaminergic neurons (Ghosh et al. 1999; Chung et al. 2001; Lotharius et al. 2005; Xie et al. 2010; Falkenburger et al. 2016). Use of human induced pluripotent stem cells have given another dimension to PD research, however, their use is still limited due to intense labor and high cost (Cooper et al. 2012; Sanchez-Danes et al. 2012; Badger et al. 2014). Reactive astrogliosis is another feature found to exist in human post-mortem tissues of PD patients. Oral administration of Ashwagandha leaf and root extract in mouse and rat model systems of PD have been shown to have antioxidant and neuroprotective properties which ultimately leads to improved motor functions (Ahmad et al. 2005; RajaSankar et al. 2009b; Rajasankar et al. 2009a). This indicates that Ashwagandha derivatives can be effective on the human brain as well. To prove this, NSC can be differentiated to dopaminergic neurons or glial cells from embryonic, fetal, adult NSCs or induced pluripotent stem cells (Kim et al. 2002; Park et al. 2005; Li et al. 2008) and can be used as an *in-vitro* model system where the effects of Ashwagandha extracts can be studied. To decipher the protective mechanism of Ashwagandha at various cellular and molecular levels, one can study the effect of acute and chronic exposure of cells to toxins like MPTP, 6-OHDA, paraquat and rotenone or after introducing the α-synuclein mutation.

16.4.1.3 Huntington's Disease

Huntington's disease (HD) is an autosomal dominant neurodegenerative disorder caused by mutation in the Huntingtin gene (Htt). The disease is characterized by the loss of neurons primarily in striatal and cortical regions resulting in abnormal motor and cognitive abilities (Roos 2010; Ross and Tabrizi 2011). The underlying cause of disease is an expansion in CAG repeat in the Huntingtin gene, resulting in a long polyglutamine chain in the encoded protein Huntingtin (Htt). Htt protein has been used as a major tool to study the HD pathogenesis *in-vitro*. Various *in vitro* cell culture systems such as SH-SY-5Y, PC-12, N2a, COS-7 and HN33 (rat hippocampal neuronal cell line) have been exploited to study the pathological features of Htt protein (Wang et al. 1999; Wyttenbach et al. 2000; Sun et al. 2001; Banez-Coronel et al. 2012). The protective effects of Ashwagandha have been studied in the rodent model of HD using 3-Nitropropionic acid (3-NP), a potent neurotoxin which mimics HD in humans. The study suggests that the root extracts of Ashwagandha reverse the 3-NP induced enzymatic, biochemical and behavioral changes, however, the molecular mechanisms still remain elusive and need further investigations. The unidentified molecular pathways of 3-NP toxicity can be assessed using human neurons or astrocytes derived from human fetal NSCs (Kumar and Kumar 2009). In addition, as there are now compelling evidence that suggest HD is a neurodevelopmental disease and hence, using NSC may also provide meaningful insights into developmental aspects of the disease mechanism.

16.4.1.4 Spinal Cord Injury

Spinal Cord Injury (SCI) occurs due to damage to the spinal cord leading to the compromised flow of sensory, motor and reflex information between the body and the brain. The neuropathological changes associated with SCI may include axonal degeneration, loss of neurons and glial cells and invasion of peripheral immune cells (Branco et al. 2007; Varma et al. 2013; Silva et al. 2014). Various *in-vitro* model systems have been effectively established to study the spinal cord injury mechanisms, as reviewed by Morrison et al (Morrison et al. 1998). Oral administration of Withanoside IV (derivatives of ashwagandha extract) or denosomin (a synthetic compound that is analogous to sominone, a metabolic product of withanoside IV), one hr after contusion injury at the L1 spinal cord, induced axonal growth in the spinal cord and helped in the recovery of hind limb motor functions (Tohda et al. 2000; Nakayama and Tohda 2007; Teshigawara et al. 2013). One of the possible mechanisms suggested by this study is the increase in the number of vimentin-secreting astrocytes at the site of injury that helps in axonal regeneration and improved motor function.

16.4.1.5 HIV Dementia

HIV-associated dementia (HAD) is neurocognitive and motor deficits associated with HIV infection of the central nervous system. It includes a wide range of neurological complications including cognitive, behavioral and motor abnormalities (Kaul 2009; Pant et al. 2012a, b; Nath 2015). Being a human specific virus, animal models of HIV infection have not been quite successful in providing the mechanistic insights into its neuropathology. Limited access to HIV-infected brain tissue at different stages of the disease confines the understanding of the exact mechanism underlying the progression of HAD. Therefore, the use of the human *in vitro* system is essential for a better understanding of the disease.

To study HIV pathogenesis in brain, several *in vitro* culture models have been developed by different laboratories such as human and rodent primary culture of fetal cortical cells, BBB model and human fetal progenitor cells derived neurons or astrocytes, as reviewed by Seth and Major (2005). These model systems have been extensively used to study the effect of HIV, various viral proteins, neurotoxins, cytokines and chemokines. Further modifications in the existing *in vitro* models will facilitate improved understanding of the HIV neuropathogenesis.

Using human fetal brain-derived neural precursor cells, our laboratory has demonstrated the effect of HIV-1 protein transactivator of transcription (Tat) on the proliferative and neurogenic properties of these cells (Mishra et al. 2010; Fatima et al. 2016). Tat exposure severely reduces both the properties which reduced further in co-morbid situations with drugs of abuse (Malik et al. 2014). The neurotoxic effects of Tat have also been assessed on mature neurons differentiated from these cells. Moreover, studies conducted using pure astrocyte and/or neuronal cultures differentiated from these cells provide relevant cellular and molecular insights into astrocyte dysfunction and indirect neurotoxicity (Tewari et al. 2015).

We plan to explore the neuroprotective effects of Ashwagandha on modulation of HIV-1 Tat induced neuronal damage. In fact the *in vitro* human fetal brain derived NSC model can be utilized to assess the neurogenic role of Ashwagandha. Using cellular and molecular approaches, we are actively pursuing experiments with hNSCs to study if HIV-1 Tat induced damage to brain cells can be minimized with the use of Ashwagandha.

16.4.2 Human Fetal Neural Stem Cell Model to Study Neurodevelopment and Synaptic Plasticity

In our experiments using human fetal neural stem cells, it has been observed that the proliferative and neurogenic abilities of these cells are enhanced upon treatment with Ashwagandha. We have gathered novel insights into mechanisms of Ashwagandha mediated augmentation of the self-renewal and differentiation capacities of hNSCs (manuscript under preparation). These findings further substantiate the scientific basis of use of Ashwagandha in neonatal, fetal and adult brain as traditional medicine.

Alcoholic extract of ashwagandha leaves has been shown to rescue the scopolamine-induced impairments in neuronal and glial plasticity in rodents (Konar et al. 2011). Synaptic plasticity is the fundamental phenomenon associated with learning and memory (Martin et al. 2000; Takeuchi et al. 2014). In addition to its role in synaptic plasticity, Ashwagandha has also been shown to improve memory deficits in AD mouse models. Hence, it would be interesting to investigate whether Ashwagandha has any possible roles in memory formation and retrieval.

16.5 Conclusion

Animal models are extensively used to understand the pathogenesis of neurodegenerative disorders at the system level. They also prove to be essential for the assessment of behavioral parameters and transplantation studies, but these models have their own limitations and drawbacks. Often, the disease mechanisms and the therapeutic strategies explored in animal models fail to translate into humans during the clinical trials, probably due to species differences. The limitations of the animal models mandated the development of human NSCs culture system, which can be a continuous source of different cell types of the brain and may be utilized to meet the demands of the neuroscience research and regenerative medicine. Human NSCs can be utilized to induce mutations of interest or can be exposed to various toxins and thereby can serve as a suitable model for neurodegenerative diseases. These cells also offer immense potential for their use as a tool for investigating the cellular and molecular mechanisms underlying the disease progression. Recent inventions of developing organoids from stem cells in 3-dimentional culture system using hydrogels have further opened new avenues for the use of stem cells in diagnosis and even prognosis of diseases as well as in drug testing and drug discovery. Human NSCs and organoids can be used for drug screening before progressing in to the clinical trials. Undoubtedly, NSCs cannot replace transgenic animal models, but it can certainly add immense value to the existing knowledge of the neurodegenerative disorders.

Because of the wide variety of experimental application, the cultures of primary human NSCs and their differentiated cell types could be a useful model to study the science of Ashwagandha. To date, most of the neuroscience studies performed using

Ashwagandha have been conducted in animal models and/or some *in-vitro* rodent or human cell lines, but not on primary human brain cells, as discussed in the chapter. Hence, human NSCs can be a valuable resource to meet the demands of the field especially for the validation of the studies before clinical trials. In addition, the identification of the role of Ashwagandha in inducing the differentiation cascade of neural progenitor towards specific cell lineage (neuron, astrocytes or oligodendrocytes) will have significant advantage in the area of neurodevelopment. This model system has robust potential to serve as an important tool for studying the role of Ashwagandha in neurodegenerative disorders at cellular and molecular level.

Acknowledgements Financial support from NBRC core funds, to Pankaj Seth laboratory is greatly acknowledged. The authors also wish to acknowledge the support of the facilities provided under the Biotechnology Information System Network (BTISNET) grant, DBT, India and Distributed Information Centre at NBRC, Manesar, India. Senior Research Fellowship to Manju Tewari and Junior Research Fellowship to Hriday Shankar Pandey from Council of Scientific and Industrial Research, New Delhi is acknowledged.

Conflict of Interests The authors declare that there is no conflict of interest regarding the publication of this book chapter.

References

Ahmad M, Saleem S, Ahmad AS, Ansari MA, Yousuf S, Hoda MN, Islam F (2005) Neuroprotective effects of Withania somnifera on 6-hydroxydopamine induced Parkinsonism in rats. Hum Exp Toxicol 24:137–147

Akerud P, Canals JM, Snyder EY, Arenas E (2001) Neuroprotection through delivery of glial cell line-derived neurotrophic factor by neural stem cells in a mouse model of Parkinson's disease. J Neurosci 21:8108–8118

Akiyama Y, Honmou O, Kato T, Uede T, Hashi K, Kocsis JD (2001) Transplantation of clonal neural precursor cells derived from adult human brain establishes functional peripheral myelin in the rat spinal cord. Exp Neurol 167:27–39

Alsanie WF, Niclis JC, Petratos S (2013) Human embryonic stem cell-derived oligodendrocytes: protocols and perspectives. Stem Cells Dev 22:2459–2476

Altman J, Das GD (1965) Autoradiographic and histological evidence of postnatal hippocampal neurogenesis in rats. J Comp Neurol 124:319–335

Alvarez-Buylla A (1992) Neurogenesis and plasticity in the CNS of adult birds. Exp Neurol 115:110–114

Atik A, Stewart T, Zhang J (2016) Alpha-synuclein as a biomarker for Parkinson's disease. Brain Pathol 26:410–418

Avior Y, Sagi I, Benvenisty N (2016) Pluripotent stem cells in disease modelling and drug discovery. Nat Rev Mol Cell Biol 17:170–182

Azari H, Rahman M, Sharififar S, Reynolds BA (2010) Isolation and expansion of the adult mouse neural stem cells using the neurosphere assay. J Vis Exp 45:2393

Badger JL, Cordero-Llana O, Hartfield EM, Wade-Martins R (2014) Parkinson's disease in a dish – using stem cells as a molecular tool. Neuropharmacology 76 Pt A:88–96

Ballatore C, Lee VM, Trojanowski JQ (2007) Tau-mediated neurodegeneration in Alzheimer's disease and related disorders. Nat Rev Neurosci 8:663–672

Banez-Coronel M, Porta S, Kagerbauer B, Mateu-Huertas E, Pantano L, Ferrer I, Guzman M, Estivill X, Marti E (2012) A pathogenic mechanism in Huntington's disease involves small CAG-repeated RNAs with neurotoxic activity. PLoS Genet 8:e1002481

Beitz JM (2014) Parkinson's disease: a review. Front Biosci (Schol Ed) 6:65–74

Benskey MJ, Perez RG, Manfredsson FP (2016) The contribution of alpha synuclein to neuronal survival and function – Implications for Parkinson's disease. J Neurochem 137:331–359

Bjugstad KB, Redmond DE Jr, Teng YD, Elsworth JD, Roth RH, Blanchard BC, Snyder EY, Sladek JR Jr (2005) Neural stem cells implanted into MPTP-treated monkeys increase the size of endogenous tyrosine hydroxylase-positive cells found in the striatum: a return to control measures. Cell Transplant 14:183–192

Branco F, Cardenas DD, Svircev JN (2007) Spinal cord injury: a comprehensive review. Phys Med Rehabil Clin N Am 18:651–679, v.

Carpenter MK, Inokuma MS, Denham J, Mujtaba T, Chiu CP, Rao MS (2001) Enrichment of neurons and neural precursors from human embryonic stem cells. Exp Neurol 172:383–397

Casarosa S, Bozzi Y, Conti L (2014) Neural stem cells: ready for therapeutic applications? Mol Cell Ther 2:31

Chen Y, Lai D (2015) Pluripotent states of human embryonic stem cells. Cell Reprogram 17:1–6

Chu K, Kim M, Park KI, Jeong SW, Park HK, Jung KH, Lee ST, Kang L, Lee K, Park DK, Kim SU, Roh JK (2004) Human neural stem cells improve sensorimotor deficits in the adult rat brain with experimental focal ischemia. Brain Res 1016:145–153

Chung KK, Zhang Y, Lim KL, Tanaka Y, Huang H, Gao J, Ross CA, Dawson VL, Dawson TM (2001) Parkin ubiquitinates the alpha-synuclein-interacting protein, synphilin-1: implications for Lewy-body formation in Parkinson disease. Nat Med 7:1144–1150

Cooper O, Seo H, Andrabi S, Guardia-Laguarta C, Graziotto J, Sundberg M, McLean JR, Carrillo-Reid L, Xie Z, Osborn T, Hargus G, Deleidi M, Lawson T, Bogetofte H, Perez-Torres E, Clark L, Moskowitz C, Mazzulli J, Chen L, Volpicelli-Daley L, Romero N, Jiang H, Uitti RJ, Huang Z, Opala G, Scarffe LA, Dawson VL, Klein C, Feng J, Ross OA, Trojanowski JQ, Lee VM, Marder K, Surmeier DJ, Wszolek ZK, Przedborski S, Krainc D, Dawson TM, Isacson O (2012) Pharmacological rescue of mitochondrial deficits in iPSC-derived neural cells from patients with familial Parkinson's disease. Sci Transl Med 4: 141ra90

Csobonyeiova M, Danisovic L, Polak S (2016) Recent advances in iPSC technologies involving cardiovascular and neurodegenerative disease modeling. Gen Physiol Biophys 35:1–12

Cummings BJ, Uchida N, Tamaki SJ, Salazar DL, Hooshmand M, Summers R, Gage FH, Anderson AJ (2005) Human neural stem cells differentiate and promote locomotor recovery in spinal cord-injured mice. Proc Natl Acad Sci U S A 102:14069–14074

De Filippis L, Binda E (2012) Concise review: self-renewal in the central nervous system: neural stem cells from embryo to adult. Stem Cells Transl Med 1:298–308

Einstein O, Karussis D, Grigoriadis N, Mizrachi-Kol R, Reinhartz E, Abramsky O, Ben-Hur T (2003) Intraventricular transplantation of neural precursor cell spheres attenuates acute experimental allergic encephalomyelitis. Mol Cell Neurosci 24:1074–1082

Falkenburger BH, Saridaki T, Dinter E (2016) Cellular models for Parkinson's disease. J Neurochem 139(Suppl 1):121–130

Fatima M, Kumari R, Schwamborn JC, Mahadevan A, Shankar SK, Raja R, Seth P (2016) Tripartite containing motif 32 modulates proliferation of human neural precursor cells in HIV-1 neurodegeneration. Cell Death Differ 23:776–786

Flax JD, Aurora S, Yang C, Simonin C, Wills AM, Billinghurst LL, Jendoubi M, Sidman RL, Wolfe JH, Kim SU, Snyder EY (1998) Engraftable human neural stem cells respond to developmental cues, replace neurons, and express foreign genes. Nat Biotechnol 16:1033–1039

Fujiwara Y, Tanaka N, Ishida O, Fujimoto Y, Murakami T, Kajihara H, Yasunaga Y, Ochi M (2004) Intravenously injected neural progenitor cells of transgenic rats can migrate to the injured spinal cord and differentiate into neurons, astrocytes and oligodendrocytes. Neurosci Lett 366:287–291

Gage FH (2000) Mammalian neural stem cells. Science 287:1433–1438

Gepstein L (2002) Derivation and potential applications of human embryonic stem cells. Circ Res 91:866–876

Ghosh SS, Swerdlow RH, Miller SW, Sheeman B, Parker WD Jr, Davis RE (1999) Use of cytoplasmic hybrid cell lines for elucidating the role of mitochondrial dysfunction in Alzheimer's disease and Parkinson's disease. Ann N Y Acad Sci 893:176–191

Gupta N, Henry RG, Strober J, Kang SM, Lim DA, Bucci M, Caverzasi E, Gaetano L, Mandelli ML, Ryan T, Perry R, Farrell J, Jeremy RJ, Ulman M, Huhn SL, Barkovich AJ, Rowitch DH (2012) Neural stem cell engraftment and myelination in the human brain. Sci Transl Med 4: 155ra137

Harris L, Zalucki O, Piper M, Heng JI (2016) Insights into the Biology and Therapeutic Applications of Neural Stem Cells. Stem Cells Int 2016:9745315

Hoffman LM, Carpenter MK (2005) Characterization and culture of human embryonic stem cells. Nat Biotechnol 23:699–708

Horgusluoglu E, Nudelman K, Nho K, Saykin AJ (2016) Adult neurogenesis and neurodegenerative diseases: A systems biology perspective. Am J Med Genet B Neuropsychiatr Genet 174:93–112

Hu BY, Weick JP, Yu J, Ma LX, Zhang XQ, Thomson JA, Zhang SC (2010) Neural differentiation of human induced pluripotent stem cells follows developmental principles but with variable potency. Proc Natl Acad Sci U S A 107:4335–4340

Ishii T, Eto K (2014) Fetal stem cell transplantation: Past, present, and future. World J Stem Cells 6:404–420

Jensen J, Hyllner J, Bjorquist P (2009) Human embryonic stem cell technologies and drug discovery. J Cell Physiol 219:513–519

Jeong SW, Chu K, Jung KH, Kim SU, Kim M, Roh JK (2003) Human neural stem cell transplantation promotes functional recovery in rats with experimental intracerebral hemorrhage. Stroke 34:2258–2263

Ji L, Zhao X, Lu W, Zhang Q, Hua Z (2016) Intracellular Abeta and its Pathological Role in Alzheimer's Disease: Lessons from Cellular to Animal Models. Curr Alzheimer Res 13:621–630

Jin X (2016) The role of neurogenesis during development and in the adult brain. Eur J Neurosci 44:2291–2299

Kaul M (2009) HIV-1 associated dementia: update on pathological mechanisms and therapeutic approaches. Curr Opin Neurol 22:315–320

Kelly S, Bliss TM, Shah AK, Sun GH, Ma M, Foo WC, Masel J, Yenari MA, Weissman IL, Uchida N, Palmer T, Steinberg GK (2004) Transplanted human fetal neural stem cells survive, migrate, and differentiate in ischemic rat cerebral cortex. Proc Natl Acad Sci U S A 101:11839–11844

Kim JH, Auerbach JM, Rodriguez-Gomez JA, Velasco I, Gavin D, Lumelsky N, Lee SH, Nguyen J, Sanchez-Pernaute R, Bankiewicz K, McKay R (2002) Dopamine neurons derived from embryonic stem cells function in an animal model of Parkinson's disease. Nature 418:50–56

Konar A, Shah N, Singh R, Saxena N, Kaul SC, Wadhwa R, Thakur MK (2011) Protective role of Ashwagandha leaf extract and its component withanone on scopolamine-induced changes in the brain and brain-derived cells. PLoS One 6:e27265

Kuboyama T, Tohda C, Komatsu K (2006) Withanoside IV and its active metabolite, sominone, attenuate Abeta(25-35)-induced neurodegeneration. Eur J Neurosci 23:1417–1426

Kuboyama T, Tohda C, Komatsu K (2014) Effects of Ashwagandha (roots of Withania somnifera) on neurodegenerative diseases. Biol Pharm Bull 37:892–897

Kulkarni SK, Dhir A (2008) Withania somnifera: an Indian ginseng. Prog Neuropsychopharmacol Biol Psychiatry 32:1093–1105

Kumar P, Kumar A (2009) Possible neuroprotective effect of Withania somnifera root extract against 3-nitropropionic acid-induced behavioral, biochemical, and mitochondrial dysfunction in an animal model of Huntington's disease. J Med Food 12:591–600

Kurapati KR, Atluri VS, Samikkannu T, Nair MP (2013) Ashwagandha (Withania somnifera) reverses beta-amyloid1-42 induced toxicity in human neuronal cells: implications in HIV-associated neurocognitive disorders (HAND). PLoS One 8:e77624

Lee HJ, Kim KS, Kim EJ, Choi HB, Lee KH, Park IH, Ko Y, Jeong SW, Kim SU (2007) Brain transplantation of immortalized human neural stem cells promotes functional recovery in mouse intracerebral hemorrhage stroke model. Stem Cells 25:1204–1212

Li X, Xu J, Bai Y, Wang X, Dai X, Liu Y, Zhang J, Zou J, Shen L, Li L (2005) Isolation and characterization of neural stem cells from human fetal striatum. Biochem Biophys Res Commun 326:425–434

Li Y, Chi XC, Li XX, Xu JC (2008) Multipotency of human neural stem cells from fetal striatum. Neuroreport 19:1679–1683

Liker MA, Petzinger GM, Nixon K, McNeill T, Jakowec MW (2003) Human neural stem cell transplantation in the MPTP-lesioned mouse. Brain Res 971:168–177

Lim DA, Alvarez-Buylla A (2016) The Adult Ventricular-Subventricular Zone (V-SVZ) and Olfactory Bulb (OB) Neurogenesis. Cold Spring Harb Perspect Biol 8:a018820

Loeffler DA, Camp DM, Bennett DA (2008) Plaque complement activation and cognitive loss in Alzheimer's disease. J Neuroinflammation 5:9

Loh YH, Agarwal S, Park IH, Urbach A, Huo H, Heffner GC, Kim K, Miller JD, Ng K, Daley GQ (2009) Generation of induced pluripotent stem cells from human blood. Blood 113:5476–5479

Lotharius J, Falsig J, van Beek J, Payne S, Dringen R, Brundin P, Leist M (2005) Progressive degeneration of human mesencephalic neuron-derived cells triggered by dopamine-dependent oxidative stress is dependent on the mixed-lineage kinase pathway. J Neurosci 25:6329–6342

Louis SA, Mak CK, Reynolds BA (2013) Methods to culture, differentiate, and characterize neural stem cells from the adult and embryonic mouse central nervous system. Methods Mol Biol 946:479–506

Lowry WE, Richter L, Yachechko R, Pyle AD, Tchieu J, Sridharan R, Clark AT, Plath K (2008) Generation of human induced pluripotent stem cells from dermal fibroblasts. Proc Natl Acad Sci U S A 105:2883–2888

Lu P, Jones LL, Snyder EY, Tuszynski MH (2003) Neural stem cells constitutively secrete neurotrophic factors and promote extensive host axonal growth after spinal cord injury. Exp Neurol 181:115–129

Malik S, Saha R, Seth P (2014) Involvement of extracellular signal-regulated kinase (ERK1/2)-p53-p21 axis in mediating neural stem/progenitor cell cycle arrest in co-morbid HIV-drug abuse exposure. J Neuroimmune Pharmacol 9:340–353

Marras C, Rochon P, Lang AE (2002) Predicting motor decline and disability in Parkinson disease: a systematic review. Arch Neurol 59:1724–1728

Marshall VS, Waknitz MA, Thomson JA (2001) Isolation and maintenance of primate embryonic stem cells. Methods Mol Biol 158:11–18

Martin SJ, Grimwood PD, Morris RG (2000) Synaptic plasticity and memory: an evaluation of the hypothesis. Annu Rev Neurosci 23:649–711

Matsui Y, Zsebo K, Hogan BL (1992) Derivation of pluripotential embryonic stem cells from murine primordial germ cells in culture. Cell 70:841–847

McBride JL, Behrstock SP, Chen EY, Jakel RJ, Siegel I, Svendsen CN, Kordower JH (2004) Human neural stem cell transplants improve motor function in a rat model of Huntington's disease. J Comp Neurol 475:211–219

McMahon SS, Albermann S, Rooney GE, Shaw G, Garcia Y, Sweeney E, Hynes J, Dockery P, O'Brien T, Windebank AJ, Allsopp TE, Barry FP (2010) Engraftment, migration and differentiation of neural stem cells in the rat spinal cord following contusion injury. Cytotherapy 12:313–325

Messam CA, Hou J, Gronostajski RM, Major EO (2003) Lineage pathway of human brain progenitor cells identified by JC virus susceptibility. Ann Neurol 53:636–646

Mishra M, Taneja M, Malik S, Khalique H, Seth P (2010) Human immunodeficiency virus type 1 Tat modulates proliferation and differentiation of human neural precursor cells: implication in NeuroAIDS. J Neurovirol 16:355–367

Molinuevo JL, Casado-Naranjo I (2014) Clinical profile of Alzheimer's disease: is the age of the patient a decisive factor? Results of the INFLUENCE study. J Alzheimers Dis 39:227–232

Monaco MC, Maric D, Bandeian A, Leibovitch E, Yang W, Major EO (2012) Progenitor-derived oligodendrocyte culture system from human fetal brain. J Vis Exp 70:4274

Morrison B 3rd, Saatman KE, Meaney DF, McIntosh TK (1998) In vitro central nervous system models of mechanically induced trauma: a review. J Neurotrauma 15:911–928

Nakayama N, Tohda C (2007) Withanoside IV improves hindlimb function by facilitating axonal growth and increase in peripheral nervous system myelin level after spinal cord injury. Neurosci Res 58:176–182

Nam H, Lee KH, Nam DH, Joo KM (2015) Adult human neural stem cell therapeutics: Current developmental status and prospect. World J Stem Cells 7:126–136

Nath A (2015) Neurologic Complications of Human Immunodeficiency Virus Infection. Continuum (Minneap Minn) 21:1557–1576

Nikol'skii NN, Gibai IA, Somova NV (2007) Human embryonic stem cells. Problems and perspectives. Tsitologiia 49:529–537

Nishino H, Hida H, Takei N, Kumazaki M, Nakajima K, Baba H (2000) Mesencephalic neural stem (progenitor) cells develop to dopaminergic neurons more strongly in dopamine-depleted striatum than in intact striatum. Exp Neurol 164:209–214

Ourednik J, Ourednik V, Lynch WP, Schachner M, Snyder EY (2002) Neural stem cells display an inherent mechanism for rescuing dysfunctional neurons. Nat Biotechnol 20:1103–1110

Pant M, Garg P, Seth P (2012a) Central Nervous System Infection by HIV-1: Special Emphasis to NeuroAIDS in India. Proc Natl Acad Sci Sect B Biol Sci 82:81–94

Pant M, Garg P, Seth P (2012b) Central Nervous System Infection by HIV-1: Special Emphasis to NeuroAIDS in India. Proc Natl Acad Sci Sect B Biol Sci 82:81–94

Park CH, Minn YK, Lee JY, Choi DH, Chang MY, Shim JW, Ko JY, Koh HC, Kang MJ, Kang JS, Rhie DJ, Lee YS, Son H, Moon SY, Kim KS, Lee SH (2005) In vitro and in vivo analyses of human embryonic stem cell-derived dopamine neurons. J Neurochem 92:1265–1276

Pluchino S, Quattrini A, Brambilla E, Gritti A, Salani G, Dina G, Galli R, Del Carro U, Amadio S, Bergami A, Furlan R, Comi G, Vescovi AL, Martino G (2003) Injection of adult neurospheres induces recovery in a chronic model of multiple sclerosis. Nature 422:688–694

Pluchino S, Zanotti L, Rossi B, Brambilla E, Ottoboni L, Salani G, Martinello M, Cattalini A, Bergami A, Furlan R, Comi G, Constantin G, Martino G (2005) Neurosphere-derived multipotent precursors promote neuroprotection by an immunomodulatory mechanism. Nature 436:266–271

Pollard SM, Benchoua A, Lowell S (2006) Neural stem cells, neurons, and glia. Methods Enzymol 418:151–169

Prakash J, Yadav SK, Chouhan S, Singh SP (2013) Neuroprotective role of Withania somnifera root extract in maneb-paraquat induced mouse model of parkinsonism. Neurochem Res 38:972–980

Rafuse VF, Soundararajan P, Leopold C, Robertson HA (2005) Neuroprotective properties of cultured neural progenitor cells are associated with the production of sonic hedgehog. Neuroscience 131:899–916

Rajasankar S, Manivasagam T, Surendran S (2009a) Ashwagandha leaf extract: a potential agent in treating oxidative damage and physiological abnormalities seen in a mouse model of Parkinson's disease. Neurosci Lett 454:11–15

RajaSankar S, Manivasagam T, Sankar V, Prakash S, Muthusamy R, Krishnamurti A, Surendran S (2009b) Withania somnifera root extract improves catecholamines and physiological abnormalities seen in a Parkinson's disease model mouse. J Ethnopharmacol 125:369–373

Reddy PH, Beal MF (2008) Amyloid beta, mitochondrial dysfunction and synaptic damage: implications for cognitive decline in aging and Alzheimer's disease. Trends Mol Med 14:45–53

Reynolds BA, Weiss S (1992) Generation of neurons and astrocytes from isolated cells of the adult mammalian central nervous system. Science 255:1707–1710

Richardson RM, Broaddus WC, Holloway KL, Fillmore HL (2005) Grafts of adult subependymal zone neuronal progenitor cells rescue hemiparkinsonian behavioral decline. Brain Res 1032:11–22

Riess P, Zhang C, Saatman KE, Laurer HL, Longhi LG, Raghupathi R, Lenzlinger PM, Lifshitz J, Boockvar J, Neugebauer E, Snyder EY, McIntosh TK (2002) Transplanted neural stem cells survive, differentiate, and improve neurological motor function after experimental traumatic brain injury. Neurosurgery 51:1043–1052. discussion 1052-1044

Rohwedel J, Sehlmeyer U, Shan J, Meister A, Wobus AM (1996) Primordial germ cell-derived mouse embryonic germ (EG) cells in vitro resemble undifferentiated stem cells with respect to differentiation capacity and cell cycle distribution. Cell Biol Int 20:579–587

Roos RA (2010) Huntington's disease: a clinical review. Orphanet J Rare Dis 5:40

Ross CA, Tabrizi SJ (2011) Huntington's disease: from molecular pathogenesis to clinical treatment. Lancet Neurol 10:83–98

Ryu JK, Kim J, Cho SJ, Hatori K, Nagai A, Choi HB, Lee MC, McLarnon JG, Kim SU (2004) Proactive transplantation of human neural stem cells prevents degeneration of striatal neurons in a rat model of Huntington disease. Neurobiol Dis 16:68–77

Saito Y, Shioya A, Sano T, Sumikura H, Murata M, Murayama S (2016) Lewy body pathology involves the olfactory cells in Parkinson's disease and related disorders. Mov Disord 31:135–138

Sanchez-Danes A, Richaud-Patin Y, Carballo-Carbajal I, Jimenez-Delgado S, Caig C, Mora S, Di Guglielmo C, Ezquerra M, Patel B, Giralt A, Canals JM, Memo M, Alberch J, Lopez-Barneo J, Vila M, Cuervo AM, Tolosa E, Consiglio A, Raya A (2012) Disease-specific phenotypes in dopamine neurons from human iPS-based models of genetic and sporadic Parkinson's disease. EMBO Mol Med 4:380–395

Sankar SR, Manivasagam T, Krishnamurti A, Ramanathan M (2007) The neuroprotective effect of Withania somnifera root extract in MPTP-intoxicated mice: an analysis of behavioral and biochemical variables. Cell Mol Biol Lett 12:473–481

Sehgal N, Gupta A, Valli RK, Joshi SD, Mills JT, Hamel E, Khanna P, Jain SC, Thakur SS, Ravindranath V (2012) Withania somnifera reverses Alzheimer's disease pathology by enhancing low-density lipoprotein receptor-related protein in liver. Proc Natl Acad Sci U S A 109:3510–3515

Seth P, Major EO (2005) Human brain derived cell culture models of HIV-1 infection. Neurotox Res 8:83–89

Shear DA, Tate MC, Archer DR, Hoffman SW, Hulce VD, Laplaca MC, Stein DG (2004) Neural progenitor cell transplants promote long-term functional recovery after traumatic brain injury. Brain Res 1026:11–22

Silva NA, Sousa N, Reis RL, Salgado AJ (2014) From basics to clinical: a comprehensive review on spinal cord injury. Prog Neurobiol 114:25–57

Singh N, Bhalla M, de Jager P, Gilca M (2011) An overview on ashwagandha: a Rasayana (rejuvenator) of Ayurveda. Afr J Tradit Complement Altern Med 8:208–213

Sohur US, Emsley JG, Mitchell BD, Macklis JD (2006) Adult neurogenesis and cellular brain repair with neural progenitors, precursors and stem cells. Philos Trans R Soc Lond B Biol Sci 361:1477–1497

Sun Y, Savanenin A, Reddy PH, Liu YF (2001) Polyglutamine-expanded huntingtin promotes sensitization of N-methyl-D-aspartate receptors via post-synaptic density 95. J Biol Chem 276:24713–24718

Svendsen CN, Caldwell MA, Shen J, ter Borg MG, Rosser AE, Tyers P, Karmiol S, Dunnett SB (1997) Long-term survival of human central nervous system progenitor cells transplanted into a rat model of Parkinson's disease. Exp Neurol 148:135–146

Takeuchi T, Duszkiewicz AJ, Morris RG (2014) The synaptic plasticity and memory hypothesis: encoding, storage and persistence. Philos Trans R Soc Lond B Biol Sci 369:20130288

Tang Z, Yu Y, Guo H, Zhou J (2002) Induction of tyrosine hydroxylase expression in rat fetal striatal precursor cells following transplantation. Neurosci Lett 324:13–16

Tang JY, Wong GH, Ng CK, Kwok DT, Lee MN, Dai DL, Lum TY (2016) Neuropsychological Profile and Dementia Symptom Recognition in Help-Seekers in a Community Early-Detection Program in Hong Kong. J Am Geriatr Soc 64:584–589

Teng YD, Lavik EB, Qu X, Park KI, Ourednik J, Zurakowski D, Langer R, Snyder EY (2002) Functional recovery following traumatic spinal cord injury mediated by a unique polymer scaffold seeded with neural stem cells. Proc Natl Acad Sci U S A 99:3024–3029
Terry RD, Masliah E, Salmon DP, Butters N, DeTeresa R, Hill R, Hansen LA, Katzman R (1991) Physical basis of cognitive alterations in Alzheimer's disease: synapse loss is the major correlate of cognitive impairment. Ann Neurol 30:572–580
Teshigawara K, Kuboyama T, Shigyo M, Nagata A, Sugimoto K, Matsuya Y, Tohda C (2013) A novel compound, denosomin, ameliorates spinal cord injury via axonal growth associated with astrocyte-secreted vimentin. Br J Pharmacol 168:903–919
Tewari M, Monika VRK, Menon M, Seth P (2015) Astrocytes mediate HIV-1 Tat-induced neuronal damage via ligand-gated ion channel P2X7R. J Neurochem 132:464–476
Thomson JA, Itskovitz-Eldor J, Shapiro SS, Waknitz MA, Swiergiel JJ, Marshall VS, Jones JM (1998) Embryonic stem cell lines derived from human blastocysts. Science 282:1145–1147
Tohda C (2008) Overcoming several neurodegenerative diseases by traditional medicines: the development of therapeutic medicines and unraveling pathophysiological mechanisms. Yakugaku Zasshi 128:1159–1167
Tohda C, Kuboyama T, Komatsu K (2000) Dendrite extension by methanol extract of Ashwagandha (roots of Withania somnifera) in SK-N-SH cells. Neuroreport 11:1981–1985
Trounson A (2005) Human embryonic stem cell derivation and directed differentiation. Ernst Schering Res Found Workshop: 27–44
Trounson A (2006) The production and directed differentiation of human embryonic stem cells. Endocr Rev 27:208–219
Varma AK, Das A, Wallace G, Barry J, Vertegel AA, Ray SK, Banik NL (2013) Spinal cord injury: a review of current therapy, future treatments, and basic science frontiers. Neurochem Res 38:895–905
Ven Murthy MR, Ranjekar PK, Ramassamy C, Deshpande M (2010) Scientific basis for the use of Indian ayurvedic medicinal plants in the treatment of neurodegenerative disorders: ashwagandha. Cent Nerv Syst Agents Med Chem 10:238–246
Vescovi AL, Reynolds BA, Fraser DD, Weiss S (1993) bFGF regulates the proliferative fate of unipotent (neuronal) and bipotent (neuronal/astroglial) EGF-generated CNS progenitor cells. Neuron 11:951–966
Wang GH, Mitsui K, Kotliarova S, Yamashita A, Nagao Y, Tokuhiro S, Iwatsubo T, Kanazawa I, Nukina N (1999) Caspase activation during apoptotic cell death induced by expanded polyglutamine in N2a cells. Neuroreport 10:2435–2438
Wang J, Gu Q, Hao J, Bai D, Liu L, Zhao X, Liu Z, Wang L, Zhou Q (2013) Generation of induced pluripotent stem cells with high efficiency from human umbilical cord blood mononuclear cells. Genomics Proteomics Bioinformatics 11:304–311
Wang T, Choi E, Monaco MC, Major EO, Medynets M, Nath A (2015) Direct induction of human neural stem cells from peripheral blood hematopoietic progenitor cells. J Vis Exp 95:52298
Whitman MC, Greer CA (2009) Adult neurogenesis and the olfactory system. Prog Neurobiol 89:162–175
Wobus AM (2001) Potential of embryonic stem cells. Mol Aspects Med 22:149–164
Wobus AM, Boheler KR (2005) Embryonic stem cells: prospects for developmental biology and cell therapy. Physiol Rev 85:635–678
Wyttenbach A, Carmichael J, Swartz J, Furlong RA, Narain Y, Rankin J, Rubinsztein DC (2000) Effects of heat shock, heat shock protein 40 (HDJ-2), and proteasome inhibition on protein aggregation in cellular models of Huntington's disease. Proc Natl Acad Sci U S A 97:2898–2903
Xie HR, Hu LS, Li GY (2010) SH-SY5Y human neuroblastoma cell line: in vitro cell model of dopaminergic neurons in Parkinson's disease. Chin Med J (Engl) 123:1086–1092
Yamanaka S (2012) Induced pluripotent stem cells: past, present, and future. Cell Stem Cell 10:678–684
Yu DX, Marchetto MC, Gage FH (2014) How to make a hippocampal dentate gyrus granule neuron. Development 141:2366–2375

Zhang SC, Wernig M, Duncan ID, Brustle O, Thomson JA (2001) In vitro differentiation of transplantable neural precursors from human embryonic stem cells. Nat Biotechnol 19:1129–1133

Zhao J, Nakamura N, Hattori M, Kuboyama T, Tohda C, Komatsu K (2002) Withanolide derivatives from the roots of Withania somnifera and their neurite outgrowth activities. Chem Pharm Bull (Tokyo) 50:760–765

Zhou T, Benda C, Dunzinger S, Huang Y, Ho JC, Yang J, Wang Y, Zhang Y, Zhuang Q, Li Y, Bao X, Tse HF, Grillari J, Grillari-Voglauer R, Pei D, Esteban MA (2012) Generation of human induced pluripotent stem cells from urine samples. Nat Protoc 7:2080–2089

Zou J, Maeder ML, Mali P, Pruett-Miller SM, Thibodeau-Beganny S, Chou BK, Chen G, Ye Z, Park IH, Daley GQ, Porteus MH, Joung JK, Cheng L (2009) Gene targeting of a disease-related gene in human induced pluripotent stem and embryonic stem cells. Cell Stem Cell 5:97–110

Chapter 17
The Potential Utility of Ashwagandha for Improving Cognitive Dysfunction in Persons with Bipolar or Other Neurocognitive Disorders

K.N. Roy Chengappa, Jessica M. Gannon, Luna Acharya, and Abhishek Rai

Abstract Bipolar Disorder is a mood disorder that can be quite severe for subgroups of patients, and often such patients manifest deficits in attention, concentration, memory, and executive functions that persist even when they are relatively well. Cognitive concerns can impair daily life functioning and impact employment, education, and interpersonal relationships. Main line treatments for bipolar disorder typically do not address cognitive impairments, and some treatments may actually worsen cognitive deficits. Treatment development for cognitive impairments in bipolar disorder is truly in its infancy; extracts of Ashwagandha (*Withania somnifera*) may hold promise as a potential treatment option. The bioactive constituents of Ashwagandha, including withanolides, indosides, Withaferin A, and others, possess anti-inflammatory-immunomodulatory, anti-oxidant, cortisol lowering, anti-stress, pro-cholinergic, anti-glutamatergic and neuroprotective properties, the very processes that are thought to underlie cognitive impairments not only in in bipolar disorder but also in other psychiatric disorders like schizophrenia and major depression. This chapter reviews this in detail and highlights the preliminary randomized controlled study assessing an extract of Ashwagandha for improving cognition in bipolar disorder. The chapter also explores the potential role of Ashwagandha extracts in the treatment of other neurocognitive disorders, and it concludes by summarizing additional future research needed to realize the potential of this ancient medicinal plant for the treatment of human ailments.

K.N. Roy Chengappa (✉) • J.M. Gannon
Western Psychiatric Institute and Clinic, University of Pittsburgh School of Medicine, Pittsburgh, PA, USA
e-mail: chengappakn@upmc.edu

L. Acharya
Icahn School of Medicine at Mount Sinai (Bronx) Program, James J. Peters VA Medical Center, Bronx, NY, USA

A. Rai
University of Pittsburgh Medical Center – Northwest, Seneca, PA, USA

S.C. Kaul, R. Wadhwa (eds.), *Science of Ashwagandha: Preventive and Therapeutic Potentials*, DOI 10.1007/978-3-319-59192-6_17

Keywords Bipolar disorder • Ashwagandha • Withanolides • Randomized controlled study • Improvement of cognitive dysfunction

17.1 Bipolar Disorder

Bipolar disorder is an episodic, often life-long mood disorder with episodes of mania classically and episodes of depression interspersed with illness free intervals (Gershon and Chengappa 2009). However, many people with this illness suffer from extended periods of low to moderate level depressive symptoms that do not fully clear up and consequently they are unable to function adequately in daily life. Rates of suicide in Bipolar Disorder are among the highest of those reported in any psychiatric illness (Goodwin and Jamison 2007). Maladaptive behaviors that accompany the illness course (examples: promiscuity, alcohol and illicit substance abuse, poor mental judgment, reckless spending and impulsivity) may lead to disastrous consequences for the person, his or her family, friends, employers and society. Although bipolar disorder has a relatively low prevalence in the general population i.e. – 1 to 2%, it ranks second among the top 19 medical conditions (physical and mental) in terms of days per year out of role (17.3 days of lost productivity annually) in a 24-country WHO study (Alonso et al. 2011).

The most recent edition of the American Psychiatric Association's book on the classification of psychiatric disorders – DSM V (APA 2013) recognizes both Bipolar I and Bipolar II as subcategories of Bipolar Disorder. Depressive episodes are common to all categories of the illness, but Bipolar I disorder is characterized by full-fledged manic episodes, which may or may not have psychotic features. Bipolar II disorder is characterized by one or more hypomanic episodes. Cyclothymia is a milder form of Bipolar Disorder where the severity and duration criteria for depressive or manic episodes are not met but such symptoms are noted on and off for at least 2 years. Other forms of Bipolar Disorder included in DSM 5 do not meet full criteria for one reason or another or are presumed to be induced by illicit substances, prescription drugs or by other medical conditions.

Medications for bipolar disorder include lithium and anti-convulsants such as valproate, lamotrigine, carbamazepine, and antipsychotic drugs like olanzapine, quetiapine, risperidone, ziprasidone, aripiprazole, asenapine, paliperidone and lurasidone (Grande et al. 2015). Some of these drugs are approved for both acute episodes of mania and/or depression and for prophylactic use to prevent future episodes. Others are approved only for acute episodes. Among the older antipsychotic drugs, chlorpromazine was approved as an anti-manic agent; however, haloperidol (in both oral and injectable formulations) was and is still used as an anti-manic agent. Even though anti-depressants are widely used in bipolar disorder they are not formally approved for bipolar depression. For severe depression and/or mania, electroconvulsive treatment is also utilized.

17.1.1 Cognitive Dysfunction in Bipolar Disorder

Up until fairly recently, the general clinical wisdom was that the severe cognitive dysfunction that often characterizes persons with schizophrenia both in the acute phase of illness and when they are relatively well did not impact people with bipolar disorder, and even if it did, it was time-limited to severe manic, mixed and depressive episodes and essentially cleared when the episode remitted. However, that view has been seriously challenged and several recent meta-analyses indicate that cognitive impairments in subgroups of patients with bipolar disorder persists well beyond major mood episodes and can impact general day to day functionally quite significantly (Martinez-Arran et al. 2004a, b; Robinson et al. 2006; Torres et al. 2007; Wingo et al. 2010; Mann-Wrobel et al. 2011). It appears that bipolar patients who experience more episodes, especially psychotic manic episodes are more likely to have persistent cognitive impairments. There is also a question of how much the cognitive impairments are driven by the illness itself versus by the medications used to treat the illness.

Adverse cognitive effects of commonly used medications in bipolar disorder are summarized by Goldberg and Chengappa (2009). In studies of cognition in bipolar patients taking antipsychotics (especially the older but also some of the newer antipsychotics), these medications were linked to cognitive impairments including mental set-shifting difficulties, slowing down of psychomotor speed, and verbal memory deficits (Goldberg 2008). Executive dysfunction (set-shifting), verbal memory deficits (Frangou et al. 2005; Altshuler et al. 2004), and psychomotor slowing (Morrens et al. 2007) have also been demonstrated. Anti-cholinergic agents (benztropine, diphenhydramine, trihexiphenydyl) used to treat extrapyramidal side effects of antipsychotics are also known to impact attention and memory. Lithium, the main stay of treatment in acute mania and maintenance treatment of bipolar disorder has mixed data with reference to cognitive effects in bipolar disorder. Some data suggests subjective memory complaints by lithium treated patients, including verbal memory and motor speed, whereas objective tests do not fully support these patient-reported complaints (Goldberg 2008). Overall in long-term treatment, neurocognitive impairment with lithium is uncommon, other than in cases of lithium toxicity (Goldberg 2008). The anti-convulsants are a broad class of agents often used as an alternative to, or in addition to, lithium and/or antipsychotic drugs in the acute or maintenance treatment of bipolar patients, and their cognitive effects vary. In the main, lamotrigine appears to have a neutral or favorable impact on cognition, divalproex has minimal to mild effects on attention and memory. Carbamazepine and oxcarbazepine are less well studied, though carbamazepine is associated with mild cognitive issues (Goldberg 2008). Topiramate is not approved for bipolar illness but is used more for its weight loss properties and potential benefits in managing alcoholism or preventing migraines, frequent co-morbidities in persons with bipolar disorder. Topiramate is associated with dosage escalation and daily dosage related cognitive concerns including word-finding difficulties, issues with verbal and non-verbal fluency, attention and concentration issues, span of memory and working memory among other side-effects (summarized in Goldberg 2008). Among anti-depressants, SSRIs (selective serotonergic reuptake inhibitors) and SNRI (serotonin and nor-epinephrine reuptake inhibitors) are rarely associated with cognitive issues, and some studies have in fact shown favorable effects for improving working memory. Tricyclic anti-depressants by virtue of their anti-cholinergic effects can exert a negative impact on learning memory. Unsurprisingly, benzodiazepines affect arousal, attention, memory and psychomotor speed.

17.1.2 Cognitive Domains That Are Impaired in People with Bipolar Disorder

Bipolar Disorder related cognitive deficits include attention, memory and executive function. For some patients the deficits are most pronounced during episodes of depression and mania but may persist when they recover (summarized by Goldberg and Burdick 2008). There is evidence that non-ill monozygotic twins or unaffected first-degree relatives of patients with bipolar illness show some of this cognitive dysfunction as well, suggesting heritability. Attention is a basic cognitive construct that involves the conscious processing of information achieved by being alert and focused, as well as concentrating on relevant information while filtering out irrelevant stimuli. Of concern is that over 26 percent of euthymic bipolar patients were at the 5th percentile or below on the Stroop Test (a measure of selective attention), as well as on a measure of vigilance (Thompson et al. 2005). Attentional concerns are compounded by mania; not surprisingly, widespread deficits on attentional and verbal learning tests manifest during a manic episode (Clark et al. 2001). With regards to memory, as summarized by Elgamal et al. 2008, verbal memory (measures of which include immediate and delayed recall, cued recall, and verbal recognition) is impaired in symptomatic and euthymic phases of the illness. On the other hand, visual memory seems to be impacted mainly during the affective episodes, as is working memory (i.e., temporary storage, manipulation of data, and retrieval). Executive functioning is essential for planning, making decisions, detecting errors and correcting them, exercising proper judgement and overcoming risk taking behavior, overcoming a habitual response and responding to a novel situation when required and appropriate, and multi-tasking. Each of these tasks can be severely impacted in manic phases of bipolar illness, and deficits in executive function may persist in subsets of patients (but not necessarily all subjects) when they recover from acute episodes (summarized by Clark and Goodwin 2008). Impaired judgement and risk taking behavior in particular may remit following manic episodes, but the capacity for sustained attention or the ability to shift attention (attentional set shifting) more often remain impaired in euthymic phases of the illness. Cognitive inflexibility and impaired problem solving may also characterize subsets of patients with bipolar depression.

17.1.3 Functional Implications of Cognitive Impairments in Bipolar Disorder?

For the person with bipolar disorder, or her family, friends and co-workers, the key issues are not what scores or percentiles are obtained on various neuropsychological cognitive tests but how this cognitive dysfunction impacts the person in her daily life. Too often, these impairments seem to have a profound impact on our patients' roles as a student, employee or employer, friend or family member. In persons with schizophrenia, another severe mental illness with broad cognitive impairments, basic life functions are seriously impacted by cognitive problems (Green 1996). Likewise, in bipolar disorder, a review concluded that in 6 out of 8 studies inferior life functioning was linked to cognitive dysfunction even after controlling for demographic, illness and mood variables (Wingo et al. 2009). In a small "n" study of acutely manic patients who had been employed prior to a manic episode, among those who experienced symptom remission, it was improvements in working memory/attention, episodic memory and executive function that robustly predicted return to work at 3 months (57% were working), and at 6 months – 71% were working. These percentages are higher than those reported in earlier studies, but an important feature of this study is that it was a prospective study and reported on the work status of patients who were employed just prior to the manic episode (Bearden et al. 2011). Torres et al. (2008) reviewed several studies that speak to the association of cognitive dysfunction and functional outcomes in persons with bipolar disorder. Using clinician rated global measures of functioning (Global Assessment of Functioning (GAF) or Global Assessment scales) wherein higher scores are associated with higher levels of psychosocial functioning, Atre-Vaidya et al. (1998), Martinez-Aran et al. (2002, 2004a, b), and Torrent et al. (2006) noted backward digit span (auditory verbal memory), verbal fluency or verbal memory, and/or attentional set shifting or trail making tests were closely associated with improved psychosocial functioning. In addition to the Bearden study (2011) noted earlier, Dickerson et al. (2004), Martinez-Aran et al. (2004a, b) and Torrent et al. (2006) also studied employment status, or occupational functioning and occupational adaptation, and further demonstrated that executive cognitions, working and verbal memory, and attentional shifting predicted occupational recovery. These research findings support the clinical observations of patients and their supports – that cognitive impairment plays a significant role in patient's recoveries and return to full role functioning following mood episodes.

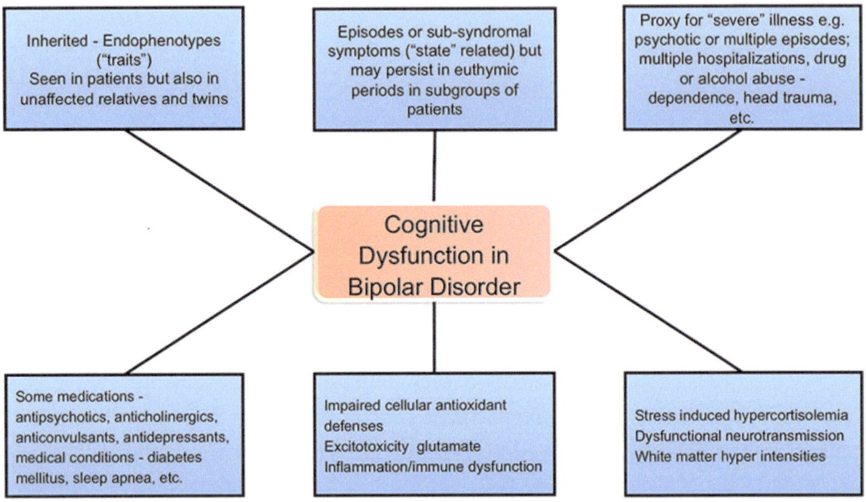

Fig. 17.1 Multiple state and trait factors, including genetics, treatment course, comorbidities, and neurobiological functioning, influence cognition in Bipolar Disorder

17.1.4 Factors and Mechanisms That May Contribute to Cognitive Impairment in Bipolar Disorder

As noted in Fig. 17.1, inherited factors may contribute to cognitive problems in people with this illness. This is not that unexpected in that bipolar disorder is a highly heritable disorder and cognitive traits, both positive and negative, may be inherited (Glahn et al. 2008). For example, unaffected relatives of bipolar patients may share some of the executive and memory problems of their affected relative (Clark et al. 2005; Ferrier et al. 2004; McIntosh et al. 2005), though not all studies have demonstrated this finding.

As alluded to in previous sections, symptoms of mania, depression, psychoses or anxiety clearly impact cognition in people with bipolar disorder, as summarized by Malhi et al. (2008). As reviewed, mania is characterized by disinhibition, poor decision-making and judgment, impaired attention, and deficits in memory and problem solving, whereas depression is accompanied by poor attention and concentration, deficits in verbal memory and verbal fluency, and executive dysfunctions. Psychosis in manic episodes is associated with deficits in working memory and executive dysfunction. Subgroups of euthymic individuals may also be impacted by cognitive inflexibility and attentional difficulties. Moreover, even lingering sub-syndromal depressive symptoms may impact cognitive abilities and impact general life functioning in persons with bipolar disorder. Clinical characteristics that indicate severity of illness, e.g. age of onset, duration of illness, numbers of mood episodes, episodes with psychotic features, ongoing sub-syndromal symptoms, frequent substance and alcohol abuse-dependence, etc. have been studied for their impact on outcomes including cognitive outcomes. Results have been mixed or negative for some variables (e.g. duration of illness, age of illness onset, or length of time euthymic) – however, an association between poorer cognitive outcomes and number of manic and especially psychotic episodes, number of hospitalizations, depressive episodes, and substance dependence and alcohol abuse (summarized by Vieta et al. 2008) has been more consistently demonstrated.

Of the many neurobiological processes that may mediate cognitive impairment in bipolar disorder, inflammatory and immunological processes have been invoked as potentially causative (Rosenblat et al. 2015). Inflammation, such as can be induced by stress or other causes can provoke a pro-inflammatory response, including elevated cytokines or proteins such as IL-6, TNF-alpha and CRP, all of which have been reported as elevated in persons with bipolar disorder. Inflammatory cascades also induce oxidative stress, which in turn can impair monoaminergic neurotransmission and impact cognition (See Fig. 17.1). There is also evidence for lower neurotrophin levels (brain derived neurotrophic factor, BDNF) in patients with bipolar disorder (Bauer et al. 2014a, b), which is also associated with cognitive impairment. Glutamate, the excitatory neurotransmitter in the brain, is reportedly over active in manic subjects (Michael et al. 2003); unchecked glutamate excitotoxicity may result in reduced neuronal viability. Post-mortem studies have shown reduced density (>40%) of glutamic acid decarboxlylase neurons in CA2/3 brain regions as well as in the dentate gyrus of the hippocampus of bipolar patients (Heckers et al. 2002). It is thus quite possible that glutamatergic excitotoxicity may contribute to cognitive dysfunction in subgroups of bipolar subjects. A highly replicated finding in MRI brain scans of people with bipolar disorder is the occurrence of white matter hyper-intensities in the periventricular and deep white matter tracts (Dupont et al. 1990, Moore et al. 2001; Tighe et al. 2012). These white matter hyper-intensities suggest neuronal insults and are associated with severe illness, cognitive dysfunction, and treatment resistance in bipolar disorder.

Stress mediated immune-inflammatory processes are detailed in Fig. 17.2. Hypercortisolemia characterizes subgroups of patients with unipolar and bipolar disorder, and may be related to stressors (Rush et al. 1996). Studies of patients with Cushing's syndrome, a condition characterized by elevated cortisol levels, indicate that these patients experience cognitive impairments (Whelan et al. 1980; Starkman et al. 2001). Also demonstrated in Cushing's syndrome patients is the finding that cortisol elevations are linked to hippocampal dysfunction, a brain structure closely associated with memory functions (Resmini et al. 2013). Furthermore, in healthy controls, hydrocortisone administration impairs working and verbal declarative memory (Lupien et al. 1999; Young et al. 1999). A review of studies of glucocorticoids and hippocampal function suggests that while studies were not consistent in demonstrating hippocampal volume loss or neuronal viability, they were more reliable with regards to loss of synaptic dendritic complexity or hippocampal dysfunction (Hibberd et al. 2000). Thus, treatments that correct elevated cortisol levels may improve neurocognitive and mood symptoms in bipolar patients (see Young et al. 2004 below).

Another endocrine abnormality that is relatively common in bipolar and unipolar patients is the occurrence of sub-clinical or overt hypothyroidism (Bauer et al.

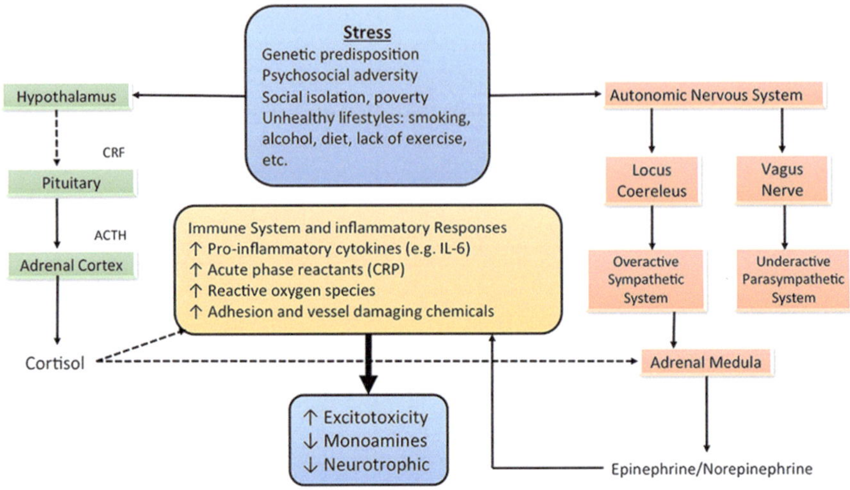

Fig. 17.2 Stress mediated immune-inflammatory processes

2014a, b). This condition is often linked to lithium treatment, with up to an estimated 20% of lithium-treated patients experiencing hypothyroidism, and with a higher incidence in women (Ozerdem et al. 2014; Bauer et al. 2014a, b). However, hypothyroidism can and does occur independently from lithium use in Bipolar Disorder. Cognitive difficulties, including slowed mentation and memory impairments are clinical features associated with subclinical and overt hypothyroidism, especially in people younger than 75 years. These deficits appear to be correlated with higher levels of thyroid stimulating hormone (TSH) (Pasqualetti et al. 2015), though not all studies are in agreement with this finding (Akintola et al. 2015). Limitations of large meta-analytic reviews of cognitive impairment and hypothyroidism are multifactorial. Most patients included in these studies were from otherwise healthy, community dwelling populations, unlike the clinical help-seeking populations, such as those with bipolar or unipolar disorder. Most community studies were also cross-sectional in nature, and definitions of subclinical hypothyroidism varied over time and often involved a single measurement of TSH. Finally, the choice of the cognitive scale (mini-mental state examination) was likely not sensitive to subtle cognitive changes.

In bipolar patients, cognitive impairments due to lithium induced overt and subclinical hypothyroidism can be ameliorated with thyroxine treatment (Tremont and Stern 1997). Ashwagandha extracts with thyroid promoting actions may thus be beneficial for thyroid induced cognitive dysfunction among bipolar patients, as discussed in the following section.

17.1.5 A General Approach to Assessing and Treating Cognitive Dysfunction in Bipolar Disorder

Patients, who describe cognitive problems, especially if these are relatively new, troublesome, and not commensurate with their age or premorbid functioning, should have their complaints taken seriously. Routine cognitive tests – performed in an office or at a patient's bedside, are good beginning to assessing such complaints. Such tests, as summarized by Goldberg and Chengappa (2009), could include: digit span forward and backward, serial seven subtraction from 100, word lists to repeat and recall later, and memory tests for recent, intermediate and remote events. If psychotropic medications have been introduced recently, and it is suspected that these agents may be responsible for cognitive issues, alternate medications thought to be more cognitively neutral may instead be prescribed. If there is suspicion of a co-occurring neurological condition (e.g. CNS lupus, multiple sclerosis, or tumors of the brain) then referral to a neurologist may be appropriate. If patients are still symptomatic, then optimization of treatment will need to occur first; that in and of itself may resolve the cognitive complaints. This may be especially true for a severe bipolar depression presenting in mid to later life, which may, like in unipolar depression, present as "pseudo-dementia". It is important to not to overlook "hidden" alcohol and other substance use disorders, as these comorbid conditions may severely impact cognitive abilities. Occupational history is also important, as exposure to CNS toxins is a possibility in certain fields. Appropriate testing for specific toxins may then be required. Common medical conditions that complicate the lives of persons with bipolar illness, including poorly controlled diabetes mellitus, and untreated obstructive sleep apnea, as well as medications used for general medical conditions that impair cognition, may all need to be carefully assessed and evaluated for impact on cognition. For some older bipolar patients, it is possible that dementia may be co-occurring and requires a more formal assessment to rule-out. If in spite of the considerations noted above, euthymic patients continue to be impacted by cognitive concerns, formal neuropsychological testing may be appropriate, if such services are readily available.

17.2 Pharmacological and Non-pharmacological Treatments for Cognitive Dysfunction

Multiple pharmacologic agents have been utilized in the treatment of impaired cognition in bipolar disorder. *Pro-cholinergic drugs* (acetylcholine esterase inhibitors such as galantamine or donepezil) and anti-glutamatergic drugs (memantine) used in Alzheimer's disease have been studied in schizophrenia and in bipolar disorder to enhance cognition (Goldberg and Chengappa 2009), but a recent review concluded the evidence is weak for their use in bipolar disorder (Veronese et al. 2016). Notably, extracts of Ashwagandha have pro-cholinergic properties, as described in the next section of this chapter. *Stimulants* like methylphenidate and methamphetamine improve cognitive issues in attention deficit hyperactivity disorders (ADHD) and have been considered for bipolar disorder, but on account of substance abuse liability and clinical concerns for precipitating mania or inducing rapid cycling, stimulants have received little study. *Dopaminergic drugs*, such as those commonly prescribed in the treatment of Parkinson's disease may be of benefit; pramiprexole was studied in an 8-week randomized controlled trial for improving cognition in bipolar disorder (Burdick et al. 2012). Even though the primary cognitive outcome was negative, when the analyses was limited to the euthymic group only (which was the original intent of the trial but due to poor recruitment, the inclusion criteria were relaxed to permit patients with sub-syndromal depressive symptoms to be enrolled), then positive results were obtained for digits backward (working memory) and Stroop Color test (processing speed), (Burdick et al. 2012). To potentially counter the clinical consequences of hypercortisolemia on mood and cognition symptoms, a *glucocorticoid receptor antagonist* – mifepristone, was studied in a small group of bipolar patients experiencing depression during a 6-week randomized cross-over double blind trial. Improvements in working and recognition memory and in verbal fluency, as well as in depressive symptom severity, were noted over placebo for the mifepristone treated group (Young et al. 2004). Recombinant *erythropoietin* infusions given intravenously (vs. saline controls), weekly for 8 weeks demonstrated improvements in sustained attention, recognition of happy faces, and processing of complex information; no improvements in verbal memory were noted (Miskowiak et al. 2014). Erythropoietin has neurotrophic properties and therefore was expected to counter lower levels of neurotrophic factors in persons with bipolar disorder; however, levels of BDNF decreased rather than increased.

Psychotherapeutic and psychosocial interventions have also been tried in treating cognitive impairment in Bipolar Disorder. Notably, *cognitive remediation* was tested against standard treatment in a randomized controlled study of 12 weeks duration in a small study of euthymic bipolar subjects. This intervention did not improve primary cognitive outcomes (Demant et al. 2015) at either 12 weeks or at the 26-week follow-up.

A review of www.clinicaltrials.gov including the search term "cognition and bipolar disorder" (accessed on June 2, 2016) indicated that several other studies exploring potential treatments are either completed, with results not yet available (e.g. *methylene blue*), or are still recruiting (e.g. computer-based cognitive remediation vs. preselected computer games, completing e-neurocognitive modules, *N-acetyl cysteine* or *minocycline* (an antibiotic with anti-inflammatory and neuroprotective properties), IV infusions of *infliximab* (a tumor necrosis factor inhibitor), etc.). Nonetheless, and at the time of writing, studies of pharmacological or non-pharmacological approaches to improve cognition in bipolar disorder remain in their infancy.

17.3 Extracts of Ashwagandha as Proposed Treatment of Cognitive Impairments in Bipolar Disorder

As noted in Fig. 17.3, the rationale for why extracts of Ashwagandha may aid in improving cognitive dysfunction in people with bipolar disorder (and potentially in other psychiatric and neuropsychiatric disorders) is related to its multi-faceted actions on target mechanisms that may underlie cognitive impairments in bipolar disorder, see also Table 17.1. As noted earlier, a recent review summarized the studies of one of the main suspect pathologies associated with cognitive impairment in bipolar disorder i.e., low-grade neuro-immune-inflammatory processes (Rosenblat et al. 2015). These pro-inflammatory cytokines and proteins lead to dysfunctional monoaminergic neurotransmission (e.g. serotonin) and also induce oxidative stress, which can, in turn, be deleterious to both neuronal function and viability. The authors opined that this mechanistic pathway might lead to both mood episodes and cognitive impairments in bipolar disorder (Rosenblat et al. 2015). In a meta-analyses, Dargel and colleagues (2015) reported CRP (C-reactive protein) levels were elevated in 730 patients with bipolar disorder compared to 888 controls suggesting an ongoing inflammatory state in the disorder; this elevation was especially marked in mania, was less so in euthymia, and was not significantly different from controls in the depressed phase. A genetic study of bipolar patients, which assessed 19 genes associated with inflammation, revealed that 52% of BPD subjects showed an inflammatory genetic signature compared to 18% of controls (Padmos et al. 2008). Moreover,

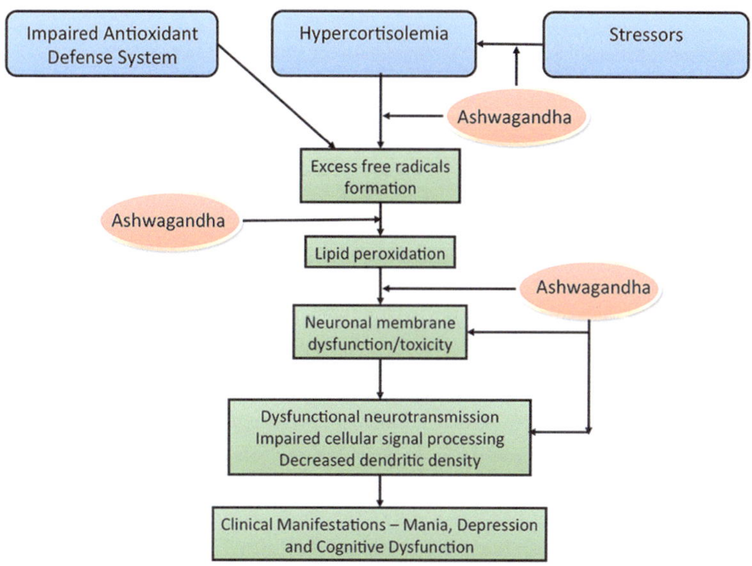

Fig. 17.3 Ashwagandha may improve cognitive symptoms of Bipolar Disorder through its effect on multiple pathways of proposed neuro-immune-inflammatory processes

Table 17.1 Why might Ashwagandha extracts, metabolites, or synthetic analogues be useful for improving cognition in bipolar and other neurocognitive disorders?

Possible targets of action	Animal data	Human data
Immune – inflammatory processes	Durg et al. (2015) (Review), Anbalagan and Sadique (1981, 1985), Begum and Sadique (1988), Aggarwal et al. (1999), Nashine et al. (1995), Jayaprakasam and Nair (2003), Khan et al. (2006), Subbaraju et al. (2006), Singh et al. (2007), Malik et al. (2007), Kour et al. (2009), and Mulabagal et al. (2009)	Kulkarni et al. (1991), Chopra et al. (2000, 2004) and Mikolai et al. (2009)
Hypercortisolemia/anti-stress/thyroid promoting/anxiolytic/weight loss	Ghosal et al. (1988), Panda and Kar (1997, 1998), Bhattacharya and Muruganandam (2003) and Khan and Ghosh (2010)	Andrade et al. (2000), Auddy et al. (2008), Cooley et al. (2009), Chandrasekhar et al. (2012), Gannon et al. (2014), Gajarmal and Shende (2015), and Choudhary et al. (2017)
Antioxidant Potential	Durg et al. (2015) (Review), Ghosal et al. (1988), Bhattacharya and Muruganandam (2003), Kumar and Kumar (2008) and Ahmed et al. (2013)	
Pro-cholinergic actions	Schliebs et al. (1997), Choudhary et al. (2004), Konar et al. (2011), and Grover et al. (2012)	
Decreased glutamatergic toxicity/neuroprotective effects/anti-dyskinestic potential	Jain et al. (2001), Naidu et al. (2006), Kumar and Kumar (2009), and Kataria et al. (2012)	
Synaptic, dendritic and axonal reconstruction	Kuboyama et al. (2005, 2006, 2014) (Review), Tohda and Joyashiki (2009), and Joyashiki et al. (2011)	
Clearance of B-amyloid, decrease in B-amyloid neurotoxicity	Sehgal et al. (2012) and Kurupati et al. (2013)	

convergent animal and human data indicate stress induced immune-inflammatory processes not only dysregulate the hypothalamo-pituitary-adrenal axis (HPA) resulting in hypercortisolemia but also disrupt normal autonomic nervous system functioning resulting in an adrenergic overdrive, (Fig. 17.2) i.e., increased norepinephrine levels (Cannon 1932; Selye 1956; Szentivanyi 1989; Chrousos 1995; Sloan et al. 2007). These biological effector molecules (cytokines, cortisol, CRP) in turn influence dopaminergic, glutamatergic, serotonergic, and adrenergic neurotransmission that are closely allied to the neurobiology and symptoms associated with mood disorders (Besedovsky et al. 1983; Sapolsky et al. 1987; Miller et al. 2009; Zalcman et al. 1994; Kronfol and Remick 2000). For instance, in animals, inflammatory cytokines induce anhedonia, social withdrawal, diminished appetite, sleep disturbances, cognitive impairments, etc. (Schiepers et al. 2005; Dantzer et al. 2008; Miller et al. 2009; Miller 2010). Examples of behavioral responses to stress in humans include deregulated sleep-wake cycles, mood instability, interpersonal conflicts, altered thinking patterns, and cognitive impairments (Rosenthal et al. 1984; Wehr et al. 1998; Frank et al. 2000; Wirz-Justice et al. 2009). In bipolar patients (BPD) experiencing stress, bi-directional behavioral and biologic responses can coalesce as circadian rhythm disturbances (Coogan and Wyse 2008) and lead to episodes of mania or depression (Wehr et al. 1987) or cognitive impairments (Rosenblat et al. 2015). These deleterious effects of effector molecules seem to occur in brain regions (e.g. hippocampus and pre-frontal cortex) that are closely associated with normal or abnormal mood states and with various cognitive abilities.

The therapeutic implications of the above referenced pathogenic mechanisms are not surprising given the evidence of anti-inflammatory drugs in improving cognition and depression. Curiously, Omega-3 fatty acids (an anti-inflammatory) have been used for treating depression in bipolar disorder with mixed results (examples: Stoll et al. 1999; Keck et al. 2006); this agent however, has not been studied for the treatment of cognitive impairment in bipolar disorder. Similarly, N-acetyl cysteine (an anti-oxidant and anti-inflammatory agent) has been targeted toward the treatment of depression in bipolar disorder rather than used as a procognitive agent (Berk et al. 2008). Non-steroidal anti-inflammatory drugs (NSAIDS) such as celecoxib have been assessed for improving bipolar depression; in a small ($n = 28$) pilot study of celecoxib, even though the time to improvement of depressive symptoms was faster (i.e. at week 1) in the celecoxib group, by the end of the study, both placebo and celecoxib group improved without significant differences between treatments (Nery et al. 2008). Again, cognition was not targeted. A search of www.clinicaltrials.gov suggests that more trials are under way. Various other treatments targeting cognitive impairments in bipolar disorder have been summarized earlier in this chapter.

Animal studies using Ashwagandha extracts (reviewed in Durg et al. 2015) indicate that bioactive constituents from Ashwagandha, including steroidal lactones such as glycowithanolides, sitoindosides, and Withaferin A, have potent brain antioxidant, neuroprotective and memory enhancing properties (Kulkarni and Dhir 2008; Ghosal et al. 1988; Panda and Kar 1997; Jain et al. 2001). A review of 8 markers of oxidative stress in bipolar disorder (superoxide dismutase, catalase, glutathione peroxidase, protein carbonyl, 3-nitrotyrosine, lipid peroxidation, nitric oxide and RNA/DNA damage) involving 971 patients with bipolar disorder and 886 controls indicated that indices of DNA/RNA damage, nitric oxide, and especially lipid peroxidation, were significantly increased among patients compared to controls (Brown et al. 2014). In a comprehensive review of rodent models of brain oxidative stress (including prolonged swimming, sleep deprivation, and chemical and physical restraint stress), pretreatment with extracts of Withania somnifera (WS) normalized levels of superoxide dismutase, catalase, glutathione peroxidase, and reduced glutathione (Durg et al. 2015). Lipid peroxidation products like malonidialdehyde and thiobarbituric acid reactive substances, which can induce neuronal dysfunction and damage if unchecked, are reduced by WS treatment. Such benefits have been reported in studies that have evaluated specific mice and rat brain regions, including the hippocampus, striatum, and forebrain (Durg et al. 2015). In an animal model of restraint stress, where nitric oxide levels can increase markedly and cause neuronal dysfunction, Ashwagandha pretreatment significantly diminished this increase in nitric acid levels (Khan and Ghosh 2010). Furthermore, chronic administration (4 weeks) of WS extracts significantly reduced lipid peroxidation, and restored glutathione, super oxide dismutase and catalase levels in a rat brain model of reserpine-induced orofacial dyskinesia and cognitive dysfunction (Naidu et al. 2006). In another rat model, poor retention of memory (in elevated plus maze task) was significantly ameliorated by a WS extract (Ghosal et al. 1988). In an immobilization stress rat model, a WS extract reduced damage to CA2 and CA3 hippocampal neurons by nearly 80% (Jain et al. 2001). Finally, in a rat model of behavioral cognitive deficits (i.e. Morris Maze Test) induced by streptozotocin and thus mediated by increased oxidative stress and cholinergic dysfunction, i.e. decreased choline acetyl transferase activity in the hippocampus, Ahmed et al. (2013) reported that pretreatment with WS extracts for 3 weeks attenuated cognitive deficits while restoring anti-oxidant status and improving cholinergic neurotransmission. Schliebs et al. (1997) reported that systemic administration of WS extracts had pro-cholinergic effects in rat brain, and Konar et al. (2011) were able to reverse scopolamine-induced anti-cholinergic actions using WS extracts. As noted earlier, glutamate excitotoxicity has been reported when bipolar patients experience mania, and unchecked glutamate excitotoxicity can result in loss of neuronal functions (cognitive problems could result) and/or neuronal and glial viability (cell death). Kataria et al. (2012) reported that a water extract of WS not only attenuated glutamate excitotoxicity in rat glioma and human neuroblastoma cell lines, but also significantly diminished protein markers of neuronal viability including heat shock proteins (HSP70) while inducing protein expression (e.g. neural cell adhesion molecule, NCAM) that is known to improve neural plasticity, learning and memory.

In animal models of inflammation or arthritis, Ashwangandha extracts exhibited significant anti-inflammatory activity, notably decreased levels of acute phase reactants and prostaglandin synthesis (Anbalagan and Sadique 1981, 1985; Begum and Sadique 1988; Aggarwal et al. 1999; Nashine et al. 1995), comparing favorably with a positive control, phenylbutazone. WS especially lowered α-2 macroglobulin levels, an acute phase reactant that is considered to be a sensitive indicator of the efficacy of anti-inflammatory agents (Anbalagan and Sadique 1985). A WS extract also inhibited NF-kB (an inflammatory signaling pathway) in mononuclear cells from healthy controls as well as in patients with rheumatoid arthritis, resulting in the suppressed production of several pro-inflammatory cytokines (Singh et al. 2007). WS withanolides also show potent Cox-2 enzyme inhibition activity (Jayaprakasam and Nair 2003; Subbaraju et al. 2006; Mulabagal et al. 2009), a mechanism closely associated with the effectiveness of several anti-inflammatory medicines.

Recent human clinical trial data exist for the use of Ashwagandha extracts in anxiety disorders, arthritis, and stress. Two randomized clinical trials (RCT) (6 and 12 weeks in duration) showed anxiolytic activity for Ashwagandha extracts in patients with anxiety disorders (Andrade et al. 2000; Cooley et al. 2009); side effects were minor and did not differ significantly from placebo. A RCT for stressed individuals in an Ayurvedic clinic used Ashwagandha extracts for 2 months and demonstrated efficacy in relieving stress symptoms as well as decreasing elevated cortisol and CRP (Auddy et al. 2008). In two RCTs for arthritis, Ashwagandha extracts combined with other herbal extracts showed efficacy in arthritic and other pain measures, with subjects' complaining of only minor side effects (Kulkarni et al. 1991; Chopra et al. 2004). A recent 8 week RCT utilizing Ashwagandha for body weight management among stressed adults showed positive results including weight loss, improved stress scores, and decreased cortisol levels compared to the placebo assigned group (Choudhary et al. 2017). Vital signs and routine laboratory measures undertaken in these studies were in the normal reference range.

On account of the anti-stress, anti-anxiety, anti-inflammatory, cortisol lowering, and cytokine and monoamine modulating properties, demonstrated through animal and human studies, Ashwagandha extracts show promise as a potential adjunctive treatment for both cognitive impairments and mood exacerbations in people with bipolar disorder.

17.3.1 The Current Clinical Evidence for Using Extracts of Ashwagandha for Enhancing Cognition in Bipolar Disorder

Based on the results of the human and animal studies summarized above, we conducted an 8-week randomized clinical trial in 60 bipolar patients who were euthymic (mentally stable) during the period before the clinical trial (4 weeks or greater, and medications had been stable for 4 weeks or longer) comparing a standardized water based extract of Ashwagandha (Sensoril) to placebo (Chengappa et al. 2013). The dosage of Sensoril was initiated at 250 mg (minimum of 8% withanolides, maximum of 2% Withaferin A, 32% of oligosaccharides) once a day during the first week and increased to 250 mg twice a day for the balance of the study. A computerized neuropsychological battery was used to assess cognition; details are available in Chengappa et al. 2013. Three cognitive tasks improved significantly in the WS treated group versus placebo; digit span backward, response time on the Flanker's test, and a measure of social cognition. Among these tests, the size of the WS treatment effect was in the moderate range for digit span backward, a measure of auditory verbal working memory. There were no statistically significant differences in any of the other cognitive tests or domains between treatments. In terms of safety, there were no serious adverse events for either placebo or WS treatments. Adverse events were relatively mild and transient, and did not lead to discontinuation from the study. Vital signs remained stable, as did EKG and laboratory findings. Weight gain was minimal (a mean of approximately 2 lbs. for the WS treated group vs. 0.6 lbs. for the placebo). Importantly, as a group, there were no worsening subjects' mood states. To our knowledge, this is the only reported study using WS in patients with bipolar disorder for cognition, and there has been no independent confirmation of the results. Limitations to the study include a small sample size, and statistical corrections for multiple comparisons were not applied. Also, the duration of the trial was relatively short (2 months) and it is not clear what the upper and lower limits of dosing might be for this standardized aqueous extract of WS (i.e. Sensoril). We reviewed thyroid indices done in this trial as a secondary paper (Gannon et al. 2014). The reason thyroid indices were assessed was that there had been a single case-report of thyrotoxicosis in an adult woman who had taken an Ashwagandha extract (van der Hooft et al. 2005), as well as two animal studies suggesting WS extracts could boost thyroid function (Panda and Kar 1998, 1999). As noted earlier, overt and subclinical hypothyroidism is associated with cognitive difficulties (especially working memory and slowing of processing time, etc.), and bipolar patients have relatively higher rates of hypothyroidism (Ozerdem et al. 2014; Bauer et al. 2014a, b). Our results hint that Ashwagandha extracts could in fact boost thyroid indices and correct sub-clinical hypothyroidism (Gannon et al. 2014). So, if we were to prospectively pre-select bipolar patients in a clinical trial with both cognitive impairments and laboratory evidence of subtle hypothyroidism, more definitive answers may be obtained regarding the benefits of WS extracts for improving cognition.

In the main study, we had expected broader improvements in cognitive domains, given the multiple effects of WS on varied targets, but obtained narrower results. So while these data are promising, more work is required to confirm the utility of WS for improving cognition in patients with bipolar disorder.

17.3.2 Ashwagandha in the Treatment of Other Neurocognitive Disorders

The immune inflammatory hypothesis has been put forward as a potential explanation for symptom exacerbations or inducing new episodes, as well as cognitive impairments, in schizophrenia (Ganguli et al. 1994, 1995; Muller and Schwarz 2006, 2010) and in major depressive disorders (Dantzer et al. 2008; Miller et al. 2009). WS extracts can potentially be evaluated for improving cognitive impairments in these two psychiatric disorders, as well as in other neuropsychiatric disorders. Please see Table 17.2. In a mouse model of Alzheimer's disease, WS extracts cleared beta-amyloid likely via hepatic mechanisms, and reversed impairments in spatial learning and working memory (Sehgal et al. 2012). In vitro and in vivo

Table 17.2 Potential utility of Ashwagandha for neurocognitive disorders

1.	Bipolar disorder
2.	Schizophrenia
3.	Major depressive disorder
4.	Attention deficit hyperactivity disorder
5.	Parkinson's disease
6.	Huntington's disease
7.	Cognitive disorders secondary to insecticides, pesticides, acaricides, etc.
8.	Cognitive disorders secondary to poorly controlled blood glucose (Diabetes Mellitus)

experiments in human neuroblastoma cell lines utilizing intracerebral injections of A beta amyloid fragments lead to dendritic and axonal degeneration in mouse brain models, which was mitigated or reversed by methanolic WS extracts (Kuboyama et al. 2005, 2006, 2014). The axonal regeneration and synaptic reconstruction was attributed to a specific WS metabolite, sominone. Sominone has been shown to improve spatial memory and axonal regeneration in another Alzheimer's mouse model (Tohda and Joyashiki 2009; Joyashiki et al. 2011). Moreover, WS extracts have cholinesterase inhibiting properties (Choudhary et al. 2004; Grover et al. 2012); drugs with such properties (e.g. donepezil, rivastigmine, and galantamine) have been approved for improving memory and behavior, but do not seem to arrest the progression of Alzheimer's disease. Therefore it follows that extracts of WS, including specific metabolites of WS like sominone, or synthetic derivatives of sominone such as denosomin may have a place in the treatment of cognitive and other behavioral disturbances in Alzheimer's disease (Kuboyama et al. 2014). HIV infection and AIDS can be complicated by neurocognitive difficulties, and Kurupati and colleagues (2013) were able to significantly mitigate beta-amyloid and HIV infection induced neurotoxicity in a human neuronal cell line using a methanol/chloroform extract of WS. Parkinson's and Huntington's diseases are characterized by significant oxidative stress, dopaminergic cell loss (Parkinson's), GABAergic abnormalities (Huntington's) at a cellular level, and by movement disorders and other neuropsychiatric symptoms and signs. In rodent models of Parkinson's (6-OH Dopa, MPTP, Maneb-paraquat) and Huntington's diseases (3-NP), WS extracts significantly reduced oxidative stress, rescued surviving dopaminergic neurons, and improved cognitive impairments and other symptoms (studies summarized in Dar et al. 2015, and Kumar and Kumar, 2008, 2009). In one Ayurvedic treatment trial of 18 patients with Parkinson's disease, 13 patients were administered cleansing plus a concoction of 4 herbal extracts (including Ashwagandha) in milk, and 5 patients were given the concoction only. Interestingly, both groups experienced improvements in motor functions and rating scale scores, and both groups experienced hyper salivation (Nagashayana et al. 2000). However, it was hard to know if the improvements were due only to Ashwagandha root extracts or another herb in the concoction, i.e. M. pruriens, which includes a significant percentage of L-dopa, an approved medication for the treatment of Parkinson's disease. It is possible that the increased rate of hyper salivation in both groups was due to the pro-cholinergic effects of Ashwagandha. Nevertheless, WS extracts may be reasonable adjunctive treatments to assess for improving cognition and other symptoms in patients with Parkinson's or Huntington's diseases. Another therapeutic area where WS extracts may be worth considering is in countering the chronic cognitive effects of anti-cholinergic pesticides, insecticides, acaricides, herbicides, and nerve gas agents (examples: organophosphorus compounds, carbamates, and others). Occupations that may be particularly exposed and vulnerable to such anti-cholinergic agents for prolonged periods (examples: farmers, gardeners, public health agency workers, employees of pest control companies, veterinarians, health care providers, others) may experience cognitive and other symptoms that may benefit from WS extracts. It is possible that WS extracts may be useful for chronic cognitive symptoms that persist following exposure to such anti-cholinergic agents rather than for acute poisoning or toxicity. In a scopolamine mouse model of central anti-cholinergic memory impairments, alcoholic extracts of WS and its purified constituent, withanone, attenuated oxidative stress and enhanced recovery of neuronal and glial cells as measured by brain-derived neurotrophic growth factor (BDNF) and other markers of cellular integrity (Konar et al. 2011). In a rat model where propoxur (a carbamate acaricide) was used to inhibit acetyl cholinesterase (AChE) and induce cognitive impairments and oxidative stress, WS extracts attenuated AChE inhibition and improved memory deficits (Yadav et al. 2010). An additional but relatively common medical ailment, type II diabetes mellitus is increasingly associated with dementia and cognitive impairment, often in the context of poorly controlled blood glucose. This cognitive impairment is often under recognized and thus neglected by patients, families, and even physicians (Cheng et al. 2012). WS extracts improve cognitive impairments in streptozotocin-induced diabetic cognitive decline in rats (Ahmed et al. 2013), and WS extracts are known to improve glucose control in skeletal myotubes and adipocytes (Gorelick et al. 2015), alloxan-induced diabetic rats, (Udayakumar et al. 2010) and in humans (Andallu and Radhika 2000). The broad anti-oxidative properties of WS may also be helpful in the treatment of diabetes. All told, WS extracts may play a role in helping control blood glucose and attenuate cognitive impairments in patients with poorly controlled diabetes.

17.4 Concluding Remarks

Modern *in vivo* and *in vitro* laboratory data and human clinical trials have begun to reveal the benefits of Ashwagandha for several cognitive disorders. Human data are still preliminary. More work is required to delineate the upper and lower dosages of Ashwagandha extracts that may be efficacious or effective for various neurocognitive disorders, whether it is safe to combine WS extracts with the main treatments used for various medical conditions, and to determine and adequate duration of treatment WS. Also, less is known as to whether or not specific extraction methods (examples: water, methanol, methanol/chloroform or others) that extrude certain constituents more than others (examples: withanolides, sitoindosides, Withaferin A, others) in specific concentrations are important for the therapeutic benefits or whether these may add to risks for human use in specific medical conditions (example: CNS disorders vs. Cancers)? Similarly, whether or not specific cultivation techniques for the Ashwagandha plant, over specific geographic regions and in certain seasons, contribute to its therapeutic potential versus risk is also less well known. For example, even though the cultivated and wild type Withania somnifera plants may look morphologically and anatomically indistinguishable, it is possible that one set of plant genes contributes to a dominant set of metabolites or secondary metabolites (sometimes referred to as a "chemotype" or "chemovar"), and this may be important for therapeutic benefits vs. risks of various WS extracts. The basic pharmacology (examples: absorption, metabolism, distribution, elimination, half-lives, toxicology and drug interactions) of various WS extracts, as well as that of specific natural metabolites and constituents (examples: withanone or sominone) or synthetic analogues of sominone (example: denosomin), needs more study. Even though emerging safety data from the long and ancient usage of Ashwagandha plant extracts (in indigenous medical systems such as Ayurveda or Unani, or in food, or in teas etc.) as well as that emerging from the limited number of human clinical trials has been reassuring, shorter and longer term efficacy and safety data in humans from modern regulatory grade randomized or pragmatic clinical trials conducted for specific medical indications is needed. Standardization of WS extracts and Good Manufacturing Practices will increase the confidence in results obtained from animal and human studies and clinical trials (Brar and Chengappa 2011). Small molecules or new chemical entities (NCEs) from WS metabolites or synthetic analogues of these metabolites (examples: withanone, sominone, denosomin) may provide standardization that is hard to come by in WS plant extracts, but whether the more benign side effect profile from the crude plant extracts is traded for greater safety concerns of these small molecules or NCEs is largely unknown at this time (Brar and Chengappa 2011). Nevertheless, plant derived treatments have contributed significantly to modern medicine (examples: aspirin, digitalis, various chemotherapeutic agents), and we optimistically predict that extracts as well as natural and synthetic constituents from the Ashwagandha plant will likely contribute significantly to the prevention and treatment of neurocognitive and other human ailments in the future.

References

Agarwal R, Diwanay S, Patki P, Patwardhan B (1999) Studies on immunomodulatory activity of Withania somnifera (Ashwagandha) extracts in experimental immune inflammation. J Ethnopharmacol 67:27–35

Ahmed ME, Javed H, Khan MM, Vaibhav K, Ahmad A, Khan A, Tabassum R, Islam F, Safhi MM, Islam F (2013) Attenuation of oxidative damage-associated cognitive decline by Withania somnifera in rat model of streptozotocin-induced cognitive impairment. Protoplasma 250:1067–1078

Akintola AA, Jansen SW, van Bodegom D, van der Grond J, Westendorp RG, de Craen AJ, van Heemst D (2015) Subclinical hypothyroidism and cognitive function in people over 60 years: a systematic review and meta-analysis. Front Aging Neurosci 11:7–150

Alonso J, Petukhova M, Vilagut G, Chatterji S, Heeringa S, Üstün TB, Alhamzawi AO, Viana MC, Angermeyer M, Bromet E, Bruffaerts R, de Girolamo G, Florescu S, Gureje O, Haro JM, Hinkov H, Hu CY, Karam EG, Kovess V, Levinson D, Medina-Mora ME, Nakamura Y, Ormel J, Posada-Villa J, Sagar R, Scott KM, Tsang A, Williams DR, Kessler RC (2011) Days out of role due to common physical and mental conditions: results from the WHO World Mental Health surveys. Mol Psychiatry 16:1234–1246

Altshuler LL, Ventura J, van Gorp WG, Green MF, Theberge DC, Mintz J (2004) Neurocognitive function in clinically stable men with bipolar I disorder or schizophrenia and normal control subjects. Biol Psychiatry 56:560–569

American Psychiatric Association (2013) Diagnostic and statistical manual of mental disorders, 5th edn (DSM-5). American Psychiatric Publishing, Washington, DC

Anbalagan K, Sadique J (1981) Influence of an Indian medicine (Ashwagandha) on acute-phase reactants in inflammation. Indian J Exp Biol 19:245–249

Anbalagan K, Sadique J (1985) Withania somnifera (ashwagandha), a rejuvenating herbal drug which controls a-2 macroglobulin synthesis during inflammation. Int J Crude Drug Res 23:177–183

Andallu B, Radhika B (2000) Hypoglycemic, diuretic and hypocholesterolemic effect of winter cherry (Withania somnifera, Dunal) root. Indian J Exp Biol 38:607–609

Andrade C, Aswath A, Chaturvedi SK, Srinivasa M, Raguram R (2000) A double-blind, placebo-controlled evaluation of the anxiolytic efficacy of an ethanolic extract of Withania Somnifera. Indian J Psychiatry 42:295–301

Atre-Vaidya N, Taylor MA, Seidenberg M, Reed R, Perrine A, Glick-Oberwise F (1998) Cognitive deficits, psychopathology, and psychosocial functioning in bipolar mood disorder. Neuropsychiatry Neuropyschol Behav Neurol 11:120–126

Auddy B, Hazra J, Mitra A, Abedon B, Ghosal S (2008) A standardized Withania somnifera Extract significantly reduces stress-related parameters in chronically stressed humans: A double-blind, randomized, placebo-controlled study. JANA 11:51–57

Bauer M, Glenn T, Pilhatsch M, Pfennig A, Whybrow PC (2014a) Gender differences in thyroid system function: relevance to bipolar disorder and its treatment. Bipolar Disord 16:58–71

Bauer IE, Pascoe MC, Wollenhaupt-Aguiar B, Kapczinski F, Soares JC (2014b) Inflammatory mediators of cognitive impairment in bipolar disorder. J Psychiatr Res 56:18–27

Bearden CE, Shih VH, Green MF, Gitlin M, Sokolski KN, Levander E, Marusak S, Hammen C, Sugar CA, Altshuler LL (2011) The impact of neurocognitive impairment on occupational recovery of clinically stable patients with bipolar disorder: A prospective study. Bipolar Disord 13:323–333

Begum V, Sadique J (1988) Long term effect of herbal drug Withania somnifera on adjuvant induced arthritis in rats. Indian J Exp Biol 26:877–882

Berk M, Copolov DL, Dean O, Lu K, Jeavons S, Schapkaitz I, Anderson-Hunt M, Bush AI (2008) N-acetylcysteine for depressive symptoms in bipolar disorder – a double-blind randomized placebo-controlled trial. Biol Psychiatry 64:468–475

Besedovsky H, del Rey A, Sorkin E, Da Prada M, Burri R, Honegger C (1983) The immune response evokes changes in brain noradrenergic neurons. Science 221:564–566

Bhattacharya SK, Muruganandam AV (2003) Adaptogenic activity of Withania somnifera: An experimental study using a rat model of chronic stress. Pharmacol Biochem Behav 75:547–555

Brar JS, Chengappa KNR (2011) Does one shoe fill all? The Randomized Controlled Clinical Trial (RCT) as a model for assessing the efficacy or effectiveness of whole systems Ayurveda or other Traditional Medicine (TM) treatments. J Ayurveda Integr Med 2:91–93

Brown NC, Andreazza AC, Young LT (2014) An updated meta-analysis of oxidative stress markers in bipolar disorder. Psychiatry Research 218:61–68

Burdick KE, Braga RJ, Nnadi CU, Shaya Y, Stearns WH, Malhotra AK (2012) A Placebo-controlled adjunctive trial of pramipexole in patients with bipolar disorder: Targeting cognitive dysfunction. J Clin Psychiatry 73:103–112

Cannon WB (1932) The wisdom of the body. WW Norton, New York

Chandrasekhar K, Kapoor J, Anishetty S (2012) A prospective, randomized double-blind, placebo-controlled study of safety and efficacy of a high-concentration full-spectrum extract of Ashwagandha root in reducing stress and anxiety in adults. Indian J Psychol Med 34:255–262
Cheng G, Huang C, Deng H, Wang H (2012) Diabetes as a risk factor for dementia and mild cognitive impairment: a meta-analysis of longitudinal studies. Intern Med J 42:484–491
Chengappa KN, Bowie CR, Schlicht PJ, Fleet D, Brar JS, Jindal R (2013) Randomized placebo-controlled adjunctive study of an extract of Withania Somnifera for Cognitive Dysfunction in Bipolar Disorder. J Clin Psychiatry 74:1076–1083
Chopra A, Lavin P, Patwardhan B, Chitre D (2000) Randomized Double-blind trial of an Ayurvedic Plant derived formulation for Treatment of Rheumatoid arthritis. Journal of Rheumatology 27:1365–1372
Chopra A, Lavin P, Patwardhan B, Chitre D (2004) A 32-week randomized, placebo-controlled clinical evaluation of RA-11, and Ayurvedic drug, on osteoarthritis of the knees. J Clin Rheumatol 10:236–245
Choudhary MI, Yousuf S, Nawaz SA, Ahmed S, Atta-ur-Rahman (2004) Cholinesterase inhibiting withanolides from Withania somnifera. Chem Pharm Bull (Tokyo) 52:1358–1361
Choudhary D, Bhattacharyya S, Joshi K (2017) Body weight management in adults under chronic stress through treatment with Ashwagandha root extract: A double-blind, randomized, placebo-controlled trial. J Evid Based Complementary and Altern Med 22:96–106
Chrousos GP (1995) The Hypothalamic-Pituitary-Adrenal Axis and Immune-Mediated Inflammation. N Engl J Med 332:1351–1363
Clark L, Goodwin G (2008) Attentional and Executive Functioning in Bipolar Disorder. In: Goldberg JF, Burdick KE (eds) Cognitive dysfunction in bipolar disorder. A guide to clinicians. American Psychiatric Publishing Inc, Arlington, pp 23–47
Clark L, Iverson SD, Goodwin GM (2001) A neuropsychological investigation of the prefrontal cortex involvement in acute mania. Am J Psychiatry 158:1605–1611
Clark L, Sarna A, Goodwin GM (2005) Impairment of executive function but not memory in first degree relatives of patients with bipolar I disorder and in euthymic patients with unipolar depression. Am J Psychiatry 162:1980–1982
Coogan AN, Wyse CA (2008) Neuroimmunology of the circadian clock. Brain Res 1232:104–112
Cooley K, Szczurko O, Perri D, Mills EJ, Bernhardt B, Zhou Q, Seely D (2009) Naturopathic care for anxiety: a randomized controlled trial ISRCTN78958974. PLoS One 4:e6628
Dantzer R, O'Connor JC, Freund GG, Johnson RW, Kelley KW (2008) From inflammation to sickness and depression: When the immune system subjugates the brain. Nat Rev Neurosci 9:46–56
Dar NJ, Hamid A, Ahmed M (2015) Pharmacologic overview of Withania somnifera, the Indian Ginseng. Cell Mol Life Sci 72:4445–4460
Dargél AA, Godin O, Kapczinski F, Kupfer DJ, Leboyer M (2015) C-Reactive Protein Alterations in Bipolar Disorder: A meta-analysis. J Clin Psychiatry 76:142–150
Demant KM, Vinberg M, Kessing LV, Miskowiak KW (2015) Effects of short-term cognitive remediation on cognitive dysfunction in partially or fully remitted individuals with bipolar disorder: Results of a randomised controlled trial. PLoS One 10:e0127955
Dickerson FB, Boronow JJ, Stallings CR, Origoni AE, Cole S, Yolken RH (2004) Association between cognitive functioning and employment status of persons with bipolar disorder. Psychol Serv 55:54–58
Dupont RM, Jernigan TL, Butters N, Delis D, Hesselink JR, Heindel W, Gillin JC (1990) Subcortical abnormalities detected in bipolar affective disorder using magnetic resonance imaging. Clinical and neuropsychological significance. Arch Gen Psychiatry 47:55–59
Durg S, Dhadde SB, Vandal R, Shivakumar BS, Charan CS (2015) Withania somnifera (Ashwagandha) in neurobehavioral disorders induced by brain oxidative stress in rodents: a systematic review and meta-analysis. J Pharmacy Pharmacol 67:879–899

Elgamal L, Sokolowska M, MacQueen G (2008) Memory deficits associated with Bipolar Disorder. In: Goldberg JF, Burdick KE (eds) Cognitive dysfunction in bipolar disorder. A guide to clinicians. American Psychiatric Publishing Inc, Arlington, pp p49–p67
Ferrier IN, Chowdhury R, Thompson JM, Watson S, Young AH (2004) Neurocognitive function in unaffected first-degree relatives of patients with bipolar disorder: a preliminary report. Bipolar Disord 6:319–322
Frangou S, Donaldson S, Hadjulis M, Landau S, Goldstein LH (2005) The Maudsley bipolar disorder project: executive dysfunction in bipolar disorder I and its clinical correlates. Biol Psychiatry 58:859–864
Frank E, Swartz HA, Kupfer DJ (2000) Interpersonal and social rhythm therapy: managing the chaos of bipolar disorder. Biol Psychiatry 48:593–604
Gajarmal AA, Shende MB (2015) A clinical evaluation of antistress activity of Ashwagandha (Withania somnifera Dunal) on employees experiencing mental stress at work place. Int J Ayur Pharma Res 3:37–45
Ganguli R, Brar JS, Rabin BS (1994) Immune abnormalities in schizophrenia: evidence for the autoimmune hypothesis. Harv Rev Psychiatry 2:70–83
Ganguli R, Brar JS, Chengappa KR, DeLeo M, Yang ZW, Shurin G, Rabin BS (1995) Mitogen-stimulated interleukin-2 production in never-medicated, first-episode schizophrenic patients. The influence of age at onset and negative symptoms. Arch Gen Psychiatry 52:668–672
Gannon JM, Forrest PE, Chengappa KNR (2014) Subtle changes in thyroid indices during a placebo-controlled study of an extract of Withania somnifera in persons with bipolar disorder. J Ayurveda Integr Med 5:241–245
Gershon S, Chengappa KNR (2009) Contemporary diagnosis and management of bipolar disorders. Handbooks in Health Care Company, Newtown
Ghosal S, Kaur R, Srivatsava RS (1988) Sitoindosides IX and X, two new withanolide glycosides from Withania Somnifera. Indian J Nat Prod 4:12–13
Glahn DC, Burdick KE, Bearden CE (2008) The Endophenotye concept. In: Goldberg JF, Burdick KE (eds) Cognitive dysfunction in bipolar disorder. A guide to clinicians. American Psychiatric Publishing Inc, Arlington, pp 69–88
Goldberg JF (2008) Adverse cognitive effects of psychotropic medications. In: Goldberg JF, Burdick KE (eds) Cognitive dysfunction in bipolar disorder. A guide to clinicians. American Psychiatric Publishing Inc, Arlington, pp 137–158
Goldberg JF, Burdick KE (eds) (2008) Cognitive dysfunction in bipolar disorder. A guide to clinicians. American Psychiatric Publishing, Inc. Arlington
Goldberg JF, Chengappa KNR (2009) Identifying and treating cognitive impairment in bipolar disorder. Bipolar Disord 11(Suppl 2):123–137
Goodwin FK, Jamison KR (2007) Manic depressive illness, 2nd edn. Oxford University Press, New York
Gorelick J, Rosenberg R, Smotrich A, Hanuš L, Bernstein N (2015) Hypoglycemic activity of withanolides and elicitated Withania somnifera. Phytochemistry 116:283–289
Grande I, Berk M, Birmaher B, Vieta E (2015) Bipolar Disorder. The Lancet 387:1561–1572
Green MF (1996) What are the functional consequences of neurocognitive deficits in schizophrenia? Am J Psychiatry 153:321–330
Grover A, Shandilya A, Agrawal V, Bisaria VS, Sundar D (2012) Computational evidence to inhibition of human acetyl cholinesterase by withanolide a for Alzheimer treatment. J Biomol Struct Dyn 29:651–562
Heckers S, Stone D, Walsh J, Shick J, Koul P, Benes FM (2002) Differential hippocampal expression of glutamic acid decarboxylase 65 and 67 messenger RNA in bipolar disorder and schizophrenia. Arch Gen Psych 59:521–529
Hibberd C, Yau JLW, Seckyl JR (2000) Glucocorticoids and the ageing hippocampus. J Anat 197:553–562
Jain S, Shukla SD, Sharma K, Bhatnagar M (2001) Neuroprotective Effects of Withania somnifera Dunn in Hippocampal Sub-regions of Female Albino Rat. Phytother Res 15:544–548

Jayaprakasam B, Nair MG (2003) Cyclooxygenase-2 enzyme inhibitory withanolides from Withania Somnifera leaves. Tetrahedron 59:841–849

Joyashiki E, Matsuya Y, Tohda C (2011) Sominone improves memory impairments and increases axonal density in Alzheimer's disease model mice, 5XFAD. Int J Neurosci 121:181–190

Kataria H, Wadhwa R, Kaul SC, Kaur G (2012) Water extract from the leaves of Withania somnifera protect RA differentiated C6 and IMR-32 cells against glutamate-induced excitotoxicity. PLoS One 7:e37080

Keck PE Jr, Mintz J, McElroy SL, Freeman MP, Suppes T, Frye MA, Altshuler LL, Kupka R, Nolen WA, Leverich GS, Denicoff KD, Grunze H, Duan N, Post RM (2006) Double-blind, randomized, placebo-controlled trials of ethyl-eicosapentanoate in the treatment of bipolar depression and rapid cycling bipolar disorder. Biol Psychiatry 60:1020–1022

Khan ZA, Ghosh AR (2010) Possible nitric oxide modulation in protective effects of withaferin A against stress induced neurobehavioral changes. J Medicinal Plants Res 4:490–495

Khan B, Ahmad SF, Bani S, Kaul A, Suri KA, Satti NK, Athar M, Qazi GN (2006) Augmentation and proliferation of T lymphocytes and Th-1cytokines by Withania somnifera in stressed mice. Int Immunopharmacol 6:1394–1403

Konar A, Shah N, Singh R, Saxena N, Kaul SC, Wadhwa R, Thakur MK (2011) Protective role of Ashwagandha leaf extract and its component withanone on scopolamine-induced changes in the brain and brain derived cells. PloS One 6:e27265

Kour K, Pandey A, Suri KA, Satti NK, Gupta KK, Bani S (2009) Restoration of stress-induced altered T cell function and corresponding cytokines patterns by Withanolide A. Intl Immunopharmacol 9:1137–1144

Kronfol Z, Remick DG (2000) Cytokines and the brain: Implications for clinical psychiatry. Am J Psychiatry 157:683–694

Kuboyama T, Tohda C, Komatsu K (2005) Neuritic regeneration and synaptic reconstruction induced by withanolide A. Br J Pharmacol 144:961–971

Kuboyama T, Tohda C, Komatsu K (2006) Withanoside IV and its active metabolite, sominone, attenuate Aβ(25-35)-induced neurodegeneration. Eur J Neurosci 23:1417–1426

Kuboyama T, Tohda C, Komatsu K (2014) Effects of Ashwagandha (Roots of Withania somnifera) on neurodegenerative diseases. Biol Pharm Bull 37:892–897

Kulkarni SK, Dhir A (2008) Withania somnifera: an Indian ginseng. Progress in Neuro-Psychopharmacol Biol Psychiatry 32:1093–1105

Kulkarni RR, Patki PS, Jog VP, Gandage SG, Patwardhan B (1991) Treatment of osteoarthritis with a herbomineral formulation: a double blind, placebo-controlled, crossover study. J Ethno Pharmacol 33:91–95

Kumar P, Kumar A (2008) Effects of root extract of Withania somnifera in 3-Nitropropionic acid-induced cognitive dysfunction and oxidative damage in rats. Int J Health Res 1:139–149

Kumar P, Kumar A (2009) Possible neuroprotective effect of Withania somnifera root extract against 3-nitropropionic acid-induced behavioral, biochemical, and mitochondrial dysfunction in an animal model of Huntington's disease. J Medicinal Food 12:591–600

Kurapati KR, Atluri VS, Samikkannu T, Nair MP (2013) Ashwagandha (Withania somnifera) reverses beta-amyloid1-42 induced toxicity in human neuronal cells: implications in HIV-associated neurocognitive disorders (HAND). PLoS One 8:e77624

Lupien SJ, Gillin CJ, Hauger RL (1999) Working memory is more sensitive than declarative memory to the acute effects of corticosteroids: a dose–response study in humans. Behav Neurosci 113:420–430

Malhi GM, Cahill CM, Mitchell P (2008) Impact of Mood, Anxiety, and Psychotic Symptoms on Cognition in Patients with Bipolar Disorder. In: Goldberg JF, Burdick KE (eds) Cognitive dysfunction in bipolar disorder. A guide to clinicians. American Psychiatric Publishing Inc, Arlington, pp 89–111

Malik F, Singh J, Khajuria A, Suri KA, Satti NK, Singh S, Kaul MK, Kumar A, Bhatia A, Qazi GN (2007) A standardized root extract of Withania somnifera and its major constituent withanolide-A elicit humoral and cell-mediated immune responses by up regulation of Th1-dominant polarization in BALB/c mice. Life Sci 80:1525–1538

Mann-Wrobel MC, Carreno JT, Dickinson D (2011) Meta-analyses of neuropsychological functioning in euthymic bipolar disorder: an update and investigation of moderator variables. Bipolar Disord 13:334–342

Martínez-Arán A, Penadés R, Vieta E, Colom F, Reinares M, Benabarre A, Salamero M, Gastó C (2002) Executive function in patients with remitted bipolar disorder and schizophrenia and its relationship with functional outcome. Psychother Psychosom 71:39–46

Martínez-Arán A, Vieta E, Colom F, Torrent C, Sánchez-Moreno J, Reinares M, Benabarre A, Goikolea JM, Brugué E, Daban C, Salamero M (2004a) Cognitive Impairment in euthymic bipolar patients: implications for clinical and functional outcomes. Bipolar Disord 6:224–232

Martínez-Arán A, Vieta E, Reinares M, Colom F, Torrent C, Sánchez-Moreno J, Benabarre A, Goikolea JM, Comes M, Salamero M (2004b) Cognitive function across manic or hypomanic, depressed, and euthymic states in bipolar disorder. Am J Psychiatry 161:262–270

McIntosh AM, Harrison LK, Forrester K, Lawrie SM, Johnstone EC (2005) Neuropsychological impairments in people with schizophrenia or bipolar disorder and their unaffected relatives. Br J Psychiatry 186:378–385

Michael N, Erfurth MN, Ohrmann P, Gossling M, Arrolt V, Heindel W, Pfliederer B (2003) Acute mania is accompanied by elevated levels of glutamate/glutamine levels within the left dorsolateral prefrontal cortex. Psychopharmacol 168:344–346

Mikolai J, Erlandsen A, Murison A, Brown KA, Gregory WL, Raman-Caplan P, Zwickey HL (2009) In vivo effects of Ashwagandha (Withania somnifera) extract on the activation of lymphocytes. J Altern Complement Med 15:423–430

Miller AH (2010) Depression and immunity: a role for T cells? Brain Behav Immun 24:1–8

Miller A, Maletic V, Raison C (2009) Inflammation and its discontents: the role of cytokines in the pathophysiology of major depression. Biol Psychiatry 65:732–741

Miskowiak KW, Ehrenreich H, Christensen EM, Kessing LV, Vinberg M (2014) Recombinant human erythropoietin to target cognitive dysfunction in bipolar disorder: a double-blind, randomized, placebo-controlled phase 2 trial. J Clin Psychiatry 75:1347–1355

Moore PB, Shepherd DJ, Eccleston D, Macmillan IC, Goswami U, McAllister VL, Ferrier IN (2001) Cerebral white matter lesions in bipolar affective disorder: relationship to outcome. Br J Psychiatry 178:172–176

Morrens M, Wezenberg E, Verkes RJ, Hulstijn W, Ruigt GS, Sabbe BG (2007) Psychomotor and memory effects of haloperidol, olanzapine, and paroxetine in healthy subjects after short term administration. J Clin Psychopharmacol 27:15–21

Mulabagal V, Subbaraju GV, Rao CV, Sivaramakrishna C, Dewitt DL, Holmes D, Sung B, Aggarwal BB, Tsay HS, Nair MG (2009) Withanolide sulfoxide from Aswagandha roots inhibits nuclear transcription factor-Kappa-B, cyclooxygenase and tumor cell proliferation. Phytother Res 22:987–992

Muller N, Schwarz M (2006) Schizophrenia as an inflammation-mediated dysbalance of glutamatergic neurotransmission. Neurotoxicity Res 10:131–148

Müller N, Krause D, Dehning S, Musil R, Schennach-Wolff R, Obermeier M, Möller HJ, Klauss V, Schwarz MJ, Riedel M (2010) Celecoxib treatment in an early stage of schizophrenia: results of a randomized, double-blind, placebo-controlled trial of celecoxib augmentation of amisulpride treatment. Schizophrenia Res 121:118–124

Nagashayana N, Sankarankutty P, Nampoothiri MR, Mohan PK, Mohanakumar KP (2000) Association of L-DOPA with recovery following Ayurveda medication in Parkinson's disease. J Neurol Sci 176:124–127

Naidu PS, Singh A, Kulkarni SK (2006) Effect of Withania somnifera root extract on reserpine-induced orofacial dyskinesia and cognitive dysfunction. Phytotherapy Res 20:140–146

Nashine K, Srivastava DN, Sahni YP (1995) Role of inflammatory mediators in anti-inflammatory activity of Withania somnifera. Indian Veterinary Medical J 19:286–288

Nery FG, Monkul ES, Hatch JP, Fonseca M, Zunta-Soares GB, Frey BN, Bowden CL, Soares JC (2008) Celecoxib as an adjunct in the treatment of depressive or mixed episodes of bipolar disorder: a double-blind, randomized placebo-controlled study. Hum Psychopharmacol 23:87–94

Özerdem A, Tunca Z, Çımrın D, Hıdıroğlu C, Ergör G (2014) Female vulnerability for thyroid function abnormality in bipolar disorder: role of lithium treatment. Bipolar Disord 16:72–82

Padmos RC, Hillegers MH, Knijff EM, Vonk R, Bouvy A, Staal FJ, de Ridder D, Kupka RW, Nolen WA, Drexhage HA (2008) A discriminating messenger RNA signature for bipolar disorder formed by an aberrant expression of inflammatory genes in monocytes. Arch Gen Psychiatry 65:395–407

Panda S, Kar A (1997) Evidence for Free radical scavenging activity of Ashwagandha root powder in mice. Indian J Physiol Pharmacol 41:424–426

Panda S, Kar A (1998) Changes in thyroid hormone concentrations after administration of Ashwagandha root extract to adult male mice. J Pharm Pharmacol 50:1065–1068

Panda S, Kar A (1999) Withania somnifera and bauhinia purpurea in the regulation of circulating thyroid hormone concentrations in female mice. J Ethnopharmacol 67:233–239

Pasqualetti G, Pagano G, Rengo G, Ferrara N, Monzani F (2015) Subclinical hypothyroidism and cognitive impairment: systematic review and meta-analysis. J Clin Endocrinol Metab 100:4240–4248

Resmini E, Santos A, Gómez-Anson B, López-Mourelo O, Pires P, Vives-Gilabert Y, Crespo I, Portella MJ, de Juan-Delago M, Webb SM (2013) Hippocampal dysfunction in cured Cushing's syndrome patients detected by (1) H-MR-spectroscopy. Clin Endocrinol 79:700–707

Robinson LJ, Thompson JM, Gallagher P, Goswami U, Young AH, Ferrier IN, Moore PB (2006) A meta-analyses of cognitive deficits in euthymic patients with bipolar disorder. J Affect Disord 93:105–111

Rosenblat JD, Brietzke E, Mansur RB, Maruschak NA, Lee Y, McIntyre RS (2015) Inflammation as a neurobiological substrate of cognitive impairment in bipolar disorder: evidence, pathophysiology and treatment implications. J Affect Disord 188:149–159

Rosenthal NE, Sack DA, Gillin JC, Lewy AJ, Goodwin FK, Davenport Y, Mueller PS, Newsome DA, Wehr TA (1984) Seasonal affective disorder. A description of the syndrome and preliminary findings with light therapy. Arch Gen Psychiatry 41:72–80

Rush AJ, Giles DE, Schlesser MA, Orsulak PJ, Parker CR Jr, Weissenburger JE, Crowley GT, Khatami M, Vasavada N (1996) The dexamethasone suppression test in patients with mood disorders. J Clin Psychiatry 57:470–484

Sapolsky R, Rivier C, Yamamoto G, Plotsky P, Vale W (1987) Interleukin-1 stimulates the secretion of hypothalamic corticotropin-releasing factor. Science 238:522–524

Schiepers OJ, Wichers MC, Maes M (2005) Cytokines and major depression. Progr Neuropsychopharmacol Biol Psychiatry 29:201–217

Schliebs R, Liebmann A, Bhattacharya SK, Kumar A, Ghosal S, Bigl V (1997) Systemic administration of defined extracts from Withania somnifera (Indian Ginseng) and Shilajit differentially affects cholinergic but not glutamatergic or GABAergic markers in rat brain. Neurochem Int 30:181–190

Sehgal N, Gupta A, Valli RK, Joshi SD, Mills JT, Hamel E, Khanna P, Jain SC, Thakur SS, Ravindranath V (2012) Withania somnifera reverses Alzheimer's disease pathology by enhancing low-density lipoprotein receptor-related protein in liver. Proc Natl Acad Sci U S A 109:3510–3515

Selye H (1956) The stress of life. McGraw-Hill, New York

Singh D, Aggarwal A, Maurya R, Naik S (2007) Withania somnifera inhibits NF-kappaB and AP-1 transcription factors in human peripheral blood and synovial fluid mononuclear cells. Phytother Res 10:905–913

Sloan RP, McCreath H, Tracey KJ, Sidney S, Liu K, Seeman T (2007) RR interval variability is inversely related to inflammatory markers: the CARDIA study. Mol Med 13:178–184

Starkman MN, Giordani B, Berent S, Schork MA, Schteingart DE (2001) Elevated cortisol levels in Cushing's disease are associated with cognitive decrements. Psychosom Med 63:985–993

Stoll AL, Severus WE, Freeman MP, Rueter S, Zboyan HA, Diamond E, Cress KK, Marangell LB (1999) Omega 3 fatty acids in bipolar disorder: a preliminary, double-blind, placebo-controlled trial. Arch Gen Psychiatry 56:407–412
Subbaraju GV, Vanisree M, Rao CV, Sivaramakrishna C, Sridhar P, Jayaprakasam B, Nair MG (2006) Ashwagandhanolide, a bioactive dimeric thiowithanolide isolated from the roots of Withania somnifera. J Nat Prod 12:1790–1792
Szentivanyi A (1989) The discovery of immune-neuroendocrine circuits in the fall of 1951. In: Hadden JW, Masek K, Nistico G (eds) Interactions among central nervous system, neuroendocrine and immune systems. Pythagora Press, Rome-Milan, pp 1–3
Thompson JM, Gallagher P, Hughes JH, Watson S, Gray JM, Ferrier IN, Young AH (2005) Cognitive Impairment in euthymic patients with bipolar affective disorder. Br J Psychiatry 186:32–40
Tighe SK, Reading SA, Rivkin P, Caffo B, Schweizer B, Pearlson G, Potash JB, Depaulo JR, Bassett SS (2012) Total white matter hyperintensity volume in bipolar disorder patients and their healthy relatives. Bipolar Disord 14:888–893
Tohda C, Joyashiki E (2009) Sominone enhances neurite outgrowth and spatial memory mediated by the neurotrophic factor receptor, RET. British J Pharmacol 157:1427–1440
Torrent C, Martínez-Arán A, Daban C, Sánchez-Moreno J, Comes M, Goikolea JM, Salamero M, Vieta E (2006) Cognitive impairment in bipolar II disorder. Br J Psychiatry 189:254–259
Torres IJ, Boudreau VG, Yatham LN (2007) Neuropsychological functioning in euthymic bipolar disorder: a meta-analysis. Acta Psychiatr Scand 116(suppl):17–26
Torres IJ, DeFreitas CM, Yatham LN (2008) Cognition and functional outcome in bipolar disorder. In: Goldberg JF, Burdick KE (eds) Cognitive dysfunction in bipolar disorder. A guide to clinicians. American Psychiatric Publishing Inc, Arlington, pp 217–234
Tremont G, Stern RA (1997) Use of thyroid hormone to diminish the cognitive side effects of psychiatric treatment. Psychopharmacol Bull 33:273–280
Udayakumar R, Kasthurirengan S, Vasudevan A, Mariashibu TS, Rayan JJ, Choi CW, Ganapathi A, Kim SC (2010) Antioxidant effect of dietary supplement Withania somnifera L. Reduce blood glucose levels in alloxan-induced diabetic rats. Plant Foods Hum Nutr 65:91–98
van der Hooft CS, Hoekstra A, Winter A, de Smet PA, Stricker BH (2005) Thyrotoxicosis following the use of Ashwagandha. Ned Tijdschr Geneeskd 149:2637–2638
Veronese N, Solmi M, Luchini C, Lu RB, Stubbs B, Zaninotto L, Correll CU (2016) Acetylcholinesterase inhibitors and memantine in bipolar disorder: a systematic review and best evidence synthesis of the efficacy and safety for multiple disease dimensions. J Affect Disord 197:268–280
Vieta E, Martinez-Aran A, Goldberg JF (2008) Cognition across the life span. In: Goldberg JF, Burdick KE (eds) Cognitive dysfunction in bipolar disorder. A guide to clinicians. American Psychiatric Publishing Inc, Arlington, pp 235–255
Wehr TA, Sack DA, Rosenthal NE (1987) Sleep reduction as a final common pathway in the genesis of mania. Am J Psychiatry 144:201–204
Wehr TA, Turner EH, Shimada JM, Lowe CH, Barker C, Leibenluft E (1998) Treatment of rapidly cycling bipolar patient by using extended bed rest and darkness to stabilize the timing and duration of sleep. Biol Psychiatry 43:822–828
Whelan TB, Schteingart DE, Starkman MN, Smith A (1980) Neuropsychological deficits in Cushing's syndrome. J Nerv Ment Dis 168:753–757
Wingo AP, Harvey PD, Baldessarini RJ (2009) Neurocognitive impairment in bipolar disorder patients: functional implications. Bipolar Disord 11:113–125
Wingo AP, Baldessarini RJ, Compton MT, Harvey PD (2010) Correlates of recovery of social functioning in type I and II bipolar disorder patients. Psychiatry Res 177:131–134
Wirz-Justice A, Bromundt V, Cajochen C (2009) Circadian disruption and psychiatric disorders: the importance of entrainment. Sleep Med Clin 4:273–284

Yadav CS, Kumar V, Suke SG, Ahmed RS, Mediratta PK, Banerjee BD (2010) Propoxur-induced acetylcholine esterase inhibition and impairment of cognitive function: attenuation by Withania somnifera. Indian J Biochem Biophys 47:117–120

Young AH, Sahakian BJ, Robbins TW, Cowen PJ (1999) The effects of chronic administration of hydrocortisone on cognitive function in normal male volunteers. Psychopharmacol (Berl) 145:260–266

Young AH, Gallagher P, Watson S, Del-Estal D, Owen BM, Ferrier IN (2004) Improvements in neurocognitive function and mood following adjunctive treatment with mifepristone in bipolar disorder. Neuropsychopharmacol 29:1538–1545

Zalcman S, Green-Johnson JM, Murray L, Nance DM, Dyck D, Anisman H, Greenberg AH (1994) Cytokine specific central monoamine alterations induced by IL-1, IL-2 and IL-6. Brain Res 643:40–49

Chapter 18
Neuroprotective Potential of *Withania somnifera* (Ashwagandha) in Neurological Conditions

Rajat Sandhir and Abhilasha Sood

Abstract *Withania somnifera*, commonly known as Ashwagandha, is a herb used in the Indian Ayurvedic medical system for centuries for its health benefits in wide range of conditions. It has been documented to enhance the functions of the nervous system and improves cognition. Medicinal properties of *Withania somnifera* are primarily attributed to the presence of active constituents known as withanolides. The multimodal beneficial effects of *WS* and its constituents stem from its property to act as an anti-oxidant, anti-inflammatory, anti-cancer, hemopoietic and rejuvenating agent. *WS* has been shown to have beneficial role in many neurological conditions that include Alzheimer's disease, Parkinson's disease, Huntington's disease, anxiety disorders, cerebral ischemia. However, the mechanisms involved in the beneficial effects of *WS* have not been well understood. The article provides information on the mechanisms involved in neuroprotective ability of *WS* in preventing various neurodegenerative conditions.

Keywords Ashwagandha • Cognition • Neurodegeneration • *Withania somnifera* • Withanolides

18.1 Introduction

Natural extracts from herbal plants have shown to offer protection in several neurodegenerative conditions, as they possess biologically active compounds of pharmacological importance. *Withania somnifera (WS)* also known as Ashwagandha, is a valuable herb that has been used in the Indian traditional system of medicine for more than 2500 years (Luxmini et al. 2015). Almost all parts of *WS* plant have been used for centuries in ayurvedic system of medicines; however dried mature roots are most commonly used to treat a variety of ailments (Durg et al. 2015). The roots of

R. Sandhir (✉) • A. Sood
Department of Biochemistry, Panjab University, Chandigarh 160014, India
e-mail: sandhir@pu.ac.in

S.C. Kaul, R. Wadhwa (eds.), *Science of Ashwagandha: Preventive and Therapeutic Potentials*, DOI 10.1007/978-3-319-59192-6_18

WS have been used as a nerve tonic and adaptogen since decades (Singh et al. 2011). *WS* belongs to the family solanaceae, which is commonly known as the nightshade family (Babu et al. 2007). *WS* is a green woody shrub, which is 200-800 cm in height and is found all over the drier parts of South East Asia that includes India, Bangladesh, Sri-Lanka, Nepal and Pakistan, as well as parts of Australia, Africa and America (Kumar et al. 2015). *WS* and its active components have been reported to have multiple beneficial effects such as, anti-inflammatory (Gupta and Singh 2014), anticancer (Rai et al. 2016), anti-stress, immunomodulatory (Tiwari et al. 2014) and adaptogenic effects (Uddin et al. 2012) on the nervous, endocrine, and cardiovascular system (Panossian and Wikman 2010; Kumar et al. 2014).

18.2 Phytochemical Composition of *Withania somnifera*

The major constituents of the *WS* roots and leaves are steroidal alkaloids and steroidal lactones known as withanolides (Senthil et al. 2015). Therapeutic potential of *WS* is attributed to antioxidant properties of its active constituents known as withanolides (Kurapati et al. 2013). Withanolides are a group of naturally occurring C-28 steroidal lactones built on an intact ergostane structure (Fig. 18.1), in which C-22 and C-26 are oxidized to form a six-membered lactone ring (Mirjalili et al. 2009).

The basic structure of withanolides is composed of 22-hydroxy-ergostan-26-oic acid-26, 22-lactone and is designated as the "withanolide skeleton" (Mirjalili et al. 2009). According to a recent report, the concentration of withanolides has been found to be in the range of 0.001% to 1.5% (dry weight basis) in the roots and leaves of *WS* (Dhar et al. 2015). Due to the structural resemblance of withanolides to the ginsenosides present in Panax ginseng, *WS* is also known as the Indian Ginseng (Singh et al. 2010). Among all the withanolides present in *WS* root extract, Withaferin-A was the first member of withanolides family to be isolated from *WS* (Palliyaguru et al. 2016). Till date, 12 alkaloids, 35 withanolides and some sitoindosides have been isolated from *WS*, which have been tested for their therapeutic effects in neurological disorders (Singh et al. 2011). In addition to withanolides, the other alkaloid constituents of *WS* root extract includes: somniferine, somnine, somniferinine, withananine, pseudo-withanine, tropine, pseudotropine, 3-a-gloyloxytropane, choline, cuscohygrine, isopelletierine and anaferine andanahydrine (Singh et al. 2010). Withaferin-A, withanolide-A, withanoside-IV and some sitoindosides are amongst the major constituents of *WS*, which are being used for their effects on nervous system disorders (Raghavan and Shah 2015; Tiwari et al. 2014).

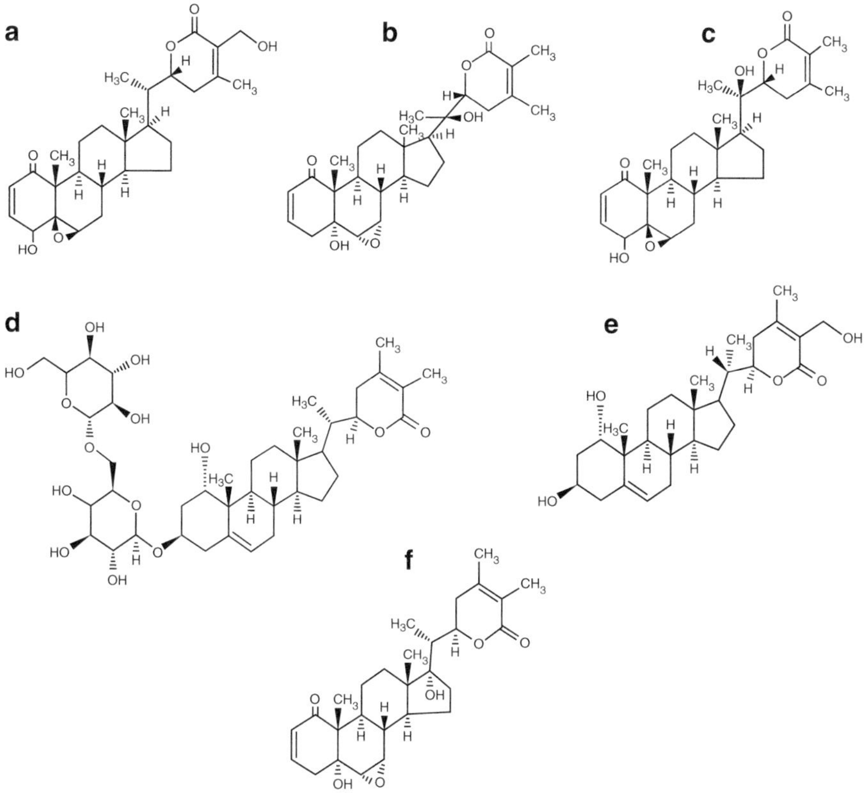

Fig. 18.1 Major Withanolides present in *Withania somnifera* root extract are Withaferin-A (a), Withanolide-A (b), Withanolide-D (c), Withanoside-IV (d), Sominone (e), Withanone (f)

18.3 Neuroprotective Activity of *Withania somnifera*

Many studies have documented the propensity of *WS* supplementation as a therapeutic strategy in treating several neurological disorders (Table 18.1). *WS* supplementation has shown to attenuate symptoms and pathology in 6-hydroxydopamine-induced model of Parkinson's disease, by its antioxidant action as evident by the attenuation of lipid peroxide formation and improved thiol levels (such as GSH and GSSG) (Ahmad et al. 2005). In an experimental model of Alzheimer's disease, *WS* root extract supplementation to the animals was shown to ameliorate behavioral deficits with enhanced amyloid beta clearance, thus protecting the neuronal cells from amyloid beta toxicity (Sehgal et al. 2012). *WS* supplementation attenuated lead-induced toxicity in glial cells by balancing the expression of glial fibrillary acidic protein (GFAP), heat shock protein (HSP70), mortalin and neural cell adhesion molecules (Kumar et al. 2014). *WS* root extract was shown to markedly salvage the degenerating cells in the hippocampus of rats subjected to immobilization stress (John 2014). Active constituents of *WS* root extract mainly

Table 18.1 Mechanisms involved in neuroprotective potential of *Withania somnifera* in various neurological disorders

Neurological condition	Proposed mechanism of action
Alzheimer's disease	Antioxidant action
	Neurite regeneration
	Inhibition of acetylcholinesterase activity
	Prevents amyloid fibril formation
	Down-regulates BACE1 and up-regulates ADAM10
	Increased amyloid clearance
	Neuritogenic properties
Parkinson's disease	Suppression of oxidative stress
	Anti-apoptotic action
	Increased tyrosine hydroxylase expression
	Preventing mitochondrial dysfunctions
Huntington's disease	Antioxidant action
	Preventing mitochondrial dysfunctions
Tardive dysknesia	Antioxidant action
	Increasing GABA levels in brain
Epilepsy	Antioxidant action
	Modulation of GABAnergic system
	Restoration of AMPA receptor function
	Modulation of NMDA receptor
Cerebral ischemia	Antioxidant action
	Anti-apoptotic action
	Upregulation of Nrf2 pathway
	Induction of Neuronal plasticity
	Increased angiogenesis
	Prevents gluatamate-induced toxicity

withanolide-A, withanolides-IV and VI were shown to promote neurite outgrowth promoting synaptic restoration and rejuvenation, in human neuroblastoma cell lines and in rat cortical neurons damaged by amyloid-β toxicity (Kuboyama et al. 2006a).

18.3.1 *Withania somnifera*: Mechanism of Action in Neuroprotection

Withanolides are the active components of the root extract of *WS*, which are primarily responsible for its wide range of biological activities (Kumar et al. 2015). Although the exact mechanism of action of *WS* is not known, the important role played by withanolides in enhancing the overall energy levels by optimizing cellular antioxidant status is well reported (Pingali et al. 2014). Withanolides are amphiphilic compounds that are reported to have antiproliferative, antimetastatic, antiangiogenic, anti-invasive and proapoptotic activities, associated with the suppression of NF-kB and NF-kB regulated gene products (Khodaei et al. 2012). Structurally, withanolides are steroidal lactones with an ergostane skeleton and the lactone ring attached to the D ring of the steroid (Mirjalili et al. 2009).

Withanolides have been shown to efficiently reconstruct neuronal networks in *in-vitro* and *in-vivo* conditions (More et al. 2012; Kataria et al. 2013). *WS* extract containing Withanolide-A, Withanoside-IV and Withanoside-VI has been shown to cause effective axonal regeneration and synaptic reconstruction (Singh et al. 2010). Recent research has shown speculated involvement of Sominone (an active metabolite of withanolides) formed as a result of its de-glycosylation by intestinal bacteria in this neuronal regeneration (Kuboyama et al. 2006b). Sominone has been reported to induce axonal regeneration and synaptic reconstruction in brain and enhance object location memory (Tohda and Joyashiki 2009). In a clinical study, *WS* supplementation was shown to increase levels of monoamine oxidase and gamma amino butyric acid (GABA), while levels of 5-hydroxytryptophan (serotonin) and glutamic acid were decreased, pointing out to its anti-excitotoxic effect (Upadhaya et al. 1990). *WS* has been reported to possess GABA agonist properties, thereby making it an anxiolytic drug involved in relieving stress and anxiety (Singh et al. 2011). *WS* supplementation has shown to improve locomotor functions and movement patterns in Parkinson's disease, which is suggested to be due to induction of catecholamine neurotransmitters (Rajasankar et al. 2009a, b). In another study, pre-supplementation with *WS* leaf extract inhibited glutamate-induced cell death and stress response, by up-regulation of HSP-70 expression (Parihar and Hemnani 2003). *WS* has been shown to possess anticonvulsant properties in acute and chronic models of epilepsy (Kulkarni and Verma 1993). *WS* supplementation has also been reported to improve neurotransmitter levels, antioxidant status and lower oxidative stress in animal model of Parkinson's disease (Rajasankar et al. 2009b). Active constituents of *WS* have been able to reverse cognitive deficits and cause reduction in cholinergic markers in Alzheimer's disease (Bhattacharya et al. 1995).

18.3.2 Role of Withania somnifera in Various Neurodegenerative Disorders

WS supplementation has been shown to confer beneficial effects in various neurodegenerative disease conditions as discussed below:

18.3.2.1 Parkinson's Disease (PD)

PD is a neurodegenerative condition, which is characterized by muscle rigidity, tremor and bradykinesia (Rascol 2009). Reduction in dopamine levels in brain regions involved in control motor functions is responsible in development of PD

(Hinterberger 1971; Piggott et al. 1999). Oxidative stress is the major contributing factor in neurodegeneration in PD brain (Jenner 2003; Olanow 2007). *WS* has been reported to confer beneficial effects in various experimental models of PD (Ahmad et al. 2005; Manjunath and Muralithara 2015; Rajasarkar et al. 2009; Prakash et al. 2014). *WS* extract has been reported to reverse neurobehavioral deficits, increased lipid peroxidation, reduced glutathione content, activities of glutathione-S-transferase, glutathione reductase, glutathione peroxidase, superoxide dismutase and catalase, catecholamine content, dopaminergic D2 receptor binding and tyrosine hydroxylase expression in 6-Hydroxydopamine-induced model of PD in a dose-dependent manner (Ahmad et al. 2005). Rajasarkar et al. (2009a) have reported oral administration of PD mouse with *WS* root extract (100 mg/kg) for 7 days or 28 days resulted in increased dopamine (DA), 3,4-dihydroxy-phenylacetic acid (DOPAC) and homovanillic acid (HVA) levels, along with normalization of thiobarbituric reactive substances levels (marker of oxidative stress) in the corpus striatum. The reduction in oxidative was accompanied by increase in glutathione levels and glutathione peroxidase activity in brain regions of PD animals. Prakash et al. (2014) further demonstrated anti-apoptotic and anti-inflammatory properties of *WS* could be significantly involved in beneficial effect in PD. Further, it was reported that supplementation with ethanolic root extract of *WS* (100 mg/kg bodyweight, i.p) for 9 weeks, inhibited the enhanced expression of inducible nitric oxide synthase (iNOS) and astroglial activation marker glial fibrillary acidic protein (GFAP) in Maneb (MB) and paraquat (PQ) induced model of PD. MB and PQ are environmental toxins that have been experimentally used to induce selective damage of dopaminergic neurons leading to the development of PD. *WS* supplementation was shown to inhibit apoptotic cell death in PD by reducing the level of the pro-apoptotic (Bax) proteins and increase in the level of anti-apoptotic (Bcl-2) proteins in the MB–PQ model. Manjunath and Muralithara (2015) have shown that neuromodulatory effect of *WS* against rotenone-induced PD model in *Drosophila melanogaster* is mediated via suppression of oxidative stress and its potential to attenuate mitochondrial dysfunctions. Supplementation of *WS* to leucine-rich-repeat-kinase-2 mutants (LRRK2) loss-of-function model of PD in drosphila resulted in increasing their lifespan, improved motor function and mitochondrial morphology (De Rose et al. 2016). These studies suggest therapeutic potential of *WS* in patients with PD.

18.3.2.2 Alzheimer's Disease

Alzheimer's disease (AD) is a progressive, neurodegenerative disease histochemically characterized by extracellular deposits of amyloid beta (Abeta) protein and intracellular neurofibrillary tangles throughout the cortical and limbic regions and is the single most common cause of dementia in the elderly (Serrano-Pozo et al. 2011). *WS* extract have been shown to reverse the behavioral deficits and pathological clues involved in experimental models of AD (Kurapati et al. 2013; Sehgal et al. 2012). In an *in vitro* study using neuronal cell lines, Kurapati et al. (2013) demonstrated *WS* mediated reversal of β-amyloid induced toxicity in SK-N-MC neuronal cells and also prevented amyloid-induced reduction in spine density, spine area, spine length and number of spines suggesting protective effect of *WS* in AD. Further, β-amyloid treated neuronal cultures had increased acetylcholinesterase activity and increased internalization of amyloid beta, which was countered on *WS* supplementation (Kurapati et al. 2013). Molecular docking studies have revealed that withanolide A has high binding affinity with human acetylcholinesterase involving Thr78, Trp81, Ser120 and His442 residues all of which fall under one or other active sites/subsites of the enzyme which may be critical to its inhibitory action on acetylcholinesterase and suggesting the likely mechanism involved in its potential therapeutic effect against AD (Grover et al. 2012).

Active constituents of *WS* (withanamides) have been reported to prevent the fibril formation and protect cells from amyloid beta toxicity by unique binding of withanamides-A and -C to the active motif of amyloid beta (25–35) (Jayaprakasam et al. 2010). Withanolide A has been shown to modulate multiple targets associated with amyloid-beta precursor protein processing and amyloid-beta protein clearance by specifically down regulating beta-secretase 1 (BACE1) and upregulating a disintegrin and metalloprotease 10 (ADAM10) (Patil et al. 2010). BACE1 is a rate-limiting enzyme in the production of Abeta from amyloid-beta precursor protein (AbetaPP), while ADAM10 is involved in non-amyloidogenic processing of AbetaPP. Shinde et al. (2015) have suggested increased efflux of amyloid peptide through activation of P-glycoprotein. In another study, active metabolite of Withanoside-IV, sominone had been reported to attenuate amyloid beta (25–35)-induced neurodegeneration by regeneration of axons, dendrites and synapses (Kuboyama et al. 2006a). In another study, oral administration with semipurified extract of the roots of *WS* (containing withanolides and withanosides) was shown to reverse behavioral deficits, plaque pathology, accumulation of β-amyloid peptides and oligomers in the brains of middle-aged and old APP/PS1 transgenic AD mice (Sehgal et al. 2012). This effect of *WS* supplementation in rapidly clearing β-amyloid peptides was shown to be related to the ability of *WS* to increase the levels of liver low-density lipoprotein receptor-related protein (LRP) in brain microvessels and the β-amyloid-degrading protease neprilysin (NEP) in liver. Liver LRP mediates endocytosis of β-amyloid peptides present in the plasma, which are then cleared through NEP and other proteases present in the liver. This ability of *WS* to induce liver LRP could be a useful tool, since the cell surface LRP in liver is required for the rapid systemic clearance of the toxic β-amyloid peptide and subsequent degradation of this peptide by proteases in the liver. Moreover, high resolution Q-TOF/MS study have been shown to demonstrate that the withanamide peaks in *WS* fruit extract crossed the blood brain barrier in the mouse after intraperitoneal administration, suggesting that an oral administration of the extract could give similar results, since the extract has lipophilic and hydrophilic functionalities and easily crosses membranes (Vareed et al. 2014). Jana et al. (2011) have shown that withanolide A promotes neuritogenic and inhibits secretase activity which may contribute towards its beneficial role in AD.

18.3.2.3 Huntington Disease

Huntington's disease (HD) is a fatal neurodegenerative disease with CAG expansion mutation in *HTT*, encoding a polyglutamine tract in the huntingtin protein resulting in selective vulnerability of striatal spiny projection neurons. The patients with HD have cognitive, motor and psychiatric disturbances. *WS* root extract has been shown to significantly improve cognitive function (assesed using morris water maze and elevated plus maze tests) and motor functions (impairment of muscle activity as assesed by rotarod and limb withdrawal tests) in 3-NP induced model of HD, which has been attributed to inhibition of oxidative stress, restoration of antioxidant status and enhancement of acetylcholinesterase enzyme activity on *WS* supplementation (Kumar and Kumar 2009).

18.3.2.4 Tardive Dyskinesia

Tardive dyskinesia is a syndrome characterized by repetitive involuntary movements, usually involving the mouth, face, tongue, limb, trunk musculature, and is one of the major side effects of long-term antipsychotic drug treatment (Mathews et al. 2005). *WS* (glycowithanolides) have been reported to have beneficial effects against neuroleptic drug, haloperidol-induced animal model of tardive dyskinesia (Bhattacharya et al. 2002). Involuntary orofacial movements (chewing movements, tongue protrusion and buccal tremors) were assessed as tardive dyskinesia parameters. *WS* (100 and 200 mg PO), administered concomitantly with haloperidol for 28 days, inhibited the induction of the neuroleptic tardive dyskinesia. The results indicated that the antioxidant effect of glycowithanolides, rather than its GABA-mimetic action, was found to be responsible for the prevention of haloperidol-induced tardive dyskinesia. In another study, chronic treatment with *WS* root extract for a period of 4 weeks to reserpine treated animals significantly and dose dependently (50 and 100 mg/kg) reduced the reserpine-induced vacuous chewing movements and tongue protrusions (Kulkarni et al. 2008).

18.3.2.5 Epilepsy

Epilepsy is a chronic disease with a heterogeneous set of symptoms that is characterized by recurrent seizures (Fisher et al. 2005). It is mainly caused due to disturbed balance of stimulatory and inhibitory neurotransmitters wherein; glutamate and GABA are known to play an important role (Fritschy 2008). Parihar and Hemnani (2003) showed that phenolic antioxidants present in *WS* mitigate the effects of kainite-induced excitotoxicity and oxidative damage in hippocampus and this might be accomplished by their antioxidative properties. *WS* has been shown to possess anticonvulsant properties in pentylenetetrazol (PTZ)-induced epilepsy through modulation of GABAnergic system (Kulkarini et al. 2008). *WS* root extract has been reported to inhibit the reuptake and increase the release of GABA in synaptosomes isolated from rat brain cortex (Homayoun et al. 2002; Dhuley 1998). Another, study suggested that the anticonvulsant mechanism of *WS* root extract is due to its binding and stimulation of benzodiazepine receptors, thereby boosting the GABAergic system in the brain (Roshanaei and Neda 2015). *WS* interferes with the GABA receptor and regulates its function by opening chloride channels that lead to increase in the seizure threshold. Thus, it can be concluded that the main mechanism of increasing the threshold of PTZ-induced seizure by *WS* root extract, is probably through GABAergic neurotransmitter system. It has also been suggested that *WS* root extract contains withanozoid and somniferin A, which possess potent anticonvulsant activity in humans and experimental animals (Isoherranen et al. 2003). *WS* has been shown to have differential activation of GABA receptor subtypes that appears to be the potential mechanism by which *WS* accomplishes its anti-epileptogenic properties. Soman et al. (2012) observed that treatment with *WS* and Withanolide A ameliorated spatial memory deficits by enhancing antioxidant system and restoring altered NMDA receptor density in temporal lobe epilepsy. In addition, Soman et al. (2013) observed that *WS* and withanolide A restored AMPA receptor function in rodent model of temporal lobe epilepsy.

18.3.2.6 Cerebral Ischemia

Stroke is a leading cause of brain injury in millions of people worldwide. In a rat model of bilateral common carotid artery occlusion, *WS* pre-supplementation (50 mg/kg) attenuated the reperfusion injury induced biochemical and histopathological alterations (Trigunayat et al. 2007). In another study, *WS* pre-supplementation was shown to attenuate oxidative stress, reduce lesion volume and restore neurological deficits in middle cerebral artery occlusion (MCAO) model of stroke (Chaudhary et al. 2003). Sood et al. (2016) observed that WS restored acetylcholinesterase activity, lipid peroxidation, thiols and attenuated MCAO induced behavioural deficits. Further, *WS* significantly reduced the cerebral infarct volume and ameliorated histopathological alterations in MCAO animals. *WS* treatment was observed to upregulate the expression of hemeoxygenase 1 (HO1) and attenuated the expression of the pro-apoptotic protein poly (ADP-ribose) polymerase-1 (PARP1) via the PARP1-AIF pathway, thus preventing the nuclear translocation of apoptosis-inducing factor (AIF) suggesting involvement of anti-apoptotic pathways and angiogenesis (Raghavan and Shah 2015). Exposure to glutamate has been implicated in loss of neural network and cell death in stroke (Kataria et al. 2012; Cimarosti and Henley 2008). Water extract from the leaves of *WS* has been reported to inhibit glutamate-induced neurotoxicity in retinoic acid differentiated rat glioma (C6) and human neuroblastoma (IMR-32) cells (Kataria et al. 2012). It was further demonstrated that pre-treatment with the *WS* extract exhibited therapeutic potential through induction of neuronal plasticity markers HSP-70, Neural cell adhesion molecule (NCAM) and its polysialylated form, PSA-NCAM that plays an important role in neuronal regeneration and repair. Baitharu et al. (2014) have shown that withanolide

A reduces neurodegeneration by restoring hypoxia induced glutathione depletion in hippocampus. Further, withanolide A increased glutathione biosynthesis in neuronal cells by upregulating glutamate-cysteine ligase (GCLC) levels through Nrf2 pathway in a corticosterone dependent manner.

18.3.2.7 Other Neurological Conditions

Gautam et al. (2013) have shown that beneficial effect of *WS* in scopolamine dementia involves upregulation of activity-regulated cytoskeletal-associated protein (Arc), a member of the immediate-early gene (IEG) family protein that plays critical role in learning and memory. An in silico study by Kumar and Patnaik (2016) has shown that anaferine, beta-sitosterol, withaferin A, withanolide A, withanolide B and withanolide D inhibit GluN2B containing NMDARs through allosteric mode suggesting beneficial effect of *WS* in treating multi-neurodegenerative diseases mediated through glutamate excitoxicity. Bhattarai et al. (2013) suggested that there are compounds in *WS* with possible glycine mimetic activities, which may be potential targets for inducing memory consolidation in hippocampal CA1 neurons. Manchanda et al. (2016) have shown that *WS* is a potential agent to suppress the acute effects of sleep loss on learning and memory impairments and may emerge as a novel supplement to control sleep deprivation-induced cognitive impairments through restoration of expression of proteins involved in synaptic plasticity, cell survival, and apoptosis in the hippocampus. Kataria et al. (2016) showed that *WS* reduced the intracranial tumor volumes *in vivo* and suppressed the tumor-promoting proteins p-nuclear factor kappa B (NF-κB), p-Akt, vascular endothelial growth factor (VEGF), heat shock protein 70 (HSP70), PSA-NCAM, and cyclin D1 in the rat model of orthotopic glioma allograft. Further, reduction in glial fibrillary acidic protein (GFAP) and upregulation of mortalin and neural cell adhesion molecule (NCAM) expression specifically in tumor-bearing tissue further indicated the anti-glioma efficacy of *WS in vivo*. Candelario et al. (2015) observed that key constituents in *WS* have an important role in the development of pharmacological treatments for neurological disorders associated with altered GABAergic signalling that include anxiety disorders, sleep disturbances, muscle spasms, and seizures. It was suggested that differential activation of GABA receptor subtypes elucidates a potential mechanism by which *WS* accomplishes its reported adaptogenic properties. Parihar et al. (2016) reported oral administration of *WS* root extract stabilized mitochondrial functions and prevented oxidative damage in the hypothalamus of diabetic rats. Baitharu et al. (2014) found that withanolide A increased glutathione biosynthesis in neuronal cells by upregulating γ-glutamylcysteinyl ligase (GCLC) levels through Nrf2 pathway in a corticosterone dependent manner. Kumar et al. (2014) observed *WS* to attenuate lipid peroxidation and accentuate antioxidant enzymes; catalase and superoxide dismutase in brain as well as liver and kidney, suggesting its ability to act as a free radical scavenger protecting cells against lead induced toxic insult. Yin et al. (2013) found GABA-mimetic actions of *WS* on substantia gelatinosa neurons of the trigeminal subnucleus caudalis in mice suggesting

WS may be effecting in modulating orofacial pain processing. Konar et al. (2011) observed that *WS* prevented downregulation of neuronal cell markers NF-H, MAP2, PSD-95, GAP-43, BDNF and glial cell marker GFAP and upregulation of DNA damage and oxidative stress following scopolamine induced effect on brain or brain derived cells. Sominone, an aglycone of withanoside IV, identified as an active metabolite after oral administration of withanoside IV was observed to increase phosphorylation of RET (a receptor for the glial cell line-derived neurotrophic factor, GDNF) in hippocampal neurons without affecting the synthesis and secretion of GDNF. The densities of axons and dendrites were increased in the hippocampus by sominone administration. *WS* has been shown to be effective in reversing haloperidol induced catalepsy, which is believed to be due its antioxidant properties (Nair et al. 2008). Kuboyama et al. (2002) found withanolide A-treated cells, had increased length of NF-H-positive processes compared with vehicle-treated cells, whereas, the length of MAP2-positive processes was increased by withanosides IV and VI suggesting that axons are predominantly extended by withanolide A, and dendrites by withanosides IV and VI.

Based on the reported literature it appears that neuroprotective potential of *Withania somnifera* primarily involves anti-oxidant, anti-inflammatory, anti-apoptotic and anti-amyloidogenic mechanisms. In addition, it promotes neurogenesis, angiogenesis and restored neurotransmitter levels thereby preventing neurodegneration. In addition it has been shown to mediate its effects through modulation of Nrf2 pathway. However, further studies would be required for translation of pre-clinical studies on *Withania somnifera* to therapeutics against neurological conditions.

References

Ahmad M, Saleem S, Ahmad AS, Ansari MA, Yousuf S, Hoda MN, Islam F (2005) Neuroprotective effects of *Withania somnifera* on 6-hydroxydopamine induced Parkinsonism in rats. Hum Exp Toxicol 2:137–147

Babu PV, Gokulakrishnan A, Dhandayuthabani R, Ameethkhan D, Kumar CV, Ahamed MI (2007) Protective effect of Withania somnifera (solanaceae) on collagen glycation and cross-linking. Comp Biochem Physiol B Biochem Mol Biol 14:308–313

Baitharu I, Jain V, Deep SN, Shroff S, Sahu JK, Naik PK, Ilavazhagan G (2014) Withanolide A prevents neurodegeneration by modulating hippocampal glutathione biosynthesis during hypoxia. PLoS One 9:e105311

Bhattacharya SK, Bhattacharya D, Sairam K, Ghosal S (2002) Effect of Withania somnifera glycowithanolides on a rat model of tardive dyskinesia. Phytomedicine 9:167–170

Bhattacharya SK, Kumar A, Ghosal S (1995) Effects of glycowithanolides from Withania somnifera on an animal model of Alzheimer's disease and perturbed central cholinergic markers of cognition in rats. Phytother Res 9:110–113

Bhattarai JP, Park SJ, Han SK (2013) Potentiation of NMDA receptors by Withania somnifera on hippocampal CA1 pyramidal neurons. Am J Chin Med 41:503–513

Candelario M, Cuellar E, Reyes-Ruiz JM, Darabedian N, Feimeng Z, Miledi R, Russo-Neustadt A, Limon A (2015) Direct evidence for GABAergic activity of Withania somnifera on mammalian ionotropic GABAA and GABAρ receptors. J Ethnopharmacol 171:264–272

Chaudhary G, Sharma U, Jagannathan NR, Gupta YK (2003) Evaluation of Withania somnifera in a middle cerebral artery occlusion model of stroke in rats. Clin Exp Pharmacol Physiol 30:399–404

Cimarosti H, Henley JM (2008) Investigating the mechanisms underlying neuronal Death in ischemia using in vitro oxygen-glucose deprivation: potential involvement of protein SUMOylation. Neuroscientist 14:626–636

De Rose F, Marotta R, Poddighe S, Talani G, Catelani T, Setzu MD, Solla P, Marrosu F, Sanna E, Kasture S, Acquas E, Liscia A (2016) Functional and Morphological Correlates in the Drosophila LRRK2 loss-of-function Model of Parkinson's Disease: Drug Effects of Withania somnifera (Dunal) Administration. PLoS One 11:e0146140

Dhar N, Razdan S, Rana S, Bhat WW, Vishwakarma R, Lattoo SK (2015) A decade of molecular understanding of withanolide biosynthesis and in vitro studies in Withania somnifera dunal: prospects and perspectives for pathway engineering. Front Plant Sci 6:1–20

Dhuley JN (1998) Effect of ashwagandha on lipid peroxidation in stress-induced animals. J Ethnopharmacol 60:173–178

Durg S, Dhadde SB, Vandal R, Shivakumar BS, Charan CS (2015) Withania somnifera (Ashwagandha) in neurobehavioural disorders induced by brain oxidative stress in rodents: a systematic review and meta-analysis. J Pharm Pharmacol 67:879–899

Fisher RS, Boas WVE, Blume W, Elger C, Genton P, Lee P, Engel J (2005) Epileptic Seizures and Epilepsy: Definitions Proposed by the International League Against Epilepsy (ILAE) and the International Bureau for Epilepsy (IBE). Epilepsia 46:470–472

Fritschy JM (2008) Epilepsy, E/I Balance and $GABA_A$ Receptor Plasticity. Front Mol Neurosci 1:5

Gautam A, Wadhwa R, Thakur MK (2013) Involvement of hippocampal Arc in amnesia and its recovery by alcoholic extract of Ashwagandha leaves. Neurobiol Learn Mem 106:177–184

Grover A, Shandilya A, Agrawal V, Bisaria VS, Sundar D (2012) Computational evidence to inhibition of human acetyl cholinesterase by withanolide a for Alzheimer treatment. J Biomol Struct Dyn 29:651–662

Gupta A, Singh S (2014) Evaluation of anti-inflammatory effect of Withania somnifera root on collagen-induced arthritis in rats. Pharm Biol 52:308–320

Hinterberger H (1971) The biochemistry of catecholamines in relation to Parkinson's disease. Aust N Z J Med 1(Suppl 1):14–18

Homayoun H, Khavandgar S, Dehpour A (2002) The role of α_2-adrenoceptors in the modulatory effects of morphine on seizure susceptibility in mice. Epilepsia 43:797–804

Isoherranen N, Yagen B, Bialer M (2003) New CNS-active drugs which are second-generation valproic acid: can they lead to the development of a magic bullet? Curr Opin Neurol 16:203–211

Jayaprakasam B, Padmanabhan K, Nair MG (2010) Withanamides in Withania somnifera fruit protect PC-12 cells from β-amyloid responsible for Alzheimer's disease. Phytother Res 24:859–863

Jana CK, Hoecker J, Woods TM, Jessen HJ, Neuburger M, Gademann K (2011) Synthesis of withanolide A, biological evaluation of its neuritogenic properties, and studies on secretase inhibition. Angew Chem Int Ed Engl 50:8407–8411

Jenner P (2003) Oxidative stress in Parkinson's disease. Ann Neurol 53:26–38

John J (2014) Therapeutic potential of withania somnifera: A report on phytopharmacological properties. IJPSR 5:2131–2148

Kataria H, Wadhwa R, Kaul SC, Kaur G (2012) Water extract from the leaves of Withania somnifera protect RA differentiated C6 and IMR-32 cells against glutamate-induced excitotoxicity. PLoS One 7:1–13

Kataria H, Wadhwa R, Kaul SC, Kaur G (2013) Withania somnifera water extract as a potential candidate for differentiation based therapy of human neuroblastomas. PLoS One 8:1–12

Kataria H, Kumar S, Chaudhary H, Kaur G (2016) Withania somnifera Suppresses Tumor Growth of Intracranial Allograft of Glioma Cells. Mol Neurobiol 53:4143–4158

Khodaei M, Jafari M, Noor M (2012) Remedial use of withanolides from Withania coagulans (Stocks) dunal. Adv Lif Sci 2:6–19

Konar A, Shah N, Singh R, Saxena N, Kaul SC, Wadhwa R, Thakur MK (2011) Protective role of Ashwagandha leaf extract and its component withanone on scopolamine-induced changes in the brain and brain-derived cells. PLoS One 6:e27265

Kuboyama T, Tohda C, Zhao J, Nakamura N, Hattori M, Komatsu K (2002) Axon- or dendrite-predominant outgrowth induced by constituents from Ashwagandha. Neuroreport 13:1715–1720

Kuboyama T, Tohda C, Komatsu K (2006a) Withanoside IV and its active metabolite, sominone, attenuate Abeta(25-35)-induced neurodegeneration. Eur J Neurosci 23:1417–1426

Kuboyama T, Hirata K, Aoki-Kinoshita KF, Kashima H, Yasuda H (2006b) A gram distribution kernel applied to glycan classification and motif extraction. Genome Inform 17:25–34

Kulkarni SK, Akula KK, Dhir A (2008) Effect of Withania somnifera dunal root extract against pentylenetetrazol seizure threshold in mice: possible involvement of GABAergic system. Indian J Exp Biol 46:465–469

Kulkarni SK, Verma A (1993) Protective effect of BR-16A (Mentat), a herbal preparation on alcohol abstinence-induced anxiety and convulsions. Indian J Exp Biol 31:435–439

Kumar A, Kumar R, Rahman MS, Iqubal MA, Anand G, Niraj PK, Ali M (2015) Phytoremedial effect of Withania somnifera against arsenic-induced testicular toxicity in charles foster rats. Avicenna J Phytomed 5:355–364

Kumar P, Kumar A (2009) Possible neuroprotective effect of Withania somnifera root extract against 3-nitropropionic acid-induced behavioral, biochemical, and mitochondrial dysfunction in an animal model of Huntington's disease. J Med Food 12:591–600

Kumar P, Singh R, Nazmi A, Lakhanpal D, Kataria H, Kaur G (2014) Glioprotective effects of Ashwagandha leaf extract against lead induced toxicity. Biomed Res Int 2014:1–15

Kumar G, Patnaik R (2016) Exploring neuroprotective potential of Withania somnifera phytochemicals by inhibition of GluN2B-containing NMDA receptors: An in silico study. Med Hypotheses 92:35–43

Kurapati KR, Atluri VS, Samikkannu T, Nair MP (2013) Ashwagandha (Withania somnifera) reverses beta-amyloid1-42 induced toxicity in human neuronal cells: implications in HIV-associated neurocognitive disorders (HAND). PLoS One 8:1–15

Luxmini KPAMK, Dharmadasa RM, Samarasinghe K, Muthukumarana PRM (2015) Comparative pharmacognostic study of different parts of Withania somnifera and its substitute ruellia tuberosa. World J Agric Res 3:28–33

Manchanda S, Mishra R, Singh R, Kaur T, Kaur G (2016) Aqueous leaf extract of Withania somnifera as a potential neuroprotective agent in sleep-deprived rats: a mechanistic study. Mol Neurobiol. [Epub ahead of print]

Manjunath MJ, Muralidhara (2015) Standardized extract of Withania somnifera (Ashwagandha) markedly offsets rotenone-induced locomotor deficits, oxidative impairments and neurotoxicity in Drosophila melanogaster. J Food Sci Technol 52:1971–1981

Mathews M, Gratz S, Adetunji B, George V, Mathews M, Basil B (2005) Antipsychotic-induced movement disorders: evaluation and treatment. Psychiatry (Edgmont) 2:36–41

Mirjalili MH, Moyano E, Bonfill M, Cusido RM, Palazon J (2009) Steroidal lactones from Withania somnifera, an ancient plant for novel medicine. Molecules 14:2373–2393

More SV, Koppula S, Kim IS, Kumar H, Kim BW, Choi DK (2012) The role of bioactive compounds on the promotion of neurite outgrowth. Molecules 17:6728–6753

Nair V, Arjuman A, Gopalakrishna HN, Nandini M (2008) Effect of Withania somnifera root extract on haloperidol-induced catalepsy in albino mice. Phytother Res 22:243–246

Olanow CW (2007) The pathogenesis of cell death in Parkinson's disease – 2007. Mov Disord 22:335–342

Palliyaguru DL, Singh SV, Kensler TW (2016) Withania somnifera: from prevention to treatment of cancer. Mol Nutr Food Res 1:1–12

Panossian A, Wikman G (2010) Effects of adaptogens on the central nervous system and the molecular mechanisms associated with their stress—protective activity. Pharmaceuticals (Basel) 3:188–224

Parihar MS, Hemnani T (2003) Phenolic antioxidants attenuate hippocampal neuronal cell damage against kainic acid induced excitotoxicity. J Biosci 28:121–128
Parihar P, Shetty R, Ghafourifar P, Parihar MS (2016) Increase in oxidative stress and mitochondrial impairment in hypothalamus of streptozotocin treated diabetic rat: antioxidative effect of Withania somnifera. Cell Mol Biol (Noisy-le-grand) 62:73–83
Patil SP, Maki S, Khedkar SA, Rigby AC, Chan C (2010) Withanolide A and asiatic acid modulate multiple targets associated with amyloid-beta precursor protein processing and amyloid-beta protein clearance. J Nat Prod 73:1196–1202
Piggott MA, Marshall EF, Thomas N, Lloyd S, Court JA, Jaros E, Perry EK (1999) Striatal dopaminergic markers in dementia with Lewy bodies, Alzheimer's and Parkinson's diseases: rostrocaudal distribution. Brain 122:1449–1468
Pingali U, Pilli R, Fatima N (2014) Effect of standardized aqueous extract of Withania somnifera on tests of cognitive and psychomotor performance in healthy human participants. Pharmacognosy Res 6:12–18
Prakash J, Chouhan S, Yadav SK, Westfall S, Rai SN, Singh SP (2014) Withania somnifera alleviates parkinsonian phenotypes by inhibiting apoptotic pathways in dopaminergic neurons. Neurochem Res 39:2527–2536
Raghavan A, Shah ZA (2015) Withania somnifera: a pre-clinical study on neuroregenerative therapy for stroke. Neural Regen Res 10:183–185
Rai M, Jogee PS, Agarkar G, dos Santos CA (2016) Anticancer activities of Withania somnifera: Current research, formulations, and future perspectives. Pharm Biol 54:189–197
Rajasankar S, Manivasagam T, Sankar V, Prakash S, Muthusamy R, Krishnamurti A, Surendran S (2009a) Withania somnifera root extract improves catecholamines and physiological abnormalities seen in a Parkinson's disease model mouse. J Ethnopharmacol 125:369–373
Rajasankar S, Manivasagam T, Surendran S (2009b) Ashwagandha leaf extract: a potential agent in treating oxidative damage and physiological abnormalities seen in a mouse model of Parkinson's disease. Neurosci Lett 454:11–15
Rascol O (2009) "Disease-modification" trials in Parkinson disease: target populations, endpoints and study design. Neurology 72:S51–S58
Roshanaei K, Neda NN (2015) Effect of withania somnifera root extract on PTZ-induced seizure threshold in mice. RJFH 10:719–723
Sehgal N, Gupta A, Valli RK, Joshi SD, Mills JT, Hamel E, Khanna P, Jain SC, Thakur SS, Ravindranath V (2012) Withania somnifera reverses Alzheimer's disease pathology by enhancing low-density lipoprotein receptor-related protein in liver. Proc Natl Acad Sci U S A 109:3510–3515
Senthil K, Thirugnanasambantham P, Oh TJ, Kim SH, Choi HK (2015) Free radical scavenging activity and comparative metabolic profiling of in vitro cultured and field grown Withania somnifera roots. PLoS One 10:1–14
Serrano-Pozo A, Frosch MP, Masliah E, Hyman BT (2011) Neuropathological alterations in Alzheimer disease. Cold Spring Harb Perspect Med 1:a006189
Shinde P, Vidyasagar N, Dhulap S, Dhulap A, Hirwani R (2015) Natural Products based P-glycoprotein Activators for Improved β-amyloid Clearance in Alzheimer's Disease: An in silico Approach. Cent Nerv Syst Agents Med Chem 16:50–59
Singh G, Sharma PK, Dudhe R, Singh S (2010) Biological activities of Withania somnifera. Ann Biol Res 1:56–63
Singh N, Bhalla M, de Jager P, Gilca M (2011) An overview on ashwagandha: a Rasayana (rejuvenator) of Ayurveda. Afr J Tradit Complement Altern Med 8(5 Suppl):208–213
Soman S, Korah PK, Jayanarayanan S, Mathew J, Paulose CS (2012) Oxidative stress induced NMDA receptor alteration leads to spatial memory deficits in temporal lobe epilepsy: ameliorative effects of Withania somnifera and Withanolide A. Neurochem Res 37:1915–1927
Soman S, Anju TR, Jayanarayanan S, Antony S, Paulose CS (2013) Impaired motor learning attributed to altered AMPA receptor function in the cerebellum of rats with temporal lobe epilepsy: ameliorating effects of *Withania somnifera* and withanolide A. Epilepsy Behav 27(3):484–491

Sood A, Kumar A, Dhawan DK, Sandhir R (2016) Propensity of *Withania somnifera* to attenuate behavioural, biochemical, and histological alterations in experimental model of stroke. Cell Mol Neurobiol 36:1123–1138

Tiwari R, Chakraborty S, Saminathan S, Dhama K, Singh SV (2014) Ashwagandha (Withania somnifera): Role in safeguarding health, immunomodulatory effects, combating infections and therapeutic applications: A review. J Biol Sci 14:77–94

Tohda C, Joyashiki E (2009) Sominone enhances neurite outgrowth and spatial memory mediated by the neurotrophic factor receptor, RET. Br J Pharmacol 157:1427–1440

Trigunayat A, Raghavendra M, Singh RK, Bhattacharya AK, Acharya SB (2007) Neuroprotective effect of Withania somnifera (WS) in cerebral ischemia-reperfusion and long-term hypoperfusion induced alterations in rats. J Nat Remedies 7:234–246

Uddin Q, Samiulla L, Singh VK, Jamil SS (2012) Phytochemical and pharmacological profile of Withania somnifera dunal: A review. J App Pharmaceu Sci 2:170–175

Upadhyaya L, Tiwari AK, Agrawal A, Dubey GP (1990) Role of an indigenous drug geriforte on blood levels of biogenic amines and its significance in the treatment of anxiety neurosis. Act Nerv Super 32:1–5

Vareed SK, Bauer AK, Nair KM, Liu Y, Jayaprakasam B, Nair MG (2014) Blood–brain barrier permeability of bioactive withanamides present in Withania somnifera fruit extract. Phytother Res 28:1260–1264

Yin H, Cho DH, Park SJ, Han SK (2013) GABA-mimetic actions of Withania somnifera on substantia gelatinosa neurons of the trigeminal subnucleus caudalis in mice. Am J Chin Med 41:1043–1051

Chapter 19
Potential Therapeutic Use of *Withania somnifera* for Treatment of Amyotrophic Lateral Sclerosis

Kallol Dutta, Vivek Swarup, and Jean-Pierre Julien

Abstract Despite continued effort for last few decades, there are no effective therapies available to treat amyotrophic lateral sclerosis (ALS). This is, in part, due to the multi-factorial nature of the disease, which makes it difficult to treat with a single drug targeted at a specific factor, pathway or mechanism. In the ensuing chapter, we discuss the potential therapeutic action of *Withania somnifera* extracts in ALS pathogenesis. *Withania somnifera*, which is commonly known as *Ashwagandha*, has been used in traditional medical systems in India, China and some middle-eastern countries for the past 3000 years. The most common use of the plant extract is as a rejuvenator, but it also possesses potent anti-inflammatory, anti-oxidant, anti-neoplastic, and immunomodulatory properties. Based on these, the plant extract has been used in numerous diseases, including neurodegenerative diseases, with various degrees of efficacy. In this chapter, we shall present the rationale for considering *Withania somnifera* and its constituents as therapeutic possibility in ALS.

Keywords *Withania somnifera* • Extracts • Withanolides • Amyotrophic lateral sclerosis • Therapy

19.1 An Introduction to Amyotrophic Lateral Sclerosis

ALS is a "relentlessly progressive" neurodegenerative disorder primarily affecting motor neurons, which supply to voluntary muscles, including lower motor neurons (LMN) in the medulla and anterior horn of the spinal cord as well as upper motor neurons (UMN) in the cerebral cortex (de Carvalho and Swash 2016). The prognosis is bleak as majority of the people diagnosed with ALS die within 3–5 years of initial symptoms due to respiratory failure and in less than 10% of the cases the

K. Dutta • V. Swarup • J.-P. Julien (✉)
CERVO Brain Research Centre, Université Laval,
2601 De la Canardière, Québec, Québec G1J 2G3, Canada
e-mail: jean-pierre.julien@fmed.ulaval.ca

S.C. Kaul, R. Wadhwa (eds.), *Science of Ashwagandha: Preventive and Therapeutic Potentials*, DOI 10.1007/978-3-319-59192-6_19

patients survive for 10 years or more. With active and aggressive management survival could be slightly enhanced, particularly among ALS patients with bulbar dysfunction (Traynor et al. 2003).

The initial clinical features of the disease are dependent on the site of onset, which in turn is highly variable. Limb involvement (weakness of legs and arms) is the most common feature of initial disease process manifesting in about 75–80% of patients. Bulbar symptoms manifesting as dysarthria (problem with speech) or dysphagia (difficulty in swallowing) are the most common ALS presentation next to limb involvement, affecting 20–25% of patients (Mitchell and Borasio 2007). However, the age of onset of the disease symptoms is thought also to influence the rate of disease progression and survival (Yokoi et al. 2016). In LMN involvement, fasciculations (brief, spontaneous twitching of muscles) may occur early on in the disease, particularly in the tongue and limbs. UMN involvements generally are hyper-reflexic and stiff whereas, reflexes may be diminished due to LMN involvement.

In North America the median incidence rate of ALS is 1.8/100000 population. Effect of gender on incidence of ALS has been extensively evaluated but the data has been variable. It was believed that men outnumber women in the ratio of 2:1 (Kahana et al. 1976, Guidetti et al. 1996) but later studies have put forward more moderate figures such as 1.7:1 (Chen et al. 2015), and even approaching unity in some studies (O'Toole et al. 2008, Roman 1996); it is now believed to be variable based on the age group to which the patients belong (Manjaly et al. 2010). ALS can either be familial (fALS) or sporadic (sALS) in nature. A meta-analysis of 38 published reports has shown that 5.1% of total cases (confidence interval 4.1 to 6.1%) can be familial (Byrne et al. 2011) thereby leaving approximately 95% of cases to be categorized as sporadic. FALS is generally defined as "the presence of ALS in either a first or second degree relative of the index case" (Leblond et al. 2014), whereas in sALS, there are no known family history. However, lack of detailed family history may lead to an overestimation of the data for sALS. Also in twin studies where at least 1 twin had ALS, it was estimated that there was 61% heritability, thereby suggesting a major genetic role even in sALS (Al-Chalabi et al. 2010). In Table 19.1 we have provided a concise list of genes that are currently known to be associated (both with high and low risk factor) with ALS along with the proteins encoded by them and their known functions.

Studies have reported that some viral infections are correlated with ALS or ALS-like symptoms. The most studied of these has been retroviral infections. Over the last decade there have been quite a few reports of patients infected with HIV demonstrating ALS-like symptoms, even though the causal relation remain uncertain (Verma and Berger 2006; Rowland 2011; Anand et al. 2014). Probably, the strongest evidence of retroviral involvement in ALS comes from a recent study involving Human endogenous retrovirus-K (HERV-K). Integration of the HERV-K genome into mammalian germ line cells, which possibly took place millions of years ago, constitute roughly 8% of the human genome. In a sub-population of ALS cases it has been reported that the HERV-K genome can reactivate and transcribe for viral proteins in brain, which were shown to be neurotoxic in nature (Li et al. 2015).

Table 19.1 Genes associated with ALS

High risk—causative genes[a]		Low risk—susceptibility risk factor	
Gene ID	Protein/function	Gene ID	Protein/function
SOD1	Cu/Zn superoxide dismutase 1, soluble/ Oxidative stress	*ANG*	Angiogenin/ ribonuclease
ALS2	Alsin/Rho guanine nucleotide exchange factors	*FIG. 4*	SAC1 lipid phosphatase domain containing (*S. cerevisiae*)/ polyphosphoinositide phosphatase
SETX	Senataxin/RNA/DNA helicase	*ATXN2*	Ataxin 2/unknown
SPG11	Spatacsin/ transmembrane protein	*DCTN1*	Dynactin/axonal transport
FUS	Fused in sarcoma/ RNA-binding protein, DNA repair, exon splicing	*CHMP2B*	Chromatin modifying protein 2B/ recycling or degradation of cell surface receptors
VAPB	Vesicle-associated membrane protein-associated protein B/ vesicular trafficking	*DPP6*	Dipeptidyl-peptidase 6/ bind specific voltage-gated potassium channels and alters their expression and biophysical properties
TARDBP	TAR DNA binding protein 43/ transcriptional repressor, splicing regulation	*VEGF*	Vascular endothelial growth Factor/ angiogenic, vascular, growth, migration and apoptosis factor
OPTN	Optineurin/ocular tension, membrane and vesicle trafficking	*UNC13A*	Unc-13 homolog A/ play important roles in neurotransmitter release at synapses.
VCP	Valosin-containing protein/ATP-binding protein, vesicle transport and fusion	*NEFH*	Neurofilament, heavy polypeptide/ intracellular transport to axons and dendrites
UBQLN2	Ubiquilin 2/ ubiquitination, degradation	*PRPH*	Peripherin/cytoskeletal protein
SIGMAR1	Sigma non-opioid intracellular receptor 1/endoplasmic reticulum chaperone	*SQSTM1*	Sequestosome 1/ scaffold protein, NFκB signaling pathway

(continued)

Table 19.1 (continued)

High risk—causative genes[a]		Low risk—susceptibility risk factor	
Gene ID	Protein/function	Gene ID	Protein/function
DAO	D-amino-acid oxidase/oxidative deamination of of D-amino acids.	*TAF15*	TATA box binding protein (TBP)-associated factor/RNA polymerase II gene transcription
PFN1	Profilin 1/actin binding protein, actin polymerization	*ELP3*	Elongator acetyltransferase complex subunit 3/ transcript elongation
hnRNPA2B1/A1	Heterogenous nuclear ribonucleoprotein/ mRNA processing, metabolism and transport	*SMN1*	Survival of motor neuron 1/ required for efficient assembly of small nuclear ribonucleoprotein complexes
C9ORF72	Chromosome 9 open reading frame 72/ regulation of endosomal trafficking	*PON1,2,3*	Paraoxonase/ organophosphate hydrolysis
		HFE	Hemochromatosis/iron absorption
		KIFAP3	Kinesin-associated protein 3/small G protein
		APEX1	APEX nuclease 1/ apurinic/apyrimidinic endonuclease
		PGRN	Progranulin/cell growth regulator
		ITPR2	INOSITOL 1,4,5-trisphosphate receptor type 2/ second messenger intracellular calcium release channel
		PLCD1	Phospholipase C delta 1/ intracellular signal transduction
		ARHGEF28	Rho guanine nucleotide exchange factor 28/unknown

(continued)

Table 19.1 (continued)

High risk—causative genes[a]		Low risk—susceptibility risk factor	
Gene ID	Protein/function	Gene ID	Protein/function
		CHCHD10	Coiled-coil-helix-coiled-coil-helix domain-containing protein 10/ involved in mitochondrial function (Bannwarth et al. 2014)
		TBK1	Serine/threonine-protein kinase TBK1/ regulation of inflammation against pathogens (Tsai et al. 2016)
		MATR3	Martin 3/RNA- and DNA-binding protein (Johnson et al. 2014)
		CCNF	cyclin F/a component of an E3 ubiquitin-protein ligase complex (Williams et al. 2016)

Adapted, modified and updated from (Leblond et al. 2014)
[a]Includes 2 unknown genes located in chromosomes 20p13 and 9q21–q22 with unknown protein/function

Dementia resulting from frontotemporal lobar degeneration (FTD) and alterations in cognition is also associated with ALS (Wilson et al. 2001). An estimated 50% of ALS cases show some degree of cognitive decline and about 15% also manifest symptoms of FTD (Ringholz et al. 2005). Pathological changes observed in FTD are gross morphological atrophy in the frontal and anterior temporal lobes, and microscopic changes, including gliosis, inclusion bodies, swollen neurons, and microvacuolation. The inclusion bodies, in approximately 40% of cases, are composed of insoluble form of the protein Tau but almost an equal number of cases were found to be Tau negative. The pathological overlap between ALS and FTD was proposed when TDP-43 was reported to be the major ubiquitinated protein found in neuronal cell bodies of both sporadic ALS patients and the most frequent pathological form of FTD negative for Tau inclusions (Arai et al. 2006; Neumann et al. 2006). Later on, the FUS protein was also detected from ubiquinated inclusion bodies in a significant number of TDP-43-negative FTD cases (Neumann et al. 2009) thereby providing yet another link between the two diseases. As both TDP-43 and FUS are involved in RNA processing /metabolism in cells, it could be hypothesized that errors in RNA processing may be central to ALS and FTD pathogenesis. Based on this, a proposed reclassification of FTD based on the main protein component accumulated is presented as a) FTLD-tau (45%), b) FTLD-TDP (45%), c) FTLD-FUS (9%), and d) FTLD-UPS (ubiquitin-proteasome system) (1%) (Sieben et al. 2012).

More recently, hexanucleotide expansion in the *C9ORF72* gene, which is a significant contributor to fALS, has also been reported to be involved in FTD (Renton et al. 2011). Apart from these, ALS-causing mutations in various other genes such as *UBQLN2, VCP, VAPB, SQSTM1, OPTN, CHMP2B and TBK1* have also been reported in some cases of FTD (Ling et al. 2013; Gijselinck et al. 2015).

The pathogenic processes involved in ALS are multifactorial and much still remains to know. A hallmark of ALS is the presence of various inclusion bodies in degenerating lower motor neurons of the brainstem, the spinal cord and in corticospinal UMN (Sasaki and Maruyama 1994) and occasionally in surrounding reactive glial cells (Miller et al. 2004). These inclusions are classified as 'Lewy body-like' or 'Skein-like' or Bunina bodies, which are cystatin C-containing eosinophilic intraneuronal inclusions, and are found in the cell bodies of motor neurons in ALS (Kawashima et al. 1998; He and Hays, 2004; Okamoto et al. 2008). These inclusions mostly comprises of mis-folded and/or ubiquinated, hyperphosphorylated or SUMOylated aggregates of different proteins such as SOD1, TDP-43 or FUS, OPTN, UBQLN2 and the translational product of intronic repeats in the gene C9ORF72 (Neumann et al. 2006; Deng et al. 2010; Dangoumau et al. 2013; Blokhuis et al. 2013). Recent studies have raised the concern that these mis-folded and aggregated proteins may be propagated in a cell-to-cell manner reminiscent of prion proteins (Grad et al. 2015a). The two most studied proteins displaying such mechanism of spread, in relation to ALS, are SOD1 and TDP-43. Mis-folded SOD1 has been shown to propagate from cell to cell both via exosome-dependent as well as independent pathways (Urushitani et al. 2006; Silverman et al. 2016) and this misfolding could be induced by other ALS-associated proteins such as FUS or TDP-43 (Pokrishevsky et al. 2016). This prion-like propagation of SOD1 could also be replicated *in vivo* using minute quantity of mis-folded and aggregated SOD1 as seed, thereby hastening disease progression (Grad et al. 2014, 2015b; Zeineddine et al. 2015). Likewise, aggregated TDP-43 (Nonaka et al. 2013) or FUS (Nomura et al. 2014) may also propagate from cell to cell based on seeded aggregation whereby mis-folded proteins recruit and initiate template-directed mis-folding of the native protein to form new aggregates and this could, in part, be by an exosome-dependent pathway.

Chaperones that are resident within the endoplasmic reticulum (ER) are involved in the sensing of intracellular mis-folded proteins and as a part of ER stress response they attempt to correct the protein folding. In ALS, presence of mis-folded proteins does understandably elicit an ER stress response. In mutant SOD1 mouse model of fALS, an ER chaperon named Protein disulphide isomerase (PDI) has been reported to be activated and colocalize with the mis-folded SOD1 (Atkin et al. 2006). The ER stress response is also present in sALS cases as post mortem spinal cord extracts showed elevated levels of ER stress sensor kinases, chaperones and apoptotic mediators. Furthermore, elevated level of PDI could be detected from the CSF of these patients and it was widely distributed in spinal motor neurons (Atkin et al. 2008). Although ER stress response is thought to be beneficial at least at the earliest stages of the disease, surprisingly mutant SOD1 mice deficient in a key ER stress response transcription factor X-box binding protein 1 (XBP-1) significantly increased the survival time of the animals. This was found to be due to enhanced clearance of mutant SOD1 aggregates by autophagy, a cellular pathway involved in lysosome-mediated protein degradation (Hetz et al. 2009).

Post-mortem analysis of brain and spinal cord from both sALS and SOD1-related fALS cases has shown elevated levels of oxidative damage to DNA, proteins and lipids (Fitzmaurice et al. 1996; Shibata et al. 2001; Shaw et al. 1995b). Further corroboration of the involvement of oxidative stress in ALS comes from the observation that nuclear erythroid 2-related factor 2 (NRF2) – antioxidant response element (ARE) signaling is dysregulated in spinal cord of both ALS patients and animal models (Sarlette et al. 2008). The NRF2-ARE signaling pathway is a major regulator of anti-oxidant response following oxidative stress. Oxidative stress, directly or indirectly, exacerbates other pathological processes in ALS. Mitochondrial vacuolation has been reported in some SOD1 mouse models of the disease at the presymptomatic stage, indicating an early role of dysfunction in disease progression (Jaarsma et al. 2000, Higgins et al. 2003). Abnormal respiratory chain function along with oxidative damage to mitochondrial lipids and proteins has been observed from ALS patient spinal cords (Wiedemann et al. 2002). This could be linked to the caspase-mediated apoptotic death of motor neurons observed in some models of ALS (Sathasivam et al. 2005).

A number of studies have reported that excitotoxicity is a major contributing factor for motor neuron injury in ALS (Van Damme et al. 2005; Foran and Trotti 2009). In some ALS patients high level of glutamate has been reported from the CSF along with a decrease in expression of its transporters (Shaw et al. 1995a; Rothstein et al. 1992). The significance of excitotoxicity in the pathogenesis of ALS can be further appreciated by the fact that currently the only therapeutic intervention, a drug named riluzole, imparts its protective effects mainly by way of inhibiting glutamate release even if that only results in prolonging the survival of patients by a few months (Cheah et al. 2010).

Insight into the pathogenic role of glia in neuroinflammation and ALS disease progression has been reported in several studies and is now thought to constitute crucial therapeutic targets (Radford et al. 2015). Microglia and astrocytes become increasingly activated as the disease progresses in both animal models and patients and this leads to the development of neuroinflammation, which could result in both positive as well as negative effects in the surrounding milieu. Recent studies demonstrating that mutant TDP-43 or truncated variety of TDP-43 could activate microglia (Zhao et al. 2015) and experiments with C9orf72 null mice that put forward the notion that C9orf72 may be necessary for proper microglial functioning (O'Rourke et al. 2016), further corroborates the importance of microglial involvement. A complex signaling cross talk between motor neurons and microglia also contribute to severity of disease or neuroprotection (Appel et al. 2011). There has also been an over whelming number of studies that have indicated the toxic role of reactive astrocytes on motor neuron survival in ALS (Pehar et al. 2005, Haidet-Phillips et al. 2011; Almad et al. 2016; Song et al. 2016). Together, the microglia and astrocytes are believed to be involved in the non-cell autonomous death of motor neurons (Clement et al. 2003; Urushtani et al. 2006; Zhao et al. 2010).

Table 19.2 Major constituents of the *Withania somnifera* plant extracts

Withanolides (Steroidal lactones)	Alkaloids (0.13 4.3%)	Other constituents
Withaferin A	Ashwagandhine	Sitoindosides VII, VIII (acylsteryl glucosides) IX and X (C-27-Glyco-withanolides)
Withanolide A-Y	Ashwagandhinine	Saccharose
5-dehydroxywithanolide-R	Somniferiene	β-sitosterol
Withasomniferin-A	Somniferinine	Hentriacontane
Withasomniferols A-C	Withanine	Scopoletin
Withasomidienone	Withaninine	Dulcitol
	Pseudowithanine	Chlorogenic acid
	Withasomnine	Glycosides
	Isopelletierine	Essential oils
	Cuscohygrine	Choline
	Isopelletierine	Amino acids (aspartic acid, glycine, tyrosine, alanine, proline, tryptophan, glutamic acid, cystine, arginine)
	Anapherine	Iron
	Anahygrine	
	Visamine	
	Nicotine (debated)	

19.2 *Withania somnifera* in Medicine

Extracts of the plant *Withania somnifera* (WS) has been used extensively in traditional medicines for centuries as a nervine tonic, aphrodisiac, and 'adaptogen' which helps the body to rejuvenate (Mishra et al. 2000). Known by many other names such as Ashwagandha, Indian Ginseng, Winter cherry, Samm Al Ferakh and Shui Qie, this plant is a perennial shrub belonging to the family Solanaceae (nightshade). Usually the roots of the plant are used for therapeutic purposes; however, extracts from the leaves, berries and seeds have also been known to be effective. WS extracts typically contains steroidal compounds such as the ergostane-type steroidal lactones, phytosterols, as well as various alkaloids, a variety of amino acids, and high amounts of iron (see Table 19.2). However, it must be noted that the actual composition and concentration may vary from plant to plant grown under different geological or environmental conditions such as seasonal temperature, duration of light-dark period, depth of tillage, time of harvesting and concentration of fertilizers (Upton et al. 2000). The seeds of plants are reported to contain another class of compounds called withanamides, which also have potent anti-oxidant properties (Nair and Jayaprakasam 2007).

Traditionally, for therapeutic purposes, WS extracts has been administered as aqueous suspensions or as suspensions in milk or honey or clarified butter (Suresh Gupta et al. 2006) and unlike other adaptogens, which tend to be stimulating,

Ashwagandha has a calming effect (Singh et al. 2011). Extracts or purified components of WS have been reported to possess a myriad of beneficial properties such as:

(a) Anti-stress – in animals exposed to chronic stress WS root extracts or purified components has been reported to alleviate pathological changes (Bhattacharya et al. 1987; Bhattacharya and Muruganandam 2003) and act as a potent anti-oxidant substance (Bhattacharya et al. 2001; Khan et al. 2015)
(b) Anti-neoplastic – *in vitro*, WS extract or its purified components have been reported to arrest the growth of and induce apoptosis is several cancerous/tumor cell lines (Jayaprakasam et al. 2003). *In vivo*, also there have been multiple studies to demonstrate the chemopreventive activity in carcinogen-induced or cancer xenograft models (Palliyaguru et al. 2016)
(c) Immunomodulatory – WS extracts acts as potent immunomodulators. They have been shown to selectively up-regulate Th1 response in mice (Bani et al. 2006) and in myelo-suppressed mice restore haematopoiesis (Ziauddin et al. 1996). In experimental models of leucopenia WS extract has been shown to restore leukocyte counts (Davis and Kuttan 1998) and increase phagocytic capacity of macrophages (Davis and Kuttan, 2000)
(d) Anti-inflammatory– WS acts mainly by inhibiting NF-κB pathway (Singh et al. 2007; Grover et al. 2010; Dey et al. 2014); beneficial results observed post treatment in inflammatory diseases such as rheumatoid arthritis (Kumar et al. 2015)

WS extracts generally have a good safety profile. There is no single systematic study available which compares acute, sub-acute, sub-chronic or chronic dosage of WS root powder, whole plant extract, or different extracts of the plant (e.g., water, alcohol, petroleum ether), or purified alkaloids, and glycosides. However, data from different studies using differing components and doses have led us to believe that WS extracts and to a certain extent its individual components, are well tolerated by experimental animals at quite high dosages without any prominent adverse side effects. In a study comparing acute toxicity of total alkaloids from the roots of WS, suspended 10% propylene glycol, reported the LD_{50} to be 465 mg/kgBW and 432 mg/kgBW in rats and mice, respectively (Malhotra et al. 1965). In a chronic study involving administration of aqueous extracts of WS to rats at a dose of 100 mg/kgBW/day for 8 months did not have any significant adverse effect (Mishra et al. 2000). In an acute toxicity study involving alcoholic extracts of WS roots, a single intraperitoneal injection at a dose of 1100 mg/kgBW did not significantly affect survival time in mice. The LD_{50} value was reported to be 1260 mg/kgBW. The same study also reported that sub-acute intraperitoneal administration of the alcoholic extract at a dose of 100 mg/kgBW for 30 days in rats also did not result in any lethality (Sharada et al. 1993). In a separate study the same group had also shown that Withaferin A when administered i.p. has a LD_{50} of 80 mg/kgBW (Sharada et al. 1996). We have used pure Withaferin A solution in DMSO-Saline (1:9) at dosages of 3 and 4 mg/kgBW administered twice weekly, in a 3 different transgenic mice model of ALS with C57BL6/J background without noticing any adverse drug-dependent side effects (Swarup et al. 2011b; Patel et al. 2015).

In a dose-related tolerability, safety and activity study of WS extract (aqueous extract, 8:1) in normal humans, a small population consisting of either sexes were treated daily in two divided doses with increase in daily dosage every 10 days for 30 days (750, 1000 and 1250 mg/day for 10 days). Most of the subjects tolerated WS without any adverse event and the formulation appeared safe for human trials (Raut et al. 2012). There have been quite a few clinical trials conducted or being conducted with WS extracts for a variety of disorders. Table 19.3 summarizes these trials and gives an idea about the range of dosages that can be considered to be within the safe zone for use as therapeutic tool.

From the above-mentioned studies and clinical trial protocols, it is fair to conclude that WS extracts are well tolerated and high dosages of 1.2–2 g on a daily basis do not result in any prominent adverse side effects in humans.

Table 19.3 Summary of clinical trials with WS extracts depicting dosage and duration

Trial Identifier	Condition	Drug used	Dosage	Duration
NCT01793935[a]	Schizophrenia	WS extract; Sensoril®	250 mg twice daily for 1st week followed by 500 mg twice daily	12 weeks
NCT00817752[a]	Stress	WS extract	3 mL/day (conc. Not defined)	5 days
NCT00761761[a, c]	Bipolar disorder	WS extract; Sensoril®	250 mg/day for 1st week; increased to 500 mg/day till completion of study	8 weeks
NCT01311180[a]	Generalized Anxiety Disorder	WS extract; Sensoril®	250 mg every morning for 7 days and then 250 mg twice a day for 7 weeks	8 weeks
CTRI/2011/091/000222[b]	Stress and anxiety	WS extract KSM-66	250 mg twice a day	8 weeks
CTRI/2011/12/002278[b]	Chronic stress	WS extract	125 or 250 or 500 mg/day	8 weeks
CTRI/2012/05/002670[b]	Cardiorespiratory endurance	WS extract KSM-66	300 mg twice daily	90 days

(continued)

Table 19.3 (continued)

Trial Identifier	Condition	Drug used	Dosage	Duration
CTRI/2013/05/003688[b]	Stage IV cancer	WS extract	500 mg four times a day	6 months
CTRI/2013/04/003537[b]	Cognitive and psychomotor performance	WS extract	250 mg twice daily	14 days
CTRI/2015/07/006045[b]	Female Sexual Dysfunction	WS extract KSM-66	300 mg twice daily	8 weeks
CTRI/2016/04/006791[b]	Oxidative stress in athletes	WS extract KSM-66	300 mg twice daily	8 weeks
CTRI/2016/05/006903[b]	Subclinical Hypothyroidism	WS extract KSM-66	300 mg twice daily	8 weeks

[a]Retrieved from https://clinicaltrials.gov/ with search keyword "WITHANIA"
[b]Retrieved from http://ctri.nic.in/ with search keyword "WITHANIA"
[c]Available Results showed no drug-associated adverse effects on subjects

19.3 *Withania somnifera* as Potential Therapeutic Option in Neurodegenerative Diseases

In this section we shall lay out the current state of knowledge regarding the use of WS extracts in various neurological disorders followed by a discussion on how they can potentially be beneficial in ALS.

19.3.1 Withania somnifera *Extracts or Constituents for Treatment of Neurodegenerative Diseases*

So far, most of the studies regarding neuroprotective effects of WS extracts have mainly been reported from the Alzheimer's disease (AD) perspective. AD and ALS, though two distinct diseases, do share some common features such as (a) in both of the diseases protein mis-folding and altered protein recycling is involved that may lead to aggregate formation and associated pathologies in neurons, (b) the mis-folded or aggregated proteins may spread to neighboring cells in a prion-like manner and (c) neuroinflammation is involved in disease mechanisms. Albeit any therapeutic role of WS extracts in AD models does not necessarily imply that it could be replicated in ALS, it would be interesting to compare its effect on the pathways or pathologies that are common to both.

AD is characterized by deposition of amyloid plaques composed of fragment clusters of beta amyloid protein and neurofibrillary tangles composed of Tau protein in the brain that leads to neuronal death. The earliest manifestations are memory deficits and other cognitive impairments that become progressively severe with aging. Neuropathological studies on post-mortem brain samples from AD patients have revealed neuritic atrophy and synaptic loss as major causes for the cognitive impairment (Dickson and Vickers 2001) and drugs/compounds that promote neurite outgrowth could be in turn beneficial for AD (An et al. 2014; Chang et al. 2015). Withanamides, extracted from the seeds of WS, have been reported to prevent neuronal death triggered by amyloid plaques *in vitro* and molecular modeling studies showed that withanamides A and C uniquely bind to the active motif of beta-amyloid (Aβ 25–35) to prevent fibril formation (Jayaprakasam et al. 2010; Kumar et al. 2012). Since these withanamides are known to be permeable across the blood-brain barrier (Vareed et al. 2014), these compounds are predicted to be promising therapeutic option. Another study involving beta-amyloid induced toxicity in human neuronal cell lines reported that a methanol-chloroform extract of WS had protective effects and neuronal spine density, spine area and spine length were maintained post treatment with the extract (Kurapati et al. 2013). Similar observations were also made from primary cortical neurons of rat and human neuroblastoma cell line using withanolide A and withanoside IV (another type of withanolide). However, it was observed that withanolide A predominantly affected axonal extensions whereas withanoside IV affected dendrite growth (Kuboyama et al. 2002, 2005). *In vivo* studies involving withanoside IV treatment (at a dosage of 10 μmol/kg/day) in Aβ (25–35)-induced AD mice model showed significantly improved memory functions and overall neuroprotection. This protective effect was attributed to a metabolite of withanoside IV called sominone (Kuboyama et al. 2006). Sominone imparts its effect via interaction with a glial cell line-derived neurotrophic factor (GDNF) receptor called RET (Tohda and Joyashiki 2009) and when directly administered to 5XFAD mouse model of AD, it resulted in significant improvement in object recognition memory and enhanced axonal densities in the frontal and parietal cortex. However, sominone administration had no effect on plaque number or glial activation in that model (Joyashiki et al. 2011).

The cholinergic hypothesis in AD, whereby it is proposed that reduced synthesis of acetylcholine (a neurotransmitter) and/or loss of cholinergic neurons play a critical role in AD, is the basis of first line of pharmacotherapy for mild to moderate AD (Birks 2006). Cholinesterase inhibitors are drugs or compounds that prevent the action of the enzyme acetylcholine esterase that are present on postsynaptic membranes of cholinergic neurons. Following release from the pre synaptic terminal and signal transduction resulting from binding to its receptor on postsynaptic membrane, acetylcholine undergoes hydrolysis catalyzed by the enzyme acetylcholine esterase to form esters and choline. However, in AD, owing to decreased availability of acetylcholine, activities of cholinergic nervous systems are impaired which has been shown to be a contributing factor for AD-associated dementia (Perry et al. 1978). Two particular types of withanolides have been reported to be potent liner-mixed type or non-competitive inhibitors of cholinesterase, and having significant calcium antagonistic activity. These compounds were also shown to be safe by human neutrophil viability assay and thus could be potential therapeutic agents in

AD-associated dementia (Choudhary et al. 2005). Withanolides are predicted to bind with multiple amino acid residues of choline esterase, all falling under one or other active sites/subsites of the enzyme, thereby confirming its inhibitory property (Grover et al. 2012). Further justification of the role of WS on declining cognitive function in AD could be extrapolated from a study carried out on healthy human participants. After 14 days of treatment with WS, a significantly improved reaction time was observed psychomotor performance tests, with no sedative effects, when compared to placebo and to baseline testing (Pingali et al. 2014). Thus it is possible that WS could be beneficial to check or improve cognitive functions in people affected with AD.

Besides the direct neuroprotective effects, WS has been shown to have a peripheral effect by dint of which its credentials as a therapeutic tool for AD has been further strengthened. On treating two different mice models of AD (B6C3-Tg (APPswe, PSEN1)85Dbo/J and APPSwInd mice (J20 line)) with a methanol-chloroform extract of WS roots, accumulation of β-amyloid peptides (Aβ) and oligomers in the brains of the transgenic mice were found to be significantly cleared off along with marked improvement in behavioral deficits. This observation was shown to be related to the early-enhanced expression of low-density lipoprotein receptor-related protein (LRP) and the Aβ-degrading protease neprilysin (NEP) in the liver and sLRP, a peripheral sink for brain Aβ, in the plasma. This was followed by an increase of LRP and NEP in brain microvasculature thereby promoting plaque clearance in the brain (Sehgal et al. 2012).

Few other studies have also reported beneficial role of WS extracts in other neurodegenerative disease models that seems to be primarily based on its anti-oxidant properties. Administration of WS extracts to 1-methyl 4-phenyl 1,2,3,6-tetrahydropy ridine-induced Parkinson's disease (PD) model in mice revealed significant improvement in the behavioral outcomes and antioxidant status that correlated with reduced oxidative stress (Sankar et al. 2007) and restored catecholamine levels that are reduced in diseased state (RajaSankar et al. 2009). In rotenone-induced PD-like Drosophila and mice models, WS extracts were also shown to alleviate disease symptoms by decreasing oxidative stress (Manjunath and Muralidhara 2013, 2015). The role of alleviation of oxidative stress for amelioration of disease symptoms in PD by WS extracts was further demonstrated in Maneb and paraquat (environmental toxins) intoxicated mouse. These two toxins selectively affect dopanimergic neurons in the CNS thereby resulting in PD-like pathology. Apart from reducing oxidative stress markers, administration of ethanolic extract of WS in this model was also found to prevent neuronal apoptosis and decrease astrocyte activation, and thereby modulated neuroinflammation (Prakash et al. 2014). In 6-Hydroxydopamine-induced model of PD, pre-treatment of animals with varying doses of WS extracts for 21 days prior to disease induction showed a dose dependent reduction of oxidative stress markers and increase of catecholamines, thereby indicating at possible long-term effects of the drug (Ahmad et al. 2005). Apart from PD, oxidative stress also plays a critical role in Huntington's disease pathology and thus anti-oxidant drugs could constitute a possible therapeutic option (Stack et al. 2008; Jang et al. 2013; Gil-Mohapel et al. 2014). In a 3-nitropropionic acid-induced Huntington's-like mouse model, administration of two different doses of WS extracts increased the level of anti-oxidant enzymes and restored ATP synthesis in a dose-dependent manner (Kumar and Kumar 2009).

Apart from neurodegenerative disease models, effect on WS on several other *in vivo* and *in vitro* models have thrown light on other possible mechanisms that could ultimately be transcribed to neurodegenerative diseases. In adult mice that were subjected to restraint stress for 30 days, administration of WS extracts inhibited neuronal nitric oxide synthase, thereby attenuating the stress-induced production of nitric oxide that normal neuronal functions. WS treatment was also found to suppress corticosterone release and activate choline acetyltransferase enzyme, which in turn increase serotonin levels in hippocampus (Bhatnagar et al. 2009). When added onto brain slice preparations, WS extract reportedly binds to synaptic/extrasynaptic GABAA receptors in hippocampal CA1 pyramidal neurons and induced remarkable inward currents, thereby maintaining the neurons in a depolarized state that may contribute to neuroprotection (Bhattarai and Han 2014). The neuroprotective effect of WS extracts on hippocampal neurons has also been reported in sleep-deprived rat models (Manchanda et al. 2016). In *in vitro* studies involving human neuroblastoma and rat glioblastoma cell lines, treatment with WS extracts or specific WS components reportedly alleviated hydrogen-peroxide mediated oxidative stress, promoted survival, maintained functional state of the cells and resulted in an up regulation of differentiation markers (Shah et al. 2015).

19.3.2 **Withania somnifera** *Extracts or Constituents to Target ALS Pathogenesis*

ALS primarily differs from AD, PD or Huntington's disease by predominantly being a motor neuron disease with pathologies occurring in the spinal cord and the brain. As a result, many of the beneficial features of WS extracts or its components observed in AD and PD may not be sufficient to explain their neuroprotective features in ALS. Promoting neurite and/or dendrite outgrowth may be beneficial to some extent in alleviating ALS-associated FTD and their anti-cholinesterase property may be beneficial at presymptomatic state of fALS (Casas et al. 2013). However, the effect of WS in relation to ALS is something specific to this disease. As described earlier, TDP-43 inclusions in spinal motor neurons are hallmarks of ALS and/or FTD. Few years back, we had established that TDP-43 binds to and acts as a co-activator of the p65 subunit of nuclear factor kappa B (NF-κB) in ALS patient samples as well as in transgenic mouse model generated by using genomic fragments that encode wild-type human TDP-43 (Swarup et al. 2011b). The involvement of NF-κB in ALS pathogenesis has been increasingly gaining attention as FUS/TLS have also been reported to act as co-activator of NF-κB (Uranishi et al. 2001) and recently loss of function mutations in TANK Binding Kinase 1 (TBK1) has been reported from ALS cases (Tsai et al. 2016; Borghero et al. 2016). TBK-1 lies at the interface between autophagy and inflammation. On one hand it regulates the activation of NF-κB by virtue of its IKB kinase-like activity while on the other hand it associates with and phosphorylates both optineurin and p62, which can, in turn, enhance inflammation.

In spinal cords of sporadic ALS cases we observed a four-fold increase in the p65 mRNA level and p65 was found to predominantly localize in neuronal nucleus. The interaction between TDP-43 and p65 was not observed either in non-ALS control spinal cords or non-transgenic animals, leading us to propose that the interaction between these two proteins are involved in pathogenic process of the disease (Swarup et al. 2011b). To validate the above-mentioned findings, Withaferin A (WA) was used as an NF-κB inhibitor and interestingly when administered to transgenic mice overexpressing WT human TDP-43, the disease pathology found to be alleviated (Swarup et al. 2011b). A marked reduction was observed in glial activation pattern in the spinal cord of the WA-treated animals as visualized by reduced expressions of Mac-2 and GFAP and reduction of microglial Cox-2 levels (Swarup et al. 2011b). Moreover, WA treatment improved motor performance of transgenic mice expressing human WT TDP-43 and it enhanced the innervation of neuromuscular junctions (NMJs) (Swarup et al. 2011b). Similarly, we have found that WA treatment for 10 weeks (i.p. at 4 mg/kg/twice a week) of transgenic mice expressing human mutant TDP-43^{G348C} ameliorated motor deficits on rotarod test when compared to TDP-43^{G348C} mice injected i.p. with 0.9% saline (Fig. 19.1).

In cultured primary neurons overexpressing WT or mutant TDP-43 species, NF-κB inhibition by WA reduced their vulnerability to glutamate-toxicity (Swarup et al. 2011b). Taken together, targeting the p65 pathway seemed to be a potential therapeutic approach to ALS. To further investigate the role of WA in ALS we next administered the drug i.p. to two other mice models of fALS- SOD1^{G93A} and SOD1^{G37R} (Patel et al. 2015). The SOD1^{G93A} is an aggressive model of the disease but WA treatment resulted in a modest prolongation of median survival time by 8–9 days. In the SOD1^{G37R} model, which is comparatively less acute, the survival was increased by about 18 days (Patel et al. 2015). Along with delay in the disease progression rate, this prolongation of survival could be attributed to a significant reduction of mis-folded SOD1 species in spinal cord of the WA-treated animals. This reduction in SOD1 mis-folding by WA treatment could be related in part to increased levels of heat shock proteins (HSPs). WA is a known proteasome inhibitor and this could in turn induce ER stress response (Khan et al. 2012) and also induce autophagy (Hahm and Singh 2013). A major role of the ER stress response is recruitment of chaperons to correct protein mis-folding. In WA-treated mice, HSP70 and HSP25 (equivalent to human HSP27) were found to be significantly increased, which could explain normal protein function and overall neuroprotection. However, the protective effect of WA was only observed when administered at early disease stage and not when drug treatment was initiated at late stage (Patel et al. 2015). Recently, we tested whole root extract of WS administered orally to pre-symptomatic SOD^{G93A} mice starting at 50 days of age. Albeit the WS extract did not significantly alter survival of SOD^{G93A} mice, it did confer neuroprotection and amelioration of motor performance (*our unpublished observations*).

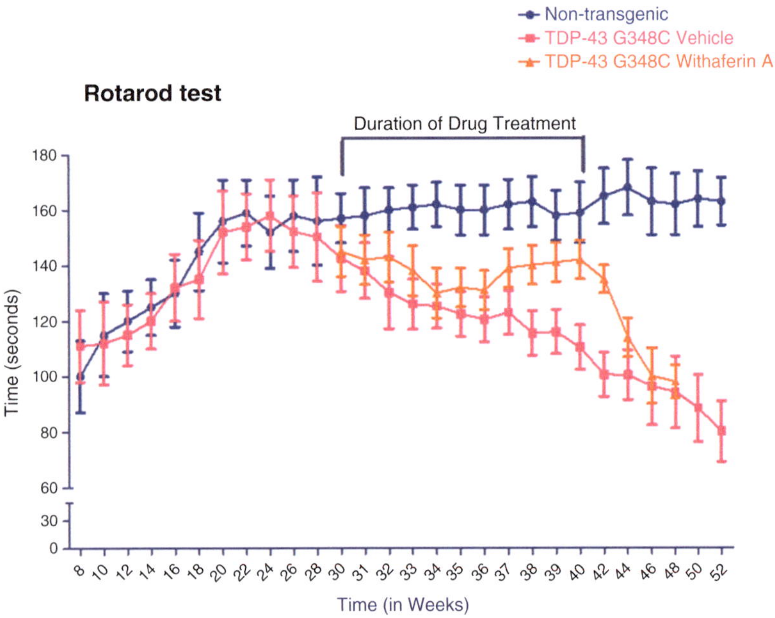

Fig. 19.1 Withaferin A ameliorated motor defects in TDP-43^{G348C} mice. Performance of TDP-43^{G348C} mice was monitored on an accelerating rotarod (4 rpm speed and 0.2 rpm/s acceleration) from 8 weeks of age onwards and compared with that of non-transgenic mice of same age. At 30 weeks, intraperitoneal Withaferin A (4 mg/kg/twice a week) or vehicle (0.9% saline) treatment was started in TDP-43^{G348C} mice that lasted for 10 weeks. Withaferin A-treated TDP-43^{G348C} mice were found to perform better on the rotarod than only vehicle-treated mice, from 6 weeks post initiation of treatment. Even after cessation of treatment at 40 weeks of age, the Withaferin A treated animals continued to perform better till 44 weeks of age. Error bars represent mean ± SEM (n = 10 each group)

Recently, we also tested the effect of oral administration of WS extract in transgenic mice expressing human mutant TDP43^{A315T}, a mouse model exhibiting motor and cognitive dysfunction with abnormal cytoplasmic TDP-43 accumulations (Swarup et al. 2011a). Our results demonstrated that 8 weeks of treatment with WS was able to ameliorate motor performance on rotarod test and cognitive function by the passive avoidance test of TDP43^{A315T} transgenic mice (Fig. 19.1). Remarkably, microscopy examination of CNS samples revealed that treatment of hTDP43^{A315T} mice with WS extract led to relocation of mis-localized cytoplasmic TDP-43 into nucleus of spinal motor neurons (Fig. 19.2) (Dutta et al. 2017).

WS and WA may confer protective or therapeutic effects through their anti-oxidant and anti-inflammatory properties. WA has been shown to induce the activation of the Keap1/Nrf2 pathway in some cells (Heyninck et al. 2016). As discussed earlier the Nrf2 pathway is an important regulator of anti-oxidant state in the body and has been found to be dysregulated in ALS and, hence it is possible that anti-oxidative role of WA in ALS could be mediated via activation of this pathway. The anti-inflammatory property WS has been attributed to multiple components but

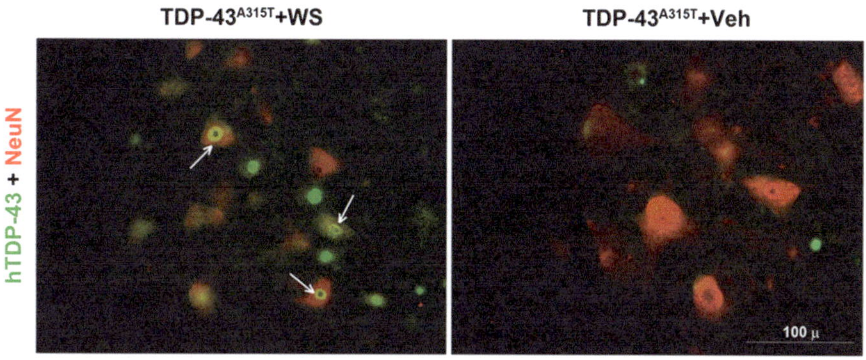

Fig. 19.2 Rescue of TDP-43 by WS extract treatment in TDP-43^{A315T} mice. TDP^{A315T} mice were treated either with *Withania somnifera* (WS) root extract (5 mg/mouse/alternate days) as a suspension in buffered saline or equal volumes of the vehicle, for 8 weeks. Post treatment, immunofluorescent staining of lumber spinal cord sections revealed that in WS-treated mice hTDP-43 was localized into the nucleus of motor neurons (white arrows), whereas in age-matched vehicle treated mice, hTDP-43 staining was present diffusedly in the cytoplasm. NeuN was used to mark neurons. Aberrant cytoplasmic presence of hTDP-43 in spinal motor neurons of our mice model is a well-characterized hallmark of the disease pathology, which is prominently observed from 8 months of age. We began our treatment when the mice were more than 14 months old and thus had mis-localized hTDP-43 in the motor neuronal cytoplasm. However, 8 weeks of WS treatment resulted in a rescue of the mis-localized hTDP-43 back into the nucleus. Image 40x; scale = 100 μ. Age of animals post 8 weeks of WS/vehicle treatment was more than 500 days

the most studied of them is WA. The NF-κB regulating role of WA has been studied in details, which could throw light on its possible inhibition of interaction with TDP-43 or suppression of inflammation in general. *In silico* studies have reported that WA interferes with the dimerization of p65 subunits of NF-κB so as to prevent their DNA-binding and thereby transcriptional activity. There is evidence that WA can interfere to dimerization of p65 by binding to residues glutamic acid and glutamine at positions 285 and 287, even though neither of these residues lay exactly in the dimerization interface (Ashkenazi et al. 2016). Another *in silico* study reported that there could be a possibility of strong intermolecular interactions of WA with IKKγ (also known as NEMO), the regulatory subunit of the IKK-kinase complex (Grover et al. 2010). The IKK-kinase complex is composed of the adapter protein IKKγ and two catalytic kinase subunits- IKKα and IKKβ. When activated IKKβ mediate the phosphorylation of the inhibitory IκB (which remains bound to inactive NF-κB, masking its nuclear localization signal) and thereby frees NF-κB. However, binding of WA with IKKγ results in hindrance to binding of IKKβ with the IKK-kinase complex and thus inhibits NF-κB activity. In addition, WA can also directly inhibit IKKβ activity by binding with a cysteine residue at position 179 (Heyninck et al. 2014). Apart from IKK-kinase complex, WA has been reported to directly alter the function of other kinases. It inhibits protein kinase C (PKC), a serine/threonine kinase (Sen et al. 2007), a kinase that can phosphorylate NF-κB at serine 311 position.

Another novel mechanism of anti-inflammation by WA could be by its inhibitory effect on High mobility group box 1 (HMGB1) protein, which is a small DNA binding protein involved in transcriptional regulation (Lee et al. 2012). In spinal cord of ALS patients high levels of HMGB1 has been reported and in transgenic mouse models of ALS, HMGB1 was found to be present in the nucleus of reactive glial cells, but not in the cytosol, thereby probably contributing to the proliferation and/or hypertrophy of these cells. However, its level in ventral motor neurons decreased with disease progression, indicating an extracellular release (Lo Coco et al. 2007). Interestingly, elevated serum autoantibody against HMGB1has been suggested to be a potent surrogate biomarker for ALS (Hwang et al. 2013). HMGB1 imparts its action by binding with the Receptor for Advanced Glycation End Products (RAGE) and concomitant to increased levels of its ligand (HMGB1) RAGE was expressed at a high level in ALS patients spinal cords (Juranek et al. 2015) and is now thought to play an important role in mediating neuroinflammation in ALS (Ray et al. 2016). WA reportedly also binds with intermediate filament proteins vimentin and glial acidic fibrillary protein (GFAP) of astrocytes. Both of these proteins are known to be elevated in ALS (Yamada et al. 1992; Swarup et al. 2011b, Patel et al. 2015). In cultured astrocytes, WA binding with these 2 intermediate filament proteins, downregulates soluble vimentin and GFAP expression to cause cell cycle G_0/G_1 arrest (Bargagna-Mohan et al. 2010) that may cause a reduction in gliosis-dependent inflammation. WA can also bind with peripherin, another intermediate filament protein (Grin et al. 2012) whose role in ALS has been discussed above.

19.4 Conclusion

Considering the involvement of multitude of different convergent or autonomous mechanisms involved in the pathogenesis of ALS, developing therapeutic approaches is a challenging prospect for this neurodegenerative disease. From the data available on the efficacy of WS extracts on various mechanisms in different models and based on our observations specifically on ALS models, we suggest that WS extracts might be considered as an experimental therapy in future human ALS clinical studies. Nonetheless, there are some drawbacks to such therapeutic approach. As with other plant extracts, it is still not clear which components of WS are responsible for what specific beneficial effects. As we have explained in this chapter, withaferin A, withanolide A, withanoside IV and withanamides seem to be the biologically active ingredients primarily responsible for the reported neuroprotective activities. However, we do not know whether the observed effects of these compounds are due to them acting alone or synergistically with other compounds present in whole extracts. Also, depending upon the huge variations in composition of WS extracts resulting from the variations in sourcing of the raw material or mechanism of extraction, a dosage with standardized composition needs to be fixed prior to administration to patients.

Thus, it can be concluded that either WS extracts or its purified components could have potential therapeutic benefits in ALS either on its own or may be as part of combination therapy with other drugs. To test this hypothesis a Phase II clinical trial has been planned using a special WS root extract and it is to be conducted by the Canadian ALS Research Network. The trial is scheduled to begin recruiting this year (2017) and will likely be executed in 10 ALS centers across Canada. Promising results obtained from this could pave the path for further trials and possibly one-day lead to the development of an effective therapeutic regime to counter ALS.

Acknowledgments This work was supported by the Canadian Institutes of Health Research. J.-P. J. holds a Canada Research Chair in neurodegeneration.

References

Ahmad M, Saleem S, Ahmad AS, Ansari MA, Yousuf S, Hoda MN, Islam F (2005) Neuroprotective effects of *Withania somnifera* on 6-hydroxydopamine induced Parkinsonism in rats. Hum Exp Toxicol 24:137–147

Al-Chalabi A, Fang F, Hanby MF, Leigh PN, Shaw CE, Ye W, Rijsdijk F (2010) An estimate of amyotrophic lateral sclerosis heritability using twin data. J Neurol Neurosurg Psychiatry 81:1324–1326

Almad AA, Doreswamy A, Gross SK, Richard JP, Huo Y, Haughey N, Maragakis NJ (2016) Connexin 43 in astrocytes contributes to motor neuron toxicity in amyotrophic lateral sclerosis. Glia 64:1154–1169

An K, Jung JH, Jeong AY, Kim HG, Jung SY, Lee K, Kim HJ, Kim SJ, Jeong TY, Son Y, Kim HS, Kim JH (2014) Neuritin can normalize neural deficits of Alzheimer/'s disease. Cell Death Dis 5:e1523

Anand KS, Wadhwa A, Garg J, Mahajan RK (2014) Amyotrophic lateral sclerosis-like presentation in a HIV-positive patient. J Int Assoc Provid AIDS Care 13:515–518

Appel SH, Zhao W, Beers DR, Henkel JS (2011) The microglial-motoneuron dialogue in ALS. Acta Myol 30:4–8

Arai T, Hasegawa M, Akiyama H, Ikeda K, Nonaka T, Mori H, Mann D, Tsuchiya K, Yoshida M, Hashizume Y, Oda T (2006) TDP-43 is a component of ubiquitin-positive tau-negative inclusions in frontotemporal lobar degeneration and amyotrophic lateral sclerosis. Biochem Biophys Res Commun 351:602–611

Ashkenazi S, Plotnikov A, Bahat A, Ben-Zeev E, Warszawski S, Dikstein R (2016) A novel allosteric mechanism of NF-kappaB dimerization and DNA binding targeted by an anti-inflammatory drug. Mol Cell Biol 36:1237–1247

Atkin JD, Farg MA, Turner BJ, Tomas D, Lysaght JA, Nunan J, Rembach A, Nagley P, Beart PM, Cheema SS, Horne MK (2006) Induction of the unfolded protein response in familial amyotrophic lateral sclerosis and association of protein-disulfide isomerase with superoxide dismutase 1. J Biol Chem 281:30152–30165

Atkin JD, Farg MA, Walker AK, Mclean C, Tomas D, Horne MK (2008) Endoplasmic reticulum stress and induction of the unfolded protein response in human sporadic amyotrophic lateral sclerosis. Neurobiol Dis 30:400–407

Bani S, Gautam M, Sheikh FA, Khan B, Satti NK, Suri KA, Qazi GN, Patwardhan B (2006) Selective Th1 up-regulating activity of *Withania somnifera* aqueous extract in an experimental system using flow cytometry. J Ethnopharmacol 107:107–115

Bannwarth S, Ait-El-Mkadem S, Chaussenot A, Genin EC, Lacas-Gervais S, Fragaki K, Berg-Alonso L, Kageyama Y, Serre V, Moore DG, Verschueren A, Rouzier C, Le Ber I, Auge G,

Cochaud C, Lespinasse F, N'guyen K, De Septenville A, Brice A, Yu-Wai-Man P, Sesaki H, Pouget J, Paquis-Flucklinger V (2014) A mitochondrial origin for frontotemporal dementia and amyotrophic lateral sclerosis through CHCHD10 involvement. Brain 137:2329–2345

Bargagna-Mohan P, Paranthan RR, Hamza A, Dimova N, Trucchi B, Srinivasan C, Elliott GI, Zhan CG, Lau DL, Zhu H, Kasahara K, Inagaki M, Cambi F, Mohan R (2010) Withaferin A targets intermediate filaments glial fibrillary acidic protein and vimentin in a model of retinal gliosis. J Biol Chem 285:7657–7669

Bhatnagar M, Sharma D, Salvi M (2009) Neuroprotective effects of *Withania somnifera* dunal.: a possible mechanism. Neurochem Res 34:1975–1983

Bhattacharya SK, Muruganandam AV (2003) Adaptogenic activity of *Withania somnifera*: an experimental study using a rat model of chronic stress. Pharmacol Biochem Behav 75:547–555

Bhattacharya SK, Goel RK, Kaur R, Ghosal S (1987) Anti-stress activity of sitoindosides VII and VIII, new acylsterylglucosides from *Withania somnifera*. Phytotherapy Res 1:32–37

Bhattacharya A, Ghosal S, Bhattacharya SK (2001) Anti-oxidant effect of *Withania somnifera* glycowithanolides in chronic footshock stress-induced perturbations of oxidative free radical scavenging enzymes and lipid peroxidation in rat frontal cortex and striatum. J Ethnopharmacol 74:1–6

Bhattarai JP, Han SK (2014) Phasic and tonic type A gamma-Aminobutryic acid receptor mediated effect of *Withania somnifera* on mice hippocampal CA1 pyramidal Neurons. J Ayurveda Integr Med 5:216–222

Birks J (2006) Cholinesterase inhibitors for Alzheimer's disease. Cochrane Database Syst Rev CD005593

Blokhuis AM, Groen EJ, Koppers M, Van Den Berg LH, Pasterkamp RJ (2013) Protein aggregation in amyotrophic lateral sclerosis. Acta Neuropathol 125:777–794

Borghero G, Pugliatti M, Marrosu F, Marrosu MG, Murru MR, Floris G, Cannas A, Occhineri P, Cau TB, Loi D, Ticca A, Traccis S, Manera U, Canosa A, Moglia C, Calvo A, Barberis M, Brunetti M, Gibbs JR, Renton AE, Errichiello E, Zoledziewska M, Mulas A, Qian Y, Din J, Pliner HA, Traynor BJ, Chiò A (2016) TBK1 is associated with ALS and ALS-FTD in Sardinian patients. Neurobiol Aging 43: 180.e1-5

Byrne S, Walsh C, Lynch C, Bede P, Elamin M, Kenna K, Mclaughlin R, Hardiman O (2011) Rate of familial amyotrophic lateral sclerosis: a systematic review and meta-analysis. J Neurol Neurosurg Psychiatry 82:623–627

Casas C, Herrando-Grabulosa M, Manzano R, Mancuso R, Osta R, Navarro X (2013) Early presymptomatic cholinergic dysfunction in a murine model of amyotrophic lateral sclerosis. Brain Behav 3:145–158

Chang PT, Talekar RS, Kung FL, Chern TR, Huang CW, Ye QQ, Yang MY, Yu CW, Lai SY, Deore RR, Lin JH, Chen CS, Chen GS, Chern JW (2015) A newly designed molecule J2326 for Alzheimer's disease disaggregates amyloid fibrils and induces neurite outgrowth. Neuropharmacology 92:146–157

Cheah BC, Vucic S, Krishnan AV, Kiernan MC (2010) Riluzole, neuroprotection and amyotrophic lateral sclerosis. Curr Med Chem 17:1942–1199

Chen L, Zhang B, Chen R, Tang L, Liu R, Yang Y, Liu X, Ye S, Zhan S, Fan D (2015) Natural history and clinical features of sporadic amyotrophic lateral sclerosis in China. J Neurol Neurosurg Psychiatry 86:1075–1081

Choudhary MI, Nawaz SA, Ul-Haq Z, Lodhi MA, Ghayur MN, Jalil S, Riaz N, Yousuf S, Malik A, Gilani AH, Ur-Rahman A (2005) Withanolides, a new class of natural cholinesterase inhibitors with calcium antagonistic properties. Biochem Biophys Res Commun 334:276–287

Clement AM, Nguyen MD, Roberts EA, Garcia ML, Boillée S, Rule M, Mcmahon AP, Doucette W, Siwek D, Ferrante RJ, Brown RH Jr, Julien JP, Goldstein LS, Cleveland DW (2003) Wild-type nonneuronal cells extend survival of SOD1 mutant motor neurons in ALS mice. Science 302:113–117

Dangoumau A, Veyrat-Durebex C, Blasco H, Praline J, Corcia P, Andres CR, Vourc'h P (2013) Protein SUMOylation, an emerging pathway in amyotrophic lateral sclerosis. Int J Neurosci 123:366–374

Davis L, Kuttan G (1998) Suppressive effect of cyclophosphamide-induced toxicity by *Withania somnifera* extract in mice. J Ethnopharmacol 62:209–214

Davis L, Kuttan G (2000) Immunomodulatory activity of *Withania somnifera*. J Ethnopharmacol 71:193–200

De Carvalho M, Swash M (2016) Lower motor neuron dysfunction in ALS. Clin Neurophysiol 127:2670–2681

Deng HX, Zhai H, Bigio EH, Yan J, Fecto F, Ajroud K, Mishra M, Ajroud-Driss S, Heller S, Sufit R, Siddique N, Mugnaini E, Siddique T (2010) FUS-immunoreactive inclusions are a common feature in sporadic and non-SOD1 familial amyotrophic lateral sclerosis. Ann Neurol 67:739–748

Dey D, Chaskar S, Athavale N, Chitre D (2014) Inhibition of LPS-induced TNF-alpha and NO production in mouse macrophage and inflammatory response in rat animal models by a novel Ayurvedic formulation, BV-9238. Phytother Res 28:1479–1485

Dickson TC, Vickers JC (2001) The morphological phenotype of beta-amyloid plaques and associated neuritic changes in Alzheimer's disease. Neuroscience 105:99–107

Dutta K, Patel P, Rahimian R, Phaneuf D, Julien J-P (2017) Withania somnifera reverses TDP-43 proteinopathy in a mouse model of ALS/FTLD. Neurotherapeutics, 14(2):447-462

Fitzmaurice PS, Shaw IC, Kleiner HE, Miller RT, Monks TJ, Lau SS, Mitchell JD, Lynch PG (1996) Evidence for DNA damage in amyotrophic lateral sclerosis. Muscle Nerve 19:797–798

Foran E, Trotti D (2009) Glutamate transporters and the excitotoxic path to motor neuron degeneration in amyotrophic lateral sclerosis. Antioxid Redox Signal 11:1587–1602

Gijselinck I, Van Mossevelde S, van der Zee J, Sieben A, Philtjens S, Heeman B, Engelborghs S, Vandenbulcke M, De Baets G, Bäumer V, Cuijt I, Van den Broeck M, Peeters K, Mattheijssens M, Rousseau F, Vandenberghe R, De Jonghe P, Cras P, De Deyn PP, Martin JJ, Cruts M, Van Broeckhoven C (2015) Loss of TBK1 is a frequent cause of frontotemporal dementia in a Belgian cohort. Neurology 85:2116–2125

Gil-Mohapel J, Brocardo PS, Christie BR (2014) The role of oxidative stress in Huntington's disease: are antioxidants good therapeutic candidates? Curr Drug Targets 15:454–468

Grad LI, Pokrishevsky E, Silverman JM, Cashman NR (2014) Exosome-dependent and independent mechanisms are involved in prion-like transmission of propagated Cu/Zn superoxide dismutase misfolding. Prion 8:331–335

Grad LI, Fernando SM, Cashman NR (2015a) From molecule to molecule and cell to cell: prion-like mechanisms in amyotrophic lateral sclerosis. Neurobiol Dis 77:257–265

Grad LI, Yerbury JJ, Turner BJ, Guest WC, Pokrishevsky E, O'neill MA, Yanai A, Silverman JM, Zeineddine R, Corcoran L, Kumita JR, Luheshi LM, Yousefi M, Coleman BM, Hill AF, Plotkin SS, Mackenzie IR, Cashman NR (2015b) Intercellular propagated misfolding of wild-type Cu/Zn superoxide dismutase occurs via exosome-dependent and -independent mechanisms. Proc Natl Acad Sci U S A 111:3620–3625

Grin B, Mahammad S, Wedig T, Cleland MM, Tsai L, Herrmann H, Goldman RD (2012) Withaferin a alters intermediate filament organization, cell shape and behavior. PLoS One 7:e39065

Grover A, Shandilya A, Punetha A, Bisaria VS, Sundar D (2010) Inhibition of the NEMO/IKKbeta association complex formation, a novel mechanism associated with the NF-kappaB activation suppression by *Withania somnifera*'s key metabolite withaferin A. BMC Genomics 11(Suppl 4):S25

Grover A, Shandilya A, Agrawal V, Bisaria VS, Sundar D (2012) Computational evidence to inhibition of human acetyl cholinesterase by withanolide a for Alzheimer treatment. J Biomol Struct Dyn 29:651–662

Guidetti D, Bondavalli M, Sabadini R, Marcello N, Vinceti M, Cavalletti S, Marbini A, Gemignani F, Colombo A, Ferrari A, Vivoli G, Solime F (1996) Epidemiological survey of amyotrophic lateral sclerosis in the province of Reggio Emilia, Italy: influence of environmental exposure to lead. Neuroepidemiology 15:301–312

Hahm ER, Singh SV (2013) Autophagy fails to alter withaferin A-mediated lethality in human breast cancer cells. Curr Cancer Drug Targets 13:640–650

Haidet-Phillips AM, Hester ME, Miranda CJ, Meyer K, Braun L, Frakes A, Song S, Likhite S, Murtha MJ, Foust KD, Rao M, Eagle A, Kammesheidt A, Christensen A, Mendell JR, Burghes AH, Kaspar BK (2011) Astrocytes from familial and sporadic ALS patients are toxic to motor neurons. Nat Biotechnol 29:824–828

He CZ, Hays AP (2004) Expression of peripherin in ubiquinated inclusions of amyotrophic lateral sclerosis. J Neurol Sci 217:47–54

Hetz C, Thielen P, Matus S, Nassif M, Court F, Kiffin R, Martinez G, Cuervo AM, Brown RH, Glimcher LH (2009) XBP-1 deficiency in the nervous system protects against amyotrophic lateral sclerosis by increasing autophagy. Genes Dev 23:2294–2306

Heyninck K, Lahtela-Kakkonen M, Van Der Veken P, Haegeman G, Vanden Berghe W (2014) Withaferin A inhibits NF-kappaB activation by targeting cysteine 179 in IKKbeta. Biochem Pharmacol 91:501–509

Heyninck K, Sabbe L, Chirumamilla CS, Szarc Vel Szic K, Vander Veken P, Lemmens KJ, Lahtela-Kakkonen M, Naulaerts S, Op De Beeck K, Laukens K, Van Camp G, Weseler AR, Bast A, Haenen GR, Haegeman G, Vanden Berghe W (2016) Withaferin A induces heme oxygenase (HO-1) expression in endothelial cells via activation of the Keap1/Nrf2 pathway. Biochem Pharmacol 109:48–61

Higgins CM, Jung C, Xu Z (2003) ALS-associated mutant SOD1G93A causes mitochondrial vacuolation by expansion of the intermembrane space and by involvement of SOD1 aggregation and peroxisomes. BMC Neurosci 4:16

Hwang CS, Liu GT, Chang MD, Liao IL, Chang HT (2013) Elevated serum autoantibody against high mobility group box 1 as a potent surrogate biomarker for amyotrophic lateral sclerosis. Neurobiol Dis 58:13–18

Jaarsma D, Haasdijk ED, Grashorn JA, Hawkins R, Van Duijn W, Verspaget HW, London J, Holstege JC (2000) Human Cu/Zn superoxide dismutase (SOD1) overexpression in mice causes mitochondrial vacuolization, axonal degeneration, and premature motoneuron death and accelerates motoneuron disease in mice expressing a familial amyotrophic lateral sclerosis mutant SOD1. Neurobiol Dis 7:623–643

Jang M, Lee MJ, Kim CS, Cho IH (2013) Korean Red Ginseng Extract Attenuates 3-Nitropropionic Acid-Induced Huntington's-Like Symptoms. Evid Based Complement Alternat Med 2013:237207

Jayaprakasam B, Zhang Y, Seeram NP, Nair MG (2003) Growth inhibition of human tumor cell lines by withanolides from *Withania somnifera* leaves. Life Sci 74:125–132

Jayaprakasam B, Padmanabhan K, Nair MG (2010) Withanamides in *Withania somnifera* fruit protect PC-12 cells from beta-amyloid responsible for Alzheimer's disease. Phytother Res 24:859–863

Johnson JO, Pioro EP, Boehringer A, Chia R, Feit H, Renton AE, Pliner HA, Abramzon Y, Marangi G, Winborn BJ, Gibbs JR, Nalls MA, Morgan S, Shoai M, Hardy J, Pittman A, Orrell RW, Malaspina A, Sidle KC, Fratta P, Harms MB, Baloh RH, Pestronk A, Weihl CC, Rogaeva E, Zinman L, Drory VE, Borghero G, Mora G, Calvo A, Rothstein JD, Drepper C, Sendtner M, Singleton AB, Taylor JP, Cookson MR, Restagno G, Sabatelli M, Bowser R, Chio A, Traynor BJ (2014) Mutations in the Matrin 3 gene cause familial amyotrophic lateral sclerosis. Nat Neurosci 17:664–666

Joyashiki E, Matsuya Y, Tohda C (2011) Sominone improves memory impairments and increases axonal density in Alzheimer's disease model mice, 5XFAD. Int J Neurosci 121:181–190

Juranek JK, Daffu GK, Wojtkiewicz J, Lacomis D, Kofler J, Schmidt AM (2015) Receptor for Advanced Glycation End Products and its Inflammatory Ligands are Upregulated in Amyotrophic Lateral Sclerosis. Front Cell Neurosci 9:485

Kahana E, Alter M, Feldman S (1976) Amyotrophic lateral sclerosis: a population study. J Neurol 212:205–213

Kawashima T, Kikuchi H, Takita M, Doh-Ura K, Ogomori K, Oda M, Iwaki T (1998) Skein-like inclusions in the neostriatum from a case of amyotrophic lateral sclerosis with dementia. Acta Neuropathol 96:541–545

Khan S, Rammeloo AW, Heikkila JJ (2012) Withaferin A induces proteasome inhibition, endoplasmic reticulum stress, the heat shock response and acquisition of thermotolerance. PLoS One 7:e50547

Khan MA, Subramaneyaan M, Arora VK, Banerjee BD, Ahmed RS (2015) Effect of *Withania somnifera* (Ashwagandha) root extract on amelioration of oxidative stress and autoantibodies production in collagen-induced arthritic rats. J Complement Integr Med 12:117–125

Kuboyama T, Tohda C, Zhao J, Nakamura N, Hattori M, Komatsu K (2002) Axon- or dendrite-predominant outgrowth induced by constituents from Ashwagandha. Neuroreport 13:1715–1720

Kuboyama T, Tohda C, Komatsu K (2005) Neuritic regeneration and synaptic reconstruction induced by withanolide A. Br J Pharmacol 144:961–971

Kuboyama T, Tohda C, Komatsu K (2006) Withanoside IV and its active metabolite, sominone, attenuate Aβ(25–35)-induced neurodegeneration. European Journal of Neuroscience 23:1417–1426

Kumar P, Kumar A (2009) Possible neuroprotective effect of *Withania somnifera* root extract against 3-nitropropionic acid-induced behavioral, biochemical, and mitochondrial dysfunction in an animal model of Huntington's disease. J Med Food 12:591–600

Kumar S, Harris RJ, Seal CJ, Okello EJ (2012) An aqueous extract of *Withania somnifera* root inhibits amyloid beta fibril formation in vitro. Phytother Res 26:113–117

Kumar G, Srivastava A, Sharma SK, Rao TD, Gupta YK (2015) Efficacy and safety evaluation of Ayurvedic treatment (Ashwagandha powder and Sidh Makardhwaj) in rheumatoid arthritis patients: a pilot prospective study. Indian J Med Res 141:100–106

Kurapati KR, Atluri VS, Samikkannu T, Nair MP (2013) Ashwagandha (*Withania somnifera*) reverses beta-amyloid1-42 induced toxicity in human neuronal cells: implications in HIV-associated neurocognitive disorders (HAND). PLoS One 8:e77624

Leblond CS, Kaneb HM, Dion PA, Rouleau GA (2014) Dissection of genetic factors associated with amyotrophic lateral sclerosis. Exp Neurol 262B:91–101

Lee W, Kim TH, Ku SK, Min KJ, Lee HS, Kwon TK, Bae JS (2012) Barrier protective effects of withaferin A in HMGB1-induced inflammatory responses in both cellular and animal models. Toxicol Appl Pharmacol 262:91–98

Li W, Lee MH, Henderson L, Tyagi R, Bachani M, Steiner J, Campanac E, Hoffman DA, Von Geldern G, Johnson K, Maric D, Morris HD, Lentz M, Pak K, Mammen A, Ostrow L, Rothstein J, Nath A (2015) Human endogenous retrovirus-K contributes to motor neuron disease. Sci Transl Med 7: 307ra153-307ra153

Ling SC, Polymenidou M, Cleveland Don W (2013) Converging mechanisms in ALS and FTD: disrupted RNA and protein homeostasis. Neuron 79:416–438

Lo Coco D, Veglianese P, Allievi E, Bendotti C (2007) Distribution and cellular localization of high mobility group box protein 1 (HMGB1) in the spinal cord of a transgenic mouse model of ALS. Neurosci Lett 412:73–77

Malhotra CL, Mehta VL, Prasad K, Das PK (1965) Studies on Withania ashwagandha, Kaul. IV. The effect of total alkaloids on the smooth muscles. Indian J Physiol Pharmacol 9:9–15

Manchanda S, Mishra R, Singh R, Kaur T, Kaur G (2016) Aqueous leaf extract of *Withania somnifera* as a potential neuroprotective agent in sleep-deprived rats: a mechanistic study. Mol Neurobiol [ePub ahead of print]

Manjaly ZR, Scott KM, Abhinav K, Wijesekera L, Ganesalingam J, Goldstein LH, Janssen A, Dougherty A, Willey E, Stanton BR, Turner MR, Ampong MA, Sakel M, Orrell RW, Howard R, Shaw CE, Leigh PN, Al-Chalabi A (2010) The sex ratio in amyotrophic lateral sclerosis: a population based study. Amyotroph Lateral Scler 11:439–442

Manjunath MJ, Muralidhara (2013) Effect of *Withania somnifera* supplementation on rotenone-induced oxidative damage in cerebellum and striatum of the male mice brain. Cent Nerv Syst Agents Med Chem 13: 43–56

Manjunath MJ, Muralidhara (2015) Standardized extract of *Withania somnifera* (Ashwagandha) markedly offsets rotenone-induced locomotor deficits, oxidative impairments and neurotoxicity in Drosophila melanogaster. J Food Sci Technol 52:1971–1981
Miller DW, Cookson MR, Dickson DW (2004) Glial cell inclusions and the pathogenesis of neurodegenerative diseases. Neuron Glia Biol 1:13–21
Mishra LC, Singh BB, Dagenais S (2000) Scientific basis for the therapeutic use of *Withania somnifera* (ashwagandha): a review. Altern Med Rev 5:334–346
Mitchell JD, Borasio GD (2007) Amyotrophic lateral sclerosis. Lancet 369:2031–2041
Nair MG, Jayaprakasam B (2007) Withanamide and withanolide compositions and method of use thereof. Google Patents
Neumann M, Sampathu DM, Kwong LK, Truax AC, Micsenyi MC, Chou TT, Bruce J, Schuck T, Grossman M, Clark CM, McCluskey LF, Miller BL, Masliah E, Mackenzie IR, Feldman H, Feiden W, Kretzschmar HA, Trojanowski JQ, Lee VM (2006) Ubiquitinated TDP-43 in frontotemporal lobar degeneration and amyotrophic lateral sclerosis. Science 314:130–133
Neumann M, Rademakers R, Roeber S, Baker M, Kretzschmar HA, Mackenzie IRA (2009) A new subtype of frontotemporal lobar degeneration with FUS pathology. Brain 132:2922–2931
Nomura T, Watanabe S, Kaneko K, Yamanaka K, Nukina N, Furukawa Y (2014) Intranuclear aggregation of mutant FUS/TLS as a molecular pathomechanism of amyotrophic lateral sclerosis. J Biol Chem 289:1192–1202
Nonaka T, Masuda-Suzukake M, Arai T, Hasegawa Y, Akatsu H, Obi T, Yoshida M, Murayama S, Mann DM, Akiyama H, Hasegawa M (2013) Prion-like properties of pathological TDP-43 aggregates from diseased brains. Cell Rep 4:124–134
O'rourke JG, Bogdanik L, Yanez A, Lall D, Wolf AJ, Muhammad AK, Ho R, Carmona S, Vit JP, Zarrow J, Kim KJ, Bell S, Harms MB, Miller TM, Dangler CA, Underhill DM, Goodridge HS, Lutz CM, Baloh RH (2016) C9orf72 is required for proper macrophage and microglial function in mice. Science 351:1324–1329
O'toole O, Traynor BJ, Brennan P, Sheehan C, Frost E, Corr B, Hardiman O (2008) Epidemiology and clinical features of amyotrophic lateral sclerosis in Ireland between 1995 and 2004. J Neurol Neurosurg Psychiatry 79:30–32
Okamoto K, Mizuno Y, Fujita Y (2008) Bunina bodies in amyotrophic lateral sclerosis. Neuropathology 28:109–115
Palliyaguru DL, Singh SV, Kensler TW (2016) *Withania somnifera*: From prevention to treatment of cancer. Mol Nutr Food Res 60:1342–1353
Patel P, Julien JP, Kriz J (2015) Early-stage treatment with Withaferin A reduces levels of misfolded superoxide dismutase 1 and extends lifespan in a mouse model of amyotrophic lateral sclerosis. Neurotherapeutics 12:217–233
Pehar M, Vargas MR, Cassina P, Barbeito AG, Beckman JS, Barbeito L (2005) Complexity of astrocyte-motor neuron interactions in amyotrophic lateral sclerosis. Neurodegener Dis 2:139–146
Perry EK, Perry RH, Blessed G, Tomlinson BE (1978) Changes in brain cholinesterases in senile dementia of alzheimer type. Neuropathol Appl Neurobiol 4:273–277
Pingali U, Pilli R, Fatima N (2014) Effect of standardized aqueous extract of *Withania somnifera* on tests of cognitive and psychomotor performance in healthy human participants. Pharmacognosy Res 6:12–18
Pokrishevsky E, Grad LI, Cashman NR (2016) TDP-43 or FUS-induced misfolded human wild-type SOD1 can propagate intercellularly in a prion-like fashion. Sci Rep 6:22155
Prakash J, Chouhan S, Yadav SK, Westfall S, Rai SN, Singh SP (2014) *Withania somnifera* alleviates parkinsonian phenotypes by inhibiting apoptotic pathways in dopaminergic neurons. Neurochem Res 39:2527–2536
Radford RA, Morsch M, Rayner SL, Cole NJ, Pountney DL, Chung RS (2015) The established and emerging roles of astrocytes and microglia in amyotrophic lateral sclerosis and frontotemporal dementia. Front Cell Neurosci 9:414

Rajasankar S, Manivasagam T, Sankar V, Prakash S, Muthusamy R, Krishnamurti A, Surendran S (2009) *Withania somnifera* root extract improves catecholamines and physiological abnormalities seen in a Parkinson's disease model mouse. J Ethnopharmacol 125:369–373

Raut AA, Rege NN, Tadvi FM, Solanki PV, Kene KR, Shirolkar SG, Pandey SN, Vaidya RA, Vaidya AB (2012) Exploratory study to evaluate tolerability, safety, and activity of Ashwagandha (*Withania somnifera*) in healthy volunteers. J Ayurveda Integr Med 3:111–114

Ray R, Juranek JK, Rai V (2016) RAGE axis in neuroinflammation, neurodegeneration and its emerging role in the pathogenesis of amyotrophic lateral sclerosis. Neurosci Biobehav Rev 62:48–55

Renton AE, Majounie E, Waite A, Simon-Sanchez J, Rollinson S, Gibbs JR, Schymick JC, Laaksovirta H, Van Swieten JC, Myllykangas L, Kalimo H, Paetau A, Abramzon Y, Remes AM, Kaganovich A, Scholz SW, Duckworth J, Ding J, Harmer DW, Hernandez DG, Johnson JO, Mok K, Ryten M, Trabzuni D, Guerreiro RJ, Orrell RW, Neal J, Murray A, Pearson J, Jansen IE, Sondervan D, Seelaar H, Blake D, Young K, Halliwell N, Callister JB, Toulson G, Richardson A, Gerhard A, Snowden J, Mann D, Neary D, Nalls MA, Peuralinna T, Jansson L, Isoviita VM, Kaivorinne AL, Holtta-Vuori M, Ikonen E, Sulkava R, Benatar M, Wuu J, Chio A, Restagno G, Borghero G, Sabatelli M, Heckerman D, Rogaeva E, Zinman L, Rothstein JD, Sendtner M, Drepper C, Eichler EE, Alkan C, Abdullaev Z, Pack SD, Dutra A, Pak E, Hardy J, Singleton A, Williams NM, Heutink P, Pickering-Brown S, Morris HR, Tienari PJ, Traynor BJ (2011) A hexanucleotide repeat expansion in C9ORF72 is the cause of chromosome 9p21-linked ALS-FTD. Neuron 72:257–268

Ringholz GM, Appel SH, Bradshaw M, Cooke NA, Mosnik DM, Schulz PE (2005) Prevalence and patterns of cognitive impairment in sporadic ALS. Neurology 65:586–590

Roman GC (1996) Neuroepidemiology of amyotrophic lateral sclerosis: clues to aetiology and pathogenesis. J Neurol Neurosurg Psychiatry 61:131–137

Rothstein JD, Martin LJ, Kuncl RW (1992) Decreased glutamate transport by the brain and spinal cord in amyotrophic lateral sclerosis. N Engl J Med 326:1464–1468

Rowland LP (2011) HIV-related neuromuscular diseases: nemaline myopathy, amyotrophic lateral sclerosis and bibrachial amyotrophic diplegia. Acta Myol 30:29–31

Sankar SR, Manivasagam T, Krishnamurti A, Ramanathan M (2007) The neuroprotective effect of *Withania somnifera* root extract in MPTP-intoxicated mice: an analysis of behavioral and biochemical variables. Cell Mol Biol Lett 12:473–481

Sarlette A, Krampfl K, Grothe C, Neuhoff N, Dengler R, Petri S (2008) Nuclear erythroid 2-related factor 2-antioxidative response element signaling pathway in motor cortex and spinal cord in amyotrophic lateral sclerosis. J Neuropathol Exp Neurol 67:1055–1062

Sasaki S, Maruyama S (1994) Immunocytochemical and ultrastructural studies of the motor cortex in amyotrophic lateral sclerosis. Acta Neuropathol 87:578–585

Sathasivam S, Grierson AJ, Shaw PJ (2005) Characterization of the caspase cascade in a cell culture model of SOD1-related familial amyotrophic lateral sclerosis: expression, activation and therapeutic effects of inhibition. Neuropathol Appl Neurobiol 31:467–485

Sehgal N, Gupta A, Valli RK, Joshi SD, Mills JT, Hamel E, Khanna P, Jain SC, Thakur SS, Ravindranath V (2012) *Withania somnifera* reverses Alzheimer's disease pathology by enhancing low-density lipoprotein receptor-related protein in liver. Proc Nat Acad Sci 109:3510–3515

Sen N, Banerjee B, Das BB, Ganguly A, Sen T, Pramanik S, Mukhopadhyay S, Majumder HK (2007) Apoptosis is induced in leishmanial cells by a novel protein kinase inhibitor withaferin A and is facilitated by apoptotic topoisomerase I-DNA complex. Cell Death Differ 14:358–367

Shah N, Singh R, Sarangi U, Saxena N, Chaudhary A, Kaur G, Kaul SC, Wadhwa R (2015) Combinations of Ashwagandha leaf extracts protect brain-derived cells against oxidative stress and induce differentiation. PLoS One 10:e0120554

Sharada AC, Solomon FE, Devi PU (1993) Toxicity of *Withania Somnifera* Root Extract in Rats and Mice. Pharm Biol 31:205–212

Sharada AC, Solomon FE, Devi PU, Udupa N, Srinivasan KK (1996) Antitumor and radiosensitizing effects of Withaferin a on Mouse Ehrlich Ascites Carcinoma in vivo. Acta Oncol 35:95–100

Shaw PJ, Forrest V, Ince PG, Richardson JP, Wastell HJ (1995a) CSF and plasma amino acid levels in motor neuron disease: elevation of CSF glutamate in a subset of patients. Neurodegeneration 4:209–216

Shaw PJ, Ince PG, Falkous G, Mantle D (1995b) Oxidative damage to protein in sporadic motor neuron disease spinal cord. Ann Neurol 38:691–695

Shibata N, Nagai R, Uchida K, Horiuchi S, Yamada S, Hirano A, Kawaguchi M, Yamamoto T, Sasaki S, Kobayashi M (2001) Morphological evidence for lipid peroxidation and protein glycoxidation in spinal cords from sporadic amyotrophic lateral sclerosis patients. Brain Res 917:97–104

Sieben A, Van Langenhove T, Engelborghs S, Martin J-J, Boon P, Cras P, De Deyn P-P, Santens P, Van Broeckhoven C, Cruts M (2012) The genetics and neuropathology of frontotemporal lobar degeneration. Acta Neuropathol 124:353–372

Silverman JM, Fernando SM, Grad LI, Hill AF, Turner BJ, Yerbury JJ, Cashman NR (2016) Disease mechanisms in ALS: misfolded SOD1 transferred through exosome-dependent and exosome-independent pathways. Cell Mol Neurobiol 36:377–381

Singh D, Aggarwal A, Maurya R, Naik S (2007) *Withania somnifera* inhibits NF-kappaB and AP-1 transcription factors in human peripheral blood and synovial fluid mononuclear cells. Phytother Res 21:905–913

Singh N, Bhalla M, De Jager P, Gilca M (2011) An overview on ashwagandha: a Rasayana (rejuvenator) of Ayurveda. Afr J Tradit Complement Altern Med 8:208–213

Song S, Miranda CJ, Braun L, Meyer K, Frakes AE, Ferraiuolo L, Likhite S, Bevan AK, Foust KD, Mcconnell MJ, Walker CM, Kaspar BK (2016) Major histocompatibility complex class I molecules protect motor neurons from astrocyte-induced toxicity in amyotrophic lateral sclerosis. Nat Med 22:397–403

Stack EC, Matson WR, Ferrante RJ (2008) Evidence of oxidant damage in Huntington's disease: translational strategies using antioxidants. Ann N Y Acad Sci 1147:79–92

Suresh Gupta M, Shivaprasad HN, Kharya MD, Rana AC (2006) Immunomodulatory activity of the ayurvedic formulation "Ashwagandha Churna". Pharm Biol 44:263–265

Swarup V, Phaneuf D, Bareil C, Robertson J, Rouleau GA, Kriz J, Julien JP (2011a) Pathological hallmarks of amyotrophic lateral sclerosis/frontotemporal lobar degeneration in transgenic mice produced with TDP-43 genomic fragments. Brain 134:2610–2626

Swarup V, Phaneuf D, Dupre N, Petri S, Strong M, Kriz J, Julien JP (2011b) Deregulation of TDP-43 in amyotrophic lateral sclerosis triggers nuclear factor kappaB-mediated pathogenic pathways. J Exp Med 208:2429–2447

Tohda C, Joyashiki E (2009) Sominone enhances neurite outgrowth and spatial memory mediated by the neurotrophic factor receptor, RET. Br J Pharmacol 157:1427–1440

Traynor BJ, Alexander M, Corr B, Frost E, Hardiman O (2003) Effect of a multidisciplinary amyotrophic lateral sclerosis (ALS) clinic on ALS survival: a population based study, 1996–2000. J Neurol Neurosurg Psychiatry 74:1258–1261

Tsai PC, Liu YC, Lin KP, Liu YT, Liao YC, Hsiao CT, Soong BW, Yip PK, Lee YC (2016) Mutational analysis of TBK1 in Taiwanese patients with amyotrophic lateral sclerosis. Neurobiol Aging 40:191.e11-6

Upton R, Graff A, Evans F, Williamson E, Sudberg S, Reich E, Bieber A, Gattu M, Spelman K, Patel NG, Shumake RL (2000) Ashwagandha root-*Withania somnifera*: analytical, quality control, and therapeutic monograph in American Herbal Pharmacopoeia and Therapeutic Compendium, ed Upton R:1–26

Uranishi H, Tetsuka T, Yamashita M, Asamitsu K, Shimizu M, Itoh M, Okamoto T (2001) Involvement of the pro-oncoprotein TLS (translocated in liposarcoma) in nuclear factor-kappa B p65-mediated transcription as a coactivator. J Biol Chem 276:13395–13401

Urushitani M, Sik A, Sakurai T, Nukina N, Takahashi R, Julien JP (2006) Chromogranin-mediated secretion of mutant superoxide dismutase proteins linked to amyotrophic lateral sclerosis. Nat Neurosci 9:108–118

Van Damme P, Dewil M, Robberecht W, Van Den Bosch L (2005) Excitotoxicity and amyotrophic lateral sclerosis. Neurodegener Dis 2:147–159

Vareed SK, Bauer AK, Nair KM, Liu Y, Jayaprakasam B, Nair MG (2014) Blood-brain barrier permeability of bioactive withanamides present in *Withania somnifera* fruit extract. Phytother Res 28:1260–1264

Verma A, Berger JR (2006) ALS syndrome in patients with HIV-1 infection. J Neurol Sci 240:59–64

Wiedemann FR, Manfredi G, Mawrin C, Beal MF, Schon EA (2002) Mitochondrial DNA and respiratory chain function in spinal cords of ALS patients. J Neurochem 80:616–625

Williams KL, Topp S, Yang S, Smith B, Fifita JA, Warraich ST, Zhang KY, Farrawell N, Vance C, Hu X, Chesi A, Leblond CS, Lee A, Rayner SL, Sundaramoorthy V, Dobson-Stone C, Molloy MP, Van Blitterswijk M, Dickson DW, Petersen RC, Graff-Radford NR, Boeve BF, Murray ME, Pottier C, Don E, Winnick C, Mccann EP, Hogan A, Daoud H, Levert A, Dion PA, Mitsui J, Ishiura H, Takahashi Y, Goto J, Kost J, Gellera C, Gkazi AS, Miller J, Stockton J, Brooks WS, Boundy K, Polak M, Munoz-Blanco JL, Esteban-Perez J, Rabano A, Hardiman O, Morrison KE, Ticozzi N, Silani V, De Belleroche J, Glass JD, Kwok JB, Guillemin GJ, Chung RS, Tsuji S, Brown RH Jr, Garcia-Redondo A, Rademakers R, Landers JE, Gitler AD, Rouleau GA, Cole NJ, Yerbury JJ, Atkin JD, Shaw CE, Nicholson GA, Blair IP (2016) CCNF mutations in amyotrophic lateral sclerosis and frontotemporal dementia. Nat Commun 7:11253

Wilson CM, Grace GM, Munoz DG, He BP, Strong MJ (2001) Cognitive impairment in sporadic ALS: a pathologic continuum underlying a multisystem disorder. Neurology 57:651–657

Yamada T, Kawamata T, Walker DG, Mcgeer PL (1992) Vimentin immunoreactivity in normal and pathological human brain tissue. Acta Neuropathol 84:157–162

Yokoi D, Atsuta N, Watanabe H, Nakamura R, Hirakawa A, Ito M, Katsuno M, Izumi Y, Morita M, Taniguchi A, Oda M, Abe K, Mizoguchi K, Kano O, Kuwabara S, Kaji R, Sobue G (2016) Age of onset differentially influences the progression of regional dysfunction in sporadic amyotrophic lateral sclerosis. J Neurol 263:1129–1136

Zeineddine R, Pundavela JF, Corcoran L, Stewart EM, Do-Ha D, Bax M, Guillemin G, Vine KL, Hatters DM, Ecroyd H, Dobson CM, Turner BJ, Ooi L, Wilson MR, Cashman NR, Yerbury JJ (2015) SOD1 protein aggregates stimulate macropinocytosis in neurons to facilitate their propagation. Mol Neurodegener 10:57

Zhao W, Beers DR, Henkel JS, Zhang W, Urushitani M, Julien JP, Appel SH (2010) Extracellular mutant SOD1 induces microglial-mediated motoneuron injury. Glia 58:231–243

ZHao W, Beers DR, Bell S, Wang J, Wen S, Baloh RH, Appel SH (2015) TDP-43 activates microglia through NF-kappaB and NLRP3 inflammasome. Exp Neurol 273:24–35

Ziauddin M, Phansalkar N, Patki P, Diwanay S, Patwardhan B (1996) Studies on the immunomodulatory effects of Ashwagandha. J Ethnopharmacol 50:69–76

Chapter 20
Neuromodulatory Role of *Withania somnifera*

Gurcharan Kaur, Taranjeet Kaur, Muskan Gupta, and Shaffi Manchanda

Abstract *Withania somnifera*, commonly known as Ashwagandha, is one of the important herbs being used in Ayurveda since times immemorial. It has been classified as a 'rasayana' herb owing to its adaptogenic and rejuvenating properties. Both root and leaf extracts of the plant have been used for the treatment of various disorders such as cancer, anxiety, inflammation and various neurological disorders due to its wide array of properties. This chapter focuses on the effects of *W. somnifera* in various CNS disorders. Various pre-clinical studies investigating the use of *W. somnifera* in modulation of neuroplasticity, anxiety, neuroinflammation and neuroprotection have been discussed in detail. A plethora of studies confirm the use of *W. somnifera* and its active phytochemicals (Withaferin-A, Withanone, Withanoside IV, Withanolide A, sitoindosides VII-X) alone or in combination as potential therapeutic agents. *W. somnifera* can be incorporated as an important dietary supplement for management of anxiety and associated cognitive and functional impairments.

Keywords *Withania somnifera* • Active phytochemicals • Combinatorial effect • Dietary supplement • Anxiety management

20.1 Ashwagandha – The Ayurvedic Boon

Ayurveda is the most ancient and the traditional medicinal system with historical roots in the Indian subcontinent (Leslie 1998). Ayurvedic therapies have been integrated in complementary and alternative medicine as preventive measures to treat wide range of chronic diseases and due to their least side effects compared to conventional drugs (Humber 2002; Patwardhan et al. 2005; Jafari et al. 2014). Moreover, ayurvedic formulations can be used in combination with other drugs without any adverse drug-drug interactions (Hopkins 2008). Modernized practices derived from traditional ayurvedic medicinal system are a type of complementary or alternative medicine (Jafari et al. 2014). High costs and increased side effects of new drugs,

G. Kaur (✉) • T. Kaur • M. Gupta • S. Manchanda
Department of Biotechnology, Guru Nanak Dev University, Amritsar 143005, Punjab, India
e-mail: kgurcharan.neuro@yahoo.com

S.C. Kaul, R. Wadhwa (eds.), *Science of Ashwagandha: Preventive and Therapeutic Potentials*, DOI 10.1007/978-3-319-59192-6_20

lack of curative treatment for several chronic diseases, and microbial resistance are some of the reasons that have contributed in directing the public interest towards complementary and alternative medicine (Humber 2002; Patwardhan et al. 2005). Herbal products are being actively used as alternative medicine because of their multi-component approach to target multiple sites for their mode of action (Bent and Ko 2004; Tachjian et al. 2010). Moreover, their formulation in single delivery system and their least side effects make them promising candidates to treat several central nervous system ailments (Borisy et al. 2003; Keith et al. 2005).

A large number of FDA approved drugs such as morphine, atropine, eugenol, reserpine etc. are derived from the plants (Prakash and Gupta 2005; reviewed by Rajendrana and Balachandarb 2013; Godara et al. 2014; Baliga et al. 2016). Several bioactive compounds have been identified till now which are being used to treat various types of abnormalities. Among these compounds, *Ginkgo biloba* for cure of dementia and cognitive impairment, withanolides and other glycolides isolated from *Withania somnifera* (WS) as anti-inflammatory, anticancer and memory enhancing compounds, curcumin isolated from *Curcuma longa* for its anti-inflammatory properties, *Lycium barbarum* for age-associated diseases are some of the compounds that have been identified to treat various ailments (Chainani-Wu 2003; Yu et al. 2005; Shah et al. 2009; Pawar et al. 2011; Gautam et al. 2013; Hashiguchi et al. 2015; Gupta and Kaur 2016; Manchanda et al. 2017).

Ashwagandha -'*Queen of Ayurveda*' is being used in traditional medicinal system. The herbal plant is widely distributed across southeast Asia and from the Mediterranean region to South Africa. In India, it is widely grown in Punjab, Rajasthan, Haryana, Mandsaur district of Madhya Pradesh, Uttar Pradesh and Gujarat (Bhatia et al. 1987; Kulkarni and Dhir 2008). The traditional use of WS is focused on its rejuvenating and life prolonging properties for various muscular, cardiac and neuronal conditions, wherein WS vitalizes and invigorates several body organs to promote general health and longevity (Singh et al. 2011). Different parts of the plant such as leaves, roots, stems, bark and even whole plant extract are being used as therapeutic interventions for the treatment of a wide array of nervous, cardiac, gastric and psychiatric conditions (Dar et al. 2015, 2016). Preclinical research and clinical studies have supported the therapeutic use of this plant for cognitive and memory-related ailments (Minhas et al. 2011; Pawar et al. 2011; Ahmed et al. 2013; Pingali et al. 2014).

20.2 Potential Benefits of Leaves of Ashwagandha

Majority of studies on Ashwagandha have explored potential beneficial effects of root-based extract or their active components rather than the leaf-based formulations. Several recent studies have explored the potential of leaves of this plant for their therapeutic value as it offers several advantages over root such as their eco-friendly and bio-friendly base. There is no need to sacrifice the plant to prepare the extract, unlike root-based preparations. Additionally, recent reports have used water-based crude formulations of leaves as compared to root based alcoholic extracts with an aim to scientifically validate the traditional use of Ashwagandha as an ayurvedic supplement. Moreover, the use of powder or water based extract is convenient, safe and easy to prepare with no need to use organic solvents for extraction procedure.

20.3 Some Pre-clinical Studies on Ashwagandha

Neuroplasticty is the ability of brain to synthesize and reorganize synaptic and neuronal connections in response to any environmental stimulus or injury. Neuroprotection is the event that leads to regeneration of damaged neurons or connections resulting in recovery of the neuronal function. Any compound that has the ability to confer neuroprotection can be classified as a neuroprotective agent. The neuroprotective and neuroplasticty inducing potential of Ashwagandha has been explored in numerous *in vitro* and *in vivo* studies. In human neuroblastoma cell line, root extract of Ashwagandha increased the percentage of the neuronal population with significant neurite outgrowth in a dose-dependent manner. The study demonstrated upregulation in the expression of synaptic and neuronal growth markers MAP2 (Microtubule-associated protein) and PSD95 (post synaptic density protein) (Tohda et al. 2000) with the Ashwagandha treatment. Further, standardized dose of withanoside IV (steroidal saponin isolated from WS root extract) improved memory in Amyloid β (25–35) injected mice model of Alzheimer's disease. Sominone was identified as one of the main metabolites of withanoside IV with significant regenerative potential. The compound was proposed to reinforce reconstruction of neuronal connections by activation of RET (receptor tyrosine kinase of glial cell line-derived neurotrophic factor) signaling and thus enhanced memory in mice (Tohda and Joyashiki 2009). Similar effects of withanoside IV were observed with injured neurons wherein it enhanced synaptogenesis and neurite outgrowth in the brain (Kuboyama et al. 2006).

Tohda and Kuboyama group (2000) investigated the effects of methanol extracts of Ashwagandha on neurite outgrowth using an *in vitro* culture system and found the neurite outgrowth-promoting activity of Ashwagandha methanolic extract in human neuroblastoma SK-N-SH cells and rat cortical neurons. Withanolide A, withanoside IV, and withanoside VI were identified as active constituents behind the neurite outgrowth promoting activity of the methanol extract (Zhao et al. 2002; Kuboyama et al. 2002). They also studied the effect of these phytochemicals on an *in vitro* axonal atrophy and synaptic degeneration model established using an active partial fragment of Aβ such as Aβ 25–35. Each of these 3 compounds induced axonal growth and synaptic densities even in the presence of Aβ 25-35 in both *in vitro* and *in vivo* model of Alzheimer's disease (Kuboyama et al. 2005, 2006). This group also investigated the effects of withanoside IV on spinal cord injured mice and found that oral administration of withanoside IV for 21 d induced axonal growth in the spinal cord and recovered hind limb motor function after 1 hr of contusion injury at the L1 spinal cord. They also found that withanoside IV gets metabolized into sominone in the intestine and is thought as an active principle for axonal growth activity (Nakayama and Tohda 2007; Tohda and Joyashiki 2009; Joyashiki et al. 2011).

A novel compound "denosomin" an analogous derivative of sominone was synthesized by Matsuya et al. (2009) and showed axonal growth activity comparable to sominone in cultured cortical neurons. Recently this group also investigated the effects of denosomin on spinal cord-injured mice and found that consecutive oral administration of denosomin for 14 d one hour after contusion injury at the L1 spinal cord facilitated axonal growth in the injured center and recovered hind limb motor function. Additionally, it was found that Vimentin-secreting GFAP-positive reactive astrocytes also increased in the injured center that helps in axonal growth and motor function recovery in spinal cord-injured mice (Teshigawara et al. 2013).

20.3.1 Ashwagandha in Neuroplasticity and Neuroprotection

A study by Konar et al. (2011) explored the potential neuroprotective role of WS in scopolamine-induced amnesia and reported that WS protects brain-derived cells against amnesia. The study revealed that WS prevented the scopolamine-induced down-regulation of expression of BDNF and GFAP (glial fibrillary acidic protein) in a dose-dependent manner. Furthermore, leaf extract of WS has been reported to effectively protect brain-derived cells against amnesia and glutamate stress by upregulating activity-regulated cytoskeletal-associated protein (Arc) (Gautam et al. 2013). Numerous neurological conditions are known to be associated with elevated levels of glutamate (excitatory neurotransmitter). In line with this, studies by Kataria et al. (2011, 2012) reported that leaf extract of Ashwagandha protects retinoic acid (RA) differentiated glioma and neuroblastoma cells against glutamate-induced excitotoxicity in a dose-dependent manner. The study demonstrated that RA differentiated cells, when exposed to glutamate, undergo loss of neural network and cell death that was accompanied by increase in the stress protein HSP70. ASH-WEX (water extract of Ashwagandha) was preventive in glutamate-induced cell death and was able to revert glutamate-induced changes in HSP70 to a large extent. Furthermore, the analyses on the neuronal plasticity marker polysialylated neural cell adhesion molecule (PSA-NCAM) and neural cell adhesion molecule (NCAM) revealed the therapeutic potential of ASH-WEX for prevention of neurodegeneration associated with glutamate-induced excitotoxicty (Kataria et al. 2012). Shah et al. (2009) investigated the anti-proliferative and differentiation potential of the alcoholic extract of Ashwagandha leaves (i-Extract), its constituents (Withaferin-A, Withanone, Withanolide A) and their combinations on glioma (C6 and YKG1) cell lines. Both extract and its active phytochemicals inhibited the proliferation and migration of glioma cells by enhancing the expression of GFAP, NCAM and Mortalin along with induction of senescence-like growth arrest and differentiation of glioma cells. Water Extract of *W. somnifera* leaves also inhibited the proliferation of human IMR-32 neuroblastoma cells by causing cell cycle arrest at G0/G1 phase and apoptosis via modulating the expression of cell cycle marker Cyclin D1, anti-apoptotic marker Bcl-xL and pAKt. It also induced the differentiation of the IMR-32 cells as indicated by morphological changes and NF200 expression. It was also accompanied by upregulated neural cell adhesion molecule expression and reduction in its polysialylation along with downregulation of MMP3 and MMP9 levels, thus elucidating its anti-metastatic activity (Kataria et al. 2013). Bioactivity-based size fractionation and NMR analysis of water extract of Ashwagandha leaves revealed the presence of triethylene glycol (TEG), Withanone as the active anticancer component (Wadhwa et al. 2013). This extract and its components also caused the selective killing of cancer cells by induction of ROS signaling (Widodo et al. 2010). These studies reported that Ashwagandha has the potential to reduce the invasiveness of brain cancer type by differentiation-based therapy.

A recent study by Kataria et al. (2016) gave the first pre-clinical evidence for tumor suppressing role of Ashwagandha in intracranial allograft of glioma cells. The study demonstrated that ASH-WEX induced the blockade of G2/M phase of cell cycle and caused the activation of multiple pro-apoptotic pathways. Oral feeding of ASH-WEX reduced the volume of intracranial tumor *in vivo* and suppressed the tumor-promoting proteins nuclear factor kappa B (NF-κB), p-Akt, vascular endothelial growth factor (VEGF), heat shock protein 70 (HSP70), PSA-NCAM, and cyclin D1 in rats. Furthermore, ASH-WEX has also been shown to prevent neurodegenerative conditions against glutamate and lead induced excitotoxicity (Kumar et al. 2014; Kataria et al. 2012).

Ashwagandha-mediated neuroprotection has been explored in various other *in vivo* model system studies. The neuroprotective role of root extract of Ashwagandha using animal model of restrain stress demonstrated the inactivation of nicotinamide adenine dinucleotide phosphate (NADPH)-d activity that was induced by stress (Bhatnagar et al. 2009). The study proposed that Ashwagandha extract mediated inhibition was not a direct effect on NADPH-d activity but by suppression of corticosterone release, and activation of choline acetyltransferase followed by a subsequent increase in serotonin levels in the hippocampus region of the brain. Therefore the mechanism of Ashwagandha mediated neuroprotection is attributed to its role in alteration of stress responsive neurotransmitters and downregulation of oxidative stress. Furthermore, Ashwagandha is known to suppress the release of glucocorticoids under chronic stress, which can be explored to treat several neurodegenerative and oxidative stress conditions. Altogether, different active components of Ashwagandha such as withanone, withanolide A, withanoside IV, VI and sominone are important therapeutic candidates for the treatment of numerous nervous system structural and behavioral abnormalities.

20.3.2 Ashwagandha: Modulator of Behavioral Plasticity

Both basic and clinical studies have explored the potential of this plant in improvement of physiological and cognitive functions of the body in wide range of neurological conditions. Ashwagandha is known for sleep inducing and memory enhancing potential among its diverse range of other biological roles. Somnifera as the species name of the plant means sleep inducing. A recent study by Manchanda et al. (2017) reported the beneficial role of Ashwagandha in preventing memory dysfunction in acute sleep deprived (SD) Wistar rats. Sleep is vital for maintenance of energy homeostasis and memory consolidation (storage of short term memory into long term memory). The study used gentle method for total sleep deprivation of rats carried out for duration of 12 hours between 6:00 AM to 6:00 PM. The study demonstrated that oral feeding of aqueous extract of Ashwagandha (ASH-WEX) for 15 consecutive days prior to SD regimen prevented memory dysfunction in rats. ASH-WEX mediates its protective role by modulation of synaptic plasticity, activation of anti-apoptosis and anti-stress pathways. Likewise, some recent reports on Parkinson's disease models have suggested that Ashwagandha root and leaf extract reverses locomotor deficits by reducing the rotenone induced oxidative and mitochondrial impairments (Prakash et al. 2013; Rajasankar et al. 2009; Manjunath and Muralidhara 2015). Oral administration of semi-purified root extract of Ashwagandha predominant in withanolides and withanosides to transgenic mice model of Alzheimer's disease has been shown to reverse behavioral deficits, and Aβ plaque pathology (Sehgal et al. 2012). The study proposed that Ashwagandha mediates its neuroprotective role by upregulation of liver LRP (lipoprotein receptor related protein) and targeted the periphery for clearance of Aβ. Enhanced expression of LRP in brain micro-vessels and the Aβ-degrading protease neprilysin (NEP) may be the attributing factors for the decreased Aβ levels in brain. The study offers a unique mechanism for the clearance of Aβ by enhancing the Aβ transport and its sequestration by plasma sLRP to reduce the AD pathology.

Methanolic extract of Ashwagandha applied in CA1 neurons of hippocampus region of brain was observed to exert a neuroprotective role in mice and may be important in process of memory (Bhattarai and Han 2014). The study reported that Ashwagandha causes activation of synaptic/extrasynaptic GABA (gamma-Aminobutyric acid) receptors type A, suggesting a link between WS mediated neuroprotection and GABA signaling. Ashwagandha has been known to promote overall health and vitality by exhibiting a pleiotropic action. A randomized study reported the use of root extract of Ashwagandha as a precognitive agent used as an adjunct with the medication in bipolar patients (Chengappa et al. 2013). The study tested patients' cognitive ability for different domains such as reaction time, social cognition and working memory and Ashwagandha extract appears to improve cognitive capacity in all three domains in bipolar patients. In addition, a clinical study by Choudhary et al. (2015) has reported the efficacy of root extract of Ashwagandha in improving cardiac endurance and physical performance in a mixed population of athletic adults, thereby improving their quality of life. Another recent clinical study has reported the efficacy of Ashwagandha root extract in improving muscle strength by exercise based evaluation of muscle strength. The study proposed that extract supplementation reduces exercise-induced muscle injury by downregulating serum creatine kinase levels (Wankhede et al. 2015).

Among the various roles of Ashwagandha, a recent clinical study has evaluated the potential of this ayurvedic plant for weight management (Choudhary et al. 2017). The study reported that root extract of Ashwagandha managed body weight in adults under chronic stress conditions by regulating their eating behavior. The study also demonstrated that weight-managing role of the root extract may be attributed to the reduced serum cortisol levels and physiological and psychological stress manifested by the Ashwagandha extract (Choudhary et al. 2017). In our study using obesity model of Wistar rats, we have reported beneficial role of leaf powder of Ashwagandha in preventing brain function impairments that were mediated by modulation of hippocampal plasticity (Manchanda and Kaur 2017). The study further demonstrated the impairment in recognition memory and locomotor coordination in obese rats, which was prevented by supplementation of the Ashwagandha leaf powder. The study also explored the molecular mechanism of these alterations and proposed that leaf powder of the plant caused upregulation of brain-derived neurotropic factor and its receptor, which were down regulated in diet-induced obesity (Manchanda and Kaur 2017). Furthermore, the activation of PI3K/Akt pathway of cell survival and plasticity was observed along with activation of immediate early genes (c-Jun and c-fos) in the obesity model that is attributable to the neuroprotective role of the leaf powder of Ashwagandha (Fig. 20.1).

20.3.3 Role of W. somnifera *in Anxiety*

Many recent studies have investigated anxiolytic properties of *W. somnifera* in laboratory settings. The plant has been reported to curb anxious behavior in both rats (Baitharu et al. 2013) and humans (Chandrasekhar et al. 2012; Khyati and Anup 2013; Pratte et al. 2014). Aqueous concentrate of roots of *W. somnifera* (commercially available from Dabur), after extraction with chloroform and spray drying, has shown anxiolytic properties in a dose-dependent manner (Bhattacharya et al. 2000). The anxiolytic property of *W. somnifera* was similar to lorazepam, a drug belonging to benzodiazepines family of drugs that have been used for the treatment of anxiety disorders. The anxiolytic effect of this extract has been attributed to the GABA-mimetic effect of glycowithanolides of the plant, which have been identified as withaferin and sitoindosides VII-X by HPTLC.

Recently, pre-supplementation of hydro-alcoholic root extract of *W. somnifera* for 30 days (300 mg/kg body weight) in rats has shown improvement in behavioral deficits caused due to middle cerebral artery occlusion (MCAO) in model of ischemic stroke (Sood et al. 2016). Ischemic stroke is a cerebrovascular event that is coupled with damage of the brain tissue due to significant deficiency of glucose and oxygen. It is caused due to reduction or complete obstruction in artery that supplies

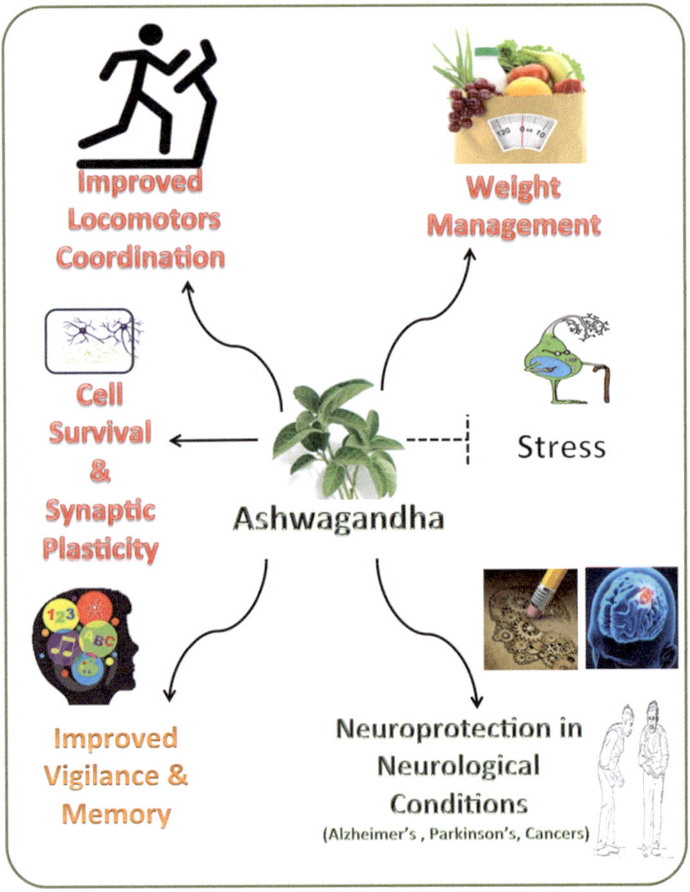

Fig. 20.1 The graphical abstract depicting pleiotropic roles of Ashwagandha in neuroplasticity mediated by modulation of multiple pathways such as synaptic plasticity, cell survival, senescence/apoptotic cell death. → indicates activation and —| indicates inhibition

blood to the brain (Rink and Khanna 2011). MCAO by intraluminal suture technique has been considered as one of the best pre-clinical models to mimic ischemic stroke in humans (Carmichael 2005). Pre-supplementation with *W. somnifera* has been shown to be instrumental in ameliorating MCAO induced anxiety-like behavior, which was correlated to the restoration of acetylcholinesterase levels after MCAO. The anxiolytic activity of *W. somnifera* has again been attributed to the presence of sitoindosides and Withaferin A in root extract of the plant, which have been reported to modulate cholinergic neurotransmission (Pingali et al. 2014). The leaf and fruit extracts of *W. somnifera* consisting of mixture of Withaferin A and sitoindosides VII-X have shown increase in cortical muscarinic acetylcholine capacity (Pingali et al. 2014).

In a recent study by Dey et al. (2016), a commercially available root extract of the plant with 2.7% (w/w) withanolides (Natural Remedies Pvt. Ltd., Bengaluru, India) has been shown to be instrumental in suppressing marble burying behavior in electric foot shock induced stress in mice. This activity has been attributed to phytochemicals other than withanolides that have been characterized in *W. somnifera* such as Withalongolide-A, an analogue of Withaferin-A (Kumar et al. 2015) and Withanamides (Mirjalili et al. 2009).

In another model of chronic stress, rats were given mild, unpredictable footshock, administered once daily for 21 days (Bhattacharya and Muruganandam 2003). Aqueous: ethanol (1:1) extract of two-year old thin roots of *W. somnifera* was further extracted with chloroform and was spray dried. The dry powder mixed in 0.3% carboxymethyl cellulose was used for the study. *W. somnifera* was administered one hour before footshock daily for 21 days. *W. somnifera* administered rats showed reduced levels of behavioral depression as evident from Porsolt's swim stress-induced behavioral despair test. Chronic stress led to significant increase in the immobility period and number of escape failures with decrease in number of avoidance response in the swin stress-induced behavioral despair test, but *W. somnifera* significantly reversed these effects in a dose-dependent manner owing to its adaptogenic properties.

Alcoholism is a major public health problem causes enormous damage to health and overall quality of life. The withdrawal of alcohol from an addict tends to cause anxiety and depression. The effect of ethanolic root extract of *W. somnifera* suspended in 0.5% (w/v) carboxy methyl cellulose has been studied in rat model of ethanol withdrawal (Gupta and Rana 2008), wherein the animals were initially provided a free access to ethanol for 15 days through liquid diet (4.5% v/v first day, 7.5% v/v second day and 9% v/v third day onwards for 15 days). From 16th day, animals were fed with ethanol-free liquid diet. Three days post-withdrawal, administration of *W. somnifera* ethanolic root reduced anxiety in the animals. The anxiolytic effect of the extract has been attributed to GABA mimetic and adaptogenic properties of the plant, which helps to downregulate $GABA_A$ receptor.

One of the anxiety related mental disorders is obsessive-compulsive disorder (OCD), which is characterized by persistent and distress causing thoughts (Eisen et al. 2004) occurring due to disturbances in serotonergic system (van Dijk et al. 2008). Chemically, treatment for OCD is the use of anti-depressants that cause inhibition of serotonin (5-hydroxytryptamine or 5-HT) reuptake (Frazier and Hensler 1994). Kaurav et al. (2012) have studied the effect of methanolic and aqueous extracts of roots of *W. somnifera* on marble burying behavior of mice that is a well-established model for OCD. The effects of both the extracts have been shown to be comparable to fluoxetine, a standard anti-OCD drug, which inhibits the uptake of 5-HT. The mode of action of both, methanolic and aqueous extracts have been proposed to be similar to that of fluoxetine. Both the extracts have been shown to facilitate serotonergic transmission at the level of post-synaptic 5-HT receptors.

Further, the anxiolytic effect of *W. somnifera* has also been documented in aquatic environment, which serves as a dump for most of the environmental pollutants. This leads to oxidative stress in aquatic animals by alteration of neurobehavioral responses. Benzo[a]pyrene (B[a]P), a prototype of polycyclic aromatic hydrocarbon, is known for causing neurotoxicity and behavioral alterations (Saunders et al. 2006). In a zebrafish model of B[a]P induced behavioral alterations, aqueous leaf extract of *W. somnifera* has shown anxiolytic effect (Mohanty et al. 2016). Zebrafish naturally display scototaxis behavior, i.e. preference for darkness (Maximino et al. 2010) and it is usually employed for the analysis of anxiety-like behavior in zebrafish. The supplementation of aqueous leaf extract of *W. somnifera* along with B[a]P for 72 hours ameliorated the anxiety-like behavior, otherwise observed in fish with B[a]P water (Mohanty et al. 2016).

Recently, we also reported the anxiolytic effect of water extract from leaves of *W. somnifera* (ASH-WEX) in acute sleep deprived female Wistar rats (Kaur et al. 2017). In the present day society, sleep deprivation due to suboptimal work schedules and unhealthy lifestyle contributes to sleep pathology, which is responsible for various neuro-psychiatric disorders. Sleep deprivation often leads to anxiety disorders and depression. Our study evaluated the effect of pre-administration of ASH-WEX on anxiety-like behavior in acute sleep deprived rats. The animals were divided into three groups – vehicle-undisturbed sleep (VUD), vehicle-sleep deprived (VSD) and ASH-WEX fed sleep deprived (WSD). VUD and VSD rats were fed with water as a vehicle and WSD rats were fed with ASH-WEX orally for 15 days and then VSD and WSD animals were sleep deprived for 12 h on 15th day. VSD rats showed anxiety-like behavior in elevated plus maze test, which has been linked to reactive gliosis and dysregulation in cytokine signalling. WSD animals showed reduction in anxiety-like behavior along with reduction in expression of pro-inflammatory cytokines TNFα, IL-1β and IL-6. ASH-WEX was also instrumental in enhancing cell survival and reducing apoptosis as witnessed from down regulation of NFκB, AP-1, cytochrome c and up regulation of Bcl-xL in WSD group.

In another independent *in vivo* study, we reported that the administration of leaf powder of *W. somnifera* was instrumental in ameliorating anxiety in high-fat diet induced obesity model. Rats were divided into four groups: Low fat diet (LFD) on regular chow, High fat diet (HFD) group on feed containing 30% fat by weight, Low fat diet plus extract (LFDE) group on regular chow supplemented with dry leaf powder of Ashwagandha – 1 mg/g body weight (ASH) and high fat diet plus extract (HFDE) group on diet containing high fat diet supplemented with ASH. The animals were kept on respective dietary regimen for 12 weeks. HFD animals showed high level of anxiety in the elevated plus maze test, which was linked to the occurrence of reactive gliosis and inflammation as marked by increase in the expression of GFAP, JNK, AP-1, Iba1, IL-1β and IL-6 as compared to LFD animals. HFDE animals showed reduction in anxiety-like behavior in elevated plus maze test along with reduced expression of GFAP, JNK, AP-1, Iba1, IL-1β and IL-6. ASH administration also led to reduction of apoptosis as witnessed from the expression of Bcl-xL.

Hence, all the above-mentioned studies scientifically validate the anxiolytic properties of *W. somnifera*, which are summarized in the graphical abstract below. The effects of *W. somnifera* have been found to be equivalent to those of the chemical drugs already being used in clinical setting. So, this plant may serve as an

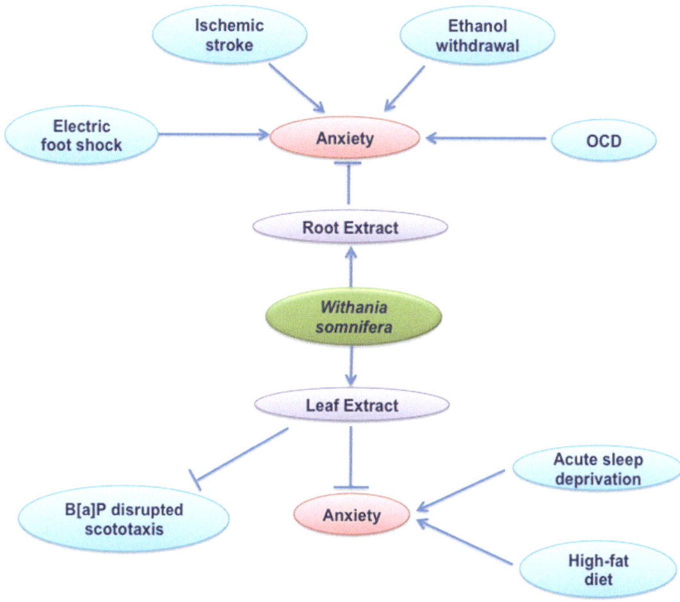

Fig. 20.2 Root and leaf extracts of *W. somnifera* possess anxiolytic properties, which has been validated under different paradigms

effective dietary supplement for management of stress and associated functional impairments (Fig. 20.2).

20.3.4 Role of W. somnifera *in Inflammation*

Different pharmacological experiments both *in vitro* and *in vivo* models have demonstrated the ability of *Withania somnifera* to exhibit anti-inflammatory, anti-oxidative, anti-microbial, anti-anxiety, immunomodulatory activities lending support to the rationale behind its traditional uses. Most of the therapeutic properties of *W. somnifera* are attributed to bioactive steroidal lactones called withanolides – a group of naturally occurring C28 steroidal lactones. Both leaves and roots are found to contain these steroidal lactones or withanolides in varied proportion.

Ashwagandha has been regarded as an adaptogen in Ayurveda for its ability to normalize physiological functions through modulating the endocrine and immune system. Many clinical and animal studies conducted to-date evident the beneficial effect of Ashwagandha and its bioactive components on immune modulation, offering promising antioxidant and anti-inflammatory responses against the neurodegenerative diseases. Anbalagan and Sadique (1981) reported that W. *somnifera* is an effective anti-inflammatory agent with higher activity as compared with hydrocortisone, a commonly prescribed anti-inflammatory drug. In another study by Anbalagan and Sadique (1984), *W. somnifera* was found to inhibit the α_2- macroglobulin, an indicator of anti-inflammatory activity in dose dependent manner in serum of rats.

Begum and Sadique (1988) reported the effect of oral administration of *Withania somnifera* root extract before the injection of inflammatory agent. Paw swelling and bony degenerative changes were seen in Freund's adjuvant induced arthritis. It was reported that this plant cause the significant reduction in both paw swelling and degenerative changes better than the reference drug, hydrocortisone. A1-hindawi et al. (1992) also found that *W. somnifera* inhibited the granuloma formation in cotton pellet implantation in rats comparable to hydrocortisone sodium succinate treatment. Sumantran et al. (2008) evaluated the effects of *W. somnifera* roots aqueous extract and glucosamine sulphate (GlcS) on the levels of nitric oxide (NO) and GAGs secreted by the knee cartilage of chronic osteoarthritis patients using validated explants model of *in vitro* cartilage damage.

Neuroinflammation is the root cause of various neurodegenerative diseases accompanied by the release of the various toxic products, free radicals, inflammatory mediators like cytokines, MMPs etc., membrane physiological disturbances, mitochondrial homeostasis dysfunctions, denaturation of proteins ultimately leading to neuronal damage. Many studies have shown that Ashwagandha exhibits multi target action against the various consequences of the neuroinflammation. It inhibited the stainless steel implant induced inflammation in adult zebrafish by reducing the inflammatory cytokine TNFα mRNA levels (Sivamani et al. 2014). The aqueous extract of *W. somnifera* roots was found to be protective against H_2O_2 and Aβ (1-42) aggregated fibril cytotoxicity in differentiated PC 12 cells (Kumar et al. 2010). It also strongly inhibited Aβ fibrils formation which is responsible for amyloid plaque formation in Alzehmier's disease in a dose dependent manner as revealed by inhibition of fibrillogenesis measured by transmission electron microscopy and thioflavin fluorescent assay (Kumar et al. 2012). Liquid chromatography- serial mass spectroscopy data revealed the presence of various withanolides including withaferin-A as the possible phytochemicals responsible for *W. somnifera* neuroprotective role (Kumar et al. 2010).

Withaferin-A from *W. somnifera* also inhibited the LPS induced expression of iNOS both at mRNA and protein levels by blocking Akt and subsequently downregulating NFκB activity in RAW 264.7 cells (Oh et al. 2008). Kaileh et al. (2007) found that withaferin-A as well as leaf extract of *Withania* inhibits NFκB activation by preventing the tumor necrosis factor induced activation of IκB Kinase β via thioalkylation – sensitive redox mechanism along with MEK1/ERK dependent Ser-181 hyperphosphorylation indicating that pure Withaferin-A and Withaferin-A enriched WS extracts can be used as novel class of NFκB inhibitors and act as promising anti-inflammatory agents for treatment of various inflammatory disorders. It also inhibited the TNFα induced intercellular adhesion molecule (ICAM)-1 and vascular cell adhesion molecule (VCAM)-1 by reducing Akt and NFκB activation in human epithelial cells (Oh and Kwon 2009). Anti-neuroinflammatory potential of withaferin-A was further supported by inhibition of LPS induced PGE_2 production and

COX-2 expression in BV-2 cells and primary microglia by reducing phosphorylation and nuclear translocation of STAT1.

Grover et al. (2010) demonstrated WA docking into active NEMO/IKKβ complex required for the formation of active IκB kinase complex, one of the essential steps for NFκB signaling pathway using semi flexible docking analysis. Results from the molecular dynamics simulations in water showed that the trajectories of the native protein and protein complexed with WA are stable for long time period of 26 ns. Withaferin-A also blocks the PAF-induced activation of NFκB as determined using reporter assays, IL8 measurements and fluorescent imaging of NFκB subunit p65 translocation in the cellular model of cystic fibrosis inflammation (Maitra et al. 2009).

Withaferin-A treatment ameliorated disease symptoms like reduction of denervated neuromuscular junctions and attenuation of neuroinflammation in TDP 43 transgenic mouse model of Amyotrophic lateral sclerosis (Swarup et al. 2011). This group further reported that early stage treatment at time of onset of the initial motor functions deficits extended the life span of ALS mice expressing either $SOD1^{G93A}$ or $SOD1^{G37R}$ mutants by reducing neuronal stress, inflammation, up regulation of Hsp25 (mouse ortholog of Hsp27) and HSP70 and decreasing misfolded SOD1 levels (Patel et al. 2015). As amplified migration and proliferation of microglial cells is the major cellular event responsible for neuroinflammation and subsequent neuron degeneration, water soluble extract of the Ashwagandha also prevents the LPS and TNFα induced increase in CCL5 (a chemoattractant for monocytes), while attenuating the increase in CCL2 expression and NFκB activation in NRK-52E cells (Grunz-Borgmann et al. 2015).

Parkinson's disease (PD) is a neurodegenerative disorder associated with impairment of balance and coordination. The midbrain and corpus striatum of PD mouse show increased levels of superoxide dismutase (Saggu et al. 1989), catalase (Chen and Fang 2005) and malondialdehyde (Sharma et al. 2008); and reduced levels of glutathione (Aoyoma et al. 2008) and glutathione peroxidase (Genc et al. 2002). Rajasankar et al. (2009) reported the effect of Ashwagandha leaf extract on the levels of these enzymes mouse model of PD. It was found that treatment with Ashwagandha leaf Extract 100 mg/kg for 7 days significantly improved these enzyme levels compared to untreated PD mouse brain. In the PD mouse, grooming, stride length, movement, rearing were found to be decreased compared to the control. In addition, narrow beam walk and foot slippery errors were increased. Treatment with Ashwagandha leaf extract improved all these physiological abnormalities. These findings suggest that *Withania somnifera* can act as potential drug in treating oxidative damage and physiological abnormalities seen in the Parkinson's disease.

Neuroprotective potential of *Withania somnifera* (WS) and its possible mechanism in MB-PQ induced mouse model of Parkinson disease has been recently reported (Prakash et al. 2013). In this study, an ethanolic root extract of *Withania* was co-treated with the MB-PQ induced mouse model of PD which was shown to significantly rescue canonical indicators of PD including compromised locomotor activity, reduced dopamine in the substantia nigra and various aspects of oxidative damage. This extract also significantly improved the MB–PQ mediated induction of a pro-apoptotic state by reducing Bax and inducing Bcl-2 protein expression, respectively. Finally, WS was found to reduce GFAP expression, pro-inflammatory marker of astrocyte activation. Altogether, this study provided evidence that WS treatment provides nigrostriatal dopaminergic neuroprotection against MB–PQ induced Parkinsonism by the modulation of oxidative stress and apoptotic machinery possibly accounting for the behavioral effects.

Further, Anti-inflammatory and anti-oxidative properties of Ashwagandha and its two withanolide constituents, namely, Withaferin-A and Withanolide A, was investigated using the murine immortalized BV-2 microglial cells (Sun et al. 2016). It was found that Ashwagandha extracts not only effectively inhibited lipopolysaccharide (LPS)-induced nitric oxide (NO) and reactive oxygen species (ROS) production in BV-2 cells, but also stimulated the nuclear factor (erythroid- derived 2)-like 2 (Nrf2) pathway, leading to induction of heme oxygenase-1 (HO-1), both in the presence and absence of LPS. Withaferin-A was found to be tenfold more effective than Withanolide A. In serum-free culture, Ashwagandha and both withanolides also suppressed LPS induced production of long thin processes (filopodia) in BV-2 microglial cells. Taken together, these results suggest an immunomodulatory role for Ashwagandha and its withanolides, and their ability to suppress oxidative and inflammatory responses in microglial cells by simultaneously down-regulating the NFκB and upregulating the Nrf2 pathways.

Our lab also investigated the anti-neuroinflammatory potential of *Withania somnifera* leaf aqueous extract (ASH-WEX) and one of its active chloroform fraction (fraction IV) in both LPS and Aβ activated murine BV-2 cell line and primary microglial cells. It was observed that ASH-WEX and FIV suppressed the morphological changes like the amoeboid morphology, expression of MHCII and microglial activation protein Iba-1 induced by LPS and β-amyloid treatment. Both the extracts were also found to control the proliferation rate of microglia as revealed by G2/M phase arrest in cell cycle analysis along with the downregulated expression of PCNA and cyclin D1. At the molecular level, ASH-WEX and FIV also inhibited the production of various pro-inflammatory cytokines TNF-α, IL-6, IL-1β, MMPs, reactive oxygen and nitrogen species etc via downregulating the expression of NFκB, AP-1 inflammatory signaling molecules as demonstrated by Sandwich ELISA based assays, immunofluorescence staining, and western blotting. Both the extracts also prevented the oxidative damage and mitochondrial dysfunction induced by LPS in both BV-2 cell line and primary microglial cells. Further, wound scratch assay showed that ASH-WEX and FIV also attenuated the migration of the microglial cells towards the site of injury. Induction of apoptosis of the activated microglia was found to be another possible mechanism behind the anti-inflammatory potential of ASH-WEX and FIV. High performance liquid chromatography data revealed the presence of Withaferin-A and Withanone in ASH-WEX and FIV as the potential phytochemicals responsible for anti-neuroinflammatory activity of Ashwagandha (Gupta and Kaur 2016) (Fig. 20.3).

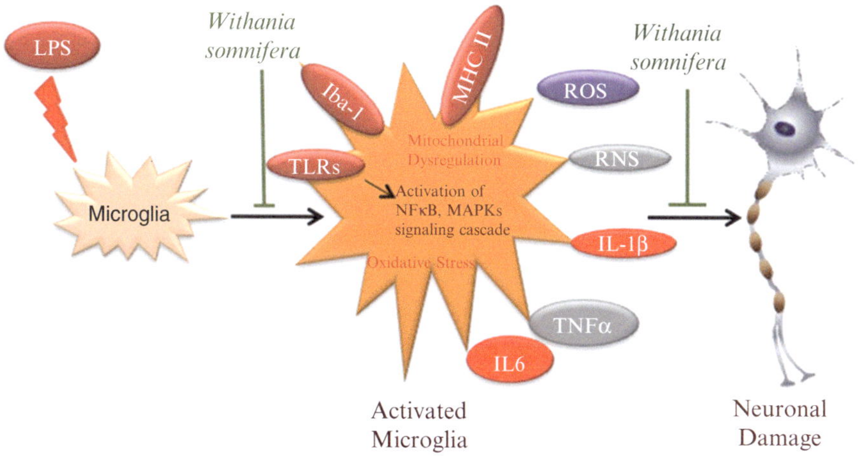

Fig. 20.3 Graphical representation of the possible inflammatory cascades or proteins involved in the anti-neuroinflammatory potential of the ASH-WEX and FIV

20.4 Conclusion

This chapter discusses the effects of *W. somnifera* in various CNS disorders. Various pre-clinical studies revealed that *W. somnifera* is a promising candidate for modulating plasticity, anxiety and neuroinflammation in various CNS afflictions that poses overall neuroprotective effects. The available scientific data supports that *W. somnifera* is a real potent rejuvenating agent due to its multiple pharmacological actions like anti-tumor, anti-stress, neuroprotective and anti-inflammatory. Therefore, it is predicted to be a useful natural drug or dietary supplement for interventions of brain pathologies including Parkinson's disease, memory loss, stress-induced disorders, tumors and others.

References

Ahmed ME, Javed H, Khan MM, Vaibhav K, Ahmad A, Khan A, Tabassum R, Islam F, Safhi MM, Islam F (2013) Attenuation of oxidative damageassociated cognitive decline by Withania somnifera in rat model of streptozotocin-induced cognitive impairment. Protoplasma 250:1067–1078

Al-Hindawi MK, Al-Khafaji SH, Abdul-Nabi MH (1992) Antigranuloma activity of Iraqi *Withania somnifera*. J Ethnopharmacol 37:113–116

Anbalagan K, Sadique J (1981) Influence of an Indian medicine (Ashwagandha) on acutephase reactants in inflammation. Indian J Exp Biol 19:245–249

Anbalagan K, Sadique J (1984) Role of prostaglandins in acute phase proteins in inflammation. Biochem Med 31:236–245

Aoyama K, Matsumura N, Watabe M, Nakaki T (2008) Oxidative stress on EAAC1 is involved in MPTP-induced glutathione depletion and motor dysfunction. Eur J Neurosci 27:20–30

Baitharu I, Jain V, Deep SN, Hota KB, Hota SK, Prasad D, Ilavazhagan G (2013) *Withania somnifera* root extract ameliorates hypobaric hypoxia induced memory impairment in rats. J Ethnopharmacol 145:431–441

Baliga MS, Rao S, Rai MP, D'souza P (2016) Radio protective effects of the Ayurvedic medicinal plant Ocimum sanctum Linn. (Holy Basil): a memoir. J Cancer Res Ther 12:20

Begum VH, Sadique J (1988) Long-term effect of herbal drug *Withania somnifera* on adjuvant-induced arthritis in rats. Indian J Exp Biol 26:877–882

Bent S, Ko R (2004) Commonly used herbal medicines in the United States: a review. Am J Med 116(7):478–485

Bhatnagar M, Sharma D, Salvi M (2009) Neuroprotective effects of *Withania somnifera* dunal.: a possible mechanism. Neurochem Res 34:1975–1983

Bhattacharya SK, Bhattacharya A, Sairam K, Ghosal S (2000) Anxiolytic-antidepressant activity of *Withania somnifera* glycowithanolides: an experimental study. Phytomedicine 7:463–469

Bhattacharya SK, Muruganandam AV (2003) Adaptogenic activity of *Withania somnifera*: an experimental study using a rat model of chronic stress. Pharmacol Biochem Behav 75:547–555

Bhattarai JP, Han SK (2014) Phasic and tonic type A γ-Aminobutryic acid receptor mediated effect of *Withania somnifera* on mice hippocampal CA1 pyramidal Neurons. J Ayurveda Integr Med 5:216–222

Bhatia P, Rattan S, Cavallius J, Clark BFC (1987) Withania somnifera (Ashawagandha), a so-called rejuvenator, inhibits growth and macromolecular synthesis on human cells. Med Sci Res 15:515–516

Borisy AA, Elliott PJ, Hurst NW, Lee MS, Lehar J, Price ER, Serbedzija G, Zimmermann GR, Foley MA, Stockwell BR, Keith CT (2003) Systematic discovery of multicomponent therapeutics. Proc Natl Acad Sci U S A 100:7977–7982

Carmichael ST (2005) Rodent models of focal stroke: size, mechanism, and purpose. NeuroRx 2:396–409

Chainani-Wu N (2003) Safety and anti-inflammatory activity of curcumin: a component of tumeric (Curcuma longa). J Altern Complement Med 9:161–168

Chandrasekhar K, Kapoor J, Anishetty S (2012) A prospective, randomized double-blind, placebo-controlled study of safety and efficacy of a high-concentration full-spectrum extract of ashwagandha root in reducing stress and anxiety in adults. Indian J Psychol Med 34:255–262

Chen Y, Fang F (2005) Ginsenoside Rg1 reduces MPTP-induced substantia nigra neuron loss by suppressing oxidative stress. Acta Pharmacol Sin 26:56–62

Chengappa KR, Bowie CR, Schlicht PJ, Fleet D, Brar JS, Jindal R (2013) Original research randomized placebo-controlled adjunctive study of an extract of *Withania somnifera* for cognitive dysfunction in bipolar disorder. J Clin Psychiatry 74:1076–1083

Choudhary B, Shetty A, Langade DG (2015) Efficacy of Ashwagandha (*Withania somnifera* [L.] Dunal) in improving cardiorespiratory endurance in healthy athletic adults. Ayu 36:63–68

Choudhary D, Bhattacharyya S, Joshi K (2017) Body weight management in adults under chronic stress through treatment with Ashwagandha root extract: a double-blind, randomized, placebo-controlled trial. J Evid Based Complement Altern Med 22:96–106

Dar NJ, Hamid A, Ahmad A (2015) Pharmacologic overview of Withania somnifera, the Indian Ginseng. Cell Mol Life Sci 72:4445–4460

Dar AP, Singh RL, Kamal AM, Dar AT (2016) Unique medicinal properties of Withania somnifera: Phytochemical constituents and protein component. Curr Pharm Des 22:535–540

Dey A, Chatterjee SS, Kumar V (2016) Low dose effects of a *Withania somnifera* extract on altered marble burying behavior in stressed mice. J Intercult Ethnopharmacol 5:274–277

Eisen JL, Phillips KA, Coles ME, Rasmussen SA (2004) Insight in obsessive-compulsive disorder and body dysmorphic disorder. Compr Psychiatry 45:10–15

Frazier A, Hensler JG (1994) Serotonin. In: Siegal JG (eds) Basic neurochemistry: molecular cellular and medical aspects, 5th edn. Raven Press, New York, pp 283–308

Gautam A, Wadhwa R, Thakur MK (2013) Involvement of hippocampal Arc in amnesia and its recovery by alcoholic extract of Ashwagandha leaves. Neurobiol Learn Mem 106:177–184

Genc S, Akhisaroglu M, Kuralay F, Genc K (2002) Erythropoietin restores glutathione peroxidase activity in 1-methyl-4-phenyl-1, 2, 5, 6-tetrahydropyridine-induced neurotoxicity in C57BL mice and stimulates murine astroglial glutathione peroxidase production *in vitro*. Neurosci Lett 321:73–76

Godara R, Katoch M, Katoch R, Yadav A, Parveen S, Vij B, Khajuria V, Singh G, Singh NK (2014) In vitro acaricidal activity of and its components, scopolamine and atropine, against Rhipicephalus (Boophilus) microplus. Sci World J 2014:1–6

Grover A, Shandilya A, Punetha A, Bisaria VS, Sundar D (2010) Inhibition of the NEMO/IKKβ association complex formation, a novel mechanism associated with the NF-κB activation suppression by *Withania somnifera*'s key metabolite withaferin A. BMC Genomics 11:S25

Grunz-Borgmann E, Mossine V, Fritsche K, Parrish AR (2015) Ashwagandha attenuates TNF-α- and LPS-induced NF-κB activation and CCL2 and CCL5 gene expression in NRK-52E cells. BMC Complement Altern Med 15:434

Gupta GL, Rana AC (2008) Effect of *Withania somnifera* Dunal in ethanol-induced anxiolysis and withdrawal anxiety in rats. Indian J Exp Biol 46:470–475

Gupta M, Kaur G (2016) Aqueous extract from the *Withania somnifera* leaves as a potential anti-neuroinflammatory agent: a mechanistic study. J Neuroinflammation 13:193

Hashiguchi M, Ohta Y, Shimizu M, Maruyama J, Mochizuki M (2015) Meta-analysis of the efficacy and safety of Ginkgo biloba extract for the treatment of dementia. J Pharm Health Care Sci 1:14

Hopkins AL (2008) Network pharmacology: the next paradigm in drug discovery. Nat Chem Biol 4:682–690

Humber JM (2002) The role of complementary and alternative medicine. JAMA 288:1655–1656

Jafari S, Abdollahi M, Saeidnia S (2014) Personalized medicine: a confluence of traditional and contemporary medicine. Altern Ther Health Med 20:31–40

Joyashiki E, Matsuya Y, Tohda C (2011) Sominone improves memory impairments and increases axonal density in Alzheimer's disease model mice, 5XFAD. Int J Neurosci 121:181–190

Kaileh M, Berghe WV, Heyerick A, Horion J, Piette J, Libert C, De Keukeleire D, Essawi T, Haegeman G (2007) Withaferin a strongly elicits IÂ B Kinase beta hyperphosphorylation concomitant with potent inhibition of its Kinase activity. J Biol Chem 282:4253–4264

Kataria H, Shah N, Kaul SC, Wadhwa R, Kaur G (2011) Water extract of ashwagandha leaves limits proliferation and migration, and induces differentiation in glioma cells. Evid Based Complement Alternat Med 2011:267614

Kataria H, Wadhwa R, Kaul SC, Kaur G (2012) Water extract from the leaves of *Withania somnifera* protect RA differentiated C6 and IMR-32 cells against glutamate-induced excitotoxicity. PLoS One 7:e37080

Kataria H, Wadhwa R, Kaul SC, Kaur G (2013) *Withania somnifera* water extract as a potential candidate for differentiation based therapy of human neuroblastomas. PLoS One 8:e55316

Kataria H, Kumar S, Chaudhary H, Kaur G (2016) *Withania somnifera* suppresses tumor growth of intracranial allograft of glioma cells. Mol Neurobiol 53:4143–4158

Kaur T, Singh H, Mishra R, Manchanda S, Gupta M, Saini V, Sharma A, Kaur G (2017) *Withania somnifera* as a potential anxiolytic and immunomodulatory agent in acute sleep deprived female Wistar rats. Mol Cell Biochem 427:91–101

Kaurav BP, Wanjari MM, Chandekar A, Chauhan NS, Upmanyu N (2012) Influence of *Withania somnifera* on obsessive compulsive disorder in mice. Asian Pac J Trop Med 5:380–384

Keith CT, Borisy AA, Stockwell BR (2005) Innovation: Multicomponent therapeutics for networked systems. Nat Rev Drug Discov 4:71–78

Konar A, Shah N, Singh R, Saxena N, Kaul SC, Wadhwa R, Thakur MK (2011) Protective role of Ashwagandha leaf extract and its component withanone on scopolamine-induced changes in the brain and brain-derived cells. PLoS One 6:e27265

Kuboyama T, Tohda C, Zhao J, Nakamura N, Hattori M, Komatsu K (2002) Axon-or dendrite-predominant outgrowth induced by constituents from Ashwagandha. Neuroreport 13:1715–1720

Kuboyama T, Tohda C, Komatsu K (2005) Neuritic regeneration and synaptic reconstruction induced by withanolide A. Br J Pharmacol 144:961–971

Kuboyama T, Tohda C, Komatsu K (2006) Withanoside IV and its active metabolite, sominone, attenuate Ab (25-35)-induced neurodegeneration. Eur J Neurosci 23:1417–1426

Kulkarni SK, Dhir A (2008) Withania somnifera: an Indian Ginseng. Prog Neuro-Psychopharmacol Biol Psychiatry 32:1093–1105

Kumar P, Singh R, Nazmi A, Lakhanpal D, Kataria H, Kaur G (2014) Glioprotective effects of Ashwagandha leaf extract against lead induced toxicity. Biomed Res Int 2014:182029

Kumar S, Seal CJ, Howes MJR, Kite GC, Okello EJ (2010) *In vitro* protective effects of *Withania somnifera* (L.) dunal root extract against hydrogen peroxide and β-amyloid (1–42) induced cytotoxicity in differentiated PC12 cells. Phytother Res 24:1567–1574

Kumar S, Harris RJ, Seal CJ, Okello EJ (2012) An aqueous extract of *Withania somnifera* root inhibits amyloid β fibril formation *in vitro*. Phytother Res 26:113–117

Kumar V, Dey A, Hadimani MB, Emeraldjavascript M (2015) Chemistry and pharmacology of *Withania somnifera*: an update. TANG 5:1–1

Leslie CM (1998) Asian medical systems: A comparative study, vol 3. Motilal Banarsidass Publisher, Delhi

Maitra R, Porter MA, Huang S, Gilmour BP (2009) Inhibition of NFκB by the natural product Withaferin A in cellular models of Cystic Fibrosis inflammation. J Inflamm 6:15

Manchanda S, Mishra R, Singh R, Kaur T, Kaur G (2017) Aqueous leaf extract of *Withania somnifera* as a potential neuroprotective agent in sleep-deprived rats: a mechanistic study. Mol Neurobiol 54:3050–3061

Manchanda S, Kaur G (2017) *Withania somnifera* leaf alleviates cognitive dysfunction by modulting hippocampal plasticity in high fat diet induced obesity model. BMC Complement Altern Med 17:136

Manjunath MJ, Muralidhara (2015) Standardized extract of *Withania somnifera* (Ashwagandha) markedly offsets rotenone-induced locomotor deficits, oxidative impairments and neurotoxicity in Drosophila melanogaster. J Food Sci Technol 52:1971–1981

Matsuya Y, Yamakawa YI, Tohda C, Teshigawara K, Yamada M, Nemoto H (2009) Synthesis of sominone and its derivatives based on an RCM strategy: discovery of a novel anti-Alzheimer's disease medicine candidate "denosomin". Org lett 11:3970–3973

Maximino C, De Brito TM, de Mattos Dias CAG, Gouveia A, Morato S (2010) Scototaxis as anxiety-like behavior in fish. Nat Protoc 5:209–216

Minhas U, Minz R, Bhatnagar A (2011) Prophylactic effect of Withania somnifera on inflammation in a non-autoimmune prone murine model of lupus. Drug Discov Ther 5:195–201

Mirjalili MH, Moyano E, Bonfill M, Cusido RM, Palazon J (2009) Steroidal lactones from *Withania somnifera*, an ancient plant for novel medicine. Molecules 14:2373–2393

Mohanty R, Das SK, Singh NR, Patri M (2016) *Withania somnifera* leaf extract ameliorates benzo [a] pyrene-induced behavioral and neuromorphological alterations by improving brain antioxidant status in zebrafish (*Danio rerio*). Zebrafish 13:188–196

Nakayama N, Tohda C (2007) Withanoside IV improves hindlimb function by facilitating axonal growth and increase in peripheral nervous system myelin level after spinal cord injury. Neurosci Res 58:176–182

Oh JH, Lee TJ, Park JW, Kwon TK (2008) Withaferin A inhibits iNOS expression and nitric oxide production by Akt inactivation and down-regulating LPS-induced activity of NF-κB in RAW 264.7 cells. Euro J Pharmacol 599:11–17

Oh JH, Kwon TK (2009) Withaferin A inhibits tumor necrosis factor-α-induced expression of cell adhesion molecules by inactivation of Akt and NF-κB in human pulmonary epithelial cells. Int Immunopharmacol 9:614–619

Patel P, Julien JP, Kriz J (2015) Early-stage treatment with Withaferin A reduces levels of misfolded superoxide dismutase 1 and extends lifespan in a mouse model of amyotrophic lateral sclerosis. Neurotherapeutics 12:217–233

Patwardhan B, Warude D, Pushpangadan P, Bhatt N (2005) Ayurveda and traditional Chinese medicine: a comparative overview. Evid Based Complement Alternat Med 2:465–473

Pawar P, Gilda S, Sharma S, Jagtap S, Paradkar A, Mahadik K, Ranjekar P, Harsulkar A (2011) Rectal gel application of Withania somnifera root extract expounds anti-inflammatory and muco-restorative activity in TNBS-induced inflammatory bowel disease. BMC Complement Altern Med 11:34

Pingali U, Pilli R, Fatima N (2014) Effect of standardized aqueous extract of *Withania somnifera* on tests of cognitive and psychomotor performance in healthy human participants. Pharmacognosy Res 6:12–18

Prakash P, Gupta N (2005) Therapeutic uses of Ocimum sanctum Linn (Tulsi) with a note on eugenol and its pharmacological actions: a short review. Indian J Physiol Pharmacol 49:125

Prakash J, Yadav SK, Chouhan S, Singh SP (2013) Neuroprotective role of *Withania somnifera* root extract in maneb–paraquat induced mouse model of parkinsonism. Neurochem Res 38:972–980

Pratte MA, Nanavati KB, Young V, Morley CP (2014) An alternative treatment for anxiety: a systematic review of human trial results reported for the Ayurvedic herb ashwagandha (*Withania somnifera*). J Altern Complement Med 20:901–908

RajaSankar S, Manivasagam T, Surendran S (2009) Ashwagandha agent in treating oxidative damage and physiological abnormalities leaf extract: a potential seen in a mouse model of Parkinson's disease. Neurosci Lett 454:11–15

Rajendrana R, Balachandarb S (2013) Review on Ocimum Tenuiflorum-a potential herb for drinking water disinfection. South Asian J Biol Sci 3:1–10

Rink C, Khanna S (2011) Significance of brain tissue oxygenation and the arachidonic acid cascade in stroke. Antioxid Redox Signal 14:1889–1903

Saggu H, Cooksey J, Dexter DA, Wells FR, Lees A, Jenner P, Marsden CD (1989) A selective increase in particulate superoxide dismutase activity in parkinsonian substantia nigra. J Neurochem 53:692–697

Saunders CR, Das SK, Ramesh A, Shockley DC, Mukherjee S (2006) Benzo (a) pyrene-induced acute neurotoxicity in the F-344 rat: role of oxidative stress. J Appl Toxicol 26:427–438

Sehgal N, Gupta A, Valli RK, Joshi SD, Mills JT, Hamel E, Khanna P, Jain SC, Thakur SS, Ravindranath V (2012) *Withania somnifera* reverses Alzheimer's disease pathology by enhancing low-density lipoprotein receptor-related protein in liver. Proc Natl Acad Sci U S A 109:3510–3515

Shah N, Kataria H, Kaul SC, Ishii T, Kaur G, Wadhwa R (2009) Effect of the alcoholic extract of Ashwagandha leaves and its components on proliferation, migration, and differentiation of glioblastoma cells: combinational approach for enhanced differentiation. Cancer Sci 100:1740–1747

Sharma A, Kaur P, Kumar B, Prabhakar S, Gill KD (2008) Plasma lipid peroxidation and antioxidant status of Parkinson's disease patients in the Indian population. Parkinsonism Relat Disord 14:52–57

Singh N, Bhalla M, Jager PD, Gilca M (2011) An overview on Ashwagandha: a Rasayana (Rejuvenator) of Ayurveda. Afr J Tradit Complement Altern Med 8:208–213

Sivamani S, Joseph B, Kar B (2014) Anti-inflammatory activity of *Withania somnifera* leaf extract in stainless steel implant induced inflammation in adult zebrafish. J Genetic Eng Biotechnol 12:1–6

Sood A, Kumar A, Dhawan DK, Sandhir R (2016) Propensity of *Withania somnifera* to attenuate behavioural, biochemical, and histological alterations in experimental model of stroke. Cell Mol Neurobiol 36:1123–1138

Khyati SS, Anup BT (2013) A randomized double blind placebo controlled study of ashwagandha on generalized anxiety disorder. Int Ayurvedic Med J 1:1–7

Sumantran VN, Chandwaskar R, Boddul S, Patwardhan B, Chopra A, Wagh UV (2008) The relationship between chondroprotective and anti-inflammatory effects of *Withania somnifera* root and glucosamine sulphate on human osteoarthritic cartilage in vitro. Phytother Res 22:1342–1348

Sun GY, Li R, Cui J, Hannink M, Gu Z, Fritsche KL, Lubahn DB, Simonyi A (2016) *Withania somnifera* and its withanolides attenuate oxidative and inflammatory responses and up-regulate antioxidant responses in BV-2 microglial cells. Neuromolecular Med 18:241–252

Swarup V, Phaneuf D, Dupre N, Petri S, Strong M, Kriz J, Julien JP (2011) Deregulation of TDP-43 in amyotrophic lateral sclerosis triggers nuclear factor κB–mediated pathogenic pathways. J Exp Med 208:2429–2447

Tachjian A, Maria V, Jahangir A (2010) Use of herbal products and potential interactions in patients with cardiovascular diseases. J Am Coll Cardiol 55:515–525

Teshigawara K, Kuboyama T, Shigyo M, Nagata A, Sugimoto K, Matsuya Y, Tohda C (2013) A novel compound, denosomin, ameliorates spinal cord injury via axonal growth associated with astrocyte-secreted vimentin. Br J Pharmacol 168:903–919

Tohda C, Kuboyama T, Komatsu K (2000) Dendrite extension by methanol extract of Ashwagandha (roots of *Withania somnifera*) in SK-N-SH cells. Neuroreport 11:1981–1985

Tohda C, Joyashiki E (2009) Sominone enhances neurite outgrowth and spatial memory mediated by the neurotrophic factor receptor, RET. Br J Pharmacol 157:1427–1440

van Dijk A, Klompmakers A, Denys D (2008) Role of serotonin in obsessive–compulsive disorder. Fut Neurol 3:589–603

Wadhwa R, Singh R, Gao R, Shah N, Widodo N, Nakamoto T, Kaul SC (2013) Water extract of ashwagandha leaves has anticancer activity: Identification of an active component and its mechanism of action. PLoS One 8:e77189

Wankhede S, Langade D, Joshi K, Sinha SR, Bhattacharyya S (2015) Examining the effect of *Withania somnifera* supplementation on muscle strength and recovery: a randomized controlled trial. J Int Soc Sports Nutr 12:43

Widodo N, Priyandoko D, Shah N, Wadhwa R, Kaul SC (2010) Selective killing of cancer cells by Ashwagandha leaf extract and its component Withanone involves ROS signaling. PLoS One 5:e13536

Yu MS, Leung SKY, Lai SW, Che CM, Zee SY, So KF, Yuen WH, Chang RCC (2005) Neuroprotective effects of anti-aging oriental medicine Lycium barbarum against B-amyloid peptide neurotoxicity. Exp Gerontol 40:716–727

Zhao J, Nakamura N, Hattori M, Kuboyama T, Tohda C, Komatsu K (2002) Withanolide derivatives from the roots of *Withania somnifera* and their neurite outgrowth activities. Chem Pharm bull 50:760–765

Chapter 21
Therapeutic Effects of Ashwagandha in Brain Aging and Clock Dysfunction

Anita Jagota and K. Kowshik

Abstract The master pacemaker, the suprachiasmatic nucleus (SCN), containing light-entrained circadian clock orchestring mammalian circadian rhythms in physiology and behavior is located in hypothalamus. SCN possessing core circadian machinery genes regulates the synthesis and release of melatonin (messanger of darkness) from the pineal gland via multisynaptic pathway. The decline in endogenous levels of this multitasking molecule with aging is associated with circadian dysfunction, neurodegeneration and brain aging as well as alterations in the endogenous defense and survival mechanisms. The age-related neurodegenerative disorders in the elderly have increased dramatically parallel to increase in longevity limiting quality of life. This is linked to need for development of effective therapeutic agents for healthy aging. Herbal extracts or formulations have been in use since ancient cultures involving herbal medicines which are of greater scientific interest as have been demonstrated to be capable of treating disease and improving health often without any significant side effects. Ashwagandha, *Withania somnifera* (WS), has been reported to have biologically active constituents such as alkaloids, steroidal compounds, glycowithanolides etc. having anti-oxidative, anti-aging and anti-inflammatory properties. The studies in our laboratory have revealed the therapeutic and adaptogenic potential of hydroalcoholic leaf extract of WS on age induced alterations in various clock gene expression and its modulators. This chapter is aimed to summarize the current knowledge on the changes of the circadian system in advanced age and the therapeutic effects of WS on brain aging and clock dysfunctions.

Keywords Antioxidant • *Withania somnifera* • Aging • Circadian rhythms • Suprachiasmatic nucleus

---

A. Jagota (✉) • K. Kowshik
Neurobiology and Molecular Chronobiology laboratory, Department of Animal Biology, School of Life Sciences, University of Hyderabad, Hyderabad 500 046, India
e-mail: ajsl@uohyd.ernet.in; anita_jagota@yahoo.com

S.C. Kaul, R. Wadhwa (eds.), *Science of Ashwagandha: Preventive and Therapeutic Potentials*, DOI 10.1007/978-3-319-59192-6_21

21.1 Biological Clocks: Time Keeping System

The organisms possess endogenous physiological timing mechanisms termed biological clocks that rhythmically modulate the functioning of cells, tissues and organs. The clocks are autonomous and provide the organism an intrinsic temporal organization, a timed pattern of change (rhythm) in physiology or behavior that is brought in phase by the process called as entrainment through environmental cues (zeitgebers); photic i.e. light-dark cycle and nonphotic such as cycles of temperature, sound, humidity, food availability, social interaction etc. (Quintero et al. 2003; Jagota 2012).

In mammals, 'master circadian pacemaker' or 'master clocks' are localized in discrete sites in central nervous system (CNS) to the bilaterally paired suprachiasmatic nucleus (SCN) in hypothalamus (Jagota et al. 2000; Reiter et al. 2010; Welsh et al. 2010). This system controls a wide variety of physiological functions including sleep-wake cycles, body temperature rhythms, secretion of hormones, locomotor activities and feeding behavior (Jagota et al. 2000; Partch et al. 2014).

The time keeping system has a tripartite organization. The afferent pathway involves the transmission of external cues to the central clock through the melanopsin containing ganglion cells of the retinohypothalamic tract (RHT) to the SCN with glutamate and pituitary adenylate cyclase – activating polypeptide (PACAP) as the major neurotransmitters in this pathway (Albrecht, 2012). The geniculohypothalamic tract (GHT) and the retino-raphe pathway also provide photic and nonphotic inputs, respectively to the SCN. The neurotransmitters involved in GHT are gamma amino butyric acid (GABA) and neuropeptide Y whereas the serotonin is present in the retino raphe tract (Jagota 2006, 2012). The SCN regulates the rhythmic synthesis and release of melatonin (N-acetyl-5-methoxytryptamine; an indole hormone (message of darkness)) with high levels at night and low during day. With the onset of darkness, tryptophan is converted into serotonin via 5-hydroxytryptophan in pinealocytes. Then N-acetylation of serotonin by arylalkylamine N-acetyltransferase (AANAT) followed by methylation of the 5-hydroxy moiety by hydroxyindole-O-methyl-transferase (HIOMT) results in melatonin synthesis. (Reiter et al. 2010; Yonei et al. 2010). Additionally we have reported recently from our laboratory daily rhythms in the various components of serotonin metabolism (Reddy and Jagota 2015). Melatonin synthesis at night is under regulation of post-transcriptional and post-translational mechanisms triggered by termination of norepinephrine release from superior cervical ganglion (SCG) terminals (Lee et al. 2009). It is then transported into the blood and cerebrospinal fluid (CSF) of the third ventricle and has been reported to act through MT1, MT2 and MT3 receptors (Dubocovich 2007), found in various parts of brain including SCN and peripheral organs such as vasculature, reproductive organs, adrenal cortex etc. Melatonin has been identified as a multitasking molecule, acting as an antioxidant and a free radical scavenger, eliminating hydroxy radicals, peroxyl radicals and peroxynitrite (Konturek et al. 2007), as well as activating several antioxidative enzymes such as superoxide dismutase or glutathione peroxidase etc. (Yonei et al. 2010; Manikonda and Jagota 2012). Both blood and CSF melatonin acts on SCN via feedback mechanisms involving MT1 and MT2 receptors for resetting the circadian pacemaker and regulating circadian processes such as sleep (Reiter et al. 2010, Yonei et al. 2010).

The involvement of positive and negative feedback loops regulate rhythmic oscillations when they move away from regular manner. This is achieved by several clock genes, *Clock*, *Per1*, *Per2*, *Per3*, *Bmal1*, *Cry1* and *Cry2* etc. which are expressed abundantly in the SCN (Dibner et al. 2010; Welsh et al. 2010). The positive regulators include *Clock*, *Bmal1* and *Rorα*, whereas the negative elements comprise of *Per*, *Cry* and *Rev-erbα*. All genes are expressed rhythmically except *Clock* that is constitutively expressed. The circadian timing machinery involves the highly interdependent transcriptional and translational feedback loops (TTFL), whose interplay results in vigorous ~24 h rhythm of gene expression. The core TTFL is coordinated by four dedicated clock proteins *viz.* BMAL1, CLOCK, PER and CRY as well as their post translational modifications involving kinases and phosphatases that regulate the localization and stability of these integral clock proteins. During the subjective day, dimers of clock proteins BMAL1-CLOCK bind to the E-box elements in the promoters of repressors *Per, Cry* as well as *Rorα*, *Rev-erbα* and several other Clock controlled genes (*Ccg*), initiating their transcription. During the subjective night, PER and CRY proteins reach critical levels in the cytoplasm and dimerizes. The protein dimer translocate back to the nucleus and disrupts the BMAL1- CLOCK heterodimer repressing their own transcription along with the transcription of *Ccg*. Repression on CLOCK-BMAL1 gets relieved as PER and CRY proteins are degraded through ubiquitin dependent pathways and the 24 h periodicity sustains. In addition to the core TTFL, there exists an auxiliary TTFL to endure the former. It consists of activator: retinoid- related orphan receptors (RORs) and repressor: REV-ERBs (Partch et al. 2014). The expression of clock gene *Bmal1* has been reported to be dependent on the expression of *Rev-erbα* where the presence of REV-ERBα inhibits *Bmal1* expression (Sato et al. 2006; Li and Zhang 2015). Further, the kinetics and phase characteristics of individual clock components are dependent upon specific promoter elements as the morning time- E/E' box and the day time D box elements mediating the expression of *Rev-erbα, Per1, Per2 and Per3*. The REV-ERB/ROR response element dictates the night time expression of *Bmal1* (Ueda et al. 2005) whereas 'combinatorial regulation' involving E/E', D box as well as the intronic RREs confer phase delay in *Cry1* expression (CT12) which is critical in keeping the biological clock on time (Ukai-Tadenuma et al. 2011). The existence of these interdependent feedback loops maintain accurate circadian timing by providing robustness against noise and environmental disturbances, and helps to generate phase alterations in circadian transcriptional output that optimally modulate gene expression influencing the physiology (Jagota 2012; Partch et al. 2014).

The post-translational modifications of clock proteins are important for ensuring the maintenance of circadian rhythms, as they can modulate the activity and turnover of major clock components (Bellet and Sassone Corsi 2010). Interestingly, the NAD^+ salvage pathway as well as direct deacetylation activity link through Silent information regulator 1 (sirtuin 1 (*Sirt1*) to the circadian rhythm machinery (Jung-Hynes et al. 2010).

21.2 Brain Aging and Clock Dysfunction

Aging has been linked to progressive physiological changes in an organism that lead to senescence, or a decline of biological functions and of the organism's ability to adapt to metabolic stress. The inflammatory cascade is exaggerated during aging due to increased expression of proinflammatory genes leading to the oxidative damage of biomolecules due to an imbalance between prooxidants (ROS) and antioxidants (Sohal and Orr 2012). The antioxidant status of an individual could be an important factor in determining age-associated pathologies, including cancer, cardiovascular disease, arthritis, and several neurodegenerative diseases. When the biological clock begins to fail, rhythms that regulate cell function and health get disrupted and this predisposes the brain to neurodegeneration. Molecular clock oscillations decline during aging and neurological diseases such as, Alzheimer's disease (AD), Parkinson's disease (PD), Huntington's disease, sleep wake and circadian rhythm pathologies such as advanced sleep wake syndrome (ASPS), delayed sleep wake syndrome (DSPS), free running syndrome (FRD) and sleep fragmentation (SF) etc. (Jagota 2012). Many researchers have reported loss of phase relationships of clock genes between the different brain regions (Cermakian et al. 2011) and significant loss of neurons in SCN of AD patients (Zhou et al. 1995). A weakened circadian output and disrupted melatonin regulation and levels were reported in PD patients by some workers (Kudo et al. 2011; Videnovic et al. 2014). In addition, *Bmal1* transcript levels were also reported to decline in leukocytes of PD patients (Cai et al. 2010). We have recently reported alterations in various clock gene expression such as *Per1, Per2, Cry1, Cry2 and Bmal1* in aging rats (Mattam and Jagota 2014) and rotenone induced PD (RIPD) rat model (Mattam and Jagota 2015). Few studies have also revealed that SIRT1 governs the central circadian function by activating transcription of *Bmal1* and *Clock* via PPARγ coactivator 1α (PGC-1α) (Asher et al. 2008; Nakahata et al. 2009). Chang and Guarente in 2011 reported a model explaining how the SIRT1 mediated circadian control in SCN decays with aging. Interestingly nuclear factor erythroid 2-like factor 2 (NRF2) had been directly involved in cellular defense against oxidative stress and SIRT1 in transcriptional activation of clock genes (Zhang et al. 2015).

In mammals, it has been observed that there are major alterations in output rhythms of the circadian clock as a result of aging. These changes are marked by a shift in the phase and decrease in amplitude (Hofman and Swaab 2006; Gibson et al. 2009). In rodents, age-related changes in circadian rhythms have been reported for body temperature, activity-wakefulness, locomotor activity patterns, drinking behavior etc. (Weinert 2000). We have reported age-induced abolition of daily rhythm of leptin (Reddy and Jagota 2014), serotonin (Jagota and Kalyani 2008, 2010) and serotonin metabolism (Reddy and Jagota 2015) in rat SCN. Additionally, we have reported the alterations in some of the core clock genes in SCN with aging (Mattam and Jagota 2014). Earlier we had reported from our laboratory age induced alterations in lipid peroxidation and antioxidant enzymes in rat liver (Manikonda and Jagota 2012).

Recently we also reported age-induced alterations in NO rhythms in various peripheral clock (Vinod and Jagota 2016). In humans, aging is associated with a phase advance in melatonin secretion and body temperature rhythms. It has been well known that, unlike young human subjects, older individuals show earlier habitual time of sleeping and waking along with disturbed sleep and impaired cognitive performance (Miyata et al. 2013). These results may indicate that the circadian systems are less able to adjust to environmental time cues with age, and as a result are more susceptible to desynchronization (Arellanes-Licea et al. 2014). The mechanisms leading to the aging-related alterations are largely unknown, but they are paralleled by changes in the neurochemical and electrophysiological output of the SCN with no change in cell number or size (Madeira et al. 1995).

There are various therapeutic approaches being developed to address aging and related disorders. Melatonin, an endogenous synchronizer of rhythms that is also a potent antioxidant possesses therapeutic effects against the age-induced alterations in circadian clock (reviewed Hardeland et al. 2012; Jagota 2012). Additionally we had also reported differential restoration of daily rhythms of lipid peroxidation and antioxidant enzymes in aging rat liver with melatonin administration (Manikonda and Jagota 2012). We had reported recently that melatonin could differentially restore the age induced alterations in the daily rhythms of some of the core clock genes, serotonin and various components of serotonin metabolism in SCN (Jagota and Kalyani 2010; Mattam and Jagota 2014; Reddy and Jagota 2015) as well as daily nitric oxide (NO) rhythms (Vinod and Jagota 2016. Additionally it proved useful when administered on RIPD rat model as reported recently from our laboratory (Mattam and Jagota 2015).

Herbal extracts or formulations have been in use since ancient cultures involving medicinal use of plants. The herbal medicines are of greater scientific interest as many international studies have demonstrated that these are capable of treating disease and improving health often without any significant side effects. A pharmaceutical drug typically uses a synthesised version of a plant's active ingredient but practitioners of herbal medicine maintain that an active ingredient can lose its impact or become less safe, if used in isolation from the rest of the plant. Various herbal formulations like curcumin, resveratrol, wolfberry extracts, ginseng, ashwagandha etc. have been extensively investigated for their anti-oxidative, anti-aging and anti-inflammatory properties (Jagota and Reddy 2007; Ho et al. 2015; Lee and Oh 2015; Wadhwa et al. 2016; Farooqui 2016; Gocmez et al. 2016).

21.3 *Withania somnifera* (Ashwagandha)

Withania somnifera (WS), known as ashwagandha or Indian winter cherry or Indian Ginseng has been used in Indian traditional medicines since documented history, classified as a rasayana herb that promotes health, enhance longevity and create a sense of well-being. It is widely distributed in the dry regions of India, Middle East, North Africa and the Mediterranean regions (Dar et al. 2015). Table 21.1 shows the

Table 21.1 Effects of various active components of *Withania somnifera* (WS) extracts on various neuropathological and physiological disorders

Extract/component	Effects	References
WS extract	Improved cognitive capacity in bipolar disorder patients	Chengappa et al. (2013)
WS Root extracts (RE)		
RE	Decrease of mean immobility time in a mice model of depression	Shah et al. (2006)
	Effects on circadian rhythms of lipid peroxidation products and antioxidants in Gentamicin induced nephrotoxicity	Jeyanthi et al. (2010)
	Physiological and behavioral alterations in sleep deprived mice	Kumar and Kalonia (2007)
	GABAergic mechanism in improving electrophysiological parameters and sleep promoting effects in sleep disturbed rats	Kumar and Kalonia (2008)
	Enhanced life span in *C. elegance*	Kumar et al. (2013)
	Neuroprotective effects in 6-OHDA induced PD rats	Ahmad et al. (2005)
	Improves catecholamine levels, physiological and motor function in MPTP model of PD mice	Rajasankar et al. (2009)
Ethanolic RE	Downregulation of apoptosis of dopaminergic neurons in MB-PQ induced PD mice	Prakash et al. (2014)
Aqueous RE	Regulation of lipid peroxidation, alteration in glutathione content, glutathione-S-transferase activity in arthritic rats	Khan et al. (2015)
	GBAergic activity of WS on $GABA_A$ and GABAρ, the main inhibitory receptors of mammalian CNS	Candelario et al. (2015)
	Countering effects on temporal lobe epilepsy associated neurodegeneration	Soman et al. (2012)
Chloroform-methanol RE	Alzheimer's disease	Sehgal et al. (2012)
WS Leaf extract (LE)		
Aqueous LE	*In vitro* free radical scavenging activity	Panchawat (2011)
Ethanolic LE		
Aqueous LE	Neuroprotective effects in sleep deprived rats	Manchanda et al. (2016)
	Therapeutic effects on neuroblastoma	Kataria et al. (2013) and Shah et al. (2015)
Alcoholic LE	Therapeutic effects on glioblastoma	Shah et al. (2009, 2015)
Alcoholic (i-extract)	Protection against Scopolamine induced amnesia	Konar et al. (2011) and Gautam et al. (2016)

(continued)

Table 21.1 (continued)

Extract/component	Effects	References
Component		
Withaferin A	Inhibitory effects on cholinesterase in treating cognitive deficits and in AD and related dementias	Schliebs et al. (1997)
	Regulation of NF-kB marking its anti-inflammatory property	Grover et al. (2010)
	Anxiety, inflammation and rheumatism	Berghe et al. (2012)
	Regeneration of both axons and dendrites in memory deficit mice with neuronal atrophy and synaptic loss	Kulkarni and Dhir (2008)
	Neuroprotection, stress resistance and prolonged life expectancy in *C. elegance*	Akhoon et al. (2016)
	Prevention of neurodegeneration during hypoxia	Baitharu et al. (2014)
Withanone	Downregulation of senescence in human fibroblasts	Widodo et al. (2009)
Withanoside IV and its metabolite somnione	Improved memory and prevention of neurodegeneration induced by amyloid beta (25-35) in mice	Kuboyama et al. (2006)

compilation of various effects of active components of WS on miscellaneous functions and pathologies.

The WS extracts have been reported to display neuroprotective, neuroregenerative and anticancer potentials (Kataria et al. 2013). The root extract of WS has been reported to possess revitalizing, life prolonging, aphrodisiac and sedative properties. In Ayurveda, it is extensively used for the treatment of disorders such as, rheumatoid arthritis, dehydration and chronic fatigue. The fruits and leaves have been related to treating ulcers and tumors. Clinical trials corroborate the benefits of WS in case of anxiety, inflammation and rheumatism (Berghe et al. 2012). The biologically active constituents found in WS were reported as alkaloids (ashwagandhin, cuscohygrine, anahygrine, topine, etc.), steroidal compounds, including ergostane type steroidallactones, withaferin A, withanolides A–Y, withasomniferin A, withasomnidienone, withasomnierose A-C, withanone, glycowithanolides etc. (Dar et al. 2015). The alcoholic extract had higher concentrations of most of these components but only trace quantities of glycowithanolides were found to be soluble in water whereas the aqueous extract had higher concentration of glycowithanolides but only trace quantities of alcohol soluble components. The nootropic potential and therapeutic effects of ashwagandha leaf extracts on various brain pathologies such as AD, PD, Neuroblastoma and Glioma had been reviewed by Wadhwa et al. 2016. It has been reported as key adaptogen due to its action on Hypothalamic-Pituitary-Adrenal (HPA) axis modulating the imbalance between neuroendocrine and immune system thus regulating the body physiology under stressful conditions (Verma and Kumar 2011). Earlier studies have reported the antioxidant properties of WS such as free radical scavenging activity (Panchawat 2011), regulation of lipid peroxidation, alteration in glutathione content, glutathione-S-transferase activity, catalase activity and decrease in nitrites levels (Khan et al. 2015). Grover et al. (2010) showed the regulation of NF-kB by Withaferin A marking its anti-inflammatory property. Clinical trials and animal research supported the use of WS for treatment of anxiety, stress, depression, insomnia, cancer, cognitive and age-associated neurodegenerative disorders (Konar et al. 2011; Sehgal et al. 2012). WS root extract treatment has been linked with improving the body weight, behavioral alterations such as reduced locomotion and anxiety in sleep deprived mice by some workers. Further, the possible involvement of GABAergic mechanism in improving electrophysiological parameters and sleep-promoting effects of WS in sleep disturbed rats has been reported. A recent study in sleep deprived rats has explored the neuroprotective effects of aqueous leaf extract of WS (Manchanda et al. 2016).

Some workers have reported the inhibitory effects of Withaferin A on cholinesterase which is of interest in treating cognitive deficits and in AD and related dementias (Schliebs et al. 1997). Withanoside IV and its metabolite somnione had been related with improving memory and preventing neurodegeneration induced by amyloid beta (25-35) in mice. In addition, in cultured rat cortical neurons damaged by amyloid beta (25-35), somnione treatment has been linked with synaptic reconstruction, axonal and dendritic regeneration potential (Kuboyama et al. 2006). Similarly, studies have investigated the potential of WS in improving the catecholamines and antioxidant enzyme activities, motor function in rodent models of PD (Ahmad et al. 2005; Rajasankar et al. 2009). In Maneb or Manganese ethylene-bis-dithiocarbamate and Paraquat (MB-PQ) induced PD mice, root extract of WS enhanced Bcl2 levels and declined Bax levels thus affecting the apoptosis of dopaminergic neurons (Prakash et al. 2014). Clinical studies have demonstrated the beneficial precognitive effects of WS on bipolar disorder (Chengappa et al. 2013). Withanolide A induces significant regeneration of both axons and dendrites in memory deficit mice with neuronal atrophy and synaptic loss (Kulkarni and Dhir 2008). Candelario et al. 2015 reported the GABAergic activity of WS on $GABA_A$ and GABAρ, the main inhibitory receptors of mammalian CNS, adding on to the mechanism of its adaptogenic properties. In addition to this, WS leaf extract downregulated senescence in human fibroblasts and the root extract could extend the life span in *C. elegance* (Widodo et al. 2009; Kumar et al. 2013). Several researchers have shown the free radical scavenging activities of various antioxidants (Table 21.2).

The 50% maximal inhibitory concentration (IC_{50}) is determined as the concentration of the tested sample resulting in a 50% reduction of the initial diphenyl-1-picryl hydrazyl (DPPH) concentration, measured from the linear regression concentration curve of the test extract (μg/ml) against the percentage of the free radical scavenging activity. Though the anti-inflammatory, anti-oxidant and neuroprotective roles of WS using aqueous or alcoholic extracts of roots and leaves have been extensively studied, we in our laboratory compared the antioxidant potential of the hydro-alcoholic extract of WS leaves and roots using DPPH free radical scavenging assay (Brand-Williams et al. 1995). The hydro-alcoholic extract of WS leaves and roots showed a scavenging activity of 75.9 and 34.8 % with DPPH IC_{50} – 659.9 and 1449.2 μg/ml respectively (Fig. 21.1). On comparison of DPPH IC_{50} val-

Table 21.2 Comparison of various antioxidants by DPPH free radical scavenging assay

Antioxidant	Activity DPPH IC_{50} (μg/ml)	References
Ascorbic acid	9.84	Kano et al. (2005)
Melatonin (Semi-aqueous)	238.08	Mishra et al. (2012)
Curcumin (Pure)	16	Jagota and Reddy (2007)
Resveratrol (Pure)	17.8	Gulcin (2010)
Licium barbarum (Wolfberry) fruit extract	267.6	Li et al. (2007)
Panax ginseng (Chinese ginseng) aqueous extract	833.3	Kim et al. (2002)
Panax quinquefolius (North American ginseng) extract	2000	Kitts et al. (2000)
WS Leaf extract (Methanolic)	101.73	Alam et al. (2012)
WS Root Extract (Methanolic)	801.93	
WS Leaf extract (Aqueous)	1455	Panchawat (2011)
WS Leaf extract (Ethanolic)	1663	
WS Leaf extract (Hydro-alcoholic)	659.9	Jagota and Kowshik (present study)
WS Root Extract (Hydro-alcoholic)	1449.2	

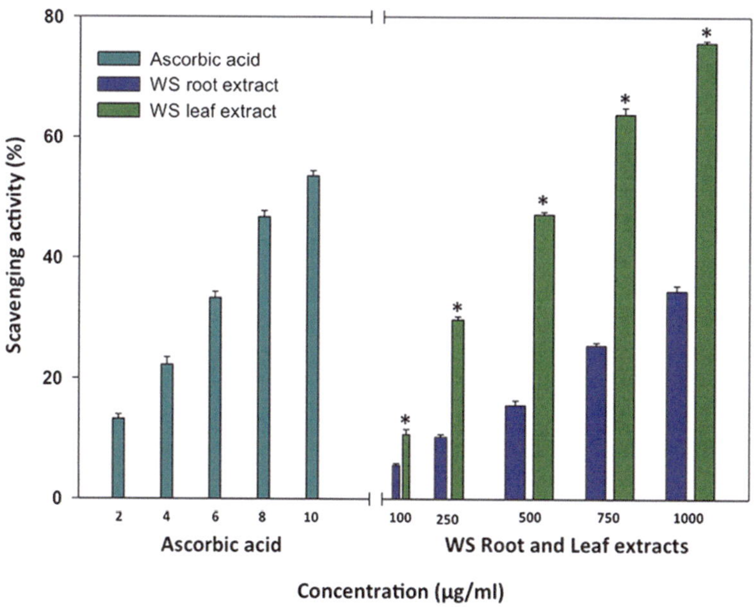

Fig. 21.1 Comparision of free radical scavenging activity of the hydro-alcoholic *Withania somnifera* root and leaf extracts with the standard antioxidant ascorbic acid using DPPH assay. Each value is mean ± SEM (n=4) (*significant difference between leaf and root extract ($p \leq 0.005$))

ues hydro-alcoholic extract of WS leaves showed significantly high antioxidant activity than aqueous and ethanolic leaf extracts and also methanolic root extract Table 21.2. Based on these results we preferred to use hydro-alcoholic extract of WS leaves due to presence of both alcohol soluble components (alkaloids (ashwagandhin, cuscohygrine, anahygrine, topine etc.), steroidal compounds, including ergostane type steroidallactones, withaferin A, withanolides A-Y, withasomniferin A, withasomnidienone, withasomnierose A-C, withanone and water soluble such as glycowithanolides etc. We have recently reported age-induced circadian disruption in clock gene expression *(rPer1*, *rPer2*, *rCry1*, *rCry2* and *rBmal1)* (Mattam and Jagota 2014) and we extended to study the therapeutic and adaptogenic potential of hydroalcoholic extract of WS on such age induced alterations in mRNA expression of these clock genes.

The effects of hydro-alcoholic leaf extract of WS suspended in 0.5% carboxy methyl cellulose (CMC) were studied by WS treatment (WST) at ZT-11 orally 100 mg/kg body weight in three age groups of male Wistar rats: 3, 12 and 24 months (m) for 15 days. Similarly age matched control animals (Vehicle treated (VT)) were administered with 0.5% CMC (1 ml/kg body weight). The time of WS leaf extract administration was 1 hour before the reported time of maximum lipid peroxidation (Manikonda and Jagota 2012). On 16th day animals of all the three age groups were decapitated and the brains were dissected out carefully at ZT-0, 6, 12 and 18. The SCN was carefully punched out from brain slices made with rat brain slicer (Zivic Instruments; Pittsburg USA) (Jagota and Reddy 2007). The expression of *rPer1*, *rPer2*, *rCry1*, *rCry2* and *rBmal1* transcripts were measured by relative qRT-PCR by the SYBR Green (Applied Biosystems, Foster, USA) detection method (Kamphuis et al. 2005; Mattam and Jagota 2014). There were no significant differences in age-matched controls and vehicle treated (VT) animals in clock gene expression levels and daily rhythms. We report here that hydro-alcoholic WS leaf extract showed differential restoratory effects on age induced alterations in clock gene expression levels and daily rhythms (Figs. 21.2 and 21.3). The peak expression of positive limb component *rBmal1* showed peak at ZT-18 (mid-subjective night) whereas peak expression of negative limb components *rPer1 and rPer2 were* at ZT-6 (mid-subjective day) however *rCry1, rCry2* were at ZT-12 (Onset of night) respectively. These peak expression time points of studied clock mRNAs are in agreement with earlier researchers in 3 m vehicle treated animals (Park and Kang 2004) as well as our previous reports (Mattam and Jagota 2014).

The administration of hydro-alcoholic leaf extract of WS resulted in restoration of the phases of *rPer1* in 12 and 24m animals in comparison to 3m VT animals with a phase advance of 12h and a phase delay of 6h respectively in comparison to the respective age matched VT group (Fig. 21.2). Further, in 24m animals WS treatment decreased the daily pulse of *rPer1* by 1.6 fold compared to the 24m VT group, restoring the pulse in comparison to 3m VT group (Fig. 21.3). In case of *rPer2*, WS treatment increased the mean 24h levels by 3.3 folds in 24m animals and decreased the daily pulse by 4.2 fold in 12m animals thus resulting in their restoration with respect to 3m VT (Fig. 21.3). In 12m animals, *rCry1* mRNA expression peaked at ZT-12 with WS administration, showing a phase advance of 6h in comparison to

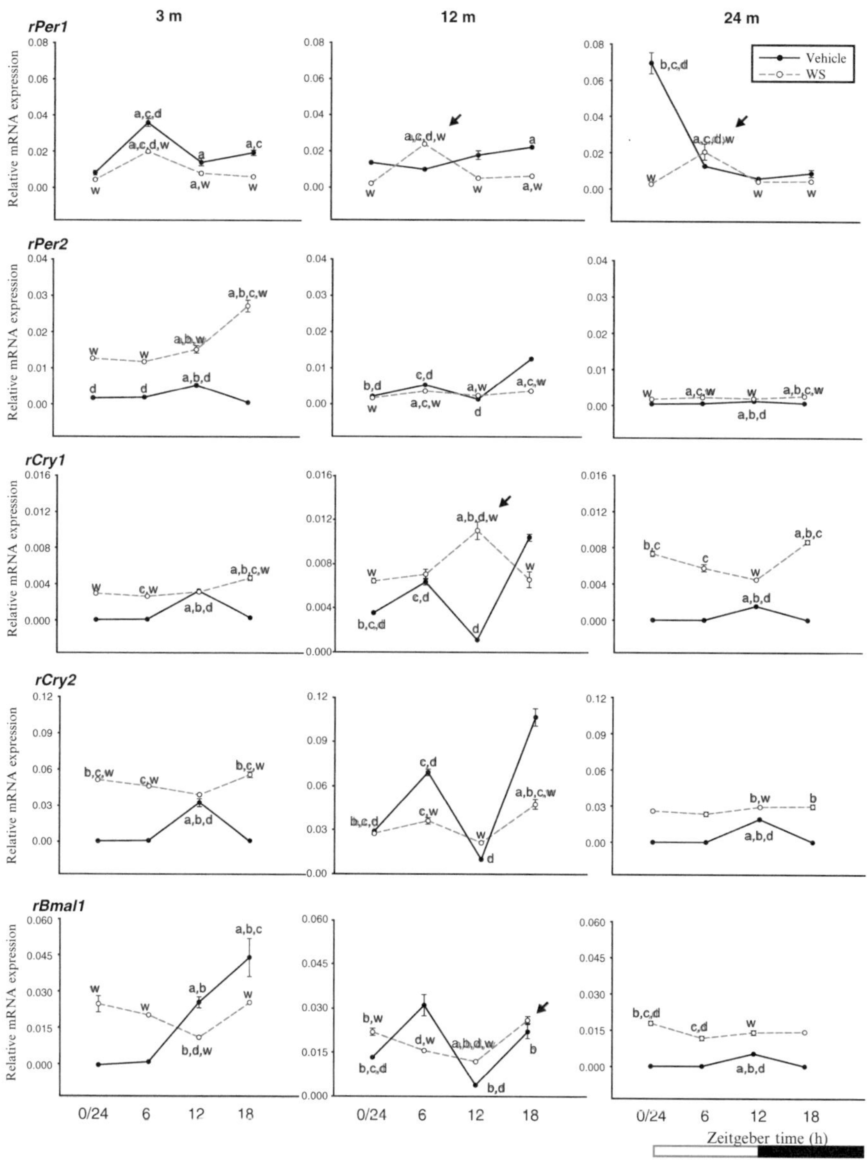

Fig. 21.2 Effect of the hydro-alcoholic *Withania somnifera* leaf extracts administration on daily rhythms of *rPer1, rPer2, rCry1, rCry2* and *rBmal1* mRNA expression in SCN in 3, 12 and 24 m old rats. Each value is mean ± SEM (n = 4), $p \leq 0.05$ and expressed as relative gene expression. $pa \leq 0.05$; $pb \leq 0.05$, $pc \leq 0.05$ and $pd \leq 0.05$ (where 'a', 'b', 'c' and 'd' refers to comparison with ZT-0, 6, 12 and 18 respectively within the group). $pw \leq 0.05$ (where 'w' refers to comparison of gene expression levels at same time point in the age matched vehicle group)

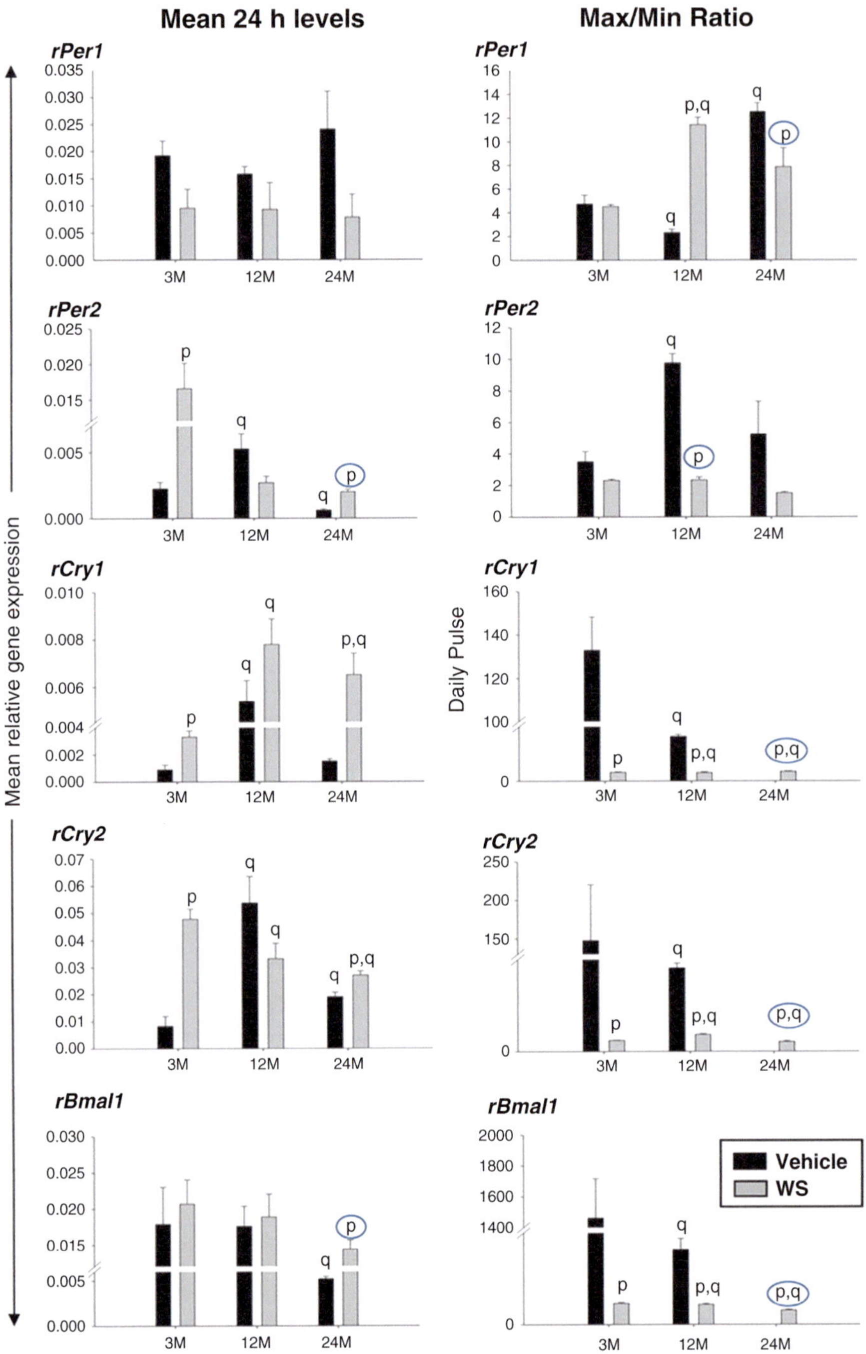

Fig. 21.3 Effect of the hydro-alcoholic *Withania somnifera* leaf extracts administration on mean 24h levels and daily pulse of *rPer1, rPer2, rCry1, rCry2* and *rBmal1* mRNA expression in SCN in 3, 12 and 24m old rats. Each value is mean ± SEM, $p \leq 0.05$ and expressed as mean relative gene expression. $pp \leq 0.05$ (where 'p' refers to comparison with age matched vehicle group). $P_q \leq 0.05$ (where 'q' refers to comparison with 3 m vehicle group)

12m VT group, restoring the phase in comparison to 3m VT group. In addition, the daily pulse of *rCry1* and *rCry2* in 24m showed a partial restoration even though these declined by 68.5 and 115.6 folds respectively in comparison to 3m VT (Figs. 21.2 and 21.3). There was a phase restoration in 12m *rBmal1* compared to 3m VT upon WS treatment, where a phase delay of 6h was observed in comparison to age matched VT (Fig. 21.2). WS treatment in 24m animals increased *rBmal1* mean 24h levels by 2.8 folds in comparison to age matched vehicle group, thus restoring the levels when compared to 3m VT group. However, *rBmal1* daily pulse showed a partial restoration when compared to the respective age matched vehicle groups and was decreased by 115.6 and 945.8 folds respectively compared to 3m VT group (Fig. 21.3). WS administration in middle-aged animals (12m VT) restored the phase of *rPer1, rCry1* and *rBmal1* compared to the young adults (3m VT). The melatonin treatment to the 12m age group had restored the phase of clock genes except *rPer1* (Mattam and Jagota 2014). The daily pulse of *rPer2* was restored in 12m WST group compared to 3 m VT group. However, we have reported previously melatonin had restored *rPer1* daily pulse in 12m (Mattam and Jagota 2014). In 24m animals, aging caused severe alterations in the phase and daily pulse of all the clock genes studied, whereas the WST restored the phase and daily pulse of *rPer1*. However, WST decreased the daily pulse of *rCry1*, *rCry2* and *rBmal1* in 24 m compared to 3 m VT. The restoration in the mRNA expression *rCry1*, *rCry2* and *rBmal1* is similar to melatonin treatment (Mattam and Jagota 2014).

Withanolide, one of the active components of WS with its steroidal structure (Das et al. 2002) might act similar to estradiol in activating signaling cascades (Konar et al. 2011). Further, Konar et al. (2011) discussed a possible mechanism of CREB phosphorylation by i-Extract resulting in attenuation of scopolamine-induced alterations in the expression of BDNF and GFAP. Thus WS leaf extract may be mediating its effects through activation of cAMP and MAPK pathway contributing to CREB phosphorylation. CREB binding to the CREs in the promoters of *Per1* and *Per2* then lead to transcriptional activation (Travnickova-Bendova et al. 2002). The decreased mean 24h levels of *rBmal1* and *rPer2* in 24m VT animals are consistent with earlier reports (Duncan et al. 2013). The decreased levels of *Bmal1, Per* and *Cry* could be correlated with increase in oxidative stress, reduced life span (Rakshit and Giebultowicz 2013) and neurodegeneration (Krishnan et al. 2012; Musiek 2015). Ashwagandha being an antioxidant and a neurogenesis inducer (Kuboyama et al. 2005; Soman et al. 2012) appeared to have upregulated the expression of *Bmal1, Per 2 and Cry1,2* transcript levels proving its role as an anti-aging component.

The pairwise correlation analysis (Fig. 21.4) showed a significant positive correlation between *rPer1* and *rPer2*; *rPer1* and *rCry2* in the light phase of 12m WS treatment, which was negative in light phase of 12m VT. Similarly, a significant negative correlation could be seen between *rCry1* and *rBmal1* in both light and dark phases. WS treatment in 24m rat SCN resulted in a positive correlation between *rPer1* and *rPer2* in both light and dark phases, which was negative in 24m VT. A significant positive correlation was found in the dark phase between *rPer2* and

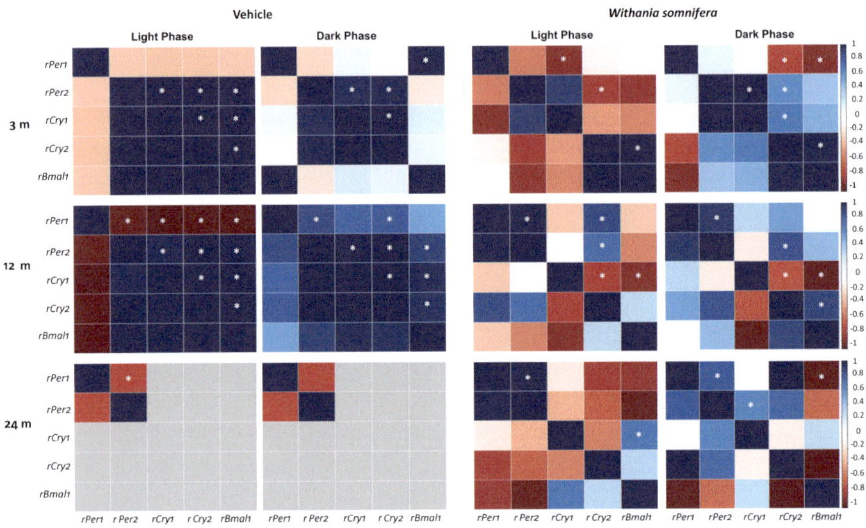

Fig. 21.4 Effect of hydro-alcoholic *Withania somnifera* leaf extracts administration on pair wise correlation between mean light (ZT-0, 6, 12) and mean dark (ZT-12, 18, 24/0) phase values of *rPer1*, *rPer2*, *rCry1*, *rCry2* and *rBmal1* in SCN of 3, 12 and 24m (m) aged rats. Every colored square represents correlation coefficient values between the parameters. Where, a positive correlation is indicated by shades of blue, negative correlation by shades of red and white indicates no correlation. Grey squares indicate that the values were not considered for correlation analysis *statistically significant value between parameters ($p \leq 0.05$)

rCry1, which was similar to that of 3m VT, however a negative correlation between *rPer1* and *rBmal1* was established in both the phases of 24m (Fig. 21.4).

In old aged animals (24m VT), a negative correlation between *rPer1* and *rPer2* in both the phases indicated the robustness of age-induced alterations. The WS treatment abolished the negative correlation thus restoring the positive correlation between *rPer1* and *rPer2*. The anti-phasic expression of *Per* and *Bmal1* genes had been demonstrated by many researchers (Yan et al. 1999). The WS treatment reestablished a significant negative correlation between *rPer1* and *rBmal1* in the dark phase of 24m animals. WST restored the correlations between *Per* and *Cry* genes as restored by melatonin reported earlier (Mattam and Jagota 2014) proving WS as one of the potent anti-aging agent. Interestingly, with melatonin administration this negative correlation was in the light phase (Mattam and Jagota 2014). Aging drastically affected the correlation between the clock genes in the dark phase by diminishing the transcript levels.

As NRF2 daily rhythms in aging rat SCN is directly involved in cellular defense against oxidative stress and SIRT1 in transcriptional activation of clock genes, their role could be pivotal in reversing age associated circadian dysfunction. It was previously reported that Rev-erbα regulates the expression of PGC-1α (Gerhart-Hines et al. 2013), which is a crucial mediator of SIRT1 through which *Bmal1* transcripts are regulated. Additionally, wide range of genes controlling cell survival and metabolism are targeted by *Rev-erbα* in addition to its critical role in transcriptional

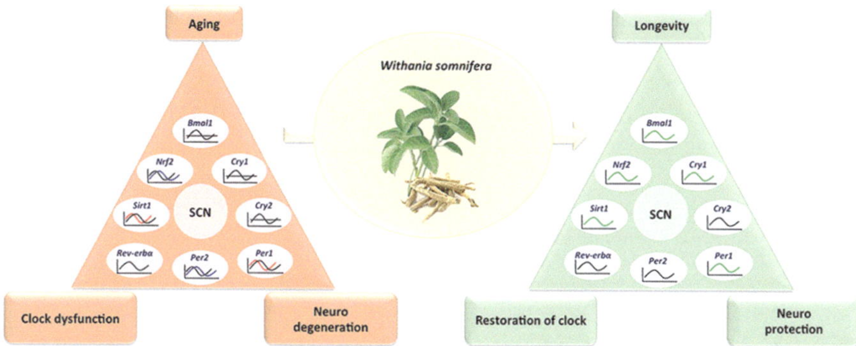

Fig. 21.5 Schematic representation of neuroprotection and longevity and restoration of clock function with hydro-alcoholic leaf extracts of *Withania somnifera*. Rhythmic; Abolished rhythm; Phase advance (Red line indicates the advance in phase); Phase delay (Blue line indicates the delay in phase); Phase restoration

repression and regulation of clock. Studies have demonstrated that REV-ERBα is the key protein by which oxidative stress and inflammation affects the biological clock (Yang et al. 2014) and also its pivotal role in enabling the organism to withstand oxidative stress maintaining the cellular homeostasis (Sengupta et al. 2016). We therefore have further studied effects of WS extract on mRNA expression of Nrf2, *Sirt*1 and *Rev-erba*, one of the key negative limb components of the molecular clock. The summary of preliminary results obtained in this study has been presented in Fig. 21.5. There was a decrease in daily pulse of mRNA expression of *Sirt1* levels with a phase advance, which was restored upon WST. In 12m animals, *Nrf2* expression however showed a phase delay, which was restored upon WST. Interestingly, aging did not alter the phase of *Rev-erba* and WS treatment also did not bring any change to the phase of this transcript. Even the mean 24h levels and the daily pulse did not show any significant change upon aging and WS treatment.

21.4 Conclusion

Brain aging has been associated with chronodisruption. Age associated complications in clock function could be addressed through various strategies including light therapy, dietary paradigms, pharmacological interventions as well as herbal formulations. Ashwagandha being one of the extensively researched medicinal herbs is a potent antioxidant and demonstrated to be highly persuasive in its action against various age related brain pathologies. Currently there is very limited information on its therapeutic role in age-induced alterations of clock function. However, our data reveals the potential of hydroalcoholic leaf extract of WS in ameliorating such alterations in some of the core clock transcript expression levels and daily rhythms.

Further, it is essential to investigate the role of this potent herb in influencing the clock components at the translational level. In addition, an extensive study of the cellular targets of the active constituents of WS and the mechanisms involved would be crucial to develop novel therapeutic strategies for improving age related clock dysfunction.

Acknowledgement The work is supported by ICMR (Ref. No. 55/7/2012-/BMS) and UPE grants to AJ. DST INSPIRE fellowship to KK is acknowledged.

References

Ahmad M, Saleem S, Ahmad AS, Ansari MA, Yousuf S, Hoda MN, Islam F (2005) Neuroprotective effects of Withania somnifera on 6-hydroxydopamine induced Parkinsonism in rats. Hum Exp Toxicol 24:137–147

Akhoon BA, Pandey S, Tiwari S, Pandey R (2016) Withanolide A offers neuroprotection, ameliorates stress resistance and prolongs the life expectancy of *Caenorhabditis elegans*. Exp Gerentol 78:47–56

Albrecht U (2012) Timing to perfection: the biology of central and peripheral circadian clocks. Neuron 74:246–260

Alam N, Hossain M, Mottalib MA, Sulaiman SA, Gan SH, Khalil MI (2012) Methanolic extracts of *Withania somnifera* leaves, fruits and roots possess antioxidant properties and antibacterial activities. BMC Complement Altern Med 12:175

Arellanes-Licea E, Caldelas I, De Ita-Pérez D, Díaz-Muñoz M (2014) The circadian timing system: a recent addition in the physiological mechanisms underlying pathological and aging processes. Aging Dis 5:406–418

Asher G, Gatfield D, Stratmann M, Reinke H, Dibner C, Kreppel F, Schibler U (2008) SIRT1 regulates circadian clock gene expression through PER2 deacetylation. Cell 134:317–328

Baitharu I, Jain V, Deep SN, Shroff S, Sahu JK, Naik PK, Ilavazhagan G (2014) Withanolide A Prevents Neurodegeneration by Modulating Hippocampal Glutathione Biosynthesis during Hypoxia. PLoS One 9:e105311

Bellet MM, Sassone-Corsi P (2010) Mammalian circadian clock and metabolism – the epigenetic link. J Cell Sci 123:3837–3848

Berghe WV, Sabbe L, Kaileh M, Haegeman G, Heyninck K (2012) Molecular insight in the multifunctional activities of Withaferin A. Biochem Pharmacol 84:1282–1291

Brand-Williams W, Cuvelier ME, Berset CLWT (1995) Use of a free radical method to evaluate antioxidant activity. Lebensm Wiss u Technol 28:25–30

Cai Y, Liu S, Sothern RB, Xu S, Chan P (2010) Expression of clock genes Per1 and Bmal1 in total leukocytes in health and Parkinson's disease. Eur J Neurol 17:550–554

Candelario M, Cuellar E, Reyes-Ruiz JM, Darabedian N, Feimeng Z, Miledi R, Russo-Neustadt A, Limon A (2015) Direct evidence for GABAergic activity of Withania somnifera on mammalian ionotropic GABA A and GABAρ receptors. J Ethnopharmacol 171:264–272

Cermakian N, Lamont EW, Boudreau P, Boivin DB (2011) Circadian clock gene expression in brain regions of Alzheimer's disease patients and control subjects. J Biol Rhythms 26:160–170

Chang H, Guarente L (2011) SIRT1 mediates central circadian control in the SCN by a mechanism that decays with aging. Cell 153:1448–1460

Chengappa KR, Bowie CR, Schlicht PJ, Fleet D, Brar JS, Jindal R (2013) Randomized placebo-controlled adjunctive study of an extract of Withania somnifera for cognitive dysfunction in bipolar disorder. J Clin Psychiatry 74:1076–1083

Dar NJ, Hamid A, Ahmad M (2015) Pharmacologic overview of *Withania somnifera*, the Indian Ginseng. Cell Mol Life Sci 72:4445–4460
Das A, Shanker G, Nath C, Pal R, Singh S, Singh HK (2002) A comparative study in rodents of standardized extracts of *Bacopa monniera* and *Ginkgo biloba*: Anticholinesterase and cognitive enhancing activities. Pharmacol Biochem Behavior 73:893–900
Dibner C, Schibler U, Albrecht U (2010) The mammalian circadian timing system: organization and coordination of central and peripheral clocks. Annu Rev Physiol 72:517–549
Dubocovich ML (2007) Melatonin receptors: role on sleep and circadian rhythm regulation. Sleep Med 8:34–42
Duncan MJ, Prochot JR, Cook DH, Tyler SJ, Franklin KM (2013) Influence of aging on Bmal1 and Per2 expression in extra-SCN oscillators in hamster brain. Brain Res 1491:44–53
Farooqui AA (2016) Effects of curcumin on neuroinflammation in animal models and in patients with Alzheimer disease. In: Therapeutic potentials of curcumin for Alzheimer disease. Springer International Publishing, Cham, pp 259–296
Gautam A, Kaul SC, Thakur MK (2016) Alcoholic extract of Ashwagandha leaves protects against amnesia by regulation of Arc function. Mol Neurobiol 53:1760–1769
Gerhart-Hines Z, Feng D, Emmett MJ, Everett LJ, Loro E, Briggs ER, Bugge A, Hou C, Ferrara C, Seale P, Pryma DA, Khurana TS, Lazar MA (2013) The nuclear receptor Rev-erbα controls circadian thermogenic plasticity. Nature 503:410–413
Gibson EM, Williams WP, Kriegsfeld LJ (2009) Aging in the circadian system: considerations for health, disease prevention and longevity. Exp Gerontol 44:51–56
Gocmez SS, Gacar N, Utkan T, Gacar G, Scarpace PJ, Tumer N (2016) Protective effects of resveratrol on aging-induced cognitive impairment in rats. Neurobiol Learn Mem 131:131–136
Grover A, Shandilya A, Punetha A, Bisaria VS, Sundar D (2010) Inhibition of the NEMO/IKKβ association complex formation, a novel mechanism associated with the NF-κB activation suppression by Withania somnifera's key metabolite withaferin A. BMC genomics 11:25
Gülçin İ (2010) Antioxidant properties of resveratrol: a structure–activity insight. Innov Food Sci Emerg Technol 11:210–218
Hardeland R, Madrid JA, Tan DX, Reiter RJ (2012) Melatonin, the circadian multioscillator system and health: the need for detailed analyses of peripheral melatonin signalling. J Pineal Res 52:139–166
Ho YS, Li XA, Hung CHL, Chang RCC (2015) Prevention of neurodegeneration for Alzheimer's disease by *Lycium barbarum*. In: Chang RC, So K (eds) *Lycium Barbarum* and human health. Springer Netherlands, London, pp p99–111
Hofman MA, Swaab DF (2006) Living by the clock: the circadian pacemaker in older people. Ageing Res Rev 5:33–51
Jagota A (2006) Suprachiasmatic nucleus: the center for circadian timing system in mammals. Proc Indian Natl Sci Acad 71:275–288
Jagota A (2012) Age-induced alterations in biological clock: therapeutic effects of melatonin. In: Thakur MK, Rattan SIS (eds) Brain aging and therapeutic interventions. Springer Netherlands, London, pp p111–p129
Jagota A, Kalyani D (2008) Daily serotonin rhythms in rat brain during postnatal development and aging. Biogerontology 9:229–234
Jagota A, Kalyani D (2010) Effect of melatonin on age induced changes in daily serotonin rhythms in suprachiasmatic nucleus of male wistar rat. Biogerontology 11:299–308
Jagota A, Reddy MY (2007) The effect of curcumin on ethanol induced changes in Suprachiasmatic nucleus (SCN) and Pineal. Cell Mol Neurobiol 27:997–1006
Jagota A, de la Iglesia HO, Schwartz WJ (2000) Morning and evening circadian oscillations in the suprachiasmatic nucleus in vitro. Nat Neurosci 3:372–376
Jeyanthi T, Subramanian P, Kumaravel P, Sivaperumal R (2010) Influence of *Withania somnifera* on circadian rhythms of lipid peroxidation products and antioxidants in Gentamicin induced nephrotoxicity. Biol Rhythm Res 41:477–486

Jung-Hynes B, Reiter RJ, Ahmad N (2010) Sirtuins, melatonin and circadian rhythms: building a bridge between aging and cancer. J Pineal Res 48:9–19

Kamphuis W, Cailotto C, Dijk F, Bergen A, Buijs RM (2005) Circadian expression of clock genes and clock-controlled genes in the rat retina. Biochem Biophys Res Commun 330:18–26

Kano M, Takayanagi T, Harada K, Makino K, Ishikawa F (2005) Antioxidative activity of anthocyanins from purple sweet potato, Ipomoera batatas cultivar Ayamurasaki. Biosci Biotechnol Biochem 69:979–988

Kataria H, Wadhwa R, Kaul SC, Kaur G (2013) *Withania somnifera* water extract as a potential candidate for differentiation based therapy of human neuroblastomas. PLoS One 8:e55316

Khan MA, Subramaneyaan M, Arora VK, Banerjee BD, Ahmed RS (2015) Effect of Withania somnifera (Ashwagandha) root extract on amelioration of oxidative stress and autoantibodies production in collagen-induced arthritic rats. J Complement Integr Med 12:117–125

Kim YK, Guo Q, Packer L (2002) Free radical scavenging activity of red ginseng aqueous extracts. Toxicology 172:149–156

Kitts DD, Wijewickreme AN, Hu C (2000) Antioxidant properties of a North American ginseng extract. Mol Cell Biochem 203:1–10

Konar A, Shah N, Singh R, Saxena N, Kaul SC, Wadhwa R, Thakur MK (2011) Protective role of Ashwagandha leaf extract and its component withanone on scopolamine-induced changes in the brain and brain-derived cells. PLoS One 6:e27265

Konturek SJ, Konturek PC, Brzozowska I, Pawlik M, Sliwowski Z, Czesnikiewicz-Guzik M, Kwiecien S, Brzozowski T, Bubenik GA, Pawlik WW (2007) Localization and biological activities of melatonin in intact and diseased gastrointestinal tract (GIT). J Physiol Pharmacol 58:381–405

Krishnan N, Rakshit K, Chow ES, Wentzell JS, Kretzschmar D, Giebultowicz JM (2012) Loss of circadian clock accelerates aging in neurodegeneration-prone mutants. Neurobiol Dis 45:1129–1135

Kuboyama T, Tohda C, Komatsu K (2005) Neuritic regeneration and synaptic reconstruction induced by withanolide A. Br J Pharmacol 144:961–971

Kuboyama T, Tohda C, Komatsu K (2006) Withanoside IV and its active metabolite, sominone, attenuate Aβ (25–35)-induced neurodegeneration. Eur J Neurosci 23:1417–1426

Kudo T, Loh DH, Truong D, Wu Y, Colwell CS (2011) Circadian dysfunction in a mouse model of Parkinson's disease. Exp neurol 232:66–75

Kulkarni SK, Dhir A (2008) *Withania somnifera*: an Indian ginseng. Prog Neuropsychopharmacol Biol Psychiatry 32:1093–1105

Kumar A, Kalonia H (2007) Protective effect of *Withania somnifera* Dunal on the behavioral and biochemical alterations in sleep-disturbed mice (Grid over water suspended method). Indian J Exp Biol 45:524–528

Kumar A, Kalonia H (2008) Effect of *Withania somnifera* on sleep-wake cycle in sleep-disturbed rats: Possible GABAergic mechanism. Indian J Pharm Sci 70:806–810

Kumar R, Gupta K, Saharia K, Pradhan D, Subramaniam JR (2013) *Withania somnifera* root extract extends lifespan of *Caenorhabditis elegans*. Ann Nurosci 20:13–16

Lee LSJ, Liu T, Chattoraj A, Zhang SL, Wang L, Lee TM and Wang MM (2009) Posttranscriptional regulation of pineal melatonin synthesis in *Octodon degus*. J Pineal Res 47:75–81

Lee Y, Oh S (2015) Administration of red ginseng ameliorates memory decline in aged mice. J Ginseng Res 39:250–256

Li S, Zhang L (2015) Circadian control of global transcription. Biomed Res Int 187809:1–8

Li XM, Li XL, Zhou AG (2007) Evaluation of antioxidant activity of the polysaccharides extracted from *Lycium barbarum* fruits in vitro. Eur Polym J 43:488–497

Madeira MD, Sousa N, Santer RM, Paula-Barbosa MM, Gundersen HJ (1995) Age and sex do not affect the volume, cell numbers, or cell size of the suprachiasmatic nucleus of the rat: an unbiased stereological study. J Comp Neurol 361:585–601

Manchanda S, Mishra R, Singh R, Kaur T, Kaur G (2016) Aqueous leaf extract of *Withania somnifera* as a potential neuroprotective agent in sleep-deprived rats: a mechanistic study. Mol Neurobiol [ePub ahead of print]

Manikonda PK, Jagota A (2012) Melatonin administration differentially affects age-induced alterations in daily rhythms of lipid peroxidation and antioxidant enzymes in male rat liver. Biogerentology 13:511–524

Mattam U, Jagota A (2014) Differential role of melatonin in restoration of age-induced alterations in daily rhythms of expression of various clock genes in suprachiasmatic nucleus of male Wistar rats. Biogerentology 15:257–268

Mattam U, Jagota A (2015) Daily rhythms of serotonin metabolism and the expression of clock genes in suprachiasmatic nucleus of rotenone-induced Parkinson's disease male Wistar rat model and effect of melatonin administration. Biogerontology 16:109–123

Mishra K, Ojha H, Chaudhury NK (2012) Estimation of antiradical properties of antioxidants using DPPH assay: a critical review and results. Food Chem 130:1036–1043

Miyata S, Noda A, Iwamoto K, Kawano N, Okuda M, Ozaki N (2013) Poor sleep quality impairs cognitive performance in older adults. J Sleep Res 22:535–541

Musiek ES (2015) Circadian clock disruption in neurodegenerative diseases: cause and effect? Front Pharmacol 6:29

Nakahata Y, Sahar S, Astarita G, Kaluzova M, Sassone-Corsi P (2009) Circadian control of the NAD^+ salvage pathway by CLOCK-SIRT1. Science 324:654–657

Panchawat S (2011) In vitro free radical scavenging activity of leaves extracts of *Withania somnifera*. Rec Res Sci Tec 3:40–43

Park K, Kang HM (2004) Cloning and circadian expression of rat Cry1. Mol Cells 18:256–260

Partch CL, Green CB, Takahashi JS (2014) Molecular architecture of the mammalian circadian clock. Trends Cell Biol 24:90–99

Prakash J, Chouhan S, Yadav SK, Westfall S, Rai SN, Singh SP (2014) *Withania somnifera* alleviates parkinsonian phenotypes by inhibiting apoptotic pathways in dopaminergic neurons. Neurochem Res 39:2527–2536

Quintero JE, Kuhlman SJ, McMahon DG (2003) The biological clock nucleus: a multiphasic oscillator network regulated by light. J Neurosci 23:8070–8076

RajaSankar S, Manivasagam T, Sankar V, Prakash S, Muthusamy R, Krishnamurti A, Surendran S (2009) *Withania somnifera* root extract improves catecholamines and physiological abnormalities seen in a Parkinson's disease model mouse. J Ethnopharmacol 125:369–373

Rakshit K, Giebultowicz JM (2013) Cryptochrome restores dampened circadian rhythms and promotes healthspan in aging Drosophila. Aging cell 12:752–762

Reddy VDK, Jagota A (2014) Effect of restricted feeding on nocturnality and daily leptin rhythms in OVLT in aged male Wistar rats. Biogerontology 15:245–256

Reddy MY, Jagota A (2015) Melatonin has differential effects on age-induced stoichiometric changes in daily chronomics of serotonin metabolism in SCN of male Wistar rats. Biogerontology 16:285–302

Reiter RJ, Tan DX, Fuentes-Broto L (2010) Melatonin: a multitasking molecule. Prog Brain Res 181:127–151

Sato TK, Yamada RG, Ukai H, Baggs JE, Miraglia LJ, Kobayashi TJ, Welsh DK, Kay SA, Ueda HR, Hogenesch JB (2006) Feedback repression is required for mammalian circadian clock function. Nat Genet 38:312–319

Schliebs R, Liebmann A, Bhattacharya SK, Kumar A, Ghosal S, Bigl V (1997) Systemic administration of defined extracts from *Withania somnifera* (Indian Ginseng) and Shilajit differentially affects cholinergic but not glutamatergic and GABAergic markers in rat brain. Neurochem Int 30:181–190

Sehgal N, Gupta A, Valli RK, Joshi SD, Mills JT, Hamel E, Khanna P, Jain SC, Thakur SS, Ravindranath V (2012) *Withania somnifera* reverses Alzheimer's disease pathology by enhancing low-density lipoprotein receptor-related protein in liver. Proc Natl Acad Sci 109:3510–3515

Sengupta S, Yang G, O'Donnell JC, Hinson MD, McCormack SE, Falk MJ, La P, Robinson MB, Williams ML, Yohannes MT, Polyak E, Nakamaru-Ogiso E, Dennery PA (2016) The circadian gene Rev-erbα improves cellular bioenergetics and provides preconditioning for protection against oxidative stress. Free Radic Biol Med 93:177–189

Shah PC, Trivedi NA, Bhatt JD, Hemavathi KG (2006) Effect of Withania somnifera on forced swimming test induced immobility in mice and its interaction with various drugs. Indian J Physiol Pharmacol 50:409–415

Shah N, Kataria H, Kaul SC, Ishii T, Kaur G, Wadhwa R (2009) Effect of the alcoholic extract of Ashwagandha leaves and its components on proliferation, migration, and differentiation of glioblastoma cells: combinational approach for enhanced differentiation. Cancer Sci 100:1740–1747

Shah N, Singh R, Sarangi U, Saxena N, Chaudhary A, Kaur G, Kaul SC, Wadhwa R (2015) Combinations of Ashwagandha leaf extracts protect brain-derived cells against oxidative stress and induce differentiation. PloS One 10:e0120554

Sohal RS, Orr WC (2012) The redox stress hypothesis of aging. Free Radic Biol Med 52:539–555

Soman S, Korah PK, Jayanarayanan S, Mathew J, Paulose CS (2012) Oxidative stress induced NMDA receptor alteration leads to spatial memory deficits in temporal lobe epilepsy: ameliorative effects of *Withania somnifera* and Withanolide A. Neurochem Res 37:1915–1927

Travnickova-Bendova Z, Cermakian N, Reppert SM, Sassone-Corsi P (2002) Bimodal regulation of mPeriod promoters by CREB-dependent signaling and CLOCK/BMAL1 activity. Proc Natl Acad Sci 99:7728–7733

Ueda HR, Hayashi S, Chen W, Sano M, Machida M, Shigeyoshi Y, Iino M, Hashimoto S (2005) System-level identification of transcriptional circuits underlying mammalian circadian clocks. Nat Genet 37:187–192

Ukai-Tadenuma M, Yamada RG, Xu H, Ripperger JA, Liu AC, Ueda HR (2011) Delay in feedback repression by cryptochrome 1 is required for circadian clock function. Cell 144:268–281

Verma SK, Kumar A (2011) Therapeutic uses of *Withania somnifera* (Ashwagandha) with a note on withanolides and its pharmacological actions. Asian J Pharma Clin Res 4:1–4

Videnovic A, Noble C, Reid KJ, Peng J, Turek FW, Marconi A, Rademaker AW, Simuni T, Zadikoff C, Zee PC (2014) Circadian melatonin rhythm and excessive daytime sleepiness in Parkinson disease. JAMA neurol 71:463–469

Vinod Ch, Jagota A (2016) Daily NO rhythms in peripheral clocks on aging in male Wistar rats: Protective effects of exogenous melatonin. Biogerontology 17:859–871

Wadhwa R, Konar A, Kaul SC (2016) Nootropic potential of Ashwagandha leaves: beyond traditional root extracts. Neurochem Int 95:109–118

Weinert D (2000) Age-dependent changes of the circadian system. Chronobiol Int 17:261–283

Welsh DK, Takahashi JS, Kay SA (2010) Suprachiasmatic nucleus: cell autonomy and network properties. Annu Rev Physiol 72:551–577

Widodo N, Shah N, Priyandoko D, Ishii T, Kaul SC, Wadhwa R (2009) Deceleration of senescence in normal human fibroblasts by withanone extracted from ashwagandha leaves. J Gerontol A Biol Sci Med Sci 64:1031–1038

Yan L, Takekida S, Shigeyoshi Y, Okamura H (1999) Per1 and Per2 gene expression in the rat suprachiasmatic nucleus: circadian profile and the compartment-specific response to light. Neuroscience 94:141–150

Yang G, Wright CJ, Hinson MD, Fernando AP, Sengupta S, Biswas C, La P, Dennery PA (2014) Oxidative stress and inflammation modulate Rev-erbα signaling in the neonatal lung and affect circadian rhythmicity. Antioxid Redox Signal 21:17–32

Yonei Y, Hattori A, Tsutsui K, Okawa M, Ishizuka B (2010) Effects of melatonin: basics studies and clinical applications. Anti-Aging Med 7:85–91

Zhang H, Davies KJ, Forman HJ (2015) Oxidative stress response and Nrf2 signaling in aging. Free Radic Biol Med 88:314–336

Zhou JN, Hofman MA, Swaab DF (1995) VIP neurons in the human SCN in relation to sex, age, and Alzheimer's disease. Neurobiol Aging 16:571–576

Part IV
Active Ingredient Enriched Ashwagandha – Biotechnologies

Chapter 22
Biotechnological Approaches in Propagation and Improvement of *Withania somnifera* (L.) Dunal

Kulwinder Kaur, Pritika Singh, Rupam Guleri, Baldev Singh, Kuldeep Kaur, Varinder Singh, and Pratap Kumar Pati

Abstract *Withania somnifera* is a high value medicinal plant of the family Solanaceae. It is known for its anti-tumour, anti-stress, anti-aging, cardio-protective, neuro-protective and anti-inflammatory properties. Its pharmaceutical properties are attributed to a wide range of secondary metabolites, such as steroidal lactones, alkaloids, glycowithanolides, flavanol glycosides, phenolics and sterols. The traditional cultivation of *W. somnifera* is limited mainly by poor seed viability and germination, low yield and inconsistency in production of secondary metabolites. The infestation with various pests and pathogens also throw a major challenge in its cultivation. Biotechnological approaches involving organ, tissue and cell culture offer potential solution to the existing problems. *In vitro* propagation helps in rapid multiplication of elite cultivars and facilitate in raising quality planting materials. Genetic manipulation and secondary metabolite engineering hold great promise for enhancement of secondary metabolites and for overall crop improvement. The present chapter briefly discuss the challenges in *W. somnifera* and present a quick overview of biotechnological advances to address these challenges. It also highlights the futuristic approaches that would lay a foundation in the conglomeration of *W. somnifera* as an ideal model medicinal plant.

Keywords *Withania somnifera* • *In vitro* propagation • Secondary metabolites • Stress • Cell suspensions • Genetic transformation

---

K. Kaur • P. Singh • R. Guleri • B. Singh • K. Kaur • V. Singh • P.K. Pati (✉)
Department of Biotechnology, Guru Nanak Dev University, Amritsar 143 005, Punjab, India
e-mail: pkpati@yahoo.com

S.C. Kaul, R. Wadhwa (eds.), *Science of Ashwagandha: Preventive and Therapeutic Potentials*, DOI 10.1007/978-3-319-59192-6_22

22.1 Introduction

Withania somnifera (L.) Dunal is a high value medicinal plant and is extensively used in Indian, African and Unani traditional system of medicine. It belongs to the family Solanaceae and is commonly referred to as Ashwagandha, Asgandh, Winter cherry or Indian ginseng. *W. somnifera* enjoys a position in the monographs on selected medicinal plants prepared by World Health Organization (WHO) (Mirjalili et al. 2009) and is among top 32 important medicinal plants chosen by National Medicinal Plant Board of India (www.nmpb.nic.in) due to its immense demand in the domestic and international markets (Prajapati et al. 2006). The genus *Withania* includes more than 23 species which are widely distributed in tropical and subtropical areas such as dry regions of Africa, South Asia and Central Asia, particularly in Bangladesh, Pakistan, Afghanistan, Sri Lanka, Egypt, Morocco, South Africa, India, Congo and Jordon (Kumar et al. 2007, 2011).

The plant is dicotyledonous, erect, woody evergreen, branched shrub that grows about 30–75 cm in height. The leaves are simple, petiolate, ovate, entire and arranged in alternate fashion (Mirjalili et al. 2009). Axil contains a cymose cluster of 5–25 inconspicuous pale green flowers. The roots are cylindrical and fleshy with light brown epidermis and white medulla. Flowers are 4–6 mm in diameter having short pedicel, gamosepalous, 5 sepals persistent and having acute linear lobes. Its corolla is gamopetalous, having 5-lobes that are recurved or spreading, acute pubescent and greenish yellow while stamens arise from the base of petals and are slender filaments. Fruit is enclosed in the green persistent calyx. It is green when unripe and turns orange-red when mature enclosing numerous small seeds (Singh and Kumar 1998). However, considerable morphogenetic variability exists in different accessions of *W. somnifera* (Kumar et al. 2007).

22.2 Economic Importance

W. somnifera is used in various forms (ointments, decoctions, powder, infusions and syrup) as herbal medicine (Kumar et al. 2007; Davis and Kuttan 2000). The extracts and different bioactive constituents are known to possess anticancer, anti-convulsant, anti-oxidative, anti-aging, anti-spasmodic, anti-inflammatory, cardio-protective, immunomodulatory, adaptogenic and neuroprotective properties (Singh et al. 2015b). The pharmacological traits of *W. somnifera* are due to the existence of a wide range of structurally diverse set of secondary metabolites in leaves and roots. Several bioactive alkaloids and phytochemicals such as tropine, psudotropine, hygrine, ashwagandhine, choline, cuscohygrine, flavanol glycosides, isopelletierine, anaferine, tropine, sitoindosides (saponins), anhygrine,withanolides, withanosomine, glycowithanolides, withanamides, sterols and phenolics are present in different parts of the plant (Misra et al. 2005, 2008; Matsuda et al. 2001; Chatterjee et al. 2010; Chaurasiya et al. 2012).

Conventionally, leaves and roots of *W. somnifera* are principally used in preparation of herbal medicines (Jayaprakasam et al. 2003; Kumar et al. 2011). Recently NMR and chromatographic techniques have been employed for metabolic profiling of leaf and root extract of *W. somnifera*. A total of 48 primary and secondary metabolites from roots and 62 from leaves have been reported. Among these, 29 metabolites are common to both the tissues. However, a noteworthy difference in the profile of withanolides has been reported in leaf and root tissues (Chatterjee et al. 2010). Withaferin A and Withanone are mainly found in leaves, whereas root is the major site of withanolide A (Singh et al. 2015b). Along with spatial variations, the distribution of secondary metabolites also shows temporal as well as quantitative and qualitative variations with respect to different chemotypes and growth conditions (Chatterjee et al. 2010; Dhar et al. 2013). Further, *W. somnifera* from different origins can be discriminated by the ratio of two groups of withanolides *viz.*4-OH and 5,6-epoxy withanolides and 5-OH and 6,7-epoxy withanolides in their leaves (Namdeo et al. 2011). Age of the plant has a significant bearing on the metabolic contents of the plant. The content of Withaferin A increases in leaves and roots with the growth of the plant from young (6 weeks old) to mature (12–18 weeks old) stage (Pal et al. 2011). However, accumulation of most of the withanolides declined at over-maturation stage (Dhar et al. 2013).

22.3 Major Challenges and Strategies for Crop Improvement in *W. somnifera*

Though *W. somnifera* is an important medicinal plant, it encounters several challenges in its mass propagation, improvement/production of secondary metabolites and protection from various biotic and abiotic stresses. The present chapter highlights some of the challenges and research endeavour to address these emerging issues.

22.3.1 Mass Propagation

W. somnifera is conventionally propagated via seeds. However, its traditional cultivation is restricted due to poor seed germination, low percentage of seed viability and seedling survival (Vakeswaran and Krishnasamy 2003), long gestation period between planting and harvesting (Rani et al. 2003) and susceptibility towards pathogens (Pati et al. 2008) and pests (Sharma and Pati 2011b). Further, effect of geographical and seasonal variations on the yield and composition of secondary metabolites in plants, low yield of withanolides under *in vivo* conditions, a narrow genetic base and presence of intraspecific chemotypic variations impose a challenge in its cultivation and improvement.

Fig. 22.1 Micropropagation in *W. somnifera*. (**a**) Initiation of aseptic cultures from nodal explant in MS medium. (**b**) Shoot proliferation in MS medium supplemented with BAP (5.0 μM). (**c**) Rooting of *in vitro* shoots in MS medium supplemented with IBA (10.0 μM). (**d**) Hardening of microshoots

Biotechnological approach such as micropropagation provides a suitable option for rapid multiplication and raising uniform quality clones (Fig. 22.1). In *W. somnifera*, micropropagation has been performed using different explants *viz.* seeds (Supe et al. 2006; Sen and Sharma 1991), shoot tips (Furmanowa et al. 2001; Ray and Jha 2001; Sivanesan 2007; Ahmad Baba et al. 2013; Sen and Sharma 1991) and nodal segments (Sabir et al. 2008; Soni et al. 2011; Fatima et al. 2011; Sivanandhan et al. 2011). Most of these studies have documented the use of Murashige and Skoog's (Murashige and Skoog 1962) medium for seed germination and micropropagation. Seed germination relies largely on the type of plant growth regulators (PGR), light and temperature. Maximum seed germination has been reported with GA at 25 °C (Khanna et al. 2013). Among different carbon sources, sucrose (2–3%) is mainly used for the proliferation of shoots and rooting (Sivanesan and Murugesan 2008).The growth and proliferation of microshoots and their rooting is significantly affected by the exogenous supply of PGRs. BAP is most effective for the formation of multiple shoots and its proliferation (Soni et al. 2011; Ray and Jha 2001; Nayak et al. 2013). However, BAP in combination with IAA (Sivanesan 2007; Sivanesan and Murugesan 2008) or IBA (Furmanowa et al. 2001; Sen and Sharma 1991) or KN (Sabir et al. 2008) have also been frequently used for multiple shoot formation and its proliferation. Plant growth medium supplemented with inorganic nutrients *viz.* $ZnSO_4$, $CuSO_4$ and polyamines showed enhanced shoot proliferation (Fatima et al. 2011; Sivanandhan et al. 2011). BAP was also found to be most suitable in liquid culture of *W. somnifera* (Sivanandhan et al. 2012). However, hyperhydricity of shoots limit the use of liquid culture system in *W. somnifera*. Microshoots have also been successfully rooted in both MS and half-strength MS medium. Among various auxins, IBA is most effective for *in vitro* rooting (Udayakumar et al. 2013; Ghimire et al. 2010). The rooted microshoots are hardened in a mixture of sand and soil (1:1) and the plantlets are established under field conditions (Kulkarni et al. 2000; Supe et al. 2006).

22.3.2 *Withanolide Biosynthesis and Their Modulation in* W. sominifera

W. somnifera has been extensively studied due to possession of a wide range of pharmaceutically important secondary metabolites. Among these, the major metabolites are withanolides that are a group of steroidal lactones (C-28) based on ergostane skeleton and constitute the prime metabolites of *W. somnifera*. Withanolides are synthesized from 24-methylene cholesterol, which itself originate from isoprenoid pathway (Sangwan et al. 2008). In plants, the biosynthesis of isoprenoid occurs *via* two independent pathways, Mevalonate (MVA) pathway and 2-C-methyl-D-erythritol-4-phosphate (MEP) pathway (Fig. 22.2). The predominant MVA pathway operates in cytosol and accounts for 75% of carbon contribution for the synthesis of withanolides, whereas, non-MVA/MEP pathway operates in the plastids and contributes to 25% of carbon for the synthesis of withanolides (Chaurasiya et al. 2012). During MVA pathway, acetyl-CoA undergoes activation to form acetoacetyl-CoA. Acetoacetyl-CoA molecule condenses with acetyl-CoA to form 3-hydroxy-3-methylglutaryl-CoA (HMG-CoA) in the presence of enzyme HMG-CoA synthase (HMGS). In an irreversible reaction catalysed by 3-hydroxy-3-methylglutaryl-CoA reductase (HMGR), 3-hydroxy-3-methylglutaryl-CoA is converted into mevalonic acid. Next two steps are phosphorylation reactions that are catalysed by mevalonate kinase and phosphomevalonate kinase that results in the formation of 5-phosphomevalonate and 5-pyrophosphomevalonate, respectively. 5-pyrophosphomevalonate leads to the formation of 3-isopentenyl pyrophosphate (IPP) with the help of enzyme mevalonate-5-pyrophosphate decarboxylase. IPP condenses with its isomer 3,3-dimethyl allyl pyrophosphate (DMAPP) to yield geranyl pyrophosphate (GPP) in a reaction catalysed by farnesyl diphosphate synthase (FPPS). Further, trans-geranyl pyrophosphate condenses with another molecule of IPP to form farnesyl pyrophosphate (FPP). In the subsequent step, two molecules of farnesyl diphosphate condense by the enzyme squalene synthase (SQS) to yield squalene, a liner 30-carbon compound. Squalene epoxidase (SE)

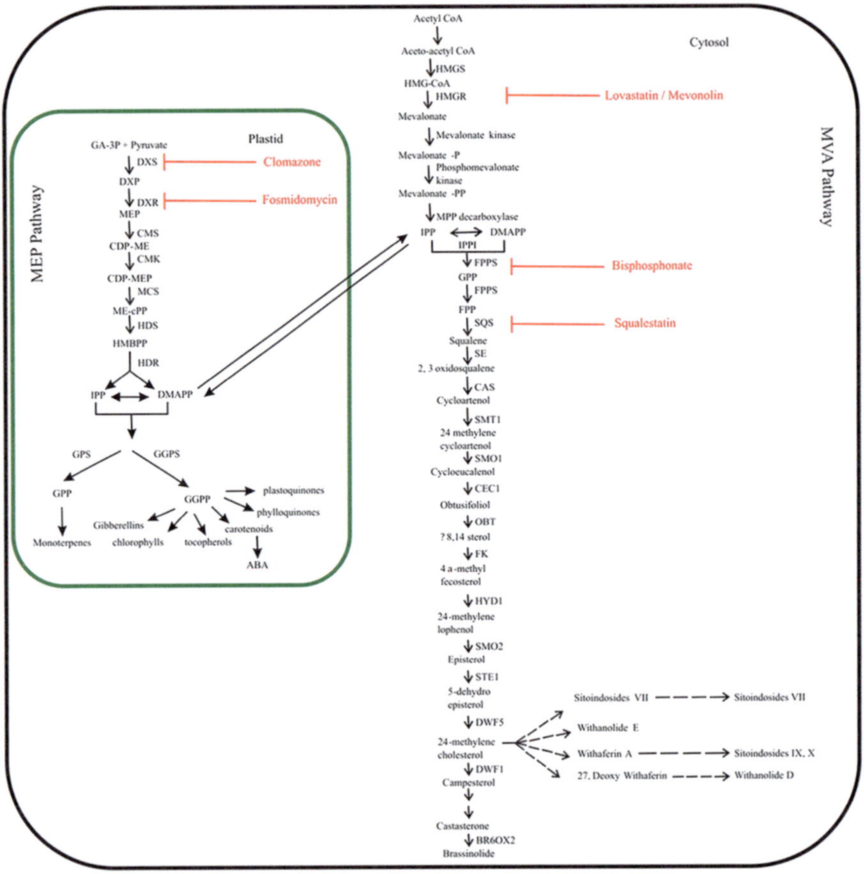

Fig. 22.2 Withanolide biosynthesis pathway and site of action of selected inhibitors. *HMG-CoA* 3-hydroxy-3-methylglutaryl-CoA, *HMGS* HMG-CoA synthase, *HMGR* 3-hydroxy-3-methylglutaryl-CoA reductase, *IPP* 3-isopentenyl pyrophosphate, *DMAPP* 3,3-dimethyl allyl pyrophosphate, *IPPI* isopentenyl pyrophosphate isomerase, *GPP* geranyl pyrophosphate, *FPPS* farnesyl diphosphate synthase, *SQS* squalene synthase, *SE* squalene epoxidase, *CAS* cycloartenol synthase, *SMTI* sterol methyltransferaseI, *SMOI* sterol-4α-methyl oxidase 1, *CECI* cycloeucalenol cycloisomerase, *OBT* obtusifoliol 14-demethylase, *FK* delta 14-sterol reductase, *HYDI* C-7,8 sterol isomerase, *DWFI* delta-24 sterol reductase, *SMO2* sterol-4a-methyl oxidase 2, *STEI* C-5 sterol desaturase, *DWF5* sterol delta-7 reductase, *BR6OX2* brassinosteroid-6-oxidase 2, *GA-3P* D-glyceraldehyde-3-phosphate, *DXP* 1-deoxy-D-xylulose-5-phosphate, *DXS* DXP synthase, *DXR* DXP reductoisomerase, *MEP* 2-methyl D-erythritol 4-phosphate, *CMS* 4-(cytidine-5-diphospho)-2-C-tmethyl-D-erythritol synthase, *CDP-ME* 4-diphospho-cytidyl-2-methyl-D-erythritol, *CMK* 4-(cytidine-5-diphospho)-2-C-methyl-D-erythritol kinase, *CDP-MEP* 2-C-methyl-D-erythritol-2-phosphate, *MCS* 2-C-methyl-D-erythritol-2,4-cyclodiphosphate synthase, *ME-cPP* – 2-C-methyl-D-erythritol-2,4-cyclodiphosphate, *HDS* hydroxyl methyl butenyl 4-diphosphate synthase, *HMBPP* hydroxyl methyl butenyl 4-diphosphate, *HDR* hydroxy methyl butenyl 4-diphosphate reductase, *GPP* geranyl diphosphate, *GPS* GPP synthase, *GGPP* geranylgeranyl diphosphate, *GGPS* geranylgeranyl diphosphate synthase, *ABA* abscisic acid

then convert squalene into 2,3-oxidosqualene, which after ring closure, yield various steroidal triterpenoidal skeletons (Singh et al. 2015b; Mirjalili et al. 2009).

MEP pathway starts with the condensation of pyruvate with D-glyceraldehyde-3-phosphate (GA-3P) to yield 1-deoxy-D-xylulose-5-phosphate (DXP) and this reaction is catalysed by DXP synthase (DXS). DXP is further converted into 2-methyl D-erythritol 4-phosphate (MEP) by the enzyme DXP reductoisomerase (DXR). 4-(cytidine-5-diphospho)-2-C-tmethyl-D-erythritol synthase (CMS) in a CTP-dependent reaction catalyse the conversion of MEP into 4-diphospho-cytidyl-2-methyl-D-erythritol (CDP-ME). The later then undergoes phosphorylation to form 2-C-methyl-D-erythritol-2-phosphate (CDP-MEP) by 4-(cytidine-5-diphospho)-2-C-methyl-D-erythritol kinase (CMK). Further, in a reaction catalysed by 2-C-methyl-D-erythritol-2,4-cyclodiphosphate synthase (MCS), CDP-MEP leads to the formation of 2-C-methyl-D-erythritol-2,4-cyclodiphosphate (ME-cPP). In the subsequent steps, ME-cPP yield hydroxyl methyl butenyl 4-diphosphate (HMBPP) in a reaction catalysed by hydroxyl methyl butenyl 4-diphosphate synthase (HDS). The enzyme hydroxy methyl butenyl 4-diphosphate reductase (HDR), converts HMBPP into a mixture of IPP and DMAPP (Hunter 2007; Rodriguez-Concepcion and Boronat 2002).

There has been a huge interest in enhancement of withanolide production in *W. somnifera*. However, the major bottlenecks in this direction is the existence of chemotypic variations (Bhatia et al. 2013), unavailability of information for genome sequence and the cumbersome process of chemical synthesis of withanolides (Grover et al. 2013). In recent years various strategies have been adapted to modulate the withanolide production in *W. somnifera*.

22.3.3 *In Vitro* Culture System and Their Potential in Modulation of Withanolide Production

In vitro strategies *viz*. organ, tissue and cell culture and genetic manipulations can potentially overcome the challenges faced to increase the production of pharmaceutically important secondary metabolites under *in vivo* conditions. In *W. somnifera*, *in vitro* culture system is used as an alternative system for production of secondary metabolites. It has high metabolic rate and active growth in a short time period, leading to increased secondary metabolite production and its accumulation. A significant increase in withanolides production has been shown in *in vitro* shoot culture system as compared to field grown plants (Sharada et al. 2007; Sabir et al. 2008).

In *in vitro* culture system, the secondary metabolite production is governed by media formulations, type and concentration of carbon source and PGR. Half strength MS medium is most efficient for both the production of withanolide A and biomass accumulation in adventitious root culture (Praveen and Murthy 2012 & Praveen and Murthy 2013). In *in vitro* shoot culture, the concentration of sucrose is directly proportional to withaferin A and withanolide D production (Sivanandhan et al. 2015b).

In most of the reports, BAP has been documented to be most suitable for withanolide production (Ray and Jha 2001; Mir et al. 2014). However, MS medium supplemented with a combination of BAP and KN resulted in higher accumulation of withanolide A (Sangwan et al. 2007). Differential accumulation of withanolides has been recorded in *in vitro* flowers and fruits. *In vitro* fruits have been shown to accumulate enhanced level of withanolide A and withanolide B, whereas, the level of withaferin A was more in *in vitro* flowers as compared to fruits. However, withanones did not show significant difference in *in vitro* fruits and flowers (Sivanandhan et al. 2015b). Further, exogenous application of elicitors such as MeJ and SA, and polyamines in the shoot culture medium led to marked improvement in secondary metabolite production (Sivanandhan et al. 2011, 2012).

Cell suspension culture holds the potential for enhancing secondary metabolites production (Mulabagal and Tsay 2004) and is also used to study cellular and molecular processes of secondary metabolite production and accumulation. The potential of cell suspension culture is affected by various factors, such as media composition, source of carbon and PGRs. MS medium is most effective for both the production of withanolide A and biomass accumulation (Nagella and Murthy 2010, 2011). As compared to glucose, fructose and maltose, sucrose was reported to be the best source of carbon for production of withanolide A (Sivanandhan et al. 2013b; Nagella and Murthy 2011). Manipulation of medium salt concentrations has important bearing in enhancing the production of withanolides. It has been observed that concentration of nitrogen and ammonia/nitrate ratio has great influence on the production of secondary metabolites. A high NO_3 and low NH_4 concentration supports both cell growth and production of withanolide (Nagella and Murthy 2011). PGRs also influence the potential of cell suspension culture for secondary metabolite production. MS medium having KN and 2,4-D led to higher accumulation of withanolide A (Nagella and Murthy 2011). In addition, elicitation acts as a promising strategy to raise the production of pharmaceutically important secondary metabolites in cell suspension cultures. Inorganic compounds *viz.* MeJ, SA and arachidonic acid, and biotic factors *viz.* components of plant cell wall and culture filtrates of *Fusarium solani*, *Alternaria alternata* and *Verticilium dahilae* act as potential elicitors (Sivanandhan et al. 2013b).

22.3.4 Enhancing Secondary Metabolite Production by Genetic Manipulation

Agrobacterium - mediated genetic transformation provides an excellent opportunity to study the function of various genes, understand their regulation and for increasing the productivity of novel secondary metabolites (Cucu et al. 2002). Several factors such as *Agrobacterium* strain, time of infection, method of co-cultivation, concentration of phytohormones, selection conditions, type and age of explant and germplasm of the plant have significant bearing on transformation efficiency

(Herrera-Estrella et al. 2004). The earliest attempt to transform *W. somnifera* was made by infecting its leaves with wild type octopine and nopaline strains of *A. tumefaciens*, resulted to shooty teratomas only (Ray and Jha 1999). The fertile transgenic plants of *W. somnifera* were developed with very low transformation efficiency (1.67%) by infecting the leaf segments with *A. tumefaciens* (LBA4404) harbouring pIG121Hm vector (Pandey et al. 2010). The leaf segments were co-cultivated with *A. tumefaciens* in MS medium fortified with 8.9 μM BAP and 8.0 μM IAA. The growth of *A. tumefaciens* was suppressed by using augmentin and transformants were selected on 50 mg/l kanamycin. The transformation efficiency was improved up to 10% by employing the techniques of vacuum infiltration and sonication. Addition of thiol compounds, sodium thiosulphate (125 mg/l), L-cysteine (100 mg/l) and dithiothreitol (75 mg/l) in co-cultivation medium were also found to be beneficial in improving the transformation efficiency of *W. somnifera* (Sivanandhan et al. 2015a).

The development of genetic transformation system offers the advantage of exploring systematic metabolic engineering by modulating the expression of the key genes of withanolide biosynthetic pathway. Overexpression of squalene synthase, a rate limiting enzyme in the biosynthesis of withanolide, resulted in significant enhancement in Withanolide A content (2.5- folds) and squalene synthase activity (4-folds) in recombinant cell lines as compared to control. Withaferin A was additionally produced in transformed cell lines which was not produced in non-transformed cells (Grover et al. 2013). Transgenic plants overexpressing *squalene synthase* gene showed increase in the expression of *squalene synthase* and total withanolide content upto 2–5 folds and 1.5–2 folds, respectively (Patel et al. 2015). Due to significant enhancement in the production of withanolides in transformants, this strategy can be explored for effective metabolic engineering.

Transgenic hairy root cultures have revolutionized the secondary metabolites production due to their capacity to synthesis the metabolites at a much faster rate. Hairy roots results from integration of T-DNA of pRi plasmid (root inducing plasmid) from *Agrobacterium rhizogenes* to the susceptible explant and their induction frequency can be enhanced by subjecting the explants to sonication (15 s) followed by heat treatment for 5 min at 41 °C (Thilip et al. 2015). Hairy root cultures are genetically stable, produce biomass in large scale and are easily maintained. Various parameters *viz.* type and age of explant, culture medium, *Agrobacterium* stain, concentration of acetosyringone, and co-cultivation period are critical in determining transformation efficiency (Sivanandhan et al. 2014; Pawar and Maheshwari 2004). In *W. somnifera*, leaves and cotyledons show higher transformation potential compared to other explants (Murthy et al. 2008; Pawar and Maheshwari 2004). MS (Murashige and Skoog) medium is best suited for hairy roots growth and biomass accumulation (Murthy et al. 2008; Saravanakumar et al. 2012). Manipulation with source of carbon and its concentration results in differential accumulation of withanolides in hairy root cultures of *W. somnifera*. Low concentration of sucrose (3%) is effective for the growth of transformed hairy roots and production of withaferin A and withanolide A (Praveen and Murthy 2012). Whereas, high concentration of sucrose (4%) results in increased production of Withaferin A in hairy roots.

Supplementing the medium with glucose (5%) leads to enhanced production of withanolide A as well as withaferin A (Doma et al. 2012). Withanolide production can also be increased in hairy root cultures by manipulating the concentration of macroelements and nitrogen source. Growth medium containing moderate concentration of NH_4 (14.38 mM) with higher concentration of NO_3 (37.60 mM) favours higher withanolide A production and biomass accumulation (Praveen and Murthy 2013).

The transformed hairy root cultures of *W. somnifera* were found to be distinct in their ability to produce a wide array of chemical compounds (Kim et al. 2002; Giri and Narasu 2000; Bandyopadhyay et al. 2007). Transformed root cultures were shown to accumulate enhanced amount of withanolide A, Withanolide D and Withaferin A (Ray et al. 1996; Ray and Jha 1999; Bandyopadhyay et al. 2007; Murthy et al. 2008). Hairy roots grown in liquid medium also showed significant enhancement in the antioxidant activity (Kumar et al. 2005). A positive correlation between elicitor treatment and withanolide production could be established in *W. somnifera* hairy roots. The optimum concentrations of MeJ and SA which leads to increased withanolide A, withaferin A and withanone production were reported to be 15 μM and 150 μM, respectively (Sivanandhan et al. 2013a). Constitutive expression of β-cryptogein (a fungal elicitor protein) in hairy root cultures of *W. somnifera* resulted in metabolic shift from withanolide biosynthesis to phenypropanoid biosynthetic pathway (Sil et al. 2015) resulting in higher amount of ferulic acid along with increase in activity of enzyme phenylalanine ammonia lyase (PAL) in cryptogein-cotransformed hairy roots.

22.3.5 *Characterization of Withanolide Biosynthetic Pathway Genes*

The major bottle neck in enhancement of pharamaceutically important secondary metabolites in *W. somnifera* has been the lack of proper information on their biosynthetic pathways. Recently, withanolide biosynthetic pathway genes such as HMGR (Akhtar et al. 2013), FPPS (Gupta et al. 2011), SQS (Bhat et al. 2012), SE (Razdan et al. 2013), DXS and DXR (Gupta et al. 2013), Cyt P450 (Rana et al. 2013), sterol glycosyltransferase (Sharma et al. 2007) and genes of oxido squalene cyclase (OSC) super-family: cycloartenol synthase (OSC/CAS), β-amyrin synthase (OSC/BS) and lupeol synthase (OSC/LS) (Dhar et al. 2014) have been identified, cloned and characterized. The expression analysis of HMGR, FPPS, SQS, DXS and DXR showed the differential expression of these genes in different chemotypes, tissue, mechanical injury and in response to elicitors (Gupta et al. 2011, 2013; Bhat et al. 2012; Akhtar et al. 2013). Also, the expression of HMGR, FPPS, SQS, DXS and DXR was enhanced in young leaves in comparison to mature leaves and roots. The high expression in leaves may attribute to higher rate of biosynthesis of withanolides in young leaves (Chaurasiya et al. 2007; Gupta et al. 2012 & Gupta et al. 2013).

Further, low expression of DXS and DXR in roots suggests that these enzymes are localized in plastids. These studies indicate that although leaves are the major site for biosynthesis of withanolides, an independent system for the biosynthesis of withanolide may exist in roots also (Sangwan et al. 2008).

Comparative expression analysis of the genes combined with the overexpression and knockout studies have facilitated the characterization of withanolide biosynthetic pathway genes and also provide clue for enhancing the accumulation of withanolides. One of the most important aspects in the study of gene expression by quantitative real-time PCR (qRT-PCR) is the selection of suitable reference genes for data normalization. Author's laboratory studied 11 candidate references genes *viz.* Actin (*ACT*), cyclophylin (*CYP*), 26S ribosomal RNA (*26S*), 18S ribosomal RNA (*18S–rRNA*), glyceral-dehyde-3-phosphate dehydrogenase (*GAPDH*), ubiquitin (*UBQ*), elongation factor-1 (*EF-1*), alpha-tubulin (*TUA*), beta-tubulin (*TUB*), Sand family protein (*SAND*) and ribosomal protein L2 (*RPL2*) to test the stability of their transcripts under various conditions. *26S, UBQ and TUB* were found to be best reference genes for drought, cold and biotic stress respectively. *T-SAND* was most stably expressed gene in tissues treated with salt and salicylic acid (SA), whereas, *18S–rRNA* was most stable in abscisic acid (ABA) treated samples (Singh et al. 2015c). This work has made significant contribution in improving the quality of q-PCR mediated gene expression data in *W. somnifera.*

Recently virus induced gene silencing (VIGS) approach has been exploited to analyse the function of the genes involved in withanolide biosynthetic pathway. VIGS of *squalene synthase* gene resulted in significant reduction in the level of sitosterol (15%), campesterol (16%), stigmasterol (26%) and total withanolides (54%) implicating a major role of *squalene synthase* in withanolide biosynthesis. Further, *WsSQS*-vigs causing down regulation of pathogenesis related (PR) proteins such as PR1 (46%), PR3 (50%), PR5 (76%) and nonexpressor of PR, NPR 1 (45%) and NPR3 (81%) resulted in reduced tolerance to biotic stress (Singh et al. 2015a). In another study, overexpression of *WsHMGR2* in tobacco led to increase in the level of sitosterol (160%), stigmasterol (123%) and campesterol (546%). However, the increase in the level of these sterols was comparative low in *WsDXR2* overexpressing tobacco, sitosterol (25%), stigmasterol (28%) and campesterol (66%). The enhanacement in cholesterol content was marginally higher in *WsDXR2* (41%) than *WsHMGR2* (38%). Further, RNAi mediated transient gene silencing approach was followed by using pART27:*WsHMGR2i* and pART27:*WsDXR2i.* A higher reduction in the level of sitosterol, stigmasterol and campesterol was observed in pART27:*WsHMGR2i* as compared to pART27:*WsDXR2i*. The reduction in the level of cholesterol was comparatively higher in pART27:*WsDXR2i*. The RNAi suppression data was further validated by using specific inhibitors such as mevinolin (HMGR inhibitor) and fosmidomycin (DXR inhibitor) and a similar trend of sterol reduction has been reported (Singh et al. 2014). These studies indicated that MVA pathway is predominant in contributing carbon flux for withanolide biosynthesis and thereby metabolic engineering involving manipulation of MVA pathway genes may prove to be more beneficial to enhance secondary metabolites production. A recent study has highlighted the importance of nitrogen in enhancing the content of sterols and withaferin A (Pal et al. 2016). Treatment with ammonium sulfate resulted in higher expression of key genes of withanolide biosynthetic pathway *viz. FPPS, SMT1, SMT2, ODM, SMO1* and *SMO2*, leading to significant increase in the content of sterols. The improvement in the yield of secondary metabolites is probably linked to the expression of WRKY transcription factors. This study suggests that identification of transcription factors involved in withanolide biosynthetic pathway would be a promising approach for modulation of withanolide biosynthetic pathway.

22.4 Biotic and Abiotic Stress

For enhancing the commercial value of *W. somnifera,* studies are being conducted for the production of biotic and abiotic stress tolerant cultivar. *W. somnifera* is exposed to a plethora of various biotic (Sharma and Pati 2011a, b, 2012a, b) and abiotic (Rout and Sahoo 2012, 2013) stresses. At present, its agroecosystem is characterized by 26 major and minor insect pests (Chaudhary 2013) and various fungal pathogens (Pati et al. 2008; Maiti et al. 2007). In field conditions, this plant is frequently attacked by various pests *viz. Henosepilachna vigintioctopunctata* (Sharma and Pati 2011b), *Oxyrachis tarandus* (Sharma and Pati 2011a), *Phenacoccus solenopsis* (Sharma and Pati 2012a), *Tetranychus utricae* (Sharma and Pati 2012b), *Meloidogyne incognita* (Sharma and Pandey 2009), *Helicoverpa armigera* and *Epilachna vigintioctopunctata* (Kumar et al. 2009) and pathogens *viz. Alternaria alternata* (Pati et al. 2008), *A. dianthicola* (Maiti et al. 2007), *Pithomyces chartarum* (Verma et al. 2008) and phytoplasma (Khan et al. 2006). Among various diseases, leaf spot disease caused by *A.alternata* is one of the most prevalent diseases of *W. somnifera* (Sharma et al. 2014). This disease is characterized by brownish to black spots on the leaves (Fig. 22.3). During severe infection, 80–90% of leaves of a single plant may get infested with *A. alternata*, causing significant biodeterioration of pharmaceutical important secondary metabolites. A respective reduction of 15.4% and 76.3% in the content of withaferin A and total withanolides has been recorded in *W. somnifera* after infestation with *A. alternata* (Pati et al. 2008). Beside these, various post-infectional changes at morphological, biochemical and physiological level due to leaf spot disease have also been reported (Sharma et al. 2011; Sharma et al. 2014). Further, maintaining the quality of herbal products has been a critical issue for pharmaceutical industries. As leaf spot disease in *W. somnifera* is caused by *A. alternata,* the fungal infected herbal products might lead to serious health implications. Hence, consistent efforts in the development of sensitive detection system for fungal contamination in the leaves of *W. somnifera* are worth pursuing (Sharma 2013).

The research on stress tolerance in *W. somnifera* is at its nascent stage mainly due to limited information about different cultivars and their potential of resistance to pests and pathogens. A comprehensive study involving traditional and advanced biotechnological approaches need to be initiated. The traditional method of developing biotic resistance in plants involves conventional screening of plants resistant

Fig. 22.3 Leaf spot disease of *W. somnifera*. (**a**) Leaf spot disease infected Plant, (**b**) severely infected leaf, (**c**) isolation of fungus, (**d**) identification of causal organism, *A. alternata*

to major pests and pathogens for identification of resistant germplasms followed by breeding to inculcate resistant trait into sensitive cultivars. However, pests and pathogens frequently overcome single gene resistance in the host, marker-assisted breeding (MAS) would facilitate pyramiding of resistant genes providing broad spectrum resistant (Singh et al. 2001). Advanced biotechnological approaches involve overexpression of plant's defence genes to confer biotic stress resistance in plants. Plants show a rapid change in its gene expression in response to attack of pathogen or abiotic stress, resulting in synthesis of specific proteins that help to combat stress. Many of these inducible proteins are pathogenesis-related (PR) proteins (Sarowar et al. 2005). To date, 17 distinct families of PR proteins (PR1 to PR17) are found in plants (Sels et al. 2008). Overexpression of some of these PR proteins in different plants has conferred resistance to a large number of pathogens (Liu et al. 1994; Datta et al. 1999; Fagoaga et al. 2001; Alexander et al. 1993; Niderman et al. 1995). Author's laboratory is actively engaged in research on PR proteins, particularly PR1 and PR5. We are characterising *WsPR1* and *WsPR5* and also trying to elucidate their mechanism of action. One another approach that has gained considerable interest in recent years is to strengthen the plant resistance to stresses by enhancing the cell wall components such as lignin. Hence, modulation of phenylpropanoid pathway leading to the production of lignin holds great promise to combat biotic stress.

Global climate changes and unethical agricultural practices create stress conditions for plants and affect their growth and development. In this context, some preliminary studies on the influence of heavy metal stress on *W. somnifera* have been conducted. Copper toxicity induced reduction in fresh weight, chlorophyll and carotenoid content and the length of root and shoot has been observed. Further, increase in lipid peroxidation and level of $O_2^{\cdot-}$ and H_2O_2 and a significant decline in the activity of antioxidant enzymes *viz.* catalase (CAT), glutathione reductase (GR), glutathione peroxidase (GPx), and superoxide dismutase (SOD) reflected the inability of *W. somnifera* to eliminate the reactive oxygen species (ROS) resulted from copper induced oxidative stress (Khatun et al. 2008). However, Copper treatment led to induction of new genes as revealed by the appearance of new proteins in sodium dodecyl sulphate-polyacrylamide gel electrophoresis (SDS-PAGE).These induced proteins are considered to be involved in the defence of Copper toxicity (Rout and Sahoo 2013). Similar morphological and physiological changes have also been reported in high concentrations of iron (Rout and Sahoo 2012).Various studies need to be conducted for producing salinity, drought, elevated temperature and heavy metal stress tolerant plants.

22.5 Conclusion

W. somnifera has generated considerable interest among researchers due to its possession of pharmaceutically important secondary metabolites. Efforts are being carried out to overcome the bottleneck in its cultivation and improvement. Various biotechnological approaches *viz.in vitro* propagation, organ and cell suspension culture and genetic manipulations are being explored to solve the challenges of *W. somnifera.* Overall a comprehensive approach involving genomics, transcriptomics, proteomics, interactomics and metabolomics along with metabolic engineering and identification of new transcription factors will pave the path to the development of an ideal model plant of *W. somnifera*, having high commercial value.

Acknowledgements We duly acknowledge the financial support received under University For Potential Excellence (UPE) program from UGC, New Delhi, Govt. of India.

References

Ahmad Baba I, Alia A, Saxena RC, Itoo A, Kumar S, Ahmad M (2013) *In vitro* propagation of *Withania Somnifera* (L.) Dunal (Ashwagandha) an endangered medicinal plant. Intl J Pharm Sci Invention 2:6–11

Akhtar N, Gupta P, Sangwan NS, Sangwan RS, Trivedi PK (2013) Cloning and functional characterization of *3-hydroxy-3-methylglutaryl coenzyme A reductase* gene from *Withania somnifera*: an important medicinal plant. Protoplasma 250:613–622

Alexander D, Goodman RM, Gut-Rella M, Glascock C, Weymann K, Friedrich L, Maddox D, Ahl-Goy P, Luntz T, Ward E (1993) Increased tolerance to two oomycete pathogens in transgenic tobacco expressing pathogenesis-related protein 1a. Proc Natl Acad Sci USA 90:7327–7331

Bandyopadhyay M, Jha S, Tepfer D (2007) Changes in morphological phenotypes and withanolide composition of Ri-transformed roots of *Withania somnifera*. Plant Cell Rep 26:599–609

Bhat WW, Lattoo SK, Razdan S, Dhar N, Rana S, Dhar RS, Khan S, Vishwakarma RA (2012) Molecular cloning, bacterial expression and promoter analysis of *squalene synthase* from *Withania somnifera* (L.) Dunal. Gene 499:25–36

Bhatia A, Bharti SK, Tewari SK, Sidhu OP, Roy R (2013) Metabolic profiling for studying chemotype variations in *Withania somnifera* (L.) Dunal fruits using GC-MS and NMR spectroscopy. Phytochemistry 93:105–115

Chatterjee S, Srivastava S, Khalid A, Singh N, Sangwan RS, Sidhu OP, Roy R, Khetrapal CL, Tuli R (2010) Comprehensive metabolic fingerprinting of *Withania somnifera* leaf and root extracts. Phytochemistry 71:1085–1094

Chaudhary V (2013) Arthropods associated with ashwagandha, (*Withania somnifera* Dunal) in semi arid region of Gujarat. Karnataka J Agric Sci 26:433–435

Chaurasiya ND, Gupta VK, Sangwan RS (2007) Leaf ontogenic phase-related dynamics of withaferin a and withanone biogenesis in ashwagandha (*Withania somnifera* Dunal.)-an important medicinal herb. J Plant Biol 50:508–513

Chaurasiya ND, Sangwan NS, Sabir F, Misra L, Sangwan RS (2012) Withanolide biosynthesis recruits both mevalonate and DOXP pathways of isoprenogenesis in Ashwagandha *Withania somnifera* L.(Dunal). Plant Cell Rep 31:1889–1897

Cucu N, Negoianu-Tenea G, Gavrila L (2002) Genetically modified medicinal plants-II. Transfer and expression of a marker kanamycin resistance gene in *Atropa belladonna* plants. Roman Biotechnol Lett 7:869–874

Datta K, Velazhahan R, Oliva N, Ona I, Mew T, Khush GS, Muthukrishnan S, Datta SK (1999) Over-expression of the cloned rice thaumatin-like protein (*PR-5*) gene in transgenic rice plants enhances environmental friendly resistance to *Rhizoctonia solani* causing sheath blight disease. Theor Appl Genet 98:1138–1145

Davis L, Kuttan G (2000) Immunomodulatory activity of *Withania somnifera*. J Ethnopharmacol 71:193–200

Dhar N, Rana S, Bhat WW, Razdan S, Pandith SA, Khan S, Dutt P, Dhar RS, Vaishnavi S, Vishwakarma R (2013) Dynamics of withanolide biosynthesis in relation to temporal expression pattern of metabolic genes in *Withania somnifera* (L.) Dunal: a comparative study in two morpho-chemovariants. Mol Biol Rep 40:7007–7016

Dhar N, Rana S, Razdan S, Bhat WW, Hussain A, Dhar RS, Vaishnavi S, Hamid A, Vishwakarma R, Lattoo SK (2014) Cloning and functional characterization of three branch point *oxidosqualene cyclases* from *Withania somnifera* (L.) dunal. J Biol Chem 289:17249–17267

Doma M, Abhayankar G, Reddy VD, Kavi Kishor PB (2012) Carbohydrate and elicitor enhanced withanolide (withaferin A and withanolide A) accumulation in hairy root cultures of *Withania somnifera* (L.) Indian J Exp Biol 50:484–490

Fagoaga C, Rodrigo I, Conejero V, Hinarejos C, Tuset JJ, Jn A, Pina JA, Navarro L, Pena L (2001) Increased tolerance to *Phytophthora citrophthora* in transgenic orange plants constitutively expressing a tomato pathogenesis related protein PR-5. Mol Breed 7:175–185

Fatima N, Ahmad N, Anis M (2011) Enhanced *in vitro* regeneration and change in photosynthetic pigments, biomass and proline content in *Withania somnifera* L.(Dunal) induced by copper and zinc ions. Plant Physiol Biochem 49:1465–1471

Ghimire BK, Seong ES, Kim EH, Lamsal K, Yu CY, Chung IM (2010) Direct shoot organogenesis from petiole and leaf discs of *Withania somnifera* (L.) Dunal. Afr J Biotechnol 9:7453–7461

Giri A, Narasu ML (2000) Transgenic hairy roots: recent trends and applications. Biotechnol Adv 18:1–22

Grover A, Samuel G, Bisaria VS, Sundar D (2013) Enhanced withanolide production by overexpression of *squalene synthase* in *Withania somnifera*. J Biosci Bioeng 115:680–685

Gupta P, Akhtar N, Tewari SK, Sangwan RS, Trivedi PK (2011) Differential expression of *farnesyl diphosphate synthase* gene from *Withania somnifera* in different chemotypes and in response to elicitors. Plant Growth Regul 65:93–100

Gupta N, Sharma P, Kumar RJS, Vishwakarma RK, Khan BM (2012) Functional characterization and differential expression studies of squalene synthase from *Withania somnifera*. Mol Biol Rep 39:8803–8812

Gupta P, Goel R, Pathak S, Srivastava A, Singh SP, Sangwan RS, Asif MH, Trivedi PK (2013) *De novo* assembly, functional annotation and comparative analysis of *Withania somnifera* leaf and root transcriptomes to identify putative genes involved in the withanolides biosynthesis. PLoS One 8:e62714

Herrera-Estrella L, Simpson J, MartÃnez-Trujillo M (2004) Transgenic plants: an historical perspective. In: Transgenic plants: methods and protocols. Humana Press, Totowa, pp 3–31

Hunter WN (2007) The non-mevalonate pathway of isoprenoid precursor biosynthesis. J Biol Chem 282:21573–21577

Jayaprakasam B, Zhang Y, Seeram NP, Nair MG (2003) Growth inhibition of human tumor cell lines by withanolides from *Withania somnifera* leaves. Life Sci 74:125–132

Khan JA, Srivastava P, Singh SK (2006) Sensitive detection of a phytoplasma associated with little leaf symptoms in *Withania somnifera*. Eur J Plant Pathol 115:401–408

Khanna PK, Kumar A, Chandra R, Verma V (2013) Germination behaviour of seeds of *Withania somnifera* (L.) Dunal: a high value medicinal plant. Physiol Mol Biol Plants 19:449–454

Khatun S, Ali MB, Hahn E-J, Paek K-Y (2008) Copper toxicity in *Withania somnifera*: growth and antioxidant enzymes responses of *in vitro* grown plants. Environ Exp Bot 64:279–285

Kim Y, Wyslouzil BE, Weathers PJ (2002) Secondary metabolism of hairy root cultures in bioreactors. In Vitro Cell Devel Biol-Plant 38:1–10

Kulkarni AA, Thengane SR, Krishnamurthy KV (2000) Direct shoot regeneration from node, internode, hypocotyl and embryo explants of *Withania somnifera*. Plant Cell Tissue Organ Cult 62:203–209

Kumar V, Murthy KNC, Bhamid S, Sudha CG, Ravishankar GA (2005) Genetically modified hairy roots of *Withania somnifera* Dunal: a potent source of rejuvenating principles. Rejuvenation Res 8:37–45

Kumar A, Kaul MK, Bhan MK, Khanna PK, Suri KA (2007) Morphological and chemical variation in 25 collections of the Indian medicinal plant, *Withania somnifera* (L.) Dunal (Solanaceae). Genet Resour Crop Evol 54:655–660

Kumar A, Singh CP, Pandey R (2009) Influence of environmental factors on the population buildup of *Helicoverpa armigera*Hübner and *Epilachna vigintioctopunctata* (Febr.) on Aswagandha (*Withania somnifera* Linn.) J Entomol Res 33:123–127

Kumar A, Mir BA, Sehgal D, Dar TH, Koul S, Kaul MK, Raina SN, Qazi GN (2011) Utility of a multidisciplinary approach for genome diagnostics of cultivated and wild germplasm resources of medicinal *Withania somnifera*, and the status of new species, *W. ashwagandha*, in the cultivated taxon. Plant Syst Evol 291:141–151

Liu D, Raghothama KG, Hasegawa PM, Bressan RA (1994) Osmotin overexpression in potato delays development of disease symptoms. Proc Natl Acad Sci U S A 91:1888–1892

Furmanowa MA, Gajdzis-Kuls D, Ruszkowska J, Czarnocki Z, Gy O, A S, Rani R, Upadhyay SN (2001) *In vitro* propagation of *Withania somnifera* and isolation of withanolides with immunosuppressive activity. Planta Med 67:146–149

Maiti CK, Sen S, Paul AK, Acharya K (2007) First report of *Alternaria dianthicola* causing leaf blight on *Withania somnifera* from India. Plant Dis 91:467–467

Matsuda H, Murakami T, Kishi A, Yoshikawa M (2001) Structures of withanosides I, II, III, IV, V, VI, and VII, new withanolide glycosides, from the roots of Indian *Withania somnifera* DUNAL. And inhibitory activity for tachyphylaxis to clonidine in isolated guinea-pig ileum. Bioorg Med Chem 9:1499–1507

Mir BA, Khazir J, Hakeem KR, Koul S, Cowan DA (2014) Enhanced production of withaferin-A in shoot cultures of *Withania somnifera* (L) Dunal. J Plant Biochem Biotechnol 23:430–434

Mirjalili MH, Moyano E, Bonfill M, Cusido RM, Palazón J (2009) Steroidal lactones from *Withania somnifera*, an ancient plant for novel medicine. Molecules 14:2373–2393

Misra L, Lal P, Sangwan RS, Sangwan NS, Uniyal GC, Tuli R (2005) Unusually sulfated and oxygenated steroids from *Withania somnifera*. Phytochemistry 66:2702–2707

Misra L, Mishra P, Pandey A, Sangwan RS, Sangwan NS, Tuli R (2008) Withanolides from *Withania somnifera* roots. Phytochemistry 69:1000–1004

Mulabagal V, Tsay H-S (2004) Plant cell cultures-an alternative and efficient source for the production of biologically important secondary metabolites. Int J Appl Sci Eng 2:29–48

Murashige T, Skoog F (1962) A revised medium for rapid growth and bio assays with tobacco tissue cultures. Physiol Plant 15:473–497

Murthy HN, Dijkstra C, Anthony P, White DA, Davey MR, Power JB, Hahn EJ, Paek KY (2008) Establishment of *Withania somnifera* hairy root cultures for the production of withanolide A. J Integr Plant Biol 50:975–981

Nagella P, Murthy HN (2010) Establishment of cell suspension cultures of *Withania somnifera* for the production of withanolide A. Bioresour Technol 101:6735–6739

Nagella P, Murthy HN (2011) Effects of macroelements and nitrogen source on biomass accumulation and withanolide-A production from cell suspension cultures of *Withania somnifera* (L.) Dunal. Plant Cell Tissue Organ Cult 104:119–124

Namdeo AG, Sharma A, Yadav KN, Gawande R, Mahadik KR, Lopez-Gresa MP, Kim HK, Choi YH, Verpoorte R (2011) Metabolic characterization of *Withania somnifera* from different regions of India using NMR spectroscopy. Planta Med 77:1958–1964

Nayak SA, Kumar S, Satapathy K, Moharana A, Behera B, Barik DP, Acharya L, Mohapatra PK, Jena PK, Naik SK (2013) *In vitro* plant regeneration from cotyledonary nodes of *Withania somnifera* (L.) Dunal and assessment of clonal fidelity using RAPD and ISSR markers. Acta Physiol Plant 35:195–203

Niderman T, Genetet I, Bruyere T, Gees R, Stintzi A, Legrand M, Fritig B, Mosinger E (1995) Pathogenesis-related PR-1 proteins are antifungal (isolation and characterization of three 14-kilodalton proteins of tomato and of a basic PR-1 of tobacco with inhibitory activity against *Phytophthora infestans*). Plant Physiol 108:17–27

Pal S, Singh S, Ashutosh KS, Madan MG, Suman PSK, Ajit KS (2011) Comparative withanolide profiles, gene isolation, and differential gene expression in the leaves and roots of *Withania somnifera*. J Hortic Sci Biotechnol 86:391–397

Pal S, Yadav AK, Singh AK, Rastogi S, Gupta MM, Verma RK, Nagegowda DA, Pal A, Shasany AK (2016) Nitrogen treatment enhances sterols and withaferin A through transcriptional activation of jasmonate pathway, WRKY transcription factors, and biosynthesis genes in *Withania somnifera* (L.) Dunal. Protoplasma: 1–11

Pandey V, Misra P, Chaturvedi P, Mishra MK, Trivedi PK, Tuli R (2010) *Agrobacterium tumefaciens*-mediated transformation of *Withania somnifera* (L.) Dunal: an important medicinal plant. Plant Cell Rep 29:133–141

Patel N, Patel P, Kendurkar SV, Thulasiram HV, Khan BM (2015) Overexpression of *squalene synthase* in *Withania somnifera* leads to enhanced withanolide biosynthesis. Plant Cell Tissue Organ Cult 122:409–420

Pati PK, Sharma M, Salar RK, Sharma A, Gupta AP, Singh B (2008) Studies on leaf spot disease of *Withania somnifera* and its impact on secondary metabolites. Indian J Microbiol 48:432–437

Pawar PK, Maheshwari VL (2004) *Agrobacterium rhizogenes* mediated hairy root induction in two medicinally important members of family Solanaceae. Indian J Biotechnol 3:414–417

Prajapati ND, Purohit SS, Sharma AK, Kumar T (2006) A handbook of medicinal plants a complete source book, reprint. Hindustan Printing Press, Agrobios, Jodhpur

Praveen N, Murthy HN (2012) Synthesis of withanolide a depends on carbon source and medium pH in hairy root cultures of *Withania somnifera*. Ind Crop Prod 35:241–243

Praveen N, Murthy HN (2013) Withanolide a production from *Withania somnifera* hairy root cultures with improved growth by altering the concentrations of macro elements and nitrogen source in the medium. Acta Physiol Plant 35:811–816

Rana S, Lattoo SK, Dhar N, Razdan S, Bhat WW, Dhar RS, Vishwakarma R (2013) *NADPH-cytochrome P450 reductase*: molecular cloning and functional characterization of two paralogs from *Withania somnifera* (L.) Dunal. PLoS One 8:e57068

Rani G, Virk GS, Nagpal A (2003) Callus induction and plantlet regeneration in *Withania somnifera* (L.) Dunal. In Vitro Cell Dev Biol Plant 39:468–474

Ray S, Jha S (1999) Withanolide synthesis in cultures of *Withania somnifera* transformed with *Agrobacterium tumefaciens*. Plant Sci 146:1–7

Ray S, Jha S (2001) Production of withaferin A in shoot cultures of *Withania somnifera*. Planta Med 67:432–436

Ray S, Ghosh B, Sen S, Jha S (1996) Withanolide production by root cultures of *Withania somnifera* transformed with *Agrobacterium rhizogenes*. Planta Med 62:571–573

Razdan S, Bhat WW, Rana S, Dhar N, Lattoo SK, Dhar RS, Vishwakarma RA (2013) Molecular characterization and promoter analysis of *squalene epoxidase* gene from *Withania somnifera* (L.) Dunal. Mol Biol Rep 40:905–916

Rodriguez-Concepcion M, Boronat A (2002) Elucidation of the methylerythritol phosphate pathway for isoprenoid biosynthesis in bacteria and plastids. A metabolic milestone achieved through genomics. Plant Physiol 130:1079–1089

Rout JR, Sahoo SL (2012) Morphological and protein profile alterations in *Withania somnifera* L. with response to iron stress. Indian J Life Sci 2:21–25

Rout JR, Sahoo SL (2013) Antioxidant enzyme gene expression in response to copper stress in *Withania somnifera* L. Plant Growth Regul 71:95–99

Sabir F, Sangwan NS, Chaurasiya ND, Misra LN, Tuli R, Sangwan RS (2008) Rapid micropropagation of *Withania somnifera* L. accessions from axillary meristems. Journal of Herbs, Spices & Medicinal Plants 13:123–133

Sangwan RS, Chaurasiya ND, Lal P, Misra L, Uniyal GC, Tuli R, Sangwan NS (2007) Withanolide a biogeneration in in vitro shoot cultures of Ashwagandha (*Withania somnifera* Dunal), a main medicinal plant in Ayurveda. Chem Pharm Bull 55:1371–1375

Sangwan RS, Chaurasiya ND, Lal P, Misra L, Tuli R, Sangwan NS (2008) Withanolide a is inherently de novo biosynthesized in roots of the medicinal plant Ashwagandha (*Withania somnifera*). Physiol Plant 133:278–287

Saravanakumar A, Aslam A, Shajahan A (2012) Development and optimization of hairy root culture systems in *Withania somnifera* (L.) Dunal for withaferin-A production. Afr J Biotechnol 11:16412

Sarowar S, Kim YJ, Kim EN, Kim KD, Hwang BK, Islam R, Shin JS (2005) Overexpression of a pepper basic *pathogenesis-related protein 1* gene in tobacco plants enhances resistance to heavy metal and pathogen stresses. Plant Cell Rep 24:216–224

Sels J, Mathys J, De Coninck BMA, Cammue BPA, De Bolle MFC (2008) Plant pathogenesis-related (PR) proteins: a focus on PR peptides. Plant Physiol Biochem 46:941–950

Sen J, Sharma AK (1991) Micropropagation of *Withania somnifera* from germinating seeds and shoot tips. Plant Cell Tissue Organ Cult 26:71–73

Sharada M, Ahuja A, Suri KA, Vij SP, Khajuria RK, Verma V, Kumar A (2007) Withanolide production by *in vitro* cultures of *Withania somnifera* and its association with differentiation. Biol Plant 51:161–164

Sharma A (2013) Study of leaf spot disease in *WIthania somnifera* (L.) Dunal employing morphological, biochemical and molecular approaches. PhD Thesis, Guru Nanak Dev University

Sharma P, Pandey R (2009) Biological control of root-knot nematode, *Meloidogyne incognita* in the medicinal plant, *Withania somnifera* and the effect of biocontrol agents on plant growth. Afr J Agric Res 4:564–567

Sharma A, Pati PK (2011a) First report of *Withania somnifera* (L.) Dunal, as a new host of cowbug (*Oxyrachis tarandus*, fab.) in plains of Punjab, northern India. World Appl Sci J 14:1344–1346

Sharma A, Pati PK (2011b) First record of 28-spotted ladybird beetle, *Henosepilachna vigintioctopunctata* (F.) infesting *Withania somnifera* (L.) Dunal in Punjab province of Northern India. Pest Technol 5:91–92

Sharma A, Pati PK (2012a) First record of Ashwagandha as a new host to the invasive mealybug (*Phenacoccus solenopsis* Tinsley) in India. Entomol News 123:59–62

Sharma A, Pati PK (2012b) First record of the carmine spider mite, *Tetranychus urticae*, infesting *Withania somnifera* in India. J Insect Sci 12:50

Sharma LK, Madina BR, Chaturvedi P, Sangwan RS, Tuli R (2007) Molecular cloning and characterization of one member of *3β-hydroxy sterol glucosyltransferase* gene family in *Withania somnifera*. Arch Biochem Biophys 460:48–55

Sharma A, Sharma I, Pati PK (2011) Post-infectional changes associated with the progression of leaf spot disease in *Withania somnifera*. J Plant Pathol 93:397–405

Sharma A, Vats SK, Pati PK (2014) Post-infectional dynamics of leaf spot disease in *Withania somnifera*. Ann Appl Biol 165:429–440

Sil B, Mukherjee C, Jha S, Mitra A (2015) Metabolic shift from withasteroid formation to phenylpropanoid accumulation in cryptogein-cotransformed hairy roots of *Withania somnifera* (L.) Dunal. Protoplasma 252:1097–1110

Singh S, Kumar S (1998) *Withania somnifera*: the Indian ginseng ashwagandha. Central Institute of Medicinal and Aromatic Plants, Lucknow

Singh S, Sidhu JS, Huang N, Vikal Y, Li Z, Brar DS, Dhaliwal HS, Khush GS (2001) Pyramiding three bacterial blight resistance genes (xa5, xa13 and Xa21) using marker-assisted selection into indica rice cultivar PR106. Theor Appl Genet 102:1011–1015

Singh S, Pal S, Shanker K, Chanotiya CS, Gupta MM, Dwivedi UN, Shasany AK (2014) Sterol partitioning by HMGR and DXR for routing intermediates toward withanolide biosynthesis. Physiol Plant 152:617–633

Singh AK, Dwivedi V, Rai A, Pal S, Reddy SGE, Rao DKV, Shasany AK, Nagegowda DA (2015a) Virus-induced gene silencing of *Withania somnifera squalene synthase*negatively regulates sterol and defence-related genes resulting in reduced withanolides and biotic stress tolerance. Plant Biotechnol J 13:1287–1299

Singh P, Guleri R, Singh V, Kaur G, Kataria H, Singh B, Kaur G, Kaul SC, Wadhwa R, Pati PK (2015b) Biotechnological interventions in *Withania somnifera* (L.) Dunal. Biotechnol Genet Eng Rev 31:1–20

Singh V, Kaul SC, Wadhwa R, Pati PK (2015c) Evaluation and selection of candidate reference genes for normalization of quantitative RT-PCR in *Withania somnifera* (L.) Dunal. PLoS One 10:e0118860

Sivanandhan G, Mariashibu TS, Arun M, Rajesh M, Kasthurirengan S, Selvaraj N, Ganapathi A (2011) The effect of polyamines on the efficiency of multiplication and rooting of *Withania somnifera* (L.) Dunal and content of some withanolides in obtained plants. Acta Physiol Plant 33:2279–2288

Sivanandhan G, Rajesh M, Arun M, Jeyaraj M, Dev GK, Arjunan A, Manickavasagam M, Muthuselvam M, Selvaraj N, Ganapathi A (2012) Effect of culture conditions, cytokinins, methyl jasmonate and salicylic acid on the biomass accumulation and production of withanolides in multiple shoot culture of *Withania somnifera* (L.) Dunal using liquid culture. Acta Physiol Plant 35:715–728

Sivanandhan G, Dev GK, Jeyaraj M, Rajesh M, Arjunan A, Muthuselvam M, Manickavasagam M, Selvaraj N, Ganapathi A (2013a) Increased production of withanolide A, withanone, and withaferin A in hairy root cultures of *Withania somnifera* (L.) Dunal elicited with methyl jasmonate and salicylic acid. Plant Cell Tissue Organ Cult 114:121–129

Sivanandhan G, Dev GK, Jeyaraj M, Rajesh M, Muthuselvam M, Selvaraj N, Manickavasagam M, Ganapathi A (2013b) A promising approach on biomass accumulation and withanolides production in cell suspension culture of *Withania somnifera* (L.) Dunal. Protoplasma 250:885–898

Sivanandhan G, Selvaraj N, Ganapathi A, Manickavasagam M (2014) An efficient hairy root culture system for *Withania somnifera* (L.) Dunal. Afr J Biotechnol 13:4141–4147

Sivanandhan G, Dev GK, Theboral J, Selvaraj N, Ganapathi A, Manickavasagam M (2015a) Sonication, vacuum infiltration and thiol compounds enhance the *Agrobacterium*-mediated transformation frequency of *Withania somnifera* (L.) Dunal. PloS One 10:e0124693

Sivanandhan G, Theboral J, Dev GK, Selvaraj N, Manickavasagam M, Ganapathi A (2015b) Effect of carbon and nitrogen sources on *in vitro* flower and fruit formation and withanolides production in *Withania somnifera* (L.) Dunal. Indian J Exp Biol 53:177–183

Sivanesan I (2007) Direct regeneration from apical bud explants of *Withania somnifera* Dunal. Indian J Biotechnol 6:125

Sivanesan I, Murugesan K (2008) An efficient regeneration from nodal expiants of *Withania* somnifera Dunal. Asian J Plant Sci 7:551–556

Soni P, Bahadur AN, Tiwari U, Kanungo VK (2011) Micropropagation of a medicinal plant *Withania somnifera* (L.) Dunal by shoot bud culture. An International quarterly journal of Life Sci 6:135–137

Supe U, Dhote F, Roymon MG (2006) In vitro plant regeneration of *Withania somnifera*. Plant Tissue Cult Biotechnol 16:111–115

Thilip C, Raju CS, Varutharaju K, Aslam A, Shajahan A (2015) Improved *Agrobacterium rhizogenes*-mediated hairy root culture system of *Withania somnifera* (L.) Dunal using sonication and heat treatment. Biotech 5:949–956

Udayakumar R, Choi CW, Kim KT, Kim SC, Kasthurirengan S, Mariashibu TS, Sahaya Rayan JJ, Ganapathi A (2013) *In vitro* plant regeneration from epicotyl explant of *Withania somnifera* (L.) Dunal. J Med Plant Res 7:43–52

Vakeswaran V, Krishnasamy V (2003) Improvement in storability of Ashwagandha (*Withania somnifera* Dunal) seeds through pre-storage treatments by triggering their physiological and biochemical properties [abstract]. Seed Technol: 203–203

Verma OP, Gupta RBL, Shivpuri A (2008) A new host for *Pithomyces chartarum*, the cause of a leaf spot disease on *Withania somnifera*. Plant Pathol 57:385–385

Chapter 23
Plant Tissue Culture Based Strategies for Production of Withanolides

Deepika Soni, Preeti Rana, Sukriti Goyal, Aditi Singh, Salma Jamal, and Abhinav Grover

Abstract *Withania somnifera*, an "Indian Ginseng", is a medicinal plant of high value having rejuvenating and health promoting effects. It possesses biologically active metabolites called withanolides that are present in the leaves and roots of the plant. Withanolide A and withaferin A are the major secondary metabolites demonstrated to possess specific therapeutic properties and pharmacological activities. Commercial withanolide production largely depends on the field plants, but there is quality concern due to different biotic and abiotic factors. *In vitro* cell culture technology can be an evident approach to an increased production of these plant secondary metabolites. Withanolides, the naturally occurring and abundant plant secondary metabolites isolated from solanaceous plants are well known to be one of the most powerful herbs in Ayurv edic medicine. Withanolides have been reported to possess diverse pharmacological functions such as being anti-tumor, anti-inflammatory, anti-oxidant, anti-microbial, cardio-protective and neuro-protective, immune-modulatory, insecticidal and more. With the plant being predominantly spread and thriving in temperate and tropical regions around the world, increasing varieties of withanolides are being characterized over the years recently and probing into their mechanism of actions and functions continues.

Keywords *Withania somnifera* • Withanolides • Pharmacological functions • Tissue culture • Production

D. Soni • A. Grover (✉)
School of Biotechnology, Jawaharlal Nehru University, New Delhi 110070, India
e-mail: abhinavgr@gmail.com

P. Rana • A. Singh
Department of Biotechnology, TERI University, Vasant Kunj, New Delhi 110070, India

S. Goyal • S. Jamal
Department of Bioscience and Biotechnology, Banasthali University, Tonk, Rajasthan 304022, India

S.C. Kaul, R. Wadhwa (eds.), *Science of Ashwagandha: Preventive and Therapeutic Potentials*, DOI 10.1007/978-3-319-59192-6_23

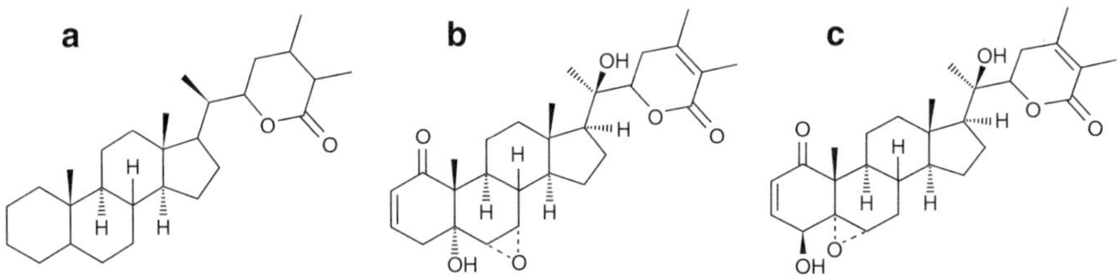

Fig. 23.1 Structure of some important withanolides: (**a**) Withanolide skeleton; (**b**) Withanolide A; (**c**) Withanolide D

23.1 Structures of Withanolides

Structurally, these compounds are composed of a steroid backbone which is attached to a lactone or its derivatives (Fig. 23.1). They are generally produced by the oxidation of steroids.

23.2 Biosynthesis of Withanolides

The withanolide biosynthesis in case of *Withania somnifera* occurs through two distinct pathways, i.e., mevalonate and the non-mevalonate pathway (Fig. 23.2). Squalene, mevalonic acid and cholesterol act as immediate precursors in the metabolic pathway of withanolides.

23.3 Hairy Root Cultures for Production of Withanolides

Praveen and Murthy (2010) carried out work on *Withania somnifera* hairy root cultures for the production of biomass and withanolide A by optimizing the growth parameters. They investigated the effects of different carbon sources with varying concentrations for higher growth and production of withanolide. The maximum accumulation of biomass and production of withanolide A was observed in case of sucrose as a carbon source (11.92 g/l DW and 11.96 mg/g DW of withanolide A). Additionally, biomass accumulation was maximized (11.92 g/l DW) at 3% sucrose concentration, whereas, withanolide A production (13.28 mg/g DW) was favored at 4% sucrose concentration. Further, the effect of pH was checked for biomass accumulation and withanolide production. Initial pH of 5.8 gave optimal biomass of hairy roots (12.1 g/l DW), while withanolide A production was highest at pH 6.0 (13.84 mg/g DW) (Table 23.1).

It has been observed that altering the concentration of the growth medium with macro nutrients and nitrogen source can enhance the production of withanolides in the hairy root cultures of *Withania somnifera* (Fig. 23.3). Maximum production of

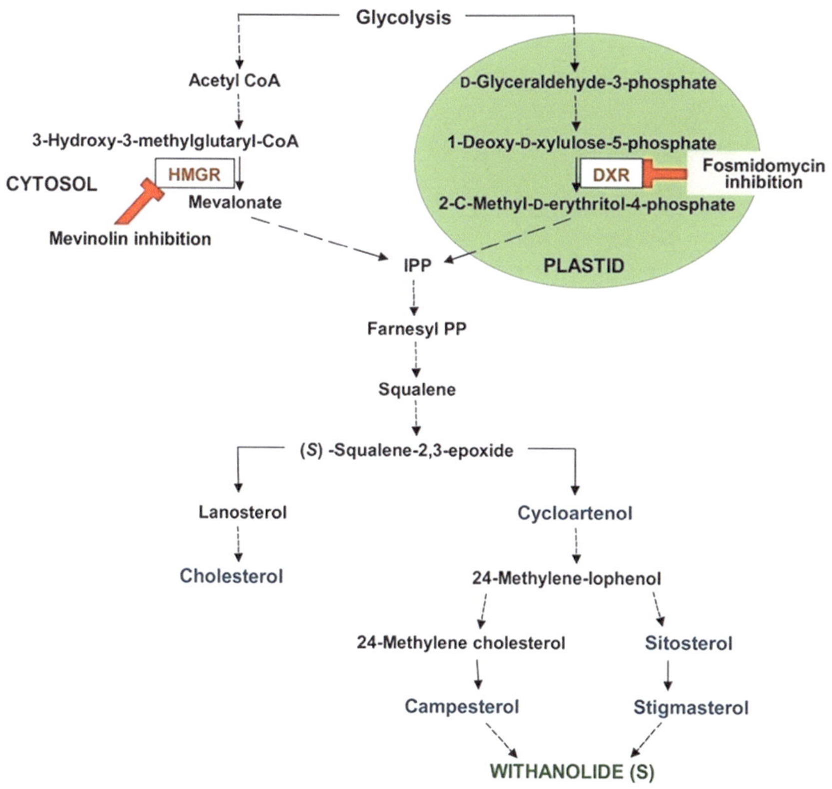

Fig. 23.2 Withanolide(s) production pathway

withanolide A was observed in case of 2.0X concentration of potassium nitrate (KNO_3) (15.27 mg/g DW). The growth of withanolide A was increased with ammonia nitrate (NH_4^+/NO_3^-) ratio having NO_3^- concentration higher than NH_4^+ (0.00/18.80 mM). At 14.38/37.60 mM (NH_4^+/NO_3^- ratio) the maximum biomass (i.e., 148.17 g/L FW and 14.79 g/L DW) growth was attained (Table 23.1) (Praveen and Murthy 2012).

Further, withanolide production has been enhanced by the elicitor treatment in hairy root cultures. Methyl jasmonate (MeJ) and salicylic acid (SA) were used as elicitors for increased biomass and withanolide production (Fig. 23.3). Methyl jasmonate at 15 μM resulted in increased withanolide A (114.38 mg/g DW; 50 fold) production with decreased biomass. In contrast, maximum withanolide A (132.44 mg/g DW; 58-fold) production has been reported with 150 μM SA with respect to control after 40 days of culture. Similarly, withaferin A (57.46 mg/g DW, 34-fold; 70.72 mg/g DW, 42-fold) and withanone (69.89 mg/g DW, 38-fold; 84.35 mg/g DW, 46-fold) production has also been increased with MeJ and SA treatment (Table 23.1). Thus, salicylic acid is an efficient elicitor for the enhanced biosysnthesis of withanolides from *W. somnifera* hairy root cultures (Sivanandhan et al. 2013a, b).

Table 23.1 Withanolide production through different plant tissue culture methods

S. No.	Source of Withanolide	Elicitor/precursor/ other factors	Time for culture	Withanolide A	Withanolide B	Withaferin A	Other withanolides	References
1	Hairy root culture (flask)	–	4 weeks	13.84 mg/g DW	–	–	–	Praveen and Murthy (2010)
2	Hairy root culture with macronutrients & nitrogen source	2.0X KNO3, $NH4^+/NO3^-$ratio of 0.00/18.80 mM	–	15.27 mg/g	–	–	–	Praveen and Murthy (2012)
3	Hairy root cultures	Salicylic acid (150 μM)	40 days (4 h elicitation)	132.44 mg/g DW	–	70.72 mg/g DW	Withanone – 84.35 mg/g DW	Sivanandhan et al. (2013a, b)
4	Adventitious root culture	–	5 weeks	8.8 ± 0.20 mg/g	–	–	–	Praveen and Murthy (2010)
5	Adventitious root culture	2% sucrose, pH 5.8	–	9.09 mg/g	–	–	–	Murthy and Praveen (2013)
6	Primary adventitious roots	Aluminium chloride-10 mg/L	4 h, 6 week	266.42 mg/g	0.225 mg/g	3.381 mg/g	0.360 mg/g withanoside V and 0.396 mg/g withanoside IV	Sivanandhan et al. (2012a, b)
7	Primary adventitious roots	Chitosan-100 mg/L	4 h, 6 week	323.85 mg/g	0.275 mg/g	4.347 mg/g	0.450 mg/g withanoside V and 0.528 mg/g withanoside IV	Sivanandhan et al. (2012a, b)
8	Adventitious root culture	Salicylic acid 150 μM	40 days (10 days elicitation)	64.65 mg/g	33.74 mg/g	17.47 mg/g	42.88 mg/g withanone, 5.34 mg/g 12-deoxy Withastramonolide, 7.23 mg/g, withanoside V, 9.45 mg/g withanoside IV	Sivanandhan et al. (2012a, b)

9	Multiple shoot cultures	–	–	–	–	0.04–0.14%	Withanolide D 0.06%	Ray and Jha (2001)
10	Multiple shoot culture	Benzyl adenine-1 ppm, kinetin 0.5 ppm	–	0.238 g/100 g	–	–	–	Sangwan et al. (2007)
11	Leaf explants infected with *A. tumefaciens strains*	–	–	–	–	0.07–0.10%	Withanolide D, 0.085–0.025%	Ray and Jha (1999)
12	Leaf & shoot explants infected with *A. tumefaciens strains*	–	4 weeks	2.05 mg/g DW	–	–	–	Grover et al. (2013)
13	Transformed root cultures	–	–	–	–	–	Withanolide D, 0.30 mg/g	Ray et al. (1996)
14	Hairy root induction on petiole explants by *A. tumefaciens*	–	–	–	–	72.3 mg/g	–	Saravanakumar and Aslam (2012)
15	Hairy root cultures transformed with *A. rhizogenes*	–	28 days	157.4 µg/g DW	–	–	–	Murthy et al. (2008)
16	Suspension culture	Salacin 750 µM	15 days	–	–	25 ± 2.9 mg/L	–	Ciddi (2006)
17	Cell suspension culture	2,4-dichloro-phenoxy acetic acid, kinetin	4 weeks	2.95 mg/g DW	–	–	–	Nagella and Murthy (2010a, b)
18	Cell suspension culture	2.0X KNO3, $NH4^+/NO3^-$ratio of 14.38/37.60 mM.	–	3.96 mg/g	–	–	–	Nagella and Murthy (2010a, b)
19	Cell suspension culture	Organic additive, seaweed extracts (*Gracilaria edulis*)	–	7.21 mg/g	4.23 mg/g	3.88 mg/g	Withanone 6.72 mg/g	Sivanandhan et al. (2013a, b)

(continued)

Table 23.1 (continued)

S. No.	Source of Withanolide	Elicitor/precursor/ other factors	Time for culture	Withanolide A	Withanolide B	Withaferin A	Other withanolides	References
20	Cell suspension culture (biotransformation)	–	–	1.181 mg/g (added) converted-16.1%	–	0.141 mg/g (added)	Withanone 1.052 mg/g (added) converted-1.3%	Sabir et al. (2011)
21	Cell suspension culture	–	–	4.98 mg/g DW	–	0.16 mg/gDW	–	Grover et al. (2013)
22	Cell suspension	*Priformospora indica* cell homogenate	15 days of elicitation	–	–	4.9 ± 0.23 mg/L	–	Ahlawat et al. (2015)
23	Cell suspension culture (shake flask & bioreactor)	Squalene (6 mM) + Chitosan (100 mg/L)	28 days	37.88 mg (shake flask)	26.53 mg (shake flask)	20.94 mg (shake flask)	160.82 mg (shake flask)	Sivanandhan et al. (2014)
				7606.75 mg (bioreactor)	4823.05 mg (bioreactor)	3732.81 mg (bioreactor)	31362.28 mg (bioreactor)	
24	Nodal explants	Polyamine 20 mg/L	4 weeks	–	6.5 times (leaf), 3.3 times (root)	1.14 times	Withanone 1.20 times	Sivanandhan et al. (2011)
25	Differentiated callus cultures	"Poshita" & "Jawahar 22"	–	–	–	1.60 mg/g & 1.40 mg/g	2.32% & 2.10%	Das et al. (2010)

Improved withanolide production has been reported through large scale cultivation of the hairy root cultures in bioreactors using elicitors like salicylic acid and methyl jasmonate (Sivanandhan et al. 2016). Enhanced production of withanolides mediated through the elicitor treatment is the result of programmed signaling involved in activation of secondary metabolite biosynthesis. The plant cell receives an intra- or extra- cellular signal of the elicitor and intitates signal transduction that results in transcription factors required for biosynthesis of secondary metabolites (Kim et al. 2009).

23.4 Production of Withanolides Using Adventitious Roots

In vitro production of withanolides has been reported in case of indirect induction of adventitious roots from the three explants namely: leaf, internode and cotyledon. Complexity of root organogenesis can be minimized using callus cultures, which can produce large quantities of metabolites (Fig. 23.3). The treatment of plant cells/ organs with elicitor is a well-established strategy for an increased production of withanolides (Smetanska 2008).

Higher biomass accumulation (108.48 g/L FW and 10.76 g/L DW) along with enhanced withanolide-A content (8.8 ± 0.20 mg/g DW) has been reported within 5 weeks of adventitious root culturing of *W. somnifera* from leaf explants. Adventitious root biomass production increased to 11-fold in suspension cultures and an increase of 21-fold in withanolide A content in contrast to leaves of the natural plant. Results indicated that the half strength MS (Murashige and Skoog) medium at 10 g/L inoculum density, supplemented with 3% sucrose is most appropriate for accumulation of biomass and production of withanolides as compared to B5 (Gamborg et al. 1968),

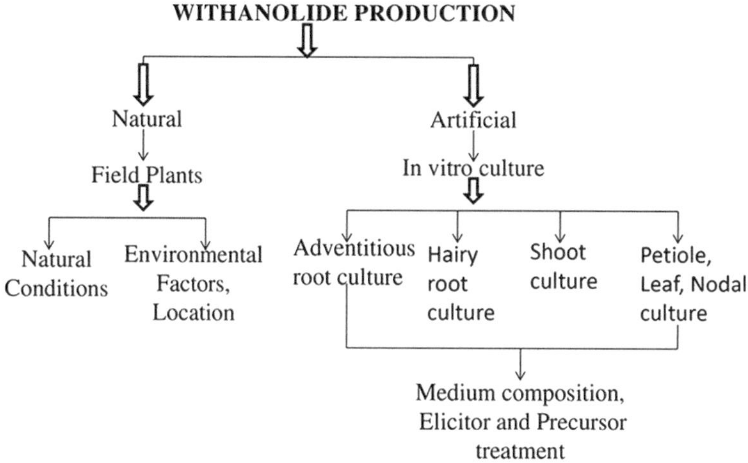

Fig. 23.3 Withanolide production by different methods in natural and artificial conditions

NN (Nitsch and Nitsch 1969) and N6 (Chu 1978). 8.27–8.65 mg/g production of withanolide has been reported in medium with above composition (Table 23.1) (Praveen and Murthy 2010). Also use of sucrose at 2% concentration as carbon source and a medium with pH range of 5.5–5.8 has been found to be important for maximum accumulation of biomass (11.33 g/L DW) and withanolide A (9.09 mg/g DW) production (Table 23.1) (Murthy and Praveen 2013).

Chitosan and aluminium chloride were used as elicitors for the increased withanolide production from callus-derived adventitious root culture. Biotic elicitor chitosan (β-1, 4-linked glucosamine), is a deacylated derivative of chitin which induces plant resistance against pathogens and thus, had been used for increased biosynthesis of secondary metabolites in different plant cell/organ cultures (El Ghaouth A et al. 1994).

Similar study with chitosan (0–200 mg/L) and aluminium chloride (0–40 mg/L) as elicitors has been reported with secondary adventitious root cultures at different exposure times (0–8 h). 10 mg/L aluminium chloride resulted in enhanced production of withanolides after 4 h exposure time on 6th week with an increase of 14-fold-withanolide A (266.42 mg/g DW), ninefold-withanolide B (0.225 mg/g DW), sevenfold- withaferin A (3.381 mg/g DW), eightfold-withanoside V (0.360 mg/g DW) and sixfold-withanoside IV (0.396 mg/g DW) compared to control. In contrast, in similar conditions, chitosan exposure at 100 mg/L resulted in higher production of withanolides with 17- fold-withanolide A (323.85 mg/g DW), 11-fold-withanolide B (0.275 mg/g DW), ninefold-withaferin A (4.347 mg/g DW), tenfold-withanoside V (0.450 mg/g DW), eightfold-withanoside IV (0.528 mg/g DW) as compared to control (Table 23.1). Maximum levels of withanolide A, withanolide B, withaferin A, withanoside V and withanoside IV (1.21, 1.22, 1.28, 1.25 and 1.33) were produced with chitosan compared to aluminium chloride elicitation. The results revealed that the elicitor concentration required for the better production of secondary metabolites is dependent on the genotype along with culture conditions including inoculum mass, age of the culture, time of exposure to the elicitor and its concentration (Sivanandhan et al. 2012a, b).

Increased withanolides production has also been reported with the adventitious root cultures using salicylic acid (SA) and methyl jasmonate (MeJ) as an elicitor in Cumbum and Kolli hills varieties of *W. somnifera* of Tamilnadu, India. Salicylic acid is more potent elicitor than methyl jasmonate for enhanced withanolide production. Salicylic acid at a concentration of 150 μM in adventitious root cultures resulted in 48-fold withanolides A (64.65 mg/g dry weight (DW), 29-fold withanolide B (33.74 mg/g DW), 20-fold withaferin A (17.47 mg/g DW), 37-fold withanone (42.88 mg/g DW) and other minor withanolides after 10 days of elicitation (Table 23.1) (Sivanandhan et al. 2012a, b).

23.5 Withanolide Production from *Withania somnifera* Shoot Cultures

Withanolide D and Withaferin A production has been reported from the shoot tip cultures (Fig. 23.3) in MS and MSSM (Murashige and Skoog basal medium + BA 1 mg/L) medium. MS medium showed an accumulation of withanolide D (0.06%) and withaferin A (0.04%). MSSM medium with 4% sucrose resulted in increased accumulation of withanolides (Withanolide D-0.08% and Withaferin A-0.16%) (Table 23.1) (Ray and Jha 2001).

Sangwan et al. used nodal segments as explants to establish multiple shoot cultures using experimental cell lines of *Withania somnifera* plants (RS-Selection-1 and RS-Selection-2). Benzyl adenine and kinetin were the two hormones whose combinations influenced the morphogenetic response and changed the levels of withanolide A biogeneration. There was a considerable escalation of Withanolide-A in the *in vitro* shoot cultures as compared to the aerial parts of field grown plants. A significant variation in the levels of withanolide A was observed (tenfold, 0.014–0.14 mg/g FW) upon changing the hormone composition. Highest withanolide A (0.238 g/100 g DW tissue) production was obtained in RS selection-1 with 1 ppm benzyl adenine and 0.5 ppm kinetin (Table 23.1) (Sangwan et al. 2007).

23.6 Withanolide Sysnthesis Using *Agrobacterium* Mediated Transformed Cultures of *Withania somnifera*

Studies have been reported on the high production of secondary metabolites using transformation experiment with *Agrobacterium tumefaciens*. Infection with the wild type octopine and nopaline strains of *Agrobacterium tumefaciens* has been demonstrated for transformation of *Withania somnifera* organ cultures. Principal withanolides found in shooty teratoma cultures were withanolide D and withaferin A and only withanolide D was found in rooty teratoma. Shooty teratomas resulted in highest production of withanolide D (0.085–0.025%) and withaferin A (0.07–0.10%). Thus, withanolide synthesis is dependent on the differentiation of transformed shoot cultures and proved to be a useful system for *in vitro* synthesis of withanolides (Ray and Jha 1999) (Fig. 23.3). Similarly, induction of withanolide A synthesis has been reported to increase upto 2.5 fold by over expression of squalene synthase in *W. somnifera* using *A. tumefaciens* resulting in 2.05 mg/g DW of withanolide A synthesis in transformed cells (Table 23.1) (Grover et al. 2013).

Withanolide D production in transformed root cultures has been reported to be higher than the untransformed cultures. The increase in withanolide was observed to 0.181 mg/L/day in transformed root cultures as compared to 0.026 mg/L/day in untransformed cultures (Table 23.1) (Ray et al. 1996).

In vitro production of the secondary metabolites of *W. somnifera* by *Agrobacterium rhizogenes* mediated hairy root production has also been reported. R1000, MTCC

2364 and MTCC 532 strains of *A. rhizogenes* were used for the hairy root induction. 50.6% transformation efficiency was observed for R 1000 strain in hairy root induction in *Withania somnifera* leaf explants using 15 s sonication and heat treatment for 5 min at 41 °C, whereas only 29.3% and 18.6% transformation efficiency was reported for MTCC 2364 and MTCC 532. Successful transfer of Ri T-DNA inside the host cells resulted in infection which was found higher for R1000 strain in comparison to other strains (Wang and Fang 1998). These studies indicated that genetic and epigenetic discrepancies seen in field-grown plants can be minimized using *in vitro* transformation of secondary metabolites (Banerjee et al. 1994; Thilip et al. 2015).

The influence of *Agrobacterium* strains, explant types and acetosyringone has been shown in the hairy root induction of *W. somnifera.* Induction of hairy roots was influenced by the petiole explants at a greater rate (64%) as compared to the intermodal segment and leaf. This could be due to the greater number of wounds bordering competent cells for transformation and regeneration by petiole explants compared to others. Further the strain of *A. rhizogenes* also influences the hairy root induction in explants as observed in R 1000 strain which influenced 64% in petiole, 42.5% in leaf and 37.7% in internodes. Addition of Acetosyringone (AS) in various steps of hairy root induction also enhanced the production of withanolides (Table 23.1). Therefore, the production of withanolides could be increased by hairy root induction in petiole explants transformed by *A. rhizogenes* (R1000) in MS half strength medium with increased transformation frequency using Acetosyringone (Saravanakumar and Aslam 2012).

Murthy et al. (2008) established hairy root cultures of *Withania somnifera* which were transformed with A. *rhizogenes*. They inoculated *A. rhizogenes* strain R1601 into explants from seedling roots, cotyledonary nodal segments, stems, young leaves, hypocotyls and cotyledons. The leaf plants and cotyledons induced formation of transformed hairy roots. Four clones of the same were established differing in their morphology. There was a fivefold increase in the biomass of cultured, transformed seedling roots in comparison to the non-transformed ones. MS-based liquid medium supplemented with sucrose (40 g/l) was optimum for the growth of transformed roots. Concentration of withanolide A in the non-transformed, cultured roots was 57.9 µg/g DW, which increased 2.7-fold more to 157.4 µg/g DW in the transformed roots (Table 23.1).

23.7 Production of Withanolides Using Cell Suspension Cultures

In order to overcome the limitations associated with the regular supply of withanolides from plant organs, cell suspension cultures have been evaluated for their ability to yield considerable amounts of the product. Withaferin A production in suspension cell cultures has been reported by (Ciddi 2006). The addition of salacin at 750 µM concentration resulted in enhanced production of withaferin A, i.e., 25 ± 2.9 mg/L in comparison to 0.47 ± 0.03 mg/L seen in controls (Table 23.1) (Ciddi 2006).

Nagella and Murthy (2010a, b) established the cell suspension cultures of *Withania somnifera* in shake flasks. They studied the effects of different media and its strength, various growth regulators (auxins, the combination of cytokinin and auxin), carbon sources and their strengths, inoculum density and pH for withanolide A production. The optimal conditions for accumulation of biomass and withanolide A production (2.95 mg/g DW), were found to be the full strength of Murashige Skoog medium, 10 g/l (FW) of inoculum, 3% w/v sucrose and an initial pH of medium was 5.8 along with culture period of 4 weeks (Table 23.1). These optimized results could be further used for scale-up process.

Withanolide A production could also be increased by changing the medium conditions and amount of macronutrients in cell suspension culture as reported by Nagella and Murthy (2010a, b). Highest accumulation of biomass (147.81 g/L FW and 14.02 g/L DW) has been reported for medium containing 0.59 concentration of ammonium nitrate (NH_4NO_3) and peak production levels of withanolide A (4.36 mg/g DW) has been reported with 2.09 X KNO_3 (Table 23.1). Also, ammonia nitrate ratio of 7.19/18.80 resulted in maximum growth (110.45 g/L FW and 9.29 gL DW) and 14.38/37.60 ratio resulted in 3.96 mg/g DW withanolide A production (Nagella and Murthy 2010a, b).

Various chemical parameters including concentrations of cytokinins or auxins along with their combinations, organic additives, carbon sources, seaweed extracts and physical parameter like agitation speed have been optimized in cell suspension culture of *W. somnifera* for maximum biomass accumulation along with sustainable production of withanolides. Sivanandhan et al. reported the maximum biomass accumulation (4.18 g DW and 16.2 g FW) and production of withnolides (withanolide A- 7.21 mg/g DW, withanolide B-4.23 mg/g DW, withaferin A-3.88 mg/g DW and withanone-6.72 mg/g DW) were attained upon treating the culture with seaweed extract of *Gracilaria edulis* at 40% level to stimulate plant cell growth and secondary metabolite production in cell suspension of *W. somnifera* (Table 23.1). Growth characteristics and withanolide production (withanolide A-5.04 mg/g DW, withanolide B-2.59 mg/g DW, withaferin A-2.36 mg/g DW and withanone-4.32 mg/g DW) has been increased with organic additive L-glutamine at 200 mg/L in combination with picrolam (1 mg/L) and KN (0.5 mg/L). Highest yield of withanolides (withanolide A 2.88 mg/g DW, withanolide B 1.48 mg/g DW) were obtained with 5% sucrose as carbon source (Sivanandhan et al. 2013a, b).

Biotransformation of withanolides is possible in *W. somnifera* cell suspension cultures. Bio conversion of withanolide A and withaferin has been reported both in the cells and the media. The 5, 6 epoxy group of withanolide A is replaced by 6, 7 epoxy group; whereas in withaferin A; 27 hydroxyl group is replaced by the 20 hydroxyl group (Table 23.1). Thus, it could be an important technique to transform more abundant withanone to valuable withanolide A (Sabir et al. 2011). In a similar study with transformed callus in cell suspension culture resulted in 4.98 and 0.16 mg/g DW of withanolide A and withaferin (Table 23.1) (Grover et al. 2013).

The biosynthesis of withaferin A can also be elicited by an endophytic fungus *Piriformospora indica.* (Ahlawat et al. 2015) have worked to show the elicitation capacity of the fungus (Fig. 23.3). Callus and cell suspension cultures of *Withania somnifera* were inoculated with variable concentrations of cell homogenate, culture filtrate and individual discs of *P. indica* at different time intervals and cultures were observed for effect on cell biomass and withaferin A production was observed using variable concentrations of culture. Elicitation potential of *P. indica* for withaferin A production was increased to 2.04 times followed by 1.78 times and 1.46 times with an addition of 3% (v/v) cell homogenate, 3% (v/v) culture filtrate and culture disc, respectively, when compared with the controls. Withaferin A production was 4.9 ± 0.23 mg/L upon addition of *P. indica* cell homogenate with an exposure time of 7 days. Under the same conditions, 4.1 ± 0.12 mg/L withaferin A was accomplished upon adding culture filtrate of the fungus (Table 23.1). The study has also proved an enhancement in the expression of genes which code for enzymes involved in withanolide biosynthetic pathway by the application of fungal cultures.

Enhanced production of withanolides had been obtained by using shake flask culture and bioreactor. The use of different elicitors and precursors further enhances the production of withanolides (Fig. 23.3). In presence of elicitors namely, aluminium chloride, cadmium chloride and chitosan after 4 h exposure to suspension culture, the biomass growth declined with chitosan and cadmium chloride whereas it remained stable with aluminum chloride. At 100 mg/L of chitosan concentration maximum amount of wtihanolides were produced on day 28 of culture with 141.39 mg total withanolides. Aluminium chloride elicitation (10 mg/L) showed increased production of withanolides with no change in total biomass. In case of precursors, squalene was found to increase the production of withanolides whereas mevalonic acid and cholesterol showed a slight increase in their production. In bioreactor, a combination of chitosan (100 mg/L) and squalene (6 mM) resulted in increased production of biomass and withanolides, than in the shake flask cultures. The maximum amount of withanolides achieved in bioreactor were withanolide A (7606.75 mg), withanolide B (4823.05 mg), withaferin A (3732.81 mg), withanone (6538.65 mg), 12 deoxy withanstramonolide (3176.63 mg), withanosides IV (2623.21 mg) and V (2861.18 mg) (Table 23.1) (Sivanandhan et al. 2014).

Higher production of withanolides has been reported by the addition of polyamines like spermidine, spermine and putrescine in nodal cultures of *Withania somnifera*. A significant increase in the amount of withanolides has been reported in *in vitro* regenerated plant leaves, stem and root compared to field grown plants. An increase of around 1.14 and 1.20 fold in the content of withaferine A and withanone respectively has been reported in the leaves of *in vitro* derived plants (Table 23.1). The roots showed a 1.10 times enhancement in withanolide A production than field grown plant roots. Withanolide B content was enhanced in the leaves (6.5 times) and roots (3.3 times) of *in vitro* plants (Table 23.1). There has been no significant change in withanolide production from the stems of both *in vitro* and field grown plants (Sivanandhan et al. 2011).

In a study with differentiated callus with multiple shoots, epicotyls and leaf resulted in the increased production amounts of withanolides and alkaloids (Fig. 23.3). Two mediums, namely 'Poshita' and 'Jawahar 22' with compositions of MS (Murashige & Skoog) medium with phytohormones like BA (6-benzyladenine), Kin (Kinetin) and IAA (Indole acetic acid) were used in different concentrations for callus culture and production of withanolides. The maximum amount of 2.32% and 2.10% withanolides and 1.60 mg/g and 1.40 mg/g withaferin A were obtained from leaf explants of Poshita and Jawahar 22 (Poshita' medium: MS + BA 3.0 mg/L + kinetin 1.0 mg/L + IAA 2.0 mg/L; 'Jawahar22' medium: MS + BA 1.0 mg/L + kinetin 3.0 mg/L + IAA 2.0 mg/L). Highest amount of withanolides A (2.42 mg/g and 2.19 mg/g) was obtained from the epicotyls explants (Poshita', medium: MS + BA 2.5 mg/L; 'Jawahar22, medium: MS + BA 1.0 mg/L + kinetin 0.5 mg/L) (Table 23.1) (Das et al. 2010).

Micropropagation of *W. somnifera* roots and stem in cultures with MS medium along with cytokinins resulted in increased productivity of the plant (Fig. 23.3). However, the micropropagated plants contain lesser amount of withanolides (Sabir et al. 2008).

23.8 Conclusion

Higher production of withanolides can be mediated through different *in vitro* cultures. The cultures include hairy root, adventitious root, multiple shoot, cell suspension and nodal from leaf, root, petiole and stem. Different factors like medium composition, carbon source, pH, amount of macronutrients, use of different elicitors at different concentrations and other inorganic and organic compounds have resulted in increased production of withanlides. MS medium with 4–5% sucrose as carbon source at pH 5.8 was found to be optimum for the different cultures. Salicylic acid (150 μM), chitosan (100 mg/L) and aluminium chloride (10 mg/L) have been used to increase the withanolide production.

References

Ahlawat S, Saxena P, Ali A, Abdin MZ (2015) Piriformospora indica elicitation of withaferin A biosynthesis and biomass accumulation in cell suspension cultures of Withania somnifera. Symbiosis 69:37–46

Banerjee S, Naqvi AA, Mandal S, Ahuja PS (1994) Transformation of Withania somnifera (L) Dunal by Agrobacterium rhizogenes: infectivity and phytochemical studies. Phytother Res 8:452–455

Chu C (1978) The N6 medium and its applications to anther culture of cereal crops. In: Proceedings of symposium plant tissue culture. Science Press, Beijing, pp 43–50

Ciddi V (2006) Withaferin A from cell cultures of Withania somnifera. Indian J Pharm Sci 68:490–492

Das A, Ghose A, Bhattacharyya A, AKD (2010) In vitro Biogeneration of alkaloids and Withanolides in Withania somnifera (L.) Dunal (Solanaceae) var. 'Poshita' and 'Jawahar 22'. Int J Plant Dev Biol 4:42–46

El Ghaouth A, Arul J, Benhamou N, Asselin A, Belanger R (1994) Effect of chitosan on cucumber plants: suppression of Pythium aphanidermatum and induction of defence reactions. Phytopathology 84:313–320

Gamborg OL, Miller RA, Ojima K (1968) Nutrient requirements of suspension cultures of soybean root cells. Exp Cell Res 50:151–158

Grover A, Samuel G, Bisaria VS, Sundar D (2013) Enhanced withanolide production by overexpression of squalene synthase in Withania somnifera. J Biosci Bioeng 115:680–685

Kim OT, Bang KH, Kim YC, Hyun DY, Kim MY, Cha SW (2009) Upregulation of ginsenoside and gene expression related to triterpene biosynthesis in ginseng hairy root cultures elicited by methyl jasmonate. Plant Cell Tissue Organ Cult 98:25–33

Murthy HN, Praveen N (2013) Carbon sources and medium pH affects the growth of Withania somnifera (L.) Dunal adventitious roots and withanolide A production. Nat Prod Res 27:185–189

Murthy HN, Dijkstra C, Anthony P, White DA, Davey MR, Power JB, Hahn EJ, Paek KY (2008) Establishment of Withania somnifera hairy root cultures for the production of withanolide A. J Integr Plant Biol 50:975–981

Nagella P, Murthy HN (2010a) Effects of macroelements and nitrogen source on biomass accumulation and withanolide-A production from cell suspension cultures of Withania somnifera (L.) Dunal. Plant Cell Tissue Organ Cult 104:119–124

Nagella P, Murthy HN (2010b) Establishment of cell suspension cultures of Withania somnifera for the production of withanolide A. Bioresour Technol 101:6735–6739

Nitsch JP, Nitsch C (1969) Haploid plants from pollen grains. Science 163:85

Praveen N, Murthy HN (2010) Production of withanolide-A from adventitious root cultures of Withania somnifera. Acta Physiol Plant 32(5):1017–1022

Praveen N, Murthy HN (2012) Withanolide A production from Withania somnifera hairy root cultures with improved growth by altering the concentrations of macro elements and nitrogen source in the medium. Acta Physiol Plant 35:811–816

Ray S, Jha S (1999) Withanolide synthesis in cultures of Withania somnifera transformed with Agrobacterium tumefaciens. Plant Sci 146:1–7

Ray S, Jha S (2001) Production of withaferin A in shoot cultures of Withania somnifera. Planta Med 67:432–436

Ray S, Ghosh B, Sen S, Jha S (1996) Withanolide production by root cultures of Withania somnifera transformed with Agrobacterium rhizogenes. Planta Med 62:571–573

Sabir F, Sangwan NS, Chaurasiya ND, Misra LN, Sangwan RS (2008) In vitro Withanolide production by Withania somnifera L. cultures. Zeitschrift für Naturforschung C 63(5–6):409–412

Sabir F, Sangwan RS, Singh J, Misra LN, Pathak N, Sangwan NS (2011) Biotransformation of withanolides by cell suspension cultures of Withania somnifera (Dunal). Plant Biotechnol Rep 5:127–134

Sangwan RS, Chaurasiya ND, Lal P, Misra L, Uniyal GC, Tuli R, Sangwan NS (2007) Withanolide A Biogeneration in *in vitro* shoot cultures of Ashwagandha (*Withania somnifera*), a main medicinal plant in Ayurveda. Chem Pharm Bull 55:1371–1375

Saravanakumar A, Aslam AAS (2012) Development and optimization of hairy root culture systems in Withania somnifera (L.) Dunal for withaferin-A production. Afr J Biotech 11:16412–16420

Sivanandhan G, Mariashibu TS, Arun M, Rajesh M, Kasthurirengan S, Selvaraj N, Ganapathi A (2011) The effect of polyamines on the efficiency of multiplication and rooting of Withania somnifera (L.) Dunal and content of some withanolides in obtained plants. Acta Physiol Plant 33:2279–2288

Sivanandhan G, Arun M, Mayavan S, Rajesh M, Jeyaraj M, Dev GK, Manickavasagam M, Selvaraj N, Ganapathi A (2012a) Optimization of elicitation conditions with methyl jasmonate and salicylic acid to improve the productivity of withanolides in the adventitious root culture of Withania somnifera (L.) Dunal. Appl Biochem Biotechnol 168:681–696

Sivanandhan G, Arun M, Mayavan S, Rajesh M, Mariashibu TS, Manickavasagam M, Selvaraj N, Ganapathi A (2012b) Chitosan enhances withanolides production in adventitious root cultures of Withania somnifera (L.) Dunal. Ind Crop Prod 37:124–129

Sivanandhan G, Kapil Dev G, Jeyaraj M, Rajesh M, Arjunan A, Muthuselvam M, Manickavasagam M, Selvaraj N, Ganapathi A (2013a) Increased production of withanolide A, withanone, and withaferin A in hairy root cultures of Withania somnifera (L.) Dunal elicited with methyl jasmonate and salicylic acid. Plant Cell Tissue Organ Cult 114:121–129

Sivanandhan G, Kapil Dev G, Jeyaraj M, Rajesh M, Muthuselvam M, Selvaraj N, Manickavasagam M, Ganapathi A (2013b) A promising approach on biomass accumulation and withanolides production in cell suspension culture of Withania somnifera (L.) Dunal. Protoplasma 250:885–898

Sivanandhan G, Selvaraj N, Ganapathi A, Manickavasagam M (2014) Enhanced biosynthesis of withanolides by elicitation and precursor feeding in cell suspension culture of Withania somnifera (L.) Dunal in shake-flask culture and bioreactor. PLoS one 9:e104005

Sivanandhan G, Selvaraj N, Ganapathi A, Manickavasagam M (2016) Elicitation approaches for Withanolide production in hairy root culture of Withania somnifera (L.) Dunal. In: Fett-Neto AG (ed) Biotechnology of plant secondary metabolism: methods and protocols. Springer, New York, pp 1–18

Smetanska I (2008) Production of secondary metabolites using plant cell cultures. In: Stahl U, Donalies UEB, Nevoigt E (eds) Food biotechnology. Springer, Berlin/Heidelberg, pp 187–228

Thilip C, Soundar Raju C, Varutharaju K, Aslam A, Shajahan A (2015) Improved Agrobacterium rhizogenes-mediated hairy root culture system of Withania somnifera (L.) Dunal using sonication and heat treatment. 3 Biotech 5:949–956

Wang G, Fang H (1998) Mechanism and technology of plant genetic engineering. Science Publisher, Beijing

Chapter 24
Establishment of Hydroponic Cultivation of Ashwagandha for Active Ingredient Enriched Leaves

Sunil C. Kaul, Sukant Garg, Kazuya Tamura, Teruo Wada, Zeenia Kaul, and Renu Wadhwa

Abstract Hydroponics, growing plants in water with appropriate nutrients and other cultivation conditions including temperature and CO_2, is a convenient way to (i) avoid the unpredictable extreme weather conditions that affect the cultivation of plants in soil, (ii) safe guard against environmental stresses including industrial pollutants and pesticides and (iii) ensure stable content of active components that are important for value of plants as either functional food or medicine. Considerable advances have been made in establishment of hydroponics cultivation conditions for a variety of leafy vegetables. However, their use has not expanded to medicinal herbs. In this chapter, we review the very first hydroponic cultivation of Ashwagandha, a medicinal herb.

Keywords Ashwagandha • Hydroponic cultivation • Light condition • Withaferin A • Withanone • Anticancer

S.C. Kaul (✉) • R. Wadhwa
DBT-AIST International Laboratory for Advanced Biomedicine (DAILAB), Biomedical Research Institute, National Institute of Advanced Industrial Science & Technology (AIST) Central 5-41, 1-1-1 Higashi, Tsukuba, Ibaraki 305-8565, Japan
e-mail: s-kaul@aist.go.jp

S. Garg
School of Integrative & Global Majors, University of Tsukuba, 1-1-1 Tennodai, Tsukuba, Ibaraki 305-8577, Japan

K. Tamura
DAI-DAN Co., Ltd., 390 Kitanagai, Miyoshi-machi, Iruma-gun, Saitama 354-0044, Japan

T. Wada
Osaka Prefecture University, 1-1 Nakakugakuencho, Sakai-city, Osaka 599-8531, Japan

Z. Kaul
Department of Molecular Virology, Immunology & Medical Genetics, The Ohio State University, Columbus, OH 43210, USA

S.C. Kaul, R. Wadhwa (eds.), *Science of Ashwagandha: Preventive and Therapeutic Potentials*, DOI 10.1007/978-3-319-59192-6_24

24.1 Hydroponic Cultivation

Hydroponics are the techniques of growing plants in water. It takes advantage of the fact that the plants can synthesize all required nutrients from inorganic ions, water, carbon dioxide and sunlight or artificial light. Upon supplementation of the essential elements in inorganic and/or organic, liquid and/or solid form in an appropriate medium, plants can grow in unnatural conditions (Nyugen et al. 2016; McDowell et al. 2013). Such capacity of plants is an extremely beneficial and valuable aspect to cope with environmental disasters, and expand their production for extensive use by human population. This is particularly important in case of seasonal plants and functional food to ensure their continued resource and active constituents. Hydroponic technology was initiated in 1970s and hydroponics are classified on the basis of nutrient and oxygen supply to the roots (Zobel et al. 1976). In the classical model of hydroponics, the roots are completely submerged into the water, saturated with nutrients and oxygen gas. If the roots are either completely or partially aerial, the system is called aeroponic. Variations such as, root growth on rockwool and clay pellets have also been reported in various types of hydroponics systems. We first provide here a basic and brief outline of these systems as follows (Fig. 24.1).

24.1.1 Flood and Drain System

In this system, the pump placed in reservoir feeds water, enriched with nutrients, to the plants placed in a tray and anchored to the stones from roots. Excessive water is drained back to the reservoir by gravitational forces through a negative loop siphon. Here, the nutrient balance and supply can be checked and adjusted according to the requirements. The system is bulky and fragile, as the plant tray has to be placed exactly above the reservoir. This system is limited because of relatively less number of plants could be grown (Fig. 24.1a).

24.1.2 Ebb and Grow System

This system is similar to the 'Flood and Drain System' except that there is a separate secondary reservoir and pump to maintain water level and pressure. Although the reservoir can be placed elsewhere to make the system less fragile, it requires excessive amount of space (Fig. 24.1b).

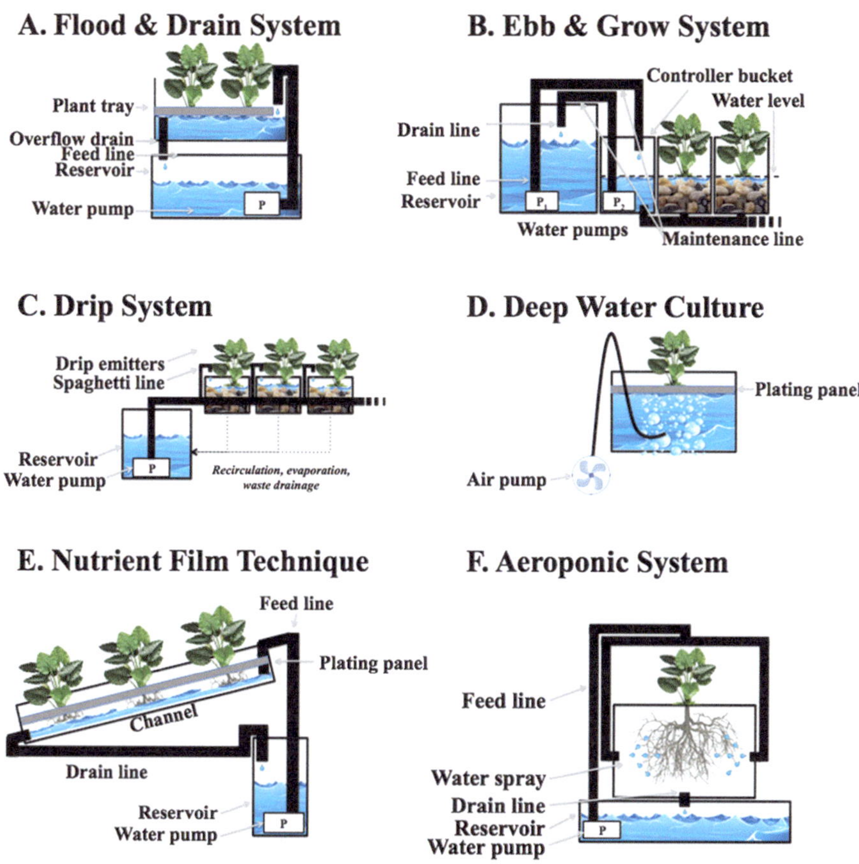

Fig. 24.1 Schematic illustration of different hydroponic cultivation systems

24.1.3 Drip System

Nutrient water is supplied slowly drop by drop via small spaghetti lines and emitters from the main channel. The principle of water drainage is though similar to *the* 'Flood and Drain' as well as 'Ebb and Grow Systems' space (Fig. 24.1c).

24.1.4 Deep Water Culture

This is the most frequently used hydroponics system. Here water is constant in the system and the plant is anchored either with the stones or some artificial fixer such as rubber fixers. Nutrient supply, cleaning and maintenance are manual. Air is supplied through an air pump that opens below the root of the plant space (Fig. 24.1d).

24.1.5 Nutrient Film Technique

In this technique, the plants are placed on an inclined tray and anchored to stones or fixers. The system uses drip system principle. It is less bulky and requires relatively less amount of water space (Fig. 24.1e).

24.1.6 Aeroponics

In this system, plants are held suspended in the air and mildly anchored to the box via the shoot-root junction. Water rich in nutrients, which is supplied through the pump in feed line, is sprayed over the root that uses oxygen from the surrounding air. Excess water is drained back to the reservoir through the drain line space (Fig. 24.1f).

While selecting a hydroponics method for cultivating a plant, it is important to choose the correct system. Regular observation and documentation is important. The choice of method, however, depends on various factors including (i) size and shape of the space available, (ii) size and number of plants to grow, (iii) maintenance and observation time allocated for cultivation and (iv) safety measures, such as power and water supply and their failures, pump specifications and failure, leaks, air pressure and humidity, emitter clogs etc.

24.2 Advantage of Hydroponics

Hydroponics has numerous advantages. The nutrient demand in the medium through the roots and supply can be monitored and controlled (Nyugen et al. 2016). It is fairly easier to document and synthesize the conditions for the growth of the plant. In the plants grown in soil, nutrients in soil bind to the soil particles and fine roots of the plant, creating an independent microenvironment. This heterogeneity adds complexity, especially when it comes to understand the process and implementation of controlled growth supplementation. Furthermore, these roots can also suffer mechanical shear and tissue damage on removal. In hydroponics, on the other hand, the root is suspended loose and suffers minimum damage while extracting the root or transfer of the plant. Nutrient bioavailability remains homogenous throughout the surface area. Conventionally when grown in soil, only the shoot of the plant is available for visualization and assessment, while in the hydroponics, the whole plant is visible and notable. Inderjit (2001) have discussed that certain agents (including biotic-the microbial, and abiotic-physical and chemical factors) present in soil have the potential to cause transformation in the components, chemicals and other nutrients essential for plant growth and characteristics in an enormous way. They may increase or decrease the content of useful ingredients in the plant or even raise the level of toxic substances. Therefore, the monitoring of soil, plant growth characteristics and their functional constituents should be a regular task requiring special expertise and resources. Hydroponics that provides stable and monitored resource of nutrients may provide an alternative choice. Furthermore, it can be a good alternative as assisted plantation where fertile land is either not sufficiently available or a good way of building up public green spaces in highly dense greenery-deficient cities (NyukHien et al. 2007). Chloramines, frequently present in the soil, interfere with metabolism and may cause severe harm to plant. Chlorine, on the other hand, not as common, has high anti-oxidant and anti-microbial action and may be more useful to the plant. Soil structure is difficult to manage or modify, but such intensive alterations are easily achievable in hydroponics structure (Mavrogianopoulos 2016). Installation of hydroponics on the walls has also been suggested (Katsoulas et al. 2017; Mazzali et al. 2013). Katsoulas et al. (2017) discussed that the hydroponics installed on the walls and roofs of buildings in hot and arid Riyadh reduced the maximum and mean air temperature inside the building from an average 36–26 °C. Benefits of hydroponics planted in the building walls were described as air cooling and thermal comfort. Lowering the inner temperature of the building may also lead to the decrease in the demand of electricity and other resources, thus reducing the overall energy cost and resources. Touliatos et al. (2016) compared the conventional hydroponics to the vertical farming to check the effect of sunlight in lettuce, and reported that the vertical farming yielded more crops per unit area. Lettuce is rapidly growing leafy vegetable, which in the reproductive process has the advantage of avoiding complexities of changes in crop biomass allocation (Morris et al. 1957; Heller et al. 2014). Its large leaves benefit it to vitally engage in metabolism, qualifying it as a fine candidate to consider sunlight-effectual plant activities. The authors here concluded with the comment that the spatial orientation of the plant also causes significant variation in the yield of crops and byproducts. We propose that the spatial orientation in accordance with direct sunlight, in addition to the hydroponics cultivation, could yield more productive germination and development of the plant. Most recently, Syu et al. (2017) have provided evidence of controlling the growth index of rice plants (*Oryzasative* L.) through hydroponically modifying the gallium concentration supply to the roots at 25/20 °C (day/night) temperature and 70–95% relative humidity under sunlight. They reported higher growth index directly in proportion to gallium concentration up to a certain limit, where further increment resulted in stunted growth. At the same time, supplementation of indium as an element to the roots in the same hydroponic system resulted in limited nutrient uptake by the plant and impeded growth suggesting that the careful formulation of the nutrient medium is an extremely important aspect of developing hydroponic plants.

24.3 Hydroponics for Functional Food and Medicinal Plants

Plants are not only the essential-most resource for human life, large number of drugs have been developed from plants. More recently, functional food for human health care and welfare has attracted much research, applications and marketing interests. Large number of plants have been experimentally tested and confirmed to possess antibacterial and antifungal properties (Pokorny 1966; Ingratta 1979; Bajaj 1998; Liu et al. 2001; Bernath 2002; Vines 2004; Canter et al. 2005; Hayden 2006; Drake et al. 2009; Rajeh et al. 2010). Roots are generally the most sensitive organs of the plant to the phytotoxic environmental stresses (Habermann et al. 2016). Although the roots enjoy the ownership of many polysaccharides, mucilages, sterols, triterpenoids and alkaloids, leaves are considered to be enriched with fresh and a larger variety of these compounds. Moreover, leaf extracts give an opportunity to not to sacrifice or damage the plant hence making the process ecological and economical. Utilizing leaves for superior medicinal activity and letting the plant grow to produce more certainly is a better choice than killing the entire plant to acquire the root with less amount of active ingredients.

It has been firmly established that plants have great capacity to tolerate the stressful conditions and match up nutrient supply-demand into equilibrium. Brazilian savanna for instance, undergoes severe stress due to the imbalance in aluminum concentration, pH, and forest fires and modeled competition for essential elements (Habermann et al. 2016). These plants develop adaptation such as coriaceous leaves, thorns and trichomes, and high intensity assembly of direct derived metabolites in the progeny (Habermann et al. 2016). Furthermore, due to the phenomenon of allelopathy wherein a plant species chemically interfere with the germination, growth and development of the other plant species (Fathima et al. 2015), the stress induced transformation have large-spread effects. Allelo-chemicals exert their mode of action via various metabolic processes viz., photosysthesis, respiration, hormonal balance, membrane permeability and reaction oxygen species production (Habermann et al. 2016). Hydroponics may offer an alternative of choice in such circumstances. Fathima et al. (2015) showed that the leaf, stem and bark extracts of the *Azadirachta indica* have the potential to allelochemically alter the uptake and metabolism of constituent chemicals in and growth of *Phaseolus vulgaris* (bean) seedlings and plant. Upon treatment in the extract range of 5% to 25%, the authors looked into stimulatory and inhibitory effect on the overall resultant content of chlorophyll, phenols, proteins and carbohydrates. Chlorophyll (a and b) and carotenoid contents increased with all the extracts, with resultant shift in position from the root-shoot axis of the carbohydrates to the cotyledon and proteins and phenols to the endosperm. To sum up, such modifications (positive and negative) are feasible; whose ease plays the key role in hydroponic system. Further to advocate the particulars, another study reported the inhibition of growth of sesame seedlings shoots due to the treatment of 1.25 mg/ml ethyl acetate that was extracted from *Blapharocalyx salicifolius* sp. (Habermann et al. 2016). Extracts of *Sapindus saponaria* delayed the germination process in lettuce and onion diaspores (Grisi et al. 2012). *Sapindus saponaria* (soapberry) is rich in saponins and has been documented to have solid allelopathic effects. Aqueous extracts from it caused a concentration dependent inhibition in the germination process and growth of seedlings, and caused morphological abnormalities.

Variations in chemotypes of Ashwagandha have been reported and have raised concerns for its use and value as a medicinal herb (Bharti et al. 2011; Namdeo et al. 2011). In view of this, we envisioned that the hydroponic cultivation might be useful to obtain a uniform resource of bioactive Ashwagandha and hence established, for the first time, its hydroponic cultivation system.

24.4 Establishment of Hydroponic Cultivation of Ashwagandha

Ashwagandha (*Withania somnifera*) is a renowned Ayurvedic herb commonly used in Indian home medicine system and is known for a variety of therapeutic and health promoting potentials attributed to bioactive components, steroidal alkaloids and lactones. In contrast to the traditional use of roots, we have been exploring bioactivities in leaves of Ashwgandha. We have reported that the leaves, as compared to the roots, contain high proportion of active Withanolides, Withaferin-A (Wi-A) and Withanone (Wi-N). The detailed description of their effects and mechanism(s) of their action have been described earlier (Shah et al. 2009; Wadhwa et al. 2013; Widodo et al. 2007, 2008, 2010). In this chapter, we illustrate the very first establishment of hydroponic cultivation of Ashwagandha in a plant factory in Saitama, Japan (Kaul et al. 2016). As illustrated in Fig. 24.2a, b , hydroponic culture chambers were designed to supply temperature-controlled air into the culture chambers (Kaul et al. 2016) and made of (i) temperature-control-able heat-insulated panels, (ii) high precision air conditioner and air duct system for air-flow, and (iii) automated ultrasonic humidifier for moisture control. CO_2 concentration was monitored with the help of sensors and regulated by supplying liquid CO_2 in the chambers. Nutrient solution was placed in the container on the lowest rack and circulated with a pump, which adjusted the concentration of the nutrients in an automated manner. Nutrients were prepared by mixing Fertilizer A and B (standard nutrients for hydroponics; 1 (ca 2.4 dS/m EC. to 1/4 unit of standard concentration were used). In order not to disturb the environment, the culture fluid supplied to culture beds (ten or more pots) was recycled (Fig. 24.2b). Levels of the nutrient cocktails in tanks was monitored by electrodes, and automatically topped up if level went below the set mark.

The pretreated seeds were sown in rockwool granule for four weeks. The seedlings were transplanted to hydroponic system (maintained at ~25 °C in light/dark cycle (16/8 h) and were grown for 6 weeks. A variety of stress treatments (as described in Table 24.1) (Kaul et al. 2016) were tested for their impact on the growth of plants. We also used Hybrid Electrode Fluorescent Lamp (HEFL) illumination (Nippon Advanced Agri Co. Ltd.) to expose the plants to four different light wavelengths (i) fluorescent, (ii) red, (ii) red:blue −1:1 and (iv) red:blue:green-1:1:1

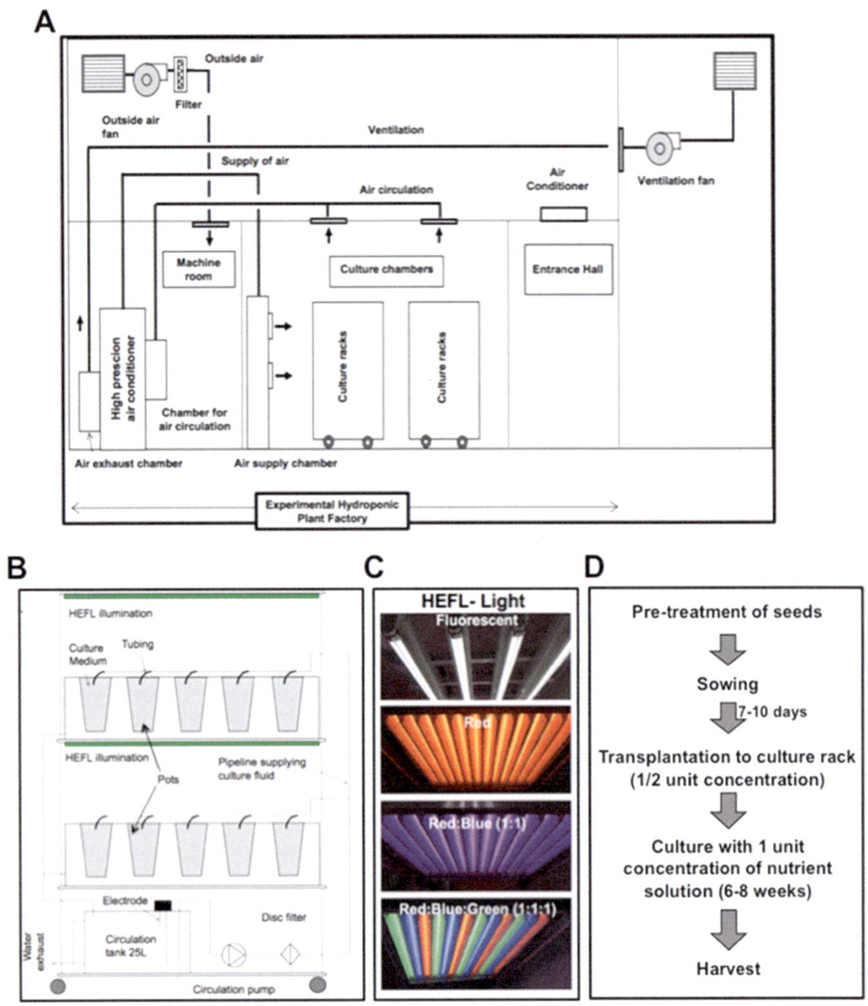

Fig. 24.2 Schematic diagram of plant factory for hydroponic cultivation of Ashwagandha (**a**). Diagrams of cultivation racks, pots and medium circulating system (**b**) and Hybrid Electrode Fluorescent Lamp (HEFL) illumination are shown (**c**). Schematic diagram showing the flow of culture conditions (**d**). (Kaul et al. 2016)

(Fig. 24.2c) (Kaul et al. 2016). In order to keep the illumination close to plants, HEFL (often used for large liquid crystal screen TV, has low energy consumption and heat generation) was used. It was also suitable for growing plants in a multiple layers and offered large cultivation area (Akihisa and Fang-Sik 2009). Culture experiments were performed as shown in the schematic diagram (Fig. 24.2d). Seeds were cleaned with water and pre-treated with gibberellin at low temperature. One week later, when the seeds showed germination they were incubated with nutrient fluid and were then transplanted on the nutrient (one unit of standard concentration)

Table 24.1 Hydroponic culture conditions (Kaul et al. 2016)

Experimental condition	Default	Remarks	EC dS/m	pH
Normal				
Fertilizer	1 U	A-type from OAT Agrio co., ltd. (EC dS/m - 2.4 and pH - 6.15)		
Temperature (light/dark)	25/25 °C			
Light source	HEFL lamp	HEFL lamp from Nippon Advanced Agri co., ltd.		
Light intensity	256 u mol/m^{2}s	Distance from light source ~5 cm		
Light/dark cycle	16/8 h			
Relative humidity	Not controlled			
Stress				
Fertilizer	Control	Enshi standard culture fluid (EC-2.4dS/m; N-16, P-4, K-8, ca-8 and mg-4 me/L)	2.47	6.37
	Modified	Modified IV standard culture fluid	2.68	6.4
	2 U	2 U culture fluid	4.62	5.97
	4 U	4 U culture fluid	8.52	5.58
	NaCl_2 U	Standard nutrient solution + NaCl 1400 ppm	5.08	6.34
	NaCl_4 U	Standard nutrient solution + NaCl 4200 ppm	9.98	6.29
	$CaCl_2$_4 U	Standard nutrient solution + $CaCl_2$ 4000 ppm	9.42	6.06
	$MgSO_4$_4 U	Standard nutrient solution + $MgSO_4/7H_2O$ 8860 ppm	6.35	6.06
Temperature (light/dark)	27/22 °C			
Light source	Fluorescent lamp			
Light intensity	150 μ mol/m^{2}s	Distance from light source ~42 cm		

filled culture rack in the culture chamber. The plants were harvested after two to 3 weeks and analyzed for the content of Withaferin-A and Withanone. Temperature of the culture chamber was adjusted to 25 °C both for light (16 h) and dark (8 h) periods. CO_2 concentration in the lighted period was controlled to ca. 1000 ppm.

As shown in Fig. 24.3, Ashwagandha leaves were successfully obtained by hydroponic cultivation. Furthermore, studies on the effect of a variety of environmental conditions including, exposure to UV, temperature, pH and nutrients (Table 24.1) on various characteristics of plants including, plant height, number of leaves, weight of shoots and roots in another experiments using deep water culture revealed no dramatic changes in these attributes suggesting that Ashwagandha is generally tolerant to a variety of environmental stresses. Some environmental stresses that caused noticeable changes either in root and leave aspects are listed here.

Fig. 24.3 Hydroponic cultivation of Ashwagandha. Time-lapse photographs of hydroponic growth of Ashwagandha at 0, 4, 6, 8 and 10 weeks. (Kaul et al. 2016)

(i) Hypertrophic roots were seen in cultivation in four units of nutrient solution
(ii) Cultivation in nutrient solution containing 4 units of NaCl caused decrease in root mass.
(iii) Exposure to UV and high temperature showed some visible alteration in growth. For example, UV-B exposure for over 30 min/day caused remarkable leaf damage and death of plants. Based on this observation, UV-B exposure

was restricted to 10 min. Plants exposed to UV-B during night caused curling of leaves. On the other hand, UV-A (16 h, during light period) exposure was well tolerated by the plants.

(iv) Biological activity investigations of the leaf extracts from UV- stressed plants showed high level of anticancer activity suggesting that in spite of the above-described phenotypic changes observed, there was no major impact on the level of anticancer bioactives.

(v) High temperature stress, light/dark (42/22 °C) was lethal.

(vi) Light/dark (37/22 °C) cycle yielded shorter plants with more lateral shoot branching that borne thick and dense green leaves. Similar to the UV-stressed, these plants also showed no difference in their cytotoxic activity in the extracts as compared to the control plants.

(vii) Hydroponically grown leaves, under a variety of conditions showed higher content of Wi-N than Wi-A. This accounted for higher anticancer activity of the extracts to cancer than to normal cells (Widodo et al. 2007; Kaul et al. 2016).

(viii) Plants cultivated in variable light conditions showed different content of Wi-A and Wi-N. Roots of the plants exposed continuously to HEFL with no dark cycle showed lowest content of Wi-A and Wi-N. Leaves from these plants showed Wi-N level similar to that of the ones cultivated under HEFL (16 h) + UV (8 h). Whereas the amount of Wi-A was significantly high in leaves from plants grown under HEFL + UV, leaves of plants cultivated under red light showed highest content of Wi-N (Wi-N:Wi-A/10:1) (Fig. 24.4). These data suggested that the

(a) Withanolide content of Ashwagandha leaves can be manipulated by their cultivation under different light conditions
(b) plants cultivated under red light possess leaves with 10-fold higher level of Wi-N as compared to Wi-A.
(c) plants cultivated under UV light possess leaves with high level of Wi-A.
(d) mixture of red and blue light may yield plants with high content of Withanone as well as Withaferin-A. In line with this prediction, the plants cultivated under Red + Blue + Green light (1:1:1 ratio) indeed showed high level of Wi-N and Wi-A (Fig. 24.4). Similar to the plants raised on land, the root extracts of hydroponically raised plants possessed less withanolides than the leaves (Fig. 24.4), and showed low cytotoxicity to cancer cells (Fig. 24.4).

Hydroponic cultivation of Ashwagandha was established. Analyses of bioactives in hydroponically raised leaves showed the high content of withanolides. Furthermore, whereas cultivation under red light yielded leaves with high content of Wi-N, UV light resulted in high level of Wi-A. These findings have predicted that the hydroponics is useful for cultivation of Ashwagandha and manipulation of its bioactive withanolides content by altering the conditions.

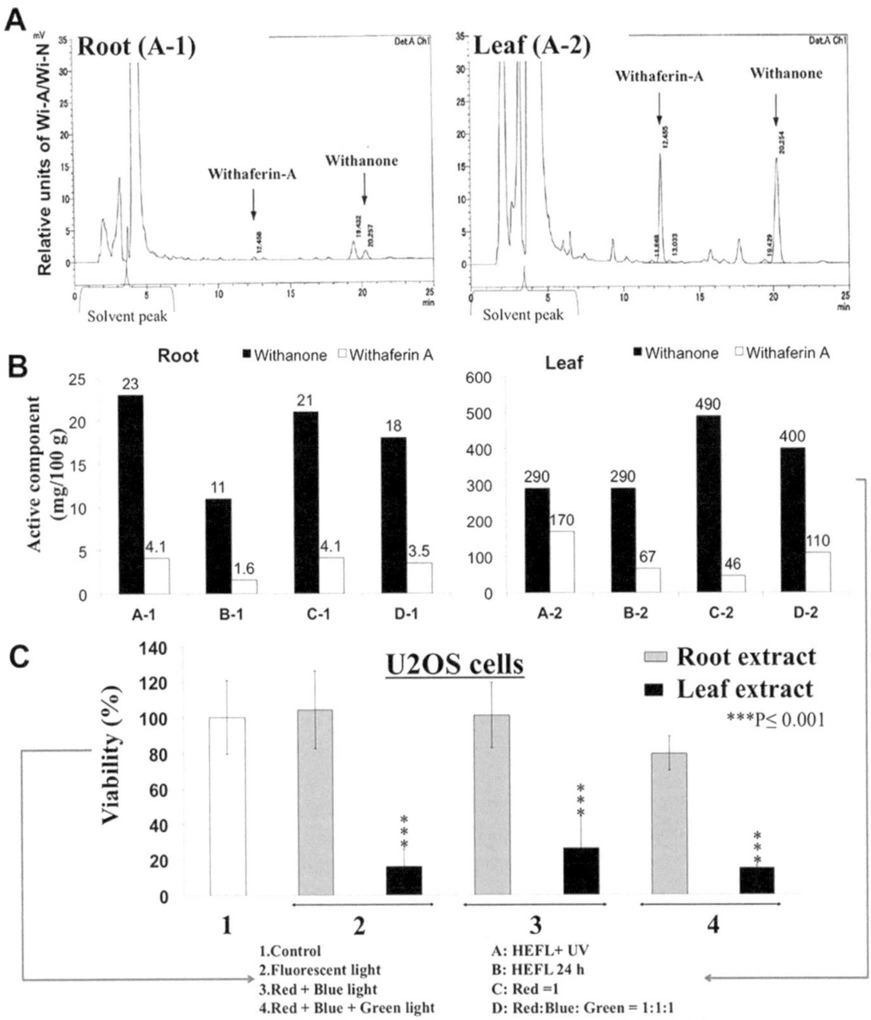

Fig. 24.4 HPLC analysis of roots (A1) and leaves (A2) from Ashwagandha plants cultivated under different light conditions (**a**). Quantitation of Withanone and Withaferin-A in root and leaves of Ashwagandha cultivated under different light conditions (**b**). Cytotoxicity of Ashwagandha root and leaf extracts to human cancer (U2OS) cells showing higher cytotoxicity of the latter (**c**). (Kaul et al. 2016)

References

Akihisa T, Fang-Sik C (2009) Development of hybrid electrode fluorescent lamp and application to plant factory. Regulation of Plant Growth Dev 44:177–181

Bajaj YPS (1998) Biotechnology for the improvement of medicinal plants. Acta Hort 457:37–45

Bernath J (2002) Strategies and recent achievements in selection of medicinal and aromatic plants. Acta Hort 576:115–128

Bharti SK, Bhatia A, Tewari SK, Sidhu OP, Roy R (2011) Application of HR-MAS NMR spectroscopy for studying chemotype variations of Withania somnifera (L.) Dunal. Magn Reson Chem 49:659–667

Canter PH, Thomas H, Ernst E (2005) Bringing medicinal plants into cultivation: opportunities and challenges for biotechnology. Trends Biotechnol 23:180–185

Drake PM, Barbi T, Sexton A, McGowan E, Stadlmann J, Navarre C, Paul MJ, Ma JK (2009) Development of rhizosecretion as a production system for recombinant proteins from hydroponic cultivated tobacco. FASEB J 23:3581–3589

Fathima M, Kumar HNK, Jagannath S (2015) Allelopathic influence of *Azadirachtaindica* leaf, stem and bark extracts on physiological responses of *Phaseolus vulgaris* L. Int J Adv Res 3:1206–1212

Grisi PU, Gualtieri SCJ, Ranal MA, Santana DG (2012) Allelopathic interference of *Sapindussaponaria* root and mature leaf aqueous extracts on diaspora germination and seeding growth of *Lactuca sativa* and *Allium cepa*. Braz J Botany 35:1–9

Habermann E, Pontes FC, Pareira VC, Imatomi M, Gualtieri SCJ (2016) Phytotoxic potential of young leaves from Blepharocalyxsalicifolius (Kunth) O. Berg (Myrtaceae). Braz J Biol 76:531–538

Hayden AL (2006) Aeroponic and hydroponic systems for medicinal herb, rhizome, and root crops. Hort Science 41:536–538

Heller H, Bar-Tal A, Assouline S, Narkis K, Suryano S, de la Forge A, Barak M, Alon H, Bruner M, Cohen S, Tsohar D (2014) The effects of containergeometry on water and heat regimes in soilless culture:lettuce as a case study. Irrig Sci 33:53–65

Inderjit (2001) Soil: environmental effects on allelochemical activity. Agron J 93:79–84

Ingratta EJ (1979) Soilless culture of greenhouse vegetables. Agdex 290/5 18. Toronto: Ontario Ministry of Agriculture

Katsoulas N, Antoniadis D, Tsirogiannis IL, Labraki E, Bartzanas T, Kittas C (2017) Microclimatic effects of planted hydroponic structures in urban environment measurements and simulations. Int J Biometeorol 61:943–956

Kaul SC, Ishida Y, Tamura K, Wada T, Iitsuka T, Garg S, Kim M, Gao R, Nakai S, Okamoto Y, Terao K, Wadhwa R (2016) Novel methods to generate active ingredients-enriched Ashwagandha leaves and extracts. PLoS One 11:e0166945

Liu J, Burdette JE, Xu H, Gu C, van Breemen RB, Bhat KP, Booth N, Constantinou AI, Pezzuto JM, Fong HH, Farnsworth NR, Bolton JL (2001) Evaluation of estrogenic activity of plant extracts for the potential treatment of menopausal symptoms. J Agric Food Chem 49:2472–2479

Mavrogianopoulos GN (2016) Irrigation dose according to substrate characteristics, in hydroponic systems. Open Agriculture 1:1–6

Mazzali U, Peron F, Romagnoni P, Pulselli RM, Bastianoni S (2013) Experimental investigation on the energy performance of livingwalls in temperate climate. Build Environ 64:57–66

McDowell SC, Akmakjian G, Sladek C, Mendoza-Cozatl D, Morrissey JB, Saini N, Mittler R, Baxter I, Salt DE, Ward JM, Schroeder JI, Guerinot ML, Harper JF (2013) Elemental concentrations in the seed of mutants and natural variants of Arabidopsis thaliana grown under varying soil conditions. PLoS One 8:e63014

Morris LG, Neale FE, Postlethwaite JD (1957) The transpiration of glasshouse crops and its relationship to the incoming solar radiation. J Agric Eng Res 2:111–122

Namdeo AG, Sharma A, Yadav KN, Gawande R, Mahadik KR, Lopez-Gresa MP, Kim HK, Choi YH, Verpoorte R (2011) Metabolic characterization of Withania somnifera from different regions of India using NMR spectroscopy. Planta Med 77:1958–1964

Nyugen NT, McInturf SA, Mendoza-Cozati DG (2016) Hydroponics: a versatile system to study nutrient allocation and plant responses to nutrient availability and exposure to toxic elements. J Vis Exp 113

NyukHien W, PuayYok T, Yu C (2007) Study of thermal performance ofextensive rooftop greenery systems in the tropical climate. Build Environ 42:25–54

Pokorny FA (1966) Pine bark as an organic amendment in the production of container plants. Georgia Agr Res 7:8–9

Rajeh MAB, Zuraini Z, Sasidharan S, Latha LY, Amutha S (2010) Assessment of *Euphorbia hirta* L. leaf, flower, stem and root extracts for their antibacterial and antifungal activity and brine shrimp lethality. Molecules 15:6008–6018

Shah N, Kataria H, Kaul SC, Ishii T, Kaur G, Wadhwa R (2009) Effect of the alcoholic extract of Ashwagandha leaves and its components onproliferation, migration, and differentiation of glioblastoma cells:combinational approach for enhanced differentiation. Cancer Sci 100:1740–1747

Syu CH, Chien PH, Huang CC, Jiang PY, Juang KW, Lee DY (2017) The growth and uptake if Ga and In of rice (*Oryzasative* L.) seedlings as affected by Ga and In concentrations in hydroponic cultures. Ecotoxicol Environ Saf 135:32–39

Touliatos D, Dodd IC, McAinsh M (2016) Vertical farming increases lettuce yield per unit area compared to conventional horizontal hydroponics. Food Energy Secur 5:184–191

Vines G (2004) Herbal harvests with a future: towards sustainable sources for medicinal plants. Plantlife International, Salisbury/Wiltshire

Wadhwa R, Singh R, Gao R, Shah N, Widodo N, Nakamoto T, Ishida Y, Terao K, Kaul SC (2013) Water extract of Ashwagandha leaves has anticancer activity: identification of an active component and its mechanism of action. PLoS One 8:e77189

Widodo N, Kaur K, Shrestha BG, Takagi Y, Ishii T, Wadhwa R, Kaul SC (2007) Selective killing of cancer cells by leaf extract of Ashwagandha:identification of a tumor-inhibitory factor and the first molecular insightsto its effect. Clin Cancer Res 13:2298–2306

Widodo N, Takagi Y, Shrestha BG, Ishii T, Kaul SC, Wadhwa R (2008) Selective killing of cancer cells by leaf extract of Ashwagandha:components, activity and pathway analyses. Cancer Lett 262:37–47

Widodo N, Priyandoko D, Shah N, Wadhwa R, Kaul SC (2010) Selective killing of cancer cells by Ashwagandha leaf extract and itscomponent Withanone involves ROS signaling. PLoS One 5:e13536

Zobel RW, Tredici PD, Torrey JG (1976) Method for growing plants aeroponically. Plant Physiol 57:344–346

MIX
Papier aus verantwortungsvollen Quellen
Paper from responsible sources
FSC® C105338

If you have any concerns about our products,
you can contact us on
ProductSafety@springernature.com

In case Publisher is established outside the EU,
the EU authorized representative is:
Springer Nature Customer Service Center GmbH
Europaplatz 3, 69115 Heidelberg, Germany

Printed by Libri Plureos GmbH
in Hamburg, Germany